T0296233

Advanced Computational Fluid and Aerodynamics

The advent of high performance computers has brought Computational Fluid Dynamics (CFD) to the forefront as a tool to analyze increasingly complex simulation scenarios in many fields. Computational aerodynamics problems are also increasingly moving towards being coupled, multi-physics and multi-scale with complex, moving geometries. The latter presents severe geometry handling and meshing challenges. Simulations also frequently use formal design optimization processes.

This book explains the evolution of CFD and provides a comprehensive overview of the plethora of tools and methods available for solving complex scenarios while exploring the future directions and possible outcomes.

Using numerous examples, illustrations and computational methods the author discusses:

- Turbulence Modeling
- Pre and Post Processing
- Coupled Solutions
- The Importance of Design Optimization
- Multi-physics Problems
- Reduced Order Models
- Large-Scale Computations and the Future of CFD

Advanced Computational Fluid and Aerodynamics is suitable for audiences engaged in computational fluid dynamics, including advanced undergraduates, researchers and industrial practitioners.

Paul G. Tucker is the Rank Professor at the University of Cambridge. He has written more than 300 journal, conference papers and technical reports. He has been a visiting a researcher at NASA and is an associate editor of the *AIAA Journal*.

Advanced Computational Fluid and Aerodynamics

PAUL G. TUCKER
University of Cambridge

Shaftesbury Road, Cambridge CB2 8EA, United Kingdom

One Liberty Plaza, 20th Floor, New York, NY 10006, USA

477 Williamstown Road, Port Melbourne, VIC 3207, Australia

314–321, 3rd Floor, Plot 3, Splendor Forum, Jasola District Centre, New Delhi – 110025, India

103 Penang Road, #05–06/07, Visioncrest Commercial, Singapore 238467

Cambridge University Press is part of Cambridge University Press & Assessment, a department of the University of Cambridge.

We share the University's mission to contribute to society through the pursuit of education, learning and research at the highest international levels of excellence.

www.cambridge.org
Information on this title: www.cambridge.org/9781107428836

First published 2016

A catalogue record for this publication is available from the British Library

Library of Congress Cataloging-in-Publication data
Tucker, Paul G.
Advanced computational fluid and aerodynamics / Paul G. Tucker.
pages cm
Includes bibliographical references.
1. Computational fluid dynamics. 2. Aerodynamics. I. Title.
TA357.5.D37T83 2016
620.1′064–dc23 2015027746

ISBN 978-1-107-07590-0 Hardback
ISBN 978-1-107-42883-6 Paperback

To my family and Rosie the Leonberger – my constant and patient companion during writing

Contents

Preface

In the past 25 years computers have become around a million times faster. This is allowing many examples where full flows or subzones involve the near-direct solution of the Navier-Stokes equations. Since these equations are remarkably exact, such simulations rival measurements. Hence, the Computational Fluid Dynamics (CFD) landscape is beginning to change dramatically. Eddy-resolving simulations should, in roughly the next 10 years, see substantial use in industry in niche areas. The use of eddy-resolving approaches moves CFD to being predictive rather than more postdictive.

CFD problems are increasingly moving towards being coupled, multi-physics and multi-scale with complex geometries. They also frequently use formal design optimization processes. This book attempts to meet this CFD evolution and give an overview of the plethora of methods available to the engineer. Unlike many other volumes, here numerical methods are restricted to just one chapter. This is partly motivated by the observation that even though a vast range of numerical methods exist, as with many other areas of CFD, such a Reynolds Averaged Navier-Stokes (RANS) and LES, just a few schemes/models see widespread use. Doubtless, many will regard this as a bold approach. However, it has enabled me to give more coverage to the areas of CFD knowledge that are needed to exploit it for aerodynamic design.

I am highly grateful to all the PhD students who have so kindly helped me with aspects of text preparation. Special thanks are due to Zaib Ali who, as ever, was a tremendous help with the text preparation. I am grateful for his careful and diligent work. Jiahuan Cui and Mahak Mahak and Richard Oriji also offered tremendous and kind help with the text preparation. I am also grateful to Richard Oriji, Hardeep Kalsi and Sanjeev Shanmuganathan for neatly drawing many of the schematics used. There are two exercises relating to writing compressible and incompressible flow solvers. Inspiration for the compressible was taken from the Cambridge University CFD course. Professor John Denton developed this course, and this inspiration is gratefully acknowledged. As stated by Confucius – I hear, I forget, I write, I remember, I do, I understand. Although time-consuming and challenging, the code-writing tasks are enlightening.

Nomenclature

The nomenclature is set out as follows. First lowercase Roman letters are given, followed by uppercase. Then Greek lowercase, followed by uppercase symbols, are given. Then superscripts and subscripts are set out. Overbars are then listed, followed by special symbols and operators. Finally the abbreviations used in the text are summarized. Please note: to save space, symbols only used once locally in the text are generally not included in this nomenclature.

Lowercase Roman

a_{ij}	Anisotropy tensor
c	Particle velocity, wave velocity, speed of sound or concentration
c'	Pseudo-acoustic speed
c_p	Specific heat capacity at constant pressure
c_v	Specific heat capacity at constant volume
d	Normal wall distance or displacement
$\tilde{d}$	Approximate wall distance function
e	Fluid internal energy due to molecular motion, fundamental charge
f_w, f_{v1}, f_{v2}, f_d	Functions in the Spalart-Allmaras turbulence model
g	Earth's acceleration due to gravity
h	Heat transfer coefficient or height
i, j, k	Array or grid point location identifiers
k	Thermal conductivity, turbulent kinetic energy, temporal weighting function component or variable to ensure that the acoustic wave speed is similar to the particle speed
k_{ij}	Coefficient in spring analogy
l	Turbulence length scale or smoothing length in SPH approach
l_μ, l_ε	Turbulence model length scales
m	Particle mass
$\dot{m}$	Mass flow rate
n	Surface normal or direction cosine
p	Static pressure, or number of stages (Chapter 4)
q	Heat flux
q_r	Radiative heat flux
$q1 \ldots q6$	Terms for transformation to curvilinear coordinate system
r	Local pressure gradient

r, θ, z	Cylindrical polar spatial coordinates
r_d	Shielding function in delayed DES
rms_ϕ	Normalised root mean square change
s	Entropy or streamwise coordinate
s_l	Laminar flame burning velocity
$\dot{s}_n$	Rate of change of species
t	Time
$t_r = t - \lvert\mathbf{x} - \mathbf{y}\rvert/c$	Retarded time
u	Displacement
u, v, w	Instantaneous x, y, z, velocity components
w	Wave number, velocity component, work done by a rotor
x, y, z	Spatial coordinates
$y_{1/2}$	Half width

Uppercase Roman

A	Area, global representation of spatial discretization, nodal coefficient, amplitude or Roe matrix element
A_μ, A_ε	Turbulence model constants
A_ω	Average cross-sectional area normal to vorticity vector
C	Courant number ($u\Delta t/\Delta x$), objective, correlation function, constant or amplitude
C_s	Smagorinsky constant
C_t	Safety factor
C_D	Drag coefficient
C_f	Skin friction coefficient
C_L	Lift coefficient
C_p	Surface pressure coefficient
D	Time step to diffusion time scale ratio or diameter scale
Da	Damköhler number or damping function
E	Young's modulus, error, flux term or energy, constant in wall function or source term in k-ε model
E_b	Black body emission
F	General force term, strong conservation flux term, speed function, switching function in Menter SST model or function in level set equation
FAR	Free air ratio
$F_{i,j}$	View factor (ratio of the radiation received by surface j to that emitted from surface i)
F_p, F_n	Forces parallel and normal to blade passages, respectively
$[F_S]$	Force matrix
F_{SST}	Delayed DES function in Menter SST framework
G	Strong conservation flux term or filter kernel/operator
GCI	Grid convergence index
Gr	Grashof number
H	Strong conservation flux term or representation of step height
I	Prolongation operator

IQ	Eddy-resolving simulation quality metric
J	Jacobian or radiosity
Kn	Knudsen number
K	Porosity, relaxation or acceleration parameter
K_n, K_{vd}	Body force model calibration constants
$[K_s]$	Stiffness matrix
$[K_f]$	Fluid system matrix
L	Length scale, linear turbulent stress component, wave operator or Laplacian
$\tilde{L}$	Free molecular path
M	Mach number
N	Number of mesh points or realizations
$\overline{NL}$	Non-linear turbulent stress component
P	Poisson's ratio or production term
$Pr = \mu c_p/k$	Prandtl number
Q	Volume flow rate or vorticity identification parameter
$Q1 \ldots Q6$	Transformation terms
R	Gas constant, radius scale, residual or energy transfer term
$\mathbf{R}$	Reynolds stress tensor
$\tilde{R}$	Universal gas constant
Re	Reynolds number
$[R]$	Coupling matrix
S	Source or strain term
S_{ij}	Mean strain rate tensor
Sc	Schmidt number
St	Strouhal number
T	Temperature or time scale
$\mathbf{T}$	Matrix of eigenvectors
T_{ij}	Lighthill stress tensor
TV	Total variation
U,	Vector of conserved variables or reference velocity
U, V, W	Velocity components aligned with transformed coordinates
U_c	Bulk or convection velocity
U_τ	Friction velocity
U_∞	Free stream velocity
V	General velocity scale or volume
Vol	Cell volume
Wf	Weighting function

Lowercase Greek

α	Grid expansion factor (Chapter 3), latency parameter in LNS model (Chapter 5), design variable, blade metal angle (Chapter 6), or weighting parameter in compact scheme or relaxation factor (Chapter 4).
β	Compressibility parameter, coefficient of thermal expansion or weighting parameter in compact scheme

$\gamma = c_p/c_v$	Ratio of specific heats, weighting parameter in compact scheme or intermittency
δ	Boundary layer thickness, grid spacing, step function or small number/perturbation
ε	Turbulence dissipation rate, smoothing parameter, strain in solid, small number, scaling parameter in level set related equations, (specified) error tolerance/level or emissivity
η	Parameter that defines time levels in discretized equations or transformed spatial variable
θ	Angle
κ	von Karman constant or weighting parameter in MUSCL scheme (see Section 4.4.3.1)
λ	Temporal discretization control parameter, Eigen values, spectral radius of matrix, viscosity coefficient ($-2\mu/3$), Lame's coefficient or wave speed (in LES filter definition)
μ	Dynamic viscosity, Lame's coefficient
μ_t	Turbulent viscosity
v	Kinematic viscosity
v_t	Turbulent kinematic viscosity
ξ, η, ζ	General, transformed coordinates
ρ	Fluid density
σ	Normal stress, Diffusion Prandtl/Schmidt number, turbulence fluctuation scale or Stefan-Boltzmann constant
τ	Transformed temporal coordinate, shear stress, pseudo time, time shift or relaxation time parameter
ϕ	General variable, flux limiter, or distribution function in lattice Boltzmann method
ψ	Stream function, internal potential
ω	Frequency (turbulence) or vorticity

Uppercase Greek

Γ	Diffusion coefficient, domain boundary or Jacobian matrix
Δ	Filter width or space shift
Δx, Δx, Δz	Grid spacings
Δt	Time-step length
Λ	Adjoint variable, spectral radius or eigenvalues
Φ	Mass fraction, general conserved variable or electric field
Ψ	Shock switch
Ω	Angular velocity or vorticity

Superscripts

eq	Equilibrium value
H	High-order component
L	Low-order component
n	Time level

new	Latest value
old	Previous value
t	Pertaining to tangential component
ΔX	Variable computed with a coarser grid spacing
$'$	Perturbation or first derivate of variable or correction in the pressure correction equation (see Section 4.7.3)
$''$	Second derivative
$+$	Dimensionless distance in wall units
$*$	Approximate value in the pressure correction equation (Section 4.7.3) or distance in wall units

Subscripts

amb	Ambient value
A	Actual value in full scale system
BD	Pertaining to backwards difference scheme
c	Convective flux
coll	Pertaining to collisions
℄	Centerline value
CVF	Pertaining to control volume face
CFD	Pertaining to CFD
DB	Pertaining to database
DES	Pertaining to the DES model
fp	Relating to a particular moving fluid particle
g	Pertaining to grid movement or flow translation
HJ	Pertaining to HJ equation
i, j, k	Array subscripts
in	Pertaining to inlet
$I1, I2$	Nodes that straddle a control volume face
k	Pertaining to turbulence kinetic energy
KEP	Pertaining to kinetic energy preserving scheme
l	Pertaining to a liquid
LES	Pertaining to LES model
max	Maximum value
min	Minimum value
M	Values in model
NB	Neighboring values
o	Reference value, or pertaining to offset
out	Pertaining to outlet
p	Pertaining to a particle or droplet
P	Central grid point
PS	Pressure surface
ref	Reference value
rel	Relative velocity component
R	Pertaining to radiation
$RANS$	Pertaining to RANS model

Roe	Pertaining to Roe scheme
s	Pertaining to solid or sand grain roughness
stat	Pertaining to stationary coordinate system
SGS	Pertaining to the subgrid scale
SS	Pertaining to suction surface
target	Target value
t	Pertaining to turbulence
up	Pertaining to point of separation
u, v, w	Pertaining to listed velocity components
v	Pertaining to a vapour or viscous flux
w, e, n, s, f, b	Geographic grid point notation for control volume face
W, E, N, S, F, B	Geographic grid point notation for grid points
x, y, z	Pertaining to the x, y and z directions, respectively
z, r, θ	Pertaining to the axial, radial and tangential directions, respectively
ϕ	Pertaining to the variable ϕ
ξ, η, ζ	General, transformed coordinates
$\Delta x, \Delta X$	Variables represented on coarse and fine grids
0	Stagnation property

Overbars

~	Dimensionless or smoothed variable
-	Averaged or filtered value
^	Relating to undivided Laplacian

Special Symbols/Operators

$N(a, b)$	Normally distributed random number operator with mean a and standard deviation b
$NS(\phi)$, $NS^s(\phi)$	Navier-Stokes and steady Navier-Stokes operator
$URANS(\phi)$	Unsteady RANS operator
$\delta(\varphi)$	Dirac delta function (see Chapter 7)
δ_{ij}	Kronecker delta ($\delta_{ij} = 1$ if $i = j$ and $\delta_{ij} = 0$ if $i \neq j$)
ε_{ijk}	Alternating third-rank unit tensor
$\mid\ \mid$	Modulus of quantity

Abbreviations

ADI	Alternating Direct Implicit
ACTRAN	ACoustic TRANsmission
ALE	Arbitrary Lagrangian-Eulerian
AUSM	Advection Upstream Splitting Method
AVPI	Pressure correction scheme variant for unsteady flows
BASIC	Beginner's All-purpose Symbolic Instruction Code
BEM	Boundary Element Method
Bi-CGSTAB	Biconjugate Gradient Stabilized Method
BREP	Boundary Representation
BTD	Balanced Tensor Diffusivity (see Section 4.8.1.4)
CAA	Computational Aeroacoustics
CAD	Computer Aided Drawing
CFD	Computational Fluid Dynamics
CGNS	CFD General Notation System
CPR	Correction Procedure via Reconstruction
CVF	Control Volume Face
CSG	Construction Solid Geometry
CVS	Control Volume Surface
DES	Detached Eddy Simulation
DFT	Discrete Fourier Transform
DG	Discontinuous Galerkin
DNS	Direct Numerical Simulation
DOE	Design of Experiment
DRAGON	Direct Replacement of Arbitrary Grid Overlapping by Non-structured
DRP	Dispersion Relation Preserving
DSM	Deterministic Stress Model
DSMC	Direct Simulation Monte-Carlo
ENO	Essentially Non-Oscillator
ERCOFTAC	European Research Community On Flow Turbulence And Combustion
FD	Finite Difference
FE	Finite Element
FFT	Fast Fourier Transform

FMM	Fast Multipole Method
FORTRAN	FORmula TRANslating System
FT	Forward Transition
FWH	Ffowcs-Williams and Hawkings
GA	Genetic Algorithm
GCI	Grid Convergence Index
GMRES	Generalized Minimum Residual
GPU	Graphical Processor Unit
HJ	Hamilton-Jacobi
HOT	High-Order Term
HPT	High-Pressure Turbine
ICE	Implicit Continuous-fluid Eulerian
IGES	International Graphics Exchange Standard
ILES	Implicit Large Eddy Simulation
KEP	Kinetic Energy Preserving
k-d	k-dimensional
LEE	Linear Euler Equation
LES	Large Eddy Simulation
LIC	Line Integral Convolution
LNS	Limited Numerical Scales
LNSE	Linear Navier-Stokes Equations
LPT	Low-Pressure Turbine
MD	Molecular Dynamics
MATLAB	MATrix LABoratory
MDICE	Multidisciplinary Computing Environment
MDO	Multidisciplinary Design Optimization
MEM	Maximum Entropy Method
MEMS	MicroElectroMechanical Systems
MILES	Monotone Integrated Large Eddy Simulation
MMS	Method of Manufactured Solutions
MRM	Multiple Reciprocity Method
MST	Mean Source Terms
MUSCL	Monotone Upstream-Centred Schemes for Conservation Laws
NACA	National Advisory Committee for Aeronautics
NAFEMS	National Finite Element Methods and Standards NLAS Non-Linear Acoustics Solver
NLDE	Non-Linear Disturbance Equation
NLES	Numerical Large Eddy Simulation
NLAS	Non-Linear Acoustics Solver
NSS	Nearest Surface Search
PANS	Partially Averaged Navier-Stokes
ParMETIS	Parallel Graph Partitioning and Fill-reducing Matrix Ordering
PPW	Points Per Wave
PISO	Pressure Implicit with Splitting of Operator

POD	Proper Orthogonal Decomposition
RANS	Reynolds Averaged Navier-Stokes
RK	Runge-Kutta Scheme
ROM	Reduced Order Model
RPM	Recursive Projection Methods
RSM	Response Surface Methods or Reynolds Stress Model
RT	Reverse Transition
SARC	SA with Rotation or/and Curvature
SAS	Scale Adaptive-Simulation
SIMPLE	Semi-Implicit Method for Pressure-Linked Equations
SIMPLER	Semi-Implicit Method for Pressure-Linked Equations-Revised
SIMPLEC	Semi-Implicit Method for Pressure Linked Equations-Consistent
SIMPLE*	Further SIMPLE (see above) scheme variant
SIMPLE2	Further SIMPLE (see above) scheme variant
SIP	Strongly Implicit Procedure
SPH	Smooth Particle Hydrodynamics
SST	Shear Stress Transport
SD	Spectral Difference
SV	Spectral Volume
TDMA	Tri-Diagonal Matrix Algorithm
T-S	Tollmien-Schlichting
TSL	Thin Shear Layer
TVD	Total Variation Diminishing
ULIC	Unsteady Line Integral Convolution
UMIST	University of Manchester Institute of Science and Technology
URANS	Unsteady Reynolds Averaged Navier-Stokes
VLES	Very Large Eddy Simulation
WALE	Wall Adapting Local Eddy-Viscosity
WENO	Weighted Essentially Non-Oscillatory

CHAPTER I

The Need and Methods for Studying Aerodynamics

The age of chivalry has gone. That of... calculator's has succeeded and the glory of Europe is extinguished forever.

E. Burke (1972)

1.1 Introduction

According to Slotnick et al. (2014), in 2025, air transport will contribute 1.5 billion tonnes of CO_2 emissions annually. Of course, there are many other fluid dynamic engineering systems and processes that contribute to environmentally damaging emissions. Hence, it is critical to study aerodynamics with the view to creating more energy-efficient systems.

Broadly speaking, the design engineer has the following three options for informing decisions regarding aerodynamics:

- Measurements
- Analytical solutions
- Computer simulations

1.2 Computational Fluid Dynamics

Since its inception around the 1960s, computer based simulation has revolutionised fluid mechanics. The key area of computer simulation in fluid mechanics is called Computational Fluid Dynamics (CFD). It is generally taken as involving the solution of the full equations governing conservation of mass, momentum and energy in a fluid, with generally little simplification to these equations. Currently, mostly it is still best for a company to take advantage of all three of the above techniques. The development of CFD has been most strongly driven by the aerospace industry. CFD's name has sometimes been partly tarnished, with suggestions that the acronym should be standing for 'Cheats Frauds and Deceivers' or 'Colourful Fluid Dynamics'. As with any experimental method, the use of CFD often requires extreme skill. CFD codes cannot be run by engineers (focused on extremely complex multidisciplinary design tasks) lacking some specialist training and without clear guidance. The modelling of turbulence has limited the accuracy of CFD, rendering the need for comparisons with test data (validation). The uncertainty in modelling turbulence has made CFD almost a postdictive method. However, with the power of modern computers, eddy-resolving techniques (the nearly exact flow equations solved to high accuracy) are beginning to emerge. Hence, we are beginning to move into a predictive era of CFD. In the future, the way in which CFD is used is likely to change substantially.

Generally with CFD simulations, the fluid flow field is divided into numerous small cells. Processes for carrying this out are described in Chapter 3. The flow-governing equations, described in Chapter 2, are solved locally for each of these cells. The solution process ensures mass, momentum and energy conservation between the cells. The geometry of the system under study can be taken from Computer-Aided Design (CAD) information (see Chapter 7), and generally geometry changes, relative to experiments, can be easily made. Key values of CFD are:

- It is an effective means for the rapid evaluation of what-if design scenarios.
- Geometry changes can be relatively easy.
- Safe for dangerous experiments.
- Can be performed at full-scale conditions (Reynolds and Mach numbers).
- Can be linked to formal design optimization procedures.
- Sensitivity analysis can be performed.
- Computer flow visualization and analysis techniques allow relatively easy analysis of flows to provide a deeper flow physics understanding.

CFD is not necessarily purely used for what we traditionally think of as design, but also sees much use in suggesting solutions to existing deficient designs through analysing them. The use of CFD, as with other modelling tools, should increase understanding of the design problem by the rigorous requirements imposed by mathematical modelling. In addition, the use of computational modelling tools can enable radical new designs (which may otherwise be disregarded due to their departure from established designs) to be assessed relatively inexpensively without the cost of manufacture and testing of a new design. Unlike most stress analysis used for solids, CFD involves solving non-linear coupled equations. Hence, CFD requires considerable rigour to be applied correctly. Next, to further contrast their merits, the traditional approaches to studying fluid flows was be described in more detail. First experimental methods will be considered, followed by analytical analysis and then CFD.

1.3 Experimental Methods

When considering experimental methods and aerodynamics, one frequently thinks of wind tunnels. With these one would aim for geometric similarity (i.e. ensuring the model is exactly to scale) and also, through matching the Reynolds number, hopefully some degree of dynamic similarity, that is, $Re_M = Re_A$, $\rho_M U_M L_M / \mu_M = \rho_A U_A L_A / \mu_A$. The implication of this, assuming a wind tunnel is being used to study the performance of a model (the 'M' subscripts identifying this) for an actual system (subscript A) that works in air (hence the fluid properties will essentially cancel), is that for a 10th-scale model, the wind tunnel fluid velocity must be 10 times that for the actual system. Hence, if we wished to study the aerodynamics of a car going at 100 kilometers per hour (kph), with a 10th-scale model, the wind tunnel air speed would need to be 1,000 kph. This speed, however, is very close to the speed of sound, and local fluid accelerations and decelerations would produce shock waves. Also,

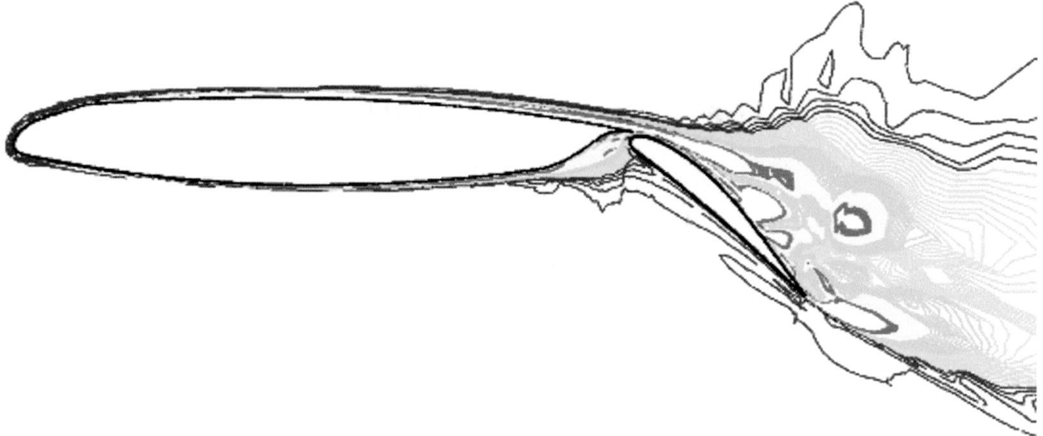

Figure 1.1. Vorticity contours for high lift configuration (from Tucker (2006) – published with kind permission of Wiley).

even if this did not occur, the air would undergo significant compression. Hence, the wind tunnel results would be inaccurate. In fact, for most model tests it is virtually impossible to get a model Reynolds number as high as that for the actual system. As well as the Mach number problem, there is also the issue of having a sufficiently large wind tunnel. Wind tunnels are very expensive to maintain and run and require highly expert staff. There are cheaper ways of achieving higher Reynolds numbers, but these have limitations and so are not discussed here.

As well as high running costs, the costs of manufacturing a scale model, in a large industrial facility, can be high. For example, a half-wing model costs around £300,000. The models need to be made from special steels. For example, the load on the mount connecting the model to the tunnel needs to withstand a force of around 25,000 N. Accuracy needs to be high, with $\pm$15–50 μm model accuracy. High-accuracy machining and testing facilities are required in addition to the actual wind tunnel and instrumentation facilities. Of course, the model needs to be instrumented with at least numerous pressure taps and these taps need to be connected to data loggers without connecting tubes impeding the air flow. The presence of an engine on a wing and the propulsive thrust from this can also dramatically influence the wind tunnel results. Hence, it is wise to include an engine simulator. These cost about £50,000 each. Thus, a wind tunnel test programme for a particular model can cost in excess of £1 million. In this context it is not difficult to see why CFD is viewed as relatively inexpensive.

With this high cost one might expect high accuracy. However, this is not always the case. For high-speed flow configurations and also high-lift configurations the wind tunnel blockage effect can significantly disrupt results. Figure 1.1 shows vorticity contours for a CFD prediction of the flow over a high-lift configuration. Around a $\pm$20% variation in predicted lift can be found when using different established CFD turbulence modelling practices. However, around a 10% variation can also be found due to the wind tunnel wall influence. This figure is ascertained through numerical

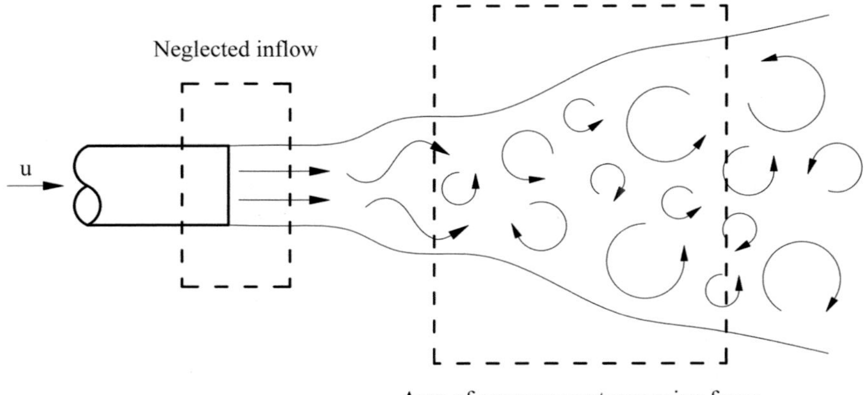

Figure 1.2. Lack of boundary condition definition for measurements.

studies. Hence, experiments do not always yield high-accuracy data. Heat transfer can be especially difficult to measure accurately. When using scaled models to study the dynamics of ships and other free-surface flows, the Froude number must also be matched. It is very rare that one can match both the Froude and Reynolds numbers.

Typically in a wind tunnel, global lift and drag will be measured along with local surface pressures. Also, to gain an idea of general flow patterns and separation points, tufts or a film of oil can be placed on surfaces. However, typically with experiments the level of data points and detail is limited relative to CFD.

A summary of the abilities of experimental methods might be that they can give accurate results at a limited number of points in space, using equipment that is often expensive to purchase and requires skill to calibrate and use successfully. Whilst time variation can be measured, the techniques often disturb the flow. A key problem with experiments is also that generally there is always some uncertainty regarding boundary conditions. This can be wind tunnel wall influences (as noted above) or can be inflow, to name but a few uncertainties. Figure 1.2 illustrates a typical problem with many experimental campaigns, but this time relating to, say, a propulsive jet. Typically the objective would be to study, say, the dynamics of the flow downstream the nozzle exit. However, this leads to the situation where most data sets for jets neglect to measure the inflow boundary conditions such as the boundary layer thickness at the inflow and the turbulence levels across the inflow. This can be a critical problem for the assessment of the CFD and indeed even with regards to the value of the measurements themselves.

Another, interesting aspect with jets is that they can be surprisingly sensitive to their surroundings, and this can greatly complicate comparisons with CFD. Essentially this is again a boundary condition issue. There are further complications with measurements, this being the aspect of repeatability. Shock locations, for example, can vary greatly between wind tunnel runs. Hence, modern measurements tend to assess repeatability. Clearly this aspect relates to uniqueness of flow solutions, and therefore a predictive CFD method should suffer from the same problem, namely a

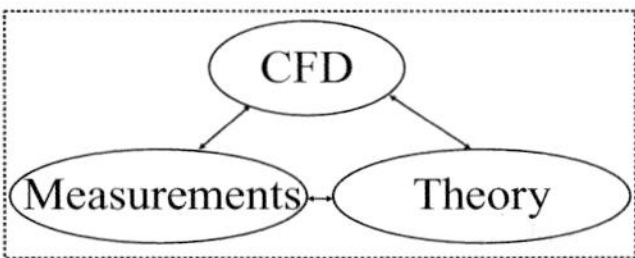

Figure 1.3. The fluid dynamics triad.

lack of repeatability as the differing runs find the different solutions, and replicate any possible physical modulations between solutions. This behaviour is certainty true for real systems.

Independent of all of this, if a number of flow situations have to be considered – say, for a number of physical geometries – experimental methods need new physical geometries to be created and then tested. This of course increases the time and expense of testing nearly linearly. Notably, however, rapid prototyping techniques are emerging, and this raises some interesting questions of how CFD and measurements might work together in the future. If high surface finish is not needed, components can be manufactured cheaply and relatively quickly using three-dimensional printers. These can even produce metallic components. This can be achieved for a range of metals (for example, stainless steel, aluminium and titanium), through the process of laser sintering. Under some circumstances, three-dimensional printing measurements might be quicker than CFD. Another interesting new area is where the CFD simulations and measurements are performed concurrently in various linked ways.

1.4 Analytical Solutions

Analytical solutions are important for gaining ballpark figures for design calculations. Generally analytical solutions essentially consist of exact solutions to simplified flow governing equations with simplified boundary conditions. Analytical solutions can be useful for verifying the soundness of CFD programs.

Notably, analytical equations can find themselves embedded in CFD programs or processes. For example, they can be embedded as models for variable distributions at the elemental level. Examples of these can be found in both temporal (see Patankar and Baliga 1978) and spatial schemes (see Patankar 1980) or for post-processing. For the latter this can be for predicting far-field sound where analytical solutions are used to connect near-field sound level data from the CFD to the far field (see Williams and Hawkings 1969). He and Oldfield (2011) use a one-dimensional analytical conduction of heat solution to connect the solid and fluid domains when performing conjugate simulations. The extent of analytical solutions is vast, and they are not outlined here, but the use of some for code assessment purposes will be found within this text. Critically, both analytical and numerical models seek to accurately solve flow-governing equations. Hence, the flow-governing equations in their wide range of forms are discussed in the next chapter.

As noted, traditionally, CFD is best used in conjunction with measurements and theory/analytical solutions in the triad shown in Figure 1.3. However, as noted

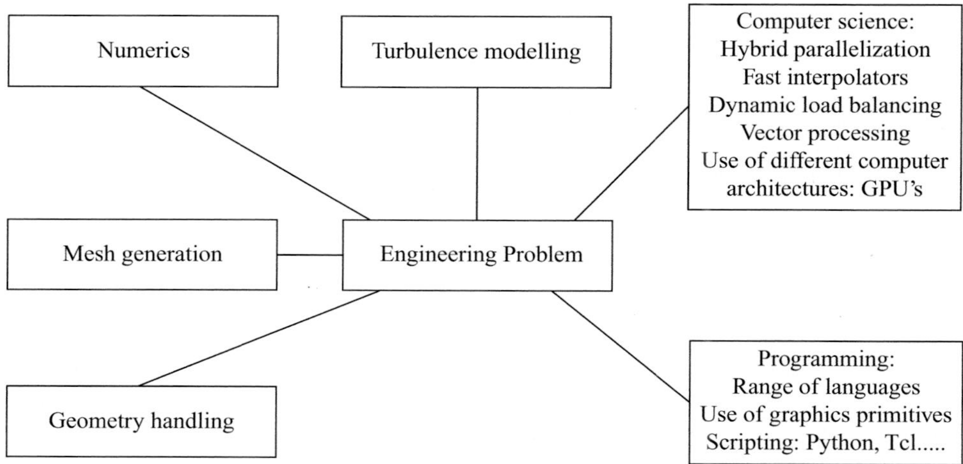

Figure 1.4. Aspects that need to be considered when bringing CFD to an engineering problem.

earlier, as simulation becomes increasingly predictive, the expectation is that the need for input from measurements is reduced.

1.5 CFD in Aerospace

Aircraft manufacturers use CFD for a wide range of aspects of design. These include flutter, fuel systems, sting corrections, cabin ventilation, cockpit ventilation, avionics cooling, ice prediction, thrust reverser design, nacelle design, auxiliary power unit inlet/outlet design, aerodynamic loads data, belly fairing design, tail design, wing design, fuselage design, power plant integration, spoiler control surfaces and flow control design such as vortex generators. Aeroengine manufacturers also use CFD for a similarly wide range of activities including even oil fire modelling.

Figure 1.4 shows aspects that need to be considered when bringing CFD to an engineering problem. As can be seen, these are wide ranging and go well beyond just the consideration of numerical methods. As outlined in Chapter 2, care needs to be exercised when choosing the most appropriate form of governing equation. This equation then needs to be integrated over grid cells. The aspect of mesh/grid generation in discussed in Chapter 3. Numerical schemes formulated around a particular grid are discussed in Chapter 4. Turbulence modelling is dealt with in Chapter 5. Geometry handling is discussed in Chapter 6. The right-hand boxes in Figure 1.4 are largely associated with computer science. This is an area of ever-increasing importance that needs to be considered as part of CFD provision. Chapter 7 discusses pre- and post-processing. Further to what is outlined in Figure 1.4, CFD is becoming an increasingly multi-physics, multi-scale, coupled system endeavour. For example, as noted by Spalart and Bogue (2003), for aircraft, the ultimate aim would be simulations of the coupled, aircraft, propulsive unit, and pilot interactions. The aeroengine manufacturers have similar pressing needs for increasingly coupled simulations.

This is especially so considering that future aircraft will need increasing integration between the airframe and the engine, ultimately with the engine blended into the aircraft. Hence, Chapter 6 looks at advanced simulation techniques. It explores the ingredients needed for many current complex engineering calculations and some future requirements. Notably, multi-scale simulations are of some importance, since the most effective future flow control techniques probably need to be effective in the very inner parts of boundary layers. This gives length scale disparities of many orders of magnitude for high-speed aerodynamic flows. Finally, Chapter 8 looks at what CFD might look like in the next 10 years and beyond.

At the end of most chapters some problems are suggested. Some of these direct the interested reader to the writing of some elementary flow solvers. One is a compressible Euler solver and the other one uses techniques related to incompressible flows. The former utilizes a more general grid structure, whilst the latter has strict cylindrical/Cartesian coordinates. The codes can be validated using analytical solutions, some of which are given in the problems section of Chapter 2. The governing equations in Chapter 2 inform the reader about the equations needed for these codes, and the questions direct them towards the task of initializing a compressible flow field and setting some boundary conditions. The meshing chapter gives the opportunity to write a very simple mesh generator for the above. The numerical methods chapter gives ideas for approaches to generating the systems of equations arsing from integrating the governing equations around the meshes and advancing them, where appropriate, through time. Chapter 6 suggests some more complex potential uses for this solver and further tasks. The pre- and post-processing chapter allows further opportunities to explore the output from the above programs. Appendices B & C provide some information to help with the code development tasks.

REFERENCES

HE, L. & OLDFIELD, M. 2011. Unsteady conjugate heat transfer modeling. *Journal of Turbomachinery*, 133, 031022.

PATANKAR, S. 1980. *Numerical heat transfer and fluid flow*, CRC Press.

PATANKAR, S. & BALIGA, B. 1978. A new finite-difference scheme for parabolic differential equations. *Numerical Heat Transfer*, 1, 27–37.

SLOTNICK, J. P., KHODADOUST, A., ALONSO, J. J., DARMOFAL, D. L., GROPP, W. D., LURIE, E. A., MAVRIPLIS, D. J. & VENKATAKRISHNAN, V. 2014. Enabling the environmentally clean air transportation of the future: a vision of computational fluid dynamics in 2030. *Philosophical Transactions of the Royal Society A: Mathematical, Physical and Engineering Sciences*, 372, 20130317.

SPALART, P. & BOGUE, D. 2003. The role of CFD in aerodynamics, off-design. *Aeronautical Journal*, 107, 323–329.

TUCKER, P. G. 2006. Turbulence modelling of problem aerospace flows. *International Journal for Numerical Methods in Fluids*, 51, 261–283.

WILLIAMS, J. F. & HAWKINGS, D. L. 1969. Sound generation by turbulence and surfaces in arbitrary motion. *Philosophical Transactions of the Royal Society of London. Series A, Mathematical and Physical Sciences*, 264, 321–342.

CHAPTER 2

Governing Equations

2.1 Introduction

Care must be taken regarding whether the chosen flow-governing equations are applicable. For most flows the Navier-Stokes equations are adequate. However, for example, Microelectromechanical systems (MEMS), which have potential for flow control, have a characteristic system length scale, L, range such that 1 mm $>$ $L > 1$ micron. This extreme range of scales creates physical modelling problems. The continuum assumption no longer fully holds and the no-slip condition at solid walls is no longer valid. Satellites in the exosphere are another system where the continuum assumption does not hold. The dimensionless parameter used to tell if the continuum assumption still holds is the Knudsen number (Kn). This is the ratio of the mean free molecular path, $\tilde{L}$ and L. Hence,

$$Kn = \frac{\tilde{L}}{L} \tag{2.1}$$

Generally the fluid can be assumed a continuum for $Kn < 0.1$. However, even then the no-slip surface boundary condition needs to be modified. For $Kn = 0$ the Navier-Stokes equations reduce to the Euler (see Section 2.2.2.1).

It can also be important to consider the Damköhler number:

$$Da = \frac{\textit{Chemical Reaction Rate}}{\textit{Convective time scale}} \tag{2.2}$$

For finite Damköhler numbers, it can be necessary to consider finite rate chemistry in the Navier-Stokes equations (see Section 2.2.2). Also, for non-Newtonian fluids, altered stress-strain relationships are required and so attention needs to be paid to such matters as well as the formulation of the Navier-Stokes equations themselves. For dynamically complex flows, the precise formulation of the Navier-Stokes equations used, when discretized, can have a strong impact on solutions.

Another aspect is the mathematical nature of the flow-governing equations. The equations presented below can be classified as elliptic, parabolic or hyperbolic. This classification is helpful in pointing to the most appropriate numerical scheme necessary to solve the governing equation. Even though, as mathematically defined, governing equations can have a single identifier, typically, in a physical sense equations can have a mixed nature. For example, an equation might be parabolic in time and elliptic or hyperbolic in space. The parabolic temporal nature reflects the fact

Figure 2.1. Isosurface of constant vorticity from numerical solution of the Navier-Stokes equations.

that future events have no impact on the past. Some equations can be formulated to accord with a physical reality so that this is also true spatially. The diffusive terms of the equations are elliptic in nature. This implies that the solution at a point is a function of information at all the surrounding boundary points. For an equation that is parabolic in space the downstream information is just dependent on upstream data. Hyperbolic equations are typically associated with high speed flows with shock waves. With these the regions of flow influence between points in a flow are more complex. Events in multiple distinct zones in a flow can disconnected in multiple directions.

Some texts refer to the combined set of the mass conservation (continuity), momentum and energy equations as the Navier-Stokes equations. Following, here generally the Navier-Stokes equations are taken to refer to just the full form of the momentum equation, including viscous-force terms. The full Navier-Stokes equations present a highly complete description of a fluid flow. In fact, nowadays, exact solutions of them are generally regarded as more accurate than measurements. As noted in Chapter 1, for the latter there are always experimental uncertainties with regards to, for example, boundary conditions and also the intrusive nature of the measurement probes. There are also more subtle issues as outlined in Chapter 1. Figure 2.1 shows instantaneous flow variable contours, from a solution of the Navier-Stokes equations, for the turbulent flow in a plane channel. The experimentally observed streak structures in the boundary layers are apparent in the solution. Unfortunately, as might be suggested from the Figure 2.1 flow complexity, analytical solutions of the Navier-Stokes equations are difficult, if not impossible, for all but the most basic, and generally steady, laminar flows. As will be shown, the Navier-Stokes equations are large non-linear equations and thus difficult to solve analytically. Hence, Figure 2.1 is a numerical solution of the Navier-Stokes and continuity equations. Figures 2.2 and 2.3 give numerical solutions for the Navier-Stokes and continuity equations for an airfoil at a high angle of attack, and the flow through

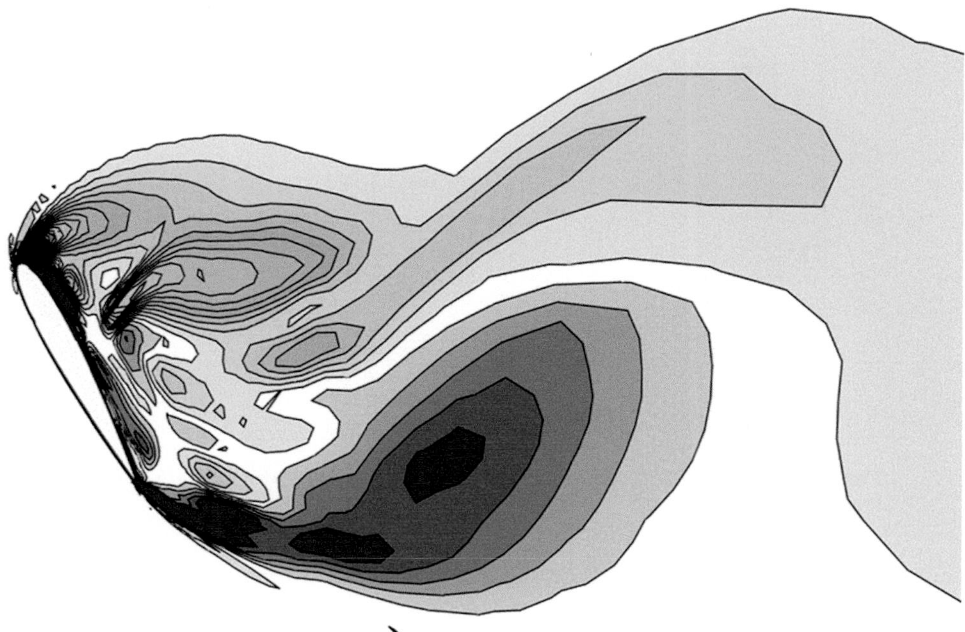

Figure 2.2. Vorticity contours from numerical solution of the Navier-Stokes equations for a NACA0012 wing at a high angle of attack. (from Tucker (2006) – published with kind permission of Wiley).

an aeroengine (bypass duct) mounted to a wing with an idealized flap. As can be seen, these are all rich in flow details.

Figure 2.4 is a flow chart showing the Navier-Stokes and related equations for increasing levels of simplification. In this chapter, these variants of the Navier-Stokes equations will be explored. First, a simple derivation of the Navier-Stokes

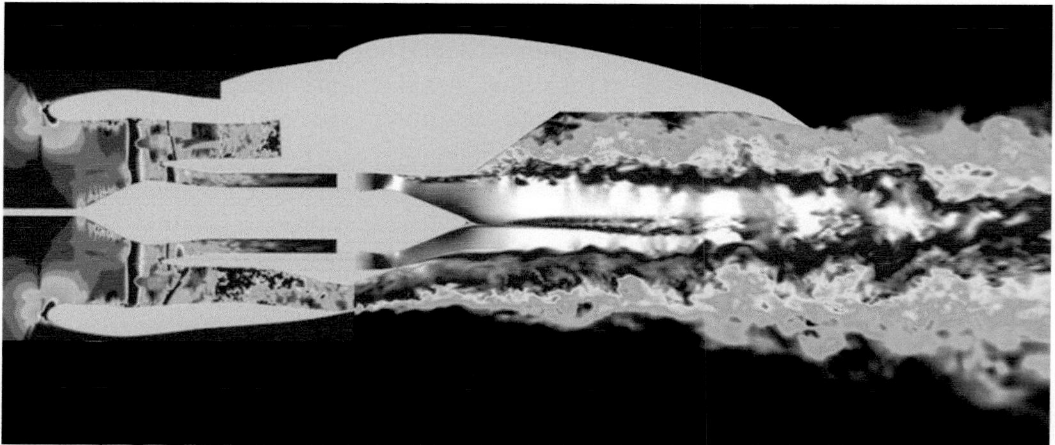

Figure 2.3. Instantaneous flow from solution of the Navier-Stokes equations for the flow though the bypass duct of a gas turbine aeroengine. (Adapted from Tyacke et al. (2015) – published with kind permission of Springer)

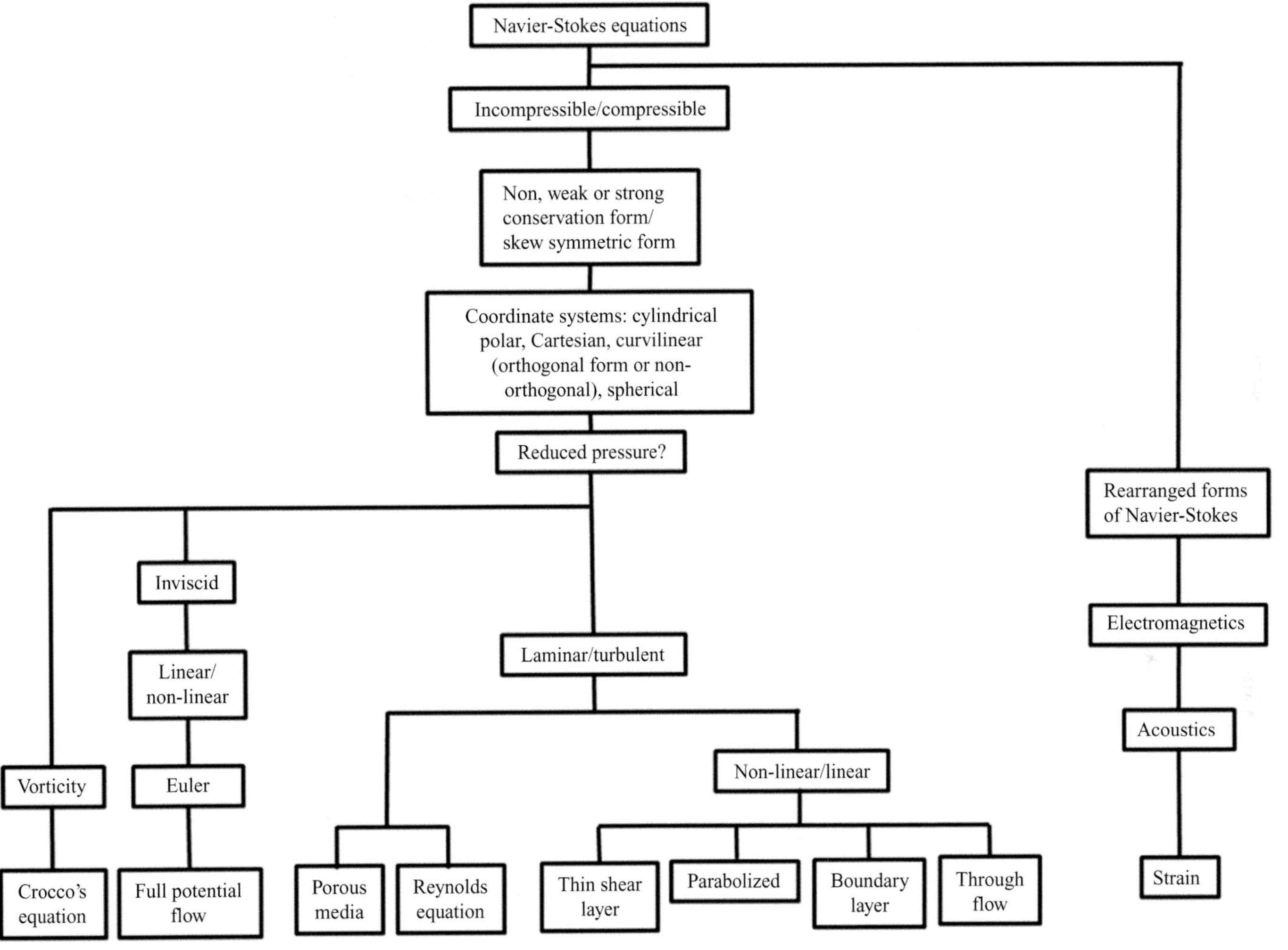

Figure 2.4. Flow chart showing the Navier-Stokes and related equations for increasing levels of simplification.

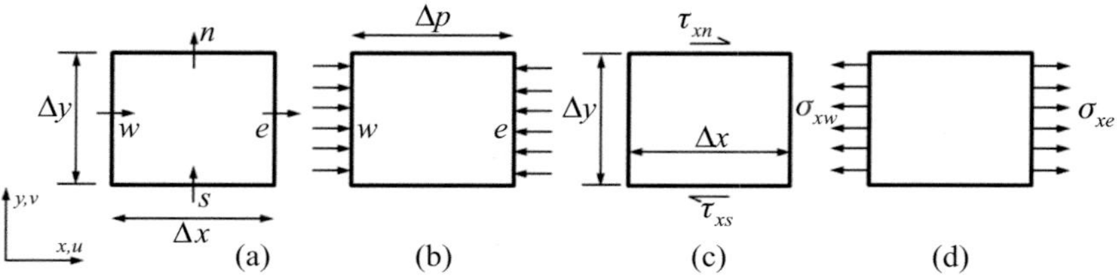

Figure 2.5. Control volume and surface forces: (a) control volume with geographical notation; (b) pressure force; (c) shear stress and (d) normal stress.

and continuity equations, inspired by Anderson (1995), will be given so that their broad nature can be better appreciated.

2.2 Derivation of Navier-Stokes and Continuity Equations

2.2.1 *Continuity Equation*

Consider fluid flowing through a control volume of height Δy and width Δx with faces marked using the geographical notation: w, e, n and s, as shown in Figure 2.5a. The fluid's x and y velocity components are u and v, respectively. If the flow is steady and we define mass flows out of the control volume as positive and into it as negative, then for conservation of mass

$$\Sigma \dot{m} = 0 \tag{2.3}$$

where $\dot{m}$ is the mass flow rate. Equation (2.3) can be obtained from the Reynolds transport theorem with the surface integral replaced by a surface summation. If the control volume is of unit depth, we can write

$$(-\rho_w u_w \Delta y) + (\rho_e u_e \Delta y) + (-\rho_s v_s \Delta x) + (\rho_n v_n \Delta x) = 0 \tag{2.4}$$

which, rearranging and dividing through by the volume per unit depth $\Delta x \Delta y$, gives

$$\frac{\rho_e u_e - \rho_w u_w}{\Delta x} + \frac{\rho_n v_n - \rho_s v_s}{\Delta y} = 0 \tag{2.5}$$

This can be more neatly expressed as

$$\frac{\Delta \rho u}{\Delta x} + \frac{\Delta \rho v}{\Delta y} = 0 \tag{2.6}$$

or, as Δx and Δy tend to zero

$$\frac{\partial \rho u}{\partial x} + \frac{\partial \rho v}{\partial y} = 0 \tag{2.7}$$

For an unsteady flow, Equation (2.7) should have a $\partial \rho / \partial t$ term added to its right-hand side. If the fluid density is constant, then we can obviously write

$$\frac{\partial u}{\partial x} + \frac{\partial v}{\partial y} = 0 \tag{2.8}$$

2.2.2 Navier-Stokes Equations

People concerned with geophysical flows tend to refer to the convective terms in the Navier-Stokes equations as the advective terms. The Navier-Stokes equations can be considered to consist of convection, diffusion and source terms. For a three-dimensional flow, the Navier-Stokes equations will consist of three equations; similarly, for a two-dimensional flow there will be two equations, and so forth. We are going to derive just one of the Navier-Stokes equations, for a two-dimensional flow. The equation we are going to derive considers just fluid forces in the x direction. Essentially, the equation expresses conservation of momentum in this direction.

Indeed the derivation will be based on the steady flow momentum equation. If we use the Reynolds transport theorem to re-express Newton's second law in a control volume form, we will get

$$F_x = \Sigma \dot{m} u \tag{2.9}$$

where momentum flow out of the system is positive and into the system is negative. Again we will apply summation to the surfaces of the control volume shown in Figure 2.5. Hence from Equation (2.9) we can write

$$F_{x,tot} = \rho_e(\Delta y)\, u_e\, u_e - \rho_w(\Delta y)\, u_w\, u_w + \rho_n(\Delta x)\, v_n\, u_n - \rho_s(\Delta x)\, v_s\, u_s \tag{2.10}$$

This can be simplified to

$$F_{x,\,tot} = \Delta y \Delta(\rho u u) + \Delta x \Delta(\rho v u) \tag{2.11}$$

Dividing through by the volume $\Delta x \Delta y$, to give forces per unit volume F', we have

$$F^{'}_{x,\,tot} = \frac{\Delta \rho u u}{\Delta x} + \frac{\Delta \rho v u}{\Delta y} \tag{2.12}$$

which as Δx and Δy tend to zero can be written as

$$F^{'}_{x,\,tot} = \frac{\partial \rho u u}{\partial x} + \frac{\partial \rho v u}{\partial y} \tag{2.13}$$

Equation (2.13) can be rewritten, using the differentiation rule for a product, as

$$F^{'}_{x,\,tot} = u\frac{\partial \rho u}{\partial x} + \rho u\frac{\partial u}{\partial x} + u\frac{\partial \rho v}{\partial y} + \rho v\frac{\partial u}{\partial y} \tag{2.14}$$

If we multiply the continuity Equation (2.7) by u and subtract it from Equation (2.14), we can then write

$$F^{'}_{x,\,tot} = \rho u\frac{\partial u}{\partial x} + \rho v\frac{\partial u}{\partial y} \tag{2.15}$$

Similar expressions to (2.13) and (2.15) can be written for the y direction. We now have to consider the net forces that give the momentum changes. These will be caused by pressure in the fluid and also normal and shear stresses. First we will consider just pressure forces.

2.2.2.1 Pressure Forces The pressure terms can be classified as source type terms in the Navier-Stokes equations. Figure 2.5b shows a control volume with the pressure that will influence the x momentum component indicated. As can be seen, the net force $F_{x,p}$, in the positive x direction, on the control volume, caused by the pressure difference $p_w - p_e(= -\Delta p)$ across it will be

$$F_{x,p} = -\Delta y \Delta p \tag{2.16}$$

After dividing by $\Delta x \Delta y$, this can be rewritten as a force per unit area

$$F'_{x,p} = -\frac{\Delta p}{\Delta x} \tag{2.17}$$

As Δx and Δy tend to zero, we can write

$$F'_{x,p} = -\frac{\partial p}{\partial x} \tag{2.18}$$

If we ignore other forces causing the fluid momentum to change and substitute F_{xp} for F_{xtot} in Equation (2.15) we will get

$$\rho u \frac{\partial u}{\partial x} + \rho v \frac{\partial u}{\partial y} = -\frac{\partial p}{\partial x} \tag{2.19}$$

A similar equation can be derived for the y direction (or by swapping u for v and x for y in Equation (2.19)) to give

$$\rho u \frac{\partial v}{\partial x} + \rho v \frac{\partial v}{\partial y} = -\frac{\partial p}{\partial y} \tag{2.20}$$

These equations are called the Euler equations after Leonhard Euler (1707–1783) (a Swiss mathematician) and for one dimensional flow ($v = 0, \partial/\partial y = 0$) reduce to

$$\rho u \frac{\partial u}{\partial x} = -\frac{\partial p}{\partial x} \tag{2.21}$$

Equation (2.21) can be integrated to give a form of the Bernoulli equation, named after Daniel Bernoulli (1700–1782)

$$p_w + \frac{1}{2}\rho u_w^2 = p_e + \frac{1}{2}\rho u_e^2 \tag{2.22}$$

where w and e represent locations at the ends of a streamline.

2.2.2.2 Shear Forces We will now consider the change in fluid momentum caused by shear forces. We will assume a convention that velocities increase in the positive coordinate directions (i.e., we have positive velocity gradients). Hence, the shear stress directions are as shown in Figure 2.5c. From this figure it can be seen that the shear force will be

$$F_{x,\tau} = \Delta x \Delta \tau_x = \Delta x(\tau_{xn} - \tau_{xs}) \tag{2.23}$$

which, again, we will write as a force per unit volume (as the limit tends to zero)

$$F'_{x,\tau} = \frac{\partial \tau_x}{\partial y} \tag{2.24}$$

We will now consider the normal stresses which cause the fluid momentum to change.

2.2.2.3 Normal Stresses Based on the convention that velocities increase as we move in the positive coordinate direction, the control volume normal stresses influencing the 'x' momentum component are indicated in Figure 2.5d Based on this convention, the velocity to the right of the w control volume face is higher than the velocity to the left. This results in a retarding suction force $\sigma_{x,e}$ acting in the direction shown. Similarly, the velocity just to the left of the right control volume face is lower than the velocity to the right of this face. This results in a suction force to the right. Therefore, the net force in the positive x direction is

$$F_{\sigma,x} = \Delta y \Delta \sigma_x = \Delta y(\sigma_{x,e} - \sigma_{x,w}) \tag{2.25}$$

Again, this is expressed in the limit as a force per unit volume

$$F_{\sigma' x} = \frac{\partial \sigma_x}{\partial x} \tag{2.26}$$

(Note that normal stress terms are often relatively small and so ignored. In general, these terms only become important when modelling shock waves where $\partial u/\partial x$ can be large.)

Therefore, from Equations (2.18), (2.24) and (2.26) we can say the total force per unit volume $F_{x,tot}$ is

$$F'_{x,tot} = -\frac{\partial p}{\partial x} + \frac{\partial \tau_x}{\partial y} + \frac{\partial \sigma_x}{\partial x} \tag{2.27}$$

Substituting this into Equation (2.15) gives

$$\rho u \frac{\partial u}{\partial x} + \rho v \frac{\partial u}{\partial y} = -\frac{\partial p}{\partial x} + \frac{\partial \tau_x}{\partial y} + \frac{\partial \sigma_x}{\partial x} \tag{2.28}$$

Equation (2.28) is the general form of the Navier-Stokes equations for the u momentum component; a similar equation can be derived for the v component.

2.2.2.4 Modelling of Viscous Stresses In the seventeenth century Newton discovered that for certain fluids shear stress is proportional to the velocity gradients. Fluids that obey this law are called Newtonian. Some fluids like blood and toothpaste are non-Newtonian. In 1845, George Gabriel Stokes obtained the following stress expressions for a *constant density* fluid (fuller expressions are given later)

$$\tau_x = \mu\left(\frac{\partial v}{\partial x} + \frac{\partial u}{\partial y}\right), \quad \sigma_x = 2\mu\frac{\partial u}{\partial x} \tag{2.29}$$

Substituting Equation (2.29) into (2.28) gives

$$\rho u \frac{\partial u}{\partial x} + \rho v \frac{\partial u}{\partial y} = -\frac{\partial p}{\partial x} + \frac{\partial}{\partial y}\left(\mu\frac{\partial v}{\partial x}\right) + \frac{\partial}{\partial y}\left(\mu\frac{\partial u}{\partial y}\right) + 2\frac{\partial}{\partial x}\left(\mu\frac{\partial u}{\partial x}\right) \tag{2.30}$$

If we assume the viscosity is constant we can write

$$\rho u \frac{\partial u}{\partial x} + \rho v \frac{\partial u}{\partial y} = -\frac{\partial p}{\partial x} + \mu \frac{\partial}{\partial y}\left(\frac{\partial v}{\partial x}\right) + \mu \frac{\partial}{\partial y}\left(\frac{\partial u}{\partial y}\right) + 2\mu \frac{\partial}{\partial x}\left(\frac{\partial u}{\partial x}\right) \tag{2.31}$$

Since, for simplicity, we have chosen to use the expressions of Stokes, for a constant density fluid, our equations will only be applicable to constant density flows. Differentiation of the continuity equation for a constant density fluid (Equation (2.8)) with respect to x, multiplied by viscosity, gives

$$0 = \mu \frac{\partial}{\partial x}\left(\frac{\partial u}{\partial x}\right) + \mu \frac{\partial}{\partial y}\left(\frac{\partial v}{\partial x}\right) \tag{2.32}$$

This can be subtracted from Equation (2.31) to give

$$\rho u \frac{\partial u}{\partial x} + \rho v \frac{\partial u}{\partial y} = -\frac{\partial p}{\partial x} + \mu \frac{\partial^2 u}{\partial x^2} + \mu \frac{\partial^2 u}{\partial y^2} \tag{2.33}$$

Similar arguments can be used to produce a momentum equation for the vertical y direction

$$\rho u \frac{\partial v}{\partial x} + \rho v \frac{\partial v}{\partial y} = -\frac{\partial p}{\partial y} + \mu \frac{\partial^2 v}{\partial x^2} + \mu \frac{\partial^2 v}{\partial y^2} \tag{2.34}$$

These are the Navier-Stokes equations for a constant density and viscosity, Newtonian fluid in a Cartesian coordinate system. These equations can be written in the following general form

$$\rho u \frac{\partial \phi}{\partial x} + \rho v \frac{\partial \phi}{\partial y} = \mu \frac{\partial^2 \phi}{\partial x^2} + \mu \frac{\partial^2 \phi}{\partial y^2} + S_\phi \tag{2.35}$$

where $\phi = u$ or v and $S_u = -\partial p/\partial x$ and $S_v = -\partial p/\partial y$. Note, the right-hand-side differential terms are elliptic. Depending on the flow conditions, the left-hand-side terms can be parabolic or hyperbolic.

2.3 Forms of the Navier-Stokes Equations

2.3.1 *Full Three-Dimensional Form of Equations*

For completeness, the Navier-Stokes and continuity equations are quoted in their full three-dimensional form, for a Cartesian (x, y, z) coordinate system where $\phi = u$, v or w. These equations apply to variable density and variable viscosity flows and are written in what is normally described as their *weak* conservative form (this aspect will be discussed further later).

$$\frac{\partial}{\partial t}(\rho\phi) + \frac{\partial}{\partial x}(\rho u\phi) + \frac{\partial}{\partial y}(\rho v\phi) + \frac{\partial}{\partial z}(\rho w\phi)$$
$$= \frac{\partial}{\partial x}\left(\Gamma_{\phi x}\frac{\partial \phi}{\partial x}\right) + \frac{\partial}{\partial y}\left(\Gamma_{\phi y}\frac{\partial \phi}{\partial y}\right) + \frac{\partial}{\partial z}\left(\Gamma_{\phi z}\frac{\partial \phi}{\partial z}\right) + S_\phi \tag{2.36}$$

Note, in the above the parabolic time derivative has also been added thus making the equation applicable to an unsteady flow. The diffusion coefficients, for the elliptic terms, become

$$\begin{aligned}\Gamma_{ux} &= 2\mu + \lambda, \Gamma_{uy} = \mu, \Gamma_{uz} = \mu \\ \Gamma_{vx} &= \mu, \Gamma_{vy} = 2\mu + \lambda, \Gamma_{vz} = \mu \\ \Gamma_{wx} &= \mu, \Gamma_{wy} = \mu, \Gamma_{wz} = 2\mu + \lambda\end{aligned} \tag{2.37}$$

where $\lambda = -2\mu/3$ (note that even after around 150 years the physical interpretation of λ is still under dispute) and the source terms are

$$\begin{aligned}S_u &= -\frac{\partial p}{\partial x} + \frac{\partial}{\partial x}\left(\lambda\frac{\partial v}{\partial y}\right) + \frac{\partial}{\partial y}\left(\mu\frac{\partial v}{\partial x}\right) + \frac{\partial}{\partial x}\left(\lambda\frac{\partial w}{\partial z}\right) + \frac{\partial}{\partial z}\left(\mu\frac{\partial w}{\partial x}\right) + \rho g_x \\ S_v &= -\frac{\partial p}{\partial y} + \frac{\partial}{\partial x}\left(\mu\frac{\partial u}{\partial y}\right) + \frac{\partial}{\partial y}\left(\lambda\frac{\partial u}{\partial x}\right) + \frac{\partial}{\partial y}\left(\lambda\frac{\partial w}{\partial z}\right) + \frac{\partial}{\partial z}\left(\mu\frac{\partial w}{\partial y}\right) + \rho g_y \\ S_w &= -\frac{\partial p}{\partial z} + \frac{\partial}{\partial x}\left(\mu\frac{\partial u}{\partial z}\right) + \frac{\partial}{\partial z}\left(\lambda\frac{\partial u}{\partial x}\right) + \frac{\partial}{\partial y}\left(\mu\frac{\partial v}{\partial z}\right) + \frac{\partial}{\partial z}\left(\lambda\frac{\partial v}{\partial y}\right) + \rho g_z\end{aligned} \tag{2.38}$$

The g terms in the above relate to components of gravity in the x, y and z directions, respectively. The continuity equation is

$$\frac{\partial \rho}{\partial t} + \frac{\partial}{\partial x}(\rho u) + \frac{\partial}{\partial y}(\rho v) + \frac{\partial}{\partial z}(\rho w) = 0 \tag{2.39}$$

Figure 2.6 gives examples of an electronics and room ventilation system made on a strictly Cartesian grid. Such grids can be well suited to efficiently solving the flow in such geometries and can be combined with trimmed or immersed boundary cell methods as discussed in Chapter 3.

2.3.2 *Concise Forms of Equations*

2.3.2.1 Tensor Forms of Equations Even after recognising that the Navier-Stokes equations can be expressed in terms of the general variable ϕ we still do not have a compact equation. The use of tensors allows (with Einstein's summation convention) us to write the Navier-Stokes equations more concisely. Equation (2.36) is expressed in tensor form as

$$\frac{\partial \rho u_i}{\partial t} + \frac{\partial \rho u_i v_j}{\partial x_j} = -\frac{\partial p}{\partial x_i} + \frac{\partial}{\partial x_j}\left[\mu\left(\frac{\partial u_i}{\partial x_j} + \frac{\partial u_j}{\partial x_i} - \frac{2}{3}\delta_{ij}\frac{\partial u_k}{\partial x_k}\right)\right] \quad (i, j, k = 1, 2, 3) \tag{2.40}$$

The Kronecker delta $\delta_{ij} = 0$ for $i \neq j$ and $\delta_{ij} = 1$ for $i = j$. The corresponding continuity equation takes the following form

$$\frac{\partial \rho}{\partial t} + \frac{\partial \rho u_i}{\partial x_i} = 0 \tag{2.41}$$

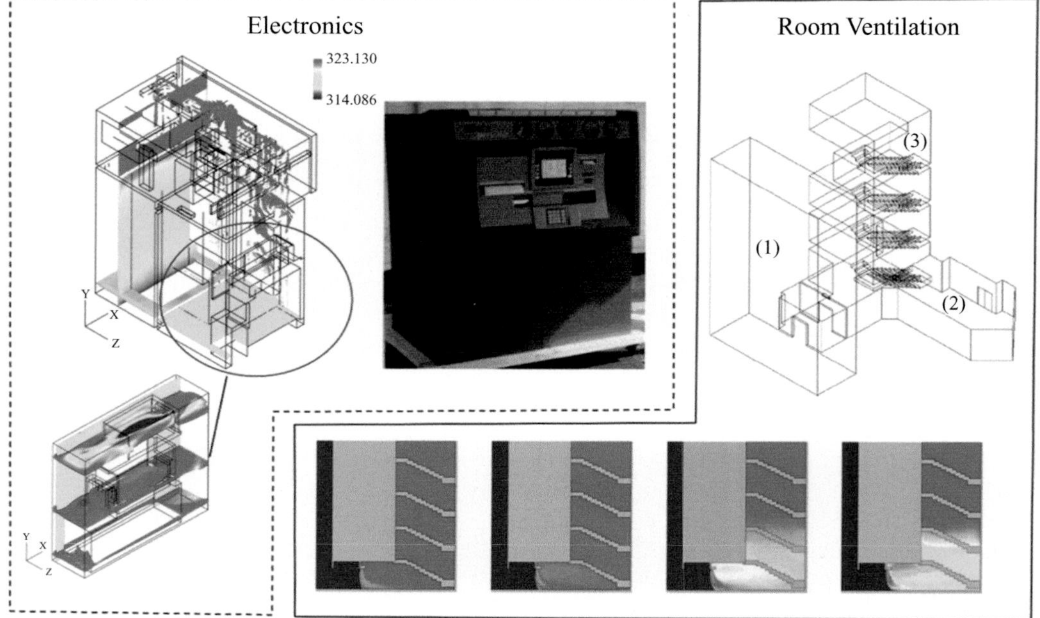

Figure 2.6. Electronics and room ventilation system CFD simulations made on a strictly Cartesian grid.

Note that the repetition of a subscript always means a summation resulting in three terms. Hence

$$\frac{\partial u_i}{\partial x_i} = \frac{\partial u_1}{\partial x_1} + \frac{\partial u_2}{\partial x_2} + \frac{\partial u_3}{\partial x_3} = \frac{\partial u}{\partial x} + \frac{\partial v}{\partial y} + \frac{\partial w}{\partial z} \tag{2.42}$$

Also, by way of another example leading into the vector forms to be discussed next

$$u_i u_i = u_1 u_1 + u_2 u_2 + u_3 u_3 = |\mathbf{u}|^2 = \mathbf{u} \cdot \mathbf{u} \tag{2.43}$$

2.3.2.2 Vector Form of Equations The use of vectors also enables further compactness of expression of the Navier-Stokes equations. Their vector form for a constant density and viscosity fluid is

$$\rho \frac{D\mathbf{u}}{Dt} = -\nabla p + \mu \nabla^2 \mathbf{u} \tag{2.44}$$

where

$$\mathbf{u} = u\mathbf{i} + v\mathbf{j} + w\mathbf{k} \tag{2.45}$$

The corresponding continuity equation is

$$\nabla \cdot \mathbf{u} = 0 \tag{2.46}$$

(i.e., the divergence of $\mathbf{u}$). For a variable density flow, the time derivative must also be included. For solving practical problems, generally, although more compact,

Table 2.1. *Transformation of velocity derivatives from Cartesian to cylindrical polar coordinates*

		i = 1	i = 2	i = 3
	j = 1	$\partial w/\partial z$	$\partial w/\partial r$	$(1/r)\,\partial w/\partial \theta$
$u^j_{,i}$	j = 2	$\partial u/\partial z$	$\partial u/\partial r$	$(1/r)\partial u/\partial \theta - v/r$
	j = 3	$\partial v/\partial z$	$\partial v/\partial r$	$(1/r)\partial v/\partial \theta - u/r$

the tensor and vector forms of the equations can sometimes be less useful. This is because before they can be used, typically the equations must be expanded from their compact to the full form.

2.3.3 *Equations in Cylindrical Polar Coordinates*

Following Morse (1980), the transformation of velocity derivatives from a Cartesian to a cylindrical polar coordinate system can be achieved using the equations given in Table 2.1 – please also see Figure 2.7. The transformations for Reynolds stress terms used when modelling turbulence (see Chapter 5) are also given by Morse.

In a cylindrical polar coordinate system the Cartesian equations become

$$\frac{\partial}{\partial t}(\rho\phi) + \frac{1}{r}\frac{\partial}{\partial r}(\rho r u\phi) + \frac{1}{r}\frac{\partial}{\partial \theta}(\rho v\phi) + \frac{\partial}{\partial z}(\rho w\phi)$$
$$= \frac{1}{r}\frac{\partial}{\partial r}\left(r\Gamma_{\phi r}\frac{\partial \phi}{\partial r}\right) + \frac{1}{r}\frac{\partial}{\partial \theta}\left(\frac{1}{r}\Gamma_{\phi\theta}\frac{\partial \phi}{\partial \theta}\right) + \frac{\partial}{\partial z}\left(\Gamma_{\phi z}\frac{\partial \phi}{\partial z}\right) + S_\phi \qquad (2.47)$$

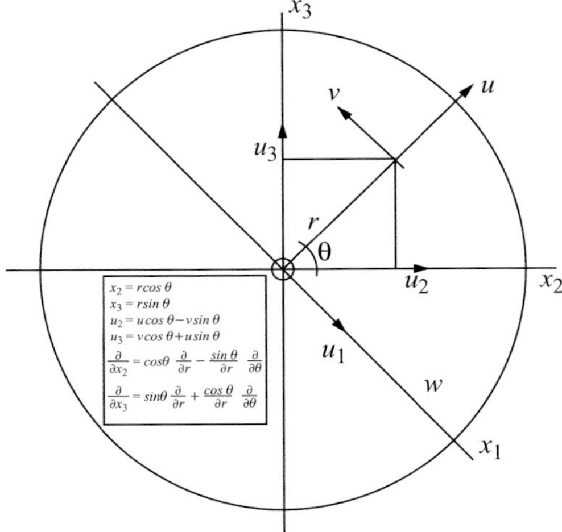

Figure 2.7. Transformation from Cartesian to cylindrical polar coordinates (Adapted from Morse 1980).

The diffusion coefficients are

$$\begin{aligned}
&\Gamma_{ur} = 2\mu + \lambda,\ \Gamma_{u\theta} = \mu,\ \Gamma_{uz} = \mu \\
&\Gamma_{vr} = \mu,\ \Gamma_{v\theta} = 2\mu + \lambda,\ \Gamma_{vz} = \mu \\
&\Gamma_{wr} = \mu,\ \Gamma_{w\theta} = \mu,\ \Gamma_{wz} = 2\mu + \lambda
\end{aligned} \tag{2.48}$$

and the corresponding source terms are

$$\begin{aligned}
S_u &= -\frac{\partial p}{\partial r} - \frac{2\mu u}{r^2} + \frac{\rho v^2}{r} + u\frac{\partial}{\partial r}\left(\frac{\lambda}{r}\right) - \frac{2\mu}{r^2}\frac{\partial v}{\partial \theta} \\
&\quad + \frac{\partial}{\partial r}\left(\frac{\lambda}{r}\frac{\partial v}{\partial \theta}\right) + \frac{\partial}{\partial \theta}\left(\mu\frac{\partial}{\partial r}\left(\frac{v}{r}\right)\right) + \frac{\partial}{\partial r}\left(\lambda\frac{\partial w}{\partial z}\right) + \frac{\partial}{\partial z}\left(\mu\frac{\partial w}{\partial r}\right) + \rho g_r \\
S_v &= -\frac{1}{r}\frac{\partial p}{\partial \theta} - \frac{\mu v}{r^2} - \frac{\rho u v}{r} - \frac{v}{r}\frac{\partial \mu}{\partial r} + \frac{2}{r^2}\frac{\partial}{\partial \theta}(\mu u) \\
&\quad + \frac{1}{r^2}\frac{\partial}{\partial r}\left(\mu r\frac{\partial u}{\partial \theta}\right) + \frac{1}{r^2}\frac{\partial}{\partial \theta}\left(\lambda\frac{\partial}{\partial r}(ru)\right) + \frac{1}{r}\frac{\partial}{\partial \theta}\left(\lambda\frac{\partial w}{\partial z}\right) + \frac{1}{r}\frac{\partial}{\partial z}\left(\mu\frac{\partial w}{\partial \theta}\right) + \rho g_\theta \\
S_w &= -\frac{\partial p}{\partial z} + \frac{1}{r}\frac{\partial}{\partial r}\left(\mu r\frac{\partial u}{\partial z}\right) + \frac{1}{r}\frac{\partial}{\partial z}\left(\lambda\frac{\partial}{\partial r}(ru)\right) + \frac{1}{r}\frac{\partial}{\partial \theta}\left(\mu\frac{\partial v}{\partial z}\right) + \frac{1}{r}\frac{\partial}{\partial z}\left(\lambda\frac{\partial v}{\partial \theta}\right) + \rho g_z
\end{aligned} \tag{2.49}$$

The continuity equation in a cylindrical polar coordinate system is

$$\frac{\partial \rho}{\partial t} + \frac{1}{r}\frac{\partial}{\partial r}(\rho r u) + \frac{1}{r}\frac{\partial}{\partial \theta}(\rho v) + \frac{\partial}{\partial z}(\rho w) = 0 \tag{2.50}$$

For equations (2.36) onwards, if the fluid density is constant, all terms involving λ vanish. If both the density and viscosity are constant, terms multiplied by viscosity in S_ϕ become zero and all $\Gamma_\phi = \mu$.

Figure 2.8 gives an example of a CFD simulation made in a cylindrical polar coordinate system for an idealized aeroengine high-pressure compressor drum. Frame (a) gives the actual geometry and idealization to it used in the simulation. The latter is given by the dashed line. Frame (b) gives velocity vectors of the fluid flow in the r-θ plane. Frame (c) gives contours of the time derivative to be discussed later.

2.3.4 *Curvilinear Form of Equations*

All the above equations can be transformed into curvilinear coordinates. Figure 2.9 shows the surface curvilinear coordinate lines for an idealized tilt rotor geometry. The flow field is superimposed on the mesh. The complete geometry can be mapped using two curvilinear grid zones. These overlay each other and hence there is an interpolation zone.

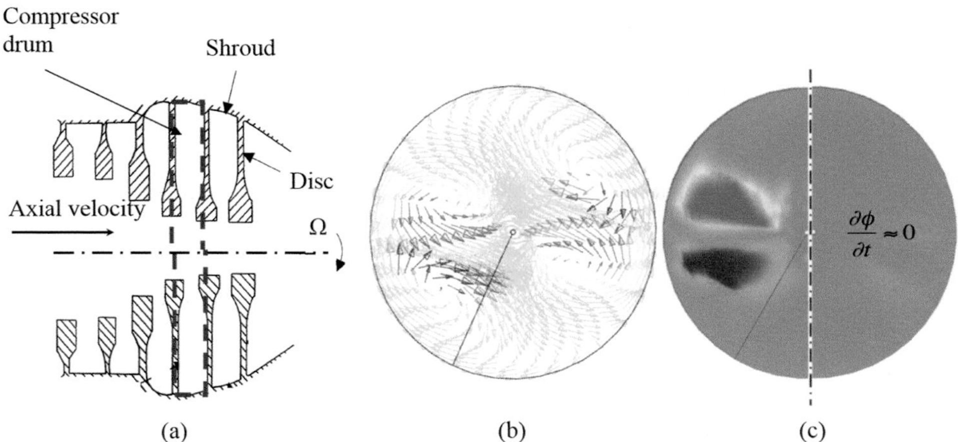

Figure 2.8. CFD simulation made in a cylindrical polar coordinate system for an idealized aeroengine high-pressure compressor drum: (a) actual geometry and idealization to it used in simulation (dashed line); (b) velocity vectors of the fluid flow and (c) contours of time derivative (left frame – coordinate system rotating with solid walls, right frame – coordinate system rotating with fluid) – Frames (b, c) published with kind permission of Elsevier.

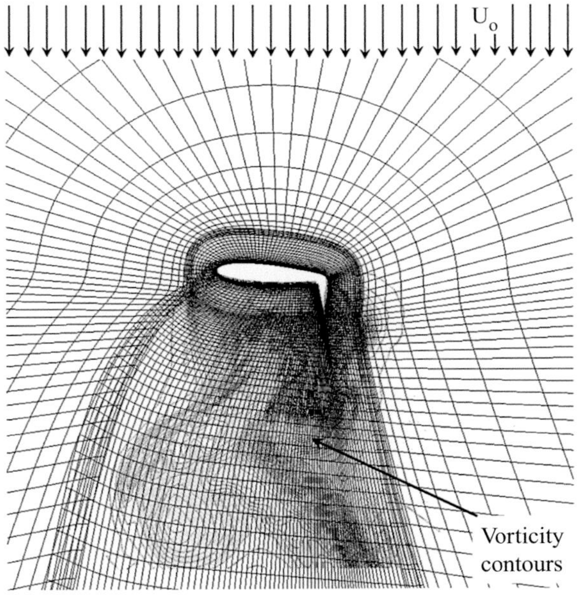

Figure 2.9. Curvilinear grid for an idealized tilt rotor geometry. (from Tucker (2013) – published with kind permission of Wiley).

The transformation can be achieved using the chain rule of differential calculus. Following Rayner (1993) we will consider transformation of the weak conservation form of Equations (2.47)–(2.50) in cylindrical polar coordinates For transformation from the r, θ, z to η, ζ, ξ system, the chain rule can be expressed as

$$\begin{aligned}
\frac{\partial \phi}{\partial r} &= \frac{\partial \phi}{\partial \eta}\frac{\partial \eta}{\partial r} + \frac{\partial \phi}{\partial \zeta}\frac{\partial \zeta}{\partial r} + \frac{\partial \phi}{\partial \xi}\frac{\partial \xi}{\partial r},\\
\frac{\partial \phi}{\partial \theta} &= \frac{\partial \phi}{\partial \eta}\frac{\partial \eta}{\partial \theta} + \frac{\partial \phi}{\partial \zeta}\frac{\partial \zeta}{\partial \theta} + \frac{\partial \phi}{\partial \xi}\frac{\partial \xi}{\partial \theta},\\
\frac{\partial \phi}{\partial z} &= \frac{\partial \phi}{\partial \eta}\frac{\partial \eta}{\partial z} + \frac{\partial \phi}{\partial \zeta}\frac{\partial \zeta}{\partial z} + \frac{\partial \phi}{\partial \xi}\frac{\partial \xi}{\partial z}.
\end{aligned} \tag{2.51}$$

Hence, in essence the transformation involves replacing the $\partial\phi/\partial r$, $\partial\phi/\partial\theta$ and $\partial\phi/\partial z$ derivatives with the right side terms in Equation (2.51). However, in doing this, derivatives with respect to r, θ and z will still remain. Ghia et al. (1976) show these derivatives are equivalent to

$$\begin{aligned}
\eta_r &= \frac{1}{J}(\theta_\zeta z_\xi - \theta_\xi z_\zeta); \quad \zeta_r = \frac{1}{J}(\theta_\xi z_\eta - \theta_\eta z_\xi); \quad \xi_r = \frac{1}{J}(\theta_\eta z_\zeta - \theta_\zeta z_\eta),\\
\eta_\theta &= \frac{1}{J}(r_\xi z_\zeta - r_\zeta z_\xi); \quad \zeta_\theta = \frac{1}{J}(r_\eta z_\xi - r_\xi z_\eta); \quad \xi_\theta = \frac{1}{J}(r_\zeta z_\eta - r_\eta z_\zeta),\\
\eta_z &= \frac{1}{J}(r_\zeta \theta_\xi - r_\xi \theta_\zeta); \quad \zeta_z = \frac{1}{J}(r_\xi \theta_\eta - r_\eta \theta_\xi); \quad \xi_z = \frac{1}{J}(r_\eta \theta_\zeta - r_\zeta \theta_\eta)
\end{aligned} \tag{2.52}$$

where the following shorthand is used: $\eta_r = \partial\eta/\partial r$ and

$$J = r_\eta(\theta_\zeta z_\xi - \theta_\xi z_\zeta) + r_\zeta(\theta_\xi z_\eta - \theta_\eta z_\xi) + r_\xi(\theta_\eta z_\zeta - \theta_\zeta z_\eta).$$

Hence Equation (2.51) can be re-expressed as

$$\begin{aligned}
\phi_r = \frac{1}{J}\phi_\theta &= \frac{1}{J}(\phi_\eta(r_\xi z_\zeta - r_\zeta z_\xi) + \phi_\zeta(r_\eta z_\xi - r_\xi z_\eta) + \phi_\xi(r_\zeta z_\eta - r_\eta z_\zeta)).\\
&(\phi_\eta(\theta_\zeta z_\xi - \theta_\xi z_\zeta) + \phi_\zeta(\theta_\xi z_\eta - \theta_\eta z_\xi) + \phi_\xi(\theta_\eta z_\zeta - \theta_\zeta z_\eta)),\\
\phi_z &= \frac{1}{J}(\phi_\eta(r_\zeta \theta_\xi - r_\xi \theta_\zeta) + \phi_\zeta(r_\xi \theta_\eta - r_\eta \theta_\xi) + \phi_\xi(r_\eta \theta_\zeta - r_\zeta \theta_\eta)).
\end{aligned} \tag{2.53}$$

Repeated application of the chain rule is used to generate transformed higher derivative terms. Substitution of Equation (2.53) and the equivalents for higher derivatives into Equation (2.47) for $\phi = u$ (i.e., radial momentum) gives the following transformed equation

$$
\begin{aligned}
&\frac{1}{J}\frac{\partial}{\partial\eta}(\rho ru^2q5+\rho uvq8+\rho ruwq2)+\frac{1}{J}\frac{\partial}{\partial\zeta}(\rho ru^2q6+\rho uvq9+\rho ruwq3)\\
&\quad+\frac{1}{J}\frac{\partial}{\partial\xi}(\rho ru^2q4+\rho uvq7+\rho ruwq1)\\
&\quad=-\frac{r}{J}\left(q5\frac{\partial p}{\partial\eta}+q6\frac{\partial p}{\partial\zeta}+q4\frac{\partial p}{\partial\xi}\right)+\frac{1}{J}\frac{\partial}{\partial\eta}\left(\frac{r\mu q5}{J}\left[\frac{4}{3}\left(q5\frac{\partial u}{\partial\eta}+q6\frac{\partial u}{\partial\zeta}+q4\frac{\partial u}{\partial\xi}\right)\right.\right.\\
&\quad-\frac{2}{3}\left(q2\frac{\partial w}{\partial\eta}+q3\frac{\partial w}{\partial\zeta}+q1\frac{\partial w}{\partial\xi}\right)-\frac{2}{3}\frac{1}{r}\left(q8\frac{\partial v}{\partial\eta}+q9\frac{\partial v}{\partial\zeta}+q7\frac{\partial v}{\partial\xi}\right)-\left.\frac{2}{3}\frac{uJ}{r}\right]\\
&\quad+\frac{r\mu q8}{J}\left[\frac{1}{r}\left(q5\frac{\partial v}{\partial\eta}+q6\frac{\partial v}{\partial\zeta}+q4\frac{\partial v}{\partial\xi}\right)+\frac{1}{r^2}\left(q8\frac{\partial u}{\partial\eta}+q9\frac{\partial u}{\partial\zeta}+q7\frac{\partial u}{\partial\xi}\right)-\frac{vJ}{r^2}\right]\\
&\quad+\left.\frac{r\mu q2}{J}\left[\left(q5\frac{\partial w}{\partial\eta}+q6\frac{\partial w}{\partial\zeta}+q4\frac{\partial w}{\partial\xi}\right)+\left(q2\frac{\partial u}{\partial\eta}+q3\frac{\partial u}{\partial\zeta}+q1\frac{\partial u}{\partial\xi}\right)\right]\right)\\
&\quad+\frac{1}{J}\frac{\partial}{\partial\zeta}\left(\frac{r\mu q6}{J}\left[\frac{4}{3}\left(q5\frac{\partial u}{\partial\eta}+q6\frac{\partial u}{\partial\zeta}+q4\frac{\partial u}{\partial\xi}\right)-\frac{2}{3}\left(q2\frac{\partial w}{\partial\eta}+q3\frac{\partial w}{\partial\zeta}+q1\frac{\partial w}{\partial\xi}\right)\right.\right.\\
&\quad-\frac{2}{3}\frac{1}{r}\left(q8\frac{\partial v}{\partial\eta}+q9\frac{\partial v}{\partial\zeta}+q7\frac{\partial v}{\partial\xi}\right)-\left.\frac{2}{3}\frac{uJ}{r}\right]\\
&\quad+\frac{r\mu q9}{J}\left[\frac{1}{r}\left(q5\frac{\partial v}{\partial\eta}+q6\frac{\partial v}{\partial\zeta}+q4\frac{\partial v}{\partial\xi}\right)+\frac{1}{r^2}\left(q8\frac{\partial u}{\partial\eta}+q9\frac{\partial u}{\partial\zeta}+q7\frac{\partial u}{\partial\xi}\right)-\frac{vJ}{r^2}\right]\\
&\quad+\left.\frac{r\mu q3}{J}\left[\left(q5\frac{\partial w}{\partial\eta}+q6\frac{\partial w}{\partial\zeta}+q4\frac{\partial w}{\partial\xi}\right)+\left(q2\frac{\partial u}{\partial\eta}+q3\frac{\partial u}{\partial\zeta}+q1\frac{\partial u}{\partial\xi}\right)\right]\right)\\
&\quad+\frac{1}{J}\frac{\partial}{\partial\xi}\left(\frac{r\mu q4}{J}\left[\frac{4}{3}\left(q5\frac{\partial u}{\partial\eta}+q6\frac{\partial u}{\partial\zeta}+q4\frac{\partial u}{\partial\xi}\right)-\frac{2}{3}\left(q2\frac{\partial w}{\partial\eta}+q3\frac{\partial w}{\partial\zeta}+q1\frac{\partial w}{\partial\xi}\right)\right.\right.\\
&\quad-\frac{2}{3}\frac{1}{r}\left(q8\frac{\partial v}{\partial\eta}+q9\frac{\partial v}{\partial\zeta}+q7\frac{\partial v}{\partial\xi}\right)-\left.\frac{2}{3}\frac{uJ}{r}\right]\\
&\quad+\frac{rq7}{J}\left[\frac{1}{r}\left(q5\frac{\partial v}{\partial\eta}+q6\frac{\partial v}{\partial\zeta}+q4\frac{\partial v}{\partial\xi}\right)+\frac{1}{r^2}\left(q8\frac{\partial u}{\partial\eta}+q9\frac{\partial u}{\partial\zeta}+q7\frac{\partial u}{\partial\xi}\right)-\frac{vJ}{r^2}\right]\\
&\quad+\left.\frac{r\mu q1}{J}\left[\left(q5\frac{\partial w}{\partial\eta}+q6\frac{\partial w}{\partial\zeta}+q4\frac{\partial w}{\partial\xi}\right)+\left(q2\frac{\partial u}{\partial\eta}+q3\frac{\partial u}{\partial\zeta}+q1\frac{\partial u}{\partial\xi}\right)\right]\right)\\
&\quad-\frac{4}{3}\frac{\mu}{rJ}\left(q8\frac{\partial v}{\partial\eta}+q9\frac{\partial v}{\partial\zeta}+q7\frac{\partial v}{\partial\xi}\right)+\frac{2}{3}\frac{\mu_{eff}}{J}\left(q2\frac{\partial w}{\partial\eta}+q3\frac{\partial w}{\partial\zeta}+q1\frac{\partial w}{\partial\xi}\right)\\
&\quad+\frac{2}{3}\frac{\mu}{J}\left(q5\frac{\partial u}{\partial\eta}+q6\frac{\partial u}{\partial\zeta}+q4\frac{\partial u}{\partial\xi}\right)-\frac{4}{3}\frac{\mu_{eff}u}{rJ}+\frac{\rho v^2}{J}
\end{aligned}
\tag{2.54}
$$

In an effort to make Equation (2.54) more compact, the following have been defined

$$q1 = r_\eta \theta_\zeta - r_\zeta \theta_\eta; \quad q2 = r_\zeta \theta_\xi - r_\xi \theta_\zeta; \quad q3 = r_\xi \theta_\eta - r_\eta \theta_\xi,$$
$$q4 = \theta_\eta z_\zeta - \theta_\zeta z_\eta; \quad q5 = \theta_\zeta z_\xi - \theta_\xi z_\zeta; \quad q6 = \theta_\xi z_\eta - \theta_\eta z_\xi,$$
$$q7 = r_\zeta z_\eta - r_\eta z_\zeta; \quad q8 = r_\xi z_\zeta - r_\zeta z_\xi; \quad q9 = r_\eta z_\xi - r_\xi z_\eta.$$

If wished, further definitions enable Equation (2.54) to be reduced to

$$
\begin{aligned}
&\frac{\partial}{\partial \eta}(\rho r u U) + \frac{\partial}{\partial \zeta}(\rho r u V) + \frac{\partial}{\partial \xi}(\rho r u W) \\
&= -rq5\frac{\partial p}{\partial \eta} - rq6\frac{\partial p}{\partial \zeta} - rq4\frac{\partial p}{\partial \xi} \\
&\quad + \frac{\partial}{\partial \eta}\left(\mu r\left[\left(Q2 + \frac{1}{3}\frac{q5^2}{J}\right)\frac{\partial u}{\partial \eta} + \left(Q6 + \frac{1}{3}\frac{q5q6}{J}\right)\frac{\partial u}{\partial \zeta} + \left(Q4 + \frac{1}{3}\frac{q4q5}{J}\right)\frac{\partial u}{\partial \xi}\right.\right. \\
&\quad + \frac{1}{3}\frac{q5q8}{rJ}\frac{\partial v}{\partial \eta} + \left(\frac{q6q8}{rJ} - \frac{2}{3}\frac{q5q9}{rJ}\right)\frac{\partial v}{\partial \zeta} + \left(\frac{q4q8}{rJ} - \frac{2}{3}\frac{q5q7}{rJ}\right)\frac{\partial v}{\partial \xi} \\
&\quad \left.\left. + \frac{1}{3}\frac{q2q5}{J}\frac{\partial w}{\partial \eta} + \left(\frac{q2q6}{J} - \frac{2}{3}\frac{q3q5}{J}\right)\frac{\partial w}{\partial \zeta} + \left(\frac{q2q4}{J} - \frac{2}{3}\frac{q1q5}{J}\right)\frac{\partial w}{\partial \xi}\right]\right) \\
&\quad + \frac{\partial}{\partial \zeta}\left(\mu r\left[\left(Q6 + \frac{1}{3}\frac{q5q6}{J}\right)\frac{\partial u}{\partial \eta} + \left(Q3 + \frac{1}{3}\frac{q6^2}{J}\right)\frac{\partial u}{\partial \zeta} + \left(Q5 + \frac{1}{3}\frac{q4q6}{J}\right)\frac{\partial u}{\partial \xi}\right.\right. \\
&\quad + \left(\frac{q5q9}{rJ} - \frac{2}{3}\frac{q6q8}{rJ}\right)\frac{\partial v}{\partial \eta} + \frac{1}{3}\frac{q6q9}{rJ}\frac{\partial v}{\partial \zeta} + \left(\frac{q4q9}{rJ} - \frac{2}{3}\frac{q6q7}{rJ}\right)\frac{\partial v}{\partial \xi} \\
&\quad \left.\left. + \left(\frac{q3q5}{J} - \frac{2}{3}\frac{q2q6}{J}\right)\frac{\partial w}{\partial \eta} + \frac{1}{3}\frac{q3q6}{J}\frac{\partial w}{\partial \zeta} + \left(\frac{q3q4}{J} - \frac{2}{3}\frac{q1q6}{J}\right)\frac{\partial w}{\partial \xi}\right]\right) \\
&\quad + \frac{\partial}{\partial \xi}\left(\mu r\left[\left(Q4 + \frac{1}{3}\frac{q4q5}{J}\right)\frac{\partial u}{\partial \eta} + \left(Q5 + \frac{1}{3}\frac{q4q6}{J}\right)\frac{\partial u}{\partial \zeta} + \left(Q1 + \frac{1}{3}\frac{q4^2}{J}\right)\frac{\partial u}{\partial \xi}\right.\right. \\
&\quad + \left(\frac{q5q7}{rJ} - \frac{2}{3}\frac{q4q8}{rJ}\right)\frac{\partial v}{\partial \eta} + \left(\frac{q6q7}{rJ} - \frac{2}{3}\frac{q4q9}{rJ}\right)\frac{\partial v}{\partial \zeta} + \frac{1}{3}\frac{q4q7}{rJ}\frac{\partial v}{\partial \xi} \\
&\quad \left.\left. + \left(\frac{q1q5}{J} - \frac{2}{3}\frac{q2q4}{J}\right)\frac{\partial w}{\partial \eta} + \left(\frac{q1q6}{J} - \frac{2}{3}\frac{q3q4}{J}\right)\frac{\partial w}{\partial \zeta} + \frac{1}{3}\frac{q1q4}{J}\frac{\partial w}{\partial \xi}\right]\right) \\
&\quad - \frac{2}{3}\left(q5\frac{\partial}{\partial \eta}(\mu u) + q6\frac{\partial}{\partial \zeta}(\mu u) + q4\frac{\partial}{\partial \xi}(\mu u)\right) \\
&\quad - q8\frac{\partial}{\partial \eta}\left(\frac{\mu v}{r}\right) - q9\frac{\partial}{\partial \zeta}\left(\frac{\mu v}{r}\right) - q7\frac{\partial}{\partial \xi}\left(\frac{\mu v}{r}\right) - \frac{4}{3}\frac{\mu}{r}\left(q8\frac{\partial v}{\partial \eta} + q9\frac{\partial v}{\partial \zeta} + q7\frac{\partial v}{\partial \xi}\right) \\
&\quad + \frac{2}{3}\mu\left(q2\frac{\partial w}{\partial \eta} + q3\frac{\partial w}{\partial \zeta} + q1\frac{\partial w}{\partial \xi} + q5\frac{\partial u}{\partial \eta} + q6\frac{\partial u}{\partial \zeta} + q4\frac{\partial u}{\partial \xi}\right) - \frac{4}{3}\frac{\mu u}{r} + \rho v^2
\end{aligned}
\tag{2.55}
$$

where U, V and W are velocity components aligned with the η, ζ, ξ coordinate lines (note that u, v and w are aligned with r, θ, z)

$$U = uq5 + \frac{v}{r}q8 + wq2,$$
$$V = uq6 + \frac{v}{r}q9 + wq3,$$
$$W = uq4 + \frac{v}{r}q7 + wq1$$

Early CFD codes directly solved for U, V and W, but results showed excessive grid sensitivity. Hence nowadays codes tend to solve for u, v and w. The following definitions are also used in Equation (2.55)

$$Q1 = \frac{1}{J}\left(q1^2 + q4^2 + \frac{q7^2}{r^2}\right); \quad Q4 = \frac{1}{J}\left(q1q2 + q4q5 + \frac{q7q8}{r^2}\right),$$
$$Q2 = \frac{1}{J}\left(q2^2 + q5^2 + \frac{q8^2}{r^2}\right); \quad Q5 = \frac{1}{J}\left(q1q3 + q4q6 + \frac{q7q9}{r^2}\right),$$
$$Q3 = \frac{1}{J}\left(q3^2 + q6^2 + \frac{q9^2}{r^2}\right); \quad Q6 = \frac{1}{J}\left(q2q3 + q5q6 + \frac{q8q9}{r^2}\right)$$

The transformed radial and tangential momentum equations, along with continuity and energy, can be found in Rayner (1993). The reduction to the Cartesian form is again, in essence, straightforward. Note, in a CFD solver, the above expanded out form of the curvilinear equations would not typically be used. They are provided to show the scale in the increase in the underlying complexity of the equations that need to be solved in the curvilinear coordinate system. This increased complexity enables the equations to be solved more simply for complex geometries. This aspect is discussed further in Chapters 3 and 4.

2.3.5 *Equations in a Rotating Frame of Reference*

The equations, in cylindrical polar or related coordinates naturally have centrifugal and Coriolis force terms. Hence, they can naturally deal with systems rotating about an axis at $r = 0$ and the complex flow physics that can result from these complex body force terms. The impact of these can be seen in Figure 2.10 (and type of flow related to that in Figure 2.8), which shows the flow in a rotating system with a temperature difference imposed between the inner and outer surfaces. A complex meandering stream arises. A range of other complex flow patterns can occur depending on the speed of rotation and the temperature difference. The equations of motion can either be written in a rotating or stationary frame of reference. The equations (2.47) to (2.55) are written for a stationary coordinate system. The transformation to the rotating frame is trivial. For rotation about the z axis, at $r = 0$, with an angular velocity Ω_g, the following transformations can be used

$$u = u_{stat},\ v = v_{stat} - \Omega_g r,\ w = w_{stat} \tag{2.56}$$

Figure 2.10. Qualitative comparison between prediction and experimental flow visualisation for a baroclinic instability in a dense fluid: (a) predicted velocity vectors; (b) experimental flow visualisation and (c) experiment visualisation at a later time. (Published, with permission, from Tucker and Long (1996).

Note that in Equation (2.56) the subscript 'stat' indicates a variable in the stationary frame. The angular velocity of the coordinate system (or grid in CFD terms), Ω_g, does not need to be the same as that of the actual system Ω. The substitution of Equation (2.56) into Equations (2.47)–(2.50) yields the transformation. The result is that the continuity and axial momentum equations do not change. However, the source terms in the radial and tangential momentum equations become

$$
\begin{aligned}
S_u = & -\frac{\partial p}{\partial r} - \frac{2\mu u}{r^2} + \frac{\rho v^2}{r} + u\frac{\partial}{\partial r}\left(\frac{\lambda}{r}\right) - \frac{2\mu}{r^2}\frac{\partial v}{\partial \theta} \\
& + \frac{\partial}{\partial r}\left(\frac{\lambda}{r}\frac{\partial v}{\partial \theta}\right) + \frac{\partial}{\partial \theta}\left(\mu\frac{\partial}{\partial r}\left(\frac{v}{r}\right)\right) + \frac{\partial}{\partial r}\left(\lambda\frac{\partial w}{\partial z}\right) \\
& + \frac{\partial}{\partial z}\left(\mu\frac{\partial w}{\partial r}\right) + \underbrace{2\rho v\Omega_g + \Omega_f^2 r}_{\text{New Terms}}\rho g_r \\
S_v = & -\frac{1}{r}\frac{\partial p}{\partial \theta} - \frac{\mu v}{r^2} - \frac{\rho u v}{r} - \frac{v}{r}\frac{\partial \mu}{\partial r} + \frac{2}{r^2}\frac{\partial}{\partial \theta}(\mu u) \\
& + \frac{1}{r^2}\frac{\partial}{\partial r}\left(\mu r\frac{\partial u}{\partial \theta}\right) + \frac{1}{r^2}\frac{\partial}{\partial \theta}\left(\lambda\frac{\partial}{\partial r}(ru)\right) + \frac{1}{r}\frac{\partial}{\partial \theta}\left(\lambda\frac{\partial w}{\partial z}\right) \\
& + \frac{1}{r}\frac{\partial}{\partial z}\left(\mu\frac{\partial w}{\partial \theta}\right) \underbrace{-2\rho u\Omega_g}_{\text{New Terms}} + \rho g_\theta
\end{aligned}
\tag{2.57}
$$

For the equations in a Cartesian coordinate system, the rotational force terms must be explicitly incorporated. For this, the following discussion closely aligns with that of Sewall (2005). The particle, p, shown in Figure 2.11, is initially located at the moving point, l. In the time period, Δt, the particle, p, has moved to p' where

$$pp' = ll' + l'p' \tag{2.58}$$

Note also that l has moved to l' in Δt and pp' gives the change in location in particle position in the absolute frame. Furthermore, $l'p'$ gives the change in particle position relative to the moving frame. Also, the change in position of the initial coordinate point location, l, is given by

$$ll' = (\mathbf{\Omega} \times \mathbf{r})\,\Delta t \tag{2.59}$$

where $\mathbf{r}$ is a position vector and $\mathbf{\Omega}$ the angular velocity of the system. The velocity vector of particle p is given by

$$\mathbf{u} = \lim_{\Delta t \to 0}\left(\frac{pp'}{\Delta t}\right) = \lim_{\Delta t \to 0}\left(\frac{l'p'}{\Delta t}\right) + (\mathbf{\Omega} \times \mathbf{r}) \tag{2.60}$$

This can be re-expressed as

$$\mathbf{u} = \underbrace{\frac{d\mathbf{r}}{dt}}_{\text{In fixed frame}} = \underbrace{\overbrace{\frac{\partial \mathbf{r}}{\partial t}}^{\text{Already accounted for}}}_{\text{In rotating frame, Origin at O}} + (\mathbf{\Omega} \times \mathbf{r}) \tag{2.61}$$

Note that Equation (2.61) is general and applicable to other vectors $\mathbf{\phi}$

$$\frac{d\mathbf{\phi}}{dt} = \left[\frac{\partial}{\partial t} + \mathbf{\Omega}\times\right]\mathbf{\phi} \tag{2.62}$$

Hence, if $\mathbf{\phi} = \mathbf{u}$, the particle acceleration can be expressed as

$$\mathbf{a} = \frac{d\mathbf{u}}{dt} = \frac{\partial \mathbf{u}}{\partial t} + \mathbf{\Omega} \times \mathbf{u} \tag{2.63}$$

Substituting Equation (2.61) into Equation (2.63), we can write the particle acceleration after using the product rule of differentiation as

$$\mathbf{a} = \frac{\partial^2 \mathbf{r}}{\partial t^2} + \underbrace{2\left(\mathbf{\Omega} \times \frac{\partial \mathbf{r}}{\partial t}\right)}_{\text{Coriolis Acceleration}} + \underbrace{\mathbf{\Omega} \times (\mathbf{\Omega} \times \mathbf{r})}_{\text{Centrifugal Acceleration}} + \underbrace{\frac{\partial \mathbf{\Omega}}{\partial t} \times \mathbf{r}}_{\text{Non-uniform coordinate system acceleration}} \tag{2.64}$$

Hence, if one is solving equations in a stationary Cartesian coordinate system and wishes to transform them to a rotating system, where the angular velocity is constant with time, one needs only to consider the following source term, scaled by density to be consistent with the forms of the Navier-Stokes equations (2.36)–(2.38)

$$\mathbf{S} = 2\rho\left(\mathbf{\Omega} \times \frac{\partial \mathbf{r}}{\partial t}\right) + \rho\mathbf{\Omega} \times (\mathbf{\Omega} \times \mathbf{r}) \tag{2.65}$$

Next, the simplified example given in Figure 2.12 will be considered relating to a 180° bend system in rotation. The Figure 2.12 system rotates about an offset r_0. Hence,

$$\mathbf{r} = (r_0 + x)\,\mathbf{i} + y\mathbf{j} + z\mathbf{k} \tag{2.66}$$

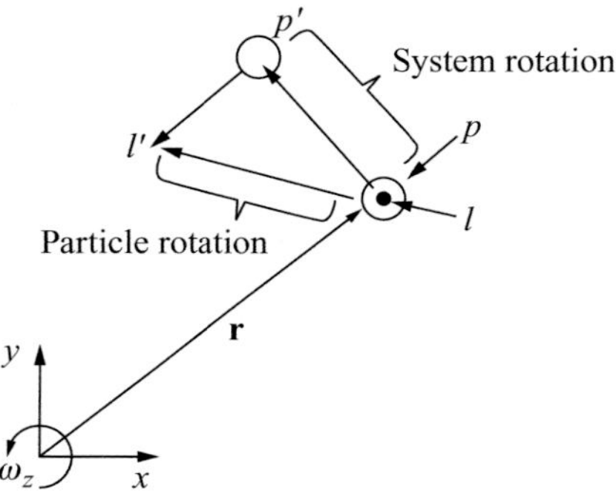

Figure 2.11. Transformation of equations in a Cartesian stationary system to a rotating frame of reference (adapted from Sewall 2005).

The velocity, now in the rotating frame, is

$$\mathbf{u} = \frac{\partial \mathbf{r}}{\partial t} = u\mathbf{i} + v\mathbf{j} + w\mathbf{k} \tag{2.67}$$

Since $\Omega_x = \Omega_y = 0$

$$\Omega = \Omega_z \mathbf{k} \tag{2.68}$$

Substituting Equations (2.66), (2.67) and (2.68) into Equation (2.64) (or Equation (2.65)) gives

$$\mathbf{a} = \left(\underbrace{-2\Omega_z v}_{\substack{\text{Coriolis}\\ \text{acceleration}}} \; - \underbrace{\Omega_z^2 (r_0 + x)}_{\substack{\text{Centrifugal}\\ \text{acceleration}}} \right) \mathbf{i} + \left(\underbrace{2\Omega_z u}_{\substack{\text{Coriolis}\\ \text{acceleration}}} \; - \underbrace{\Omega_z^2 y}_{\substack{\text{Centrifugal}\\ \text{acceleration}}} \right) \mathbf{j} \tag{2.69}$$

Hence in Equation (2.69) the Coriolis acceleration comprises $-2\Omega_z v\mathbf{i} + 2\Omega_z u\mathbf{j}$ and the centrifugal $\mathbf{a} = \Omega_z^2(r_0 + x)\mathbf{i} + \Omega_z^2 y\mathbf{j}$. Equation (2.69) implies the following force/source terms

$$S_u = 2\rho\Omega_z v + \rho\Omega_z^2(r_0 + x) \tag{2.70}$$

$$S_v = -2\rho\Omega_z u + \rho\Omega_z^2 y \tag{2.71}$$

Typically, for an aeroengine, for which the ribbed passage geometry shown in Figure 2.12 is intended to be an idealization for $r_0 \gg y$. Hence, it follows that

$$S_v = -2\rho\Omega_z u \tag{2.72}$$

The Boussinesq approximation will be discussed in more detail, but based on this, the Equation (2.70) second right-hand side term can be expressed as

$$\rho\Omega_z^2(r_0 + x) = (\rho_{ref} + \Delta\rho)\Omega_z^2(r_0 + x) \tag{2.73}$$

In Equation (2.73), ρ_{ref} is the reference density and $\Delta\rho \simeq -\beta\rho_{ref}(T - T_{ref})$ where T_{ref} is a reference temperature and $\beta = 1/T_{ref}$. Hence, the equation can be rewritten as

$$\rho\Omega_z^2(r_0 + x) = \rho_{ref}(1 - \beta(T - T_{ref}))\Omega_z^2(r_0 + x) \tag{2.74}$$

Following Sewall, when the Boussinesq approximation is used, $\rho = \rho_{ref}$ in the Coriolis terms in Equations (2.70) and (2.71). Note that Equation (2.74) can be split as $\rho_{ref}\Omega_z^2(r_0 + x) - \rho_{ref}\beta(T - T_{ref})\Omega_z^2(r_0 + x)$ and the first group of terms absorbed into the pressure, as discussed in the next section.

2.3.5.1 Reduced Pressures Sometimes it can be important to solve for reduced or modified pressures. For rotating systems, the reduced pressure

$$p = p_{stat} - \frac{1}{2}\rho\Omega_g^2 r^2 \tag{2.75}$$

can be defined. The radial momentum source term for the governing equations in cylindrical polar coordinates then becomes

$$\begin{aligned} S_u = &-\frac{\partial p}{\partial r} - \frac{2\mu u}{r^2} + \frac{\rho v^2}{r} + u\frac{\partial}{\partial r}\left(\frac{\lambda}{r}\right) - \frac{2\mu}{r^2}\frac{\partial v}{\partial \theta} + \frac{\partial}{\partial r}\left(\frac{\lambda}{r}\frac{\partial v}{\partial \theta}\right) \\ &+ \frac{\partial}{\partial \theta}\left(\mu\frac{\partial}{\partial r}\left(\frac{v}{r}\right)\right) + \frac{\partial}{\partial r}\left(\lambda\frac{\partial w}{\partial z}\right) + \frac{\partial}{\partial z}\left(\mu\frac{\partial w}{\partial r}\right) + \underbrace{2\rho v\Omega_g}_{\text{New Terms}} + \rho g_r \end{aligned} \tag{2.76}$$

Notably, for flows involving dense and deep fluids, it can also be helpful to axially reduce the pressure using

$$p = p^{old} - \rho g_z z \tag{2.77}$$

where p^{old} is the old non-reduced pressure. Then in Equation (2.49) for the axial momentum the source term now becomes

$$\begin{aligned} S_w = &-\frac{\partial p}{\partial z} + \frac{1}{r}\frac{\partial}{\partial r}\left(\mu r\frac{\partial u}{\partial z}\right) + \frac{1}{r}\frac{\partial}{\partial z}\left(\lambda\frac{\partial}{\partial r}(ru)\right) \\ &+ \frac{1}{r}\frac{\partial}{\partial \theta}\left(\mu\frac{\partial v}{\partial z}\right) + \frac{1}{r}\frac{\partial}{\partial z}\left(\lambda\frac{\partial v}{\partial \theta}\right) \end{aligned} \tag{2.78}$$

Therefore, the gravitational term vanishes, absorbed in the pressure. The prediction of the baroclinic instability flow in the rotating dense fluid shown in Figure 2.10 needs both the rotationally and gravitationally reduced pressure to keep machine round-off errors from contaminating solutions. The flow has both cyclones (vortices rotating in the same direction as the system rotation) and anti-cyclones (vortices rotating in the opposite direction to the system rotation). Between these vortices is a stream analogous to the 'jet streams' found in the earth's atmosphere. Note that, as can be seen from the two experimental frames in the figure, the actual flow has a different angular velocity to that of the system. The flow is not dissimilar to that in the high-pressure compressor drum discussed earlier in terms of its topology and the difference between the system and flow angular velocities.

Looking at the impact of the choice of coordinate system, Figure 2.13 shows the predicted flow field, represented using the stream function. The main figure shows

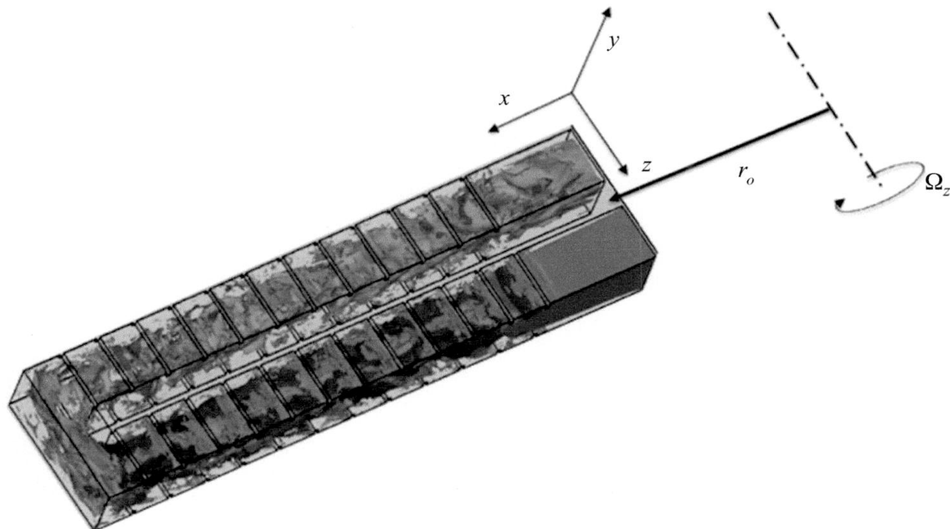

Figure 2.12. Schematic of idealized rotating ribbed passage representative of a passage found in the turbine of a gas turbine aeroengine.

the flow in a rotating frame of reference and the upper right-hand inset shows the flow in the stationary frame. The numerical errors in the stationary frame (false diffusion – see Chapter 4, Section 4.2.6.3) wrongly renders the flow axisymmetric.

Sometimes, mostly for computational convenience, flows are assumed periodic. For example, a fully developed channel flow could be modelled using just a section of channel with periodic streamwise boundary conditions. However, the pressure field will not be periodic since a pressure gradient is needed to drive the flow. To

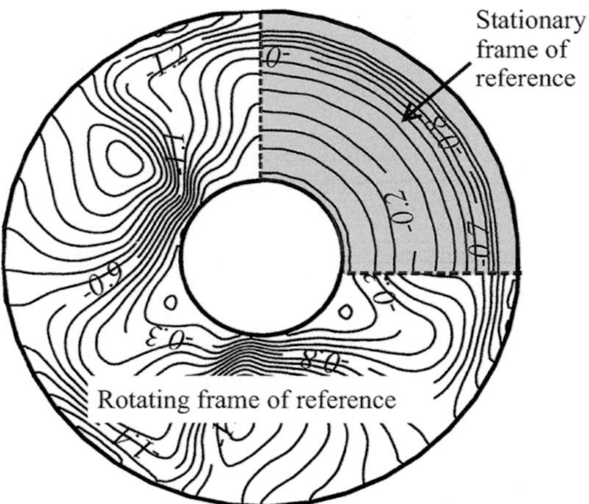

Figure 2.13. Predicted dimensionless stream function field for baroclinic instability flow in rotating and stationary frames. (adapted from Tucker and Long 1996).

make the pressure field periodic in, for example, x, the following equation could be used

$$p = p^{old} - \rho f_x x \tag{2.79}$$

where p^{old} is the non-periodic pressure field and f_x is a flow forcing term that will drive the flow.

2.4 Energy and Temperature Equations

For heat transfer problems and high-speed compressible flows it can be necessary to consider temperature or energy. The convection-diffusion transport equations (2.36) and (2.47), as expressed for the general variable, ϕ, can be directly solved for temperature where $\phi = T$. The diffusion coefficients then become

$$\Gamma_{Tr} = k/c_p, \Gamma_{T\theta} = k/c_p, \Gamma_{Tz} = k/c_p \tag{2.80}$$

where k is thermal conductivity and c_p is the specific heat capacity. If there is substantial frictional heating then the following source terms are needed in Cartesian

$$\begin{aligned} S_T = {} & \frac{2\mu}{c_p}\left[\left(\frac{\partial v}{\partial y}\right)^2 + \left(\frac{\partial w}{\partial z}\right)^2 + \left(\frac{\partial u}{\partial x}\right)^2\right] \\ & + \frac{\mu}{c_p}\left[\left(\frac{\partial v}{\partial x} + \frac{\partial u}{\partial y}\right)^2 + \left(\frac{\partial w}{\partial x} + \frac{\partial u}{\partial z}\right)^2 + \left(\frac{\partial v}{\partial z} + \frac{\partial w}{\partial y}\right)^2\right] \\ & + \frac{\lambda}{c_p}\left[\left(\frac{\partial v}{\partial y} + \frac{\partial w}{\partial z} + \frac{\partial u}{\partial x}\right)^2\right] \end{aligned} \tag{2.81}$$

and cylindrical polar coordinates

$$\begin{aligned} S_T = {} & \frac{2\mu}{c_p}\left[\left(\frac{\partial u}{\partial r}\right)^2 + \left(\frac{1}{r}\frac{\partial v}{\partial \theta} + \frac{u}{r}\right)^2 + \left(\frac{\partial w}{\partial z}\right)^2\right] \\ & + \frac{\mu}{c_p}\left[\left(\frac{\partial u}{\partial z} + \frac{\partial w}{\partial r}\right)^2 + \left(\frac{\partial v}{\partial z} + \frac{1}{r}\frac{\partial w}{\partial \theta}\right)^2 + \left(\frac{1}{r}\frac{\partial u}{\partial \theta} + \frac{\partial v}{\partial r} - \frac{v}{r}\right)^2\right] \\ & + \frac{\lambda}{c_p}\left[\left(\frac{1}{r}\frac{\partial}{\partial r}(ru) + \frac{1}{r}\frac{\partial v}{\partial \theta} + \frac{\partial w}{\partial z}\right)^2\right] \end{aligned} \tag{2.82}$$

Corresponding source terms are needed for compressive heating (see Section 2.6.3). However, for higher speed flows it is generally considered better to solve for stagnation enthalpy

$$\phi = h_0 = c_p T + \frac{1}{2}(u^2 + v^2 + w^2) \tag{2.83}$$

For a two-dimensional type of flow in cylindrical polar coordinates (axisymmetric where all $\partial/\partial\theta$ terms are zero – see Section 2.8.1), the source term takes the following form

$$S_{\phi,h_0} = \frac{\partial}{\partial x}(\mu - \Gamma_h)\frac{\partial}{\partial x}\frac{1}{2}(u^2 + v^2 + w^2) + \frac{1}{r}\frac{\partial}{\partial r}(\mu - \Gamma_h)\frac{\partial}{\partial r}\frac{1}{2}(u^2 + v^2 + w^2)$$
$$+ \frac{\partial}{\partial x}\left[\mu\frac{\partial}{\partial x}\frac{1}{2}(w^2) - \frac{2}{3}\mu w\left(\frac{\partial w}{\partial x} + \frac{\partial u}{\partial r} + \frac{u}{r}\right) + \mu_{eff} v\frac{\partial w}{\partial r}\right]$$
$$+ \frac{1}{r}\frac{\partial}{\partial r}\left[(r\mu)\frac{\partial}{\partial r}\frac{1}{2}(u^2) - \frac{2}{3}\mu u\left(\frac{\partial w}{\partial x} + \frac{\partial u}{\partial r} + \frac{u}{r}\right) + \mu u\frac{\partial u}{\partial x} - \mu\frac{v^2}{r}\right] \quad (2.84)$$

The transformation to Cartesian coordinates is readily achieved using the data in Table 2.1. Also, the extension to three-dimensions is straightforward. Forms of the enthalpy equation are outlined further later.

2.5 Turbulent Flow Equations

For turbulent flow, the substitution $\phi = \Phi + \phi'$ must be made in the Navier-Stokes equations. They must then be time averaged to yield equations for the mean velocity, Φ. The average level of the fluctuations, ϕ', must be modelled. This is mostly achieved using the Boussinesq approximation. The whole process, which yields the so-called Reynolds Averaged Navier-Stokes (RANS) equations, is outlined in Chapter 5. However, it is sufficient to say here that when the Boussinesq approximation (not to be confused with the other Boussinesq approximation noted earlier and to be discussed later for buoyancy) is used, the viscosity in the Navier-Stokes equations is replaced by the effective viscosity, $\mu_{eff} = \mu + \mu_t$, where μ_t is called the eddy viscosity. For turbulent flow, alterations to variables being solved for, source terms, diffusion terms and the viscosity given in Table 2.2 must be made to the equations (2.36)–(2.49) relating to the general variable, ϕ. Note that k is now the turbulence kinetic energy. This is discussed further in Chapter 5. For Reynolds stress turbulence models, the situation is very different and extended source terms are needed. This is also discussed further in Chapter 5. Note that for pressure generally the $2\rho k/3$ is 'ignored', it being effectively lumped in with a modified pressure. If this is the case, then one wishes to recover the true pressure, as say a post processing operation, then this kinetic energy based component must be later subtracted off.

2.6 Substantial Derivative and Convective Term Forms

2.6.1 *The Substantial Derivative*

When computing flows, the nature of the substantial derivative needs to be appreciated. For a Cartesian coordinate system, this is defined as (strictly, for the substantial derivative, the vector $\boldsymbol{U}$ should be outside the dot product)

$$\left(\frac{\partial\phi}{\partial t}\right)_{fp} = \left(\frac{\partial\phi}{\partial t}\right)_g + \nabla\cdot(\phi\mathbf{U}) = \left(\frac{\partial\phi}{\partial t}\right)_g + u\frac{\partial\phi}{\partial x} + v\frac{\partial\phi}{\partial y} + w\frac{\partial\phi}{\partial z} \quad (2.85)$$

Table 2.2. *Alteration of equations for turbulent flow*

	ϕ	Diffusion coefficients	Source terms	Viscosity μ
Momentum	–	Altered by viscosity substitution given in final column	Pressure is directly replaced by $p + \frac{2\rho k}{3}$, implying additional source term derivatives	$\mu_{eff} = \mu + \mu_t$
Temperature	–	$\Gamma_{T,eff} = \frac{\mu}{\mathrm{Pr}} + \frac{\mu_t}{\mathrm{Pr}_t}$	–	$\mu_{eff} = \mu + \mu_t$
Stagnation enthalpy	$c_p T + \frac{1}{2}(u^2 + v^2 + w^2) + k$	$\Gamma_{h,eff} = \frac{\mu}{\mathrm{Pr}} + \frac{\mu_t}{\mathrm{Pr}_t}$, $\Gamma_{k,eff} = \frac{\mu}{Sc} + \frac{\mu_t}{Sc_t}$	$S_{\phi,h} = S_{\phi,h,l} + \frac{\partial}{\partial x}\left[\frac{2}{3}\rho k w\right]$ $+\frac{1}{r}\frac{\partial}{\partial r}\left[\frac{2}{3}\rho k u\right]$ $+\frac{\partial}{\partial x}(\Gamma_{k,eff} - \Gamma_{h,eff})\frac{\partial k}{\partial x}$ $+\frac{1}{r}\frac{\partial}{\partial r}(\Gamma_{k,eff} - \Gamma_{h,eff})\frac{\partial k}{\partial r}$	$\mu_{eff} = \mu + \mu_t$
General variable (say concentration – c)	–	$\Gamma_{c,eff} = \frac{\mu}{Sc} + \frac{\mu_t}{Sc_c}$	–	$\mu_{eff} = \mu + \mu_t$

The substantial derivative, $(\partial\phi/\partial t)_{fp}$, is more usually expressed as $D\phi/Dt$. However, here, following Anderson (1995), the subscript *fp* is used, enforcing its meaning, this being the time rate of change in ϕ when a particular fluid particle is locked onto and followed. The $(\partial\phi/\partial t)_g$ term is the time rate of change at a fixed coordinate location. In a CFD context, this is the temporal change of ϕ relative to the grid. Generally, for flows classified as unsteady, it is meant $(\partial\phi/\partial t)_g \neq 0$. Even for steady flows, $(\partial\phi/\partial t)_{fp} \neq 0$.

For certain systems (classified as exhibiting unsteady flow), the fluid structure may stay exactly or much the same, but simply translate (at a velocity U_g) or rotate (at an angular velocity Ω_g) in space relative to the coordinate system/grid. The mechanism for this translation or rotation could be fluid dynamic in origin or alternatively the result of boundary movement. When viewed in a moving frame of reference, the flow can be made steady ($(\partial\phi/\partial t)_g = 0$).

Essentially, the substantial derivative is the left-hand side, as traditionally presented, of the Navier-Stokes equations. These are Galilean invariant (i.e., in a moving frame of reference their form does not change). Hence, when solved in a moving frame, the form of the substantial derivative also does not change. However, the convective velocities must be defined relative to the moving frame of reference. Therefore, for a coordinate system moving just horizontally at U_g, the Cartesian substantial derivative becomes

$$\left(\frac{\partial\phi}{\partial t}\right)_{fp} = \left(\frac{\partial\phi}{\partial t}\right)_g + (u - U_g)\frac{\partial\phi}{\partial x} + v\frac{\partial\phi}{\partial y} + w\frac{\partial\phi}{\partial z} \tag{2.86}$$

For cylindrical rotating systems, it is convenient to consider the following substantial derivative form

$$\left(\frac{\partial\phi}{\partial t}\right)_{fp} = \left(\frac{\partial\phi}{\partial t}\right)_g + (v - \Omega_g r)\frac{1}{r}\frac{\partial\phi}{\partial\theta} + \frac{u}{r}\frac{\partial r\phi}{\partial r} + w\frac{\partial\phi}{\partial z} \tag{2.87}$$

where u, v and w are the velocities in r, θ and z directions respectively, and Ω_g is the grid/coordinate system angular velocity. Figure 2.8 gives an example of a CFD simulation for an idealized aeroengine high-pressure compressor drum. Frame (c) gives contours of the time derivative $(\partial\phi/\partial t)_g$. The left frame gives contours for the coordinate system rotating with the same angular velocity as the system's solid walls. The right frame coordinate system is rotating with the main body of the fluid, which largely stays structurally the same but temporally drifts in an angular sense. Solving in the latter frame makes the flow more steady. The same is true for the flow in Figure 2.10. Equations (2.86) and (2.87) can be expressed more generally as

$$\left(\frac{\partial\phi}{\partial t}\right)_{fp} = \left(\frac{\partial\phi}{\partial t}\right)_g + (\mathbf{u} - \mathbf{U}_g)\cdot\nabla\phi \tag{2.88}$$

Incidentally, some equations related to CFD, such as those for particle transport and the vortex method (see Khatir 2000), can be solved in a completely Lagrangian

Table 2.3. *Various forms of convective term forms in the Navier-Stokes equations – from Tucker (2013), published with the kind permission of Springer*

Title	Tensor form	Source
Non-conservation	$u_j \dfrac{\partial u_i}{\partial x_j}$	–
Conservation	$\dfrac{\partial (u_j u_j)}{\partial x_i}$	–
Skew-symmetric	$\dfrac{1}{2}\left[\dfrac{\partial (u_j u_j)}{\partial x_i} + u_j \dfrac{\partial u_i}{\partial x_j}\right]$	Arakawa (1966)
Rotational form	$u_j \left(\dfrac{\partial u_i}{\partial x_j} - \dfrac{\partial u_j}{\partial x_i}\right) + \dfrac{1}{2}\dfrac{\partial (u_j u_j)}{\partial x_i}$	Deardorff (1970)

frame which follows fluid points. The 'substantial derivative' then appears as

$$\left(\frac{\partial \phi}{\partial t}\right)_{fp} = \left(\frac{\partial \phi}{\partial t}\right)_g \tag{2.89}$$

This can be appreciated for a flow with $u = w = 0$. When $u = U_g$ (see Equation (2.86)), only the $(\partial\phi/\partial t)_g$ term remains in Equation (2.85) and unsteady solution techniques must be used.

2.6.2 *Different Convective Term Forms*

The flow-governing differential equations can be converted into algebraic equations that can be typed into a computer program. This can be done using several approaches. The most popular are the finite-volume, finite-difference and finite-element methods. Of these, the first is the most widely used. Since, essentially, the Navier-Stokes equations were derived herein using a finite volume, it is perhaps not surprising that an effective solution approach for them is to reverse the processes to recover algebraic equations. As we have already noted, the same equation can be written in different forms. For example, it was noted that when all the convective term velocity components are inside the convective term derivatives we have what is called the *weak conservation* form of the Navier-Stokes equations. A property of the weak conservation form of the equations is that, in the absence of shock waves, when solved using the finite volume technique the solution will always conserve momentum. When equations of the form of (2.35) are used, this is not the case. Even though the conservation error will decrease with decreasing size of the finite volume, ideally the non-conservation form of the equations should not be used in a CFD solver. Some well-known convective term forms are given in Table 2.3.

It is widely reported that the discretization of skew-symmetric form results (with central differences) in substantially less aliasing error (discussed in Chapter 4). On the other hand, the rotational form is observed to show substantial dissipation in the numerical framework of Horiuti and Itami (1998).

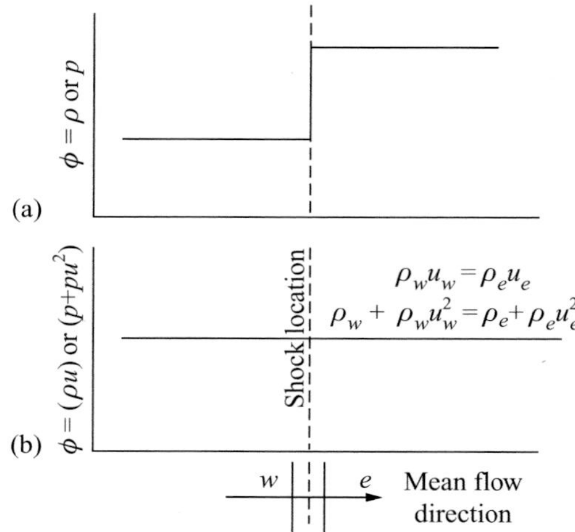

Figure 2.14. Variable distributions through a shock.

2.6.3 *Strong Conservation Form of Equations*

If there are shocks present in the flow, even the weak conservation form of the equations will become problematic. This is because the derivative of the momentum flux, $\rho u_i v_j$, (see Equation 2.40) cannot be defined through a shock feature. Figure 2.14 illustrates variable distributions through a shock. The subscript w locations are to the west or left of the shock. Similarly, the subscript e represents the east or right of the shock. As can be seen from Frame (a), ρ and p are discontinuous through the shock. However, ρu and $p + \rho u^2$ are continuous and so the gradients can be numerically defined. Hence, it is better to formulate the equations in terms of such continuous gradient variables. Therefore, for flows with shocks, it is necessary to use the Navier-Stokes equations in the *strong conservation* form. For high-speed flows, there are significant density changes and high shear stresses. The fluid has significant temperature gradients and it is important to also include the thermal energy equation. In strong conservation form, the continuity, Navier-Stokes and energy equations (having the key role of dealing with thermal energy) can be written together as

$$\frac{\partial U}{\partial t} + \frac{\partial F}{\partial x} + \frac{\partial G}{\partial y} + \frac{\partial H}{\partial z} = S_\phi \tag{2.90}$$

where U, F, G and H are column vectors that have the following form

$$U = \begin{bmatrix} \rho \\ \rho u \\ \rho v \\ \rho w \\ \rho\left(e + |\mathbf{u}|^2/2\right) \end{bmatrix}$$

$$F = \begin{bmatrix} \rho u \\ \rho uu + p + \sigma_x \\ \rho vu - \tau_{xy} \\ \rho wu - \tau_{xz} \\ \rho\left(e + |\mathbf{u}|^2/2\right)u + pu - k\frac{\partial T}{\partial x} - u\sigma_x - v\tau_{xy} - w\tau_{xz} \end{bmatrix} \tag{2.91}$$

$$G = \begin{bmatrix} \rho v \\ \rho uv - \tau_{yx} \\ \rho vv + p - \sigma_y \\ \rho wv - \tau_{yz} \\ \rho\left(e + |\mathbf{u}|^2/2\right)v + pv - k\frac{\partial T}{\partial y} - u\tau_{yx} - v\sigma_y - w\tau_{yz} \end{bmatrix}$$

$$H = \begin{bmatrix} \rho w \\ \rho uw - \tau_{zx} \\ \rho vw - \tau_{yz} \\ \rho ww + p - \sigma_z \\ \rho\left(e + |\mathbf{u}|^2/2\right)w + pw - k\frac{\partial T}{\partial z} - u\tau_{zx} - v\tau_{zy} - w\sigma_z \end{bmatrix}$$

To include body forces and internal heat generation, the following column vector can be included in Equation (2.90)

$$S_\phi = \begin{bmatrix} 0 \\ \rho g_x + ... \\ \rho g_y + ... \\ \rho g_x + ... \\ \rho \mathbf{g}\cdot\mathbf{u} + \dot{q} \end{bmatrix}$$

Note that $\dot{q}$ is a heat source. The first row of the vectors U, F, G and H, when added, expresses mass conservation. The second expresses conservation of momentum in the x direction. Similarly, the third and fourth express momentum conservation in the y and z directions, respectively. The last row expresses conservation of total energy. It includes terms that model compressive and frictional heating. In equations (2.91) e is the fluid's internal energy due to random molecular motion. Note that for the viscous stresses, the convention is that the first subscript refers to the plane on which the stress acts and the second to the direction in which the stress acts. Hence τ_{yx} corresponds to stress gained from Newton's basic viscosity law for, say, Poiseuille flow along a horizontal channel with mean flow in the x direction.

2.6.3.1 Stokes Equations The stresses in Equation (2.91) can be evaluated using the following equations derived by Stokes in 1845. Two-dimensional reduced forms

Table 2.4. *Recovery of full equations from tensors, Equation (2.94)*

i	1	2	3
$j=1$	$\sigma_{11} = \lambda\left[\frac{\partial v_1}{\partial x_1} + \frac{\partial v_2}{\partial x_2} + \frac{\partial v_3}{\partial x_3}\right] + \mu\left(\frac{\partial v_1}{\partial x_1} + \frac{\partial v_1}{\partial x_1}\right)$	$\sigma_{12} = \mu\left(\frac{\partial v_2}{\partial x_1} + \frac{\partial v_1}{\partial x_2}\right)$	$\sigma_{13} = \mu\left(\frac{\partial v_1}{\partial x_3} + \frac{\partial v_3}{\partial x_1}\right)$
$j=2$	Same as 1, 2	$\sigma_{ij} = \lambda\delta_{ij}\frac{\partial v_k}{\partial x_k} + \mu\left(\frac{\partial v_i}{\partial x_j} + \frac{\partial v_j}{\partial x_i}\right)$	$\sigma_{ij} = \lambda\delta_{ij}\frac{\partial v_k}{\partial x_k} + \mu\left(\frac{\partial v_i}{\partial x_j} + \frac{\partial v_j}{\partial x_i}\right)$
$j=3$	Same as 1, 3	Same as 2, 3	$\sigma_{ij} = \lambda\delta_{ij}\frac{\partial v_k}{\partial x_k} + \mu\left(\frac{\partial v_i}{\partial x_j} + \frac{\partial v_j}{\partial x_i}\right)$

of these equations were presented earlier.

$$
\begin{aligned}
\sigma_x &= \lambda\left(\frac{\partial u}{\partial x} + \frac{\partial v}{\partial y} + \frac{\partial w}{\partial z}\right) + 2\mu\frac{\partial u}{\partial x} \\
\sigma_y &= \lambda\left(\frac{\partial u}{\partial x} + \frac{\partial v}{\partial y} + \frac{\partial w}{\partial z}\right) + 2\mu\frac{\partial v}{\partial y} \\
\sigma_z &= \lambda\left(\frac{\partial u}{\partial x} + \frac{\partial v}{\partial y} + \frac{\partial w}{\partial z}\right) + 2\mu\frac{\partial w}{\partial z}
\end{aligned} \tag{2.92}
$$

$$
\begin{aligned}
\tau_{xy} &= \tau_{yx} = \mu\left(\frac{\partial v}{\partial x} + \frac{\partial u}{\partial y}\right) \\
\tau_{yz} &= \tau_{zy} = \mu\left(\frac{\partial w}{\partial y} + \frac{\partial v}{\partial z}\right) \\
\tau_{zx} &= \tau_{xz} = \mu\left(\frac{\partial u}{\partial z} + \frac{\partial w}{\partial x}\right)
\end{aligned} \tag{2.93}
$$

Of course, Equation (2.92) can be more compactly written using vectors to express the bracketed terms as $\nabla \cdot \mathbf{u}$. Alternatively, they can be written in the following tensor form

$$
\sigma_{ij} = \lambda\delta_{ij}\frac{\partial v_k}{\partial x_k} + \mu\left(\frac{\partial v_i}{\partial x_j} + \frac{\partial v_j}{\partial x_i}\right), \quad (i, j, k = 1, 2, 3) \tag{2.94}
$$

and hence written out fully as shown in Table 2.4. Table rows and columns are labelled with the potential i and j values. Equation (2.94) is then copied into each of the table entries. The subscripts are then replaced with their respective table i and j values. Note that since k is a repeated subscript, this is summed from 1 to 3.

When writing a computer program, these tensors can easily be recovered by placing, for example, Equation (2.94) in two 'do loops' (i.e., do $i = 1, 3$ and do $j = 1, 3$). However, this could have greater computational overhead than entering them in directly.

To further write Equation (2.94) in a more compact fashion, the strain rate S_{ij} is defined as

$$
S_{ij} = \frac{1}{2}\left(\frac{\partial v_i}{\partial x_j} + \frac{\partial v_j}{\partial x_i}\right) \tag{2.95}
$$

This allows Equation (2.94) to be re-written as

$$\sigma_{ij} = 2\mu \left[S_{ij} - \frac{1}{3} \frac{\partial v_k}{\partial x_k} \delta_{ij} \right] \tag{2.96}$$

2.7 Real Gas Form of Equations

All of the above equations relate to perfect gasses. However, for real gases, chemical and thermodynamic processes play a role. If the gas is not in chemical equilibrium, but in thermodynamic equilibrium (see Blazek 2001), then the following equations need to be used in combination with Equation (2.90)

$$U = \begin{bmatrix} \rho \\ \rho u \\ \rho v \\ \rho w \\ \rho \left(e + |\mathbf{u}|^2/2 \right) \\ \rho \Phi_1 \\ \vdots \\ \rho \Phi_{N-1} \end{bmatrix} \tag{2.97}$$

$$F = \begin{bmatrix} \rho u \\ \rho u u + p + \sigma_x \\ \rho v u - \tau_{xy} \\ \rho w u - \tau_{xz} \\ \rho \left(e + |\mathbf{u}|^2/2 \right) u + p u - k\frac{\partial T}{\partial x} - u\sigma_x - v\tau_{xy} - w\tau_{xz} + \rho \sum_{n=1}^{N} h_n \Gamma_n \frac{\partial \Phi_n}{\partial x} \\ \rho \Phi_1 u + h_n \Gamma_n \frac{\partial \Phi_n}{\partial x} \\ \vdots \\ \rho \Phi_{N-1} u + h_n \Gamma_{N-1} \frac{\partial \Phi_{N-1}}{\partial x} \end{bmatrix} \tag{2.98}$$

$$G = \begin{bmatrix} \rho v \\ \rho u v - \tau_{yx} \\ \rho v v + p - \sigma_y \\ \rho w v - \tau_{yz} \\ \rho \left(e + |\mathbf{u}|^2/2 \right) v + p v - k\frac{\partial T}{\partial y} - u\tau_{yx} - v\sigma_y - w\tau_{yz} + \rho \sum_{n=1}^{N} h_n \Gamma_n \frac{\partial \Phi_n}{\partial y} \\ \rho \Phi_1 v + h_n \Gamma_n \frac{\partial \Phi_n}{\partial y} \\ \vdots \\ \rho \Phi_{N-1} v + h_n \Gamma_{N-1} \frac{\partial \Phi_{N-1}}{\partial y} \end{bmatrix} \tag{2.99}$$

$$H = \begin{bmatrix} \rho w \\ \rho u w - \tau_{zx} \\ \rho v w - \tau_{yz} \\ \rho w w + p - \sigma_z \\ \rho\left(e + {}^{|\mathbf{u}|^2}\!/_2\right) w + p w - k\frac{\partial T}{\partial z} - u\tau_{zx} - v\tau_{zy} - w\sigma_z + \rho \sum_{n=1}^{N} h_n \Gamma_n \frac{\partial \Phi_n}{\partial z} \\ \rho \Phi_1 w + h_n \Gamma_n \frac{\partial \Phi_n}{\partial z} \\ \vdots \\ \rho \Phi_{N-1} w + h_n \Gamma_{N-1} \frac{\partial \Phi_{N-1}}{\partial z} \end{bmatrix} \tag{2.100}$$

$$S_\phi = \begin{bmatrix} 0 \\ \rho g_x + \ldots. \\ \rho g_y + \ldots. \\ \rho g_x + \ldots. \\ \rho \mathbf{g} \cdot \mathbf{u} + \dot{q} \\ \dot{s}_1 \\ \vdots \\ \dot{s}_{N-1} \end{bmatrix} \tag{2.101}$$

Note that in these equations, ρ is the total density of the mixture and Φ_n is a mass fraction. The total mixture density is equal to the sum of the densities of the separate species $\rho\Phi_n$. Also, Γ_n is the diffusivity of the particular component and h_n is its enthalpy. In addition, $\dot{s}_n$ is the rate of change of species, n, from chemical reactions. This parameter in itself is evaluated via a substantial set of chemical equations outlined in Blazek (2001) and the references therein.

Solution of these real gas equations can be of some importance when modelling steam turbines. When the steam is wet, this presents an even bigger challenge. Then the solution of droplet transport equations is necessary (Chapter 6, Section 6.4.1).

2.8 Reduced Forms of Navier-Stokes Equations

As has been shown, the full Navier-Stokes equations are lengthy and so to reduce cost and complexity simplifications can be made. Such simplifications were especially necessary in the early days of CFD when computing power was limited. The Euler equations as reduced forms of the Navier-Stokes equations were noted previously. Further simplifications based on mathematical and physical considerations are now outlined. Practical applications of these simplified equations are also considered.

2.8.1 *Axisymmetric Form of Equations*

Useful engineering problems can sometimes be solved with the aforementioned cylindrical polar coordinate equations reduced to their axisymmetric form.

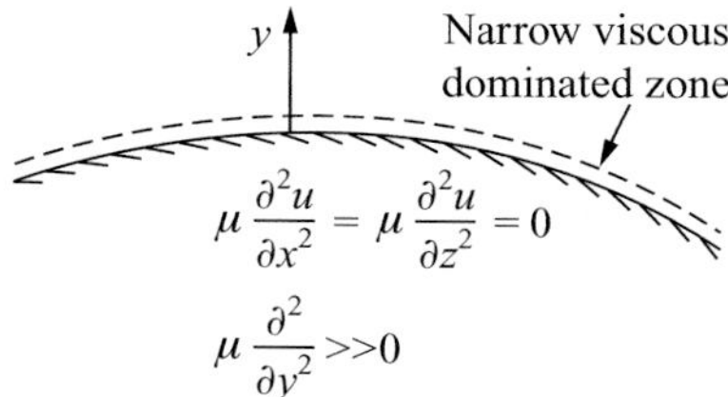

Figure 2.15. Schematic for thin shear layer form of the Navier-Stokes equations.

For an axisymmetric flow, there are no flow variable variations in the θ direction. When this is the case, all derivatives with respect to θ vanish. The flow under such circumstances is in a sense 'two-dimensional.' However, for a rotating system the tangential velocity can be a physically dominant term (it governs the centrifugal force) and then the momentum equations for all three velocity components must be solved for – as one would do for a three-dimensional flow.

2.8.2 *Orthogonal Form of Equations*

For Equation (2.55), for example, if the transformed coordinate system is constructed so that η, ζ, ξ coordinate lines are orthogonal (i.e., at right angles), then the size of the equation drops dramatically to a size comparable to the original non-transformed equation. For example, $\partial/\partial z$ terms in the original, non-transformed equations simply become $\partial/L\partial\xi$, where L is a simple geometric scaling factor. Also, $\partial/\partial\theta = \partial/L\partial\varsigma$, where L is a different scaling factor, and so forth.

2.8.3 *Thin Shear Layer Form of Equations*

To allow further simplification, aerospace solvers can use the equations in the Thin Shear Layer (TSL) form. With this assumption (see Figure 2.15), only derivatives in the wall normal direction are retained in the viscous stress and analogous heat flux terms. This acknowledges the key role of just one component of shear stress in a high Reynolds number ($Re > 1 \times 10^5$) boundary layer type flow, and indeed the wakes, with very little viscous diffusion in the streamwise direction. For high-speed flows, both of these zones tend to be highly localized. An example of a widely used solver using this approximation in the NASA CFL3D solver, see Krist et al. (1998). With appropriate use, highly accurate simulations can result. Equation (2.33) can be expressed in the thin shear layer form as

$$\rho u \frac{\partial \phi}{\partial x} + \rho v \frac{\partial \phi}{\partial y} = \mu \frac{\partial^2 \phi}{\partial y^2} + S_\phi \qquad (2.102)$$

Equation (2.102) assumes that the flow is two-dimensional and directed in the x-coordinate direction. It is worth noting that the thin shear layer equations are similar to the boundary layer equations (discussed later) but can be regarded as a slightly higher fidelity version of them. However, the boundary layer equations implicitly assume that the wall normal pressure gradient is zero and that the momentum equation in this direction drops out. For high-speed aerodynamic flows, the cells can have high aspect ratios – reaching 1000 with the long cell side parallel to the key flow direction. Hence, such a mesh implies that it is only the wall normal gradients that are expected to be substantial.

2.8.4 *The Boundary Layer Equations*

The Navier-Stokes equations can be more formally reduced to approximate equations suitable for the boundary layer using a rough order of magnitude analysis. To do this, an x location is considered where $x/\delta \gg 1$ and it is assumed $u \sim U_\infty$ and $\partial u/\partial x \sim U_\infty/x$. Also, $\partial/\partial y \sim 1/\delta$. From the continuity equation, $\partial v/\partial y \sim U_\infty/x$. Hence, $v = U_\infty \delta/x$. Therefore, the terms in the non-conservation form of the Navier Stokes equations have the following orders of magnitude

$$u\frac{\partial u}{\partial x} \sim \frac{U_\infty^2}{x}, \quad v\frac{\partial u}{\partial y} \sim \frac{U_\infty^2 \delta}{x\delta}, \quad \frac{\partial^2 u}{\partial x^2} \sim \frac{U_\infty}{x^2}, \quad \frac{\partial^2 u}{\partial y^2} \sim \frac{U_\infty}{\delta^2} \tag{2.103}$$

Considering the Bernouilli equation ($\Delta p = \Delta \rho u^2/2$) gives $\partial p/\partial x \sim U_\infty^2/x$. Since $x \gg \delta$, the smallest term is $\partial^2 u/\partial x^2$. Hence, for a boundary layer, the x momentum component of the Navier Stokes equations reduces to

$$\rho u\frac{\partial u}{\partial x} + \rho v\frac{\partial u}{\partial y} = -\frac{\partial p}{\partial x} + \mu\frac{\partial^2 u}{\partial y^2} \tag{2.104}$$

For the wall normal y momentum equation, we have

$$u\frac{\partial v}{\partial x} \sim \frac{U_\infty^2 \delta}{x^2}, \quad v\frac{\partial v}{\partial y} \sim \frac{U_\infty^2 \delta^2}{x^2 \delta}, \quad \frac{\partial p}{\partial y} \sim \frac{U_\infty^2}{\delta}, \quad \frac{\partial^2 v}{\partial x^2} \sim \frac{U_\infty \delta}{x^3}, \quad \frac{\partial^2 v}{\partial y^2} \sim \frac{U_\infty \delta}{x\delta^2} \tag{2.105}$$

This suggests $\partial p/\partial y$ is the greatly dominant term (relative to $\partial p/\partial x$) and that $\partial p/\partial y = 0$ is a reasonable approximation, that is, there is no significant pressure gradient across the boundary layer. Therefore, Equation (2.104) reduces to

$$\rho u\frac{\partial u}{\partial x} + \rho v\frac{\partial u}{\partial y} = -\frac{\partial p_\infty}{\partial x} + \mu\frac{\partial^2 u}{\partial y^2} \tag{2.106}$$

for a two-dimensional boundary layer (see Figure 2.16). In a three-dimensional flow system, it follows that Equation (2.106) becomes

$$\rho u\frac{\partial \phi}{\partial x} + \rho v\frac{\partial \phi}{\partial y} + \rho w\frac{\partial \phi}{\partial z} = \mu\frac{\partial^2 \phi}{\partial y^2} + S_\phi(x, z) \tag{2.107}$$

where

$$\begin{aligned} S_u &= -\frac{\partial p_\infty(x, z)}{\partial x} \\ S_w &= -\frac{\partial p_\infty(x, z)}{\partial z} \\ S_v &= 0 \end{aligned} \tag{2.108}$$

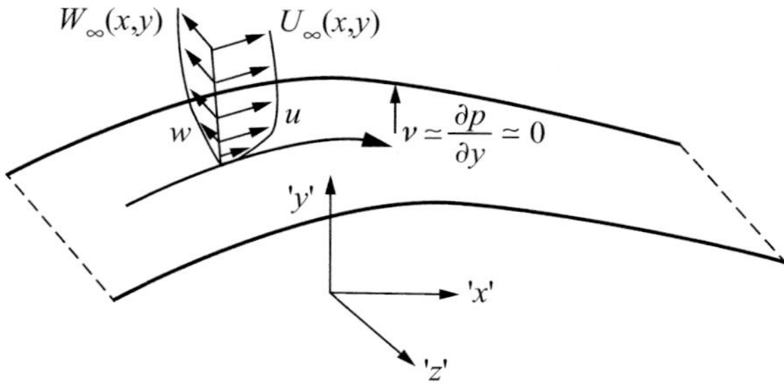

Figure 2.16. Schematic relating to the boundary layer equations.

This is identical to the thin shear layer form of the equations except that the source terms are based on the pressure external to the boundary layer. In practice, this pressure is evaluated for an inviscid flow equation solution. The wall pressure is taken as the boundary layer edge pressure. The key element is that (as shown by Prandtl) for high-speed, aerodynamic flows, the boundary layers, δ, are typically thin (scaling roughly as $\delta/L = (\rho/U_\infty L\mu)^{1/2}$). This allows the inviscid equations and the boundary layer equations to be solved separately. For thick boundary layers, an iterative coupling can be needed. Note that Equation (2.107) is solved with the following boundary conditions

$$
\begin{aligned}
u &= U_\infty\,(x, z) \\
w &= W_\infty\,(x, z)
\end{aligned}
\tag{2.109}
$$

Boundary layer equations can be solved in modern CFD. For example, see Piomelli and Balaras (2002), where the boundary layer equations are solved for a separate near wall mesh and patched to a mesh in the main body of the flow. In the main body of the flow, the 'high-fidelity' LES (Large Eddy Simulation) equations discussed in Chapter 5 are solved. For the practical application of the coupling of the boundary layer equations to the reduced order potential flow equations for highly complex, aerodynamics geometries with transonic flow, see Tinoco and Chen (1984). These two extremes in hybrid modelling are illustrated in the Figure 2.17 schematic.

As would be expected, the boundary layer equation can be written in a multitude of forms with the convective term taking on the non-conservation, weak and strong conservation forms, and so forth.

2.8.5 *Parabolised Equations*

The *parabolised Navier-Stokes* equations are used for channel or duct flows that do not have recirculations. A schematic relating to the parabolised Navier-Stokes equations is shown in Figure 2.18. The predominant flow direction is in the x direction. To help illustrate this, Equation (2.35) is first expressed in a three-dimensional form as

$$
\rho u \frac{\partial \phi}{\partial x} + \rho v \frac{\partial \phi}{\partial y} + \rho w \frac{\partial \phi}{\partial z} = \mu \frac{\partial^2 \phi}{\partial x^2} + \mu \frac{\partial^2 \phi}{\partial y^2} + \mu \frac{\partial^2 \phi}{\partial z^2} + S_\phi \tag{2.110}
$$

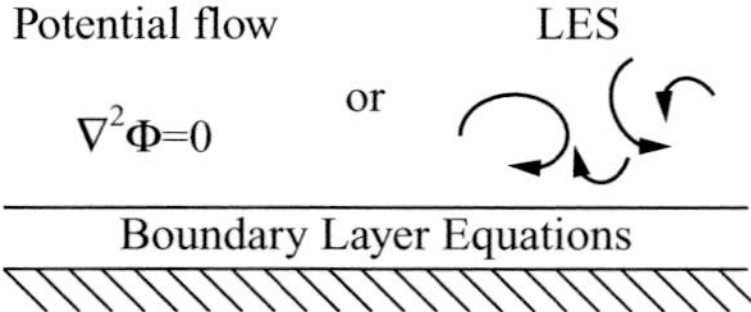

Figure 2.17. Combination of boundary layer equations with both low-order potential flow equations and also 'high-order' LES equations.

Then to form the parabolized Navier-Stokes equations, the streamwise second derivate (streamwise diffusion) for the viscous terms is removed to give

$$\rho u\frac{\partial \phi}{\partial x}+\rho v\frac{\partial \phi}{\partial y}+\rho w\frac{\partial \phi}{\partial z}=\mu\frac{\partial^2 \phi}{\partial y^2}+\mu\frac{\partial^2 \phi}{\partial z^2}+S_\phi \tag{2.111}$$

A similar treatment is needed for the energy and any turbulence modelling equations. The parabolised equations are reminiscent of the thin shear layer equations discussed previously. Critically, they are parabolic in the x-coordinate direction. Hence, x is like a time derivate and computationally efficient, time marching related approaches can be used to solve the equations. Clearly care is needed when deciding if use of this simplification is valid. A relatively widely used computer program that has successfully applied this approach is GENMIX (see Spalding 1977).

2.9 Further Reduced Equation Forms

2.9.1 *Stream Function Vorticity Equation*

As has been shown, solution of the incompressible Navier-Stokes equations typically involves solving for u, v, w and p or related products. These are sometimes referred to as the primitive variables. The equation for p can be produced by substituting the Navier-Stokes equations for u, v, and w into the continuity equation and rearranging, collecting the pressure gradient terms on the left-hand side of the equation. As might be expected, the full pressure equation is potentially substantial. The scale of the Navier-Stokes equations, the need to define a pressure field, and the number of variables involved has lead researchers to seek reduced, potentially easier to solve, equation sets. For example, for two-dimensional, incompressible, iso-viscous flow, in a Cartesian x-y system, this can be achieved by differentiating

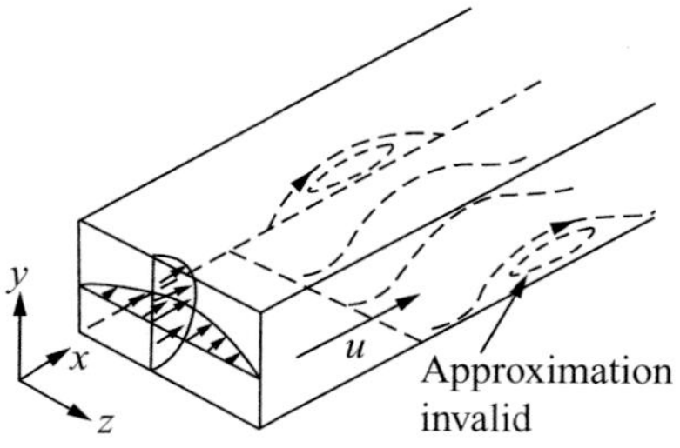

Figure 2.18. Schematic relating to the parabolised form of the Navier-Stokes equations.

the u momentum equation with respect to y and the v with respect to x. Subtraction of the two resulting equations gives the following single vorticity equation

$$\rho\frac{\partial\omega}{\partial t}+\rho u\frac{\partial\omega}{\partial x}+\rho v\frac{\partial\omega}{\partial y}=\mu\left(\frac{\partial^2\omega}{\partial x^2}+\frac{\partial^2\omega}{\partial y^2}\right) \tag{2.112}$$

The stream function (a variable that automatically satisfies the continuity equation) is defined using the following: $\partial\psi/\partial x=-v$ and $\partial\psi/\partial y=u$. Similarly, vorticity can be expressed as $\omega=\partial v/\partial x-\partial u/\partial y$. Combination of the stream function and vorticity definitions gives

$$\frac{\partial^2\psi}{\partial x^2}+\frac{\partial^2\psi}{\partial y^2}=-\omega \tag{2.113}$$

The approach, in two-dimensions, has the attraction that it involves solving for just two variables instead of the usual three (u, v and p). The velocity field is recovered by differentiation of the stream function field. Drawbacks are that, the boundary condition specification can be awkward. Furthermore, in three dimensions (see Chapter 4), there are other complications and drawbacks.

Historically, especially before the advent of powerful modern computers, even further reduced equation sets than the stream function vorticity equations have been considered, such as the stream function equation (or related velocity potential equation) – Equation (2.113) with $\omega=0$. This is a linear equation. Therefore different solutions for ψ can be summed to yield valid solutions. Certain CFD solvers solve stream function type equations away from solid surfaces, coupling these solutions with near surface boundary layer equations. An example of this is the Boeing TRANAIR code. The code can produce rapid solutions and this allows design optimization to be carried out relatively quickly. Hence, an engineer can specify a target lift and drag for a design, and then specify design constrains such as the wing weight and other manufacturing aspects. In the light of the constraints and target quantities, a wing geometry can be produced. Chapter 6 discusses such design optimization processes.

2.9.2 *Throughflow Equations*

The throughflow model is an example of simplified inviscid flow-governing equations. These absorb empirical correlations for viscous loss and other physical modelling aspects. The throughflow model is specifically intended for studying turbomachinery flows. These largely consist of series of rotating and stationary blade row pairs. Essentially a curvilinear coordinate system is adopted. However, the ξ direction coordinate lines, shown in Figure 2.19, evolve with the flow. The ξ coordinate is the flow aligned streamline – called the meridional direction. The η coordinate is the quasi-orthogonal to this and these quasi-orthogonal lines stay fixed and must be coincident with blade leading and trailing edges. The equations are solved in an axisymmetric form. Hence, the tangential direction, θ, derivatives are ignored. Force fields F_ξ, F_η and F_θ are imposed to replicate blade forces (see Chapter 6, plausible

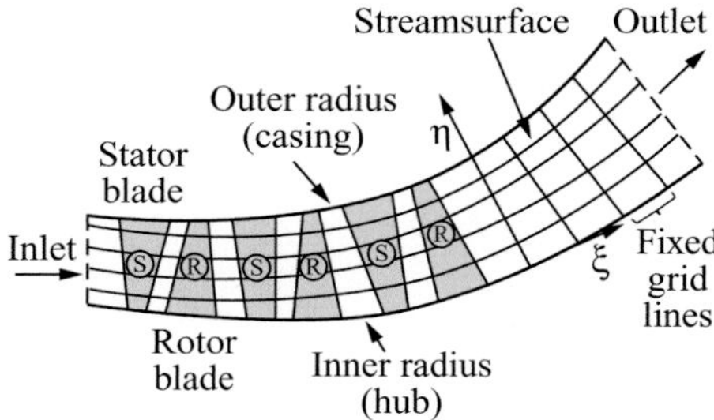

Figure 2.19. Grid system for a throughflow model.

forms of these). Following Pullan (2008), the broad form of the equation, which expresses conservation of momentum in the η direction, is

$$\rho u_\xi \frac{\partial \phi}{\partial \xi}(\sin(\varepsilon + \varepsilon')) + \frac{\phi^2}{r_c}(\cos(\varepsilon + \varepsilon')) = -\frac{\partial p}{\partial \eta} + \frac{u_\theta^2}{r}\cos\beta + F_\eta \qquad (2.114)$$

Some of the terms used in this equation are identified in Figure 2.20. Note the variable being solved for $\phi = u_\xi$. Also, r_c is the radius of curvature of the ξ streamline and ε is the local angle of this streamline. $F_\eta = 0$ outside blade rows. Also, generally, this force will be small – the blade leading and trailing edge geometry is quite parallel to the η coordinate directed lines. Note that β is a blade angle of attack (see Pullan). Also, u_θ is algebraically specified. In blade passages, it is a simple function of the blade metal angle, blade angular velocity and u_ξ. Outside blade zones, from conservation of angular momentum, ru_θ is constant. This allows the tangential velocity to be easily evaluated. The stagnation enthalpy is readily calculated, since, from the Euler work equation

$$h_0 - \Omega r u_\theta = \text{constant} \qquad (2.115)$$

Absorbing the entropy Equation (2.161), discussed further later, to remove pressure (as with Crocco's form of the Navier-Stokes equations – also discussed further later) and making use of stagnation enthalpy and that $\cos\varepsilon = dr/d\eta$, the following equation

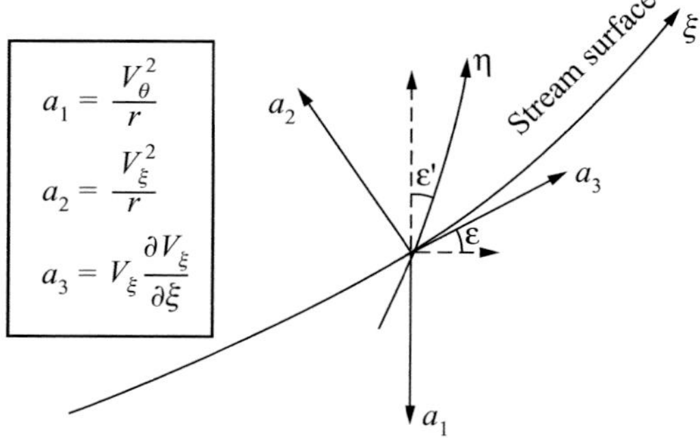

Figure 2.20. Schematic showing terms used in the throughflow model.

can be derived (see Pullan)

$$\frac{1}{2}\rho\frac{\partial(\phi^2)}{\partial\eta} - \rho u_\xi \frac{\partial\phi}{\partial\xi}(\sin(\varepsilon+\varepsilon')) - \frac{\phi^2}{r_c}(\cos(\varepsilon+\varepsilon'))$$
$$= \underbrace{\frac{\partial h_0}{\partial\eta}}_{h_0-\text{ algebraically specified}} \underbrace{-T\frac{\partial s}{\partial\eta}}_{\text{Empirical correlations}} - \underbrace{\frac{1}{2r^2}\frac{\partial(ru_\theta)^2}{\partial\eta}}_{u_\theta-\text{ specified from simple algebraic equations}} - F_\eta \tag{2.116}$$

This equation is solved with a simple integral (over the blade radial direction) mass conservation equation. This enables design or analysis of the flow in turbomachinery blade passages. The integration over η is

$$\frac{1}{2}\rho\frac{\partial(\phi^2)}{\partial\eta} = A(\phi) \tag{2.117}$$

where

$$A(\phi) = \rho u_\xi \frac{\partial\phi}{\partial\xi}(\sin(\varepsilon+\varepsilon')) + \frac{\phi^2}{r_c}(\cos(\varepsilon+\varepsilon'))$$
$$+\frac{\partial h_0}{\partial\eta} - T\frac{\partial s}{\partial\eta} - \frac{1}{2r^2}\frac{\partial(ru_\theta)^2}{\partial\eta} - F_\eta \tag{2.118}$$

Quite specific solution procedures are required for the throughflow equations. These are not outlined here. The process, in fact, involves such a large degree of empiricism that the numerical schemes used and their accuracy is of secondary importance. Notably, high fidelity CFD (even eddy-resolving simulations) could be used to refine the empirical models used. This would be important if applying the approach for radically new designs where the experimental correlation data used for model closure might not be applicable.

2.9.3 *Dimensionless Form of Equations*

To assist with making approximate solutions, it can be convenient to write the Navier-Stokes equations in a dimensionless form. It is debatable if this form of the equations is best discretized as part of a numerical solution (there are advantages when using the artificial compressibility approach to recover pressure outlined in Chapter 4 – the ad hoc compressibility parameter seems to have less variability). However, generally the preference is to leave the Navier-Stokes equations in their dimensional form in CFD programs. Non-dimensionalization can be achieved by defining the following dimensionless variables

$$\tilde{u} = \frac{u}{U}, \quad \tilde{v} = \frac{v}{U}, \quad \tilde{p} = \frac{p}{p_o}, \quad \tilde{x} = \frac{x}{L}, \quad \tilde{y} = \frac{y}{L}, \quad \tilde{t} = \frac{t}{t_o} \tag{2.119}$$

Equation (2.34) is now re-written, but this time including the gravity and unsteady flow terms. Saving space, two dimensions are stayed with and the v equation just considered. Extension to three dimensions and also the u equation is obviously straightforward:

$$\rho\frac{\partial v}{\partial t}+\rho u\frac{\partial v}{\partial x}+\rho v\frac{\partial v}{\partial y}=-\frac{\partial p}{\partial y}-\rho g+\mu\frac{\partial^2 v}{\partial x^2}+\mu\frac{\partial^2 v}{\partial y^2} \tag{2.120}$$

We will now substitute equations (2.119) into (2.120), that is, replace u in Equation (2.120) with $u = U\tilde{u}$, and so forth. Since the reference velocity, U, pressure, p_o, length scale, L, and time period, t_o, are assumed constants, these are taken outside derivatives. Hence, after dividing both sides by $\rho U^2/L$, the following is gained

$$\left\langle\frac{L}{t_o U}\right\rangle\frac{\partial\tilde{\phi}}{\partial\tilde{t}}+\tilde{u}\frac{\partial\tilde{\phi}}{\partial\tilde{x}}+\tilde{v}\frac{\partial\tilde{\phi}}{\partial\tilde{y}}=-\left\langle\frac{p_o}{\rho U^2}\right\rangle\frac{\partial\tilde{p}}{\partial\tilde{y}}-\left\langle\frac{gL}{U^2}\right\rangle+\left\langle\frac{\mu}{\rho UL}\right\rangle\left(\frac{\partial^2\tilde{\phi}}{\partial\tilde{x}^2}+\frac{\partial^2\tilde{\phi}}{\partial\tilde{y}^2}\right) \tag{2.121}$$

where $\tilde{v} = \tilde{\phi}$, the $\tilde{\phi}$ being used as a reminder of the common form of the equations. However, for $\tilde{u} = \tilde{\phi}$ the pressure gradient would be with respect to x and generally, unless a less usual geometry orientation is used, $g = 0$.

The terms in the $\langle\rangle$ brackets are themselves also dimensionless. For, example, the first bracket on the left-hand side is a Strouhal number. The $\langle p_o/\rho U^2\rangle$ is the Euler number – the ratio of pressure to inertia forces (remember this is basically what the Euler equations we derived earlier relate to, i.e., the pressure-inertia force balance). The $\langle gL/U^2\rangle$ term is the inverse of the square of the Froude number. The final term $\langle\mu/\rho UL\rangle$ is the inverse of the Reynolds number. At very high Reynolds numbers (i.e., low $\langle\mu/\rho UL\rangle$), it can be seen that (2.121) tends to the Euler equations (i.e., the Navier-Stokes equations given previously but with no viscous force terms).

2.9.4 *Viscous Force Dominated Flows*

The other extreme, to the Euler equations, is very low Reynolds numbers, sometimes called creeping flows. With low Reynolds number flows, if we ignore the gravity and unsteady term, Equation (2.121) reduces to

$$0=-\left\langle\frac{p_o}{\rho U^2}\right\rangle\frac{\partial\tilde{p}}{\partial\tilde{y}}+\left\langle\frac{\mu}{\rho UL}\right\rangle\left(\frac{\partial^2\tilde{\phi}}{\partial\tilde{x}^2}+\frac{\partial^2\tilde{\phi}}{\partial\tilde{y}^2}\right) \tag{2.122}$$

An example of a creeping flow would be the drainage of fluid through, say, a porous soil medium. Re-dimensionalizing Equation (2.122) but for the x component of velocity gives

$$0=-\frac{\partial p}{\partial x}+\mu\left(\frac{\partial^2 u}{\partial\tilde{x}^2}+\frac{\partial^2 u}{\partial\tilde{y}^2}\right) \tag{2.123}$$

Simple scaling analysis then gives

$$0=-\frac{\partial p}{\partial x}+\frac{\mu}{L^2}u \tag{2.124}$$

where $\partial p/\partial x$ is the pressure difference in the x direction. Alternatively, the previous equation can be written as

$$0 = -\frac{\partial p}{\partial x} - \frac{\mu}{K}u \tag{2.125}$$

where K is a porosity coefficient. This is called Darcy's law and is analogous to the law of resistance for the flow of electricity, Fourier's law and also Fick's law for the diffusion of mass. Similarly, for the other directions we can write

$$0 = -\frac{\partial p}{\partial y} - \frac{\mu}{K}v, \; 0 = -\frac{\partial p}{\partial z} - \frac{\mu}{K}w \tag{2.126}$$

The above can be used for an isotropic media. For an anisotropic media, the following can be written

$$\begin{aligned} u &= \frac{Q_x}{A_x} = -\frac{K_{xx}}{\mu}\frac{\partial p}{\partial x} \\ v &= \frac{Q_y}{A_y} = -\frac{K_{yx}}{\mu}\frac{\partial p}{\partial y} \\ w &= \frac{Q_z}{A_z} = -\frac{K_{zx}}{\mu}\frac{\partial p}{\partial z} \end{aligned} \tag{2.127}$$

where Q are volume flow rates and the A terms are projected area components. More fully, the following can be written

$$\mu \begin{pmatrix} u & v & w \end{pmatrix} = -\begin{pmatrix} K_{11} & K_{12} & K_{13} \\ K_{21} & K_{22} & K_{23} \\ K_{31} & K_{32} & K_{33} \end{pmatrix} \begin{pmatrix} \partial p/\partial x \\ \partial p/\partial y \\ \partial p/\partial z \end{pmatrix} \tag{2.128}$$

or in tensor form

$$\mu v_i = K_{ij}\frac{\partial p}{\partial x_j} \tag{2.129}$$

2.9.5 *Reynolds Equation*

With hydrodynamic bearings, the flow takes place between two surfaces which are close together. Two key typical geometries are shown in Figure 2.21. Frame (a) shows what is called a slider bearing. Frame (b) shows the type of bearing used for the 'big end' in a car engine and also ground based turbines. In both frames, L has been greatly exaggerated. Both systems can essentially be considered as channel flows. The small L results in a relatively small Reynolds number. Perhaps more importantly, the small L means the other coordinate directions can be considered as extensive. This means derivatives with respect to x and z can generally be neglected. Neglecting these derivatives, in the continuity equation, suggests v (wall normal velocity) is small and that inertial terms can be neglected. The combined result is that, again, the convection terms in the Navier-Stokes equations vanish. Under these low Reynolds number circumstances, equations of the type (2.122) (i.e., equations with no convective terms) can give accurate design solutions.

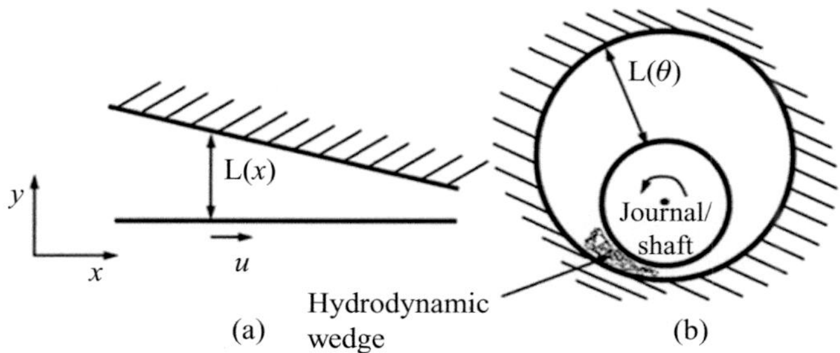

Figure 2.21. Hydrodynamic journal bearing geometries: (a) slider bearing and (b) journal bearing.

If we ignore the convective, gravity and unsteady flow terms (remembering $v \to 0$), and assume the geometry is as in Figure 2.21a, the Navier-Stokes equations reduce to

$$\frac{\partial^2 u}{\partial y^2} = -\frac{1}{\mu}\frac{\partial p}{\partial x}, \quad \frac{\partial^2 w}{\partial y^2} = -\frac{1}{\mu}\frac{\partial p}{\partial z} \tag{2.130}$$

After some integration with manipulations and approximations (see Pinkus and Sternlicht 1961), the Reynolds equation can be expressed as

$$\frac{\partial}{\partial x}\left(\frac{L^3}{\mu}\frac{\partial p}{\partial x}\right) + \frac{\partial}{\partial z}\left(\frac{L^3}{\mu}\frac{\partial p}{\partial z}\right) = 6U\frac{\partial L}{\partial x} + 12\frac{\partial L}{\partial t} \tag{2.131}$$

The last right-hand side term models the consequences of geometrical movement and U is the bearing surface velocity. Equation (2.131) in its polar coordinate form is

$$\frac{\partial}{\partial z}\left(\frac{L^3}{\mu}\frac{\partial p}{\partial z}\right) + \frac{1}{r}\frac{\partial}{\partial \theta}\left(\frac{L^3}{\mu r}\frac{\partial p}{\partial \theta}\right) = 6U\frac{\partial L}{r\partial \theta} + 12\frac{\partial L}{\partial t} \tag{2.132}$$

The last term comes into play when the shaft orbits in the bearing housing. This is quite common in reciprocating combustion engines. Due to accuracy and cost reasons, the Reynolds equations are frequently used in computer packages for bearing design. The Reynolds equation assumptions are summarised as follows:

1) Since the mass of the fluid is negligible, so are body forces;
2) Since the y or r direction extent is small relative to the other directions, the pressure and lubricant viscosity variation in these directions is negligible;
3) Fluid inertia forces are negligible and
4) The fluid is Newtonian.

It is not until high speeds are reached or if the frequency of shaft oscillation/orbiting becomes large that it is necessary to include inertial terms and solve the full Navier-Stokes equations (this is an unusual requirement). Normally, for bearings the equations are solved in a curvilinear coordinate form.

2.9.6 *Other Forms of the Flow-Governing Equations*

Many further forms and simplifications of the flow-governing equations can be written. For example, there is Crocco's equation, which hybridizes thermodynamic and fluid dynamic equations. It is essentially a vorticity transport equation. However, the flow driving pressure gradient is expressed in terms of the entropy gradient. This is achieved using Equation (2.161), discussed later. Also, all the equations so far can be written in linearized forms where they are decomposed into a mean, $\overline{\phi}$, and a component fluctuation about the mean, ϕ'. This is achieved through the substitution

$$\phi = \overline{\phi} + \phi' \tag{2.133}$$

The solution of linearized equations can be useful in gaining an estimate of unsteady temporal information such as the magnitude of acoustic waves or unsteady surface pressure loads. Indeed, non-linear disturbance equation forms of the governing equations can also be formulated (see Batten et al. 2004). Hence, there are numerous potential forms of the governing equations and those presented here are perhaps the most generally used for practical problems. Notably, the Navier-Stokes equations can also be reorganized into equations governing acoustics (see Goldstein 2003), electromagnetics (Morgan et al. 2000) and even the behaviour of solid structures. As noted, acoustics equations can be generated from simplifying the Navier-Stokes equations and there is a further extensive hierarchy of acoustics equations. These will be outlined in Chapter 6.

2.10 Equations in Solids

As outlined by Spalding (1993) and Spalding and Artemov (2009) solids/ structural equations can be arranged in a form that can be dealt with in CFD solvers. Here we restate our general convection-diffusion transport for the general variable ϕ in its compact vector form

$$\frac{\partial}{\partial t}(\rho\phi) + \nabla \cdot (\rho u\phi - \Gamma\nabla\phi) = S_{\phi} \tag{2.134}$$

Immediately, if we ignore the convection term (i.e., $\mathbf{u} = 0$), we get the conduction of heat equation where, $\phi = T$ and $\Gamma = k/c_p$. Thus, in CFD, making conjugate solutions, where the fluid flow and conduction in solids equations are solved together, is relatively easy. Just the velocity needs to be set to zero in the solid zone (potentially in a crude fashion through a large viscosity) and the appropriate thermal constant set in the solid. Care is needed with the averaging of variables at control volume faces to ensure thermal energy conservation.

Stress and strain in solids is more complicated than heat conduction. Strain is characterized by points in a domain being displaced in the x, y and z directions. The key connection with the fluid flow equations is that velocities are displacements per unit time. As noted by Newton, stresses in fluids are proportional to velocity gradients, the constant of proportionality being the fluid's viscosity. This is again akin to the observations of Hook, that the stresses in solids are proportional to the gradients of displacement. This time, the constant of proportionality is Young's modulus, E. A key difference between the fluid flow and structural processes is the Poisson ratio effect. With this, a positive gradient of displacement in one direction can cause negative gradients of displacement in the other two directions. To deal with this, additional terms are necessary. However, it seems evident that the fluid flow equations, for conservation of momentum, can be connected to static force balance equations. First we evaluate equations for displacements (velocities per unit time) and then put these into equations to give stresses. To formally derive the equations, following Spalding, we write the force balance equation for a solid as

$$\frac{\partial \sigma_{xx}}{\partial x} + \frac{\partial \sigma_{xy}}{\partial y} + \frac{\partial \sigma_{xz}}{\partial z} + F_x = 0 \tag{2.135}$$

Note that, to aid simplification, this equation just considers the force balance in the x direction. Note also that F_x is a body force and σ_{ik} are stress tensor components. For small displacements, stresses are related to strains by Hooke's law:

$$\sigma_{ik} = 2\mu\varepsilon_{ik} + \delta_{ik}(\lambda D - \Gamma\epsilon_t), i, k = x, y, z \tag{2.136}$$

where

$$D = \varepsilon_{xx} + \varepsilon_{yy} + \varepsilon_{zz} \tag{2.137}$$

and

$$\varepsilon_{ik} = \frac{1}{2}\left(\frac{\partial u_i}{\partial x_k} + \frac{\partial u_k}{\partial x_i}\right) \tag{2.138}$$

In the preceding, u_i are the displacements. Also, μ and λ are Lame's coefficients which are associated with Young's modulus and Poisson's ratio, P, as

$$\mu = \frac{E}{2(1+P)}, \quad \lambda = \frac{EP}{(1+P)(1-2P)}, \quad \Gamma = \frac{E}{1-2P} \tag{2.139}$$

Note,

$$\varepsilon_t = \beta(T - T_o) \tag{2.140}$$

where β is the coefficient of thermal expansion and T_o a reference temperature at which zero deformation is expected. If the material properties are uniform, then from the previous, the following equation can be written

$$\mu\nabla \cdot (\nabla u) + \mathbf{F} - \nabla p = 0 \tag{2.141}$$

where

$$p = -(\lambda + \mu)D + \Gamma\varepsilon_t \tag{2.142}$$

Hence, we have an equation akin to the momentum equations solved in CFD but without the convective terms. The equation for the displacement component in the

x direction (i.e., u) is written as

$$\left\{\frac{\partial}{\partial x}\left[(2\mu+\lambda)\frac{\partial u}{\partial x}\right]+\frac{\partial}{\partial y}\left[\mu\frac{\partial u}{\partial y}\right]+\frac{\partial}{\partial z}\left[\mu\frac{\partial u}{\partial z}\right]\right\}+\frac{\partial}{\partial x}[\lambda(\varepsilon_{yy}+\varepsilon_{zz})-\Gamma\varepsilon_t]$$
$$+\frac{\partial}{\partial y}\left[\mu\frac{\partial v}{\partial x}\right]+\frac{\partial}{\partial z}\left[\mu\frac{\partial w}{\partial x}\right]+F_x=0 \qquad (2.143)$$

Similar equations would need to be solved for the other two displacements. Based on these displacements, stresses can be readily evaluated.

2.11 Auxiliary Equations

2.11.1 *Boussinesq Approximation and Buoyancy*

Solution of the energy equation is not just necessary for high-speed flows. It is also needed when we wish to explore how fluid flow influences heat transfer. Alternatively, if buoyancy forces play a key role in driving the flow, then again the energy equation must be included.

If the density and viscosity changes related to temperature are small, then the energy equation can be solved independently from the momentum and continuity equations. However, if this is not the case, then the equations are coupled and their solution is more challenging. For buoyancy driven flows, the Boussinesq approximation can be used. Using this, the fact that density varies in the convective terms of momentum equations is ignored. This can be useful, but caution is required that it is a valid simplification. The density in the convective terms is left at some reference value, say, ρ_o. The buoyancy force is then treated as a source term body force. This takes the following form

$$\text{Body Force} \rightarrow S=\beta\rho_o g(T-T_o)=\frac{\rho_o g(T-T_o)}{T_o} \qquad (2.144)$$

The parameter β is discussed later. In terms of a CFD solver, this approximation, when appropriate, can reduce computational cost. Implicit in the assumption is that density changes are just a function of temperature.

The Grashof number gives the ratio of buoyancy to viscous forces. For a totally buoyancy driven flow, it is the equivalent to the Reynolds number. The Grashof number can be defined as

$$Gr=\frac{\text{Buoyancy Forces}}{\text{Viscous Forces}}=\frac{\rho^2 g\beta\Delta T L^3}{\mu^2} \qquad (2.145)$$

where L is a characteristic length scale. For a boundary layer, ideally this would be δ (the boundary layer thickness) but as with the Reynolds number, since δ is proportional to distance along a surface, the latter is more convenient and can be used. The ΔT is the difference between the surface and ambient temperatures. The parameter, β, is the coefficient of volume[1] expansion (at constant pressure). This

[1] Remember that density is the inverse of specific volume.

can be expressed as

$$\beta = -\frac{1}{\rho}\frac{\partial \rho}{\partial T}\bigg|_p \tag{2.146}$$

For a perfect gas, $\beta = 1/T$ (see Equation 2.144), where T is the ambient temperature. If the ratio of Gr/Re^2 is low (<0.01), buoyancy forces can be neglected. Hence the momentum and energy equations can be decoupled, density changes in the momentum equations being ignored.

2.11.2 *Thermodynamic Equations*

The general CFD solution of flow problems needs some additional equations. When studying fluids moving at high speeds, thermal modelling can be important and many of the additional equations in this section relate to this aspect. The first obvious relation is the equation of state for a perfect gas,

$$\rho = \frac{p}{RT} \tag{2.147}$$

where R is the perfect gas constant. The equation is a useful means of evaluating density. Its combination with the differential form of the continuity equation (with a time derivative) yields a useful means for the evaluation of pressure – providing that the flow is fast enough to give sufficient density variations. The following equation relating to the isentropic (reversible adiabatic process) expansion of a perfect gas is also frequently used

$$\rho^\gamma = \rho^\gamma_{ref}\frac{p}{p_{ref}} \tag{2.148}$$

where the *ref* subscripts indicate reference states and

$$\gamma = \frac{c_p}{c_v} \tag{2.149}$$

This parameter is called the ratio of specific heats, where c_v is the specific heat capacity at constant volume. Note that $R = c_p - c_v$. Related to Equation (2.148), the following relationship can also be derived

$$\frac{p}{p_{ref}} = \left(\frac{T}{T_{ref}}\right)^{\gamma/\gamma-1} \tag{2.150}$$

Assuming isentropic flow, the speed of sound, c, can be expressed as

$$c^2 = \frac{\partial p}{\partial \rho} = \gamma RT \tag{2.151}$$

The important dimensionless parameter for high-speed flows, noted in Chapter 1, is the Mach number, M. This is the ratio of the fluid's speed to the speed of sound, that is:

$$M = \frac{|\mathbf{u}|}{c} \tag{2.152}$$

The fluid's internal energy, already noted in Section 2.6.3 (often represented by the symbol u in many thermodynamics texts), can, using Joules law ($u = f(T)$ only), be evaluated as

$$e = c_v T \tag{2.153}$$

The constant volume specific heat can also be evaluated from

$$c_v = \frac{R}{\gamma - 1} \tag{2.154}$$

Using this and the equation of state for a perfect gas allows us to write

$$e = \frac{1}{\gamma - 1} \frac{p}{\rho} \tag{2.155}$$

Equation (2.155) is not restricted to just constant volume processes. The enthalpy of a perfect gas can be expressed as

$$h = c_p T = e + \frac{p}{\rho} \tag{2.156}$$

When considering compressible flows, the energy equation must also be solved. Generally, in aerospace orientated CFD codes, the following *total energy* is solved for

$$E = c_v T + \frac{|\mathbf{u}|}{2} \tag{2.157}$$

However, as indicated previously, the energy equation can also be formulated so that the solution variable is temperature, enthalpy or stagnation enthalpy. The stagnation enthalpy is a popular choice and can be expressed as

$$h_0 = E + \frac{p}{\rho} \tag{2.158}$$

Solving for temperature can have the advantage that boundary conditions are easily conceptualized. The specification of stagnation enthalpy is less simple. The stagnation temperature, T_0, is expressed as

$$T_0 = T + \frac{|\mathbf{u}|}{2c_p} = T\left(1 + \frac{\gamma - 1}{2} M^2\right) \tag{2.159}$$

Note that Equation (2.150) and related equations can be used with stagnation variables. For example, combining Equations (2.159) and (2.150), we have

$$\frac{p_0}{p_{ref}} = \left(\frac{T_0}{T_{ref}}\right)^{\gamma/\gamma-1} = \left(1 + \frac{\gamma - 1}{2} M^2\right)^{\gamma/\gamma-1} \tag{2.160}$$

Entropy can be evaluated from

$$Tds = de + pd\left(\frac{1}{\rho}\right) = dh - \frac{dp}{\rho} \tag{2.161}$$

Heat flux can be recovered from temperature using Fourier's law

$$q = k\frac{dT}{dn} \tag{2.162}$$

where n is a surface normal coordinate and k is thermal conductivity. Hence, the boundary condition for the energy equation cast in terms of temperature for an adiabatic (thermally insulated) wall is again straightforward – the gradient of temperature is set to zero.

2.11.3 *Fluid Properties*

When there are substantial temperature variations throughout a fluid region, then the fluid properties can change appreciably and for accuracy it is important to take this into account. Fluid viscosity is a very weak function of pressure. A popular equation for relating dynamic viscosity to temperature for air is Sutherland's law. This is given as

$$\mu = \frac{(1.46 \times 10^{-6})T^{3/2}}{110 + T} \tag{2.163}$$

where T is needed in degrees K. Sutherland's law is incorporated into many CFD solvers. However, using curve fitting (for $300 < T < 1000$) suggests the following are more accurate (to within $\pm 1\%$, i.e., half the error of that for Sutherland's law) constants (see Morse 1980)

$$\mu = \frac{(1.513 \times 10^{-6})T^{3/2}}{126 + T} \tag{2.164}$$

The fluid's thermal conductivity can be evaluated from the following expression

$$k = \frac{c_p \mu}{\mathrm{Pr}} \tag{2.165}$$

where Pr is the Prandtl number (the ratio of thermal to momentum diffusivity which characterises the ratio of the thermal to the velocity boundary layer thicknesses). The Prandtl number is generally treated as a constant and varies little with temperature. This can be seen from the following fits to established data by Morse (1980)

$$\begin{aligned} \mathrm{Pr} &= 0.680 + 3 \times 10^{-7}(600 - T)^2, \quad T < 600\ \mathrm{K} \\ \mathrm{Pr} &= 0.680 + 5 \times 10^{-5}(T - 600), \quad T > 600\ \mathrm{K} \end{aligned} \tag{2.166}$$

Typically, $\mathrm{Pr} = 0.7$ is used in CFD simulations. For solving passive scalar transport and also for the transport of variables in the turbulence models (see Chapter 5), turbulent Prandtl or Schmidt numbers are required. The standard turbulent Prandtl number is $\mathrm{Pr}_t = 0.9$. However, there can be substantial variations in published values for turbulent Schmidt/Prandtl numbers.

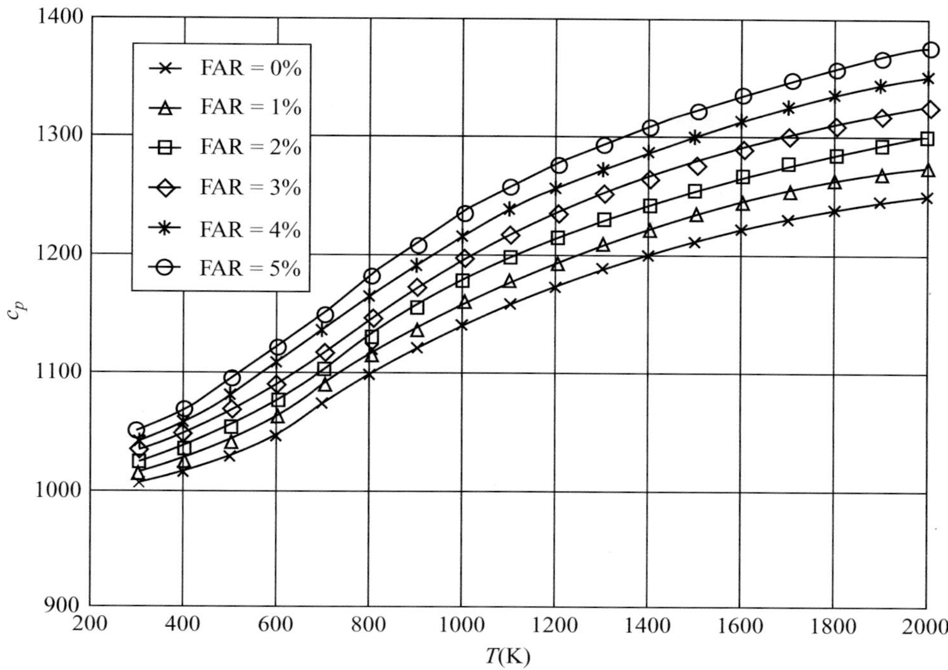

Figure 2.22. Variation of c_p with temperature and Fuel Air Ratio (FAR = Φ) for a turbine with a mixture of air. (adapted from Northall 2006).

For the thermal conductivity of air, the following curve, fit to the established data of Morse (1980), can be used

$$k = \frac{0.0025T^{3/2}}{202 + T} \tag{2.167}$$

The uncertainty in this expression is again $\pm 1\%$. For turbulent flows, an effective turbulent thermal conductivity can be evaluated based on an equation analogous to (2.165) using the turbulent Prandtl or Schmidt number combined with the eddy viscosity.

2.11.3.1 Real Gases The equations in Sections 2.11.2 and 2.11.3 relate to perfect gasses. However, for real gases, chemical and thermodynamic processes play a role and Equation (2.147) does not hold. This can occur for steam turbines, combustion and hypersonic flows. If the gases are in chemical and thermodynamic equilibrium, the task is relatively simple. There is a unique equation of state. Look up tables (with interpolation) for fluid properties (viscosity, temperature and pressure) can be used. This approach is used by Northall (2006). As noted by Northall, for a gas turbine, the temperature variation from the inlet of the high-pressure turbine to the exit of the intermediate pressure turbine can be around 1880 K to 1200 K. Also, the fuel air ratio can drop from 0.0281 to 0.0244. As Figure 2.22 shows, this results in a c_p variation of around 7%. This corresponds to a variation in γ from 1.281 to 1.306.

Note that FAR = Φ in the figure is the fuel air ratio. The turbine system is choked. Hence, the mass flow can be evaluated from

$$\dot{m} = \frac{A\,p_0}{\sqrt{c_p T_0}} \frac{\gamma}{\sqrt{\gamma - 1}} \left(\frac{\gamma + 1}{2}\right)^{-(\gamma+1)/2(\gamma-1)} \tag{2.168}$$

This implies that the change in ratio of specific heats corresponds to a 0.7% increase in mass flow rate. This machine 'capacity' is important in the design process. However, in terms of the CFD analysis, Northall observes only a maximum difference of 1.5% in computed efficiency between using a fixed c_p and the real gas model.

Note that the variation in c_p has implications for the basic equations given in Section 2.11.2. For example, the use of $h = c_p\ T$ as a simple integral to $dh = c_p dT$ is not appropriate. Instead, the following needs to be used

$$h = h_{ref} + \int_{T_{ref}}^{T} C_p(T, \Phi) dT \tag{2.169}$$

Also, integration of the entropy Equation (2.161) is complicated, thus the following equation is required

$$s_2 - s_1 = \int_{T_1}^{T2} \frac{C_p(T, \Phi) dT}{T} - R \ln(p_2/p_1) \tag{2.170}$$

A convective-diffusion transport equation of the form given earlier (i.e., $\phi = \Phi$) can be used to estimate the fuel air ratio. Northall also considers specifying simple algebraic one-dimensional distributions for through the engine Φ.

If the gas is not in chemical equilibrium, then the situation is more complex and equations (2.97)–(2.101) need to be used. The pressure for each chemical species can be expressed as

$$p_n = \rho \Phi_n \frac{\tilde{R}}{m_m} T \tag{2.171}$$

where $\tilde{R}$ is the universal gas constant and m_m is the species molecular weight. Using Equation (2.171) with Dalton's law

$$p = \sum_{n=1}^{N} p_n \tag{2.172}$$

we can write

$$p = \rho \tilde{R} T \sum_{n=1}^{N} \frac{\Phi_n}{m_n} \tag{2.173}$$

The following can also be written (see Blazek 2001)

$$p = \sum_{n=1}^{N} \left[\Phi_n \left(h f_n^{\theta} + \int_{T_{ref}}^{T} c_{p,n} dT \right) \right] - \frac{p}{\rho} \tag{2.174}$$

where T_{ref} is a reference temperature and $h f_n^{\theta}$ is the enthalpy of reformation.

2.11.3.2 Liquids Water is another widely used working fluid. For this, the following approximate equations can be used:

$$\rho = 1.043 \times 10^3[1 - 3.07 \times 10^{-3}\Delta T - 7.83 \times 10^{-6}(\Delta T)^2]\,\mathrm{kg/m^3} \quad (2.175)$$

where the kinematic viscosity is expressed as

$$\nu = 1.62 \times 10^{-6}[1 - 2.79 \times 10^{-2}\Delta T - 6.73 \times 10^{-4}(\Delta T)^2]\,\mathrm{m^2/s} \quad (2.176)$$

and

$$\frac{k}{c_p} = 1.345 \times 10^{-4}[1 + 2.33 \times 10^{-3}\Delta T]\,\mathrm{kg/ms} \quad (2.177)$$

Note that $\Delta T = T - T_{ref}$ and $T_{ref} = 295$ K.

For multiphase flows, the specification of fluid properties can be complicated. For a liquid-vapour mix in a cavitation zone Tucker and Keogh (1996), use the following expressions

$$\rho = (1 - \Phi)\,\rho_v + \Phi\rho_l \quad (2.178)$$

where Φ is vapour fraction and the subscripts l and v refer to the liquid and vapour components. Similarly

$$\phi^{-1} = (1 - \Phi)\,\phi_v^{-1} + \Phi\phi_l^{-1} \quad (2.179)$$

where $\phi = k, c_v$ or μ.

2.12 Source Terms

A range of source terms can be applied to all the differential transport equations in this chapter. For example, as noted previously, there are gravitational, buoyancy and rotational forces and so forth. A will be seen later, body forces can be applied that force the flow to behave as if metal surfaces are present but without their explicit meshing (see Gong 1999). Source terms, with carefully calibrated coefficients, can be used to represent the presence of fans and grills (see Chapter 6) with carefully calibrated coefficients. They can be used to replicate internal heat generation from, say, electricity. More specialist source terms can also be needed in CFD. For electro osmotic flow, the Navier-Stokes equations can be used through simply adding an extra source term, which drives the fluid. The external electric field, Φ, is modelled using Laplace's equation

$$\frac{\partial^2 \Phi}{\partial {x_i}^2} = 0 \quad (2.180)$$

Internal potential, ψ, can be approximated using the so called Poisson-Boltzmann equation

$$\frac{\partial^2 \psi}{\partial {x_i}^2} = -\frac{2n_0 ze}{\varepsilon\varepsilon_0}\sinh\left(\frac{ze\psi}{k_b T}\right) \quad (2.181)$$

The electrical force is characterized by adding the following source to the Navier-Stokes equations

$$S_\phi = -\rho_E \left(\frac{\partial(\psi + \Phi)}{\partial x_i} \right) \tag{2.182}$$

where $\rho_E = -2n_0 ze \sinh(ze\psi/k_b T)$. Note also that now e is the fundamental charge, k_b is the Bolzmann's constant, z is the valence of ions, ε is the dielectric constant of the electrolyte solution, ε_0 is the permittivity of a vacuum and ζ is the zeta potential. Source terms can be used to characterize momentum transfer between a particulate phase and a fluid. Also, they can be used to characterize the exchange of thermal energy between a droplet phase and the fluid (see Chapter 6). Source terms can also be used to characterize small scale features such a plasma actuators (see Fujii 2014) and vortex generators (see Bender et al. 1999, Schlichting 1979, Wendt 2001 and Dudek 2006).

Problems

(1) Fluid of constant density and viscosity flows between two infinitely long plates separated by a vertical distance h. At $y = 0$ the plate is stationary and at $y = h$ the plate moves in the x direction with a velocity U. Solve the two-dimensional Navier-Stokes and continuity equations to give the pressure and velocity distribution between the plates showing that the former is constant and the latter has a linear variation.

(2) Show that for a compressible, constant viscosity fluid, the flow-governing equations can be expressed as

$$\begin{aligned}
&\rho\frac{\partial u}{\partial t} + \rho u\frac{\partial u}{\partial x} + \rho v\frac{\partial u}{\partial y} + \rho w\frac{\partial u}{\partial z} \\
&\quad = -\frac{\partial p}{\partial x} + \mu\left(\frac{\partial^2 u}{\partial x^2} + \frac{\partial^2 u}{\partial y^2} + \frac{\partial^2 u}{\partial z^2}\right) + \frac{1}{3}\mu\frac{\partial \nabla\cdot\mathbf{v}}{\partial x} \\
&\rho\frac{\partial v}{\partial t} + \rho u\frac{\partial v}{\partial x} + \rho v\frac{\partial v}{\partial y} + \rho w\frac{\partial v}{\partial z} \\
&\quad = -\frac{\partial p}{\partial y} + \mu\left(\frac{\partial^2 v}{\partial x^2} + \frac{\partial^2 v}{\partial y^2} + \frac{\partial^2 v}{\partial z^2}\right) + \frac{1}{3}\mu\frac{\partial \nabla\cdot\mathbf{v}}{\partial y} \\
&\rho\frac{\partial w}{\partial t} + \rho u\frac{\partial w}{\partial x} + \rho v\frac{\partial w}{\partial y} + \rho w\frac{\partial w}{\partial z} \\
&\quad = -\frac{\partial p}{\partial z} + \mu\left(\frac{\partial^2 w}{\partial x^2} + \frac{\partial^2 w}{\partial y^2} + \frac{\partial^2 w}{\partial z^2}\right) + \frac{1}{3}\mu\frac{\partial \nabla\cdot\mathbf{v}}{\partial z}
\end{aligned} \tag{2.183}$$

where

$$\nabla\cdot\mathbf{v} = \frac{\partial u}{\partial x} + \frac{\partial v}{\partial y} + \frac{\partial w}{\partial z} \tag{2.184}$$

(3) Consider the case of flow between two infinitely long cylinders of inner radius a and outer radius b, where the axial velocity $w = 0$. Using the Navier-Stokes equations in cylindrical polar coordinates for a constant density and constant viscosity flow and the corresponding continuity equation, show that these equations reduce to

$$0 = -\frac{\partial p}{\partial r} + \frac{\rho v^2}{r}$$
$$0 = \frac{1}{r}\frac{\partial}{\partial r}\left(r\frac{\partial v}{\partial r}\right) - \frac{v}{r^2} \tag{2.185}$$

Show that if the inner cylinder rotates with an angular velocity Ω,

$$v = \frac{\Omega a^2}{r}\left(\frac{r^2 - b^2}{a^2 - b^2}\right) \tag{2.186}$$

Note that the axial, radial and tangential velocity components are represented by w, u and v, respectively.

(4) Using the incompressible Navier-Stokes equations in cylindrical polar coordinates, show that

$$w = 2W\left(1 - \left(\frac{r}{R}\right)^2\right) \tag{2.187}$$

for the fully developed flow through a pipe of radius R. Note that the fluid enters the pipe with a bulk average velocity W and there is no swirl ($v = 0$).

(5) Show that when Equation (2.47) is transformed to a rotating coordinate system, the following extra terms arise

$$S_{u,rotating} = S_{u,stationary} + \underbrace{2\rho v\Omega_g + \Omega_g^2 r}_{New\ Terms},\ S_{v,rotating} = S_{v,stationary} - \underbrace{2\rho u\Omega_g}_{New\ Term} \tag{2.188}$$

(6) Reduce Equation (2.47) to the steady, axisymmetric, non-conservation form for tangential velocity:

$$\rho u\frac{\partial v}{\partial r} + \rho w\frac{\partial v}{\partial z} = \frac{1}{r}\frac{\partial}{\partial r}\left(r\Gamma_{\phi r}\frac{\partial v}{\partial r}\right) + \frac{\partial}{\partial z}\left(\Gamma_{\phi z}\frac{\partial v}{\partial z}\right) - \rho\frac{uv}{r} - \mu\frac{v}{r^2} \tag{2.189}$$

Show that after multiplying through by r and rearranging, the following can be written

$$\rho u\frac{\partial (rv)}{\partial r} + \rho w\frac{\partial (rv)}{\partial z} = \frac{1}{r}\frac{\partial}{\partial r}\left(r\Gamma_{\phi r}\frac{\partial (rv)}{\partial r}\right) + \frac{\partial}{\partial z}\left(\Gamma_{\phi z}\frac{\partial (rv)}{\partial z}\right) - 2\frac{\mu}{r}\frac{\partial (rv)}{\partial r} \tag{2.190}$$

Through combining the first and last right-hand side terms in the previous equation, show that the following source term free tangential momentum

Table 2.5. *Behaviour of various forms of the tangential momentum equation as $r \to 0$*

Term	1	2	3	4	5	6
Eqn (2.189)	r^1	0	∞	0	r	∞
Eqn (2.190)	r^2	$rv = 0\|_{r=0}$	$4c_1$	$rv = 0\|_{r=0}$	$-4c_1$	NA
Eqn (2.191)	r^2	$rv = 0\|_{r=0}$	r^1	$rv = 0\|_{r=0}$	NA	NA
Eqn (2.192)	r^1	0	∞	0	∞	r^0

equation can be derived

$$\rho u \frac{\partial (rv)}{\partial r} + \rho w \frac{\partial (rv)}{\partial z} = \frac{1}{r}\frac{\partial}{\partial r}\left(r^3 \Gamma_{\phi r} \frac{\partial}{\partial r}(v/r)\right) + \frac{\partial}{\partial z}\left(\Gamma_{\phi z}\frac{\partial (rv)}{\partial z}\right) \tag{2.191}$$

By dividing terms in Equation (2.189) by r, show

$$\rho u \frac{\partial (v/r)}{\partial r} + \rho w \frac{\partial (v/r)}{\partial z} = \frac{1}{r}\frac{\partial}{\partial r}\left(r \Gamma_{\phi r} \frac{\partial (v/r)}{\partial r}\right) + \frac{\partial}{\partial z}\left(\Gamma_{\phi z}\frac{\partial (v/r)}{\partial z}\right)$$
$$-2\frac{\mu}{r}\frac{\partial (v/r)}{\partial r} - 2\rho\frac{uv}{r^2} \tag{2.192}$$

Comments: For a problem where $r \to 0$, Equation (2.192) could be problematic. Equations (2.190) and (2.191) do not have the Coriolis source term. This involves the radial velocity component in the tangential momentum equation. This gives rise to equation coupling issues. This aspect is discussed in Chapter 4. Equation (2.191) has a non-standard radial diffusion term. However, it is source term free. This simplifies ensuring the conservation of tangential momentum and hence convergence.

(7) Analyze the properties of Equations (2.189–2.192) above when $r \to 0$. Note that at the axis, $v = 0$. Assume: $v = c_1 r + c_2 r^2 + c_3 r^3 + \ldots$, $u = b_1 r + b_2 r^2 + b_3 r^3 + \ldots$. Show that as $r \to 0$, the states given in Table 2.5 arise.

Comments: For Equation (2.189), terms 3 and 6 tend to infinity. This is non-ideal. However, their difference tends to a constant of $3c_2$. For Equation (2.190), terms 3 and 4 balance at the axis. Equation (2.192) is potentially problematic, with terms tending to infinity.

(8) Using the transformations from Cartesian to cylindrical polar coordinates given in Table 2.1, transform the Navier-Stokes equations in Cartesian coordinates to cylindrical polar coordinates and verify the transformation with the equations given in Section 2.3.3.

(9) Using the following equations (based on the auxiliary equations in Section 2.11.2), generate, through an iterative process, an initial estimate of the isentropic velocity field u and v along with ρu, ρv, ρE and E with the Figure 2.23 system in mind.

$$\frac{T_{out}}{T_{0,in}} = \left(\frac{p_{out}}{p_{0,in}}\right)^{\frac{\gamma-1}{\gamma}}, \tag{2.193}$$

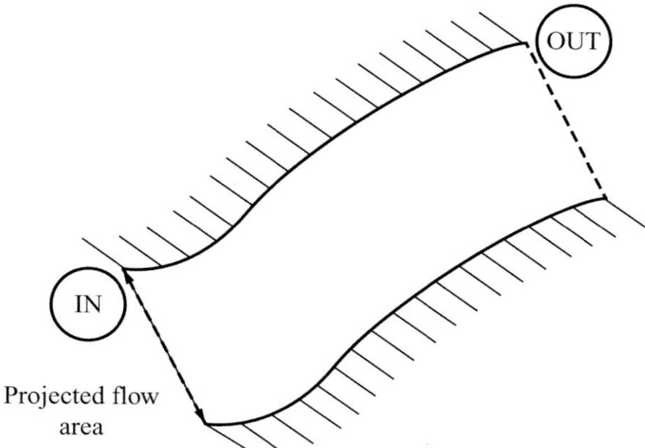

Figure 2.23. Schematic of an idealized duct system.

Steady Flow Energy Equation (SFEE) $-\frac{|\mathbf{u}|^2}{2} + c_p T =$ constant

$$p = \rho R T, \tag{2.194}$$

$$\frac{T_{lim}}{T_0} = \frac{1}{1 + 1/2\,(\gamma - 1)\,M^2} \tag{2.195}$$

In this set, $M = 1$ (for subsonic flow to impose an iterative hard limit on temperature) and T_{lim} is a temperature limit (see later).

$$\frac{\rho}{\rho_{in}} = \left(\frac{T}{T_{in}}\right)^{\frac{1}{\gamma-1}}, \tag{2.196}$$

$$u = |\mathbf{u}|\cos\theta, \qquad v = |\mathbf{u}|\sin\theta \tag{2.197}$$

Note that θ is the flow angle relative to the horizontal. Assume that the stagnation pressure and temperature at the duct inlet are known (but arbitrary) along with the duct static outlet pressure p_{out}. Make sure these values are consistent with subsonic but compressible flow. The local duct area, A, facing normal to the flow direction is also known. Use MATLAB or a computer programming language for the iterative process. The piece of program will be needed later if you wish to write a simple compressible flow solver.

The process needed is as follows: Assume a duct cross-sectional area; make an initial guess of the density and velocity at the exit by assuming isentropic flow between the inlet stagnation pressure and temperature and the exit static pressure; estimate mass flow based on quantities at the outlet. Next estimate the velocity and density at a set number of axial locations. First assume that the density is constant and equal to the exit density and that the flow does not vary in the cross-stream direction. Use the estimated velocity to calculate the static temperature, assuming that the stagnation temperature is constant. Check that this temperature is not less than T_{lim}, implying supersonic flow. Next use this temperature and the isentropic flow assumption to obtain a

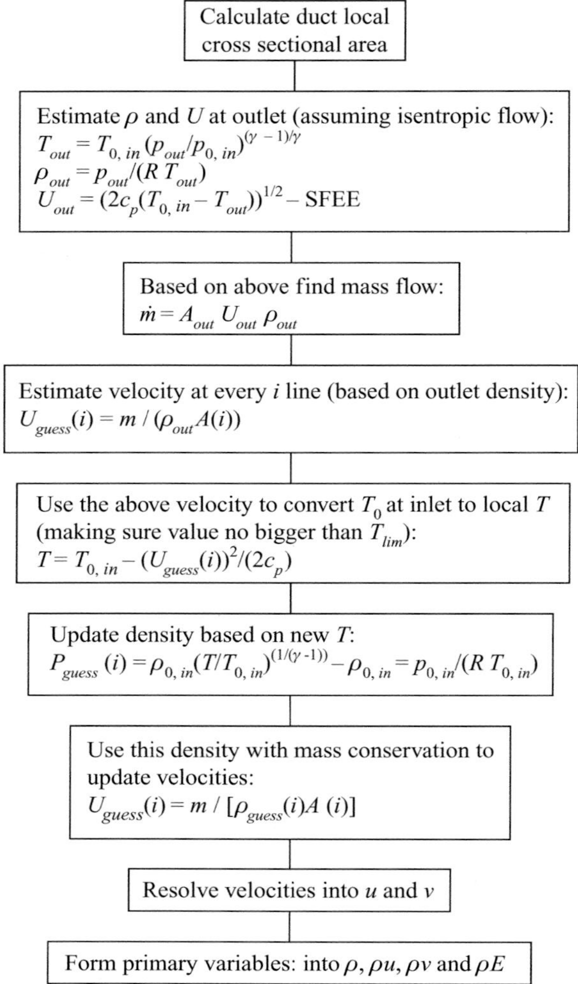

Figure 2.24. Process for gaining initial, compressible flow solution guess.

better estimate of ρ. Use this ρ and continuity considerations to obtain a better estimate of the velocity, $|\mathbf{u}|$. Finally, resolve $|\mathbf{u}|$ and assemble the variables asked for. The process is summarized in Figure 2.24.

(10) For a given inlet stagnation pressure, temperature and flow angle α, using the following equations

$$\rho_0 = \frac{p_0}{RT_0} \tag{2.198}$$

$$T = T_0\left(\frac{\rho}{\rho_0}\right)^{\gamma-1} \tag{2.199}$$

$$|\mathbf{u}| = \sqrt{2c_p\left(T_0 - T\right)} \tag{2.200}$$

$$E = c_v T + \frac{1}{2}|\mathbf{u}|^2 \tag{2.201}$$

$$u = |\mathbf{u}| \cos\alpha \tag{2.202}$$

$$v = |\mathbf{u}| \sin\alpha \tag{2.203}$$

estimate ρ, ρu, ρv and ρE at the duct inlet in Question 9. Note: Assume that the density at the boundary is known, being equal to an adjacent value in the interior of the domain.

REFERENCES

ANDERSON, J. D. 1995. *Computational fluid dynamics*, Springer.

ARAKAWA, A. 1966. Computational design for long-term numerical integration of the equations of fluid motion: Two-dimensional incompressible flow. Part I. *Journal of Computational Physics*, 1, 119–143.

BATTEN, P., RIBALDONE, E., CASELLA, M. & CHAKRAVARTHY, S. 2004. Towards a generalized non-linear acoustics solver. 10th AIAA/CEAS Aeroacoustics Conference. American Institute of Aeronautics and Astronautics, AIAA Paper Number AIAA-2004-3001.

BENDER, E., ANDERSON, B. & Y, P. 1999. Vortex generator modeling for Navier-Stokes codes. Third ASME/JSME Joint Fluids Engineering Conference, Paper No. FEDSM99-6919.

BLAZEK, J. 2001. *Computational Fluid Dynamics: Principles and Applications*, Elsevier.

DEARDORFF, J. W. 1970. A numerical study of three-dimensional turbulent channel flow at large Reynolds numbers. *Journal of Fluid Mechanics*, 41, 453–480.

DUDEK, J. C. 2006. Empirical model for vane-type vortex generators in a Navier-Stokes code. *AIAA Journal*, 44, 1779–1789.

FUJII, K. 2014. High-performance computing-based exploration of flow control with micro devices. *Philosophical Transactions of the Royal Society A: Mathematical, Physical and Engineering Sciences*, 372, 20130326.

GHIA, K., GHIA, U. & STUDERUS, C. 1976. Analytical formulation of three-dimensional laminar viscous flow through turbine cascades using surface-oriented coordinates. American Society of Mechanical Engineers, Gas Turbine and Fluids Engineering Conference, New Orleans, La. ASME Paper No. 76-FE-22.

GOLDSTEIN, M. E. 2003. A generalized acoustic analogy. *Journal of Fluid Mechanics*, 488, 315–333.

GONG, Y. 1999. *A computational model for rotating stall and inlet distortions in multistage compressors*. PhD Thesis, Massachusetts Institute of Technology.

HORIUTI, K. & ITAMI, T. 1998. Truncation error analysis of the rotational form for the convective terms in the Navier–Stokes equation. *Journal of Computational Physics*, 145, 671–692.

KHATIR, Z. 2000. *Discrete vortex modelling of near-wall flow structure in turbulent boundary layers*. PhD Thesis, The University of Warwick.

KRIST, S. L., BIEDRON, R. T. & RUMSEY, C. L. 1998. *CFL3D user's manual (version 5.0)*, NASA TM-1998-208444.

MORGAN, K., HASSAN, O., PEGG, N. E. & WEATHERILL, N. P. 2000. The simulation of electromagnetic scattering in piecewise homogeneous media using unstructured grids. *Computational Mechanics*, 25, 438–447.

MORSE, A. P. 1980. *Axisymmetric free shear flows with and without swirl.* PhD Thesis, Imperial College London (University of London).

NORTHALL, J. D. 2006. the influence of variable gas properties on turbomachinery computational fluid dynamics. *Journal of Turbomachinery*, 128, 632–638.

PINKUS, O. & STERNLICHT, B. 1961. *Theory of hydrodynamic lubrication*, McGraw-Hill.

PIOMELLI, U. & BALARAS, E. 2002. Wall-layer models for large-eddy simulations. *Annual Review of Fluid Mechanics*, 34, 349–374.

PULLAN, G. 2008. *Introduction to numerical methods for predicting turbomachinery flows*, Cambridge University Turbomachinery Course Notes.

RAYNER, D. 1993. *A numerical study into the heat transfer beneath the stator blade of an axial compressor.* D. Phil Thesis, University of Sussex.

SCHLICHTING, H. 1979. *Boundary-layer theory*, McGraw-Hill.

SEWALL, E. A. 2005. *Large eddy simulations of flow and heat transfer in the developing and 180 bend regions of ribbed gas turbine blade internal cooling ducts with rotation–effect of coriolis and centrifugal buoyancy forces.* D. Phil, Virginia Polytechnic Institute and State University.

SPALDING, D. B. 1977. *GENMIX: A general computer program for two-dimensional parabolic phenomena*, Pergamon Press Oxford.

SPALDING, D. B. 1993. Simulation of fluid flow, heat transfer and solid deformation simultaneously. NAFEMS Conference No. 4.

SPALDING, D. B. & ARTEMOV, V. I. 2009. Numerical modelling of problems involving hydrodynamics, Heat Transfer and Elasticity; Thermal Stresses arising in Air-Cooled Gas-Turbine Blades. *Proceedings of the XVII School-Seminar of Young Scientists and Specialists under Supervision of A. I. Leontiev. 'Problems of Heat and Mass Transfer and Gas Dynamics in Aerospace Technology.'*

TINOCO, E. & CHEN, A. 1984. Transonic CFD applications to engine/airframe integration. 22nd Aerospace Sciences Meeting. American Institute of Aeronautics and Astronautics, AIAA Paper No. AIAA-84-0381 .

TUCKER, P. G. 2006. Turbulence modelling of problem aerospace flows. *International Journal for Numerical Methods in Fluids*, 51, 261–283.

TUCKER, P. G. 2013. *Unsteady computational fluid dynamics in aeronautics*, Springer.

TUCKER, P. G. & KEOGH, P. S. 1996. On the dynamic thermal state in a hydrodynamic bearing with a whirling journal using cfd techniques. *Journal of Tribology*, 118, 356–363.

TUCKER, P. & LONG, C. 1996. Semi-implicit second order finite volume jet stream computations. *International Journal of Numerical Methods for Heat & Fluid Flow*, 6, 39–50.

TYACKE, J. C., NAQAVI, I. T. & TUCKER, P. G. 2015. *Body force modelling of internal geometry for jet noise prediction.* Advances in Simulation of Wing and Nacelle Stall (Radespiel, R., Niehuis, R., Kroll, N., Behrends, K. (Eds.)), Notes on Numerical Fluid Mechanics and Multidisciplinary Design, Springer, 131, 97–109.

WENDT, B. J. 2001. Initial circulation and peak vorticity behavior of vortices shed from airfoil vortex generators. NASA/CR-2001-211144.

CHAPTER 3

Mesh Generation

3.1 Introduction

Unless the boundary element method (see Brebbia and Wrobel 1980), mesh less methods such as lattice-Boltzman (see Derksen and Van den Akker 1999), smooth particle hydrodynamics (Marongiu et al. 2007, 2010) or vortex methods (see Chapter 4) are used, then a volume mesh is needed. Indeed, 'meshless' methods can need a background mesh, but more for data structure purposes. There are wide ranges of meshing techniques and some of these are discussed here. Securing a mesh for numerical simulation can be an extremely time consuming task. Notably, it is hard to assess, ahead of making a simulation, if a mesh is adequate and hence mesh development can be an iterative process. Currently, the process of subsequent mesh refinement tends to be highly manual. Although adaptive refinement has been developed for some years, this approach is seldom used. A critical element of the mesh refinement process is finding an adequate heuristic that detects the mesh zones that need improvement. Such aspects are briefly discussed in Chapter 7. As with most other areas of CFD, there are a wide range of methods and concepts. Figure 3.1 attempts to summarize these. In this chapter, an overview of the concepts in this figure will be given. First, different mesh types will be described along with their applicability. The care that is needed to consider how a particular mesh will work with a particular solver is outlined. The mesh generation techniques are given along with methods for grid control. Parameters for quantifying grid quality are described. Optimal mesh forms for turbulent flow computations are discussed. Both the fields of RANS and eddy-resolving simulations are addressed along with hybrids of these. Finally, adaptive and moving meshes are considered.

3.2 Mesh Types, Applicability and Solver Compatibility

3.2.1 *Basic Mesh Types*

The mesh or grid types used in CFD can crudely be broadly classified as: structured Cartesian (or related variants such as cylindrical polar coordinates); modified structured Cartesian (trimmed cell, immersed boundary); structured curvilinear and unstructured.

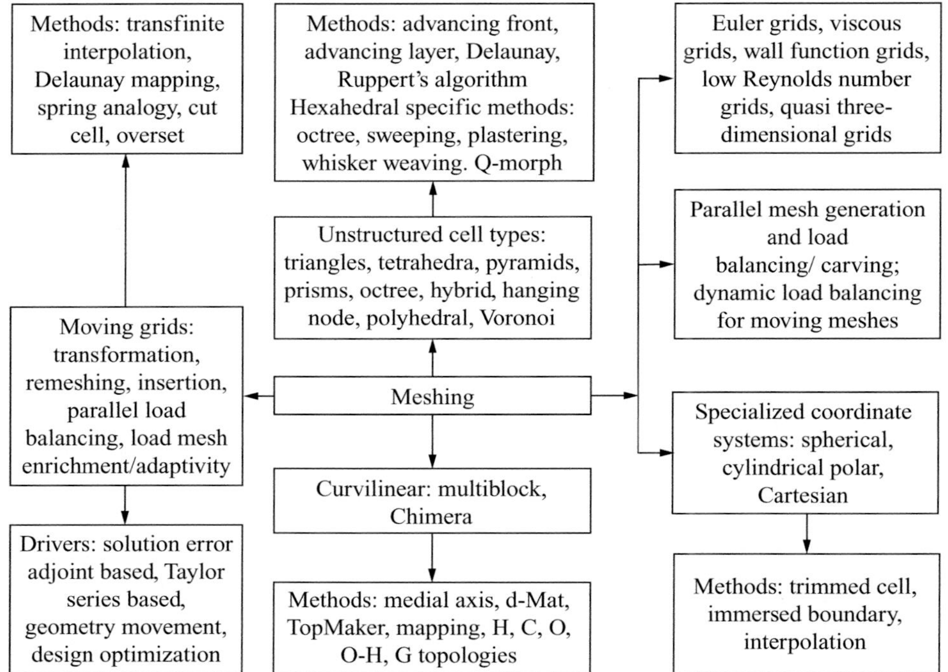

Figure 3.1. Mesh types, methods and concepts.

Figure 3.2, gives examples of these grids for a divergent duct. The grids in frames (I–III), although amenable to solution in structured flow solvers, can of course be used in unstructured. Notably, with unstructured grids (see Frame (IV)), the element labelling does not need to follow a clear order (see Figure 3.3b). However, for computational performance purposes, it can be helpful to have elements in close spatial proximity in close memory proximity (see Chapter 8). For density control purposes, the grid can be graded as shown in Figure 3.3a. This aspect and methods for locally enhancing the grid density are discussed later. For unstructured grids, the cells or building blocks are frequently triangular (for two-dimensional flows) or tetrahedral, pyramids or prisms (for three-dimensional flows). Some cell types are shown in Figure 3.4.

3.2.2 *Body Fitted Grid Types*

Broadly, body fitted or curvilinear grids can be classified into three types. The form of these is shown in Figure 3.5. The grids to the right of the figure are for the physical planes (as discussed in Chapter 4, and noted previously, this is a special, transformed and more regular plane in which complex forms of the Navier-Stokes equations are solved) and to the left are for the computational. Frame (a), shows an 'O' type grid, where edges a-b and b-c of the computational grid are wrapped round the object and abutted together. Frame (b), shows a 'C' type grid. Again the

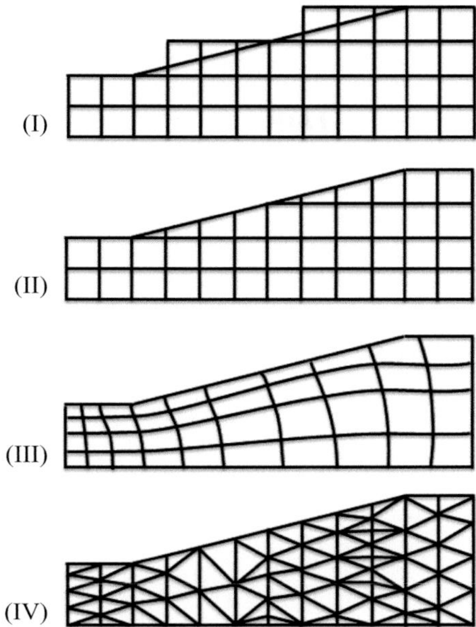

Figure 3.2. Typical grid forms for a divergent duct: (I) strictly Cartesian; (II) strictly Cartesian with boundary modifications; (III) curvilinear and (IV) unstructured (adapted from Tucker and Mosquera (2001) – Published with the kind permission of NAFEMS).

computational plane is wrapped around an object. However, this time, edges a-f and d-e form the downstream boundary. Such grids are useful for wings with sharp trailing edges. Frame (c) shows an 'H' type grid. This is formed by making an incision along e-f in the computational plane and prizing this region apart. For aerodynamic computations attempting to map airfoils, the 'C' grid is generally better than the 'H'. Indeed nowadays, the latter type grid is seldom used for this purpose. For complex domains, several curvilinear grids can be combined. This is called a multiblock

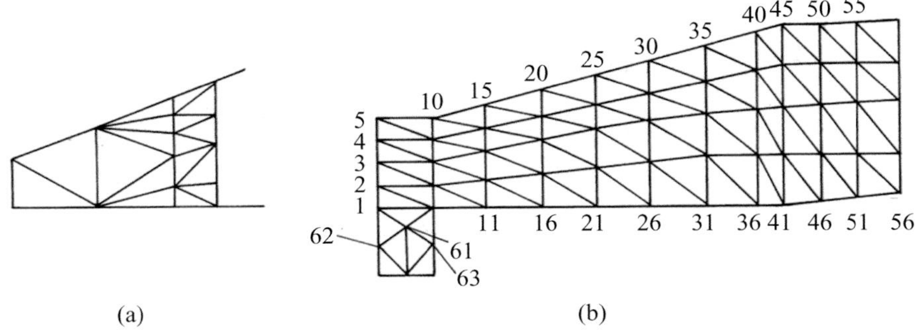

Figure 3.3. Use of unstructured meshes: (a) element grading and (b) nodal numbering (from Tucker and Mosquera (2001) – Published with the kind permission of NAFEMS).

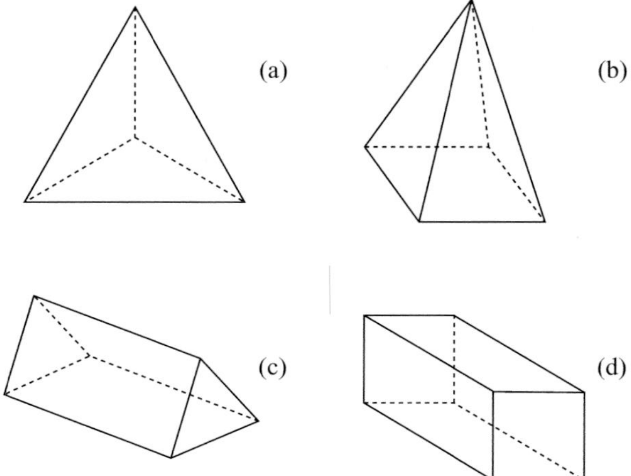

Figure 3.4. Different cell types used in unstructured flow solvers: (a) tetrahedral; (b) pyramid; (c) prism and hexahedral.

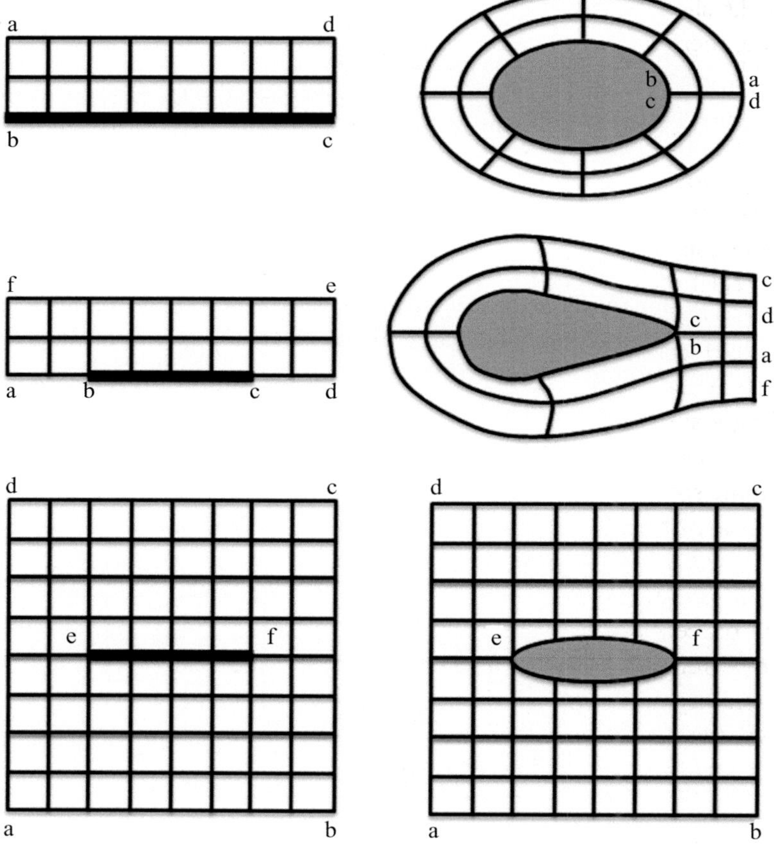

Figure 3.5. Typical curvilinear grid forms: top frame – 'O' grid, middle frame – 'C' grid and lower frame – 'H' grid (Adapted from Tucker and Mosquera (2001) – Published with the kind permission of NAFEMS).

technique. Multiblock mesh generation is difficult to automate. Such meshes are typically complex combinations of the elements shown in Figure 3.5.

More specialized meshes are:

1. Unstructured Cartesian (Octree) potentially with trimmed cells or surface layers;
2. Hybrid with structured grid generated curvilinear surface cells mixed with unstructured off surface cells;
3. Chimera where structured blocks are combined in an overset almost unstructured fashion and
4. DRAGON (Direct Replacement of Arbitrary Grid Overlapping by Non-structured) grids (see Zheng and Liou 2003).

DRAGON grids are similar to Chimera. However, they remove the overlapping area and fill the void with unstructured elements. Hence, the majority of the mesh is created from regular 'structured' elements such as quadrilaterals having good accuracy (see Section 3.6.1).

Figure 3.6 shows octree based meshes. In Frame (b), near the surface, although hard to see, there is also a surface mesh that appears akin to a curvilinear mesh. Hence, there are ranges of potential hybridizations of mesh and the mesh structure can vary dramatically with the nature of the governing equations and hence the flow being solved.

Although not fully a meshing strategy, the Immersed Boundary Method (IBM) (see Preece 2008) is an important approach that should be noted here. With IBM, typically Cartesian cells are used. The fluid-solid interface passes through these cells and no especial effort is made to ensure that the cells are boundary conforming. The flow governing equations are then modified to ensure that nodal values of the solution variables that evolve are such that they are consistent with the desired solid wall boundary conditions. For example, for a viscous flow, the nodal values surrounding a solid wall would evolve to be consistent with there being both the no-slip and impermeability conditions at the solid surface. This can either be achieved via a dynamic source/forcing term in the flow governing equations or through the more direct use of interpolation formula. As shown by Tucker and Pan (2000), the IBM approach can be hybridized with the trim cell technique (the cell faces are local modified on essentially Cartesian meshes), the different approaches being applied depending on cell topology. For highly complex geometries, the IBM approach can be powerful and efficient. With it, the mesh generation task changes to cell labelling and organising geometric information relating to surface proximity. The latter allows forcing/interpolation expressions to be generated to ensure the correct surface aerodynamic boundary conditions are evolved.

3.2.3 *Applicability of Different Mesh Types*

Curvilinear grids are frequently used for more streamlined aerodynamic flows such as airfoils. However, even with simple two-dimensional sections, when a slat

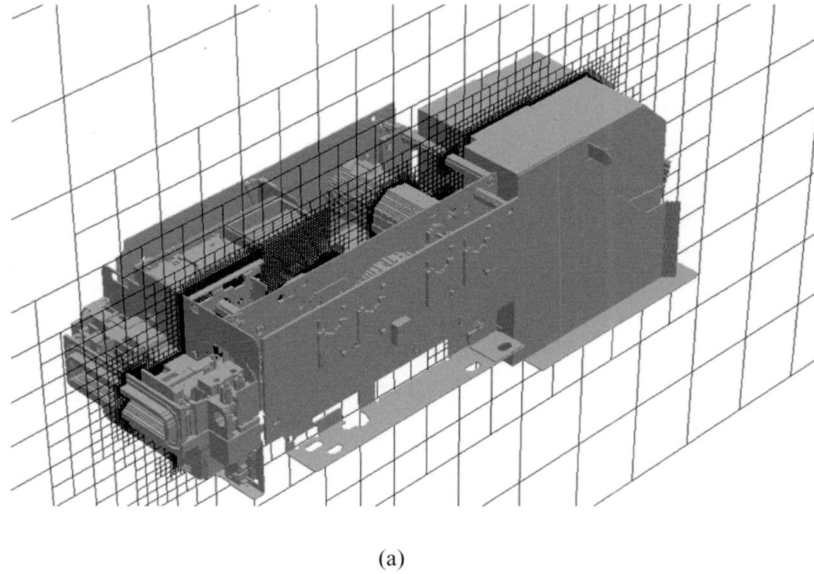

(a)

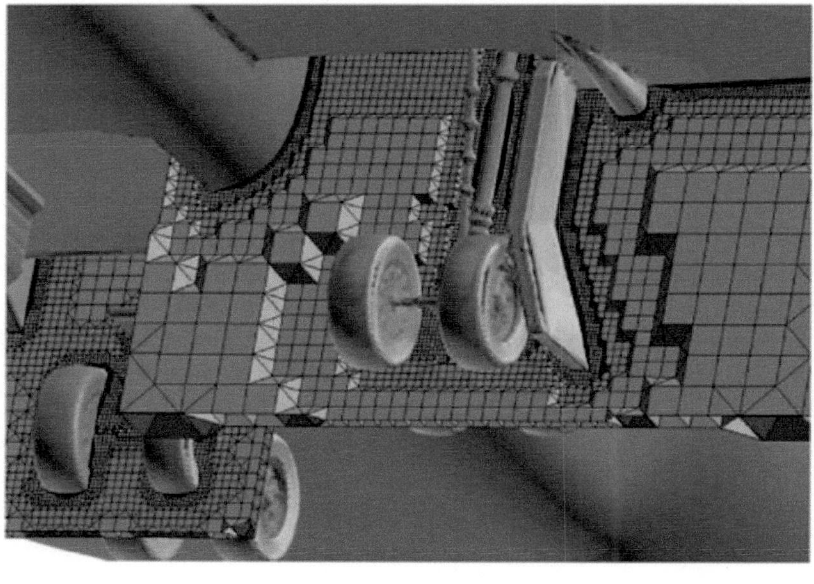

(b)

Figure 3.6. Unstructured, octree based, Cartesian grid examples: (a) mechatronic card reader and (b) aircraft landing gear (Frame (b) from Xia et al (2010) – Published with the kind permission of Elsevier).

and flap are included, the range of flow complexity can greatly expand as shown in Figure 3.7.

The arising mesh structure can become complex, see for example in Figure 3.8. The different shades represent different mesh blocks.

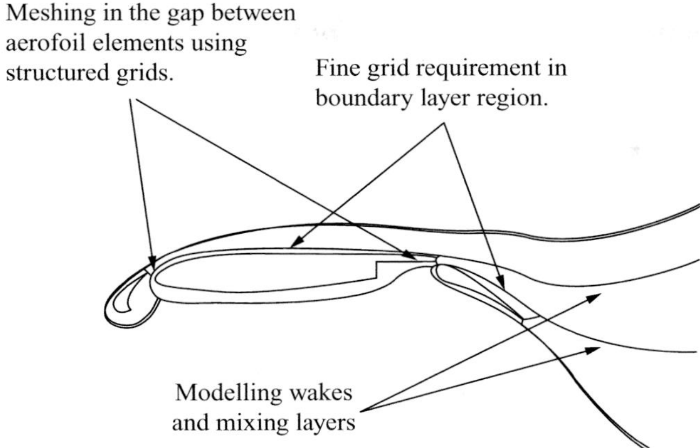

Figure 3.7. Wing-slat-flap configuration and flow physics to be captured.

Indeed, as shown in Figure 3.9, even a wing with a deployed flap can necessitate a relatively complex blocking topology. This can be greatly simplified if an overset or Chimera mesh is used. With this approach, the separate components are considered in isolation and optimal meshes for the isolated components are constructed. The meshes are then overlaid, a small overlap left and unnecessary cells deleted – this process has not been represented in Figure 3.9. Since the points will generally not match at the overlapping interface, some interpolation is necessary and

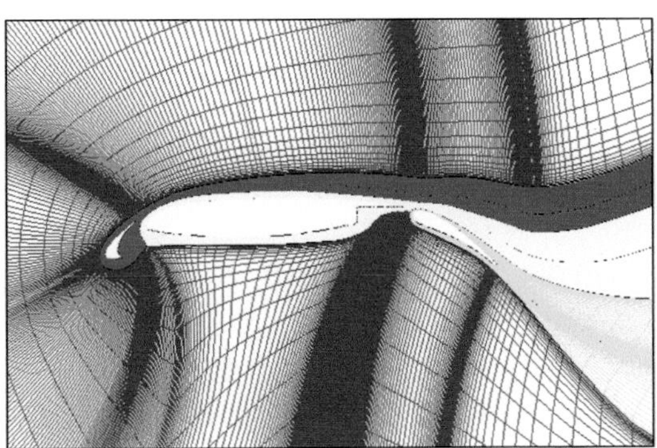

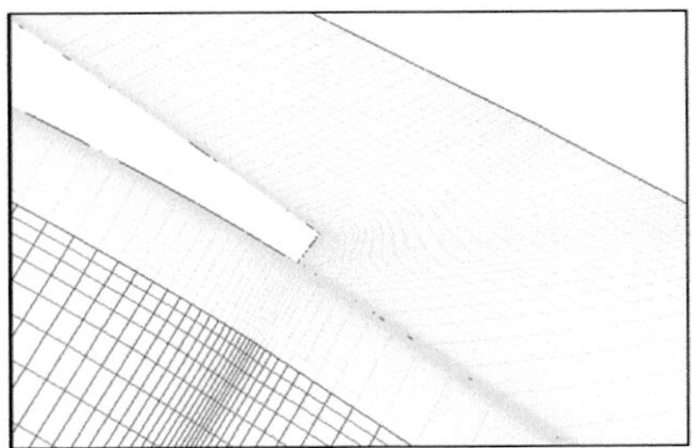

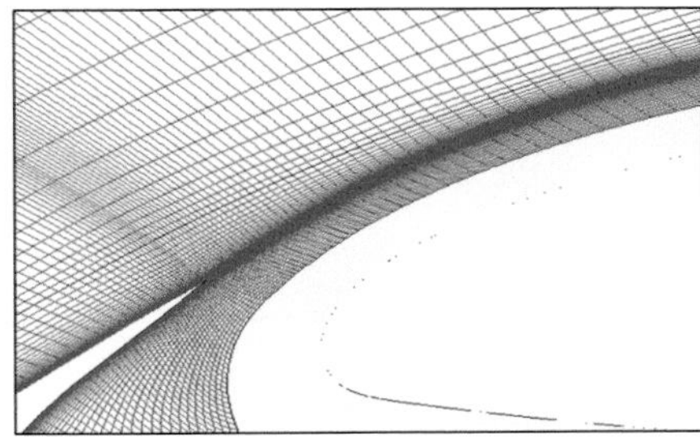

Figure 3.8. Typical curvilinear mesh for a wing-slat-flap configuration.

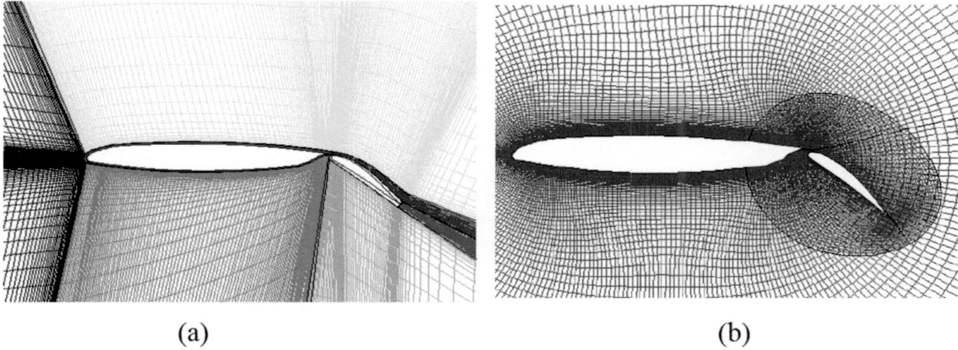

Figure 3.9. Meshing for wing-flap geometry: (a) basic curvilinear mesh and (b) overset or Chimera mesh.

this can give rise to conservation errors. This leaves the question of which has the biggest impact, that is, the low quality global grid that does not have localized zones with incomplete conservation or the globally higher quality grid with the interface zones at which local accuracy can be compromised. Chimera grids offer the possibility of modelling moving body flows most readily. For example, in Figure 3.9b, the impact of flap oscillations could be readily explored. The flap mesh would undergo a solid body rotation. The cut-out zone and interpolation stencils would need reconfiguring.

An example of a hybrid unstructured grid (strategy b), applicable to high Reynolds number aerodynamics flows (see Smith (1996)) is shown in Figure 3.10. As can be seen, the surface grid has a relatively high aspect ratio. Hence, the high near wall

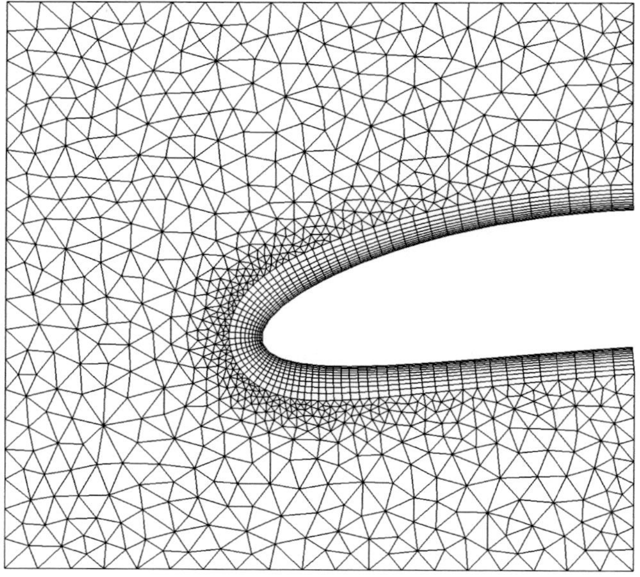

Figure 3.10. Hybrid grid example.

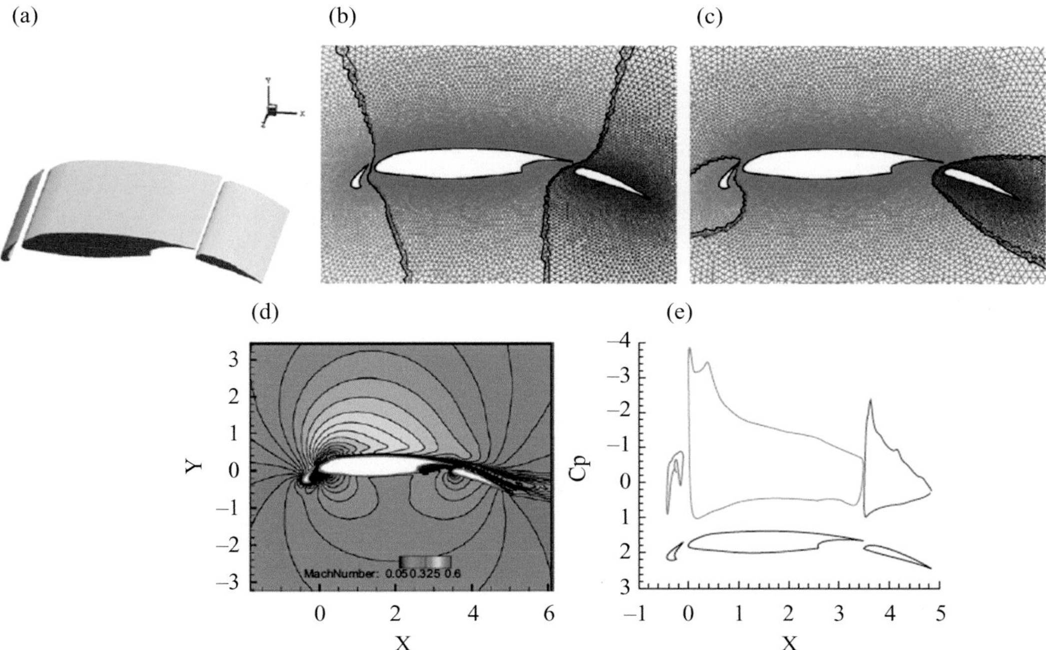

Figure 3.11. Use of unstructured overset grid for a slat-wing-flap geometry: (a) geometry, (b, c) unstructured overset meshes, (d) flow field and (e) surface pressure coefficient (From Xia et al. (2010) – Published with the kind permission of Elsevier).

velocity gradients can be resolved. Away from walls, where the variable gradients are milder, computationally less efficient triangulated cells are used. The computational efficiency of such cells is explored later.

For extremely complex geometries that potentially have body relative movement, the unstructured and overset grid methodologies can be combined. For example, Figure 3.11a shows a wing section with slat and flap. In Figure 3.11b, two-dimensional unstructured overset meshes for the slat, wing and flap are shown, and the interface between each domain is defined by medial axes (see Section 3.3.2.1). An inviscid flow solution (at Ma $= 0.3$, 5° angle of attack) obtained on the overset mesh is shown in frame (c). The Mach number contours and surface pressure coefficient are shown in frames (d) and (e), respectively. The use of unstructured overset mesh is powerful. For example, Nakahashi and Togashi (2000) use the approach to study a rocket being released from a launch vehicle.

For highly complex geometries, such as modelling cities, under bonnet flows in cars, flight vehicles in formation and so forth, unstructured octree type meshes are probably most appropriate. Such flow tend to have little classical boundary layer content, the geometries being non-streamlined. This gives rise to much separated flow. Hence, resolution of fine near wall boundary layers is of little relevance. Figure 3.6b, discussed earlier, gives a mesh of this type for an aircraft with deployed landing gear. Some specialized treatment of the surface cells is needed. To a first order

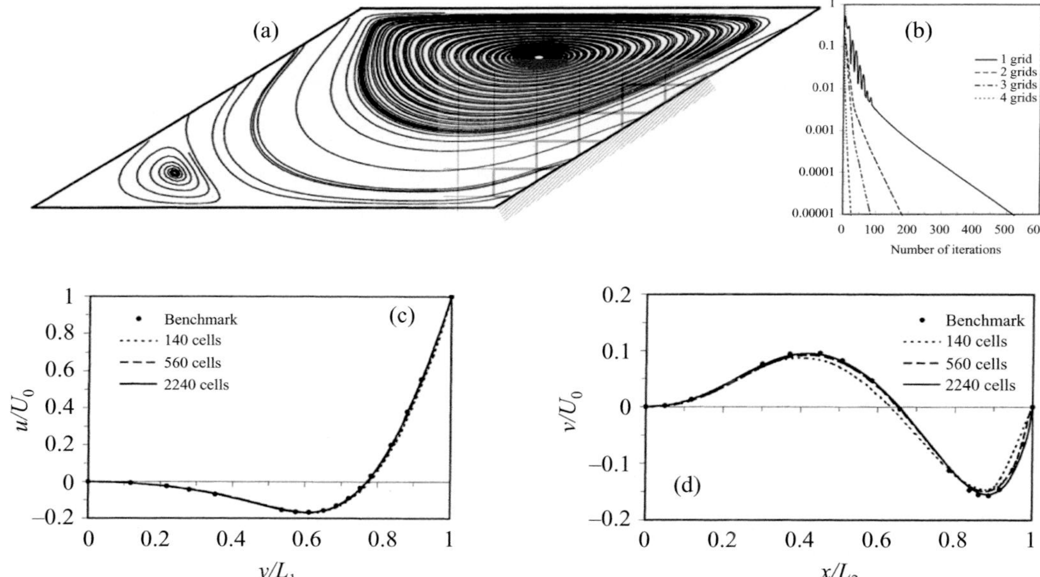

Figure 3.12. Use of hybrid trimmed-cell-IBM method for an inclined driven cavity: (a) flow streamlines (inset coarsened mesh topology), (b) multigrid convergence, (c) axial velocity profile and (d) vertical velocity profile (adapted from Tucker and Pan (2000) – Published with the kind permission of Elsevier).

approximation, the cells can be left as castellated. For higher accuracy, they can be trimmed to match the surface. Alternatively, see Dawes et al. (2009) – as indicated earlier, a surface layer mesh can be inserted as used in Figure 3.6b.

The IBM is also well suited to highly complex geometries and where there is substantial separated flow and hence wake drag. Although the method can be made accurate for capturing skin friction on attached boundary layers, for high-speed flow the approach is very inefficient. This is because it does not enable the generation of the high aspect ratio cells needed for high-speed boundary layers. To an extent, this problem can be alleviated through the use of wall functions as discussed in Chapter 5. The method is well suited to eddy-resolving simulations (again see Chapter 5). For such approaches, high numerical accuracy is needed. The hexahedral cells provide this. Also, the regular grid structure supports the use of simple high-order schemes. Figure 3.12 illustrates the use of a hybrid trimmed-cell-IBM method for an inclined driven cavity. The results are from Tucker and Pan (2000). Frame (a) shows flow streamlines, the inset being a schematic of a much coarsened (for illustrative purposes) mesh topology. Frame (b) shows the multigrid convergence acceleration. Frames (c) and (d) show mid-plane axial and vertical velocity profiles. As can be seen, accurate results can be gained and the geometric multigrid convergence acceleration (discussed in Chapter 4) is still effective.

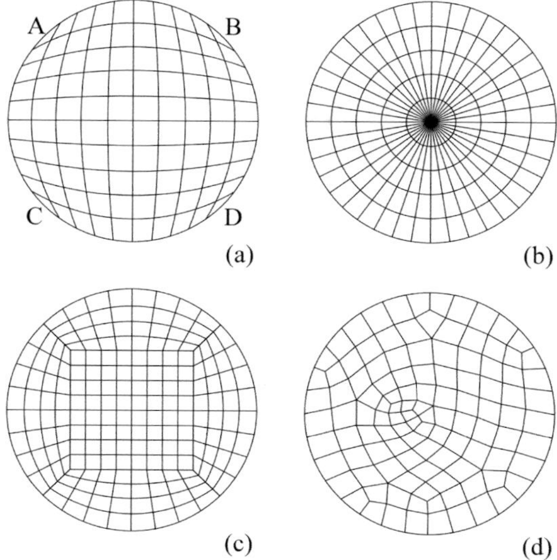

Figure 3.13. Four meshing approaches that could be applied to a circle: (a) 'H' grid, (b) 'O' grid, (c) butterfly grid and (d) paved grid (from Tucker and Mosquera (2001) – Published with the kind permission of NAFEMS).

3.2.4 *Mesh-Solver Compatibility*

It needs to be carefully considered how the particular solver will use the mesh supplied to it. For example, the cells generated might directly be used as control volumes (i.e., the full lines in Figure 3.44). Alternatively, the centres of the edges can be connected to the centroids of the cells to form the dashed control volumes in Figure 3.44. This is called the median dual cell vertex approach. Other possible variants of constructing median dual cells are described in the next chapter. However, it is sufficient to say here that clearly very different control volumes can arise from the same mesh. For the median dual mesh, the wall zone can be problematic. A half control volume is left at boundaries, which needs special treatment. Although the cell shape might broadly be theoretically compatible with the flow solver being used, this does not mean that plausible results will arise.

When creating structured CFD grids, geometries that cannot be decomposed into simple rectangles and blocks can create problems. For example, the meshing of a circle shown in Figure 3.13 gives different possible approaches to this problem. The grid in Frame (a) (called an 'H' grid) considers the circle to be a rectangle with the four sides AB, BD, DC and CA. Creating the grid is simplified but the cell quality is poor. Cells become very distorted near the 'corners' A, B, C and D. The circle circumference is discretised and the angle at the corners is nearly 180°. Some solvers handle this well; others will not work at all. The grid ('O' grid), shown in Frame (b), gives better cell quality throughout the grid, but the cells at the centre meet at a singularity, the circle centre. Again some solvers will accept this and others will not. The butterfly grid shown in Frame (c) decomposes the circle into five rectangular

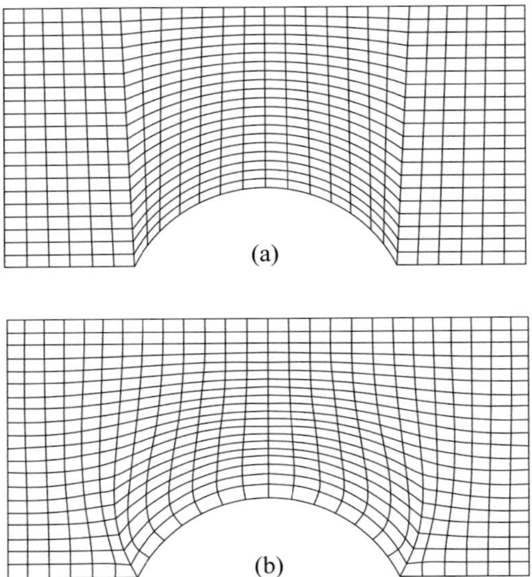

Figure 3.14. Algebraic and differential equation generated grids: (a) algebraic and (b) differential (from Tucker and Mosquera (2001) – Published with the kind permission of NAFEMS).

areas (topologically). The central square is surrounded by four four-sided regions, one on each face. A good quality grid is created which most solvers can handle well. The paved grid shown in Frame (d) is an irregular arrangement of four-sided cells to create an unstructured grid and can normally only by handled by specific solvers. Hence, it is important to consider the mesh and solver together and generate a mesh that is compatible with the flow solver and also the nature of the flow physics to be resolved. These aspects are discussed further later.

3.3 Structured Mesh Generation Techniques

3.3.1 *Basic Body Fitted Mesh Generation*

3.3.1.1 Algebraic The two most common body fitted grid generation techniques are algebraic and differential equation based. The former method involves the use of relatively simple algebraic interpolation expressions. Its advantages are that it is fast and can be used to provide an initial guess to other more refined grid generation techniques. Its disadvantages are that it will propagate the consequences of boundary singularities (corners) into the domain. Also grid points can be generated that fall outside the physical domain (over-spill). However, this can be corrected for. Figure 3.14a gives an example of an algebraically generated grid for a geometry with a hump. The most widely used algebraic grid generation technique is transfinite interpolation. This is described next.

3.3.1.2 Transfinite Interpolation Transfinite interpolation involves use of the Lagrange (polynomial) weighting functions: $1-\xi$, ξ, $1-\eta$, η. Figure 3.15 shows

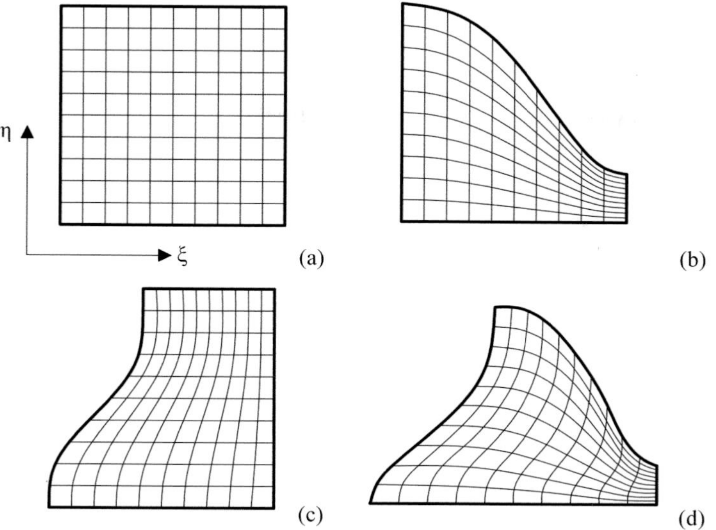

Figure 3.15. Various forms of transfinite interpolated grids: (a) computational grid, (b) η interpolated grid, (c) ξ interpolated grid and (d) ξ-η interpolated grid (adapted from Persson).

various forms of transfinite interpolated grids. Frame (a) shows the computational plane grid. Frame (b) shows a grid generated using weighting functions just based on η. These functions, in interpolation equations, are given as

$$
\begin{aligned}
x(\xi,\eta) &= (1-\eta)\,xb_s(\xi) + \eta\, xb_n(\xi) \\
y(\xi,\eta) &= (1-\eta)\,yb_s(\xi) + \eta\, yb_n(\xi)
\end{aligned} \tag{3.1}
$$

Note, xb and yb are boundary locations the subscripts giving their geographic locations (w = west, e = east and so on). Frame (c) shows an alternative, ξ, interpolated grid

$$
\begin{aligned}
x(\xi,\eta) &= (1-\xi)\,xb_w(\eta) + \xi xb_e(\eta) \\
y(\xi,\eta) &= (1-\xi)\,yb_w(\eta) + \xi\, yb_e(\eta)
\end{aligned} \tag{3.2}
$$

Combination of these allows construction of the two-dimensional interpolation expressions

$$
\begin{aligned}
x(\xi,\eta) = {} & (1-\eta)xb_s(\xi) + \eta\, xb_n(\xi) + (1-\xi)xb_w(\eta) + \xi xb_e(\eta) \\
& - \{\xi\eta\, x_n(1) + \xi(1-\eta)x_s(1) + \eta(1-\xi)x_n(0) \\
& + (1-\eta)(1-\xi)\, x_s(0)\}
\end{aligned} \tag{3.3}
$$

and

$$
\begin{aligned}
y(\xi,\eta) = {} & (1-\eta)yb_s(\xi) + \eta\, yb_n(\xi) + (1-\xi)yb_w(\eta) + \xi yb_e(\eta) \\
& - \{\xi\eta\, y_n(1) + \xi(1-\eta)y_s(1) + \eta(1-\xi)y_n(0) \\
& + (1-\eta)(1-\xi)\, y_s(0)\}
\end{aligned} \tag{3.4}
$$

The following corner identities are involved in equations (3.3) and (3.4)

$$\left(x_s(0), y_s(0)\right) = \left(x_w(0), y_w(0)\right)$$
$$\left(x_w(1), y_w(1)\right) = \left(x_n(0), y_n(0)\right)$$
$$\left(x_n(1), y_n(1)\right) = \left(x_e(1), y_e(1)\right)$$
$$\left(x_e(0), y_e(0)\right) = \left(x_s(1), y_s(1)\right)$$

Frame (d) shows an ξ-η interpolated grid resulting from the application of equations (3.3) and (3.4)

3.3.1.3 Differential Equation Mesh Generation The most common differential mesh generation technique uses an elliptic partial differential equation. The process of elliptic grid generation is based around the Poisson equation (i.e., the conduction of heat equation with sources). Hence, the expected isosurfaces of temperatures and the isosurfaces of gradients normal to these, defining the heat flux, would be expected to be smooth. Therefore, elliptic grid generation should produce smooth grids with good Jacobians (see Equation 3.39 later). Figure 3.14b gives an example of an elliptically generated mesh. Theoretically, a valid Jacobian (a term needed in the numerical discretization – see Chapter 4) is guaranteed. However, when the *discrete* differential equation is subject to numerical solution, this aspect is less certain. Nonetheless, high quality grids are expected that would be well suited to eddy-resolving simulations.

The boundary conditions for the elliptic equations are the physical grid point locations at the boundaries (i.e., x and y values in two-dimensions). If interior grid points in the x-y plane were known, these could be mapped to a simple square computational ξ-η plane. This would be achieved using the transform Jacobian (see Chapters 2 and 4). The interior transformed plane grid points could be gained from solving the following equations

$$\begin{aligned}\frac{\partial^2 \xi}{\partial x^2} + \frac{\partial^2 \xi}{\partial y^2} &= S_1(\xi, \eta)\\ \frac{\partial^2 \eta}{\partial x^2} + \frac{\partial^2 \eta}{\partial y^2} &= S_2(\xi, \eta)\end{aligned} \tag{3.5}$$

The source terms S_1 and S_2 enable grid control. Equation (3.5) can be transformed to solve for x and y as

$$\alpha \frac{\partial^2 \phi}{\partial \xi^2} + \beta \frac{\partial^2 \phi}{\partial \eta^2} = S \tag{3.6}$$

where $\phi = x$ or y and the source has the following form

$$S = -I^2 \left(S_1 \frac{\partial \phi}{\partial \xi} + S_2 \frac{\partial \phi}{\partial \eta} \right) + 2\left(x_\xi x_\eta + y_\xi y_\eta\right) \frac{\partial}{\partial \xi}\left(\frac{\partial \phi}{\partial \eta}\right) \tag{3.7}$$

where

$$\alpha = \left(\frac{\partial x}{\partial \eta}\right)^2 + \left(\frac{\partial y}{\partial \eta}\right)^2 \tag{3.8}$$

$$\beta = \frac{\partial x}{\partial \xi}\frac{\partial x}{\partial \eta} + \frac{\partial y}{\partial \xi}\frac{\partial y}{\partial \eta} \tag{3.9}$$

and

$$I = \frac{\partial x}{\partial \xi}\frac{\partial y}{\partial \eta} - \frac{\partial x}{\partial \eta}\frac{\partial y}{\partial \xi} \tag{3.10}$$

Hence, essentially Equation (3.6) is solved for the grid point locations x and y subject to Dirichlet boundary conditions. The latter are the surface grid point coordinates.

Defining the grid control functions S_1 and S_2 is complex and discussed in Pletcher et al. (2013). If it is desired to have surface normal near wall grid points, further action is required. For moving grids, the approach would be slow. Even though grid control can be challenging, with care, high quality grids can be generated. Also, boundary condition discontinuities are not propagated into the domain.

Other well-known methods of differential grid generation are the parabolic and hyperbolic methods. The former essentially tries to treat the equation (3.5) based elliptic procedure in a parabolic manner. Clearly, there is a compatibility issue. If j is the subscript moving away from a surface, then the derivatives in the elliptic operators need information at $j-1, j$ and $j+1$. Data at $j-1$ and j will be available and allow a parabolic process, but data at $j+1$ will not be available. However, $j+1$ data associated with the outer boundary can be used or data based on, say, a crude algebraic grid generation. Furthermore, once a parabolic sweep has been made, this could be refined using $j+1$ information based on the previous parabolic sweep. Hence, what is gained, with this parabolic approach, is a non-converged elliptic grid generation solution. Hyperbolic grid generation equations can also be made. Clearly for this, and also parabolic grid generation, the intent is to use these methods for external flows such as airfoils or for generating just surface layers. The hyperbolic approach will propagate boundary condition discontinuities into the core domain. However, it is fast. For three-dimensional systems, iterative solution of the Poisson equation is relatively expensive (the computational cost can be reduced by using an algebraic initial guess). Also, for external flows, a fictitious far field boundary condition must be defined. For such systems, hyperbolic and parabolic differential equation techniques are faster.

3.3.2 *Multiblock Meshes*

Although the structured, multiblock meshes can be challenging to generate, they have many attractions. Their structured nature allows for greater computational speed and their shape gives rise to substantial accuracy improvements (see Section 3.6.1). This is especially so when it is considered that high-order schemes can be relatively easily implemented on such grids. For a complex geometry, the breaking

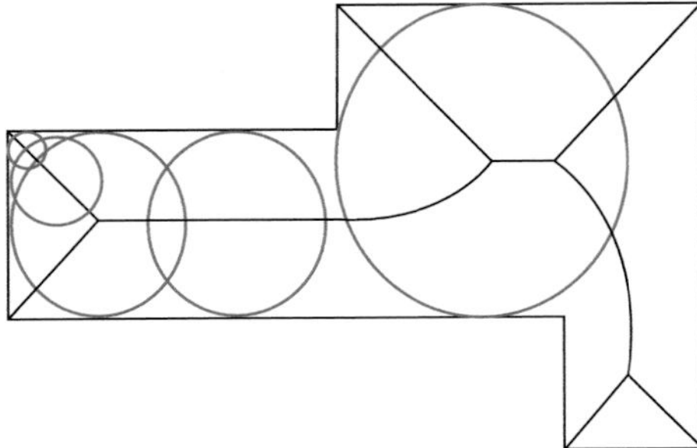

Figure 3.16. Example of generating the medial axis through locus of the maximal circle.

down of the domain into blocks, readily tessellated by a structured hexahedral, can present a substantial challenge. To carry out the process manually requires a lot of experience, skill and time – potentially several weeks. Hence, there is a need for a degree of automation.

3.3.2.1 Medial Axis The most widely known method for automatically generating sub-block zones is through the use of the medial axis. This can be generated by rotating a circle in a two-dimensional geometry or a sphere for three dimensions, tracing the locus of the centre of the circle/sphere. This will automatically theoretically generate the medial axis. Figure 3.16 shows a two-dimensional example of this process. There is a range of approaches to actually generating the medial axis in practice. For example, the axis corresponds to the maximal wall distance in the geometry. Hence, for example, the level-set (wall distance) equation can be solved (on say a coarse triangulated grid that can easily be automatically generated or a Cartesian cut cell mesh). The maximal wall distance can be extracted as the maximum level-set value. There are then a range of extraction processes and means of ensuring a thin, unambiguous surface is gained. For example, Xia et al. (2011) use pixel thinning for solutions based on a Cartesian background mesh, and Xia et al. (2010) use a Voronoi diagram for surface thinning. The other approach is to use long triangulated shards between the boundaries (constrained Delaunay triangulation of points distributed on the boundary) and extract the centroids of these (see Tam et al. 1991, Price et al. 1995, 1997). This will directly give the maximal wall distance or level-set. Indeed, this is a macro version of the approach of Xia et al. (2010).

The level-set equation, in its boundary value formulation, models fronts propagating from surfaces. This can be seen in Figure 3.17. Notably, in concave geometry zones there is a collision of fronts and these are analogous to compression shocks. These shocks correspond to the medial axis location. Similarly for convex geometry zones, there are expansion shock analogous zones (i.e., fronts departing from each other). All these zones can be readily detected through classical shock detection techniques, used in CFD, and can be potentially useful when attempting to capture the medial axis (compression shock zones) and other features useful for defining multiblock meshes.

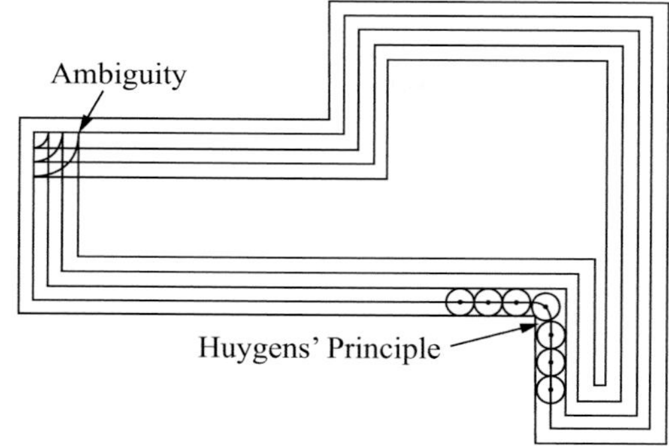

Figure 3.17. Front propagation implicit in level-set equation.

The raw medial axis (or Medial Axis Transform – MAT) shown in Figure 3.16 will not define a useful hexahedral mesh. To facilitate meshing, rules need to be introduced. A useful and widely used rule (R1) is that triangular domains are further sub-divided into quadrilateral zones. This is done by connecting the centre of the triangle to the midpoints of the edges, thus allowing simpler quadrilateral meshing. The edge centre and centroid locations do not need to be rigid locations but serve as an obvious nominal position. Two further useful rules (R2) and (R3) are given by Rigby (2004). R2 is to extend hanging features (features which are not connected to anything) to the nearest point on the geometry. There are no such features in Figure 3.16. For R3, expansion features (see Figure 3.17) are connected to the nearest 'medial vertex'; that is, if that point lies within sight of the fan caused by the expansion feature. If not, it is connected to the nearest point on the shock feature topology. Using these rules and the medial axis, the mesh show in Figure 3.18 can be produced.

However, the use of the medial axes, even with rules, generally has strong limitations. For example, Figure 3.19 shows the medial axis for a jet-wing-flap configuration. As can be seen, for this external flow system, the medial axis is no use for providing a framework for developing a multiblock hexahedral mesh. There are alternative approaches to the medial axis and these are outlined next.

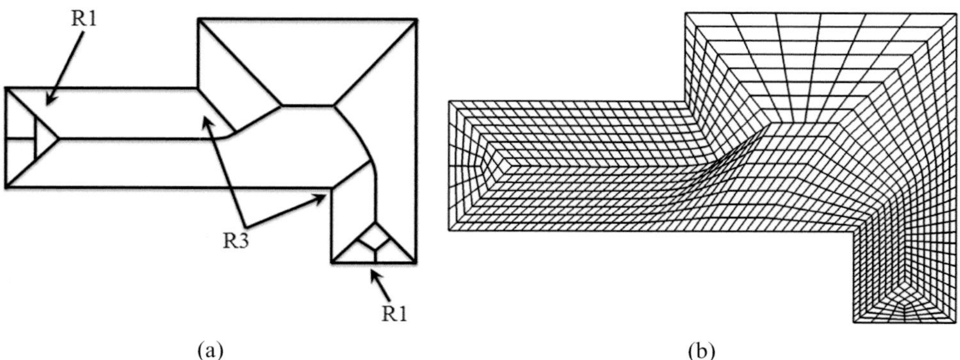

Figure 3.18. Mesh boundaries and hexahedral mesh generated from medial axis with rules: (a) mesh boundaries and (b) final mesh.

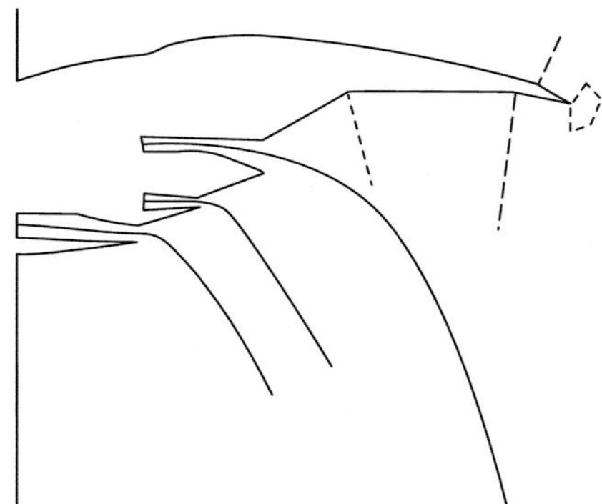

Figure 3.19. Medial axis for a jet-wing-flap configuration.

3.3.2.2 TopMaker Figure 3.20 outlines the hybrid TopMaker approach of Rigby (2004). This combines use of the medial axis and vertices. The left-hand frame in Figure 3.20 shows the medial axis with the medial vertices identified by the circular symbols. The middle frame shows how, through the use of intuitively obvious rules, the medial vertices are converted to block boundaries. These block boundaries are then further supplemented by medial axis data. For details of the full procedure and rules used see Rigby (2004).

3.3.2.3 Cartesian Fitting Figure 3.21 illustrates the creation of a block topology for a different geometry, using the so-called *Cartesian fitting* approach of Malcevic (2011). Frame (a) gives the geometry. This is a seal type geometry as found in turbomachinery. As shown in Frame (b), this is morphed to a Cartesian geometry. Then, as shown in Frame (c) this geometry has a reasonably obvious intermediate blocking. This blocking can then be morphed back to the original geometry to give finally the blocking shown in Frame (d).

Figure 3.22 shows the blocking topologies that would be gained when using the MAT and TopMaker approaches for the geometry shown in Figure 3.21a. Frame (a) gives the medial axis blocking and Frame (b) that for TopMaker. As can be seen, very different blocking styles can be gained depending on the approach used. Note that in all of these blockings, additional rules are required akin to those noted previously. These rules are discussed in detail in Rigby (2004) and Ali and Tucker (2013).

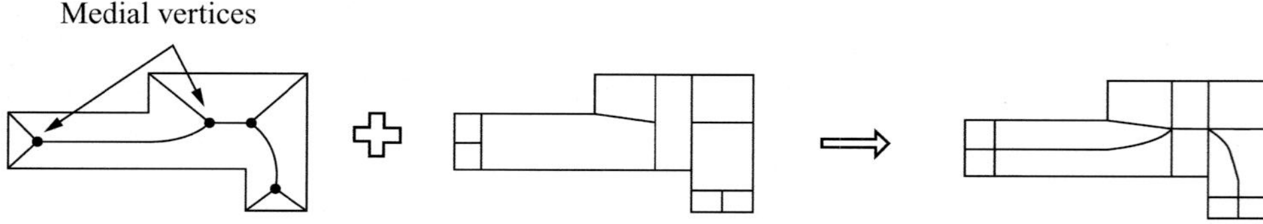

Figure 3.20. The hybrid TopMaker approach of Rigby (2004) combining use of the medial axes and vertices.

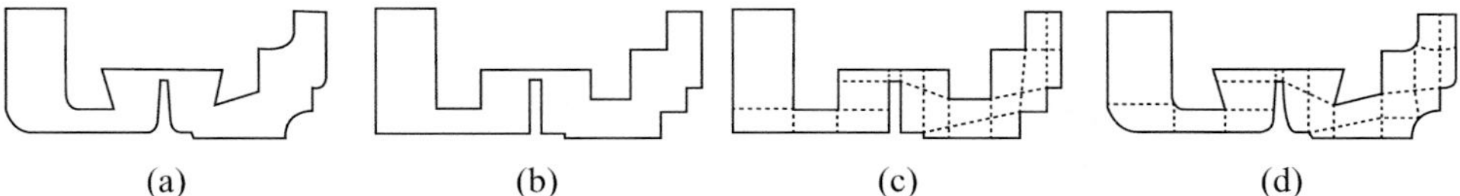

Figure 3.21. Creating a block topology using the Cartesian fitting approach of Malcevic (2011): (a) original geometry, (b) Cartesian fitted geometry, (c) intermediate blocking and (d) final blocking (Figures from Malcevic (2011) with kind permission from the author).

3.3.3 *Further Examples Showing Application of Automatic Blocking Approaches*

Next, some examples of the approaches just given will be shown for some alternative geometries. First, another seal geometry is considered – a labyrinth seal. Figure 3.23 compares different blocking approaches and resulting meshes. Frame (a) gives the medial axis blocking and mesh. Similarly, Frames (b), (c) and (d) are for TopMaker, Cartesian fitting and manually generated meshes. The latter is generated by an expert in CFD but who is not perhaps a meshing specialist. Notably, the manually generated grid is quite similar to that of TopMaker. Also, there are substantial variations in the resulting grids.

Figure 3.24 compares different blocking approaches and resulting meshes for an idealized aeroengine intake in crosswinds. The crosswind goes from left to right. The outer boundaries are chosen to match the stream tube of the crosswind as it becomes redirected around the intake lip. Figure 3.25 shows contours and streamlines of variable fields for the intake. Frame (a) gives velocity contours and streamlines of the velocity field around the lip. Frame (b) shows the adjoint counterpart for the eddy viscosity variable used in the turbulence model. The zone of sensitivity around the lip can be partially inferred from the darker (signifying higher values) upstream contours in these plots.

In Figure 3.24, (a) shows the medial axis, (b) TopMaker, (c) Cartesian fitting and (d) the manually generated meshes. Again substantial differences can be observed but it is hard to say which is the best mesh/blocking topology. This is especially so for a potential design variable of interest such as mass flow, loss, lift or drag coefficient. Adjustments in local grid refinement can be used to assess local errors, but these can be misleading. The numerical errors effectively propagate with the flow. Hence, a substantial solution error at a downstream location can be caused by deficiencies in the upstream grid. A more reliable test of grid quality is to study adjoint variables. The precise mathematical basis for these is discussed later and so this is not dwelt upon just now.

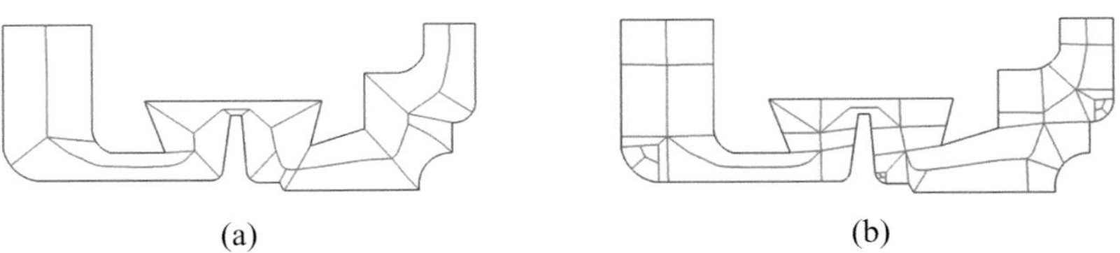

Figure 3.22. Medial axis and TopMaker blockings: (a) medial axis and (b) TopMaker.

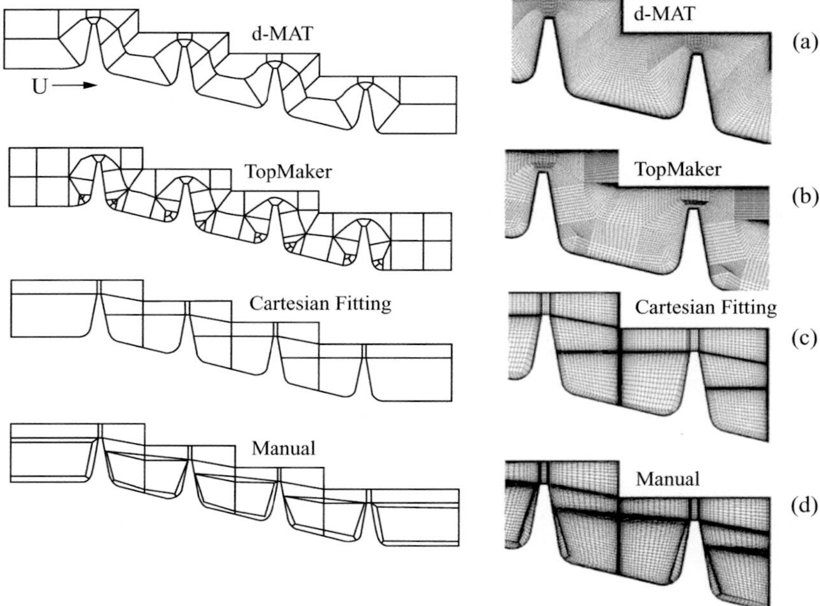

Figure 3.23. Comparison of different blocking approaches and resulting meshes for a labyrinth seal: (a) medial axis, (b) TopMaker, (c) Cartesian fitting and (d) manually generated (from Ali and Tucker (2013) – Published with the kind permission of Springer).

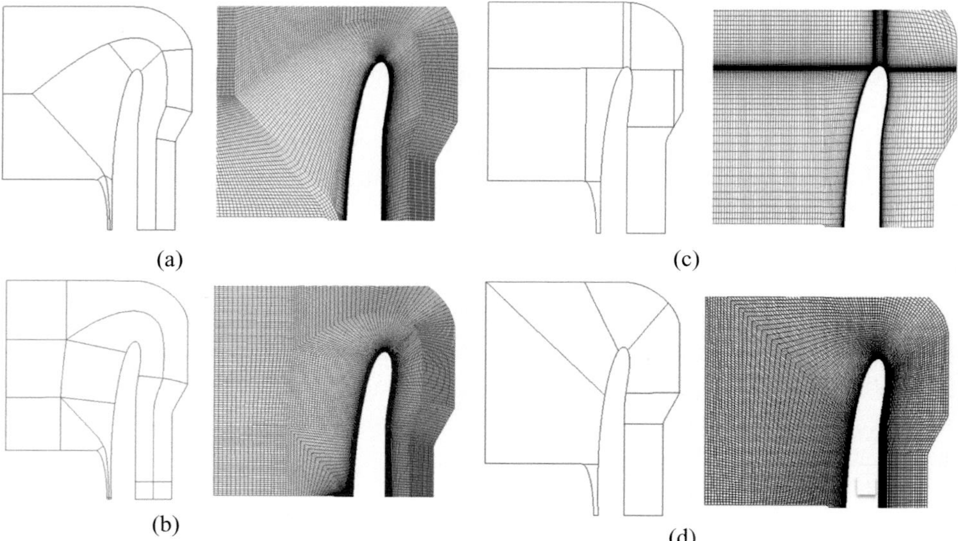

Figure 3.24. Comparison of different blocking approaches and resulting meshes for an idealized aeroengine intake in crosswinds: (a) medial axis, (b) TopMaker, (c) Cartesian fitting and (d) manually generated (from Ali and Tucker (2013) – Published with the kind permission of Springer).

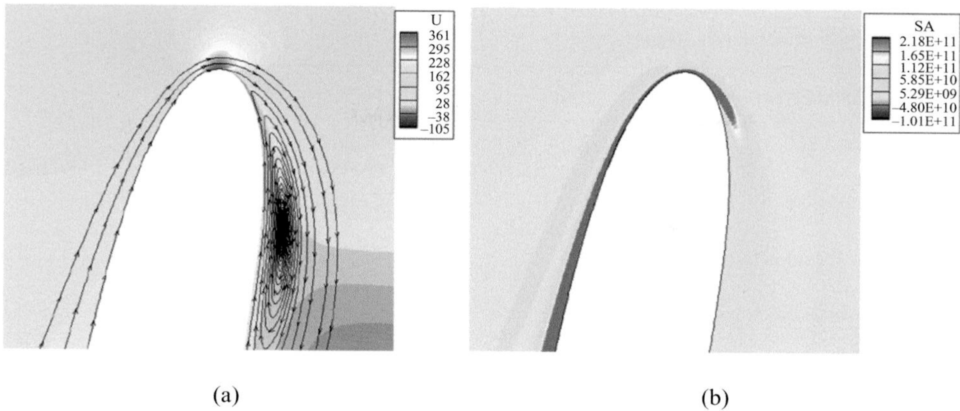

Figure 3.25. Variable fields for intake lip case: (a) velocity field around lip and (b) adjoint counterpart for eddy viscosity variable in turbulence model (from Ali and Tucker (2013) – Published with the kind permission of Springer).

Table 3.1 shows the mesh cell count, N, and the normalized total error, TE_N, estimates for the block topologies. The error in cell-wise, integrated, total pressure loss, based on the adjoint procedure is described later. It is normalized by the error value for the MAT grid simulation, making use of the Spalart-Allmaras (SA) turbulence model outlined in Chapter 5. The results indicate that the medial axis based meshes provide the most accurate values of the objective function (total pressure loss) relative to the Cartesian fitting and the manual block topologies. One of the reasons for this is that for a similar cell count, the medial axis based approach generates a mesh which has a more uniform cell size distribution in the main flow path than the Cartesian fitting approach and it also has better flow alignment. It is interesting to note that the SA model gives greater accuracy relative to the more advanced k-ω SST model (see Chapter 5) for the same grid. This is perhaps not so surprising. The SA model is specifically designed to give sensible near wall behaviour on relatively coarse meshes. This is because the key solution variable is designed to be a linear function in the log law zone.

Table 3.1. *Error in total pressure loss for labyrinth seal for different mesh topologies*

Blocking Type	Model	N	TE_N
d-MAT	SA	36,972	1.00
	$k-\omega$ SST	36,972	1.11
TopMaker	SA	35,507	1.13
	$k-\omega$ SST	35,507	1.17
Cartesian fitting	SA	41,240	2.41
	$k-\omega$ SST	41,240	2.53
Manual blocking	SA	41,882	2.47
	$k-\omega$ SST	41,882	2.55

Table 3.2. *Error in total pressure loss for engine intake with different mesh topologies*

Blocking Type	Model	N	TE_N
d-MAT	SA	26,564	1.00
	$k - \omega$ SST	26,564	1.14
TopMaker	SA	29,333	1.37
	$k - \omega$ SST	29,333	1.55
Cartesian fitting	SA	27,536	112
	$k - \omega$ SST	27,536	128
Manual blocking	SA	26,712	20
	$k - \omega$ SST	26,712	22

Table 3.2 compares the cell count and the normalized estimated error for the various methods. Due to the same reasons described for the labyrinth seal case, the MAT approach outperforms the others. The table again shows the slightly better performance of the SA model for each blocking. Further analysis (see Ali and Tucker 2013) indicates that the medial axis based approach performs well because it again has a more uniform cell size distribution and better flow alignment. Notably, for industry, where designs are relatively incremental, and only small changes in form take place between different designs, use can be made of standard blocking templates. These can be optimized using, for example, an adjoint based assessment procedure akin to the one described previously.

3.3.4 *Hybrid Blocking Approaches*

As has been noted, the medial axis does not provide any useful information for the blocking of external flows such as the wing-jet geometry. However, the level-set contours, for which the maximal values provide the MAT, can be used to generate a surface mesh zone (if the maximal level set/distance is given a specified upper limit). This is shown in Figure 3.26. Frame (a) shows contours of the level set/wall

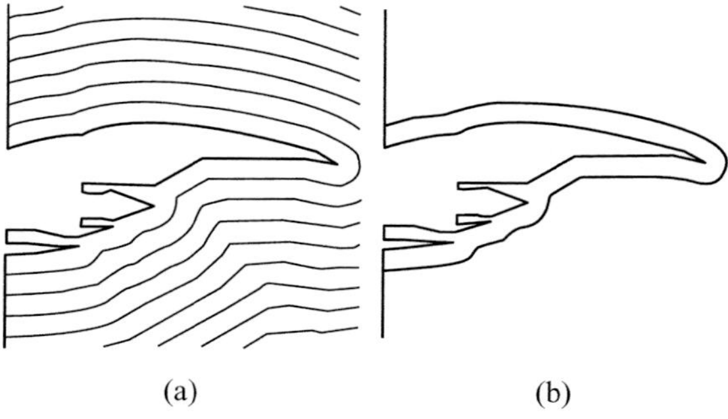

Figure 3.26. External flow geometry and distance contours for the level-set equation: (a) multiple contours for increasing distance and (b) near surface contour (from Ali and Tucker (2013) – Published with the kind permission of Springer).

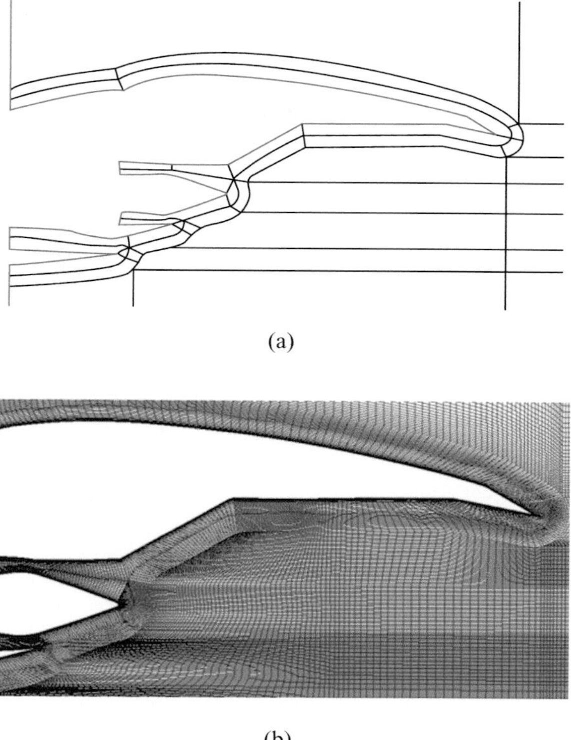

(a)

(b)

Figure 3.27. Hybrid meshing strategy for jet-wing-flap geometry: (a) hybrid blocking and resulting mesh (from Ali and Tucker (2013) – Published with the kind permission of Springer).

distance for a large extent of the domain. Frame (b) shows a surface local level set.

The near surface zone in Figure 3.26 can then be combined with the Cartesian fitting approach to generate a valid blocking. This is shown in Figure 3.27, where Frame (a) gives the hybrid blocking and Frame (b) the mesh. This example is for a two-dimensional geometry. However, the extension to three dimensions is practical. The use of the surface level set performs the very simple process of geometry defeaturing. The latter eases the meshing task and is discussed in Chapter 7.

As a final example of hybrid blocking, Figure 3.28 shows application of this approach to a tandem airfoil. Frame (a) gives the medial axis and Frame (b) the level sets of wall distances. Frame (c) shows the hybrid blocking and (d) the resulting mesh.

3.4 Unstructured Mesh Generation Techniques

Generally, only triangles and tetrahedrals are amenable to direct automatic mesh generation techniques. However, the latter can be combined to give hexahedrals. Basic automatic meshing strategies are described next.

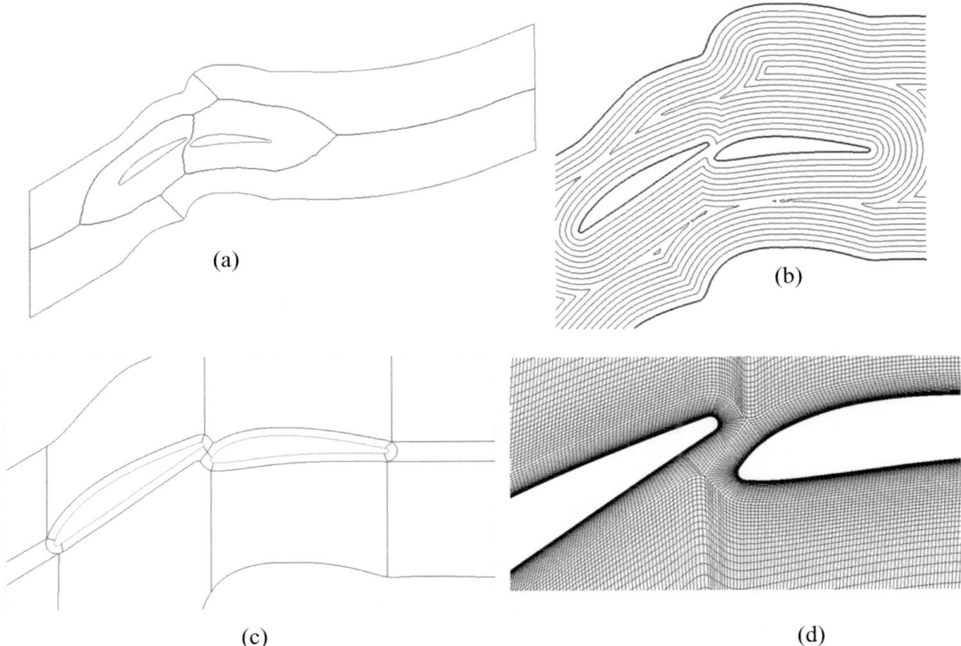

Figure 3.28. Example of hybrid meshing for tandem airfoils: (a) medial axis, (b) level sets of wall distance, (c) hybrid blocking and (d) resulting mesh (from Ali and Tucker (2013) – Published with the kind permission of Springer).

3.4.1 *Grid Based Methods*

With grid-based methods, in two dimensions, an essentially uniform grid is generated over a rectangular domain. A complex geometry can then be stamped out of this. Cut surface cells are modified to match the complex domain boundary. Figure 3.29 gives a schematic showing the grid based mesh generation process. The upper frame shows the background grid and the lower the modified mesh stamped out from this. The method has the advantage that it is fast. However, a key disadvantage is the poor quality of the generated boundary cells.

3.4.2 *Advancing Front Method*

In two dimensions, the advancing front method involves dividing the domain boundary into a series of line segments called the front. The method when meshing a square is shown in Figure 3.30. One of the front's segments is chosen to form the edge of an initial triangle. (In three dimensions, the process could involve tetrahedrons.) This initial front segment is then ignored, leaving a modified front. An element is then added to a line segment on the modified front and the procedure repeated until the domain is filled with triangles. The method has the advantages that it is conceptually

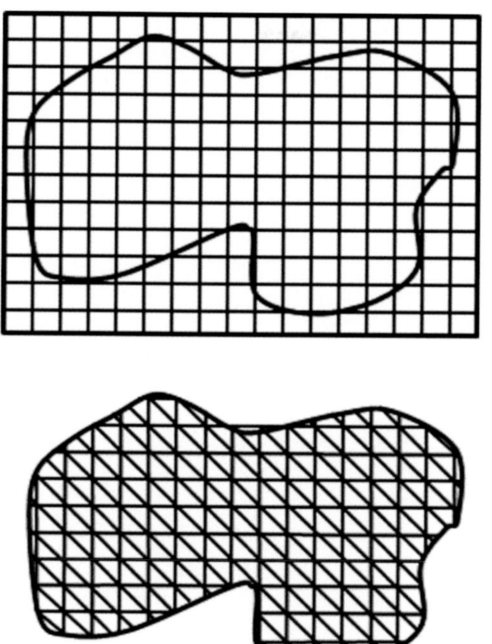

Figure 3.29. Schematic showing the grid based mesh generation process: upper frame background grid and lower frame mesh stamped out from this and modified.

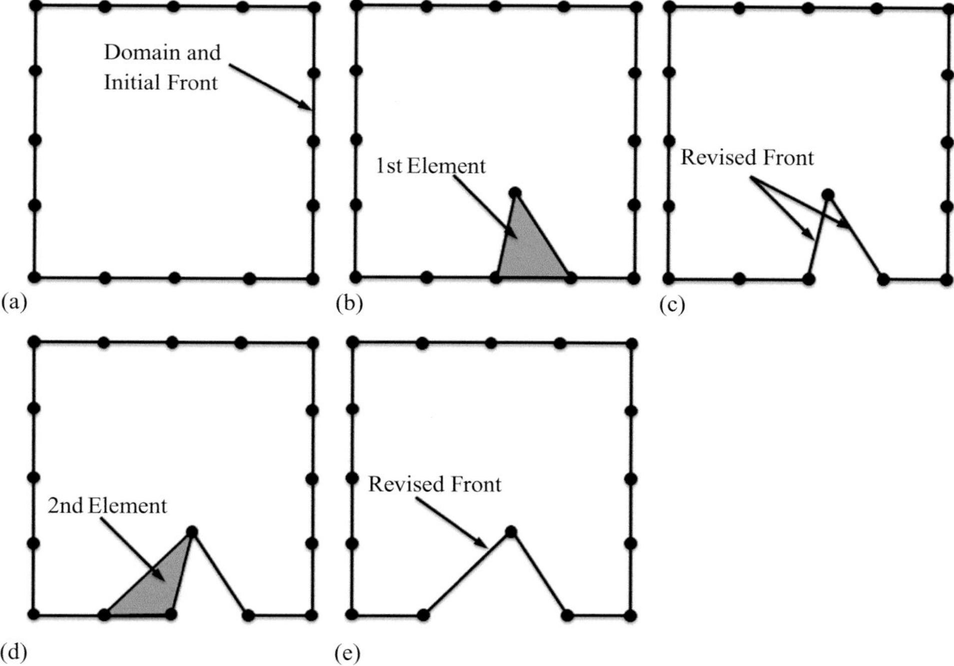

Figure 3.30. The advancing front method when meshing a square (adapted form Tucker and Mosquera (2001) – Published with the kind permission of NAFEMS).

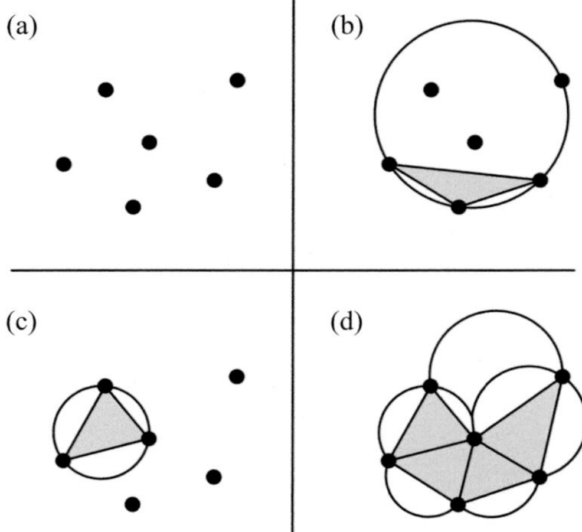

Figure 3.31. The Delaunay triangulation process: (a) point cloud, (b) invalid circumcircle, (c) valid circumcircle and resulting triangle and (d) mesh evolution process.

simple and naturally boundary conforming. However, in three dimensions, the mesh can be sensitive to the boundary surface parameterisation. A background grid that allows cell size, stretching and stretching direction to be defined can be used (see Lohner and Parikh 1988). For a uniform mesh, the background grid could consist of just one cell. Use of the background grid cell topology information is made to guide the process and advance the front. Further aspects of grid control are discussed later. Note, paved meshes use an advancing front process.

The advancing front method can be slow. Also, substantial overhead can be involved with grid control and ensuring that the grid density matches user requirements. According to Lohner (1996) the complexity of the advancing front method is O (N log (N)), where N is the number of elements. The advancing front process is inherently non-parallel. However, sufficiently decoupled/separated mesh zones can be processed in parallel.

3.4.3 *Delaunay Triangulation*

The Delaunay triangulation (refinement) technique allows greater speed. The approach seeks to ensure that no input point is inside the circumcircle of any triangle. Figure 3.31 shows the Delaunay triangulation process which underpins this grid generation method. Frame (a) shows the point cloud and Frame (b) shows an invalid circumcircle. Frame (c) shows a valid circumcircle and the resulting triangle. Frame (d) shows the mesh evolution process involving a set of valid circumcircles. To extend to three dimensions, the notion of spheres rather than circles is used. If well coded, the method is relatively fast. A key aspect of efficient Delaunay triangulation is quickly testing if a point encroaches on a triangulation. For two-dimensional triangles, one approach for testing if a point, s, encroaches (is within the circumcircle) a triangle p-q-r is to calculate the determinant

$$\begin{vmatrix} p_x & p_y & p_x^2+p_y^2 & 1 \\ q_x & q_y & q_x^2+q_y^2 & 1 \\ r_x & r_y & r_x^2+r_y^2 & 1 \\ s_x & s_y & s_x^2+s_y^2 & 1 \end{vmatrix} = \begin{vmatrix} p_x-s_x & p_y-s_y & (p_x^2-s_x^2)+(p_y^2-s_y^2) \\ q_x-s_x & q_y-s_y & (q_x^2-s_x^2)+(q_y^2-s_y^2) \\ r_x-s_x & r_y-s_y & (r_x^2-s_x^2)+(r_y^2-s_y^2) \end{vmatrix} > 0 \quad (3.11)$$

with encroachment the determinant is positive if the points are placed in a counter-clockwise order. Sua and Scott-Drysdale (1997) discuss a range of algorithms for Delaunay triangulation and their performances. The elements generated using Delaunay triangulation tend to be isotropic – the process attempting to maximise the minimum angle of triangles. A useful reference showing variants of the Delaunay method is Barth (1994). Key early algorithms for improving the quality of Delaunay meshes are that of Ruppert's (1995) point insertion. For an alternative approach see Chew (1993), Chew (1989). The relatively isotropic nature of the Delaunay triangulation can give numerical accuracy and hence such meshes are compatible with quasi-DNS type simulations. However poor elements can be produced. Hence, at the end of the automatic grid generation stage, checking and rework is required.

3.4.4 *Unstructured Quadrilateral/Hexahedral Meshes*

Quadrilateral/hexahedral meshes have the desirable property that for the same edge length they have fewer cells relative to other grid topologies and often better general numerical properties. Two-dimensional paving works well. However, the equivalent, three-dimensional plastering process, for complex geometries, can fail due to voids or poor cell quality (see Carey 2002). Hence, Carey devised the alternative approach of converting tetrahedron into hexahedron.

3.4.4.1 'Hexing the Tet' Figure 3.32, based on Carey (2002), shows the basis of converting a tetrahedron into a hexahedron. Frame (a) shows the tetrahedron A-B-C-D. The process involves connecting edge midpoints to face centroids. For example, consider tetrahedron face A-B-C. The centroid is P. The edge midpoints are E, F and G. The centroid to midpoint lines break the triangular face A-B-C into four quadrilaterals. A point X is inserted, ideally, optimally placed for grid quality. Focusing on the quadrilateral face B-E-P-F, and combining this with other centroid to mid-edge constructions along with point X, allows the formation of the hexahedron shown in Frame (b).

A notable aspect of the procedure is that it needs to be born in mind that the hexahedron mesh is a refinement to the tetrahedron mesh. Hence, the latter should be generated with a coarser resolution in mind. As pointed out by Carey (2002),

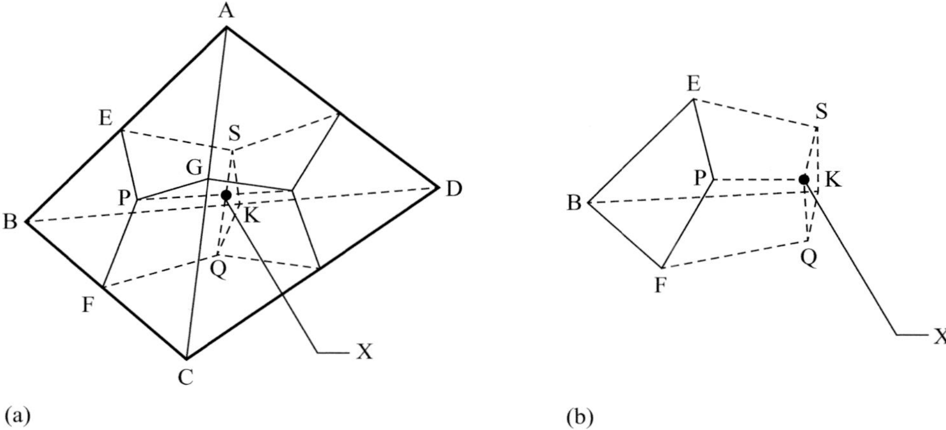

Figure 3.32. The basis of converting a tetrahedron into a hexahedron – 'Hexing the Tet' – (based on Carey 2002).

for a two-grid geometric multigrid convergence acceleration procedure (see Section 3.10.1.4), this grid generation method naturally provides a coarser grid.

3.4.4.2 Octree Meshes The term 'octree' refers to a type of data structure where a parent node has eight children. The two-dimensional equivalent is the quadtree structure with a parent having four children. Figure 3.33 indicates the quadtree data structure and how this links with a mesh. Refinement continues until the geometry

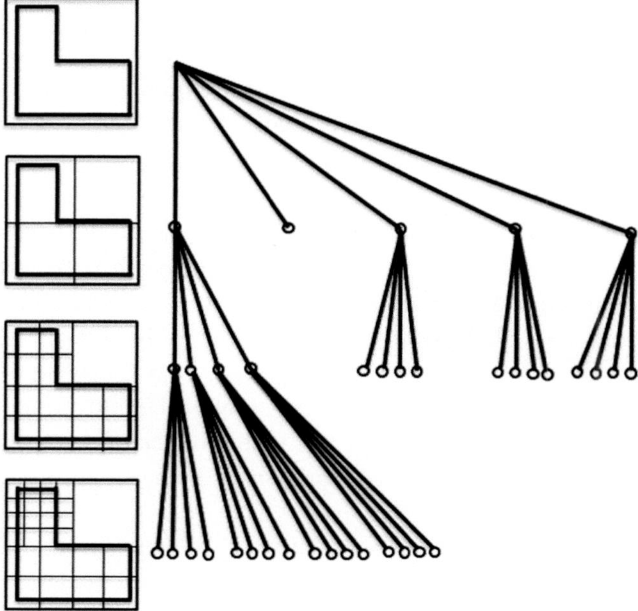

Figure 3.33. Quadtree data structure and grid generation.

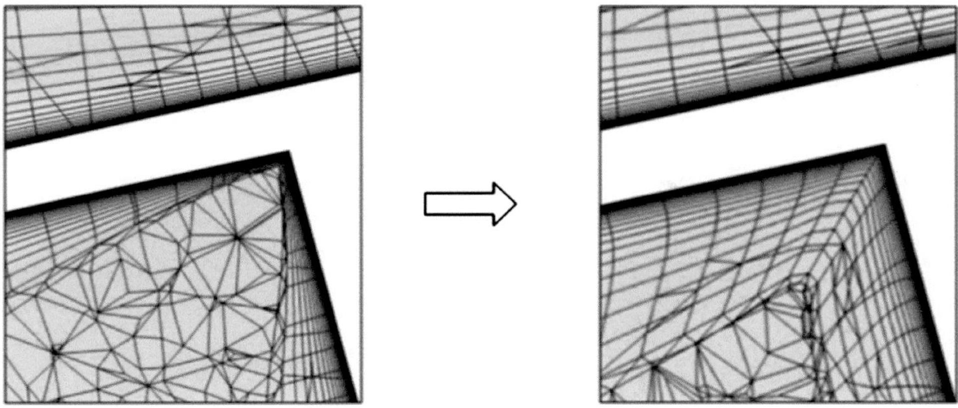

Figure 3.34. Hybrid grid generation in a convex corner: (a) incursion of unstructured cells into corner and (b) corrected mesh with adapted surface 'normals' (from Masi 2014).

(or even flow features) are well resolved. The extension to the three-dimensional quadtree structure is conceptually straightforward, the quadrilaterals being replaced by cuboids. Octree grids are typically fast to generate. Also, the clear data structure around which they are based can be exploited in the solver to define face area properties and hence speed up solver performance. The downside is the lack of boundary conformity and the inability to generate high aspect ratio cells. The cuboid elements generated with octree meshes are highly compatible with LES. There is no ambiguity with regard to the modelled turbulence length (filter) scale – most of the wide range of definitions giving the cuboid edge length as the modelled length scale. Also, the cuboid cells will give good numerical accuracy (see Section 3.6.1), which again is important for eddy-resolving simulations.

3.4.5 *Hybridization*

As shown by El-hamalawi (2004) and Frey et al. (1998), the advancing front and Delaunay triangulation techniques can be hybridized, points based on the advancing front technique being Delaunay triangulated.

Generally for Navier-Stokes type meshes, the majority of the mesh points are located in the boundary layers. Typically with hybrid meshes a flow aligned, structured near wall mesh is generated, using standard structured grid generation techniques. This is connected to an unstructured grid for the off surface, inviscid flow zones. Where surfaces are close, connecting the two zones is problematic. Close surfaces occur in areas of extreme curvature – especially corners. The structured grid generation will involve projection of surface normal. In corners these will cross over. The degenerate cells can be deleted. However, this gives rise to incursion of the unstructured grid into the boundary layer. Figure 3.34a shows this. Frame (b) shows a grid generated using a process that avoids the incursion of the unstructured cells into the corners. This incursion aspect is explored further later.

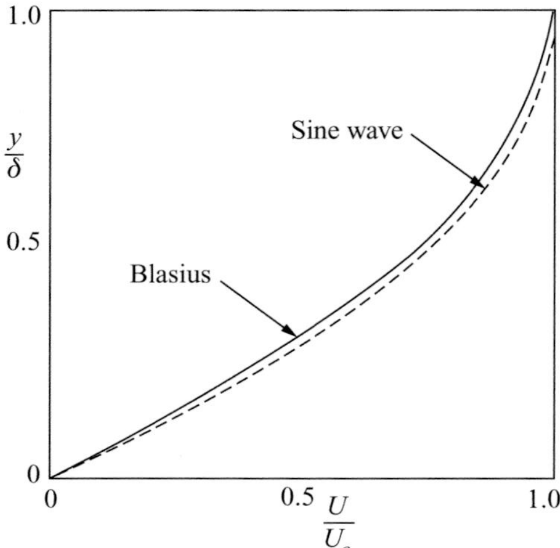

Figure 3.35. Comparison of Blasius and sine function velocity profiles (Adapted from Tucker and Mosquera (2001) – Published with the kind permission of NAFEMS).

3.5 Mesh Control

3.5.1 *Control of Near Wall Grid Distributions*

The nature of near wall grid distributions was touched on earlier. Here this is further elaborated on and the forms of typical expressions used to control near wall grid distributions briefly noted. A sine function makes a reasonable approximation to a laminar velocity profile. Figure 3.35 compares a sine function velocity profile with the exact Blasius profile solution.

If the streamwise velocity $u = f(\sin y)$, then $du/dy = f(\cos y)$, suggesting that a cosine based grid expansion function is about optimal for low Reynolds number laminar flows. The following cosine based Chebyshev polynomial (Lonsdale and Welsch 1984) can be used for such flows

$$y_j = L\left[1 - \cos(Wf)\right], \quad Wf = \xi\frac{\pi}{2}, \ \xi = \frac{j}{n_y}, \ j = 1, 2 \ldots\ldots\ldots n_y \tag{3.12}$$

where n_y is the number of grid nodes and L is the extent of the expanded grid zone. For a channel flow,

$$y_j = \frac{1}{2}\left(L_1 + L_2\right) + \frac{1}{2}(L_1 + L_2)\cos(Wf), \quad Wf = \xi\pi, \ \xi = \frac{j}{n_y}, \ j = 1, 2 \ldots\ldots n_y \tag{3.13}$$

where L_1 and L_2 identify the locations of the channel boundaries. Figure 3.36 shows various grid distributions. The horizontal shaded surfaces are solid. Frame (a) shows a cosine distribution arising essentially from the use of Equation (3.12).

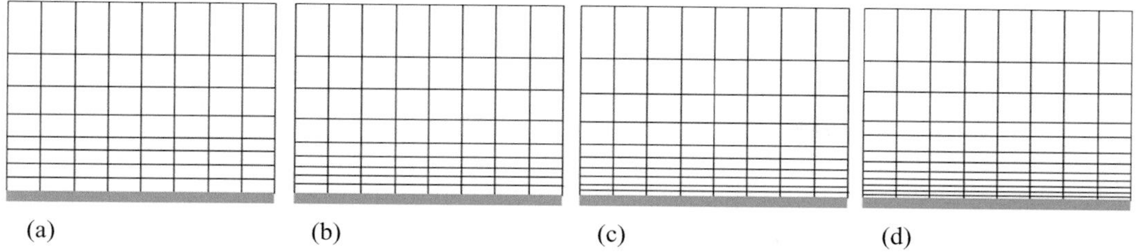

Figure 3.36. Grid distributions for a basic boundary layer: (a) cosine grid distribution, (b, c, d) geometric grid distributions with $\alpha = 1.1, 1.2$ and 1.3, respectively (Adapted from Tucker and Mosquera (2001) – Published with the kind permission of NAFEMS).

For DNS/LES of turbulent boundary layers and free shear layers, respectively, the use of the following hyperbolic expressions is preferable

$$\begin{aligned} y_i &= \frac{\tanh(\alpha Wf)}{\tanh\alpha}, \quad Wf = \xi\pi, \ \xi = \frac{j}{n_y}, \ j = 1, 2, \ldots\ldots\ldots, n_y \\ y_i &= \frac{\sinh(\alpha Wf)}{\sinh\alpha}, \quad Wf = \xi\pi, \ \xi = \frac{j}{n_y}, \ j = 1, 2, \ldots\ldots\ldots, n_y \end{aligned} \tag{3.14}$$

where α is a grid control parameter. For turbulent flow, a geometric expansion is a popular and simple expression. With this, moving away from a wall, grid spacings Δy_{i+1} are a constant multiple of previous values Δy_i. Therefore, the following can be written

$$\Delta y_{j+1} = \alpha\, \Delta\, y_j \tag{3.15}$$

where $1 \leq \alpha \leq 1.3$. For geometric expansions, the $\alpha \leq 1.3$ criterion avoids significant discretization errors (see Roache 1976). Figure 3.36b–d correspond to grid expansion factors of 1.1, 1.2 and 1.3, respectively. As can be seen, the latter factor causes substantial grid stretching. When modelling turbulent flows with wall functions, it is not necessary to commit large numbers of grid nodes close to walls, and so more uniform grids can be used. It is recommended that, where possible, prior to meshing, established correlations are used to estimate boundary layer thickness values. This will enable sensible estimates to be made of near wall grid distributions. Interestingly, some codes can adapt grids based on near wall y^+ values. More details of mesh generation with turbulence in mind are given later.

3.5.2 *Use of Sources*

For unstructured grids, the use of background sources for grid control is relatively common. The general idea is that around a source, the grid spacing would be finer and grow with distance from the source. The procedure of Lohner (1996) is outlined next for line, point and surface sources. First the line source shown in Figure 3.37a is considered. Vanishing the extent of this line source can generate a point source.

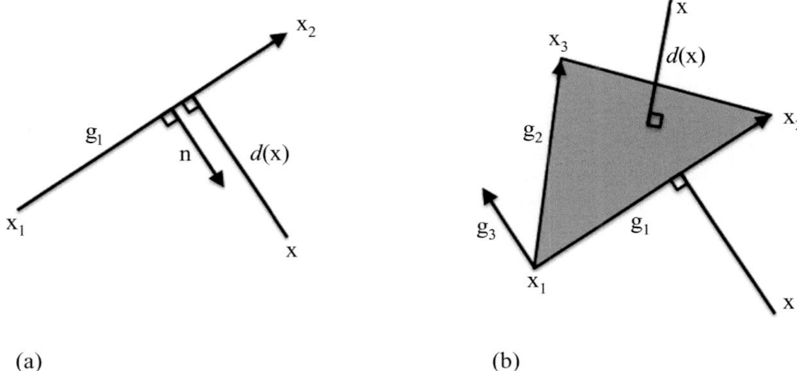

Figure 3.37. Line and area sources for mesh control from Lohner (1996): (a) line source and (b) area source.

The line source has endpoints at $\mathbf{x}_1$ and $\mathbf{x}_2$ and is defined by the vector $\mathbf{g}_1$. The normal to this vector is $\mathbf{n}$. The nearest distance $d(\mathbf{x})$ of point $\mathbf{x}$ to the line source is what is sought. The location of point $\mathbf{x}$ relative to the line source can be expressed as

$$\mathbf{x} = \mathbf{x}_1 + \xi \mathbf{g}_1 + \alpha n \tag{3.16}$$

Hence, $\mathbf{x}$ is characterized by a proportion lying along the line source and then a normal distance from the line source, where

$$\xi = \frac{(\mathbf{x} - \mathbf{x}_1)\Delta \mathbf{g}_1}{\mathbf{g}_1 \Delta \mathbf{g}_1} \tag{3.17}$$

This is further constrained to 'be on the line' using

$$\xi' = \max(0,\ \min(1, \xi)) \tag{3.18}$$

The desired distance is then given as

$$d(\mathbf{x}) = |\mathbf{x} - \mathbf{x}_1 - \xi \mathbf{g}_1| \tag{3.19}$$

For a surface source, a triangulated zone can be used as shown in Figure 3.37b. Then, as shown by Lohner (1996), the following extension to Equation (3.16) can be used

$$\mathbf{x} = \mathbf{x}_1 + \xi \mathbf{g}_1 + \eta \mathbf{g}_2 + \gamma \mathbf{g}_3 \tag{3.20}$$

where

$$\mathbf{g}_3 = \frac{\mathbf{g}_1 \times \mathbf{g}_2}{|\mathbf{g}_1 \times \mathbf{g}_2|} \tag{3.21}$$

$$\xi = (\mathbf{x} - \mathbf{x}_1)\,\Delta \mathbf{g}^1, \quad \eta = (\mathbf{x} - \mathbf{x}_1)\,\Delta \mathbf{g}^2, \quad \zeta = 1 - \xi - \eta \tag{3.22}$$

and $\mathbf{g}^1$ and $\mathbf{g}^2$ are as defined by Lohner (1996). Again, extending Equation (3.18), the following surface constraints

$$0 \le \xi, \eta, \zeta \le 1 \tag{3.23}$$

Table 3.3. *Potential grid spacing expressions based on proximity to source*

Approach	Expression	Input	Comments
Power law	$\delta(\mathbf{x}) = \delta_o[1+\rho^{\gamma}]$	$\delta_0, d_0, d_{1,}\gamma$	$1.0 < \gamma < 2.0$
Exponential function	$\delta(\mathbf{x}) = \delta_o c^{\gamma\rho}$	$\delta_0, d_0, d_{1,}\gamma$	Expensive to repeatedly compute
Polynomial	$\delta(\mathbf{x}) = \delta_o\left[1+\sum_{i=1}^{n} a_i\rho^i\right]$	$\delta_0, d_0, d_{1,}a_i$	Cheaper than exponential, $n < 3$ generally sufficient

are placed on the previous. When these constraints are violated, at edges, line sources can be used to represent behaviour associated with the edge. If Equation (3.23) is satisfied, then the closest distance of the point $\mathbf{x}$ to the surface is given by

$$d(\mathbf{x}) = \left|(1-\xi-\eta)\,\mathbf{x}_1 + \xi\mathbf{x}_2 + \eta\mathbf{x}_3 - \mathbf{x}\right| \tag{3.24}$$

Once the nearest distance from the source to the point has been evaluated, some estimate of the grid spacing, $\delta(\mathbf{x})$, based on $d(\mathbf{x})$ must be made. To ensure a constant fine mesh resolution in some proximity of extent d_0 to the source, the following transformed intermediate variable can be used

$$\rho = \max\left[0, \frac{d(\mathbf{x}) - d_0}{d_1}\right] \tag{3.25}$$

where d_1 is a scaling factor. Table 3.3 gives potential grid spacing expressions based on the proximity to the source.

Note that the same point in space can be influenced by multiple sources and hence Lohner recommends use of the following equation

$$\delta(x) = \min(\delta_1, \delta_2, \ldots\ldots, \delta_n) \tag{3.26}$$

where the minimum spacing scale implied by, say, m sources taken at a particular point. Notably, for every local operation, all the source and conversion functions to change source distances into grid spacings need to be revaluated. This potentially gives rise to considerable expense. Some geometric parameters can be pre-computed and stored rather than calculated on the fly. For complex geometries, the amount of necessary sources that need to be described can be substantial and hence problematic since this is a manual Graphical User Interface (GUI) based process.

The use of sources can be combined with the use of a background grid and near wall grid expansions (such as the geometric, Chebyshev) as discussed previously. Then the grid spacing can be defined as

$$\delta = \min[\delta_s, \delta_{BG}, \delta_{EXP}] \tag{3.27}$$

where δ_s corresponds to the spacing defined by the source and $\delta_{BG}, \delta_{EXP}$ are the spacings defined by the background grid and surface associated expansion. The latter can be stored on the CAD file (see Lohner 1996), and for extremely complex geometries this could considerably streamline the grid generation process.

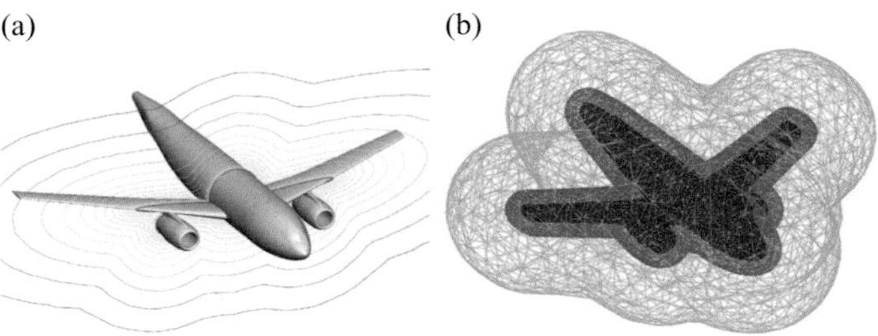

Figure 3.38. Solutions to level-set equation for a DLR-F6 wing-body configuration: (a) contours of ϕ for $F(\mathbf{x}) = 1$ and (b) level-set isosurfaces at 5, 10 and 20 (from Xia et al. 2010 – Published with the kind permission of Elsevier).

3.5.3 *Use of Level Sets*

Level sets, as indicated previously, are ideal for identifying computational inter-faces. The boundary value formulation of the level-set equation is given as

$$F\,|\nabla\phi| = 1 + \varepsilon\nabla^2\phi \tag{3.28}$$

where the dependent variable, ϕ, describes first arrival times of propagating fronts from boundaries, and $F(\mathbf{x})$ is the local speed function of these fronts. To gain numerical (entropy) solutions, $\varepsilon \to 0$. For $F(\mathbf{x}) = 1$, ϕ is equal to the distance, d, from a surface.

For bodies with relative movement, the use of overset grids often proves robust and efficient. Then the key issue becomes how to define the inter-grid boundaries. Nakahashi and Togashi (2000) use a local level-set d solution to determine the classification of vertices and the hole cutting. In Figure 3.11b, the interface between each domain is defined by medial axes. Notably, with the level-set method $F(\mathbf{x})$ can be defined in an advancing front and the different interfaces are achieved through the control of this speed function.

As noted earlier, convex zones can be problematic for mesh generation. Such a zone can be observed in the wing body junction zone for a DLR-F6 wing-body configuration in Figure 3.38. This figure shows a level-set equation solution. Frame (a) gives contours of ϕ for $F(\mathbf{x}) = 1$ and (b) gives level-set isosurfaces at 5, 10 and 20. In Figure 3.39, a cross section of the DLR-F6 wing-body configuration is shown, along with meshes separately propagated from the fuselage and wing surfaces. As shown in Frame (a), the overset grid stops propagating once the medial axis is passed. In Frame (b), this happens in a similar way, but the extrusion stops before reaching the medial boundary to leave an unfilled region for triangular elements (i.e., a DRAGON grid). The level-set based medial axis can also be 'biased'. For example, in Figure 3.39c, the medial axis is biased towards the fuselage, giving control to the mesh generation process.

As noted earlier, near wall advancing structured grid generation for hybrid meshes is also important for many applications. Hyperbolic structured mesh generators are appropriate for extruding surface meshes into near wall regions, leaving

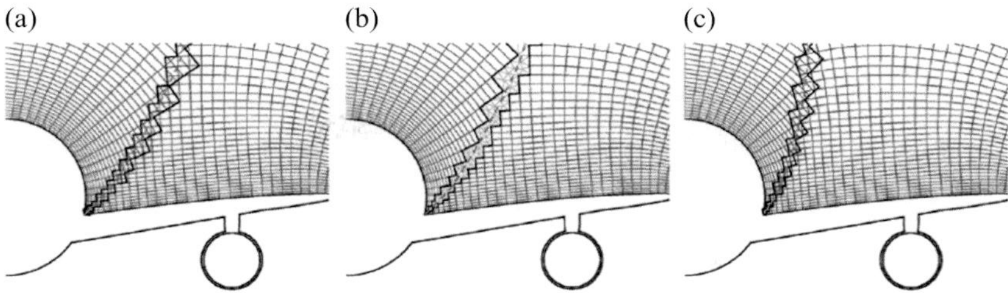

Figure 3.39. Collision boundary detection using medial axis generated from a level-set equation: (a) overset grids, (b) hybrid grids and (c) biased medial axis (from Xia et al. (2010) – Published with the kind permission of Elsevier).

the remaining domain for unstructured tetrahedral elements. Hence, the hyperbolic level-set Equation (3.28) is suitable for this task.

In Figure 3.40, the level-set d solution automatically gives mesh lines parallel to the boundary and ∇d forms the mesh lines orthogonal to the boundary and level sets. With different speed functions (see Osher and Sethian 1988 and Sethian 1994), dependent on various quantities (e.g., curvature κ), various representations of the boundary curvature as the level set moves away from the boundary can be produced. Indeed the curvature formula, used in Frame (b), is

$$F(\kappa) = 1 - \varepsilon\kappa \tag{3.29}$$

where the curvature, κ, can be expressed as

$$\kappa = \nabla \cdot \frac{\nabla\phi}{|\nabla\phi|} \tag{3.30}$$

This reduces the influence of the boundary curvature. It smooths the concave feature. This is preferable in many cases. In essence the surface normals are adjusted to delay their collision, preventing the problem shown in Figure 3.34.

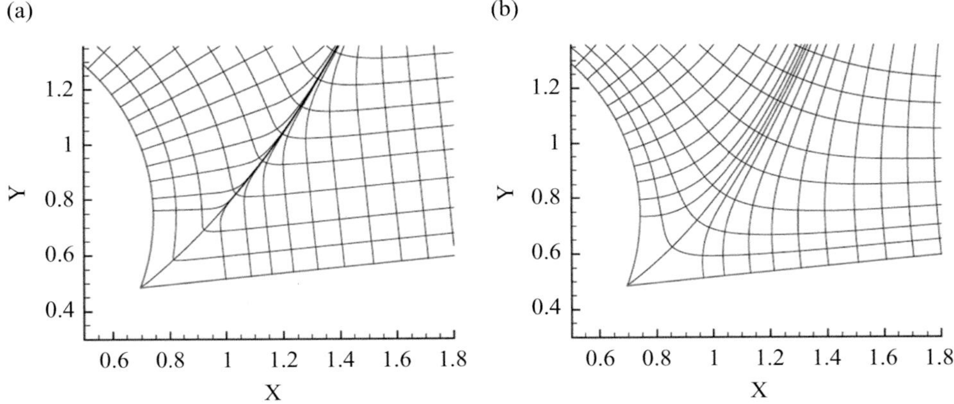

Figure 3.40. Near wall advancing layer meshes using level sets from curvature dependent speed functions: (a) $F = 1$ and (b) $F = 1 - 0.5\kappa$ (from Xia et al. (2010) – Published with the kind permission of Elsevier).

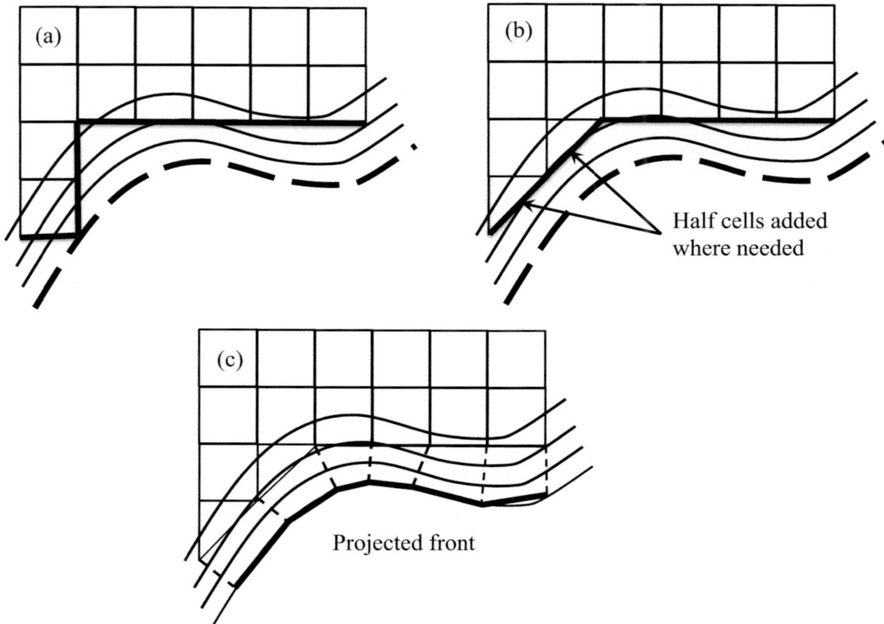

Figure 3.41. Patching a surface mesh on a core Octree mesh (based on Dawes et al. 2009).

Figure 3.41 shows a level set based process for patching a surface mesh on a core Octree mesh from Dawes et al. (2009). Cut cells are removed, exposing a relatively irregular front of cells. Half-cells can be included to improve this (see top right-hand frame). The exposed front is projected to the surface using the gradient of the level-set field to guide this. The level-set isosurfaces will define the near surface boundary conforming mesh. This level-set steered process is used in the commercial BOXER system.

3.5.3.1 Gradient Limiting Level Set Equation Persson (2006) devised the gradient limiting Hamilton-Jacobi equation (closely related to the level-set Equation (3.28)) for use with approaches like the advancing front mesh generation approach. This equation is given as

$$\frac{\partial \phi}{\partial t} + |\nabla \phi| = \min\left(|\nabla \phi|, \alpha\right) \tag{3.31}$$

where $\phi = \delta$ is a cell size scale. The factor α is a local ratio of cell size, such that $\delta(x) = \delta_o + \alpha(x - x_o)$, and δ_o is an initial spacing at a location x_o. The gradient limiting Hamilton-Jacobi is solved with the initial condition

$$\delta\,(x, t = 0) = \delta_o(x) \tag{3.32}$$

When $|\nabla\delta| \leq \alpha$, $\partial\delta/\partial t = 0$ and so δ will not change. On the other hand, if $|\nabla\delta| > \alpha$, then the grid expansion will be limited, the condition $|\nabla\delta| = \alpha$ being locally enforced. Obviously, the steady state equation is

$$|\nabla\delta| = \min(|\nabla\delta|, \alpha) \tag{3.33}$$

and thus $|\nabla\delta| \leq \alpha$. Grid size control can be enforced through both the initial and boundary conditions. With δ specified at boundaries, δ values will climb from the boundary values – information propagates outwards, the equation being hyperbolic. The other means of grid control is through initial values. If these satisfy the requirement $|\nabla\delta| \leq \alpha$ then values will not change. The following initialization is proposed by Persson (2006).

$$\delta_o(\mathbf{x}) = \min(\delta_{curv}(\mathbf{x}), \delta_{lfs}(\mathbf{x}), \delta_{ext}(\mathbf{x})) \tag{3.34}$$

where δ_{curv} is based around the inverse of surface curvature. The idea is that around highly curved surfaces more grid will be needed. Simple functions can be used to correct local grid point values so that they reflect the boundary curvature (see Persson 2006). The functions are of the following form

$$\delta_{curv} = f\left(\frac{1}{\kappa(\mathbf{x})}, d(\mathbf{x})\right) \tag{3.35}$$

where d is the nearest wall distance. The distance, d, as pointed out in Chapter 7, can be evaluated from the Hamilton-Jacobi/eikonal equation. It is necessary to approximately correct for the fact the boundary curvature is sought and not local curvature given by Equation (3.30). Hence, to approximately correct the local curvature to boundary curvature, the boundary proximity is needed. Equation (3.34) also makes use of δ_{lfs}. This is a local feature scale (e.g., the distance between two adjacent boundaries). To estimate this, Persson (2006) makes use of the medial axis.

$$\delta_{lfs}(\mathbf{x}) \propto d(\mathbf{x}) + d_{MA}(\mathbf{x}) \tag{3.36}$$

where $d_{MA}(\mathbf{x})$ is the distance from the field point being considered to the medial axis. Note that $\delta_{lfs}(\mathbf{x})$ could also be readily evaluated from summing the two roots of the Poisson wall distance equation given in Chapter 7. Equation (3.34) also has the non-geometric scale $\delta_{ext}(\mathbf{x})$, which can be used for grid adaptation and other purposes.

3.5.4 *Mesh Smoothing and Loss of Grid Control*

3.5.4.1 Smoothing Especially when grids adapt to local flow features, grid smoothing can be required. This can be done by iteratively averaging neighbouring grid node location values as

$$x_j \leftarrow \frac{1}{n_i}\sum_{j=1}^{n_i} x_j \tag{3.37}$$

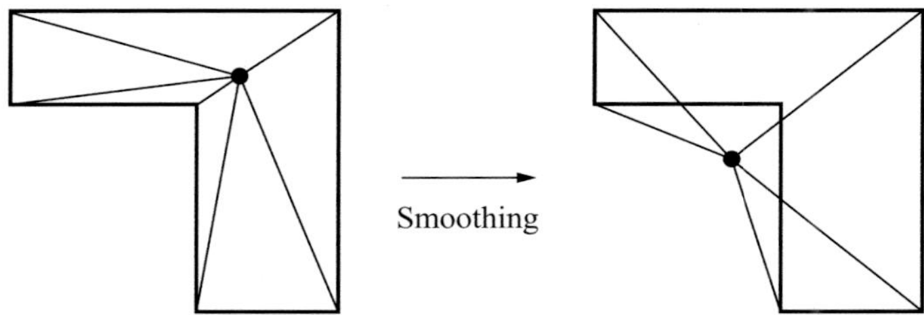

Figure 3.42. Grid spilling out of domain from excessive smoothing.

where n_i is the number of neighbours. However, care is required. For example around sharp convex features the smoothing can result in the grid spilling out of the domain. This is shown in Figure 3.42.

As noted by Lohner (1996), it can be important to smooth surface normal to ensure grid quality – the ideal being to have precise surface normal near the surface with smoothed normals away from it. Lohner proposes a Hermitian polynomial based blending between the non-smoothed normal $\boldsymbol{n}_0$ and the smoothed $\boldsymbol{n}_1$. This takes the following form

$$\mathbf{x} = \mathbf{x}_0 + \xi\delta\mathbf{n}_0 + \xi \cdot (2 - \xi) \cdot \xi\delta(\mathbf{n}_1 - \mathbf{n}_0) \tag{3.38}$$

where δ is the boundary layer thickness, $\mathbf{x}_0$ are points on the surface and ξ is a dimensionless boundary layer point distribution ranging between zero and one, and geometrically expanding. Evidently, sometimes a better point distribution can be produced by replacing $\xi \cdot (2 - \xi)$ with the following: $\eta \cdot (2 - \eta)$, where $\eta = \xi^p$ and $p = 0.5$.

3.5.4.2 Loss of Grid Control As noted by Lohner (1996), it is possible to lose grid control through the loss of numerical precision. The unstructured grid generation process can involve multiple products of grid spacings and this, for highly stretched grids, can give rise to rounding errors. For example, as noted by Lohner, for a Boeing 747 simulation, the domain would typically be of the order of 10^3 units, corresponding to ten body lengths. The minimum and maximum grid spacings will be 1×10^{-5} and 20 units with an element to volume ratio of the order of 10^{-12}. Using 64-bit computation gives 10^{-16} accuracy However, element distortion, loss of shape and surface singularities can produce rounding errors, giving the potential for a loss of grid control.

3.6 Mesh Quality and Features

3.6.1 *Impact of Cell Shape*

Figure 3.43 shows the impact of cell shape on the decay of a viscous instability wave. The wave is a sub-critical Tollmien-Schlichting. The precise details are not important here. The key point is that the analytical solution (to the Orr-Sommerfeld

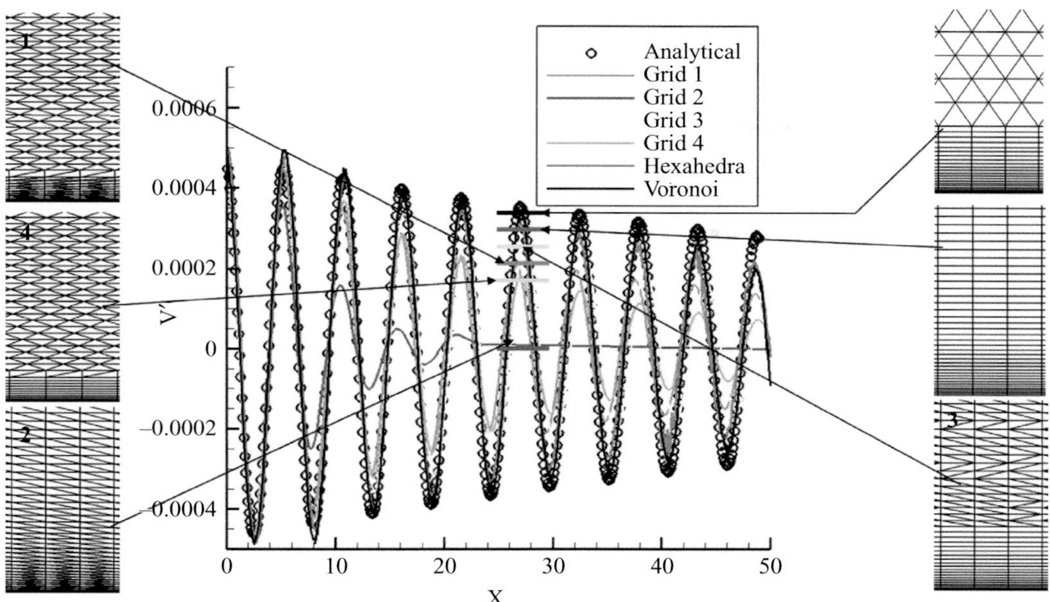

Figure 3.43. Impact of cell shape on the decay of an instability wave (from Tucker (2008) – Published with the kind permission of Elsevier).

equation – see Chapter 7) is given by the symbols. Clearly the cell shape has a substantial solution impact.

Figure 3.44 explores the traits of the cells with the most extreme performances shown in Figure 3.43. Frame (a) shows the 'Finite Element' cell 2. Frame (b) shows the Voronoi mesh and (c) the stretched Voronoi mesh. The solver used constructs the actual solution control volume by connecting centroids to midpoints of edges. Hence, the actual solution control volumes are shown by the dashed lines. The successful Voronoi control volume has the same trait as the successful hexahedral cells, this trait being that the grid points straddling the control volume face form a line orthogonal to the face. The cell type 2 does not have this orthogonality trait. However, this can be remedied, as discussed in Chapter 3, through changing the rules used to construct the control volume. The Voronoi mesh looks attractive. However, as shown in Figure 3.44c, it is difficult to generate the high aspect ratio cells necessary for high-speed shear flows. To overcome this, in Figure 3.43, a near wall hexahedral mesh is used. This also ensures similar cell counts are used (indeed the Voronoi mesh has far fewer edges than all the other meshes) and hence that the results can be sensibly contrasted. Unfortunately, as can be seen from Figure 3.43, it is hard to hybridize Voronoi meshes and smoothly connect them to high aspect ratio near wall cells.

3.6.2 *Basic Mesh Features*

Before discussing mesh quality, it is worth reviewing some very basic mesh features. Firstly, the mesh needs sufficient density – generally, for high accuracy, cells must be small. However, for computational economy, sizes should vary over

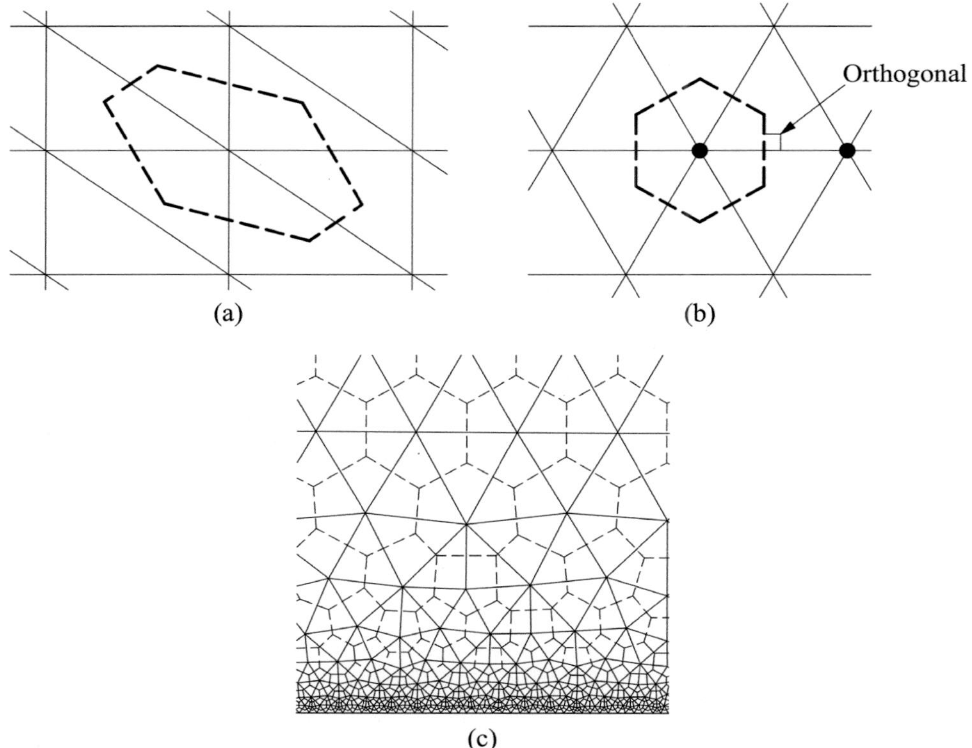

Figure 3.44. Examples of potentially bad and good cell shapes: (a) Finite Element cell 2, (b) Voronoi mesh and (c) stretched Voronoi mesh (from Tucker (2008) – Published with the kind permission of Elsevier).

the domain. Where there are steep flow gradients or complex geometry to resolve, the grid density should be increased appropriately. To avoid diffusion and dispersion errors, adjacent cell size changes should be small (i.e., the mesh should be smooth). The mesh should also have a structure that is appropriate to the flow being modelled. For highly complex geometries triangles/tetrahedrals are attractive. However, in boundary layers, these give poor accuracy. Therefore, in this region, a common practice is to use prism shaped elements. Outside the boundary layer, these are connected to, say, tetrahedral elements.

Key basic traits to aim for when generating a mesh are as follows: (1) boundary conformity (i.e., adequate resolution of the geometry), (2) adequate density to resolve high gradient regions, (3) elements that are as equiangular as possible, (4) where possible, orthogonality at boundaries, (5) smoothness, (6) cells that have an underlying discretization of sufficient accuracy, (7) if possible, move at a velocity that will minimise average cell Reynolds numbers, (8) be as flow aligned as possible, (9) have cell shapes that are consistent with the flow physics and (10) be consistent with turbulence model implementation. With regards to (8), Knupp (1995) developed an adaptive approach, where the mesh moved to enforce flow alignment. In relation to (4), the use of tetrahedral elements at boundaries should be avoided. For accuracy, hexahedral or prism cells, at boundaries, should have an included angle of around 90°.

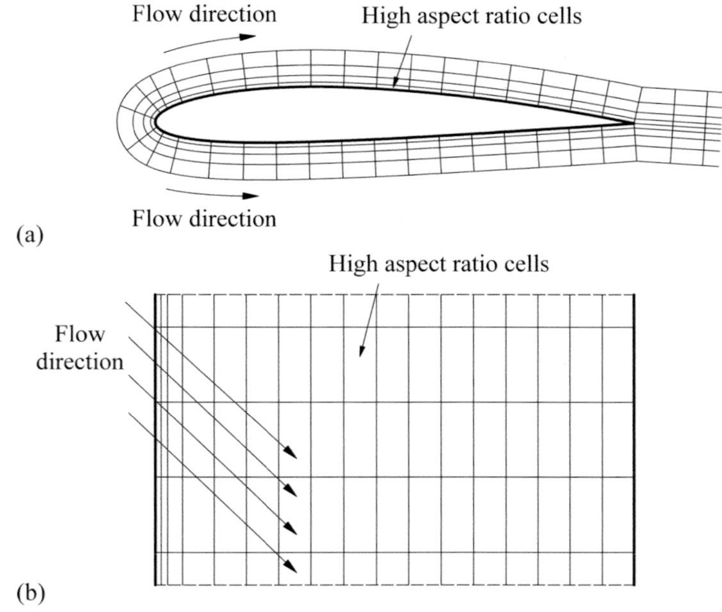

Figure 3.45. Use of high aspect ratio grids: (a) desirable and (b) undesirable (Adapted from Tucker and Mosquera (2001) – Published with the kind permission of NAFEMS).

In some situations, for economy, periodic boundary conditions can be required. A classic example is the modelling of blade sets in turbines and compressors. When using such boundary conditions, it is important to ensure the grid points at either side of the periodic interface precisely match. This is particularly so for unstructured solvers where a search is needed to check for periodic neighbours to a specified tolerance. Figure 3.45 is used to illustrate points (8) and (9) above. Frame (a) shows a flow-aligned grid with high aspect ratio cells – this ratio can be around 1000. For high-speed flows, these cells efficiently capture the high, normal to surface, variable gradients. Nodes are not wasted in the streamwise direction, where gradients are less severe. In Frame (b), the flow is shown to be moving obliquely to the grid lines. This can create significant numerical errors in the form of false diffusion (the solution produced is that for a fluid with a higher viscosity than intended). Things a mesh should not have, from Tucker and Mosquera (2001) are: (1) holes, (2) self-intersections, (3) for finite element triangular cells, a maximum interior angle of greater than 180° (this is for theoretical correctness and hence accuracy), (4) for curvilinear grids, away from boundaries, a maximum interior angle of roughly greater than 135° and (5) where it is not needed, excessive resolution.

3.6.3 *Grid Generator Diagnostics and Quality*

For quadrilateral and hexahedral meshes, the Jacobian can be a useful measure for quantifying quality. This can be expressed as

$$J_{ij} = \frac{\partial x_i}{\partial \xi_j} \tag{3.39}$$

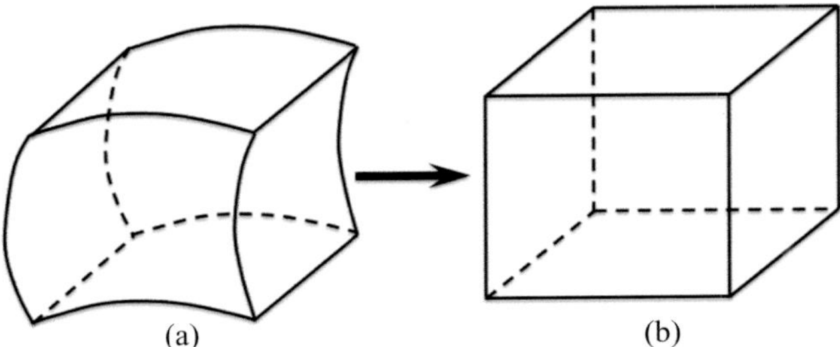

Figure 3.46. Physical and transformed mesh cells/coordinates.

where x_i corresponds to a physical coordinate system in which the element resides (see left-hand frame of Figure 3.46) and ξ_j corresponds to a transformed coordinate system into which the Frame (a) element in Figure 3.46 is mapped. The element in this system is a shown in Figure 3.46b. Clearly the more extensive the difference between the Frame (a) and (b) elements, the greater the need for J to be active. Hence, it will quantify the degree of element deformity. As pointed out by Knupp (2008), to transform the Laplacian equation $\partial^2\phi/\partial x_i^2$ to curvilinear equations we have $1/\det(J)\cdot\partial(J^{-t})_{ij}/\partial\xi_j\cdot\partial\phi/\partial\xi_j$, where J^{-t} is the transpose of the matrix. Clearly, the transformed equation must be elliptic like its parent and hence for this to be so J^{-t} must be elliptic. For this to be guaranteed, $\det(J) > 0$ over the domain (Knupp 2008). A locally negative Jacobian determinant will correspond to a negative volume. According to Pletcher et al. (2013) a measure of smoothness, I_S, for a two-dimensional system is

$$I_S = (\nabla\xi)^2 + (\nabla\eta)^2 \tag{3.40}$$

and for orthogonality

$$I_O = (\nabla\xi\cdot\nabla\eta)^2(J^{-1})^3 \tag{3.41}$$

To simply explore the impact of the grid on accuracy, it can be useful to consider one-dimensional, non-uniform grids. When discretizing $\partial\phi/\partial x = (\partial\xi/\partial x)(\partial\phi/\partial\xi)$ for a second-order central difference, on such a grid, the leading truncation error term is as follows (see Knupp 2008)

$$\frac{(\Delta\xi)^2}{6}\left(\frac{\partial x}{\partial\xi}\right)^{-1}\frac{\partial\phi^3}{\partial\xi^3} = \frac{(\Delta\xi)^2}{6}\left(\left(\frac{\partial x}{\partial\xi}\right)^2\frac{\partial\phi^3}{\partial x^3} + 3\frac{\partial x^2}{\partial\xi^2}\frac{\partial\phi^2}{\partial x^2} + \left(\frac{\partial x}{\partial\xi}\right)^{-1}\frac{\partial x^3}{\partial\xi^3}\frac{\partial\phi}{\partial x}\right) \tag{3.42}$$

Hence, as can be seen, although the error goes at $C\Delta x^2$ where C is a constant, the form of C is complex, being a function of the first, second and third derivatives of the grid transform and also the actual variable being solved for. Clearly, with the latter aspect it would be hard to make a full assessment of grid quality prior to making a simulation. Also, clearly, when moving to three-dimensional grids the situation becomes even more complex.

3.6.4 *Some Basic Grid Error Estimates*

3.6.4.1 Non-flow Aligned Grids For two-dimensional flows, that move obliquely to orthogonal grid lines false diffusion can be roughly estimated (when first order upwinding is used) using the following expression of Davis and Mallinson (1972)

$$\Gamma_{false} = \frac{\rho U \Delta x \Delta y \sin 2\theta}{4(\Delta x \sin^3\theta + \Delta y \sin^3\theta)} \tag{3.43}$$

where U is a uniform fluid velocity, θ is the angle of the flow to the grid and Δx and Δy are the horizontal and vertical grid spacings. When Equation (3.43) is applied to the momentum equations, Γ_{false} can be considered as an erroneous amount by which the fluid viscosity has effectively risen. This rise will cause smearing of profiles. The expression shows that false diffusion can be reduced with grid refinement or aligning the grid to the flow. For simple flows, body fitted grids can frequently be made flow aligned.

3.6.4.2 Grid Stretching The derivative $\partial\phi/\partial x$ can be represented on a non-uniform grid using the following second order, centred, finite difference expression

$$\left.\frac{\partial\phi}{\partial x}\right|_i = \frac{\dfrac{(\phi_{i+1}-\phi_i)\Delta x_i}{\Delta x_{i+1}} + \dfrac{(\phi_i-\phi_{i-1})\Delta x_{i+1}}{\Delta x_i}}{\Delta x_{i+1}+\Delta x_i} \tag{3.44}$$

where $\Delta x_i = x_i - x_{i-1}$ and $\Delta x_{i+1} = x_{i+1} - x_i$. Substituting Equation (3.15) into Equation (3.44), including truncated terms (see Fletcher 1997), gives

$$\left.\frac{\partial\phi}{\partial x}\right|_i + \left[\frac{\Delta x_i^2}{6}\alpha\frac{\partial^3\phi}{\partial x^3} + \frac{\Delta x_i^3}{24}\alpha(\alpha-1)\frac{\partial^4\phi}{\partial x^4} + HOT\right]_i = \frac{\dfrac{(\phi_{i+1}-\phi_i)\Delta x_i}{\alpha\Delta x_i} + \dfrac{(\phi_i-\phi_{i-1})\alpha\Delta x_i}{\Delta x_i}}{(1+\alpha)\Delta x_i} \tag{3.45}$$

The parameters in square brackets are error terms. As can be seen, the solution error increases with α. If the grid is uniform ($\alpha = 1$), the diffusive error term ($\partial^4\phi/\partial x^4$) vanishes. According to Fletcher, for a finite element method (Galerkin discretization with a linear shape function – See Appendix A), we can write

$$\left.\frac{\partial\phi}{\partial x}\right|_i + \left[\frac{\Delta x_i}{2}(\alpha-1)\frac{\partial^2\phi}{\partial x^2} + \frac{\Delta x_i^2}{6}\frac{\alpha^3+1}{\alpha+1}\frac{\partial^3\phi}{\partial x^3} + HOT\right]_i = \frac{\phi_{i+1}-\phi_{i-1}}{(1+\alpha)\Delta x_i} \tag{3.46}$$

As can be seen, for this discretization, the false diffusion term ($\partial^2\phi/\partial x^4$) increases significantly with α. Clearly, ensuring changes in grid spacing are gradual is a matter of some seriousness, and sensitivity with respect to this will be code dependent. Therefore, it important to know the behaviour of the solver you are using. A numerical solution technique that can deal with non-smooth changes in grid spacing is given by Chew (1984), the solver needing significant modification.

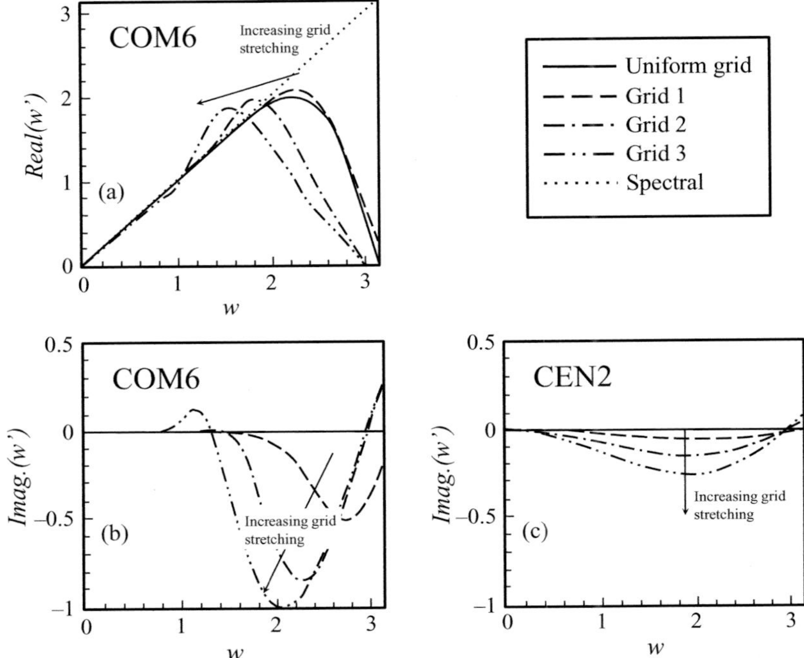

Figure 3.47. Dispersion and dissipation errors for compact and finite difference schemes on tanh grids: (a) dispersion error for sixth order compact scheme, (b) dissipation error for sixth order compact scheme and (c) dissipation error for second order central difference scheme (adapted from Chung and Tucker (2003) – Published with the kind permission of the American Institute of Aeronautics and Astronautics).

It is straightforward, using Fourier based modified wave number analysis, to study the accuracy of numerical differentiation and hence the impact of grid stretching. The property of interest can be expressed as $\phi(x) = e^{iwx}$, where w is the wave number. On differentiation this becomes $d\phi(x)/dx = iwe^{iwx} = iw\phi(x)$. The discrete scheme will give $iw'\phi(x)$, where w' is the less accurate, modified wave number. We could quantify the error in the wave number as $Error = (w' - w)/w$. Using the previous analysis, it is possible to estimate the ratio of, say, a flow scale or wave length to the grid spacing Δx. Hence, the points needed per wave length (PPW) for a specified error can be calculated. Chung and Tucker (2003) carry out this analysis for non-uniform hyperbolic sine and tangent grids. The ratio of largest to the smallest grid spacings is 100. This is consistent with typical eddy-resolving simulations. Hyperbolic sine grids are relatively popular for free shear layers and tanh boundary layers. Figure 3.47 plots dispersion and dissipation errors for compact and finite difference schemes on tanh grids. The thinner lines are for a finer grid. This has 256 grid nodes. The coarser grid has eight times fewer points with 32 nodes. The dotted line gives the exact, spectral scheme result. The other lines are for different levels of grid stretching and are as identified in the figure caption. Figure 3.47a shows how the wave number overshoots for a compact (see Chapter 4 for a description of this

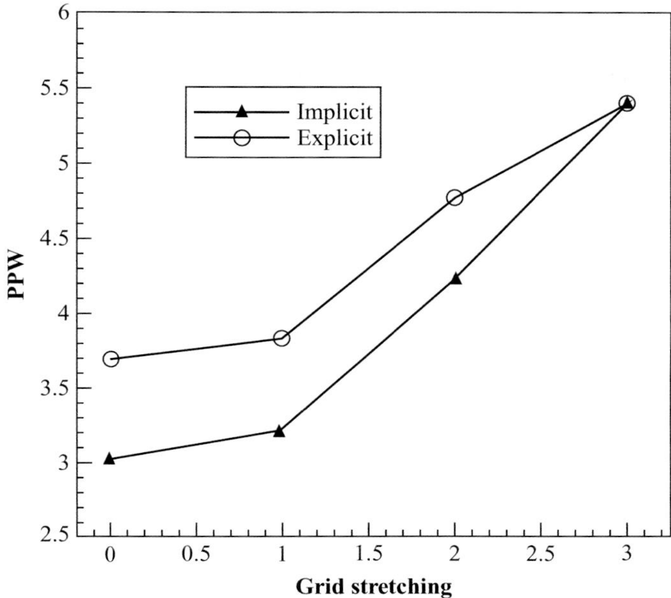

Figure 3.48. Points per wave length (PPW) (for a 10% dispersion error) with grid stretching for a fourth order standard finite difference scheme (explicit) and compact scheme (implicit) (from Tucker (2013) – Published with the kind permission of Springer).

more advanced high-order scheme) sixth order scheme when the grid becomes more stretched.

However, the key thing that happens for a non-uniform grid is that there is a finite imaginary component to w', representing, consistently with the previous discussion, dissipation. This is zero for uniform grids. As can be seen from Figure 3.47b, as the grid is stretched, for a compact scheme, the dissipation error (negative w' zone) becomes large. However, as Figure 3.47c shows, for the centred scheme (second order) this error is much smaller. Hence, since finite difference schemes are more economical, there is an attraction in staying with them for industrial flows.

This is further illustrated in Figure 3.48. This plots the PPW for a 10% dispersion error. The results are for a stretched sinh grid and a fourth order standard finite difference scheme (explicit) and compact scheme (implicit). The largest grid stretching, in the plot, is consistent with typical expansion ratios for eddy-resolving simulations. Figure 3.48 shows that for practical problems, where the grid needs stretching, the compact and standard finite difference schemes have similar accuracy. However, the latter needs less computational effort. Hence, again, clearly there is the need to understand how a particular grid will work with a particular solver. In this case, the more advanced but expensive procedure will have worse accuracy for non-smooth grids than the simpler scheme. High grid quality is not always easy to guarantee for complex geometry industrial simulations.

3.7 Grid Distributions in more Practical Flows

Discussion on more general aspects of grid distributions and also some basic theory on their impact, particularly with regard to grid stretching has been given. Now, some matters relating to grid distributions for more specific practical problems are outlined.

3.7.1 *Leading and Trailing Edge Grid Distributions*

For airfoils, leading edges can have strong accelerations and decelerations and hence fine meshes can be required in this zone. The trailing edges are more complex. There can be a tendency for flows to hang on to the trailing edge surface too long – especially if there is a large, rounded trailing edge. Then it can be necessary to blunt the trailing edge, add a cusp or even coarsen the mesh (see Denton 2010). This matter is discussed further in Chapter 7.

3.7.2 *Free Shear Layer Grid Distributions*

Unlike with boundary layers, where the shear layer location is clearer for free flows, care can be needed to ensure that the mesh follows the shear layers. This will typically be an iterative process. Hence, a preliminary flow solution is needed and then potentially repeated mesh adjustments to ensure alignment of the mesh with the shear layers. An example for a jet with an angled nozzle outlet is shown in Figure 3.49. This nozzle is relatively simple. For real engines, coaxial nozzles are used. Hence there are two shear layers. The typical grid structure for such a nozzle is shown in Figure 3.50.

The meshing for a co-flowing nozzle can be further complicated by the presence of chevrons. These make the shear layer trajectory and thickness vary with the azimuthal coordinate. Furthermore, the geometry of the chevrons can azimuthally vary.

Another aspect is that quite often it is required to gain acoustic information from simulations. This is especially so for the propulsive jets from aeroengines. Figure 3.51 shows the acoustic waves emanating from the turbulent field in an eddy-resolving simulation for a jet. Typically for a second-order standard CFD solver that does not use specialized numerical treatments, at least 10 grid points per wave length are needed to resolve an acoustic wave length. Using this and the speed of sound it is possible to estimate the acoustic frequencies supported by the grid and hence to see if these are sufficient for design requirements. Typically, this is a tough grid resolution challenge, and grid densities necessary to reach the frequencies ideally needed for engine designers are many years away.

The situation is complicated further by the fact that the geometry for a real jet nozzle is quite complex. For example, the nozzle must be mounted to the engine by a pylon. For modern high bypass ratio engines, the propulsive jet can also be quite close to the wing and flap. Hence, it can also be necessary to mesh these components.

Figure 3.49. Mesh alignment for free shear layers.

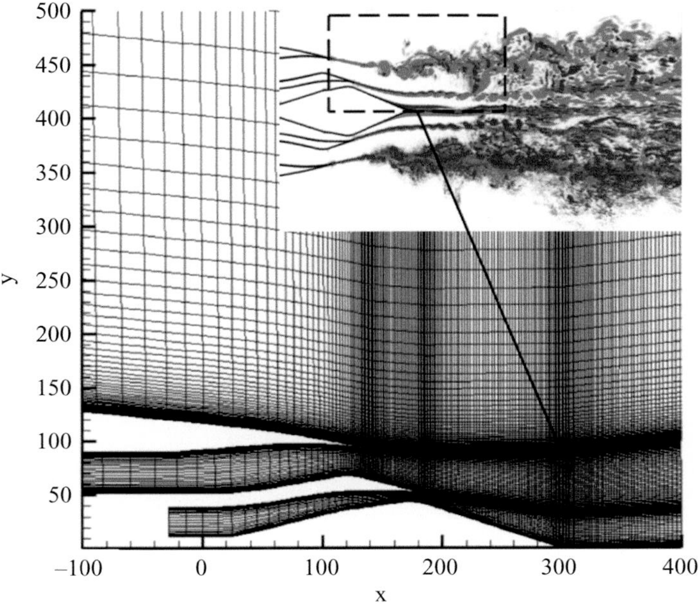

Figure 3.50. Twin shear layers in a jet with co-flow.

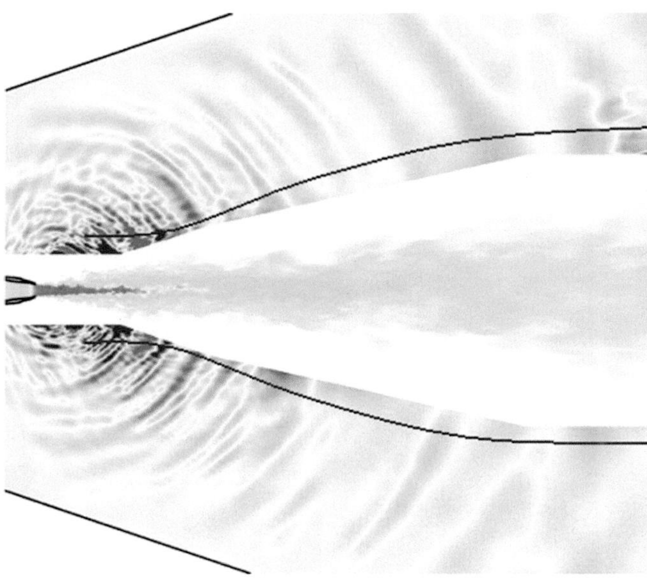

Figure 3.51. Resolution of acoustic waves on grid.

Figure 3.27 demonstrated efforts at semi-automated blocking for this system. Figure 3.52a shows a manually generated three-dimensional blocking, and Frame (b) shows the associated flow field. The presence of the pylon and wing gives rise to a complex, three-dimensional shear layer. This, with the restrictions imposed by the blocking structure, make it a great challenge to ensure adequate shear layer resolution.

Figure 3.53 shows the impact of different near axis meshes on eddy-resolving simulations for a jet. The figure plots the radial variation of the radial Reynolds stress

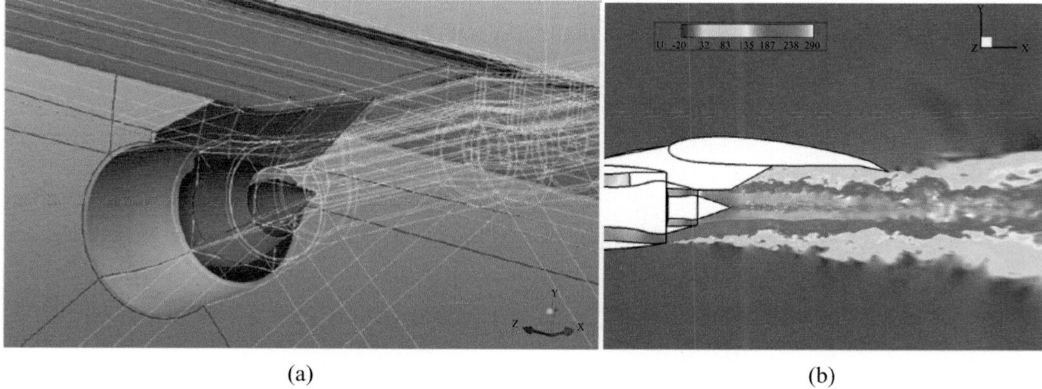

(a) (b)

Figure 3.52. Three-dimensional blocking for a jet-wing flap configuration and flow solution: (a) blocking and (b) eddy-resolving flow simulation.

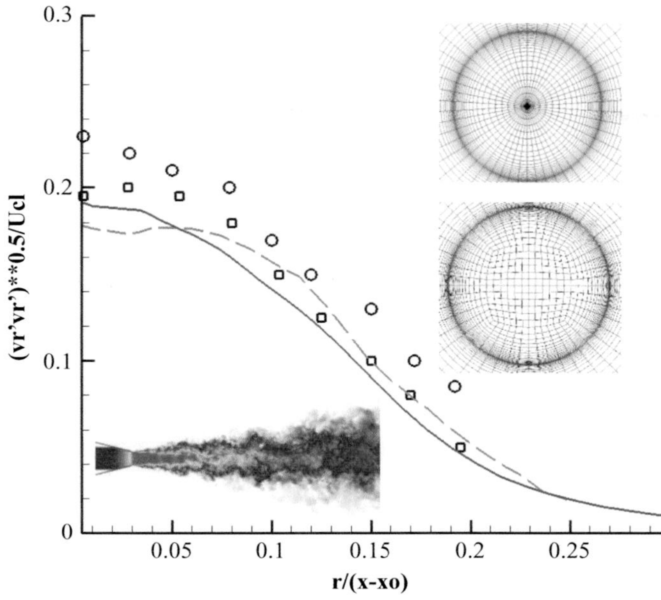

Figure 3.53. Impact of different near axis meshes on eddy-resolving simulations for a jet (dashed line is for 'O' grid and full line for butterfly grid).

component in the downstream self-similar region of the jet. The geometry is circular. Hence, an O-mesh could be used or the butterfly type grid described previously. The numerical singularity, arising from use of the 'O' grid, at the axis is best avoided. This is especially so with an explicit solver where extreme numerical stiffness will arise. For many structured solvers, the singularity at the axis presents further problems with regards to unambiguously defining velocity components at the axis. There is a range of specialist strategies to overcome this, but these are not considered here. As can be seen from Figure 3.53, the near axis mesh structure (the rest of the mesh is kept the same) has had a global impact on the flow. The differences in near axis flow structure have given rise to some flow redistribution. Although at the axis the problem arises of the strong convergence of grid azimuthal points, at larger radii the opposite problem can arise with the tangential grid spacing becoming too coarse. Towards the far field this is not such a great problem since typically with eddy-resolving simulations it is undesirable to resolve unsteady flow features at boundaries. These will give rise to boundary reflection problems. However, in summary, for jets it can be useful to use grid embedding strategies to control azimuthal grid spacings. Such a strategy is explored by Andersson et al. (2005).

Obviously, the solution grid should be sufficiently fine so that discretization approximations become small. The acceptability of error levels, clearly, depends on the solutions context. Generally, solutions should be run on at least two grids, one having a grid density which is twice the others. This aspect is discussed further in Chapter 7.

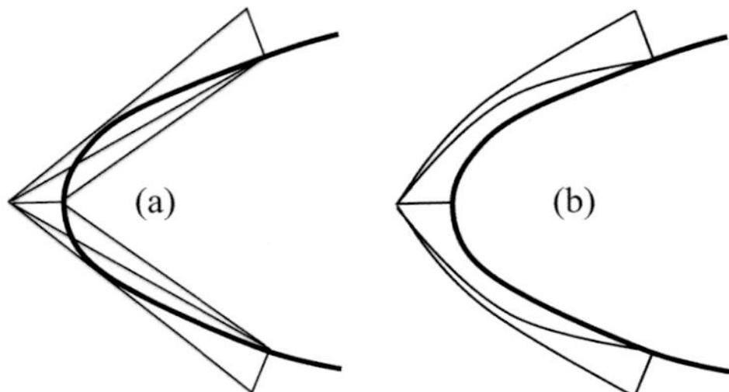

Figure 3.54. The need for high-order mesh generators: (a) non-curved internal element face crossing boundary and (b) curved internal face not crossing the boundary.

3.8 High-Order Meshing

High-order methods are attractive. Large-scale industrially relevant simulations can involve billions of cells. The use of high-order methods potentially reduces the necessary mesh count and also the volume of information that needs to be exchanged when using a domain decomposed solution with many processors. However, high-order elements with their internal collocations imply that the cell boundaries must be curved (and these also constitute extra data to transfer). Currently there are no commercial grid generators that can generate high-order scheme meshes. However, Gmsh is public domain software for high-order meshing (see Geuzaine and Remacle 2009).

The need for curved boundaries can be seen in Figure 3.54 based on Wang (2014). It shows how a straight internal element face from a standard mesh generator can cross a boundary. Figure 3.55 shows the importance of using a high-order boundary representation. Frame (a) shows the mesh structure for flow over a sphere. Frame (b) shows the low-order boundary representation and (c) the high–order boundary representation. The plots are taken from Sun et al. (2007). As can be seen, from Frame (b), the low-order boundary representation gives unacceptable error.

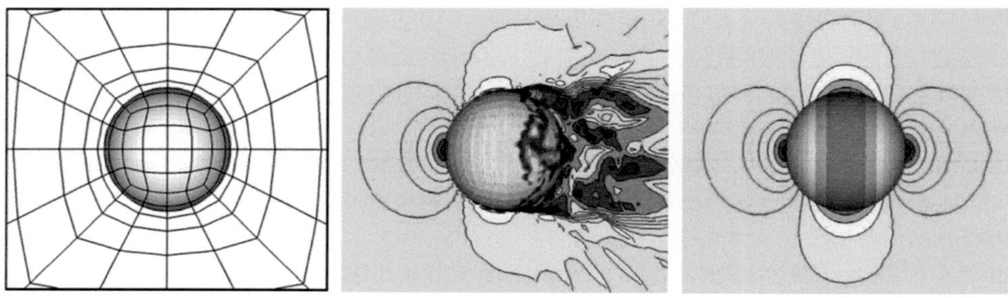

Figure 3.55. The importance of using a high-order boundary representation: (a) mesh structure, (b) low-order boundary representation and (c) high-order boundary representation (from Sun et al. (2007) – Published with the kind permission from Global Science Press).

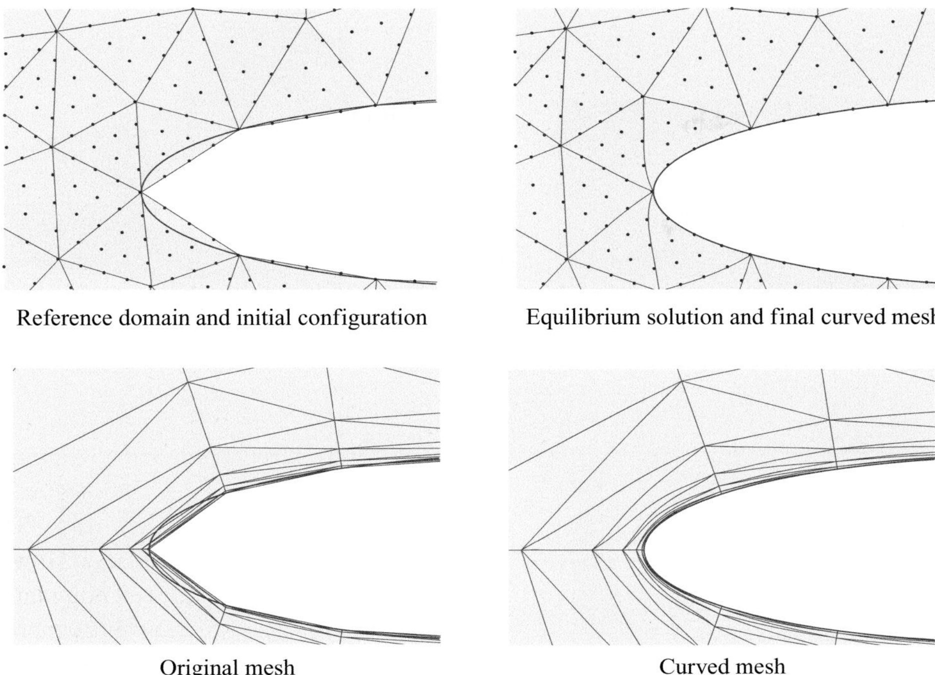

Figure 3.56. Examples of the use of a high-order mesh around the leading edge of an airfoil. The left-hand frames are for the unmodified 'low-order' mesh. The upper frames are for an isotropic mesh and the lower anisotropic (from Persson and Peraire (2009) – Published with kind permission of the authors).

A 'work around', to generate high-order element meshes using standard grid generation packages, is to first make a much finer mesh and then remove grid content. Alternatively, Persson and Peraire (2009) make a standard low-order scheme mesh. The boundary points are then projected to the actual curved boundary. Concepts from solid mechanics are then used to move the interior grid points relative to the revised boundary points. In this way, a valid mesh that does not have negative volumes is produced. An example of a mesh produced using this procedure is shown in Figure 3.56.

Figure 3.57 shows the use of high-order meshing on a Falcon aircraft. Frame (a) gives the original low-order mesh. Frame (b) gives a high-order (curved) mesh. Cells with a normalized Jacobian of less than 0.5 are identified. Frame (c) plots a histogram of cell quality when just boundary cells are curved and (d) plots cell quality when whole mesh is treated. Note that a scaled Jacobian of less than zero corresponds to cell inversion. The plots are taken from Persson and Peraire (2009). The improved mesh quality arising from using the solid mechanics based approach to propagate the impact of boundary cell distortion through the domain is clear.

The lack of high-order solvers reduces the potential for the emergence of high-order mesh generators. The lack of the latter also limits the development potential for the former.

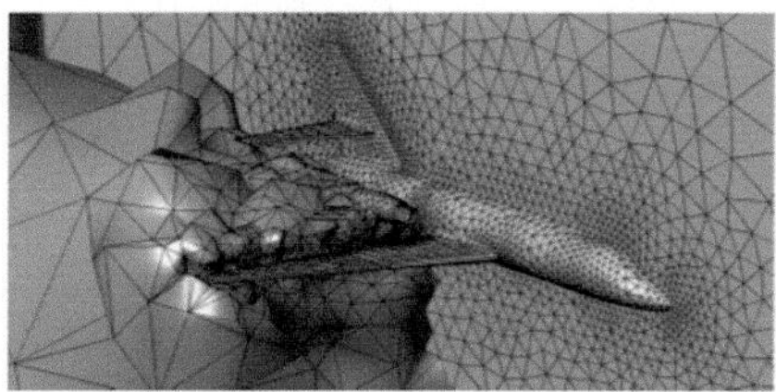

(a) Original Mesh

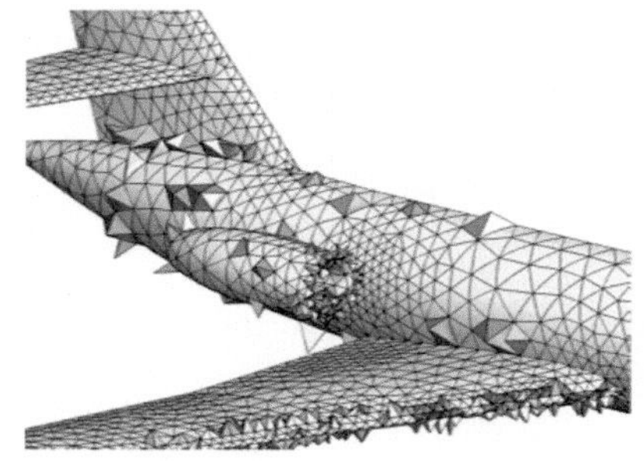

(b) Curved mesh with elements with scaled Jacobians <0.5 marked

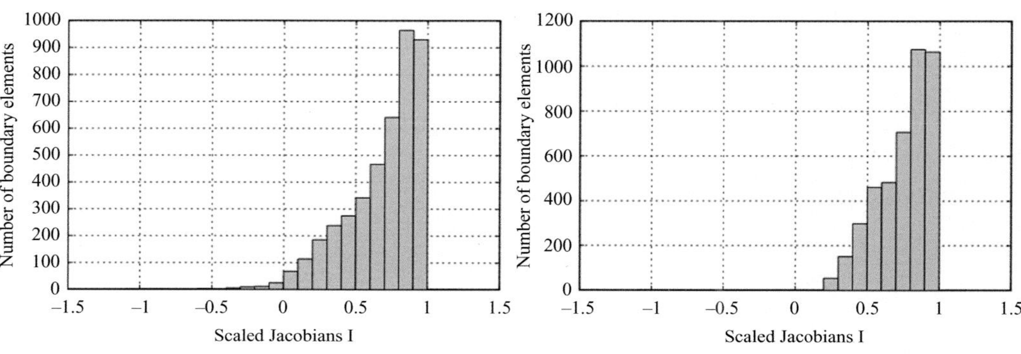

(c) Scaled Jacobians, local curving only

(d) Scaled Jacobians, elasticity equilibrium

Figure 3.57. The use of high-order meshing on a Falcon aircraft: (a) original mesh, (b) high-order (curved) mesh, (c) cell quality when just boundary cells are curved and (d) cell quality when whole mesh is treated. Note, a normalized Jacobian < 0 corresponds to cell inversion (from Persson and Peraire (2009) – Published with kind permission of the authors).

3.9 Meshing for Turbulence

For Euler and potential flow solutions, there is no need to capture the boundary layer profile and hence there is no need for heavy near wall grid clustering. Laminar flow needs grid clustering but as noted previously, the Reynolds numbers will generally be low. Hence the grid clustering requirements will not be too severe.

Modelling turbulent flow gives rise to greater complexity with regards to grid structure. Obviously the turbulent boundary layer has several zones. These are the viscous sublayer, buffer layer, log-law layer and the law of the wake zone (see Chapter 5 for more details). These areas have differing grid resolution requirements. Crucially, these requirements greatly depend on the turbulence modelling strategy that is used. Also, even though there might be the presence of solid walls, there is no guarantee that this wall supports a classical boundary layer. For example, the boundary layer can be modified by the flow physics external to the wall.

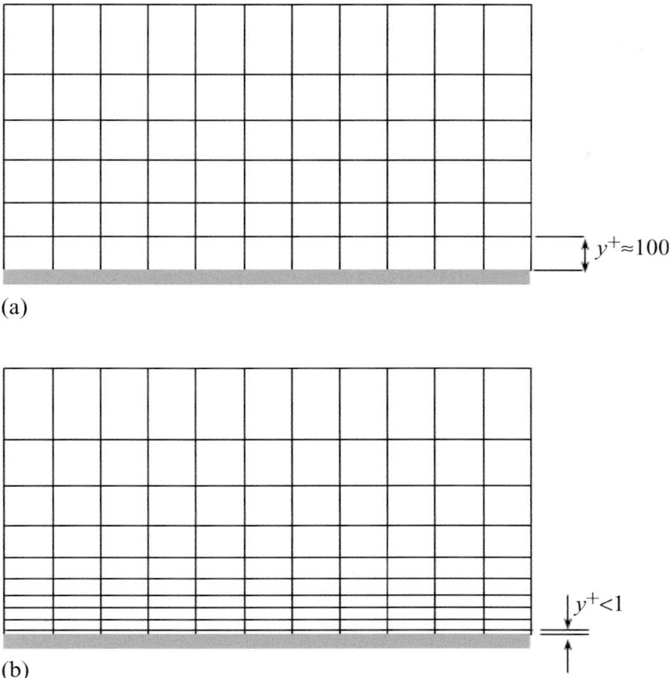

Figure 3.58. Potential forms of high and low Reynolds number turbulence model grids: (a) high Reynolds number model and (b) low Reynolds number turbulence model (from Tucker and Mosquera (2001) – Published with the kind permission of NAFEMS).

The turbulence modelling strategy will greatly influence the near wall meshing requirements.

3.9.1 *RANS Meshes*

As will be discussed in Chapter 5, RANS based turbulence models can be broadly classified into both high and low Reynolds number forms. With the latter, extremely small near wall grid spacings can be needed. Indeed, for the low Reynolds number k-ε model (an extremely popular model for internal flows), this can tend to the requirements needed for a direct numerical simulation. On the other hand, generally, the expectation is that for high Reynolds number formulations of turbulence models, the first off wall grid node will sit in around the middle of the log law region. However, nowadays adaptive wall functions see relatively widespread use. With these, the grid design with regards to the first off wall grid node location is not so critical. This is because the modelling function will adapt to the location of the first of wall grid node. Figure 3.58 gives potential grid forms for both high (a) and low (b) turbulence model simulations.

Returning to low Reynolds number turbulence models, simple algebraic turbulence models (like the mixing length with an algebraic equation for the mixing length based on normal wall distance) are more forgiving with regard to grid when in their low Reynolds number form. Hence, unlike for the k-ε model, for which the first off-wall grid location must be well below $y^+ < 1$ (this is a dimensionless wall distance defined in Chapter 5), for the Spalart-Allmaras and k-l models – both strongly connected to the algebraic mixing length model – first off-wall grid nodes at $y^+ =$ 2–3 can be acceptable.

3.9.2 *Eddy Resolving Simulation Grids*

The key challenge with eddy-resolving simulations is, of course, that since the turbulence being resolved is three-dimensional, flow that can be treated as two-dimensional with RANS needs to be modelled as three-dimensional. Hence, for a two-dimensional airfoil flow, the grid needs to be extruded in the spanwise direction. However, since the boundary layer is populated by fine elongated streaks (see Chapter 5) the number of grid nodes in this direction can be substantial. Also, these streaks need axial resolution. Hence again this places resolution constraints.

Chapter 5 (see Table 5.2) gives the near grid resolution requirements in the three coordinate directions for DNS, LES and hybrid RANS-LES. DNS, with the need to resolve all the way down to the Kolmogorov scales (the smallest in a turbulent flow) has the severest grid requirements. As can be seen from the tabulated data in Chapter 5, wall resolved LES requirements tend to those of DNS. Wall modelled LES, basically, uses a wall function to deal with the log wall region. Therefore, the streak zone, extending to around the outer part of the buffer layer, is modelled. Thus, the grid requirements are considerably less and the grid structure needed is quite different.

Table 5.1 gives the number of grid nodes typically found necessary for resolving the outer part of the boundary layer for eddy-resolving simulations. All the above-tabulated data is based on tests or estimates for second-order flow solvers. The outer layer estimates are also for a cube of extent equal to the boundary layer thickness. Next, grid design, specifically for LES, is discussed.

3.9.2.1 LES Grids Enforcing the wall resolved LES requirements in Table 5.2 can need a complex grid design. For example, if the boundary layer has developed from a small thickness, the leading edge mesh requirements can be severe. What is more, the optimal near wall grid design, to resolve the streaks, would suggest the need for relatively high aspect ratio cells. Hence, the need to connect the isotropic grid requirements in the outer part of the boundary layer to the highly anisotropic requirements in the inner part of the boundary layer. This gives some strong meshing challenges. Indeed, as noted by Chapman et al. (1975), hanging nodes would be required. Figure 3.59, through a crude schematic, illustrates the disparity of scales of the mesh in different flow zones. Notably, just very roughly every fifth grid node is shown. Importantly in the grid requirements estimates of Chapman and others, such ideal, hard to generate meshes, as shown in the figure, are assumed. Note the requirement for extreme changes in grid resolution with streamwise direction.

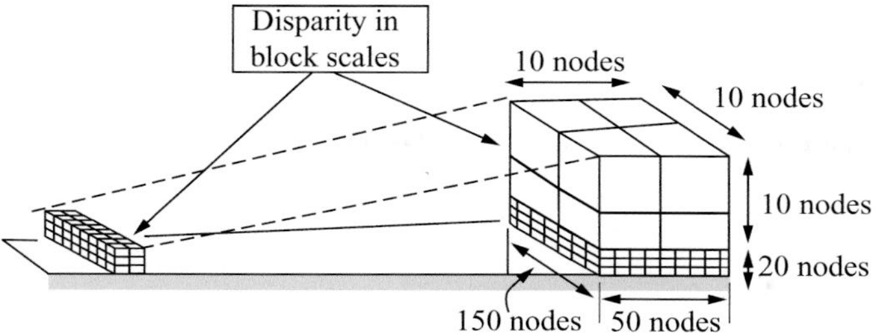

Figure 3.59. Sketch of complexity of generating a wall resolved LES mesh (roughly every fifth grid node omitted).

Without hanging nodes, this could propagate downstream and gives rise to the need for impractically large meshes.

To gain consistency with these requirements, the use of embedded, hanging node grids has been made (see Addad et al. 2008a).

3.9.2.2 Hybrid LES-RANS Grids The use of hybrid RANS-LES presents another range of meshing options. For the classical DES (Detached Eddy Simulation) approach, the general intent is that the boundary layer is fully modelled in RANS mode and any detached zones treated as LES. Notably, the switch over is defined as where the nearest wall distance $d = C_{DES}\,\Delta$. The parameter, C_{DES}, is a constant. The modelled turbulence related scale, Δ, has a range of definitions. These are given in Chapter 5. However, the key thing is that $\Delta$is generally defined by grid spacings. Hence, it is the grid that controls the RANS-LES interface location. This gives the mesh designer very strong control over the turbulence modelling. For example, Figure 3.60 gives the location of the RANS-LES interface for DES on a double delta configuration. Frame (a) gives the interface location in wall units (i.e., expressed as a y^+ or d^+) and (b) gives the surface mesh and contours of d. The figure is taken from Tucker (2006). As can be seen, the mesh imposes a substantial variation in the RANS-LES location. Note that for illustrative purposes, an artificial value of C_{DES} is used. Hence, in practical simulation terms, the interface, as a result of this artificial value, would be too close to the wall.

A rigid interface can be used. For example, the interface can be set at a fixed wall distance corresponding to, say, an average y^+ value. This hard interface then places the interface location under the control of the turbulence modeller. However, attempting to define an interface at a fixed y^+ value will also give rise to interface irregularities (discussed further in Chapter 5). Hence, the generation of a smooth interface that does not give rise to flow passing in and out of RANS and LES zones, or other problems, is a tough challenge. A range of approaches have been devised to remove the burden of careful grid design from the user, letting the turbulence model pick the best strategy, depending on the grid resolution relative to the local flow. For example, there is the LNS (Local Numerical Scales) approach of Batten et al. (2002), where a so-called latency parameter is used. This scales the eddy viscosity so that LES is used in zones where the grid is sufficiently fine and RANS is used in coarse

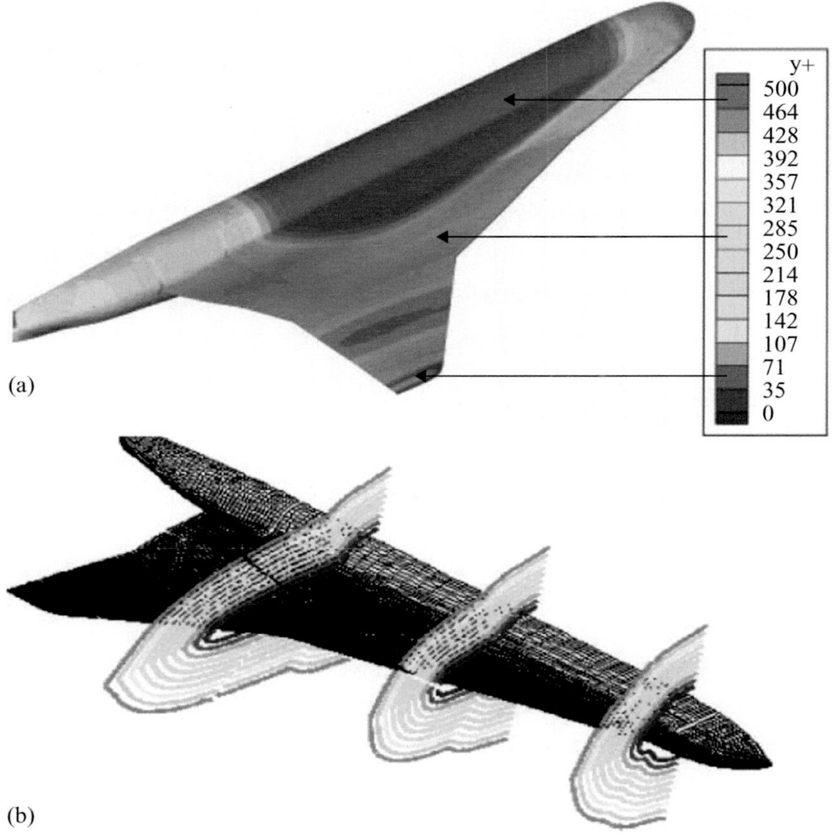

Figure 3.60. Location of RANS-LES interface for DES on a double delta configuration: (a) interface location in wall units and (b) surface mesh and contours of nearest surface distance, *d* (from Tucker (2006) – Published with the kind permission of Wiley).

grid zones. However, this can give rise to numerous zones where RANS patches are upstream of LES. This results in no resolved turbulence entering the downstream LES zones and hence, without subsequent grid re-design or other measures, poor accuracy.

An alternative turbulence modelling method, discussed in Chapter 5, is the NLDE (Non-Linear Disturbance Equation) approach. With this, an initial RANS solution is made. Then non-linear, large-scale disturbances around this are solved for. Figure 3.61 shows the relative grid requirements for these differing approaches.

The Zonal DES (ZDES) approach of Deck (2012) presents a further range of meshing challenges. With this approach, for zones with mild curvature and non-severe pressure gradients, where the RANS model will struggle to predict separation, an especially fine grid is embedded. Figure 3.62 shows a crude schematic illustrating this concept. The embedded zone is intended to allow the simulation to just locally be run in LES mode. This offers the best possibility of the separation point

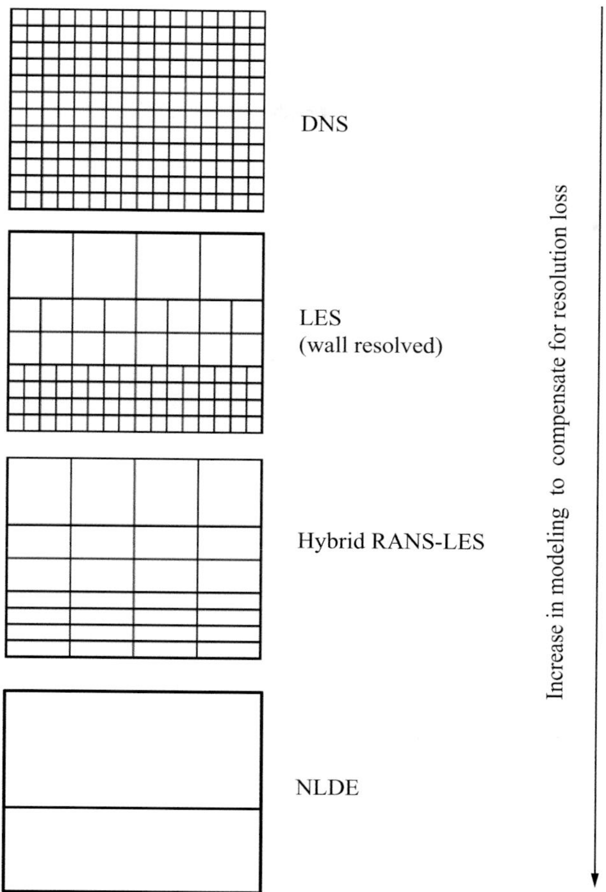

Figure 3.61. Grid requirements for different eddy-resolving approaches (adapted form Batten et al. 2002).

being predicted accurately. The ZDES approach also allows for extensive zones to be treated in fully RANS mode, with a two-dimensional grid connected to three-dimensional eddy resolved grids. Hence, as can be seen from this discussion, the turbulence modelling strategy and grid design are highly coupled elements.

Figure 3.63, extended from Chapter 2, further illustrates this point, showing a simple grid design for DES based on the work of Spalart (2001). The grid relates to a highly idealized V22 tilt rotor. For the flight configuration shown, the rotor is giving a downwash on the wing. The flap is orientated so as not to obstruct the downwash. The mesh consists of three zones. Zone I has a coarse C-mesh. In the zone occupied by this mesh, the fluid mostly has the uniform free stream velocity. Hence, this is an Euler type zone as noted by Spalart. There is then a viscous O-mesh around the airfoil section – Zone II. Finally, there is a wake block – Zone III – this being an 'H' grid block. To capture boundary layers, Zone II is heavily clustered near walls. However, examination of the eddy viscosity contours given in the top right-hand

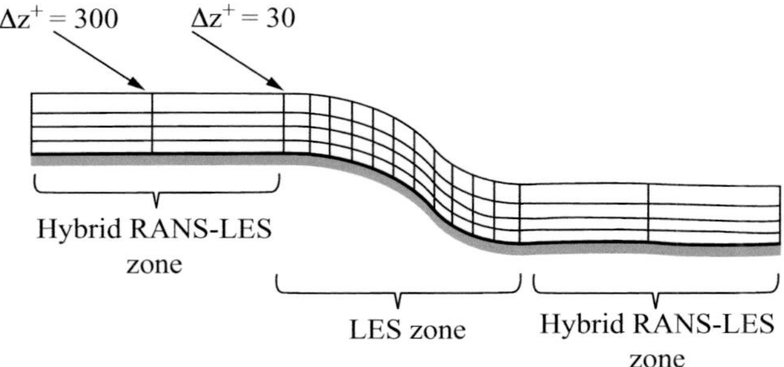

Figure 3.62. Example of a ZDES surface mesh zone.

figure insert shows that there is very little classical boundary layer content. On the top of the airfoil, the surface is subjected to strong impingement. The lower surface is in the wake zone. Indeed, the only zone approaching that of a classical boundary layer is the flap surface. This will have a developing boundary layer. The wake block is relatively fine. This is because it needs to resolve the integral scales in the wake. Each of these scales needs around 10 grid points.

This might all suggest that mesh generation can be driven by engineering insight. Although this is largely true, as shown by Larsson and Wang (2014), such an approach

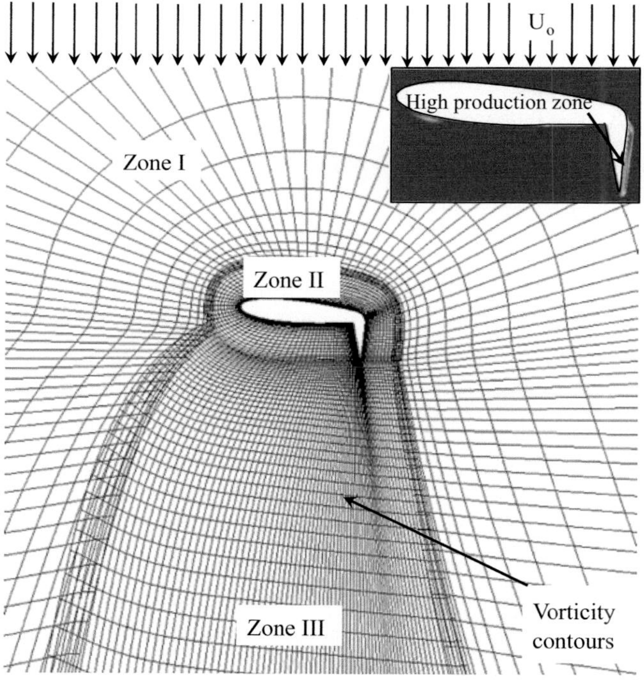

Figure 3.63. Schematic of the grid design for a simple DES (adapted from Tucker 2006).

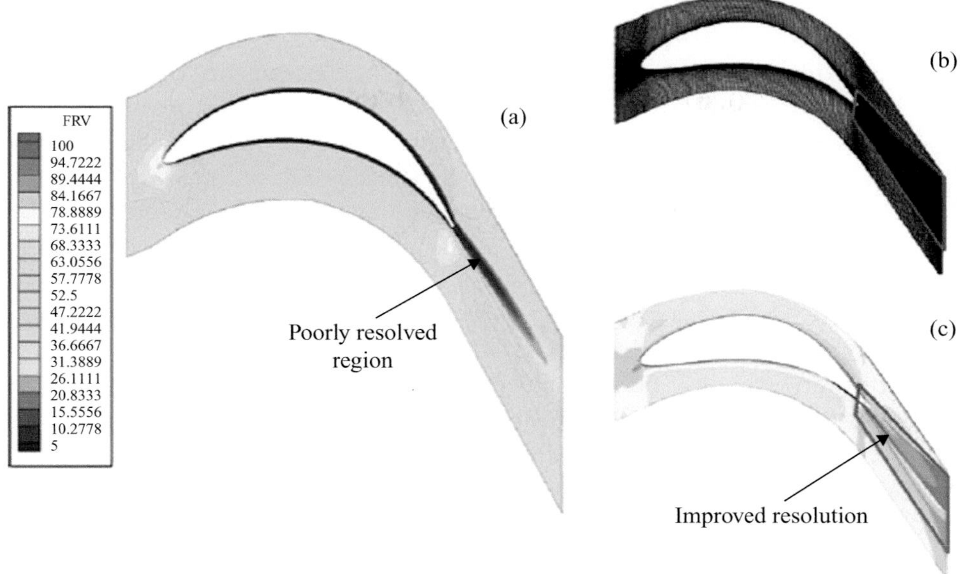

Figure 3.64. Use of RANS constructed energy spectrum to determine eddy-resolving grid resolution: (a) contours of the fraction of resolved turbulence energy based on an energy spectrum constructed from RANS information, (b) finer embedded wake block and (c) revised RANS based estimates of the percentage of resolved scales (from Tucker 2013) – Published with the kind permission of Springer).

can fail. Then it is necessary to have meshes that adapt based on more formal mathematical error estimates. Such approaches are, to an extent, discussed later.

3.9.3 *Use of Correlations and Preliminary RANS to Inform Grid Design*

Clearly, with all of the above, surface shear stress and boundary layer thickness estimates are necessary to correctly mesh a boundary layer. These can initially be estimated based on analytical expressions. For example, those for flat plate boundary layers and so forth. However, for rotating flows, the nature of the boundary layer can vary considerably. For example, on rotating disks, non-entraining Ekman layers can be generated. On rotating cylindrical surfaces, Stewartson layers can be produced. If there is substantial buoyancy, the nature of the boundary layer flow and hence meshing requirements can again greatly change. Hence, it is first necessary to make the best initial estimate considering the known physics and then perform a simulation. Based on this evidence, the grid can be further refined. However, it needs to be born in mind that too coarse a grid will underestimate the wall shear stress and hence the severity of the meshing requirements.

For eddy-resolving simulations, it can also be helpful to perform a preliminary RANS to guide grid design. For example Figure 3.64a shows contours of the fraction of resolved turbulence energy based on an energy spectrum constructed based on RANS information. The procedure is outlined in Tyacke et al. (2013). From Frame

(a) it can be identified that the wake zone is poorly resolved. Hence, in Frame (b) a finer wake block is embedded. Frame (c) shows revised RANS based estimates, of the percentage of resolved scales. With regards to the boundary layer zones, the wall shear stress from the RANS can be used. These can be combined with the inner layer mesh requirement estimates, given in Chapter 5, to estimate the near wall grid spacings. Similarly boundary layer thickness estimates can be extracted from the RANS. Then the outer layer grid requirement estimates given in Chapter 5 can be used to inform the grid design.

The wake zones are more challenging. These are areas where the predictive accuracy of RANS is generally poor – particularly if the length scale estimates are based directly on turbulence model variables. For simple bluff bodies it is known that the integral scales will be of the object's size. Hence, since approximately 10-grid points would be needed to resolve such a scale, an estimate to the grid requirements can be easily made. Wake widths extracted from RANS may well be a more reliable estimate to grid requirements, in more complex wakes, than using scales implied by the RANS model variables. Indeed, in some RANS models, no length scale data is directly available (see, for example, the Spalart and Allmaras (1992) model).

As noted by Piomelli (2014), a useful objective for eddy-resolving simulations is to fix the ratio of the integral scales to the filter width, Δ, typically defined by grid spacing. Then it is roughly ensured that the percentage of resolved energy is constant over the domain. Based on dimensional analysis, the following standard RANS expression can be used to estimate the integral scales

$$L_{RANS} = C\frac{(\overline{u_i'u_i'})^{3/2}}{\epsilon} = 2C\frac{k^{3/2}}{\varepsilon} \tag{3.47}$$

where ε is the rate of dissipation of turbulence, k is the turbulence kinetic energy and $\overline{u_i'u_i'}$ are the Reynolds stresses. These can all be extracted from different types of RANS models (see Chapter 5). To get the large scales, $C \approx 1$. The approach of Piomelli seeks to generate a mesh such that L_{RANS}/Δ is approximately constant throughout the domain.

3.9.3.1 Taylor Microscale based LES Grid Design Addad et al. (2008b) use estimates of the Taylor microscales (an intermediate length scale, where viscosity has a significant impact and above which the inertial subrange is expected) to design LES grids. This scale can be expressed as

$$\lambda_{ii,k} = \sqrt{\frac{2\overline{u_i'u_i'}}{\left(\overline{\frac{\partial u_i'}{\partial x_k}\frac{\partial u_i'}{\partial x_k}}\right)}} \tag{3.48}$$

where k is now the coordinate direction and the subscripts ii refer to the velocity components. The Taylor microscales are, in fact, a combination of the Kolmogorov, η, and integral scales, L, such that $\lambda \sim (L)^{1/3}\eta^{2/3}$. Based on idealizations from homogenous turbulence, the Taylor microscales can be approximated for a RANS framework as

$$\lambda_{RANS,i} = \sqrt{15\frac{\overline{u_i'u_i'}\nu}{\epsilon}} \tag{3.49}$$

where the subscript *RANS* indicates a RANS model based estimate. Also, ν is the kinematic viscosity.

Using DNS data, Addad and colleagues deduce that Taylor microscale estimates are reasonable. The aim for LES is that the grid spacing is such that the filter sits within the inertial subrange. Use of the Taylor microscale ensures that, where possible, this requirement is well satisfied. However, for high Reynolds number flows, some limiting, based on L, is needed such that

$$\Delta x_k = \max\left[\lambda_{ii,k}, \frac{L_{RANS}}{10}\right] \tag{3.50}$$

This limiting prevents excessively small length scales being generated. When λ is based on a RANS model, the following limit is also proposed

$$\lambda_{ii,k} = \max(\lambda_{RANS,k}, \eta/5) \tag{3.51}$$

where $\eta = (\nu^3/\epsilon)^{1/4}$.

The approach of Addad et al. (2008b) suggests quite cuboid cells that might potentially be approximated using the octree mesh generation process. The Taylor microscale approach suggests that in the wall normal direction, current grids based on more empirical data are nearly five times too fine. They also suggest that they are two times too coarse in the streamwise direction. The grid form suggested by the preceding procedure is not that dissimilar to the grid shown in Figure 3.59.

3.10 Moving Meshes and Adaptation

3.10.1 *Adaptation*

In CFD simulations it can sometimes be important to have meshes that adapt with flow features or geometry. Various mesh adaptation approaches and their advantages and disadvantages are summarized in Table 3.4. With mesh enrichment, an initial coarse mesh can be enriched to resolve flow features.

Adaptive solutions, where mesh spacings are automatically adjusted during computations through potentially a combination of enrichment, movement and remeshing to minimise errors, are attractive.

3.10.1.1 Structured Grid, Flow Feature Based Adaptation Grid adaptation techniques that just involve the redistribution of grid points without insertion are best suited to structured grid solvers being relatively easy to implement in this framework. The curvilinear grid transforms discussed in Chapter 4 can be used as part of a solution procedure to cluster grid points in areas of high gradients (see Anderson

Table 3.4. *Mesh adaptation strategies and their advantages and disadvantages (from Tucker (2013) – published with the kind permission of Springer)*

Meshing Strategy	Advantages	Disadvantages
Enrichment	Most practical for steady flows	Memory intensive in 3D; Distorted elements can be generated; De-refinement beyond initial mesh challenging
Movement	Efficient for modest boundary movements	Need to define sufficient initial points; Not easy to guarantee valid mesh
Remeshing	Can deal with extensive boundary movements	Time consuming in 3D

1995). A crude adaptive grid strategy for a fixed number of grid points is to define relationships such that

$$\begin{aligned}\Delta x &= C_x \Delta\xi\, f\left(\frac{1}{|\partial\phi/\partial x|}\right)\\ \Delta y &= C_y\, \Delta\eta\, f\left(\frac{1}{|\partial\phi/\partial y|}\right)\end{aligned} \tag{3.52}$$

where C_x and C_y are arbitrary scale factors. The transformed grid and hence $\Delta\xi$ and $\Delta\eta$ are fixed. Grid refinement could be based on, for example, $\phi = p$, T or ρ. In regions of high ϕ gradient, as can be seen from Equations (3.52), the grid spacing in the physical plane will be made closer. The final form of the adaptive grid can in itself be helpful in interpreting the physical nature of the predicted flow.

Figure 3.65 gives a simple example of grid redistribution for the supersonic flow over wedge with an angle of 5.27°. With an inlet pressure of 23.5 N/m^2 (M = 1.6), an oblique shock is formed at the compression corner and it reflects back from the top wall. The shock deflection is perfectly designed to impinge on the expansion corner of the wedge and suppress the growing expansion fan. Similar shock systems are employed by supersonic jet engine intakes. In Figure 3.65, the Euler equations are solved. The solver used is a simple overlapping grid/cell vertex type to be described in the next chapter.

The Frame (a) flow also shows an initial mesh. Frame (b) shows a first refinement to the mesh and Frame (c) shows a second refinement. The refinements are based on looking at the smoothed density field. Multiple smoothing sweeps are needed. Also, the grid refinement is based around projecting the density field rather than directly evaluating density gradients. It is clear from the right-hand frames that the shock structure is sharper for the adapted meshes. The inlet M = 1.6 and wedge angle of 5.727° should (from analytical data) result in a shock angle of 45.0° relative to horizontal and a post-shock Mach number of 1.40. The downstream shock should then have an angle of 48.1° relative to horizontal and post-shock Mach number of 1.19. The predictions agree well with these expected analytical values.

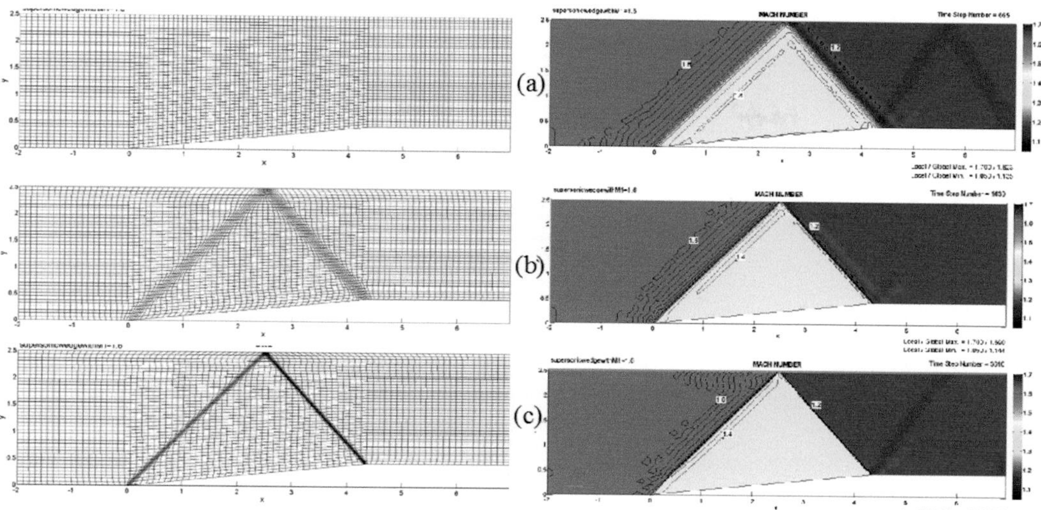

Figure 3.65. Two stages of adaptive mesh refinement for a transonic Euler solution (from Sampson 2013).

Figure 3.66 shows contours of total pressure, indicating the losses incurred by the shocks. The overall loss created by the shock system is only of the order of 0.5%, the oblique shocks being nearly isentropic. The oblique shocks perform the majority of the compression (e.g., here the compression is from M = 1.6 to 1.19) before a much-weakened normal shock finally takes the flow to a subsonic speed. Interestingly, if using the adjoint procedure (see Section 3.10.2) to adapt meshes to meet a global design constraint, then the adjoint procedure suggests that less grid should be used around the shock. Hence, the shock is being treated as a crude jump condition.

3.10.1.2 Structured Grid, Grid Feature Based Adaptation Notably, the grid can also be adapted or internally optimized to minimise grid centred objective functions. These could be, for example, smoothness, I_S, orthogonality, I_O, and a measure of

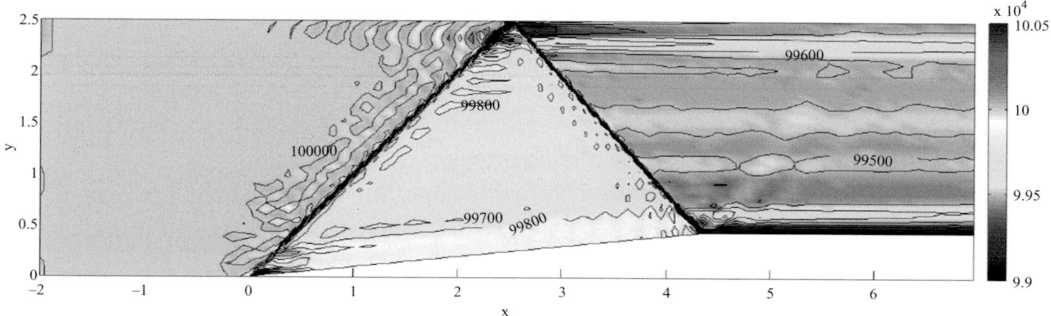

Figure 3.66. Variation of total pressure over supersonic wedge (from Sampson 2013).

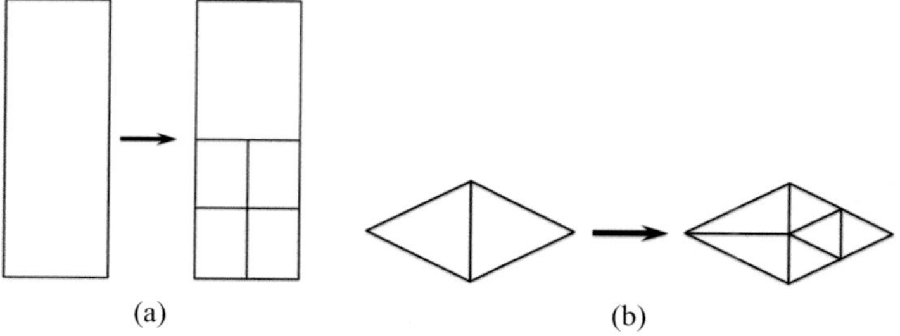

Figure 3.67. The use of hanging nodes and the subdivision of triangles: (a) hanging nodes and (b) subdivision of triangles into additional triangles (from Tucker and Mosquera (2001) – Published with the kind permission of NAFEMS).

volume, I_V. These objectives can be weighted as follows to form a linear combination of a total objective

$$I_T = I_S + Wf_S \int I_S \, dV + Wf_O \int I_O \, dV + Wf_V \int I_V \, dV \tag{3.53}$$

where I_S and I_O are defined by Equations (3.40) and (3.41) respectively. Also, $I_V = IWf$ (see Equation 3.10). I_T can be directly minimized by setting up Euler-Lagrange equations using a variational approach. This Euler-Lagrange approach is a classical optimization procedure that will allow minimization of a function. The equation system is large and the process perhaps best carried out using a symbolic mathematics package. By way of an example Pletcher et al. (2013) just consider I_S and show that the application of the Euler-Lagrange approach to Equation (3.41) alone simply yields the standard elliptic grid generation equations given earlier. Full solution of an optimization system based on Equation (3.53) is computationally expensive. However, for eddy-resolving simulations, high quality grids are necessary. Then, as a percentage of the overall computational cost, when considering the potential benefits arising from the high quality grids, the use of this multi-objective optimization/grid adaptation approach has some attractions.

3.10.1.3 Unstructured Grid Refinement and De-refinement When grid point insertion is used with quadrilateral grids, hanging nodes are produced as shown in Figure 3.67a. As discussed in Chapter 4, for cell centred, unstructured codes these can be naturally dealt with. For cell vertex, median dual cell methods, such hanging nodes cannot be accommodated, for reasons outlined in Chapter 4. Hence, the numerical scheme places limits on the form of the adaptivity.

Unstructured solvers lend themselves best to adaptation with grid point insertion. For example, for triangular meshes, triangles can be subdivided to give just triangles as shown in Figure 3.67b.

The use of unstructured grid enrichment, involving what is called adaptive grid embedding, is given by Davis and Dannenhoffer (1994) and Coelho et al. (1991). The application of the enrichment approach to time dependent shock hydrodynamics is

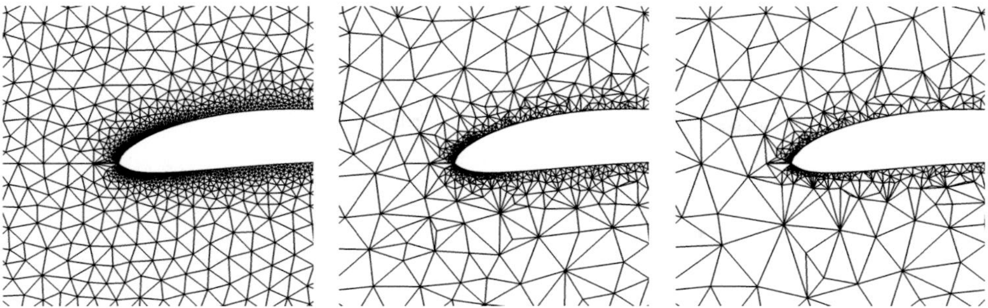

Figure 3.68. Shortest edge removal.

shown by Berger and Colella (1989). Importantly, multilevel (or multigrid) methods, discussed in Chapter 4, lend themselves (see Brandt 1980) to grid adaptation and embedding (see Thompson and Ferziger 1989).

Especially with LES in mind, it would be important to be sure that mesh movement/changes do not introduce spurious motions or mass sources and sinks. Time averaging of results on a moving grid is also challenging. In addition, the continual changing of the LES filter width raises theoretical questions. A key, general difficulty is that multiple grid refinements can generate distorted elements and mesh quality is especially important for LES. Also, in three dimensions, local mesh refinement – in a grid generation context – can be expensive. As noted, large changes in grid spacings can generate substantial numerical error. The gradient limiting Hamilton-Jacobi equation is useful in this context for ensuring that the grid expansion when using adaptive grids is bounded. Persson (2006) presents an example of the use of this approach when using adaptive meshing to capture a shock wave.

3.10.1.4 Mesh De-refinement and Multigrid Convergence Acceleration As well as locally refining grids, to keep simulation costs under control, it can be necessary to coarsen grids. If it is wished to de-refine, the origins of the original refinement must be stored and this can be expensive. De-refinement is also key to the use of multigrid convergence acceleration. With this and also grid sequencing (where a crude initial guess is made on a coarse mesh and then interpolated down to successively finer meshes), it can be necessary to generate coarse meshes based on a fine base mesh. To coarsen the mesh there are a range of approaches. For example, as shown in Moinier (1999), shortest edges can be removed. Use of this approach is shown in Figure 3.68. Alternatively, as shown by Mavriplis (1999), cell agglomeration can be used. The cell growth process for this is shown in Figure 3.69.

According to Brandt (1977), a two to one coarsening ratio is the most effective for multigrid convergence acceleration. For turbulent flows, Shyy et al. (1993) recommend that first off-wall grid node positions are fixed at all grid levels. For complex geometries, the implementation of geometric multigrid can be limited by the fact that as the base grid is successively coarsened the resulting grid cannot support the geometry.

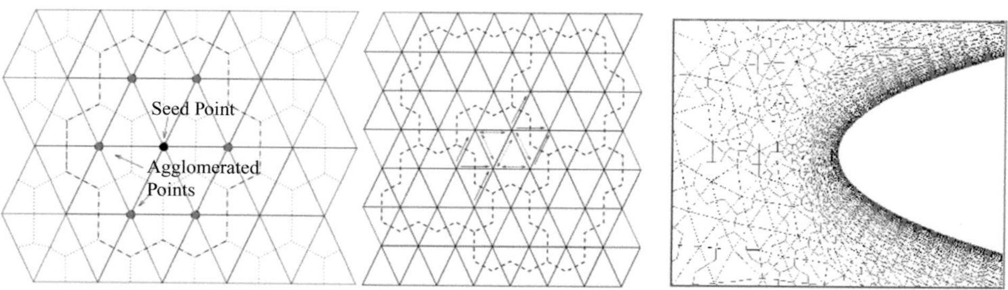

Figure 3.69. Cell agglomeration (Mavriplis (1999) – Published with the kind permission of the author).

3.10.2 *Adjoint Analysis*

As has been shown, it can be important to assess the impact of local grid quality on a global potential design variable of interest, such as mass flow, or a loss, lift or drag coefficient. Adjustments in local grid refinements can be used, in combination with the Taylor series expansion, such as discussed in Chapter 7, to assess the local discretization truncation error. Such estimates simply relate local changes in the solution variables to changes in mesh. However, these estimates can be misleading. Numerical errors propagate with the flow and hence cannot be evaluated locally. A substantial solution error at a downstream location can be caused by deficiencies in the upstream grid. A more reliable test of grid quality is to study adjoint variables. Here the procedure of Venditti and Darmofal (2002) is described, which was exploited earlier for assessing the accuracy of multiblock grids. With this procedure, two grids with spacings of ΔX and Δx are considered. The latter is the finer and is considered as a uniform refinement to the grid with the spacings of ΔX. The coarser grid is assumed to give an approximate solution. It is expected that it captures at least the basic flow features but does not have the accuracy of the grid with the spacing Δx. The finer grid provides the necessary level of accuracy but is expensive to compute. The residual vectors (essentially the flow equations set to be solved) for the discretized system on the coarse and fine grids are represented by $A_{\Delta X}(\phi_{\Delta X})$ and $A_{\Delta x}(\phi_{\Delta x})$, where ϕ is the (primal) solution to the equations. The process has the accuracy of an objective function, $C(\phi)$, in mind – such as drag or lift and so forth. Using a Taylor series, we can write an estimate for an exact objective $C_{\Delta x}(\phi_{\Delta x})$ about an estimate $C_{\Delta x}(\phi_{\Delta x}^{\Delta X})$ as

$$C_{\Delta x}(\phi_{\Delta x}) \approx C_{\Delta x}\left(\phi_{\Delta x}^{\Delta X}\right) + \left[\frac{\partial C}{\partial \phi}\right]_{\phi_{\Delta x}^{\Delta X}} \left(\phi_{\Delta x} - \phi_{\Delta x}^{\Delta X}\right) \tag{3.54}$$

where $\phi_{\Delta x}^{\Delta X}$ represents a variable computed with a coarse step of ΔX and represented on a finer grid of Δx. The transfer to the finer grid can be expressed as

$$\phi_{\Delta x}^{\Delta X} = I_{\Delta x}^{\Delta X} \phi_{\Delta X} \tag{3.55}$$

where $I_{\Delta x}^{\Delta X}$ represents a prolongation operator that can involve linear or higher order interpolations. Similar, to Equation (3.54) the residual vector, A, can be expressed as

$$A_{\Delta x}(\phi_{\Delta x}) \approx A_{\Delta x}\left(\phi_{\Delta x}^{\Delta X}\right) + \left[\frac{\partial A}{\partial \phi}\right]_{\phi_{\Delta x}^{\Delta X}} \left(\phi_{\Delta x} - \phi_{\Delta x}^{\Delta X}\right) \tag{3.56}$$

which can be rearranged as

$$\phi_{\Delta x} - \phi_{\Delta x}^{\Delta X} = \left[A_{\Delta x}(\phi_{\Delta x}) - A_{\Delta x}\left(\phi_{\Delta x}^{\Delta X}\right)\right] \left[\frac{\partial A}{\partial \phi}\right]_{\phi_{\Delta x}^{\Delta X}}^{-1} \tag{3.57}$$

Note that since, when the governing equations are satisfied,

$$A_{\Delta x}(\phi_{\Delta x}) = 0 \tag{3.58}$$

Equations (3.57) and (3.54) can be combined to give

$$C_{\Delta x}(\phi_{\Delta x}) = C_{\Delta x}\left(\phi_{\Delta x}^{\Delta X}\right) + \left[\frac{\partial C}{\partial \phi}\right]_{\phi_{\Delta x}^{\Delta X}} \left[\frac{\partial A}{\partial \phi}\right]_{\phi_{\Delta x}^{\Delta X}}^{-1} A_{\Delta x}\left(\phi_{\Delta x}^{\Delta X}\right) \tag{3.59}$$

The furthest right-hand group of terms in Equation (3.59) are an estimate of the solution error

$$E = -\left[\frac{\partial C}{\partial \phi}\right]_{\phi_{\Delta x}^{\Delta X}} \left[\frac{\partial A}{\partial \phi}\right]_{\phi_{\Delta x}^{\Delta X}}^{-1} A_{\Delta x}\left(\phi_{\Delta x}^{\Delta X}\right) \tag{3.60}$$

Alternatively, Equation (3.60) could be viewed as a solution correction. For computational efficiency and convenience, the following adjoint variable is defined

$$\Lambda_{\phi,\Delta x}^{T}\left(\phi_{\Delta x}^{\Delta X}\right) = -\left[\frac{\partial C}{\partial \phi}\right]_{\phi_{\Delta x}^{\Delta X}} \left[\frac{\partial A}{\partial \phi}\right]_{\phi_{\Delta x}^{\Delta X}}^{-1} \tag{3.61}$$

or, rearranging

$$\left[\frac{\partial A_{\Delta x}}{\partial \phi_{\Delta x}}\right]_{\phi_{\Delta x}^{\Delta X}}^{T} \Lambda_{\phi,\Delta x}\left(\phi_{\Delta x}^{\Delta X}\right) = -\left[\frac{\partial C}{\partial \phi}\right]_{\phi_{\Delta x}^{\Delta X}}^{T} \tag{3.62}$$

However, Equation (3.62) needs solution on the fine mesh and this we seek to avoid. Hence, it is rewritten as

$$\left[\frac{\partial A}{\partial \phi}\right]_{\phi_{\Delta X}}^{T} \Lambda_{\phi,\Delta x} = -\left[\frac{\partial C}{\partial \phi}\right]_{\phi_{\Delta X}}^{T} \tag{3.63}$$

This is a simultaneous equation set. Once $\Lambda_{\phi,\Delta X}$ has been computed, this field is interpolated onto the fine mesh via a projection operator

$$\Lambda_{\phi_{\Delta x}^{\Delta X}} = I_{\Delta x}^{\Delta X} \Lambda_{\phi,\Delta X} \tag{3.64}$$

The error, E, is computed from

$$E = \Lambda_{\phi_{\Delta x}^{\Delta X}}^{T} A_{\Delta x}\left(\phi_{\Delta x}^{\Delta X}\right) \tag{3.65}$$

Hence, the error is a function of the residual weighted by the adjoint variable. This process is summarized in Figure 3.70.

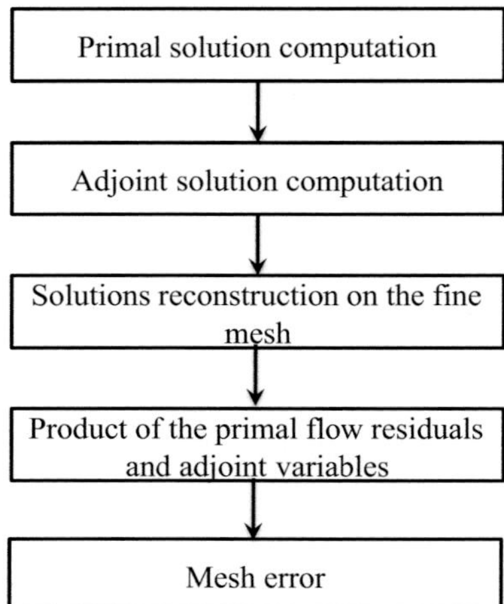

Figure 3.70. Summary of adjoint error estimation for grid.

3.10.3 *Space-Time Meshes and Adaptation*

With space-time methods, a mesh that extends into the temporal domain is used. Based on the temporal truncation error, this mesh can be adapted in time (see Mani and Mavriplis 2010, Bell and Surana 1994)) or in space also. A key attraction of this approach is that regions of high unsteadiness activity can be treated with smaller time steps than the rest of the domain. The governing equations, in say a finite volume method, are integrated (flows are summed) around temporal as well as spatial edges. A typical mesh, with a single spatial dimension, is shown in Figure 3.71. With time varying spatial mesh deformations, the space-time domain mesh could look as shown in Figure 3.72.

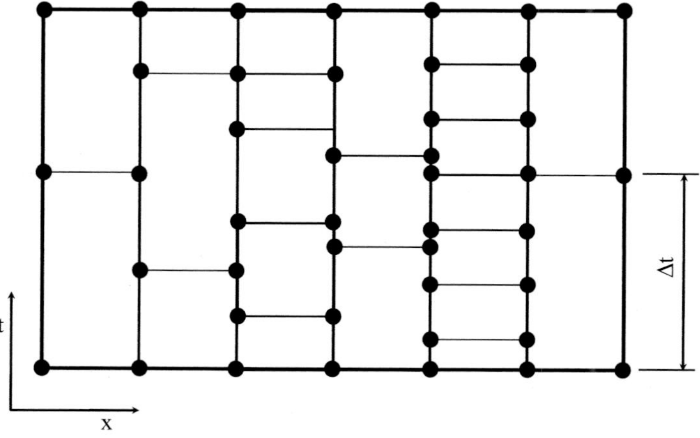

Figure 3.71. A typical space-time mesh, with a single spatial dimension (from Tucker (2013) – Published with the kind permission of Springer).

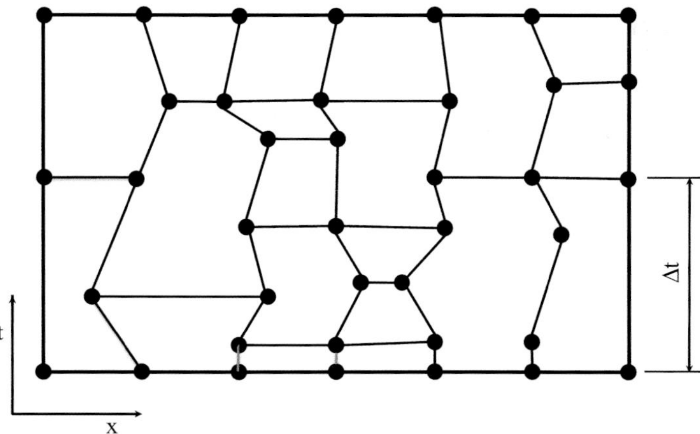

Figure 3.72. Mesh, with a single, time-adapting, spatial dimension (from Tucker (2013) – Published with the kind permission of Springer).

Figure 3.73 shows the form of a space-time element and hence the surfaces about which the integration (described in Chapter 4) of the governing equations needs to be performed.

3.10.4 *Boundary Movement*

Just moving the mesh to account for simple boundary movements is relatively straightforward. This is especially so if they involve structured meshes. However, when there are large surface movements and complex mesh topologies it is not easy to guarantee a physically sensible mesh. Then full re-meshing can be needed.

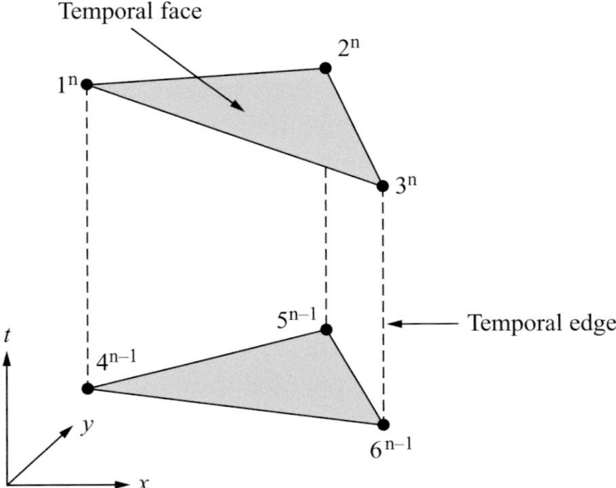

Figure 3.73. A space-time element at two time locations (shaded zones are the areas of the temporal faces) (from Tucker (2013) – Published with the kind permission of Springer).

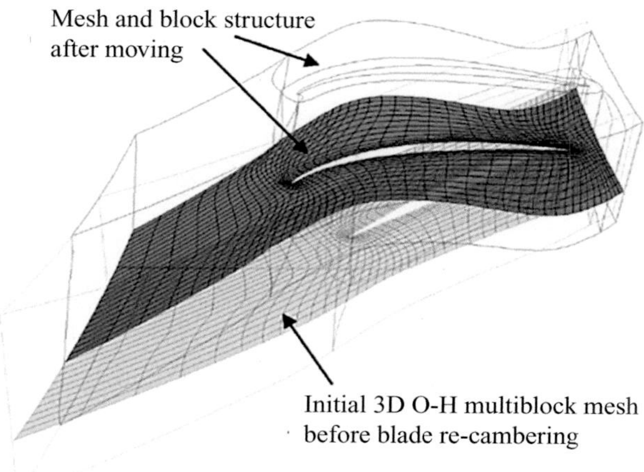

Figure 3.74. Mesh movement using simple algebraic interpolations (from Page et al. (2013) – Published with the kind permission of the American Society of Mechanical Engineers).

However, this is expensive and will also involve interpolation of the solution on the old grid zone to the new. Ideally a moving boundary scheme needs to combine the advantages of the three methods (noted earlier), that is, mesh movement, local enrichment and also zone marking and complete re-meshing. Such a strategy is discussed later. Next, key mesh movement strategies are outlined.

3.10.4.1 Mesh Movement Algorithms

3.10.4.1.1 Algebraic Interpolations For structured multiblock meshes, perturbations on existing grids can be performed. Algebraic shearing is widely used (see Potsdam and Guruswamy 2001). With this, surface displacements are distributed along grid lines of constant index. Grid line rotation operations can also be implemented. The displacement is weighted so that it tends to zero towards the outer boundary. Such a method is not readily extendable to multiblock grids where the outer boundary is likely to be another grid block. Figure 3.74 shows the use of a simple algebraic interpolation based technique for an aerodynamic section. As can be seen, reasonable grid quality can be (robustly) maintained for substantial grid displacements.

3.10.4.1.2 Delaunay Mapping Figure 3.75 shows the movement of a hybrid viscous mesh for an airfoil using a Delaunay graph. Frame (a) shows the moving airfoil geometry. Frame (b) shows the Delaunay graph. This is essentially triangular shards that extend, in this case, to the four corners of a rectangular domain. Note that the corners are not shown in the figure. Frame (c) shows the deformed mesh. These images are taken from Liu et al. (2006). With the approach of Liu and colleagues, the Delaunay graph of the boundary points is stored. There is a one-to-one mapping between this and interior grid points. Hence, once the boundary has moved and the

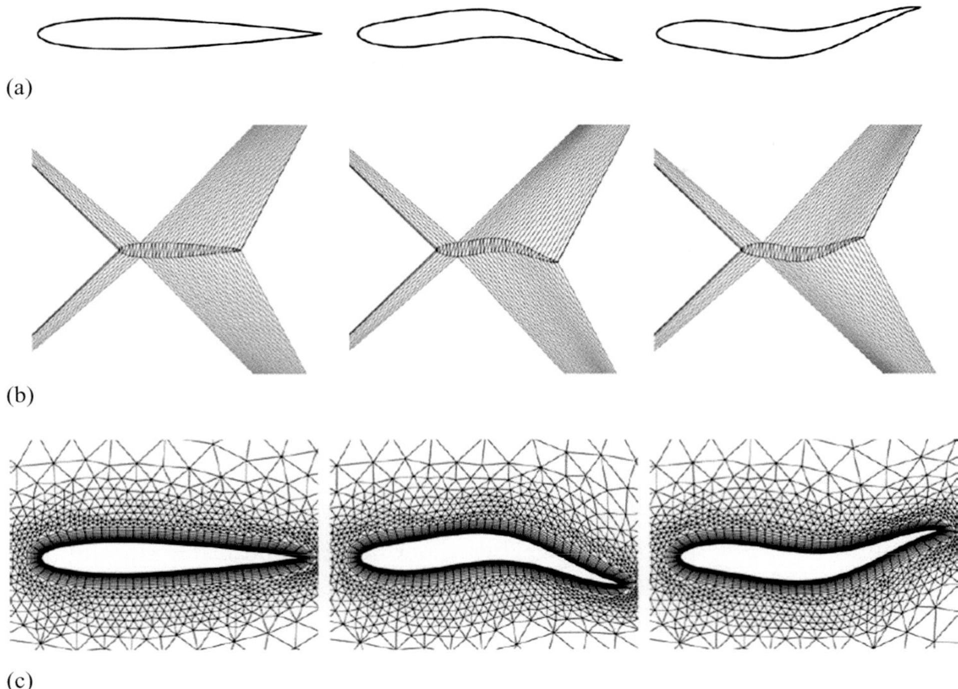

Figure 3.75. Movement of a hybrid viscous mesh for an airfoil using a Delaunay graph: (a) moving airfoil geometry, (b) Delaunay graph and (c) the moved mesh (from Liu et al. (2006) – Copyright Elsevier).

boundary point Delaunay graph has been adjusted, through mapping the mesh can be updated. With boundary rotation, the Delaunay graph can become invalid. Since more costly searches take place at the pre-processing stage, the method is especially efficient.

3.10.4.1.3 Spring Analogy The most widely used approach is to assume that the edges connecting nodes behave like springs that respond to small boundary displacements in either a linear or non-linear fashion. This is achieved by solving the following internal equilibrium equation

$$\sum_j k_{ij}\left[L_{ij}^n - L_{ij}^{n-1}\right] = 0 \tag{3.66}$$

where subscripts i and j are the indices for the two ends of a grid line. The spring coefficient, k_{ij}, can be taken to be the inverse of the edge length or some adjacent volume-based scale. Also, L_{ij}^n and L_{ij}^{n-1} give edge lengths before and after movement respectively. For highly stretched near wall grids, negative volumes can arise. In this zone the spring stiffness can be locally increased. This will result in the near wall cells moving as a solid in the surface vicinity. The spring analogy process can take around 10–15% of the total computing time (see Hassan et al. 2007b).

Table 3.5. *Advantages and disadvantages of different mesh movement approaches (adapted from Tucker (2013) – Published with the kind permission of Springer)*

Approach	Advantages	Disadvantages
Interpolation/Algebraic operations	Lacks generality	Fast
Delaunay Mapping	Can be applied to unstructured meshes; fast; allows large mesh movements	With boundary rotation the approach can become invalid
Spring analogy	Robust, general and provides good quality meshes	Computationally expensive; memory intensive
Overset grids with relative movement	Allows extreme mesh movements	Grid interface definition can be awkward; lack of interface conservation

3.10.4.1.4 Non-deforming Meshes that Account for Movement For bodies with relative movement, the use of overset grids often proves robust and efficient. The overset grids can either be structured or unstructured. Component meshes can slide relative to each other. A key issue with such grids is how to define the inter-grid boundaries. Nakahashi and Togashi (2000) use the local level set to determine the classification of vertices and the hole cutting. Notably, the level-set/eikonal is most suitable for this purpose.

For flows involving adjacent regions rotating at different angular velocities, sliding meshes are frequently used. A classic example is in a gas turbine engine, where a set of stationary blades are positioned upstream of a set of rotating blades. The stationary and rotating blades are best solved with stationary and rotating meshes, respectively. To enable interfacing of the two meshes, some interpolation and book-keeping is required. Sliding meshes can also be used to reduce numerical stiffness and discretization errors (see Tucker 1997).

It seems worth briefly noting here the Cartesian cut cell approach (see for example Yang et al. 1997). This can naturally deal with moving objects by time dependently changing the cell cutting associated with an object moving through a Cartesian background mesh. The approach has a range of limitations. For example, like all cut cell techniques, it is non-ideal for cases where accurate skin friction is critical. However, like many approaches it has its niche. Table 3.5 summarizes advantages and disadvantages of different basic mesh movement approaches.

3.10.4.2 Hybrid Mesh Movement Approach Ideally, a moving boundary scheme needs to combine advantages of the three methods noted in Section 3.10.4.1 (i.e., mesh movement, local enrichment and also zone marking and complete re-meshing).

Figure 3.76 gives a flow chart showing how such a meshing strategy, proposed by Hassan et al. (2007b), works. As can be seen from the chart, the initial mesh is generated and its quality assessed. The latter could be based on some of the element quality heuristics noted previously. In the time loop, the following happens: surface coordinates are updated; based on these the volume mesh is updated and then the

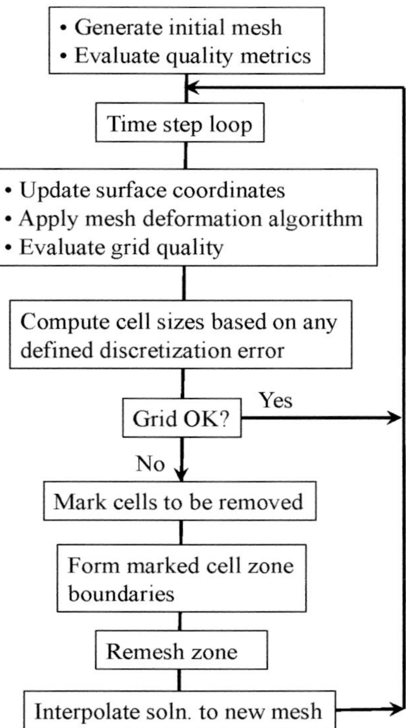

Figure 3.76. Flow chart showing a typical moving meshing strategy (Procedure of Hassan et al. (2007b) from Tucker (2013) – Published with the kind permission of Springer).

quality of this new volume mesh is tested. Based on the flow solution and the current grid spacings, grid spacings necessary to keep the solution error below a specified level are calculated. If the new grid is satisfactory, the new time step is started. If it is not, the elements that failed tests are marked and grouped to form zone boundaries or holes. The elements in the zone boundary region are deleted and the hole re-meshed. Then the solution is interpolated onto the new grid. The new time step is recomputed. Note that where possible, for viscous moving boundary cases the surface grid is held fixed.

An example of modelling an F16 stores release using the above hybrid simulation strategy is given in Hassan et al. (2007a, 2007b). With this simulation the payload trajectory is not prescribed but its 6° of freedom solved for. Impressive agreement is found with measured trajectories. These simulations can be seen in Figure 6.32, in Chapter 6. Hassan et al. (2007a, 2007b) also show a shuttle booster simulation with a prescribed component motion. The shuttle moves back and then rotates through 15°. The surface mesh is fixed. This simulation also involves zonal re-meshing.

It is important to stress that the mesh generation is these simulations is performed in parallel. Dynamic parallel mesh partitioning takes place. Nonetheless, the volume meshing takes around half the computational time. Ideally any zones to be re-meshed must be contained completely on one processor (i.e., a zone marked for re-meshing should not be spread across multiple processors).

3.11 Conclusions

A brief overview of some different mesh types along with their generation and control has been given. Some time was spent exploring how to automatically generate hexahedral meshes. Such meshes have advantageous properties. Basic mesh quality metrics were outlined.

Mesh solver compatibility was identified as an important aspect. It needs to be appreciated how a particular solver deals with particular types of cells. The grid and the CFD program must be seen as a team that needs to work together. The cell topology, as partially shown, essentially introduces different differential terms into the solver. The impact of such terms can become especially evident when performing eddy-resolving simulations.

The important aspect of designing the mesh to deal with different types of turbulent flow was given. Considerable attention was given to ideal mesh topologies for eddy-resolving simulations. There seems to be much more work that needs to be carried out in this area to ensure optimal meshes and guidance with respect to their generation. This is especially important for such simulations where the mesh densities are intrinsically high and waste must be avoided. Moving adaptive meshes are attractive, but these pose substantial challenges for eddy-*resolving* simulations. However, they could be important for more complex eddy-resolving flow fields where user insights could be fallible. Parallel mesh generation is an important aspect for the future but for time adapting meshes presents a challenge, dynamic load balancing being needed. Higher order elements might be a critical component for future large-scale CFD simulations. Such schemes offer the potential for greater accuracy at lower mesh counts and hence avoid data transfer bottlenecks when performing massively parallel simulations.

Problems

1. Verify the transformation of Equations (3.5) into Equation (3.6).
2. Represent Equation (3.5) in a polar system and show that for $S_1(\xi, \eta) = 0$ and $S_2(\xi, \eta) = 1/\eta$ for a uniform transformed plane, a uniform physical plane grid mapping arises.
3. Write a program to algebraically generate the Figure 3.65 unadapted mesh. First make the mesh uniform (using Equations (3.1) with $\eta =$ float $(j-1)/$ float $(nj-1)$ where j is the vertical grid index, float expresses conversion of an integer to a real number and nj is the number of vertical grid nodes) and then adapt the mesh near walls (see Equation (3.15)) so that it is suitable for a turbulent flow solution with a low Reynolds number turbulence model. Write the mesh in the Plot3D format outlined in Chapter 7. This will allow you to view it in standard CFD post-processing software as outlined in Chapter 7.
4. Verify that half the cross product of the cell diagonal vectors **VD1** and **VD2** shown in Figure 3.77 gives the cell area

$$A = \frac{1}{2}[\mathbf{VD1} \times \mathbf{VD2}] \tag{3.67}$$

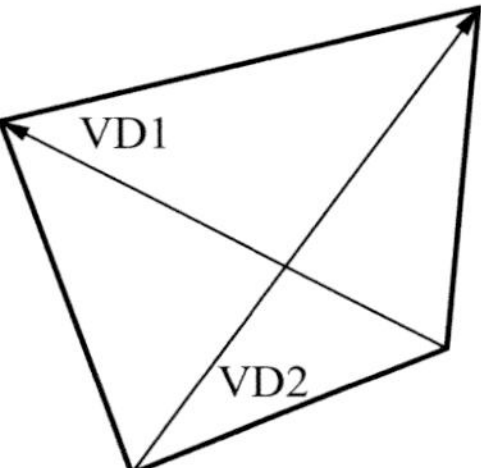

Figure 3.77. Cross product of diagonals to work out cell area.

5. Work out the areas of the cells in the meshes generated in Question 3 and check that they are sensible.
6. Based on the Figure 3.78 nomenclature in Frames (a) and (b), work out mesh face projected areas as seen by flow in the x and y directions. Give the faces a sign convention (say, flow component into the cell has a positive sign and out negative). Hence, if mass flow, for example, were integrated/summed around the control volume for a flow with constant velocities (and density) in both the x and y directions, mass flow would be observed to be conserved.

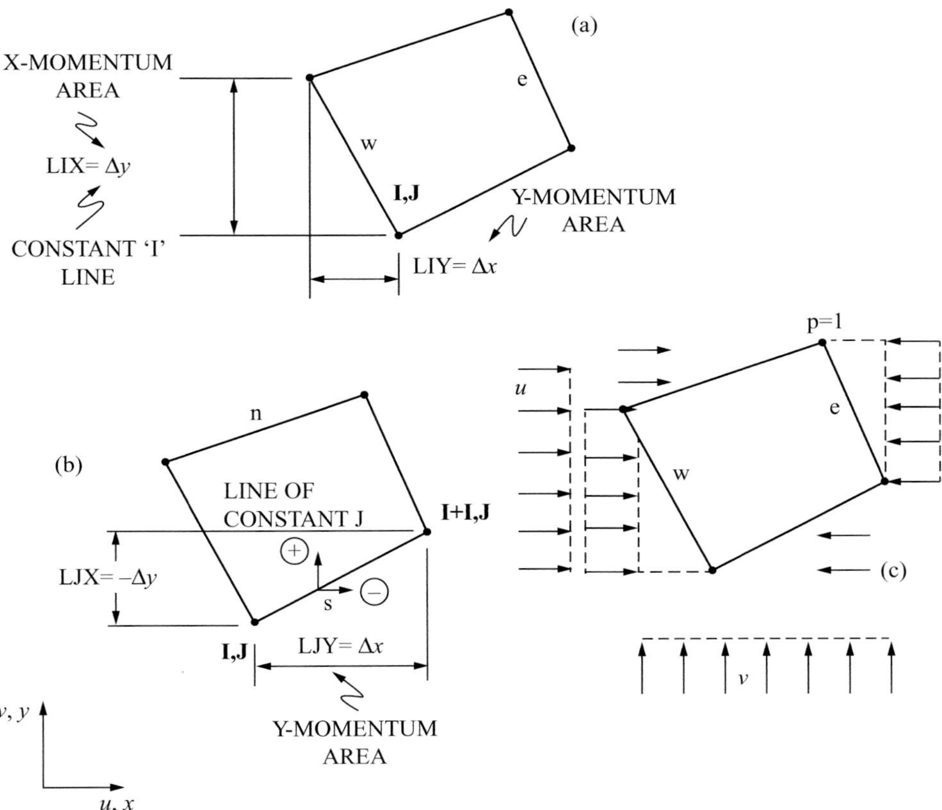

Figure 3.78. Control volume face areas and verification of their evaluation: (a) west face areas, (b) south face areas and (c) integration test to check/verify area evaluation.

7. Assuming that the fluid has a uniform velocity $u = 1$, $v = 1$ and constant density (or alternatively that it is stationary and has a constant unit pressure over faces) – see Figure 3.78c – integrate/sum the face areas for different control volume shapes to check for a zero volume flow (or net force). Note that this requires that a consistent sign convention is employed.
8. Write a program to elliptically generate the unadapted Figure 3.65 mesh for a wedge angle of 5.27°.
9. Alter your Euler solver, written as part of Chapter 4, to adapt the mesh (keeping a fixed number of grid points) for the case given in Figure 3.65.

REFERENCES

ADDAD, Y., PROSSER, R., LAURENCE, D. R. P., MOREAU, S. & MENDONCA, F. 2008a. On the use of embedded meshes in the LES of external flows. *Flow, Turbulence and Combustion*, 80, 393–403.

ADDAD, Y., GAITONDE, U., LAURECE, D. & ROLFO, S. 2008b. Optimal unstructured meshing for large eddy simulations, *Quality and Reliability of Large-Eddy Simulations*, 93–103, Springer.

ALI, Z. & TUCKER, P. G. 2013. Multiblock structured mesh generation for turbomachinery flows, Proc. 22nd International Meshing Roundtable, Orlando, Florida, 13–16 Oct 2013. Editors: Sarrate J, Staten M. Springer. pp. 165–182.

ANDERSON, J. D. 1995. *Computational fluid dynamics: The basics with applications*, McGraw-Hill.

ANDERSSON, N., ERIKSSON, L.-E. & DAVIDSON, L. 2005. LES prediction of flow and acoustic field of a coaxial jet, 11th AIAA/CEAS Aeroacoustics Conference, 23–25 May, Monterey, California, AIAA Paper No. AIAA-2005-2884.

BARTH, T. J., 1994. Aspects of unstructured grids and finite-volume solvers for the Euler and Navier-Stokes equations, *VKI Series 1994-04, Computational Fluid Dynamics*.

BATTEN P., GOLDBERG U. & CHAKRAVARTHY S. 2002. LNS – An approach towards embedded LES, 40th Aerospace Sciences Meeting and Exhibit, Reno/NV, AIAA Paper No. AIAA-2002-0427.

BELL, C. B. & SURANA, K. S. 1994. A space-time coupled p-version least-squares finite element formulation for unsteady fluid dynamics problems. *International Journal of Numerical Methods in Engineering*, 37, 3545–3569.

BERGER, M. J. & COLELLA, P. 1989. Local adaptive mesh refinement for shock hydrodynamics. *Journal of Computational Physics*, 82, 64–89.

BRANDT, A. 1977. Multilevel adaptive solutions to boundary value problems. *Mathematics of computation*, 31(138), 333–390.

BRANDT, A. 1980. Multilevel adaptive computations in fluid dynamics. *AIAA Journal*, 18(10), 1164.42172.

BREBBIA, C. A. & WROBEL, L. C. 1980. Steady and unsteady potential problems using the boundary element method. *Recent Advances in Numerical Methods in Fluids*, Vol. 1, 1–25, Pineridge Press Ltd.

CAREY, G. F. 2002 (March). Hexing the tet. *Communications in Numerical Methods in Engineering*, 18(3), 223–227.

CASTREJON, A. & SPALDING, D. B. 1988. An experimental and theoretical study of transient free-convection flow between horizontal concentric cylinders. *International Journal of Heat and Mass Transfer*, 31(2), 273–284.

CHAPMAN, D. R., MARK, H. & PIRTLE, M. W. 1975. Computers vs. wind tunnels for aerodynamic flow simulations. *Astronaut. Aeronaut*, 13, 12–35.

CHESSHIRE, G. & HENSHAW, W. D., 1990. Composite overlapping meshes for the solution of partial differential equations. *Journal of Computational Physics*, 90, 1–64.

CHEW, J. W. 1984. Development of a computer program for the prediction of flow in a rotating cavity. *International Journal for Numerical Methods in Fluids*, 4, 667–683.

CHEW, L. P. 1989. Constrained Delaunay Triangulations. *Algorithmica*, Vol. 4, 97–108, Springer.

CHEW, L. P. 1993. Guaranteed-quality mesh generation for curved surfaces. *Proceedings of the Ninth Annual Symposium on Computational Geometry*, 274–280.

CHUNG, Y. M. & TUCKER, P. G. 2003. Accuracy of higher-order finite difference schemes on non-uniform grids. *AIAA Journal*, 41(8), 1609–1611.

COELHO, P., PEREIRA, J. C. F. & CAVALHO, M. G. 1991. Calculation of laminar recirculating flows using a local non-staggered grid refinement system. *International Journal for Numerical Methods in Fluids*, 12, 535–557.

DAVIS, G. DE VAHL & MALLINSON, G. D. 1972. False diffusion in numerical fluid mechanics. *University of New South Wales, School of Mechanical and Industrial Engineering, Report 1972/FMT/1.*

DAVIS, R. L. & DANNENHOFFER, J. F. 1994. Three-dimensional adaptive grid-embedding Euler technique. *AIAA Journal*, 32(6), 1167–1174.

DAWES, W. N., KELLAR, W. P. & HARVEY, S. A. 2009. Using level sets as the basis for a scalable, parallel geometry engine and mesh generation system. 47th AIAA Aerospace Sciences Meeting and Exhibit, AIAA Paper No. AIAA-2009-0372.

DECK S. 2012. Recent improvements in the zonal detached eddy simulation (ZDES) formulation. *Theoretical and Computational Fluid Dynamics*, 26, 523–550.

DEMIRDZIC, I. & PERIC, M. 1988. Space conservation law in finite volume calculations of fluid flow. *International Journal for Numerical Methods in Fluids*, 8, 1037–1050.

DENTON, J. D. 2010. Some limitations of turbomachinery CFD. *Proceedings of ASME Turbo Expo 2010: Power for Land, Sea and Air*, GT2010-22540.

DERKSEN, J. & VAN DEN AKKER, H. 1999. Large eddy simulations of stirred tank flow. *Engineering Turbulence Modelling and Experiments*, 257–266, Elsevier Science Ltd.

DIMITRIADIS, K. P. & LESCHZINER, M. A. 1989. Multilevel convergence acceleration for viscous and turbulent transonic flows computed with cell-vertex method. Proceedings of the Fourth Copper Mountain Conference on Multigrid Methods, Copper Mountain, Colorado, *1–15*.

EL-HAMALAWI, A. 2004. A 2D combined advancing front-Delaunay mesh generation scheme. *Finite Elements in Analysis and Design*, 40, 967–989.

FERZIGER, J. H. & PERIC, M. 1977. *Computational methods for fluid dynamics*, Springer.

FLETCHER, C. A. J. 1997. *Computational techniques for fluid dynamics: Fundamental and general techniques*, Vol. 1, Springer.

FREY, P. J., BOROUCHAKI, H. & GEORGE, P.-L. 1998. 3D Delaunay mesh generation coupled with advancing-front approach. *Computer Methods in Applied Mechanics and Engineering*, 157, (1–2), 115–131.

GAITONDE, A. L. & FIDDES, S. P. 1993. A three-dimensional moving mesh method for the calculation of unsteady transonic flows. *Recent Developments and Applications in Aeronautical CFD*, p. 13

GEUZAINE, C. & REMACLE, J.-F. 2009. Gmsh: a three-dimensional finite element mesh generator with built-in pre- and post-processing facilities. *International Journal for Numerical Methods in Engineering*, 79(11), 1309–1331.

HARLOW, F. H. & WELSCH, J. E. 1965. Numerical calculation of time dependent viscous incompressible flow with free surface physics. *The Physics of Fluids*, 8(12), 2182–2189.

HASSAN O., MORGAN K. & WEATHERILL N. 2007a. Unstructured mesh methods for the solution of the unsteady compressible flow equations with moving boundary components. *Philos. Trans. R. Soc. A, Math. Phys. Eng. Sci.*, 365 (1859), 2531–2552.

HASSAN, O., S., RENSEN, K. A., MORGAN, K. & WEATHERILL, N. P. 2007b. A method for time accurate turbulent compressible fluid flow simulation with moving boundary components employing local remeshing. *International Journal for Numerical Methods in Fluids*, 53(8), 1243–1266.

HUJEIRAT, A. & RANNACHER, R. 1998. A method for computing compressible, highly stratified flows in astrophysics based on operator splitting. *International Journal for Numerical Methods in Fluids*, 28, 1–22.

JIAO, X. & WANG, D. 2011. Reconstructing high-order surfaces for meshing, *Engineering with Computers*, 28(4), 361–373, Springer.

JUN, L. 1986. *Computer Modelling of flows with a free surface*. PhD Thesis, University of London.

KNUPP, P. 1995. Mesh generation using vector fields, *Journal of Computational Physics*, 119, 142–148.

KNUPP, P. M. 2008. Remarks on mesh quality. 45th AIAA Aerospace Sciences Meeting and Exhibit, 7–10 January, 2007, Reno, NV, AIAA Paper No. AIAA-2008-933

LOHNER, R. & PARIKH, P. 1988. Generation of three-dimensional unstructured grids by the advancing-front method. *International Journal for Numerical Methods in Fluids*, Vol. 8, 1135–1149.

LOHHNER, R. 1996. Progress in grid generation via the advancing front technique. *Engineering with Computers*, Vol. 12, 186–210, Springer.

LARSSON J. & WANG Q. 2014. The prospect of using LES and DES in engineering design and the research required to get there, *Philosophical Transactions of the Royal Society (Series A: Mathematical, Physical and Engineering Sciences*, A 372, 20130329. (doi:10.1098/rsta.2013.0329).

LIU, X., QIN, N. & XIA, H. 2006. Fast dynamic grid deformation based on Delaunay graph mapping. *Journal of Computational Physics*, 211(2), 405–423.

LONSDALE, G. & WELSCH, J. E. 1984. The pressure correction method and the use of a multigrid technique for laminar source sink flow between co-rotating discs. *Numerical Analysis Report No. 95*, Department of Mathematics, The University of Manchester.

MALCEVIC, I. 2011. Automated blocking for structured CFD gridding with an application to turbomachinery secondary flows. 20th AIAA Computational Fluid Dynamics Conference, Honolulu, Hawaii, AIAA Paper Number AIAA-2011-3049.

MANI, K. & MAVRIPLIS, D. J. 2010. Spatially non-uniform time-step adaptation for functional outputs in unsteady flow problems. 48th AIAA Aerospace Sciences Meeting, AIAA Paper Number AIAA-2010-121.

MARONGIU, J. C., LEBOEUF, F. & PARKINSON E. 2007. Numerical simulation of the flow in a Pelton turbine using the meshless method smoothed particle hydrodynamics: a new simple solid boundary treatment. *Proc. Inst. Mech. Eng. A, Journal of Power Energy*, 221(6), 849–856.

MARONGIU, J. C., LEBOEUF, F., CARO, J. Ë. & PARKINSON E. 2010. Free surface flows simulations in Pelton turbines using an hybrid SPH-ALE method. *Journal of Hydraulic Research*, 48(S1), 40–49.

MASI, A. 2014. *CFD modeling of tail planes*, 1st year PhD Report, School of Engineering, Cambridge University.

MAVRIPLIS, D. J. 1999. Directional agglomeration multigrid techniques for high-Reynolds-number viscous flows. *AIAA Journal*, 37(10), 1222–1230.

MIYATA, H. & NISHIMURA, S. 1985. Finite difference simulation of non-linear ship waves. *Journal of Fluid Mechanics*, 157, 327–357.

MOINIER P. 1999. *Algorithm developments for an unstructured viscous flow solver*. PhD thesis, University of Oxford.

NAKAHASHI, K. & TOGASHI, F. 2000. Inter-grid boundary definition method for overset unstructured grid approach. *AIAA Journal*, 38(11), 2077–2084.

OSHER, S. J. & SETHIAN, J. A. 1988. Fronts propagating with curvature dependent speed: algorithms based on Hamilton–Jacobi formulations. *J Comput Phys*, 79, 12–49.

PAGE, J. H., HIELD, P. & TUCKER P. G. 2013. Inverse design of 3D multistage transonic fans at dual operating points. *Proceedings of ASME Turbo Expo*. ASME Paper GT2013-95062.

PERSSON, P-O & PERAIRE, J. 2009. Curved mesh generation and mesh refinement using lagrangian solid mechanics. *Proceedings of 47th AIAA Aerospace Sciences Meeting and Exhibit*, AIAA-2009-949.

PERSSON, P-O. 2006. Mesh size functions for implicit geometries and PDE-based gradient limiting. *Engineering with Computers*, 22(2), 95–109.

PIOMELLI, U. 2014. Large-eddy simulations in 2030 and beyond. *Philosophical Transactions of the Royal Society (Series A: Mathematical, Physical and Engineering Sciences)*, A 372, 20130320. (doi:10.1098/rsta.2013.0320).

PLETCHER, R. H., TANNEHILL, J. C. & ANDERSON A. A. 2013. *Computational fluid mechanics and heat transfer*, CRC Press.

POTSDAM, M. A. & GURUSWAMY, G. P. 2001. A parallel multiblock mesh movement scheme for complex aeroelastic applications. 39th Aerospace Sciences Meeting and Exhibit. January AIAA Paper No AIAA-2001-716.

PREECE, A. 2008. An investigation into methods to aid the simulation of turbulent separation control. PhD Thesis, School of Engineering, The University of Warwick.

PRICE, M. A. & ARMSTRONG, C. G. 1997. Hexahedral mesh generation by medial surface subdivision: Part II solids with flat and concave edges. *International Journal for Numerical Methods in Engineering*, 40(1), 111–136.

PRICE, M. A., ARMSTRONG, C. G. & SABIN, M. A. 1995. Hexahedral mesh generation by medial surface subdivision: Part I solids with convex edges. *International Journal for Numerical Methods in Engineering*, 38(19), 3335–3359.

RIGBY, D. L. 2004. *Topmaker: A technique for automatic multi-block topology generation using the medial axis*, NASA/CR-213044.

ROACHE, P. J. 1976. *Computational fluid dynamics*, Hermosa.

ROACHE, P. J. 1994. Perspective: a method for uniform reporting of grid refinement studies. *Journal of Fluids Engineering*, 116, 405–413.

RUGE, J. & STUBEN, K. 1986. *Algebraic Multigrid*, 210, Arbeitspapiere der GMD.

RUPPERT, J. 1995. A Delaunay Refinement Algorithm for Quality 2-dimensional Mesh Generation. *Journal of Algorithms* 18(3), 548–585.

SAITO, H. & SCRIVEN, L. E. 1981. Study of coating flow by finite element method. *Journal of Computational Physics*, 42, 53–73.

SAMPSON, F. W. 2013. *Final year 4A2, CFD, report*, School of Engineering, Cambridge University.

SETHIAN, J. A. 1994. Curvature flow and entropy conditions applied to grid generation. *Journal of Computational Physics*, 115(2), 440–454.

SHYY, W., SUN, C., CHEN, M. & CHANG, K. C. 1993. Multigrid computation for turbulent recirculating flows in complex geometries. *Numerical. Heat Transfer*, A23, 79–98.

SMITH, R. J. 1996. *Automatic grid generation for compressible Navier-Stokes solvers in aerodynamic design for complex geometries*. Doctor of Engineering Thesis, University of Manchester Institute of Science and Technology (UMIST).

SPALART, P. R. & ALLMARAS, S. R. 1992. A one equation turbulence model for aerodynamic flows. *Recherche Aérospatiale*, 1, 5–21.

SPALART, P. R. 2001. *Young-person's guide to detached-eddy simulation grids*. NASA/CR-2001-211032.

SUA, P. & SCOT-DRYSDALE, R. L. 1997 (April). A comparison of sequential delaunay triangulation algorithms, *Computational Geometry*, 7(5–6), 361–385.

SUN, Y., WANG, Z. J. & LIU, Y. 2007. High-order multidomain spectral difference method for the Navier-Stokes equations on unstructured hexahedral grids. *Communications in Computational Physics*, 2(2), 310–333.

TAM, T. K. H. & ARMSTRONG, C. G. 1991. 2D finite element mesh generation by medial axis subdivision. *Advances in Engineering Software and Workstations*, 13(5), 313–324.

TANNER, R. I., NICKELL, R. E. & BILGER, R. W. 1975. Finite element methods for the solution of some incompressible non-Newtonian fluid mechanics problems with free surfaces. *Computer Methods in Applied Mechanics and Engineering*, 6, 154.4270.

THOMPSON, M. C. & FERZIGER, J. H. 1989. A multigrid adaptive method for incompressible flows. *Journal of Computational Physics*, 82, 94–121.

TUCKER, P. G. 1997. Numerical precision and dissipation errors in rotating flows. *International Journal for Numerical Methods in Heat Fluid Flow*, 7(7), 647–658.

TUCKER, P. G. 2006. Turbulence modelling of problem aerospace flows. *International Journal for Numerical Methods in Fluids*, 51, 261–283.

TUCKER, P. G. 2013. *Unsteady computational fluid dynamics in aeronautics*, Springer.

TUCKER, P. G. & MOSQUERA, A. 2001. *Introduction to grid and mesh generation for CFD*, Published by NAFEMS (Ref. R0079).

TUCKER, P. G. & PAN, Z. 2000. A Cartesian cut cell method for incompressible viscous flow. *Applied Mathematical Modeling*, 24, 591–606.

TYACKE, J., TUCKER, P. G., JEFFERSON-LOVEDAY, R. J., VADLAMANI, R., WATSON, R., NAQAVI, I. & YANG X. 2013. Les for turbines: methodologies, cost and future outlooks, Proceedings of ASME Turbo Expo 2013, San Antonio, Texas, 3–7 June. ASME Paper No. GT2013-94416.

VENDITTI, D. A. & DARMOFAL, D. L. 2002. Grid adaptation for functional outputs: Application to two-dimensional inviscid flows. *Journal of Computational Physics*, 176(1), 40–69.

WANG, Z. J. 2014. High-order CFD tools for aircraft design, *Proceedings Royal Society*, A 372, 20130318. (doi:10.1098/rsta.2013.0318).

XIA, H. & TUCKER, P. G. 2010. Finite volume distance field and its application to medial axis transforms. *International Journal of Numerical Methods in Engineering*, 82(1), 114–134.

XIA, H. & TUCKER, P. G. 2011. Fast equal and biased distance fields for medial axis transform with meshing in mind. *Applied Mathematical Modelling*, 35, 5804–5819.

XIA, H., TUCKER, P. G. & DAWES, W. N. 2010. Level sets for CFD in aerospace engineering. *Progress in Aerospace Sciences – Invited Paper*, 46(7), 274–283.

XIA, H., TUCKER, P. G. & COUGHLIN, G. 2012. Novel applications of BEM based level set approach. *International Journal of Boundary Element Methods*, 36, 907–912.

YANG, G., CAUSON, D. M., INGRAM, D. M., SAUNDERS, R. & BATTEN, P. 1997. Cartesian cut cell method for compressible flows, Part B: Moving body problems. *The Aeronautical Journal,* Paper No. 2120, February, 57–65.

ZHENG, Y. & LIOU, M-S. 2003. A novel approach of three-dimensional hybrid grid methodology: part I. Grid generation. *Computer Methods in Applied Mechanics and Engineering,* 192, 4147–4171.

CHAPTER 4

Numerical Methods

4.1 Introduction

Through the method of lines it is possible to regard the spatial and temporal integrations as separate processes. Essentially, through this method, the spatial discretization can be considered to comprise an operator $A(\Phi)$ and the temporal term equated to this. Hence, the following can be written

$$\frac{\partial \Phi}{\partial t} = A(\Phi) \tag{4.1}$$

For improved accuracy it can be important to exploit the intrinsic link between spatial and temporal discretization. Schemes that exploit this link typically extend spatial schemes through the inclusion of Estimated Streaming Terms (see Leith 1965 and Leonard 1979) or balancing terms such as with the Lax-Wendroff scheme. This link is only briefly considered here. For further discussion, see Tucker (2001, 2013).

Initially, in this chapter, through the method of lines, the spatial discretization is considered in isolation. Then finally the temporal is addressed. The two key methods for solving the fluid flow equations are the finite difference and finite volume methods. Hence, this chapter mostly focuses on these approaches. The finite element method does not see extensive use in fluid flow solvers (note that certain classes of finite element method overlap with the finite volume). Hence, that approach is not dealt with here. However, Appendix A provides a very simple and brief example of the use of the finite element method to solve diffusion terms. Such finite element discretizations have, in the past, been combined with the finite volume method for the convective terms. For a full discussion on the use of the finite element method in fluid flow see Zienkiewicz et al. (2005). For completeness, some non-volume based grid approaches are considered. These are important for niche applications. When considering fluid flow problems, the evaluation of the pressure field provides a substantial complication. Hence, a section on this is provided along with one on effectively solving the large simultaneous equation sets arising when computing large-scale problems.

The finite difference method is discussed first and then the finite volume. However, it is worth noting that essentially the finite volume method is used to derive the fluid flow equations (see Chapter 2). Hence, this seems the most compatible method for the solution of the flow governing equations. As industrial problems grow in scale, the use of higher order schemes may become of greater importance. Hence,

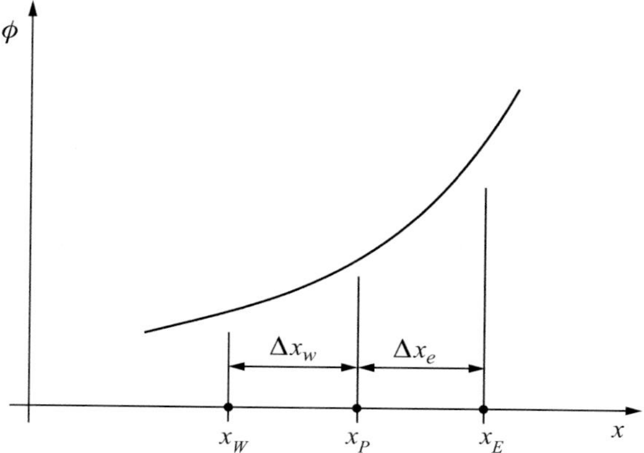

Figure 4.1. Geographic based nomenclature for the finite difference method.

such schemes are also briefly discussed here. Then some meshless methods are considered. Next, discussion on the solution of the pressure field and simultaneous equations is provided. Finally, the temporal discretization is considered.

PART A – SPATIAL PROCEDURES

4.2 Finite Difference, Related Methods and Numerical Traits

The finite difference method is relatively conceptually simple. It has two key bases: the geometric means of formulating finite difference expressions and the more formal Taylor series based approach. To start with we derive some finite difference equations for uniform grids. These will then be extended to non-uniform grids. Initially, the finite difference equations will be produced using a geometric approach. This essentially invokes the basic definition of a finite difference. A shortcoming of this approach is that although direct, there is no information on the resulting scheme's accuracy. Use of the Taylor series overcomes this.

4.2.1 *Geometric Approach*

By assuming a uniform grid, $\Delta x = \Delta x_w = \Delta x_e$, and from viewing Figure 4.1 we can produce the following expressions:

backwards difference

$$\frac{\phi_P - \phi_W}{\Delta x} = \left.\frac{d\phi}{dx}\right|_P \tag{4.2}$$

forwards difference

$$\frac{\phi_E - \phi_P}{\Delta x} = \left.\frac{d\phi}{dx}\right|_P \tag{4.3}$$

central difference

$$\frac{\phi_E - \phi_W}{2\Delta x} = \left.\frac{d\phi}{dx}\right|_P \tag{4.4}$$

And similarly

$$\left.\frac{d^2\phi}{dx^2}\right|_P = \left[\left(\frac{\phi_E - \phi_P}{\Delta x}\right) - \left(\frac{\phi_P - \phi_W}{\Delta x}\right)\right] = \frac{\phi_E - 2\phi_P + \phi_W}{\Delta x^2} \tag{4.5}$$

4.2.2 *Taylor Series Approach*

We will now derive the same finite difference expressions using the Taylor series approach. First we write the Taylor series for some function $\phi(x)$.

$$\phi(x) = \phi(x_P) + (x - x_P)\left.\frac{d\phi}{dx}\right|_P + \frac{(x - x_p)^2}{2!}\left.\frac{d^2\phi}{dx^2}\right|_P + \frac{(x - x_P)^3}{x!}\left.\frac{d^3\phi}{dx^2}\right|_P \tag{4.6}$$

If $x = x_E$ in Equation (4.6), then we can write

$$\phi_E = \phi_P + \Delta x\left.\frac{d\phi}{dx}\right|_P + \frac{(\Delta x)^2}{2}\left.\frac{d^2\phi}{dx^2}\right|_P + \frac{\Delta x^3}{3!}\left.\frac{d^3\phi}{dx^3}\right|_P \tag{4.7}$$

Similarly, if $\Delta x = \Delta x_W$ (i.e., at the moment we assume a uniform grid), then

$$\phi_W = \phi_p - \Delta x\left.\frac{d\phi}{dx}\right|_P + \frac{(\Delta x)^2}{2!}\left.\frac{d^2\phi}{dx^2}\right|_P - \frac{\Delta x^3}{3!}\left.\frac{d^3\phi}{dx^3}\right|_P \tag{4.8}$$

Equation (4.7) (ignoring terms multiplied by $(\Delta x)^2$ and higher powers) gives a forwards difference (cf. Equation (4.3))

$$\frac{\phi_E - \phi_P}{\Delta x} = \left.\frac{d\phi}{dx}\right|_P + HOT \tag{4.9}$$

where HOT involves $(\Delta x)^1$ and so the expression is generally referred to as being first order accurate. Equation (4.8) (ignoring terms multiplied by $(\Delta x)^2$ and higher powers) gives a backwards difference (cf. Equation (4.2))

$$\frac{\phi_P - \phi_W}{\Delta x} = \left.\frac{d\phi}{dx}\right|_P + HOT \tag{4.10}$$

Again, for the reasons just stated this is described as being first order accurate. If we subtract Equation (4.7) from (4.8) we get

$$\phi_W - \phi_E = -2\Delta x\left.\frac{d\phi}{dx}\right|_P - \frac{2\Delta x^3}{3!}\left.\frac{d^3\phi}{dx^3}\right|_P \tag{4.11}$$

If we then ignore the $(\Delta x)^3$ multiplied term and rearrange (cf. Equation (4.4)), we get

$$\left.\frac{d\phi}{dx}\right|_P = \frac{\phi_E - \phi_W}{2\Delta x} + HOT \tag{4.12}$$

This is a central difference where $HOT \propto \Delta x^2$. The power of 2 in the HOT would generally mean that Equation (4.12) is referred to as a second order approximation. However, according to Leonard (1979), "The 'order' of a numerical scheme is classically defined to be the highest degree (e.g. $\phi = A + Bx + Cx^2 + Dx^3$ highest degree $= 3$) of a polynomial for which the algorithm is exact." Therefore, another way of looking at the accuracy/order of a scheme is to consider the polynomial $\phi = A + Bx + Cx^2 + Dx^3$, and its subsequent differentiation, $d\phi/dx = B + 2Cx + 3Dx^2$, $d^2\phi/dx^2 = 2C + 6Dx$, $d^3\phi/dx^3 = 6D$. The term that has been ignored is '$6D$', that is, the discretization is exact for a solution of the form

$$\phi = A + Bx + Cx^2 \tag{4.13}$$

Therefore, there seems to be no inconsistency and the finite difference approximation is second order accurate. Now we will add Equations (4.7) and (4.8):

$$\phi_E + \phi_W = 2\phi_P + (\Delta x)^2 \left.\frac{d^2\phi}{dx^2}\right|_P + \frac{2(\Delta x)^4}{4!}\left.\frac{d^4\phi}{dx^4}\right|_P \tag{4.14}$$

Ignoring $(2(\Delta x)^4/4!)[d^4\phi/dx^4]|_P$ and rearranging gives (cf. Equation (4.5))

$$\left.\frac{d^2\phi}{dx^2}\right|_P = \frac{\phi_E + \phi_W - 2\phi_P}{\Delta x^2} + HOT \tag{4.15}$$

where $HOT \propto \Delta x^2$. Hence we might conclude that we have a second order scheme. Again we can consider a polynomial of the form $\phi = A + Bx + Cx^2 + Dx^3 + Ex^4$, and its subsequent differentiation: $d\phi/dx = B + 2Cx + 3Dx^2 + 4Cx^3$, $d^2\phi/dx^2 = 2C + 6Dx + 12Ex^2$, $d^3\phi/dx^3 = 6D + 24Ex$, $d^4\phi/dx^4 = 24E$(i.e. E is ignored or x^4). This means the scheme would be exact if the solution was a polynomial of order 3 and so the scheme could be viewed as third order. Clearly it can be seen that we need to be careful about what we mean when we refer to the order of a numerical scheme. We will now derive non-uniform grid expressions. With regards to order, we will generally base it on, unless stated otherwise, the power that the grid spacing is raised to in the numerator of the leading Taylor series truncated term.

4.2.2.1 Non-uniform Grid Second Order Centred Representation of $\partial\phi/\partial x$ From the Taylor series (Equation (4.6)), we can write

$$\phi_E = \phi_P + \Delta x_e \left.\frac{d\phi}{dx}\right|_P + \frac{\Delta x_e^2}{2!}\left.\frac{d^2\phi}{dx^2}\right|_P + \frac{\Delta x_e^3}{3!}\left.\frac{d^3\phi}{dx^3}\right|_P \tag{4.16}$$

and

$$\phi_W = \phi_P - \Delta x_w \left.\frac{d\phi}{dx}\right|_P + \frac{\Delta x_w^2}{2!}\left.\frac{d^2\phi}{dx^2}\right|_P - \frac{\Delta x_w^3}{3!}\left.\frac{d^3\phi}{dx^3}\right|_P \tag{4.17}$$

Subtraction of Equation (4.16) from (4.17) gives (if terms multiplied by $(\Delta x)^3$ are ignored)

$$\frac{d\phi}{dx} = \frac{\Delta x_w \phi_E}{\Delta x_e(\Delta x_e + \Delta x_w)} - \frac{(\Delta x_w - \Delta x_e)\phi_P}{\Delta x_e \Delta x_w} - \frac{\Delta x_e \phi_W}{\Delta x_w(\Delta x_e + \Delta x_w)} + HOT \tag{4.18}$$

where $HOT \propto \Delta x^2$ and so is a second order approximation. More memorably, Equation (4.18) can be written as a linearly weighted interpolation of $d\phi/dx$ at the midpoint locations w and e

$$\frac{d\phi}{dx} = \frac{\left[\frac{(\phi_E - \phi_P)\Delta x_w}{\Delta x_e} - \frac{(\phi_W - \phi_P)\Delta x_e}{\Delta x_w}\right]}{\Delta x_w + \Delta x_e} \tag{4.19}$$

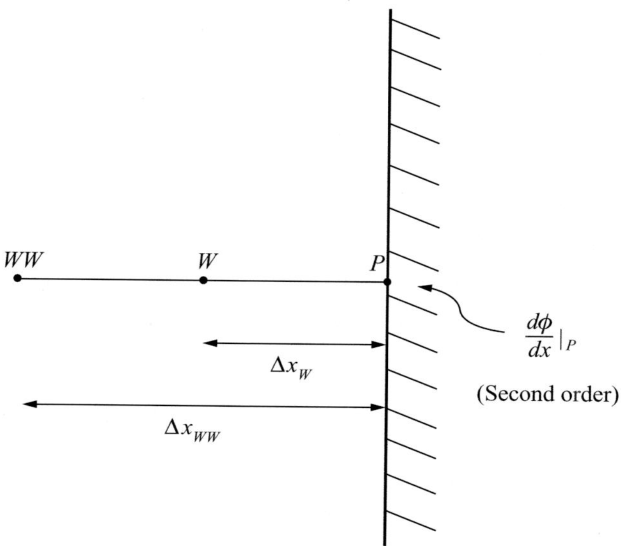

Figure 4.2. Formulation of second order backwards differences.

4.2.2.2 Second Order Backwards Difference (Non-uniform Grid) From the Taylor series and viewing Figure 4.2, we have

$$\phi_W = \phi_P - \frac{\Delta x_w}{1!}\frac{d\phi}{dx}\bigg|_P + \frac{(\Delta x_w)^2}{2!}\frac{d^2\phi}{dx^2}\bigg|_P - \frac{(\Delta x_w)^3}{3!}\frac{d^3\phi}{dx^3}\bigg|_P \tag{4.20}$$

$$\phi_{WW} = \phi_P - \frac{\Delta x_{ww}}{1!}\frac{d\phi}{dx}\bigg|_P + \frac{(\Delta x_{ww})^2}{2!}\frac{d^2\phi}{dx^2}\bigg|_P - \frac{(\Delta x_{ww})^3}{3!}\frac{d^3\phi}{dx^3}\bigg|_P \tag{4.21}$$

Multiplying (4.20) by $(\Delta x_{ww})^2$ and (4.21) by $(\Delta x_w)^2$ and subtracting gives

$$\frac{d\phi}{dx}\bigg|_P (\Delta x_w \Delta x_{ww}^2 - \Delta x_w^2 \Delta x_{ww}) = (\phi_W - \phi_P)\Delta^2 x_{ww} - (\phi_{WW} - \phi_P)\Delta x_w^2 \tag{4.22}$$

which can be rewritten as

$$\frac{d\phi}{dx}\bigg|_P = \frac{(-\phi_W + \phi_P)\Delta x_{ww}^2 - (-\phi_{WW} + \phi_P)\Delta x_w^2}{\Delta x_w \Delta x_{ww}(\Delta x_{ww} - \Delta x_w)} \tag{4.23}$$

Clearly the previous finite difference equations can be directly inserted into the exact differential flow equations to produce discrete equations to be used in a CFD program.

4.2.3 *Compact Schemes*

It is straightforward to generate higher order finite difference schemes. However, these give rise to large stencils which are problematic at boundaries. Also, with most simulations nowadays being performed in parallel mode with domain decomposition, if the order is maintained at the block boundaries, the data that needs to be transferred is large. This gives a substantial solution overhead. In this situation the

Table 4.1. *Coefficients for different compact schemes (from Tucker (2013) – Published with the kind permission of Springer)*

Derivative	Order	Wf_W	Wf_P	Wf_E	α	β	γ
ϕ_i'	4	¼	1	¼	3/2	0	0
ϕ_i''	4	1/10	1	1/10	6/5	0	0
ϕ_i'	6	1/3	1	1/3	14/9	1/9	0
ϕ_i''	6	2/11	1	2/11	12/11	3/11	0

compact (or Pade) schemes are in principle attractive. These schemes allow use of a relatively small finite difference stencil to gain high-order accuracy. On the other hand, the approach would typically involve (for implicit solution) a costly matrix inversion. This cost negates the interface advantage noted. For a function ϕ, a compact scheme for a first derivative on a uniform grid can be expressed as

$$Wf_W\phi_W' + Wf_P\phi_P' + Wf_E\phi_E' = \alpha\frac{\phi_E - \phi_W}{2\Delta x} + \beta\frac{\phi_{EE} - \phi_{WW}}{2\Delta x} \tag{4.24}$$

where Wf, α and β are constants such as those given in Table 4.1. Hence, it is not just the nodal values that are unknowns but also the derivatives, ϕ', at these nodal points. Note that the subscripts WW and EE correspond to points on an enlarged stencil adjacent to the existing W and E points. The coefficients in the equation can be evaluated from Taylor series expansion about node P and some manipulation. Equation (4.24) can be expressed in matrix form. On a uniform grid, the compact scheme will typically have better accuracy than a finite difference scheme of equivalent order. For second derivatives, the following equation can be used

$$\begin{aligned}Wf_W\phi_W'' + Wf_P\phi_P'' + Wf_E\phi_E'' &= \alpha\frac{\phi_E - 2\phi_P + \phi_W}{\Delta x^2}\\ &+\beta\frac{\phi_{EE} - 2\phi_P + \phi_{WW}}{4\Delta x^2} + \gamma\frac{\phi_{EEE} - 2\phi_P + \phi_{WWW}}{9\Delta x^2}\end{aligned} \tag{4.25}$$

Again, the subscripts WWW and EEE correspond to further enlarged stencil components. Table 4.1 gives the coefficients for different fourth and sixth order accuracy compact schemes. These coefficients can be modified to improve certain numerical properties (see, for example, Laizet and Lamballais 2009, Lele 1992, Kim and Lee 1996, Tang and Baeder 1998).

Next, a simple, essentially finite difference discretization is presented that is based around a finite volume. Hence, the method is hybrid and connects with the later discussed finite volume discretization.

4.2.4 *Simple Finite Difference–Finite Volume Discretization*

As shown in Chapter 2, the Navier-Stokes equations, in conservation form, for a two-dimensional, constant kinematic viscosity (ν), steady flow can be expressed as

$$\frac{\partial u\phi}{\partial x} + \frac{\partial v\phi}{\partial y} = \nu\left(\frac{\partial^2\phi}{\partial x^2} + \frac{\partial^2\phi}{\partial y^2}\right) + S_\phi \tag{4.26}$$

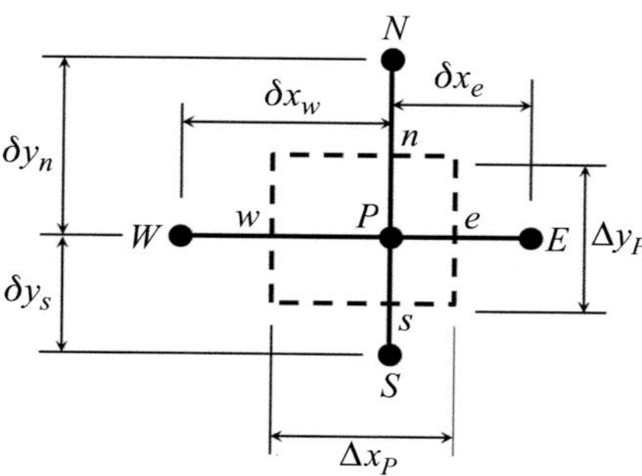

Figure 4.3. Control volume and notation for a basic finite difference/finite volume discretization.

where $\phi = u$ or v and $S_u = -\partial p/\rho\partial x$ and $S_v = -\partial p/\rho\partial y$. Equation (4.26) will be discretized about the Figure 4.3 non-uniform control volume. Discretizing Equation (4.26) about this control volume gives

$$\left(\frac{u_e\phi_e - u_w\phi_w}{\Delta x_p}\right) + \left(\frac{v_n\phi_n - v_s\phi_s}{\Delta y_p}\right) = \nu\frac{\left(\frac{\partial\phi}{\partial x}\big|_e - \frac{\partial\phi}{\partial x}\big|_w\right.}{\Delta x_p} + \frac{\left.\frac{\partial\phi}{\partial y}\big|_n - \frac{\partial\phi}{\partial y}\big|_s\right)}{\Delta y_p} + S_\phi \quad (4.27)$$

which can be re written as

$$\left(\frac{u_e\phi_e - u_w\phi_w}{\Delta x_p}\right) + \left(\frac{v_n\phi_n - v_s\phi_s}{\Delta y_p}\right)$$
$$= \nu\left[\left(\frac{\phi_E - \phi_P}{\delta x_e \Delta x_p}\right) - \left(\frac{\phi_P - \phi_W}{\delta x_w \Delta x_p}\right) + \left(\frac{\phi_N - \phi_P}{\delta y_n \Delta y_p}\right) - \left(\frac{\phi_P - \phi_S}{\delta y_s \Delta y_p}\right)\right] + S_\phi \quad (4.28)$$

We need expressions for the control volume ϕ values in terms of grid point values. This extensive subject is discussed further later. However, the use of second order central differencing, in fact, would correspond to the following interpolations

$$\phi_w = \frac{\phi_W + \phi_P}{2}, \quad \phi_e = \frac{\phi_E + \phi_P}{2}, \quad \phi_n = \frac{\phi_N + \phi_P}{2}, \quad \phi_s = \frac{\phi_S + \phi_P}{2} \quad (4.29)$$

Substituting (4.29) into (4.28) gives after minor rearrangement

$$\left(\frac{(u_e - u_w)}{2\Delta x_p} + \frac{(v_n - v_s)}{2\Delta y_p} + \frac{\nu}{\delta x_e \Delta x_p} + \frac{\nu}{\delta x_w \Delta x_p} + \frac{\nu}{\delta y_n \Delta y_p} + \frac{\nu}{\delta y_s \Delta y_p}\right)\phi_P$$
$$= \left(\frac{-u_e}{2\Delta x_p} + \frac{\nu}{\delta x_e \Delta x_p}\right)\phi_E + \left(\frac{u_w}{2\Delta x_p} + \frac{\nu}{\delta x_w \Delta x_p}\right)\phi_W \quad (4.30)$$
$$+ \left(\frac{-v_n}{2\Delta y_p} + \frac{\nu}{\delta y_n \Delta y_p}\right)\phi_N + \left(\frac{v_s}{2\Delta y_p} + \frac{\nu}{\delta y_s \Delta y_p}\right)\phi_S + S_\phi$$

Defining

$$A_E = -\frac{u_e}{2\Delta x_p} + \frac{\nu}{\delta x_e \Delta x_p}, \quad A_W = \frac{u_w}{2\Delta x_p} + \frac{\nu}{\delta x_w \Delta x_p} \quad (4.31)$$

$$A_N = -\frac{v_n}{2\Delta y_p} + \frac{\nu}{\delta y_n \Delta y_p}, \quad A_S = \frac{v_s}{2\Delta y_p} + \frac{\nu}{\delta y_s \Delta y_p} \tag{4.32}$$

Equation (4.30) can be re written as

$$\left(\frac{(u_e - u_w)}{2\Delta x_p} + \frac{(v_n - v_s)}{2\Delta y_p} + \frac{\nu}{\delta x_e \Delta x_p} + \frac{\nu}{\delta x_w \Delta x_p} + \frac{\nu}{\delta y_n \Delta y_p} + \frac{\nu}{\delta y_s \Delta y_p} \right) \phi_P \\ = A_E \phi_E + A_W \phi_W + A_N \phi_N + A_S \phi_S + S_\phi \tag{4.33}$$

Discretizing the continuity equation in a form that is consistent with the Navier-Stokes equations as defined previously (i.e., $\partial u/\partial x + \partial v/\partial y = 0$) gives

$$\left(\frac{u_e - u_w}{\Delta x_p} \right) + \left(\frac{v_n - v_s}{\Delta y_p} \right) = 0 \tag{4.34}$$

Subtracting (4.34) times ϕ_P from the left-hand side of Equation (4.33) enables the following to be written

$$A_P \phi_P = A_W \phi_W + A_E \phi_E + A_N \phi_N + A_S \phi_S + S_\phi \tag{4.35}$$

where

$$A_P = A_W + A_E + A_N + A_S$$

and

$$S_u = -\frac{1}{\rho} \left(\frac{p_w - p_e}{\Delta x_p} \right), \quad S_v = -\frac{1}{\rho} \left(\frac{p_n - p_s}{\Delta y_p} \right)$$

Hence, as can be seen, a set of equations is gained that can be readily solved in a matrix form. Alternatively, all the matrix coefficients can be lumped together to form the global term $A(\phi)$ in Equation (4.1) and the system marched through a pseudo-time. The rearrangement involving the continuity equation and ϕ_P, to ensure that the diagonal coefficient is equal to the sum of the neighbouring coefficients, gives physical realism (see Patankar 1980). Also, it is an important physical step to producing a matrix equation that will converge with basic iterative solvers. It should be noted that, except for low cell Reynolds numbers, the central difference method becomes unstable. Under these circumstances, to control the growth of oscillations in the solution, a discrete smoother (typically of the form $(\Delta x)^2 \nabla^4 \phi$) could be added in Equation (4.35). The use of smoothing in this fashion is discussed later.

An observation that can be made from the finite difference based derivations is that this method is not that well suited to mapping complex geometries. Immersed boundary methods (see Preece 2008) and trimmed cell methods can be used as discussed in Chapter 3. With the former, interpolation or forcing terms are used to ensure that nodes inside the solid have finite values that are consistent with grid zones on the surface having the impermeability and no-slip boundary conditions. Trimmed cell treatments ensure that the local finite difference stencil is adjusted such that the surface shape is resolved. However, with high-speed aerodynamic flows, at design conditions typically the vast majority of the mesh is needed in the boundary layer and is also extremely close to the surface. Also, high aspect ratio cells are optimal for RANS simulations (see Chapter 3). The basic finite difference method described here, even with boundary modifications, does not meet these needs well. In this situation, the Navier-Stokes equations are best first transformed into a curvilinear system. Then the transformed equations can be discretized using finite difference related approaches (see sections 4.2.1–4.2.4) in a regular computational plane. This approach, for single domains, also naturally facilities the use of high-order finite difference expressions. Use of a body fitted curvilinear coordinate system is discussed next.

4.2.5 *Body Fitted Grids*

Rather than focusing on Cartesian or cylindrical grids that have limited geometric applicability the governing equations, as shown in Chapter 2, can be expressed in a general curvilinear coordinate system. Using the chain rule of differential calculus it is possible to map between the curvilinear mesh in the physical, (x, y, z, t) space and a regular Cartesian looking mesh in the transformed $(\xi, \eta, \varsigma, \tau)$ plane. The transformed plane is sometimes described as the computational plane, this being where the differential equation (expressed in terms of the transformed derivatives) is solved. In the computational plane the Cartesian nature of the grid is very amenable to the application of the finite difference related methods described above. The transformation (see Rayner 1992) simply involves making the following chain rule substitutions in the governing (x, y, z, t) equations, where $\xi = \xi(x, y, z, t)$, $\eta = \eta(x, y, z, t)$, $\varsigma = \varsigma(x, y, z, t)$ and $\tau = \tau(t)$

$$\begin{aligned}
\frac{\partial}{\partial x} &= \left(\frac{\partial \xi}{\partial x}\right)\frac{\partial}{\partial \xi} + \left(\frac{\partial \eta}{\partial x}\right)\frac{\partial}{\partial \eta} + \left(\frac{\partial \varsigma}{\partial x}\right)\frac{\partial}{\partial \varsigma} + \left(\frac{\partial \tau}{\partial x}\right)\frac{\partial}{\partial \tau} \\
\frac{\partial}{\partial y} &= \left(\frac{\partial \xi}{\partial y}\right)\frac{\partial}{\partial \xi} + \left(\frac{\partial \eta}{\partial y}\right)\frac{\partial}{\partial \eta} + \left(\frac{\partial \varsigma}{\partial y}\right)\frac{\partial}{\partial \varsigma} + \left(\frac{\partial \tau}{\partial y}\right)\frac{\partial}{\partial \tau} \\
\frac{\partial}{\partial z} &= \left(\frac{\partial \xi}{\partial z}\right)\frac{\partial}{\partial \xi} + \left(\frac{\partial \eta}{\partial z}\right)\frac{\partial}{\partial \eta} + \left(\frac{\partial \varsigma}{\partial z}\right)\frac{\partial}{\partial \varsigma} + \left(\frac{\partial \tau}{\partial z}\right)\frac{\partial}{\partial \tau}
\end{aligned} \tag{4.36}$$

The bracketed derivatives in Equation (4.36) are called metrics. In the following, variables treated as fixed in the partial differentiations contained in Equation (4.36) are indicated using subscripts

$$\left(\frac{\partial}{\partial x}\right)_{yzt}\left(\frac{\partial}{\partial y}\right)_{xzt}\left(\frac{\partial}{\partial z}\right)_{xyt}\left(\frac{\partial}{\partial t}\right)_{xyz}\left(\frac{\partial}{\partial \xi}\right)_{\eta\varsigma\tau}\left(\frac{\partial}{\partial \eta}\right)_{\xi\varsigma\tau}\left(\frac{\partial}{\partial \varsigma}\right)_{\xi\eta\tau}\left(\frac{\partial}{\partial \tau}\right)_{\xi\eta\varsigma} \tag{4.37}$$

An example of the form of the transformed equations is given in Chapter 2. However, they can in fact be expressed more compactly than shown in Chapter 2. See, for example, Jefferson-Loveday (2008). The equations will not be further outlined here. Notably, for time varying grids the temporal derivate also needs to be transformed. The use of a transformed time coordinate is also useful for modelling periodic flows in turbomachinery (see Section 6.3.2) when the geometry involved does not have natural periodicity (see Giles 1988, 1991). This process is known as time-inclining.

A key issue is the formation of the metric terms. Firstly, these should be expressed in a form consistent with the main finite differences used in the discretization. For highly non-orthogonal grids, their form can have a strong impact on the solution accuracy (see Visbal and Gaitonde 2002), as will non-smooth grid stretching (Gamet et al. 1999). Visbal and Gaitonde (2002) contrast standard, for example,

$$\begin{aligned} \xi_x &= y_\eta z_\varsigma - y_\varsigma z_\eta \\ \eta_x &= y_\varsigma z_\xi - y_\xi z_\varsigma \\ \zeta_x &= y_\xi z_\eta - y_\eta z_\xi \end{aligned} \tag{4.38}$$

and the conservation metric form of Thomas and Lombard (1979)

$$\begin{aligned} \widehat{\xi_x} &= (y_\eta z)_\varsigma - (y_\varsigma z)_\eta \\ \widehat{\eta_x} &= (y_\varsigma z)_\xi - (y_\xi z)_\varsigma \\ \widehat{\varsigma_x} &= (y_\xi z)_\eta - (y_\eta z)_\xi \end{aligned} \tag{4.39}$$

On skewed grids, they found the latter more accurate. Notably, although the underlying finite difference basis of curvilinear grids naturally allows the use of high order schemes, the metric terms also need to be represented to a consistent order. This aspect is often neglected in high-order curvilinear coordinate codes. What is more, as noted in Chapter 3, multiple connected curvilinear grids are used to map more complex engineering geometries. To maintain a high order in such situations the grids must substantially overlap. The number of grid nodes that overlap needs to increase with the numerical order. This gives a substantial data transfer overhead and the domain decoupling a slowing of iterative convergence. The decomposition also presents some serious grid generation challenges. These were outlined in Section 3.3 where are range of approaches are described to enable the relatively automated generation of curvilinear grids.

4.2.6 *Numerical Considerations*

4.2.6.1 Algorithm Consistency We will now consider the use of Equations (4.12) and (4.15) to model the simple transport equation, $\rho u \partial\phi/\partial x = \Gamma\partial^2\phi/\partial x^2$. If we used these equations, since for both we noted $HOT \propto (\Delta x)^2$, we might conclude that the algorithm is consistent (i.e., both components are second order). However, as noted by Leonard (1979), when defining the order based on the highest order of polynomial, for which the scheme is exact, we can see some potential inconsistency. Equation (4.15) is third order in this polynomial sense and (4.12) is second order. Hence, for convection-dominated flows, it could be argued that we will have a second order method and for diffusion-dominated cases the method is third order. Whatever the definition for the order that we use, after applying Equations (4.12) and (4.15), we should strictly conclude (in a slightly conservative sense) that we have a second order scheme. The lowest order spatial component is generally rightly assumed to dictate the scheme's accuracy. However, it is not unusual to use reduced order schemes just at boundaries (as can be inferred from the Equation (4.23) derivation, high-order schemes are less easy to implement at boundaries). In this non-ideal situation, the order of the algorithm is frequently taken as equal to that of the higher order scheme used in the domain interior. Obviously, having the order of the scheme at the boundary being of a lower order to that in the interior is not ideal. The impact of this inconsistency will be flow physics dependent.

4.2.6.2 The Modified Equation In Section 4.2.2 it could be seen that using the Taylor series enabled us to study the order of the discretization of derivatives. However, it also allows us to study some of the mathematical traits of the final governing equation when all the finite difference components have been combined. This is often referred to as *modified equation analysis*. Essentially, the idea is to explore what the exact differential equation is that the discrete approximate equation in fact represents. This process is important and powerful. Typically, the discrete equation will represent the intended original differential equation but also have other differential terms. These might well be unhelpful error terms, but also can be helpful as in the ILES (Implicit LES) approach to be discussed later. With this the error, terms are turned to our advantage being used for modelling purposes.

The modified equation can again be formed by replacing terms in the discretized equation with their Taylor series expansion. The process can be seen by considering the basic one-dimensional transport equation, $u\partial\phi/\partial x$. Discretizing this equation using, this time, the first order backwards difference Equation (4.10) and staying with the geographic notation directly gives

$$\frac{u(\phi_P - \phi_W)}{\Delta x_P} = 0 \tag{4.40}$$

Taylor series expansion for ϕ_W yields

$$\phi_W = \phi_P - \left(\frac{\partial\phi}{\partial x}\right)_P \Delta x_P + \left(\frac{\partial^2\phi}{\partial x^2}\right)_P \frac{(\Delta x_P)^2}{2!} - \left(\frac{\partial^3\phi}{\partial x^3}\right)_P \frac{(\Delta x_P)^3}{3!} + HOT \tag{4.41}$$

Replacing ϕ_W in Equation (4.40) with its Taylor expansion gives the following, dropping the P subscript because all terms are now for this point,

$$u\frac{\partial\phi}{\partial x} = C_0\frac{\partial^2\phi}{\partial x^2} + C_1\frac{\partial^3\phi}{\partial x^3} + HOT \tag{4.42}$$

where $C_0 = u\Delta x/2$ and $C_1 = -u\Delta x^2/6$.

Including the time derivative, $\partial\phi/\partial t$, in the original equation ($u\partial\phi/\partial x$) to give $\partial\phi/\partial t + u\partial\phi/\partial x$ and discretizing using a forwards difference (called the explicit Euler time scheme – see Section 4.8), yields $C_0 = u\Delta x(1 - u\Delta t/\Delta x)/2$ and $C_1 = u(\Delta x)^2(3u\Delta t/\Delta x - 2(u\Delta t/\Delta x)^2 - 1)/6$. For $\Delta t = 0$, C_0 and C_1 are the same as for the steady analysis. However, for finite Δt, some interaction can be observed between the spatial and temporal discretization. Hence, as noted earlier, although through the methods of lines we have decoupled these processes, the two aspects do interact.

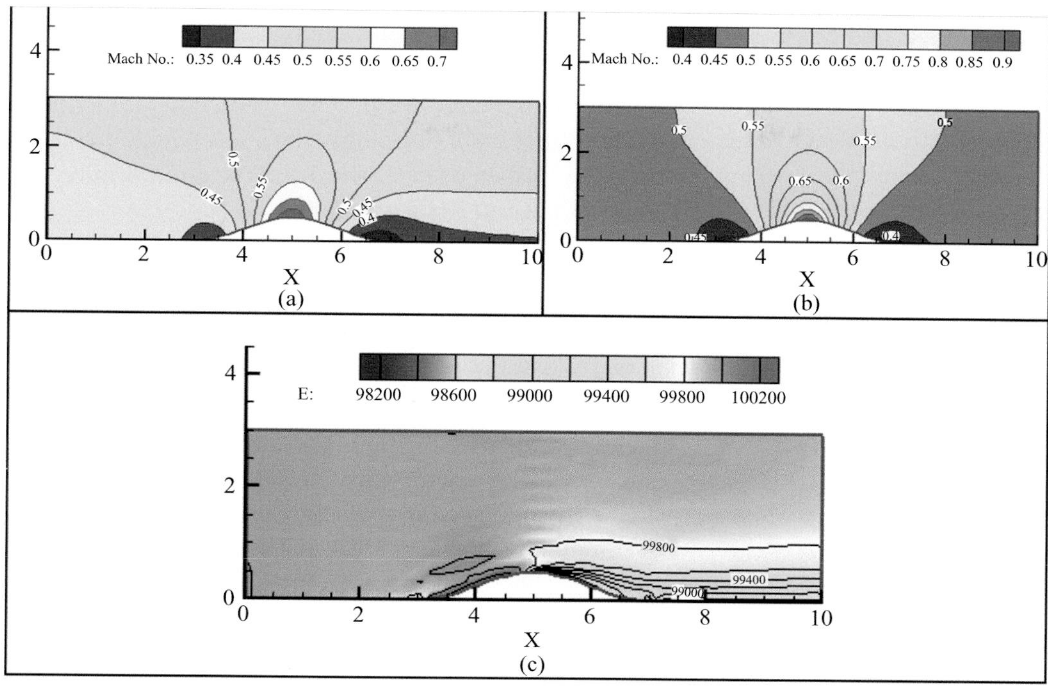

Figure 4.4. Subsonic flow over a bump from solution of the Euler equation: (a) Mach number contours for high numerical dissipation, (b) Mach number contours for relatively low level of numerical smoothing and (c) contours of total pressure loss.[1]

4.2.6.3 Dissipation and Dispersion As can be seen, from Equation (4.42), the additional diffusion term $C_0\partial^2\phi/\partial x^2$ can arise. This term mostly gives rise to what is called *false or numerical diffusion*. Depending on the magnitude of $u\Delta t/\Delta x$ (known as the Courant number and being the ratio of the distance a fluid particle would travel in a time step relative to the grid size), the false diffusion term can be positive (diffusive) or negative (anti-diffusive). The latter causes numerical instability – some level of diffusivity is necessary for numerical convergence. On the other hand, excessive erroneous diffusion will cause smearing of flow profiles and general distortion of the flow field. Figure 4.4 shows Euler solutions for subsonic flow over a bump. Frame (a) gives Mach number contours for high numerical diffusion. Ideally, for a perfect solution of the Euler equation, the flow should be symmetrical about a vertical line centred on the bump. In Frame (a), this is not the case, the lack of symmetry being caused by excessive numerical diffusion. Frame (b) shows Mach number contours for a relatively low level of numerical diffusion. This improves the flow symmetry. Frame (c) gives contours of total pressure loss. This shows that the false diffusion generates, where there are substantial velocity gradients, a loss of energy in the flow.

The $C_1\partial^3\phi/\partial x^2$ term, in the modified equation analysis, produces what is called a *dispersion error*. This error causes 'wiggles' (pointwise oscillations) to appear in

[1] Figure 4.4 was kindly provided by Dr V. Nagabhushana Rao.

solutions and phase errors (waves will propagate at the wrong rate). Higher order terms (*HOT* in Equation (4.42)) also appear in the modified equation. Those with even order derivatives will cause dissipative behaviour (i.e., will damp flow gradients and unsteadiness). Odd ordered derivatives will cause dispersive behaviour, giving wiggles and phase errors. The omitted leading term, arising from the discretization, generally suggests the most striking behaviour of the scheme.

4.2.6.4 Stability and Aliasing As a numerical solution progresses when integrated through time or a pseudo-time, instabilities can grow. Hirt (1968) gives a heuristic stability theory based around the modified equation approach and making stability inferences from this. The stability of equations systems can also be studied using the approach of von Neumann. This is Fourier series based. However, it is only suitable for differential equations with simple boundary conditions and constant coefficients. Hence, such a simplified procedure yields information that is only suggestive of the performance of a numerical scheme when applied to a non-linear equation in three dimensions. Glass and Rodi (1982) and Gresho et al. (1984) present a simplification to the von Neumann stability analysis based on just selecting individual Fourier series components.

The shortest wavelength that can be resolved by a grid is $2\Delta x$. Energy from wavelengths that cannot be resolved by the grid can erroneously become combined with longer wavelengths. This is called aliasing, which ultimately results in solution divergence. Specialized schemes can be designed to reduce aliasing errors. For example, Laizet and Lamballais (2009) tune scheme coefficients to make the scheme more dissipative at high wave numbers. The skew-symmetric form of the Navier-Stokes equations, given in Chapter 2, exhibits substantially less aliasing when discretized. Hence, discretization of this form of the flow governing equation is popular for eddy-resolving simulations. On the other hand, Ducros et al. (2000) find that in finite volume form the second order central difference discretization naturally provides a skew-symmetry. Hence, when considering or designing a numerical scheme it can be important to pay some attention to the scheme's aliasing properties. This is especially so if integrations are to be made over long time periods.

4.2.6.5 Resolution of Schemes Clearly not just the order of the scheme is important but also the dissipation and dispersion traits. The latter two aspects are sometimes associated with the scheme's 'resolution'. The Dispersion Relation Preserving (DRP) schemes look at both the order and resolution of schemes. They seek a best compromise between these two aspects, to yield the potentially most efficient scheme. Hence, we might approximate a first derivate as

$$\frac{\partial \phi}{\partial x} = \frac{\cdots + Wf_W \phi_W + Wf_E \phi_E + \cdots}{\Delta x}. \tag{4.43}$$

With DRP (see Tam and Webb 1993, Lockard et al. 1995), the coefficients in the above are altered from what they would be for a standard high-order scheme, so that the scheme's resolution is improved at the sacrifice of some formal order. Table 4.2 summarizes some DRP scheme coefficients.

Table 4.2. *DRP scheme coefficients from Broeckhoven et al. (2007)*

Source	$Wf_W = Wf_E$	$Wf_{WW} = Wf_{EE}$	$Wf_{WWW} = Wf_{EEE}$
Tam and Webb (1993)	0.79926643	−0.18941314	0.02651995
Tam and Shen (1993)	0.77088238	−0.16670590	0.20843142

Next, the finite volume method is discussed along with more advanced convective discretizations beyond those offered by the finite difference method. In this situation, the derivation of the modified equation and other analysis is more challenging. This can be seen, for example, in the work of Grinstein et al. (2011).

4.3 Finite Volume Method

Much of the basic flux integration procedures for high-speed flows with the finite volume method are outlined in detail in Blazek (2005). The initial discussion on the finite volume method below largely follows this text. For a detailed outline of a general unstructured compressible flow solver format, see Moinier (1999). For the finite volume method, the form of $A(\Phi)$ in Equation (4.1) is

$$A(\Phi) = -\frac{1}{V}\left[\oint_{\partial V}(\mathbf{F}_c - \mathbf{F}_V)dL - \int_V \mathbf{S}dV\right] \tag{4.44}$$

where $\boldsymbol{S}$ is a source term, V is a cell volume and L represents the surface area of the volume. Also, $\boldsymbol{F}_c$ and $\boldsymbol{F}_V$ correspond to convective and viscous fluxes, respectively. The critical thing to note is that a surface integral is involved. In discrete form this can be expressed as

$$A(\Phi) \approx -\frac{1}{V_{i,j,k}}\left[\sum_{m=1}^{N}(\mathbf{F}_c - \mathbf{F}_V)_m \Delta L_m - \mathbf{S}_{i,j,k}V_{i,j,k}\right] \tag{4.45}$$

where N, is the number of control volume faces and the subscripts i, j and k are grid point location references. For an unstructured solver, a single index would be used and a topology array would define all the faces associated with a particular cell. As can be seen, the numerical solution of various differential equations, such as the Navier-Stokes and Euler ($\mathbf{F}_V = 0$) can effectively be configured to entail the integration (summing) of fluxes around control volumes. This integration is not surprising – the derivation of the Navier-Stokes and Euler equations in Chapter 2 consisted of summing flows (mass, momentum and enthalpy) around control volumes.

With the finite volume method, there is a range of potential control volume configurations. These are summarized, for basic quadrilaterals, in Figure 4.5. The control volumes in Frames (a) and (b) are more for historical interest. Both essentially involve overlapping grids. With Frame (a) the variable is evaluated at the cell centres. However, ultimately, the cell centre values are stored at the cell corners and the scheme is classified as a cell vertex method. The vertex storage is achieved through an averaging operator (from cell centred data) discussed further later. In Frame (b) the grids are staggered. This procedure originated from incompressible

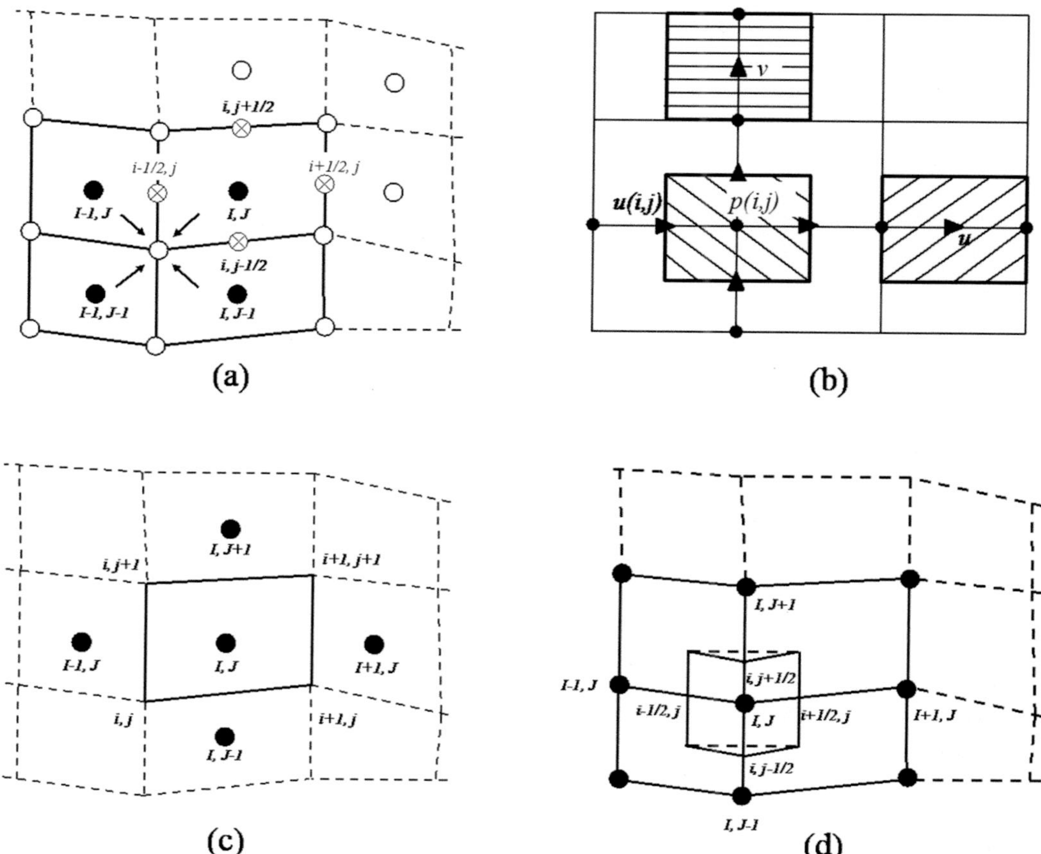

Figure 4.5. Potential quadrilateral control volume structures for the finite volume method: (a) cell vertex scheme with overlapping control volumes, (b) staggered control volumes, (c) cell centred scheme and (d) cell vertex scheme with (median dual control volume) mesh.

CFD methods and will be discussed further later. The pressure is evaluated at the cell centres. The staggered momentum volumes are such that the pressure, driving the flow, is located precisely at their control volume faces. This gives rise to good energy conserving numerical properties. However, the coding complexity is substantially increased. Frame (c) gives a classic cell centred methodology. The variables are again, as with Frame (a), stored at the cell centre. Notably, with the cell centred method, the control volume exactly corresponds with the mesh lines. The next alternative is the *cell vertex approach*, shown in Frame (d), with typically the *median dual control volume*. With this approach, control volumes are constructed around the imported mesh. There are two main alternatives for doing this. One is that cell centroids are simply connected as indicated by the dashed lines (centroid dual). Note that these overlay the full lines on the vertical edges. The other, more popular, option is that the cell centroids are connected to edge midpoints – see full lines in Frame (d). This is called the median dual. Figure 4.6 gives an example of the

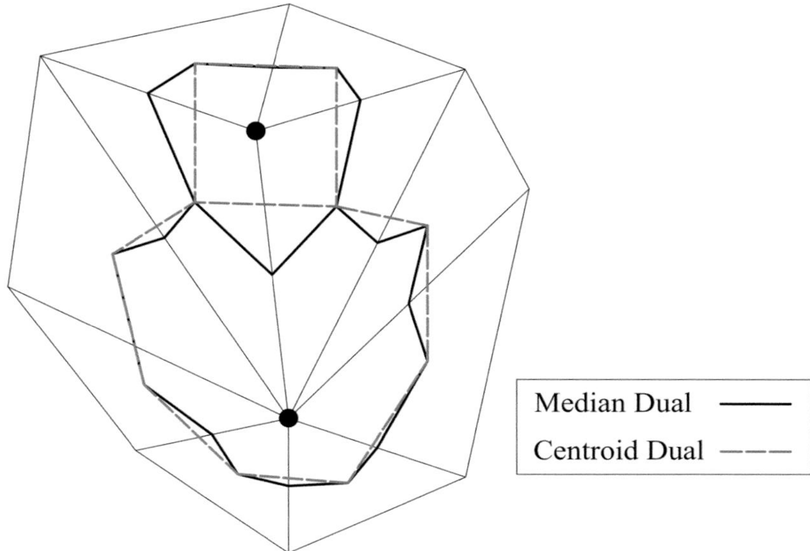

Figure 4.6. Control volumes arising using the unstructured method with a triangulated mesh and dual control volumes.

Frame (d) control volume constructions for an unstructured triangulated grid. The disadvantage of connecting centroids to face midpoints is that the control volume faces will have multiple normals. This is manifested as unusual control volume faces for the median dual. As shown in Chapter 3, this can have substantial accuracy penalties. However, connecting the cell centroids to cell edge midpoints prevents cell edges spilling out of the domain or incursions into the domain. The latter is shown in Frame (a) of Figure 4.7. Frame (b) shows how the incursion is remedied through connecting centroids to edge midpoints. Another key thing to note from this figure is that the cell vertex method gives rise to half-control volumes near walls.

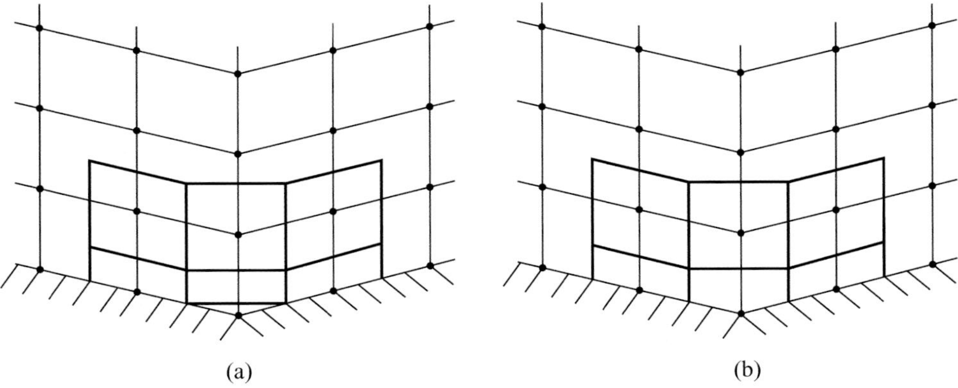

Figure 4.7. Potential near wall half-control volumes for the classical cell vertex method: (a) control volume formed by the simple connection of centroids and (b) control volumes formed by connecting centroids to the midpoints of edges.

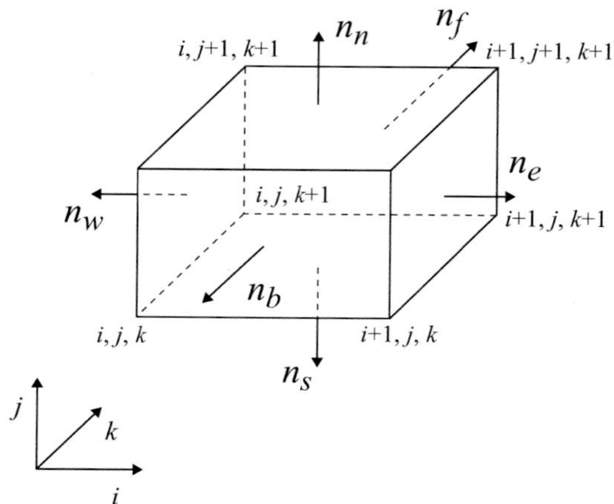

Figure 4.8. Three-dimensional hexahedral control volume.

All of these methods can be extended to three dimensions. Figure 4.8 gives a three-dimensional control volume and a potential nomenclature. Unless all the straight lines that form the faces are planar, a unique face normal cannot be defined. This is required for the discretization (see Section 4.3.2). However, for smooth grids an averaged normal gives acceptable accuracy.

4.3.1 *Advantages and Disadvantages of Differing Control Volumes*

The near wall, half-control volumes, shown in Figure 4.7, are a key disadvantage of the cell vertex method. The coding near walls can become complex, with different subroutines required to specifically form fluxes for these near wall control volumes. The half-control volume means that the residual for the control volume resides on the wall. Cleary, the wall is a non-ideal place to have numerical errors and discretization ambiguities. The implementation of periodic boundary conditions is also complicated. For unsteady flows, the cell vertex approach ideally needs a mass matrix (see Section 4.8). Also, the dual control volume of the cell vertex scheme can be problematic around trailing edges.

As noted, a key disadvantage of staggered grids is code implementation. The cell vertex method of Frame (a) has a serious limitation arising from averaging data to the vertex from the four surrounding cells. This averaging results in the fluxes at the internal faces cancelling out. The result of this is that the solution cell (effective control volume) is four times larger in area, in two dimensions, than expected. Clearly, the situation is worse in three dimensions. Also, the Frame (a) cell vertex approach does not allow implementation of the more modern advanced convective term treatments to be discussed later.

On balance, relative to the classic cell vertex, median dual, control volume method, for simplicity of coding (i.e. avoiding the need for half-control volumes at walls and a mass matrix), the cell centred method would seem best. However, in

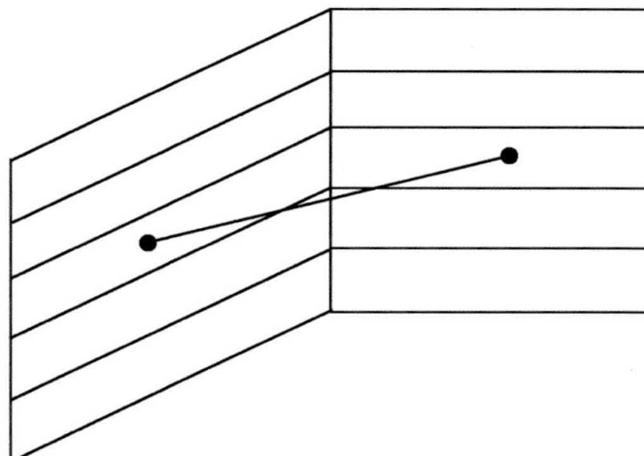

Figure 4.9. Potential control volume interpolation for skewed cells when using the cell centred finite volume method.

the domain interior, for meshes of reasonable quality and steady flow, there is little numerical difference between the cell-centred and cell vertex methods. For badly formed cells, though, all methods can show problematic behaviour. For example, as shown in Figure 4.9, with the cell centred method, the cell face averaging procedure can become inaccurate for skewed control volumes.

With regards to unstructured (non-hexahedral) grids, as can be inferred from Figure 4.6, the cell centred scheme gives rise to considerably more control volumes and hence degrees of freedom (around a factor of three for a typical mixed element mesh) relative to the cell vertex, median dual, control volume method. On the other hand, the control volume, for the latter, has a greater number of fluxes/greater connectivity and this might suggest greater accuracy. However, there is no clear-cut evidence suggesting superiority of either approach. A key problem with the cell vertex, median dual, control volume scheme is that the fluxes are generally assumed to be orthogonal to the cell faces. As can be seen from Figure 3.44, in Chapter 3, and also in the lower control volume of Figure 4.10, this can be far from the case. As can be seen from Figure 3.43, in Chapter 3, which looks at the propagation of a Tollmien-Schlichting wave, this lack of orthogonality results in dramatic amounts of numerical dissipation. The grid structure shown in Figure 4.10 is not uncommon in boundary layers. A potential remedy is to formulate the control volume using the containment dual control volume (see Hu and Nicolaides 1992)). This procedure, based around the generation of a bounding circle, or sphere, in three dimensions, is illustrated in Figure 4.11. The result of using this containment circle can be seen from the upper control volume in Figure 4.10.

As well as accuracy, there is the aspect of computational cost to consider. The cell centred method involves summations over cells faces. The cell vertex, median dual, control volume method involves summations over edges. For a tetrahedral mesh there are around two times as many faces to loop over. Since computational operations for the cell vertex and cell centred methods are very similar, for each face/edge integration for the same mesh, the cell centred approach is around twice the computational cost. For hexahedral meshes the cost disparity goes away. Also,

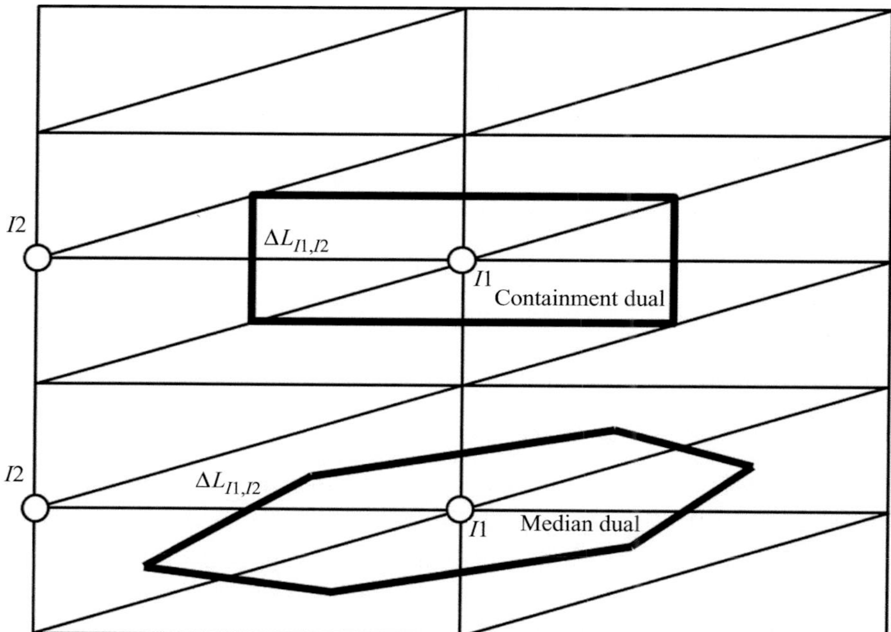

Figure 4.10. Potential control volumes arising from the use of the median dual cell vertex method (lower mesh usual control volume construction procedure and upper mesh control volume structure arising from use of containment circle).

for meshes with prism layers, often used to deal with boundary layer zones, where there is a heavy mesh concentration, the disparity considerably lessens. Both the cell vertex and cell centred meshes require the storage of two integers (as markers – see Section 4.4) and three real numbers (face vectors). However, the cell centred method requires an additional six real numbers to store vectors to faces. Although the storage requirements vary considerably depending on the cell type, the cell centred

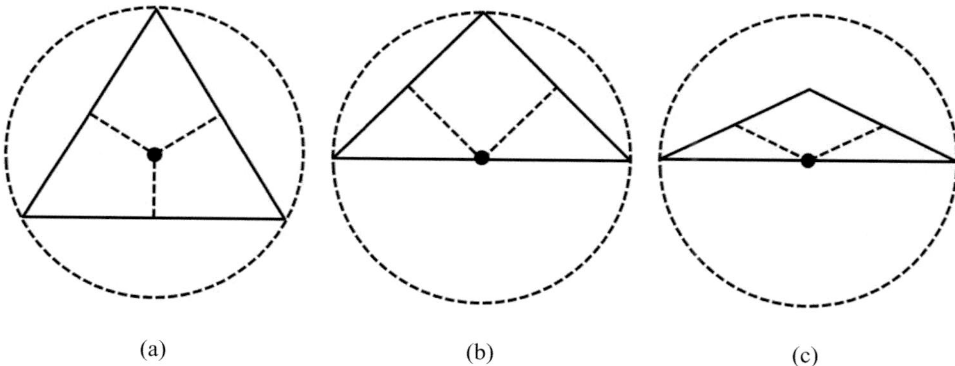

Figure 4.11. The use of a containment circle to improve control volume quality (adapted from Blazek 2005).

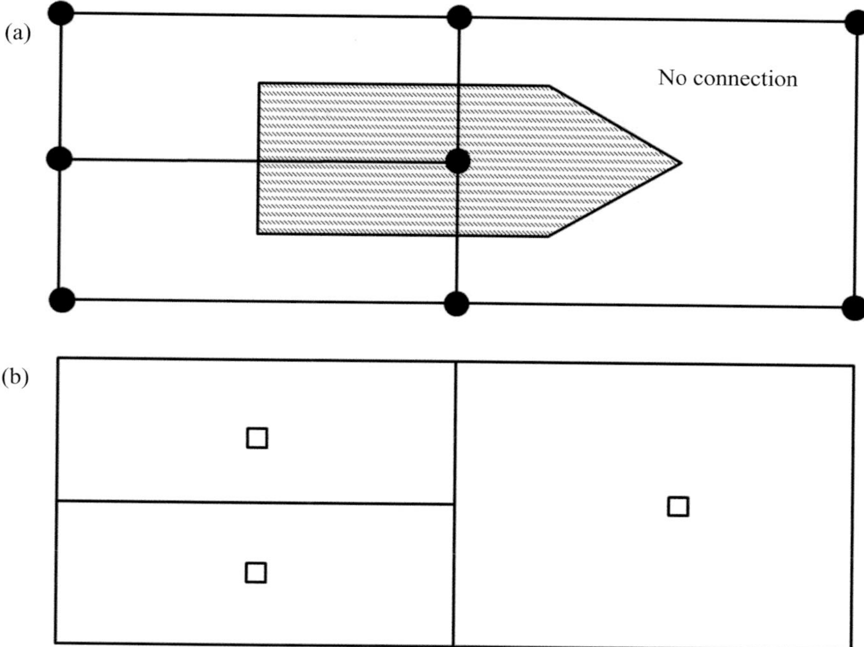

Figure 4.12. Hanging nodes and the cell centred and cell vertex with median dual control volume meshes: (a) cell vertex method with median dual control volume and (b) cell centred method.

method typically needs over twice the storage of the cell vertex, median dual, control volume approach. Notably, the cell centred method will naturally handle the non-conforming interface shown in Figure 4.12. Such interfaces are problematic for the median dual control volume method and hampers the use of adaptive meshing. This can be seen from Figure 4.12. Frame (a) shows that, for the cell vertex method, there is no data flow from the left to the right-hand grid zone. This is not the case for the cell centred configuration shown in Frame (b).

4.3.2 *Control Volume Face Integration*

As noted previously, the discretization involves the summing of fluxes over the cell edges. To do this, an estimate of solution variables, stored at surrounding nodes, must be represented at the control volume faces. There are a range of methods for this which will be discussed later. Essentially interpolation is used. This can take a wide range of forms with substantially differing levels of complexity. Then, to estimate the flow of quantities into the control volume, the face normal and area is needed. Evaluation of this information is discussed next. Note that in Chapter 3, this process was essentially considered as one of the problem exercises. However, the process was not formally discussed. The following discussion is largely in two dimensions. The key required information for the flux integrations is the face normal and area, ΔA. In two dimensions, the area is a length, L, per unit depth. Hence we can write

$$\mathbf{L}_m = \mathbf{n}_m \Delta L = \begin{bmatrix} L_{x,m} \\ L_{y,m} \end{bmatrix} \tag{4.46}$$

where the subscript m relates to the face number. The face vectors for the control volumes shown in Figure 4.5 can consequently be expressed as

$$\begin{aligned} \mathbf{L}_w &= \begin{bmatrix} y_{i,j} - y_{i,j+1} \\ x_{i,j+1} - x_{i,j} \end{bmatrix}, \quad \mathbf{L}_e = \begin{bmatrix} y_{i+1,j+1} - y_{i+1,j} \\ x_{i+1,j} - x_{i+1,j+1} \end{bmatrix} \\ \mathbf{L}_n &= \begin{bmatrix} y_{i,j+1} - y_{i+1,j+1} \\ x_{i+1,j+1} - x_{i,j+1} \end{bmatrix}, \quad \mathbf{L}_s = \begin{bmatrix} y_{i+1,j} - y_{i,j} \\ x_{i,j} - x_{i+1,j} \end{bmatrix} \end{aligned} \tag{4.47}$$

where the lower case i, j subscripts refer to the vertices of the control volume. From rearranging Equation (4.46), the face normal vector can be expressed as

$$\mathbf{n}_m = \frac{\mathbf{L}_m}{\Delta L} \tag{4.48}$$

and

$$\Delta L_m = |\mathbf{L}_m| = \sqrt{L_{x,m}^2 + L_{y,m}^2} \tag{4.49}$$

where $L_{x,m}$ and $L_{y,m}$ are edge extents when viewed in the x and y directions. These terms only need to be stored for the north and west faces. The values evaluated for the same faces on the adjacent vertical and right-hand cells can then be used/stored with the signs reversed. Note that since this derivation is based around control volume faces, it can be readily extended to, for example, triangles. However, clearly the i, j referencing is inappropriate and should be based on, say, a clockwise numbering system with a single subscript. The similar process can be carried out for a three-dimensional control volume. Hence, for the west face, in Figure 4.8, the following can be defined

$$\begin{aligned} \Delta x_1 &= x_{i,j+1,k+1} - x_{i,j,k} \\ \Delta x_2 &= x_{i,j+1,k} - x_{i,j,k+1} \\ \Delta y_1 &= y_{i,j+1,k+1} - y_{i,j,k} \\ \Delta y_2 &= y_{i,j+1,k} - y_{i,j,k+1} \\ \Delta z_1 &= z_{i,j+1,k+1} - z_{i,j,k} \\ \Delta z_2 &= z_{i,j+1,k} - z_{i,j,k+1} \end{aligned} \tag{4.50}$$

The face vector is then expressed as

$$\mathbf{L}_w = \frac{1}{2} \begin{bmatrix} \Delta z_1 \Delta y_2 - \Delta y_1 \Delta z_2 \\ \Delta x_1 \Delta z_2 - \Delta z_1 \Delta x_2 \\ \Delta y_1 \Delta x_2 - \Delta x_1 \Delta y_2 \end{bmatrix} \tag{4.51}$$

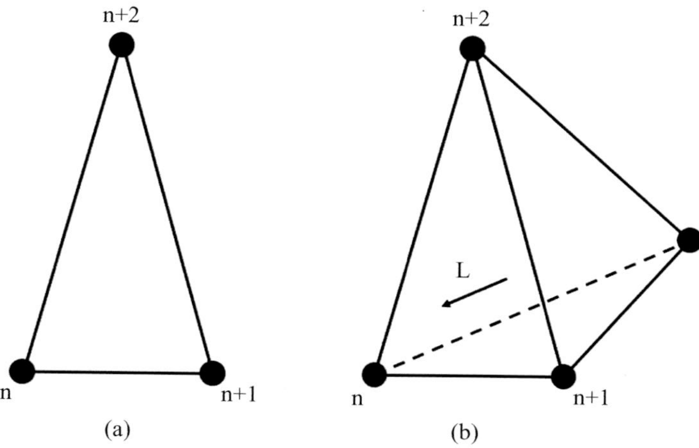

Figure 4.13. Triangular and tetrahedral elements: (a) triangular and (b) tetrahedral.

where $\mathbf{L}_w = \mathbf{n}_w \Delta L_w$. This time, only, $\mathbf{L}_w$, $\mathbf{L}_s$ and $\mathbf{L}_f$ need to be stored. The equivalent values for an adjacent cell, with the signs reversed, allows the full control volume integration. The edge normals are again calculated with Equation (4.48), but this time with

$$\Delta \mathbf{L}_m = \left|\mathbf{L}_m\right| = \sqrt{L_{x,m}^2 + L_{y,m}^2 + L_{z,m}^2} \tag{4.52}$$

This procedure is only approximate if the face is non-planar, but for most grids is expected to give acceptable accuracy. For a triangular face, as shown in Figure 4.13b, the face vector can be exactly computed using the following equations

$$\begin{aligned}
\Delta xy_1 &= (x_n - x_{n+1})(y_n + y_{n+1}) \\
\Delta xy_2 &= (x_{n+1} - x_{n+2})(y_{n+1} + y_{n+2}) \\
\Delta xy_3 &= (x_{n+2} - x_n)(y_{n+2} + y_n) \\
\Delta yz_1 &= (y_n - y_{n+1})(z_n + z_{n+1}) \\
\Delta yz_2 &= (y_{n+1} - y_{n+2})(z_{n+1} + z_{n+2}) \\
\Delta yz_3 &= (y_{n+2} - y_n)(z_{n+2} + z_n) \\
\Delta zx_1 &= (z_n - z_{n+1})(x_n + x_{n+1}) \\
\Delta zx_2 &= (z_{n+1} - z_{n+2})(x_{n+1} + x_{n+2}) \\
\Delta zx_3 &= (z_{n+2} - z_n)(x_{n+2} + x_n)
\end{aligned} \tag{4.53}$$

Then the face vector can be expressed as

$$\mathbf{L}_w = \frac{1}{2}\begin{bmatrix} \Delta yz_1 + \Delta yz_2 + \Delta yz_3 \\ \Delta zx_1 + \Delta zx_2 + \Delta zx_3 \\ \Delta xy_1 + \Delta xy_2 + \Delta xy_3 \end{bmatrix} \tag{4.54}$$

Note that for a median dual control volume strategy, each 'face integration' can involve multiple faces (eight for a tetrahedral based grid). This is very evident in Figure 4.6. The multiple normal must then be averaged.

4.3.3 Areas/Volumes

Note that as part of the discretization the element area, or, in three-dimensions, volume, is needed. The area evaluation for a hexahedral was outlined in Chapter 3. For a triangular cell (see Figure 4.13a), the area can be expressed as

$$A = \frac{1}{2}\begin{bmatrix} (x_n - x_{n+1})(y_n + y_{n+1}) + \\ (x_{n+1} - x_{n+2})(y_{n+1} + y_{n+2}) + \\ (x_{n+2} - x_n)(y_{n+2} + y_n) \end{bmatrix} \tag{4.55}$$

where n is an edge node number.

As noted, the cell volume also needs to be computed and a range of procedures are available for this purpose (see Hirsch 2007). Using the divergence theorem, the following equation for volume can be derived

$$V = \frac{1}{3}\sum_{m=1}^{N} [\mathbf{r} \cdot \mathbf{L}]_m \tag{4.56}$$

where **r** is the midpoint of the control volume face and N is the number of faces. It is exact for a volume with planar quadrilateral faces or triangles.

4.3.4 Viscous Flux Evaluation

Although not trivial, the integration of $\mathbf{F}_v$ around the control volume is considerably simpler than $\mathbf{F}_c$. Indeed, for a four-sided cell, finite differences could be used, instead of the finite volume integration. This would give a mixed type discretization. For irregular cells, a transformation into a regular computational plane would be needed. There is also the possibility of using an approximate) Galerkin finite element type discretization – see Appendix A. However, sticking with the finite volume method, there are a range of alternatives. All the popular methods are based around use of the Green's (divergence) theorem. As shown in Figure 4.14, relating to a cell centred scheme, a potential option is to define auxiliary control volumes. In Figure 4.14, this has been constructed by the connection of mid-edges. The usual Green's theorem can be used to integrate around the control volume, to gain a gradient estimate at the main 'west' control volume face. This is marked 'X' in Figure 4.14. For the x gradient of a general variable, ϕ, the following can be written

$$\frac{\partial \phi}{\partial x} = \frac{1}{V}\int_{\partial V} \phi dL_x \approx \frac{1}{V}\sum_{m=1}^{N} \phi_m dL_x \tag{4.57}$$

where L and V refer to the auxiliary cell. Clearly for the west and east faces of the auxiliary control volume, ϕ_m is directly available as $\phi_{I-1,J}$ and $\phi_{I,J}$ for the above Equation (4.57) summation. The north face value of ϕ_m can be estimated as $\phi_{i,j+1} = (\phi_{I,J} + \phi_{I-1,J} + \phi_{I-1,J+1} + \phi_{I,J+1})/4$, and a similar expression written for the south face. Extension to three dimensions is in essence straightforward. The other face gradients can be assembled in a similar way through defining auxiliary control volumes. The final integration, around the main control volume, gives the

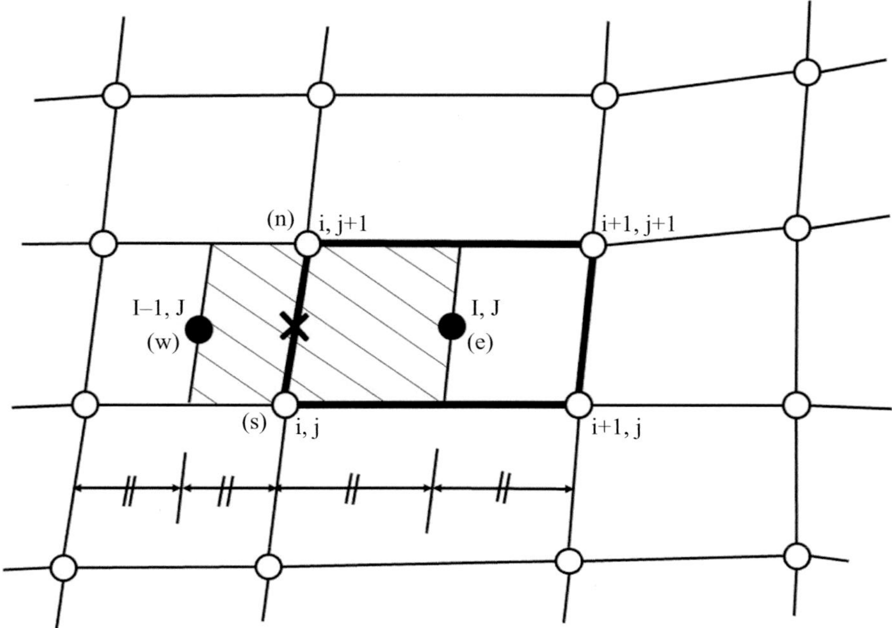

Figure 4.14. Auxiliary integration volume to form viscous flux, $\mathbf{F}_v$.

second order derivate needed for the diffusion term. Care needs to be taken to avoid odd-even decoupling, but relative to the problems associated with the convective term discretization, the issues are minor. Cleary, a smooth grid enables the procedure to give the most plausible estimates. An analogous procedure can be used for a cell vertex method. For the thin shear layer approximation, evaluation of the irrelevant gradient terms can be neglected.

The use of the staggered control volumes just described is expensive in terms of both computer operations and storage. A more general procedure is to apply Green's theorem to the main solution control volumes. The gradients at the control volume centres are then simply averaged to the control volume faces. The averaged flux can be expressed as

$$\overline{\nabla\Phi_{I2I1}} = \frac{1}{2}(\nabla\Phi_{I2} + \nabla\Phi_{I1}) \tag{4.58}$$

However, being an average of two central differences, it will not damp high frequency oscillations. This is especially so in boundary layers, where the viscous fluxes are dominant and any dissipation associated with the convective fluxes will be inadequate for damping purposes. To alleviate this, the component of $\nabla\Phi$ along, say, the edge (in a median dual cell vertex method) is replaced by a simple difference along the edge. Hence, there is the following modification

$$\nabla\Phi_{I2I1} = \overline{\nabla\Phi_{I2I1}} - \left(\overline{\nabla\Phi_{I2I1}} \cdot \delta s_{I2I1} - \frac{(\Phi_{I2} - \Phi_{I1})}{|x_{I2} - x_{I1}|}\right) \delta s_{I2I1} \tag{4.59}$$

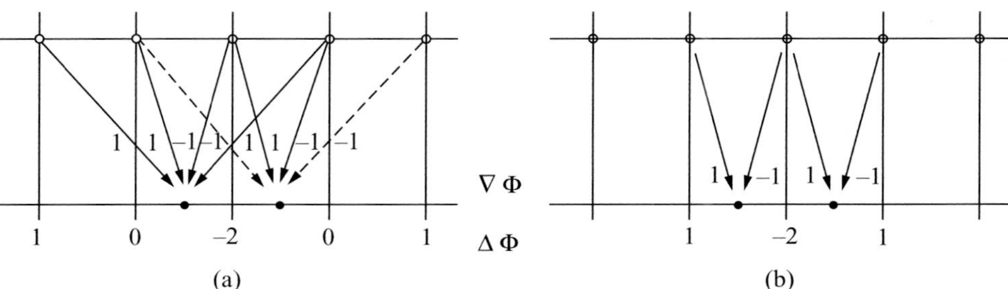

Figure 4.15. One-dimensional stencil of $\Delta\Phi$ when using: (a) central differencing and (b) a simple difference along one edge (from Moinier 1999).

where

$$\delta s_{I2I1} = \frac{x_{I2} - x_{I1}}{|x_{I2} - x_{I1}|} \tag{4.60}$$

In a boundary layer, the strong gradients along the shortest wall normal edges will be effectively emphasised. This will damp out high frequency oscillations. These can be especially problematic when using multigrid convergence acceleration – discussed later.

Figure 4.15 illustrates the stencil of the Laplacian in one dimension when using central differencing and a simple difference along one edge. The former is shown in Frame (a) and the latter in Frame (b). The figure is taken from Moinier (1999) and best considered with Figure 4.16. The key point to note is that the application of the Frame (a) stencil from Figure 4.15 will not damp the oscillation shown in Figure 4.16 – the stencil will not see the oscillation, even though a stencil of five points is being used. However, the simple, three-point Frame (b) stencil of Figure 4.15 will. Hence, a scheme without the Equation (4.59) correction will struggle to converge. This approach can be extended also to a cell centred discretization, the derivative being applied along the connection between the centroids of adjacent cells. This approach needs no extra storage – thus making it attractive.

Gauss-Green's theorem-based approaches tend to become inaccurate on mixed element meshes. An alternative method is the least squares approach. This is based on using a first order, Taylor series approximation for each edge (for cell vertex) or control volume face for the cell centred method. For the cell vertex, median dual, control volume method, the following can be written for an edge

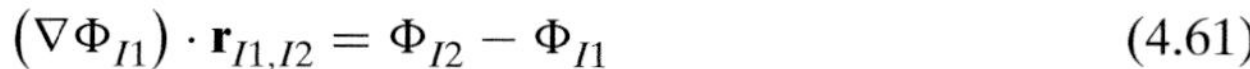

$$\left(\nabla\Phi_{I1}\right) \cdot \mathbf{r}_{I1,I2} = \Phi_{I2} - \Phi_{I1} \tag{4.61}$$

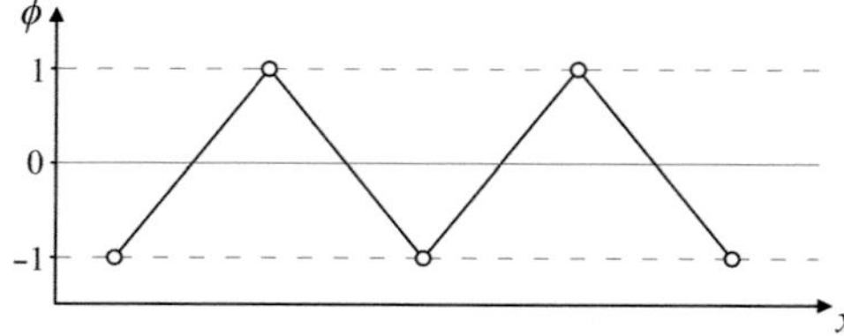

Figure 4.16. Pointwise oscillation in solution variable (from Moinier 1999).

where $\mathbf{r}_{I1,I2}$ is the vector connecting the nodes that straddle the edge under consideration. For all the edges connected to the focal node point $I1$, the following matrix system can be written

$$\begin{bmatrix} \Delta x_{I1,1} & \Delta y_{I1,1} & \Delta z_{I1,1} \\ \Delta x_{I1,2} & \Delta y_{I1,2} & \Delta z_{I1,2} \\ \vdots & \vdots & \vdots \\ \Delta x_{I1,I2} & \Delta y_{I1,I2} & \Delta z_{I1,I2} \\ \vdots & \vdots & \vdots \\ \Delta x_{I1,N} & \Delta y_{I1,N} & \Delta z_{I1,N} \end{bmatrix} \begin{bmatrix} (\partial\Phi/\partial x) \\ (\partial\Phi/\partial y) \\ (\partial\Phi/\partial z) \end{bmatrix}_{I1} = \begin{bmatrix} Wf\,(\Phi_1 - \Phi_{I1}) \\ Wf\,(\Phi_1 - \Phi_{I1}) \\ \vdots \\ Wf\,(\Phi_{I2} - \Phi_{I1}) \\ \vdots \\ Wf\,(\Phi_N - \Phi_{I1}) \end{bmatrix} \tag{4.62}$$

where Wf is a weighting function but typically $Wf = 1$. The nomenclature $\Delta\phi_{I1,I2}$ is shorthand for $\phi_{I2} - \phi_{I1}$. This matrix is a system of the form $[\mathbf{A}][\mathbf{x}] = [\mathbf{S}]$. The idea is to invert $[\mathbf{A}]$, to evaluate $[\mathbf{x}]$ for the necessary gradients. The only difficulty is that the $[\mathbf{A}]$ is ill-conditioned. Hence, it needs to be solved with care, making a process that appears in essence quite simple a more specialist task – especially if the approach is resilient to a wide range of grids and grid qualities. The method is most attractive for mixed element grids, where the Gauss-Green approach becomes problematic. Linear reconstruction has a similar cost to use of the Gauss-Green theorem but requires extra storage of pre-computed terms.

4.4 Convective Flux Evaluation

Once the geometric face information is known, then the mass, momentum or, say, enthalpy flow through the faces can be evaluated. For mass flow, clearly the density at the control volume face is needed along with the velocity. For momentum flow the same data is needed. The obvious simple strategy would be to independently interpolate these values, ϕ, from surrounding nodes to the control volume face, form the flux $F_c(\Phi)$ and scale this by area. Alternatively, $F_c(\Phi)$ could be evaluated at node points and then interpolated to the control volume face. The simplest expedient, for the cell centred scheme, would be to have the flux at the control volume face, $I - 1/2$, expressed as

$$(F_c)_{I-1/2} = \frac{1}{2}(F_c(\Phi)_{I-1} + F_c(\Phi)_I) \tag{4.63}$$

Then, to complete the integration, the following must be evaluated

$$(F_c\Delta L)_{I-1/2,J} = \frac{1}{2}(F_c(\Phi)_{I-1,J} + F_c(\Phi)_{I,J})\Delta L_{I-1/2,J} \tag{4.64}$$

Note that the uppercase subscripts indicate control volume nodal values are being used. Hence, for mass flux, $F_c(\Phi)$ is the product of the normal velocity at the face and the face density. The area multiplication then gives the mass flow, or mass flow per unit depth in two dimensions. For momentum, the mass flux is multiplied by the velocity component associated with the component of momentum being considered.

For the energy equation, typically the mass flux would be multiplied by enthalpy and then face area to gain an enthalpy flow. For the cell vertex scheme, with the dual mesh, again, Equations (4.63) and (4.64) are directly applicable. For the cell vertex scheme, with the overlapping control volumes, the interpolation has the more distinctive form

$$(F_c\Delta L)_{i,j} = \frac{1}{2}(F_c(\Phi)_{i,j} + F_c(\Phi)_{i,j+1})\Delta L_{i,j} \tag{4.65}$$

Note that the lower case subscripts indicate that the cell vertex values are being used. These are directly available. This interpolation relates to a west face, i, j, corresponding to the node at the lower left-hand corner of the face. The interpolation at first appears convenient but substantially limits the implementation of more advanced cell face interpolation techniques. It needs to be reiterated here that with the cell vertex scheme, with overlapping control volumes, cell centred data is summed and stored at cell vertices. For a non-uniform grid, ideally the interpolation should be volume based. Alternatively (see Section 4.4.3), it could have some form of upwind biasing. Unfortunately, the most obvious volume weighting gives rise to instabilities. Hence generally, an equal weighting is given to surrounding node values. For example, in the approach as classically used, the residual is summed as

$$A_{i,j} = A_{I-1,J} + A_{I+1,J} + A_{I,J-1} + A_{I,J+1} \tag{4.66}$$

Then the following is solved

$$\frac{\Delta\Phi}{\Delta t} = -\frac{1}{V}A_{ij}(\Phi) \tag{4.67}$$

to update the solution variable Φ (or rather the vector of conserved variables: $\rho, \rho u, \rho v, \rho w, \rho E$). Note that a critical problem, evident from deeper analysis of Equation (4.66), and noted earlier, is that the fluxes at the internal faces cancel. Hence, essentially the solution is being made on a super-cell of the four (in two dimensions) surrounding cells. Hence, the volume in Equation (4.67) needs to be the sum of these four cells. For steady flows, this precise volume is not that important. An alternative strategy (see Denton 1992) is to apply the update Equation (4.67) to the smaller sub-cells to get an update for Φ. Then the following equation can be applied

$$\Phi_{i,j} = \left[\frac{\Phi_{I-1,J} + \Phi_{I+1,J} + \Phi_{I,J-1} + \Phi_{I,J+1}}{4}\right] \tag{4.68}$$

With this approach the volume in Equation (4.67) would be for the smaller sub volume.

Clearly, for a structured, quadrilateral grid, the flux summation to form the total residual, $A(\Phi)$, is relatively straightforward. For an unstructured procedure it is not so simple. Clearly, the summation is still over the control volume faces. Hence pointers are needed to identify the cells that share the same face. Then the code would essentially look like

Do $m = 1, N$
$I1 =$ pointer to cell 1
$I2 =$ pointer to cell 2

$$(F\Delta L)_{I1,I2} = \frac{1}{2}(F(\Phi)_{I1} + F(\Phi)_{I2})\Delta L_{I1,i2} + \text{Smoother}$$

$$A(\Phi)_{I1} = A(\Phi)_{I1} + (F\Delta L)_{I1,I2}$$
$$A(\Phi)_{I1} = A(\Phi)_{I1} - (F\Delta L)_{I1,I2}$$

End do

The key point here is the change of sign. A flux that contributes to one cell will be a deficit to the adjacent cell. The procedure is virtually the same for the median dual cell vertex method, except that the summation is based around edge lines rather than cell faces.

4.4.1 *Numerical Smoothing*

If the flow speed is such that the local cell Reynolds number, or more generally, the Peclet number ($\rho u\phi/\Gamma$, where Γ is a diffusion coefficient and u is cell velocity) is around unity, then the interpolation is straightforward. Second order accuracy can be gained by taking half the surrounding (straddling) node values, as in the Equations (4.6.3)–(4.6.5). The difficulty occurs at high Peclet numbers. Then, the numerical scheme will become unstable and also exhibit dispersion. To overcome this, the grid needs to be refined or some form of increased smoothing added, thus increasing the effective value of Γ. Clearly, this is undesirable and hence needs to be done judiciously.

A crude way of increasing smoothing would be to evaluate the fluxes with Equations (4.64) or (4.65). Then, smoothing could subsequently be applied. Simple smoothing can be achieved by taking a weighted average of surrounding node values. For a uniform grid, a smoothing operator could be defined based on the Laplacian given as follows (Note that L is now used, temporarily, to represent the Laplacian and not edge length).

$$L(\Phi) = \frac{1}{4}\begin{bmatrix} & 1 & \\ 1 & -4 & 1 \\ & 1 & \end{bmatrix}\Phi \tag{4.69}$$

This gives, for $L\,(\Phi) = 0$,

$$\tilde{\Phi}_{i,j} = \varepsilon_{\Phi}\left[\frac{\Phi_{i-1,j} + \Phi_{i+1,j} + \Phi_{i,j-1} + \Phi_{i,j+1}}{4}\right] \tag{4.70}$$

where $\tilde{\Phi}_{i,j}$ is the smoothed value and $\varepsilon_{\Phi} = 1$. Alternatively, $\varepsilon_{\Phi} \approx \Gamma_{\Phi}$ to represent some form of artificial diffusion scaling parameter. As just noted, a crude way of applying the smoothing is to apply it as a post processing stage (Denton 1992). Then the following equation can be used

$$\bar{\Phi}_{i,j}^{new} = (1 - \varepsilon_{\Phi})\Phi + \varepsilon_{\Phi}\tilde{\Phi}_{i,j}^{old} \tag{4.71}$$

where $\bar{\Phi}_{i,j}^{new}$ is the final smoothed value and $\Phi_{i,j}^{old}$ is the unsmoothed value. Note that for an Euler solution, smoothed values at walls can be needed. Then the following equation can be used, where *i,j* corresponds to a lower wall node.

$$\tilde{\Phi}_{i,j} = \varepsilon_{\phi}\left[\frac{1}{3}\Phi_{i-1,j} + \frac{1}{3}\Phi_{i+1,j} + \frac{2}{3}\Phi_{i,j+1} - \frac{1}{3}\Phi_{i,j+2}\right] \tag{4.72}$$

Equation (4.72) relates to a linear extrapolation to the wall from two interior points. This value is then averaged with the average of the wall values adjacent to the node under consideration.

Use of the discrete equivalent of a Laplacian would reduce the effective cell Peclet number. Indeed, more sophisticated variants of Equation (4.70), with judiciously selected cell weighting functions (i.e., not just ¼) are essentially used in some modern ILES methods (see Chapter 5).

The more usual procedure would be to explicitly include this smoothing element in the flux interpolation. It is also more usual to have a smoothing of higher order. This ensures that the numerical error, from the smoothing, vanishes quickly with grid refinement. Hence, the scheme is of higher order. The flux terms, at a control volume face, F_{CVF}, would generally take the general form

$$F_{CVF} = \frac{1}{2}(F(\Phi)_{I2} + F(\Phi)_{I1}) + \varepsilon_{\Phi}[\hat{L}(\Phi)_{I2} - \hat{L}(\Phi)_{I1}] \tag{4.73}$$

where $I1$ and $I2$ are the nodes adjacent to the control volume being considered. Also, the hat is used to represent an undivided Laplacian. The key thing to note is that since the Laplacian is included in the flux integration around the control volume. Hence, ultimately, the process will give a fourth order smoother. Around shocks, fourth order smoothing is not suitable. It will induce Gibbs phenomena. In such zones, a globally second order smoother is needed (i.e., a lower order scheme component). Dealing with shocks is discussed later, as are more general forms of ε_{Φ}. The Laplacian Equation (4.70) is only exact on a uniform Cartesian grid. However, ranges of functions, some quite arbitrary, scale the Laplacian term utilized in the smoother. Hence, there is no need for the Laplacian to be that exact, especially if an approximate Laplacian that saves computational cost can be used (see Moinier 1999, Holmes et al. 1986 and Frink 1994).

4.4.1.1 Pseudo Laplacian Following Moinier (1999), the pseudo-Laplacian can be formulated in the following simple form based on Green's theorem, where the L values with no hat correspond to edge lengths or areas.

$$\hat{L}(\Phi)_{I1} = \frac{1}{N}\sum_{m=1}^{N}\frac{1}{2}(\Phi_{I2} + \Phi_{I1})\mathbf{n}_{I1,I2}\Delta L_{I1,I2} \tag{4.74}$$

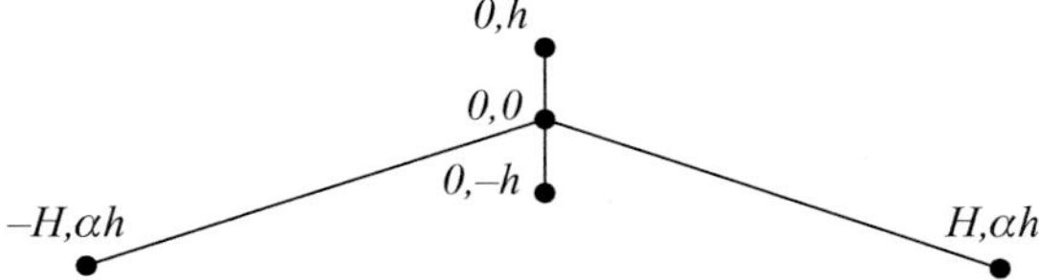

Figure 4.17. Difficulty for highly stretched grids when using approximate Laplacians (from Moinier 1999).

This equation (which must also include where necessary boundary contribution terms) has been used with success for structured type grids and can work for unstructured. However, it is obviously not guaranteed to give a zero pseudo-Laplacian for a linear variable distribution. Hence, the following modification is helpful for (second order) accuracy on non-smooth grids

$$\hat{L}_{I1}^{lp}(\Phi) = \hat{L}_{I1}(\Phi) - \nabla\Phi_{I1} \cdot \hat{L}_{I1}(x) \tag{4.75}$$

where $\hat{L}_{I1}(x) = (\hat{L}_{I1}x, \hat{L}_{I1}y, \hat{L}_{I1}z)^T$. The final right-hand side group of terms, for a linear function, is designed to cancel the function (4.74), provided $\nabla\Phi_{I1}$ is exact for the linear function. The following approximation for $\nabla\Phi_{I1}$ can be used

$$\nabla\Phi_{I1} = \sum \frac{1}{2}(\Phi_{I2} + \Phi_{I1})\mathbf{n}_{I2I1}\Delta L_{I2I1} = \sum \frac{1}{2}(\Phi_{I2} - \Phi_{I1})\mathbf{n}_{I2I1}\Delta L_{I2I1} \tag{4.76}$$

However, for highly stretched grids, even the linear preserving option is not adequate – for example, as pointed out by Moinier, around the leading edge of an airfoil $H/h \sim 1000$ and $\alpha \sim 100$ where the nomenclature can be identified in Figure 4.17.

The finite difference approximation to $\nabla\Phi_{I1}$ is given as

$$\Phi_x = \begin{bmatrix} & 0 & \\ -\frac{1}{2H} & 0 & \frac{1}{2H} \\ & 0 & \end{bmatrix}\Phi, \quad \Phi_y = \begin{bmatrix} & \frac{1}{2h} & \\ 0 & 0 & 0 \\ & -\frac{1}{2h} & \end{bmatrix}\Phi \tag{4.77}$$

The pseudo Laplacian, L, can be represented using Equation (4.69). Since $L(x) = (o, 2\alpha h)$,

$$\nabla\Phi \cdot L(x) = \begin{bmatrix} & \alpha & \\ 0 & 0 & 0 \\ & -\alpha & \end{bmatrix}\Phi \tag{4.78}$$

The problem is that since $\alpha \sim 100$, the magnitudes of the Equation (4.78) terms can differ substantially from those in Equation (4.69). Hence, Equation (4.75) becomes ill behaved. There is loss of diagonal dominance. This can result in a loss of numerical stability. To circumvent this problem, Moinier propose the following anisotropic scaling in the pseudo-Laplacian

$$\widehat{L}_{I2}(\Phi) = \left(\sum_{m=1}^{N} \frac{1}{|\mathbf{x}_{I1} - \mathbf{x}_{I2}|}\right)^{-1} \sum_{m=1}^{N} \frac{(\Phi_{I1} - \Phi_{I2})}{|\mathbf{x}_{I1} - \mathbf{x}_{I2}|} \tag{4.79}$$

Then a repeat of the analysis (see Moinier) gives

$$\widehat{L}(\Phi) = \frac{Hh}{2(H+h)} \begin{bmatrix} & \frac{1}{h} & \\ \frac{1}{H} & -2\frac{H+h}{Hh} & \frac{1}{H} \\ & \frac{1}{h} & \end{bmatrix} \Phi \tag{4.80}$$

with

$$\widehat{L}\,(\mathbf{x}) = \frac{Hh}{2\,(H+h)} \begin{pmatrix} 0 \\ \frac{2\alpha h}{H} \end{pmatrix} \tag{4.81}$$

and

$$\nabla\Phi \cdot \widehat{L}(\mathbf{x}) = \frac{Hh}{2\,(H+h)} \begin{bmatrix} & \frac{\alpha}{H} & \\ 0 & 0 & 0 \\ & \frac{-\alpha}{H} & \end{bmatrix} \Phi \tag{4.82}$$

The instability only happens for $|\alpha|\, h > H$, which is a much less severe restriction. Figure 4.18, taken from Moinier (1999), shows the impact of using the linear preserving Laplacian for an RAE2822 airfoil. Frame (a) gives the mesh, Frame (b) gives Mach number contours for basic Laplacian and Frame (c) gives Mach number contours for the linear preserving Laplacian. The latter clearly has less dispersion error in the solution contours.

Holmes and Connell (1986) and Frink (1994)) use the more sophisticated, distance weighted form of pseudo-Laplacian:

$$\hat{L}(\Phi_{I1}) = \sum_{m=1}^{N} Wf_{I1,I2}(\Phi_{I2} - \Phi_{I1}) \tag{4.83}$$

where again, N is the number of adjacent nodes/control volumes and $I1$ and $I2$ correspond to the nodes that straddle a control volume face. The weighting function, Wf, takes the form

$$Wf_{I1,I2} = 1 + \Delta Wf_{I1,I2} \tag{4.84}$$

The weighting functions are based on solving an optimization problem using Lagrange multipliers (see Frink 1994), where

$$\Delta Wf_{I1,I2} = \lambda_{x,I1}(x_{I2} - x_{I1}) + \lambda_{y,I1}(y_{I2} - y_{I1}) + \lambda_{z,I1}(z_{I2} - z_{I1}) \tag{4.85}$$

The concept is that it is best if the weighting functions are as close to unity as possible. The optimization hence seeks to minimize $\Delta Wf_{I1,\,I2}$ such that for a linear distribution the pseudo-Laplacian is zero. In Equation (4.8.5) x, y and z are nodes and the Lagrange multipliers are

$$\begin{aligned} \lambda_x &= \frac{R_x C_{11} + R_y C_{12} + R_z C_{13}}{b} \\ \lambda_y &= \frac{R_x C_{21} + R_y C_{22} + R_z C_{23}}{b} \\ \lambda_z &= \frac{R_x C_{31} + R_y C_{32} + R_z C_{33}}{b} \end{aligned} \tag{4.86}$$

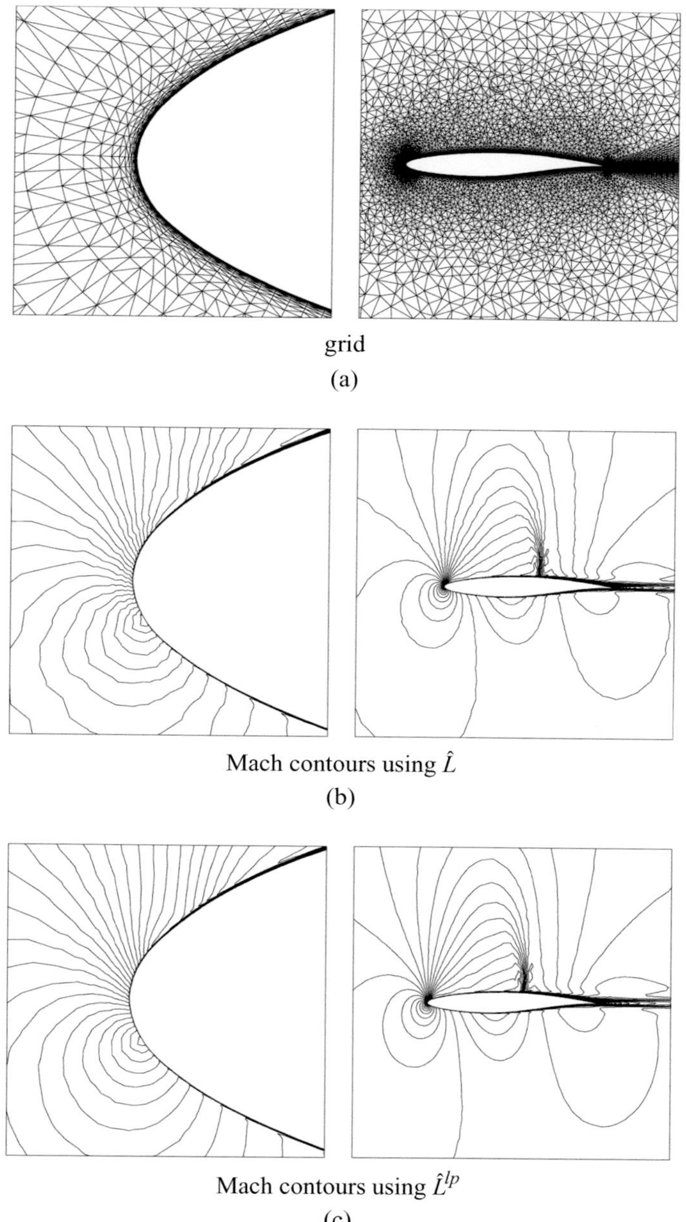

Figure 4.18. Impact of using the linear preserving Laplacian for RAE2822 airfoil: (a) mesh, (b) Mach number contours for basic Laplacian and (c) Mach number contours for linear preserving Laplacian (from Moinier 1999).

Also,

$$b = I_{xx}I_{yy}I_{zz} - I_{xx}I_{yz}^2 - I_{yy}I_{xz}^2 - I_{zz}I_{xy}^2 + 2I_{xy}I_{yz}I_{yz} \tag{4.87}$$

$$\begin{aligned}
C_{11} &= I_{yy}I_{zz} - I_{yz}^2 \\
C_{12} &= I_{xz}I_{yz} - I_{xy}I_{zz} \\
C_{13} &= I_{xy}I_{yz} - I_{xz}I_{yy} \\
C_{21} &= I_{xz}I_{yz} - I_{xy}I_{zz} \\
C_{22} &= I_{xx}I_{zz} - I_{xz}^2 \\
C_{23} &= I_{xy}I_{xz} - I_{xx}I_{yz} \\
C_{31} &= I_{xy}I_{yz} - I_{xz}I_{yy} \\
C_{32} &= I_{xz}I_{xy} - I_{xx}I_{yz} \\
C_{33} &= I_{xx}I_{yy} - I_{xy}^2
\end{aligned} \tag{4.88}$$

The first order moments can be expressed as

$$\begin{aligned}
R_{x,I1} &= \sum_{I2=1}^{N} (x_{I2} - x_{I1}) \\
R_{y,I1} &= \sum_{I2=1}^{N} (y_{I2} - y_{I1}) \\
R_{z,I1} &= \sum_{I2=1}^{N} (z_{I2} - z_{I1})
\end{aligned} \tag{4.89}$$

The second order moments are

$$\begin{aligned}
I_{xx,I1} &= \sum_{I2=1}^{N} (x_{I2} - x_{I1})^2 \\
I_{yy,I1} &= \sum_{I2=1}^{N} (y_{I2} - y_{I1})^2 \\
I_{zz,I1} &= \sum_{I2=1}^{N} (z_{I2} - z_{I1})^2 \\
I_{xy,I1} &= \sum_{I2=1}^{N} (x_{I2} - x_{I1})(y_{I2} - y_{I1}) \\
I_{xz,I1} &= \sum_{I2=1}^{N} (x_{I2} - x_{I1})(z_{I2} - z_{I1}) \\
I_{yz,I1} &= \sum_{I2=1}^{N} (y_{I2} - y_{I1})(z_{I2} - z_{I1})
\end{aligned} \tag{4.90}$$

For a linear function, on any grid, Equations (4.83)–(4.90) gives the Laplacian as equal to zero. For poor quality grids, it can yield a negative, anti-diffusive Laplacian. This will be seriously problematic for stability. Hence, clipping can be needed.

Next there is the question of the form of the parameter ε_Φ in Equation (4.73). This could be a constant but this is too primitive. Hence, it is typically some scaled (using ε_A) value of the spectral radius of the convective flux Jacobian $\Lambda_{I1,I2}$. Therefore, the following equation is used

$$\varepsilon_\Phi = \varepsilon_A \Lambda_{I1,I2} \tag{4.91}$$

where

$$\Lambda = \frac{1}{N}\sum_{m=1}^{N}(|\mathbf{U}_m| - \mathbf{c}_m)\mathbf{\Delta L}_m \tag{4.92}$$

and $|\mathbf{U}_m|$ is the velocity in the direction normal to the control volume face, while c is the speed of sound. The actual face value is needed and this is averaged as

$$\Lambda_{I1,I2} = \frac{\Lambda_{I1} + \Lambda_{I2}}{2} \tag{4.93}$$

4.4.2 *Modelling Shocks*

Returning to Equation (4.73), around shocks, a lower order smoothing is required. Hence, typically the flux takes the very broad form

$$F_{CVF} = \frac{1}{2}(F(\Phi)_{I2} + F(\Phi)_{I1}) + \varepsilon_\Phi\big[(1-\Psi)\big[\hat{L}(\Phi)_{I2} - \hat{L}(\Phi)_{I1}\big] - \Psi\big[\Phi_{I2} - \Phi_{I1}\big]\big] \tag{4.94}$$

The key term is Ψ, which is the shock switch. The idea is that this is zero away from shocks and unity in the shock zone. A typical form is

$$\Psi = \min\left(\varepsilon\left|\frac{p_j - p_i}{p_j + p_i}\right|^2, 1\right) \tag{4.95}$$

(see Moinier 1999). This equation does not account for edge lengths and this information can also be included. The parameter ε is a constant. Typically, the form of (4.94) is more complex than shown including 'max' functions and also distance weightings to ensure optimal numerical behaviour. Notably, if a standard compressible flow solver is used for an eddy-resolving simulation, the shock smoother will be very active. The small-scale flow components will be interpreted as shocks. However, the second order Laplacian that the shock modelling includes is not an unwelcome term for eddy-resolving simulations. As will be seen later, most turbulence models (including LES models) increase the coefficients, multiplying the diffusion terms in the governing equations. Hence, any added bias towards including more second order diffusion is more welcome. The fourth order smoother can be rather damaging for LES. Limiters can be used that are specifically designed to ensure the correct blend of the fourth and second order smoothers are used to ensure good performance when making eddy-resolving simulations. The use of such limiters relates to the area of ILES and Monotone Integrated Large Eddy Simulation (MILES).

For classical LES, the presence of any form of smoothing is unwelcome and ideally the shock smoothing component should just be active where there is a shock. The difficulty is that for LES, as noted previously, the oscillations at the small scale, generated by the presence of eddies, can erroneously trigger the shock switch. Hence, special shock switches have been designed. The Ducros et al. (1999) switch takes the form

$$\Psi = \frac{(\nabla \cdot \mathbf{u})^2}{(\nabla \cdot \mathbf{u})^2 + (\omega)^2 + \varepsilon} \tag{4.96}$$

where ε is a small number to prevent division through by zero, and ω is the resolved vorticity ($\nabla \times \mathbf{u}$). Bisek et al. (2013), when making compact scheme LES, use the Swanson and Turkel (1992) pressure gradient detector

$$\Psi = \frac{|p_{i+1} - 2p_i + p_{i-1}|}{(1-\omega)(|p_{i+1} - p_{i-1}|) + \omega(p_{i+1} + 2p_i + p_{i-1})}, \begin{cases} \Psi > 0.05, & \text{Roe scheme} \\ \Psi \leq 0.05, & \text{centered scheme} \end{cases} \tag{4.97}$$

which is used to switch between a diffusive upwind (Roe 1981; see Section 4.4.3.1) scheme and a centred scheme.

4.4.3 *Interpolation, its Limiting, Biasing and Control*

When implementing the Section 4.4.1 procedure of using a smoother in the flux term, in actual coding terms the flux would be discretized as a simple averaging (analogous to central differencing). Then Equation (4.44) would be modified to include the smoothing as source terms. Hence, Equation (4.44) would look as

$$\frac{\partial \Phi}{\partial t} = A(\Phi) \approx -\frac{1}{V_{i,j,k}} \left[\sum_{m=1}^{N} (\mathbf{F_c} - \mathbf{F_v})_m \Delta L_m - \mathbf{S}_{i,j,k} V_{i,j,k} + f_1(\Phi)\tilde{\nabla}^2 \Phi + f_2(\Phi)\tilde{\nabla}^4 \Phi \right] \tag{4.98}$$

where the tildes in $f_1(\Phi)\tilde{\nabla}^2\Phi$ (for shocks) and $f_2(\Phi)\tilde{\nabla}^4\Phi$ (away from shocks) indicate that these are approximate discrete operators.

This procedure involves a centred interpolation with the addition of smoothing terms. Instead, interpolations can be used that are biased, based on the information flow directions. Such procedures are termed upwinding methods. The basic idea is that when considering the transport of a variable ϕ, the upwind value of ϕ will have more predominance at the control volume face than the value downwind. Acoustic wave information (pressure) can also be biased in this way. The idea of upwinding can also be extended so that the upwind value is biased based on values taken upstream, along a streamline (see Hassan et al. 1983).

The use of streamlines brings in a Lagrangian element and there is a class of upwinding that is described as Lagrangian-Eulerian (see Staniforth and Cote 1991). However, the standard forms of upwinding are more grid point based. Most upwind schemes just involve lines that straddle the control volume face and project along lines that connect these grid points. However, skewed upwind schemes that involve other grid points have been formulated. Then, the next factor is how many grid points upstream and downstream of the control volume face are used, the level of biasing and the information used in the biasing. With a structured solver it is easy to use a large extent of grid points either upstream or downstream of the control volume face. However, when using domain decomposition for parallel processing, this needs a larger amount of data transfer and hence will slow the solution process.

As with the field of general interpolation, the range of potential interpolation functions is substantial and can even include the use of splines (Nasser and Leschziner 1985), Hermitian interpolation (Glass and Rodi 1982, Tang and Baeder 1998), analytical flow solutions (see Spalding 1972 and Chen et al. 1981)), biased quadratic interpolations (Leonard 1979) (involving a quadratic fit to two upstream and one downstream point) and spectral representations at the element level (see Patera 1984). There is a very substantial body of literature on the convective term discretization. There are also two distinct bodies. These relate to the field of compressible flow and incompressible flow.

4.4.3.1 Compressible Flow Procedures With regards to compressible flow classically, one needs to consider the propagation of acoustic, vortical and entropy waves. The latter are not actual waves but correspond to the convection of hot spots in the flow. The 'waves' noted have different speeds and directions. The vortical and entropy 'waves' convect with the flow mean speed. The acoustic waves speeds correspond to the sum of the sound and flow speeds. They have a right running component of $u + c$ and a left running $u - c$ for a left to right flow. More advanced upwind related convective schemes attempt to account for the different directions of these information flows. For incompressible flow solvers generally the focus is simpler, the speed of sound is infinite and the vortical mode convection (and perhaps entropy) is the major focus.

A key distinction with control volume interpolations is whether the flux is interpolated to the control volume face or the variables within the flux. With compressible flow, to strongly enforce momentum conservation (note the strong conservation form of the governing equations in Chapter 2) normally the flux is interpolated. However, the kinetic energy conserving scheme of Jameson (2008a, 2008b) interpolates the variables. This means that momentum conservation is lost but that kinetic energy is conserved. However, for eddy-resolving simulations the latter is paramount.

A further distinction to make is if the interpolation focuses on making a single interpolation to a control volume face from grid points that straddle the face. A popular alternative to this is to make two interpolations to the face based on information to the left and right of the face. Then a subsequent centred interpolation at the face is made, based on this data. Hence, in Equation (4.73) rather than use data at nodes $I1$ and $I2$, left and right state data could be used instead. This gives greater stability when Equation (4.94) is used with a Runge-Kutta temporal integration. The simplest, just first order, approximation is

$$(\mathbf{\Phi}_L)_{i+\frac{1}{2}} = \mathbf{\Phi}_i$$
$$(\mathbf{\Phi}_R)_{i+\frac{1}{2}} = \mathbf{\Phi}_{i+1} \tag{4.99}$$

The following are well known classes of upwind scheme for compressible flow: flux-difference splitting, flux-vector splitting and total variation diminishing. Flux-difference splitting is by far the most popular. There are a range of flux-vector splitting schemes dating back to the 1980s. The most popular is perhaps the AUSM (Advection Upstream Splitting Method) of Liou and Steffen (1993). These have found some success when making eddy-resolving simulations in high-speed flows. With AUSM, the flux vector is decomposed into a convective and acoustic (pressure) part. These are then treated separately using upwind schemes. A key advantage of flux-vector splitting schemes is that, unlike the flux-difference splitting, they can be readily extended to real gas flows. The basis for the flux-difference splitting scheme is the work of Godunov (1959) who solved the Riemann (shock tube) problem for discontinuous states at the interface between two control volumes with left and right states. The Roe (1981) flux-difference splitting scheme is a cheaper, approximate solution to the exact Riemann problem which involves the one-dimensional Euler equations. Using flux-difference splitting, Equation (4.94) can be made into a type of upwind scheme. To do this, ε_Φ is replaced by the so-called Roe matrix, a convective flux Jacobian:

$$\varepsilon_\Phi = \overline{\mathbf{A}}_{Roe} = \frac{\partial \mathbf{F_c}}{\partial \mathbf{\Phi}} \tag{4.100}$$

The Roe scheme decomposes the flux difference across the control volume face into a sum of wave components. Note that $(\mathbf{F}_c)_R - (\mathbf{F}_c)_L = (\overline{\mathbf{A}}_{Roe})_{CVF}(\Phi_R - \Phi_L)$.The Roe matrix is based on specially averaged variables (Roe averaged). The Roe matrix can be written as the product of a matrix of left ($\mathbf{T}$) and right ($\mathbf{T}^{-1}$) eigenvectors and a diagonal matrix of Eigenvalues (Λ) $\overline{\mathbf{A}}_{Roe} = \mathbf{T}\Lambda\mathbf{T}^{-1}$. This allows the following to be written, $-(\mathbf{F}_c)_R - (\mathbf{F}_c)_L = (\overline{\mathbf{A}}_{Roe})_{CVF}(\Phi_R - \Phi_L) = \mathbf{T}\Lambda(\mathbf{Am}_R - \mathbf{Am}_L)$, where $\mathbf{Am}$ are wave amplitudes and Λ are wave speeds. Indeed

$$\Lambda = \begin{pmatrix} u \\ u \\ u \\ u+c \\ u-c \end{pmatrix} \tag{4.101}$$

where u is the fluid velocity and c is the speed of sound. A key difficulty with this is that when $u \to 0$ there can be a large disparity in eigenvalues. Indeed, the eigenvalue multiplying pressure can become large and the damping of pressure variations can damp velocity variations and hence give highly dissipated results. This will be shown later. Also, at low speeds extreme numerical stiffness can occur.

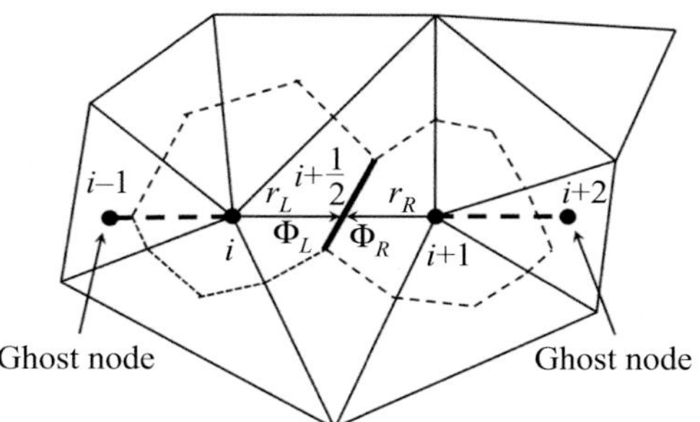

Figure 4.19. Application of MUSCL interpolation on unstructured grids through use of ghost nodes.

Total Variation Diminishing (TVD) basis can be appreciated by considering for a solution, over a domain of length L, a variable called the *total variation*, $TV = \int_0^L |\partial\phi/\partial x| dx$. Ideally, TV should not increase with time. Based on this correct physical behaviour, TVD schemes are constructed so that

$$\int_0^L \left|\frac{\partial \phi^{n+1}}{\partial x}\right| dx \leq \int_0^L \left|\frac{\partial \phi^{n}}{\partial x}\right| dx \qquad (4.102)$$

where n represents different time levels. Enforcing the TVD constraint (introduced by Harten 1983) removes erroneous solution oscillations. TVD schemes are monotonicity preserving. This means that they do not produce any new extrema in a domain of L. Also, all local minima and maxima do not decrease or increase, respectively.

An interpolation of considerable popularity is the MUSCL (Monotone Upstream Centred Scheme for Conservation Laws) interpolation of Van Leer (1977, 1979). With this, Equation (4.99) is extended beyond first order as

$$\begin{aligned}(\mathbf{\Phi}_L)_{i+\frac{1}{2}} &= \mathbf{\Phi}_i + \tfrac{1}{4}[(1-\kappa)\Delta_- + (1+\kappa)\Delta_+]_i \\ (\mathbf{\Phi}_R)_{i+\frac{1}{2}} &= \mathbf{\Phi}_{i+1} + \tfrac{1}{4}[(1-\kappa)\Delta_+ + (1+\kappa)\Delta_-]_{i+1}\end{aligned} \qquad (4.103)$$

where

$$\begin{aligned}\Delta_+ &\equiv \mathbf{\Phi}_{i+1} - \mathbf{\Phi}_i \\ \Delta_- &\equiv \mathbf{\Phi}_i - \mathbf{\Phi}_{i-1}\end{aligned} \qquad (4.104)$$

Equations (4.103) and (4.104), for different k values, gives a family of second order schemes. For example, $\kappa = 1$ gives the second order central difference scheme. On the other hand, $\kappa = -1$ gives a second order upwind scheme. Also, $\kappa = -1/3$ gives a third order upwind biased interpolation.

This MUSCL interpolation can also be applied with an unstructured solver. Figure 4.19, keeping with the above, structured grid nomenclature, shows how this is done. The line connecting the nodes that straddle the control volume face is projected to ghost nodes. Data at the nodes that surround the ghost node are interpolated to the ghost nodes. This allows the usual procedure to be applied. The process is applicable to both cell centred median dual and cell vertex solvers. It can be problematic at boundaries, since the ghost cell is outside the domain. The approach also needs extra bookkeeping and storage. It can be necessary to place limiters on the interpolation expressions. This is especially so for flows with shocks.

4.4.3.2 Piecewise Reconstructions Based on the Taylor series, both linear (Barth and Jespersen 1989) and quadratic reconstructions are possible of the following form

$$\begin{aligned} \Phi_L &= \Phi_i + \phi_i[\nabla\Phi \cdot \mathbf{r}_L] + HOT_L \\ \Phi_R &= \Phi_{i+1} + \phi_{i+1}[\nabla\Phi \cdot \mathbf{r}_R] + HOT_R \end{aligned} \tag{4.105}$$

where the node to control volume face distance vectors are indicated in Figure 4.19. The ϕ terms are here used to represent any potential flux limiter implemented. These are discussed later. Also, $\nabla\Phi = [\partial\Phi/\partial x,\ \partial\Phi/\partial y,\ \partial\Phi/\partial z]^T$. For a quadratic reconstruction, the first higher order neglected term in the above, $HOT_{\mathrm{L/R}}$ (also needs multiplying by a flux limiter construct), can be expressed as

$$\begin{aligned} HOT_L &= \frac{\mathbf{r}_L^T \mathbf{H}_I \mathbf{r}_L}{2} \\ HOT_R &= \frac{\mathbf{r}_R^T \mathbf{H}_{I+1} \mathbf{r}_R}{2} \end{aligned} \tag{4.106}$$

(see Delanaye and Essers 1997). The Hessian matrix is as

$$\mathbf{H}_I = \begin{bmatrix} \partial^2_{xx}\Phi & \partial^2_{xy}\Phi & \partial^2_{xz}\Phi \\ \partial^2_{xy}\Phi & \partial^2_{yy}\Phi & \partial^2_{yz}\Phi \\ \partial^2_{xz}\Phi & \partial^2_{yz}\Phi & \partial^2_{zz}\Phi \end{bmatrix}_I \tag{4.107}$$

where $\partial^2_{ij}\Phi = \partial^2\Phi/\partial ij$. These gradients need to be evaluated through some means. The Gauss-Green theorem, along with other methods, can be used to facilitate this. Note that the gradients in the Hessian can be evaluated to first order accuracy whilst still giving a second to third order accuracy depending on the grid non-uniformity level. Linear reconstruction on triangular and tetrahedral cells is relatively simple, since the gradient information does not need to be explicitly calculated (see Frink et al. 1991).

4.4.3.3 Flux Limiters In highly convecting flows with sharp velocity gradients, higher order spatial interpolation schemes can result in local oscillations. Therefore, these spatial schemes are sometimes bounded in some way using flux limiters or schemes like the TVD. A range of requirements can be posed which are not outlined here. However, the least is that maxima do not increase and minima do not decrease and that no new local extrema are produced as the solution progresses. Such a scheme that meets these requirements is monotonicity preserving.

Table 4.3. *A sample of some potential flux limiters (from Watson 2014)*

Flux Limiter	$\phi(r)$
Minmod	$\max(0, \min(1, r))$
Ospre	$\frac{1.5(r^2+r)}{r^2+r+1}$
Superbee	$\max(0, \min(2r, 1), \min(r, 2))$
Drikakis's Third Order	$1 - \left(1 + \frac{2r}{1+r^2}\right)\left(1 - \frac{2r}{1+r^2}\right)$
UMIST	$\max(0, \min(2r, 0.25 + 0.75r, 0.75 + 0.25r, 2))$
Van Leer (1974)	$\frac{r+\lvert r\rvert}{1+\lvert r\rvert}$

The flux limiter can be incorporated in different ways. For example, looking back at Equation (4.94), the total flux

$$F_{CVF} = \frac{1}{2}(F(\Phi)_{I2} + F(\Phi)_{I1}) + \varepsilon_{\Phi}[(1 - \Psi)F^H + \Psi F^L] \tag{4.108}$$

could be broken down and slightly modified as a low order component

$$F^L = \frac{1}{2}(F(\Phi)_{I2} + F(\Phi)_{I1}) + \varepsilon_{\Phi}\Psi[\Phi_{I2} - \Phi_{I1}] \tag{4.109}$$

and the high order component

$$F^H = \frac{1}{2}(F(\Phi)_{I2} + F(\Phi)_{I1}) + \varepsilon_{\Phi}[\hat{L}(\Phi)_{I2} - \hat{L}(\Phi)_{I1}] \tag{4.110}$$

The high order component uses, as would be expected, the larger stencil. Flux limiters can be expressed as a function of the local pressure gradient as

$$r = \frac{p_{i+1} - p_i}{p_i - p_{i-1}} = \frac{\nabla p_{i+\frac{1}{2}}}{\nabla p_{i-\frac{1}{2}}} \tag{4.111}$$

The parameter r is then used in the Table 4.3 functions.

There are a wide range of flux limiters. Those in the table include the minmod and Superbee (see Roe 1986), Ospre (see Waterson and Deconinck 1995), Third Order (see Zoltak and Drikakis 1998)), University of Manchester Institute of Science and Technology (UMIST) (see Lien and Leschziner 1994 and Van Leer 1974). Figure 4.20 plots the shapes of various flux limiters based on local pressure ratio, r (from Watson 2014).

The high order and low order fluxes are then combined as follows with the flux limiter

$$F_{CVF} = \phi(r)F^H + (1 - \phi(r))F^L \tag{4.112}$$

A key requirement is that in discontinuous/highly non-monotonic zones, $\phi(r) \to 0$. Hence, then, in Equation (4.112), the low order flux is used. Note that this flux component emphasizes the second order Laplacian component that is associated with turbulence modelling (see Chapter 5). Flux limiters can also be incorporated into the MUSCL scheme to limit the $\Delta^{+/-}$ variations and also in the other more advanced reconstructions above (see Blazek 2005). Again, an r parameter is needed but this is not just restricted to the use of pressure.

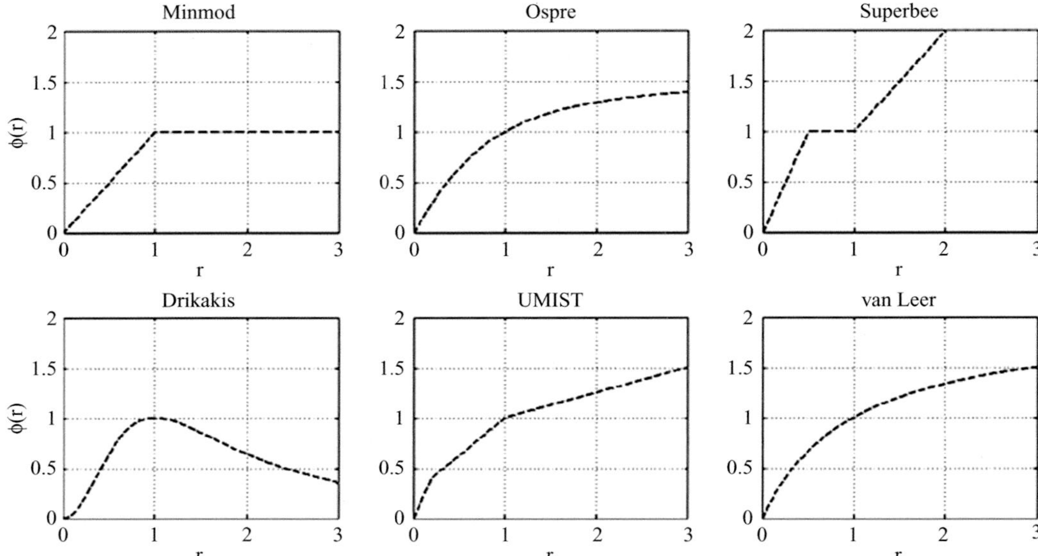

Figure 4.20. Shapes of various flux limiters based on local pressure ratio, r (from Watson 2014).

Figure 4.21 gives the predicted turbulence spectrum for homogenous, isotropic, decaying turbulence for a range of flux limiters when performing ILES using Equation (4.112) that blends the high and low order fluxes. The simulations are taken from Watson (2014). They are made using a compressible, unstructured flow solver, with hexahedral cells on a 64^3-node grid. As can be seen, the Drikakis limiter

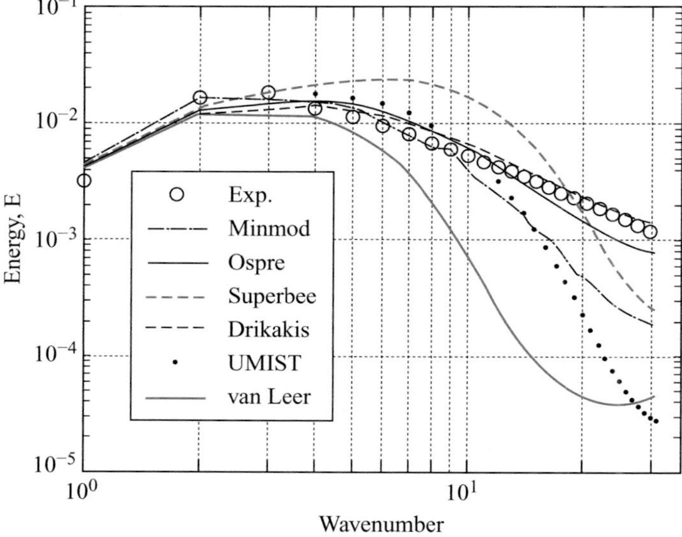

Figure 4.21. Predicted turbulence spectrum for homogenous, isotropic, decaying turbulence for a range of flux limiters when performing ILES (from Watson 2014).

(see Zoltak and Drikakis 1998), designed to replicate an LES model, acts as viable surrogate for an LES model. What is more, ILES will allow shocks to be dealt with and hence ILES has, rightly, found some popularity for simulation of compressible flows with shocks. A notable feature of limiters suited to eddy-resolving simulations is that they make the second order Laplacian smoothing contribution to the (modified) equations dominant. In the small-scale features that represent turbulence needing dissipation, the fourth order smoother becomes (through the multiple differentiation of sharp gradients) very substantial. Hence, it is an unwelcome component that has no connection with the physical modelling of turbulence.

4.4.3.4 Kinetic Energy Conservation Essentially, when discretizing the flow governing equations it is possible to ensure that momentum, kinetic energy or global total energy of the flow is conserved but not all three. The use of staggered grid variables, as discussed further later, ensures the conservation of kinetic energy. Particularly for eddy-resolving simulations, kinetic energy conservation is important. The use of staggered grids makes the program structure complex. Hence, it is preferable to avoid this, especially for unstructured flow solvers. As shown previously, for a classical unstructured scheme with a Roe matrix scaled smoother, the flux takes the form

$$F_{Roe} = \frac{1}{2}(\rho_{I1}u_{I1}\phi_{I1} + \rho_{I2}u_{I2}\phi_{I2}) - \frac{1}{2}\varepsilon|\bar{A}_{Roe}|[\hat{L}(\Phi)_{I2} - \hat{L}(\Phi)_{I1}] \tag{4.113}$$

Jameson (2008a, 2008b) proposed that for kinetic energy conservation, the first right-hand group of terms should be multipled out as

$$F_{KEP} = \frac{1}{2}(\rho_{I1} + \rho_{I2})\frac{1}{2}(u_{I1} + u_{I2})\frac{1}{2}(\phi_{I1} + \phi_{I2}) \tag{4.114}$$

Some form of smoothing could also be included. However, the kinetic energy conservation subtantially bounds and hence improves the solution stability. Therefore, considerably less smoothing can be used. However, a key problem is dealing with shocks in high-speed flows. Clearly, the kinetic energy scheme is not well suited to flows around shocks. A potential remedy to this is given where a blending is used between the classic KEP and Roe schemes:

$$F = \alpha F_{Roe} + (1 - \alpha)F_{KEP} \tag{4.115}$$

where α is a Mach number based blending parameter that, say, makes use of a hyperbolic tangent function as shown in Figure 4.22. Figure 4.23 shows the use of a Roe based scheme in the standard form to model homogenous decaying turbulence for different unstructured cell types. The cell types involved are hexahedra, prisms and tetrahedra. Roughly speaking, the prisms and tetrahedra are formed around the hexahedral grid's nodes. As is evident from the figure, both of the former have poor accuracy. However, the simple alteration to the way the basic control volume averaging is performed, to conserve kinetic energy, makes a dramatic difference. This can clearly be seen from Figure 4.24.

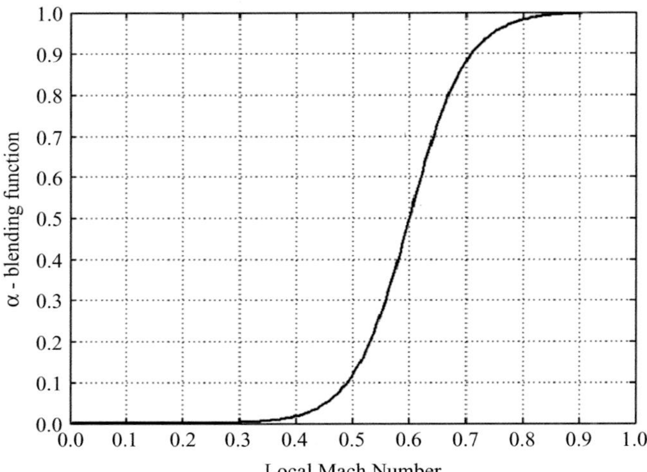

Figure 4.22. Variation of function to blend between kinetic energy conserving and standard Roe scheme with local Mach number.

4.4.3.5 Smoothing Controls for Eddy Resolving Simulations For eddy-resolving simulations the extent of the upwinding component or smoother appended to a central difference scheme can undergo a special form of control. Hence, in Equation (4.94), ε_{Φ} can be pre-multiplied by another parameter, $\varepsilon_{\Phi,1}$. The latter can be a static user-defined function. For example, Shur et al. (2003) define a target zone where it is

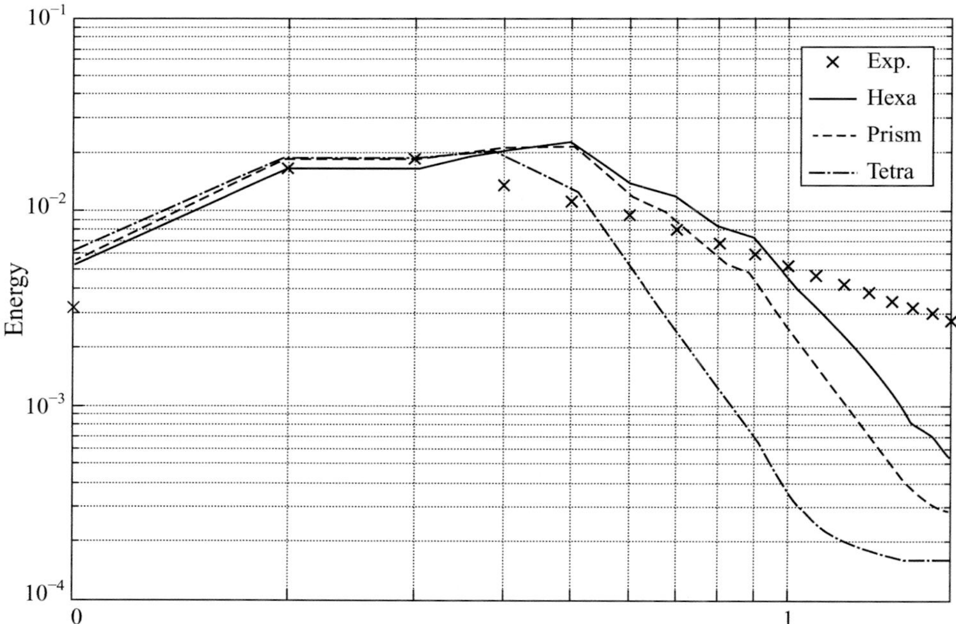

Figure 4.23. Turbulence kinetic energy spectrum for homogenous decaying turbulence with a more standard Roe based scheme for different grid topologies.

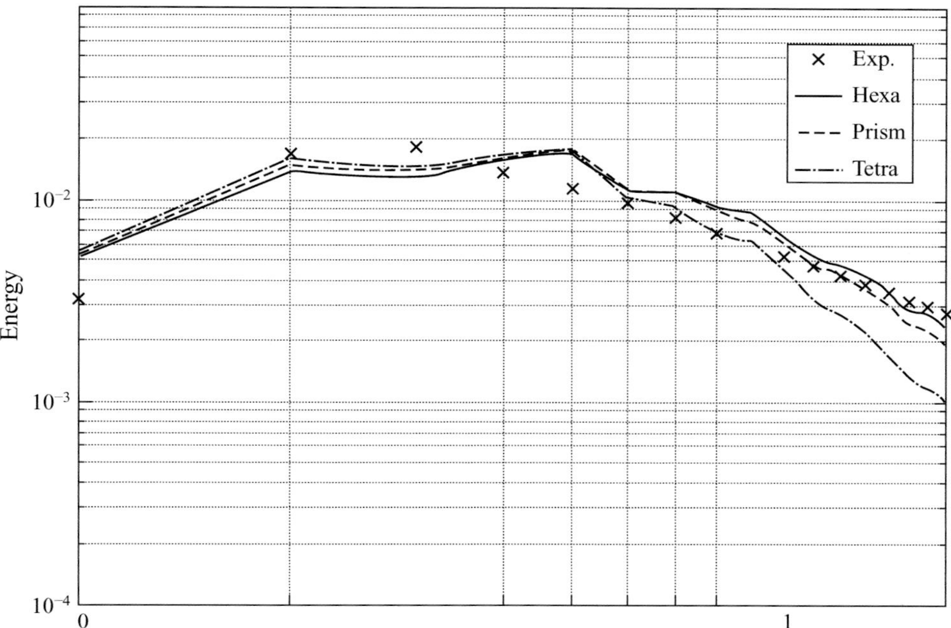

Figure 4.24. Turbulence kinetic energy spectrum for homogenous decaying turbulence with kinetic energy conserving scheme for different grid topologies.

necessary that eddies are resolved accurately (See Figure 7.22, Frame (a)). To achieve this, it is important that the level of dissipation is minimal. Hence, the neutrally dissipative central difference scheme is ideal. Thus in the target zone, $\varepsilon_{\Phi,1}$ should be set as low as possible but so that excessive dispersion error is avoided. Then away from the target zone, towards boundaries, $\varepsilon_{\Phi,1}$ can be elevated. This will prevent numerical reflections at boundaries.

A simple and effective approach for adaptively controlling $\varepsilon_{\Phi,1}$ is due to Mary and Sagaut (2002). Basically, the approach seeks to eradicate dispersion error. If associated oscillations are detected, $\varepsilon_{\Phi,1}$ is increased; otherwise it is decreased. Hence, the minimum level of dispersion is sought. The method is based around a structured grid. A line of stencil points is used, and the coexistence of a maximum and minimum indicates the presence of dispersion and thus the need to increase $\varepsilon_{\Phi,1}$. This is achieved using the following

$$\varepsilon_{\Phi,1}^{new} = \begin{cases} \min\left[\left(\varepsilon_{\Phi,1}^{old} + \Delta_\varepsilon\right), \varepsilon_{\Phi,1}^{\max}\right] & \text{if wiggle is detected} \\ \max\left[\left(\varepsilon_{\Phi,1}^{old} - \Delta_\varepsilon\right), \varepsilon_{\Phi,1}^{\min}\right] & \text{if wiggle is not detected} \end{cases} \tag{4.116}$$

Here Δ_ε, $\varepsilon_{\Phi,1}^{\max}$ and $\varepsilon_{\Phi,1}^{\min}$ correspond to the increment in $\varepsilon_{\Phi,1}$, the maximum allowable $\varepsilon_{\Phi,1}$ and the minimum allowable $\varepsilon_{\Phi,1}$. A parameter is also needed to control how frequently $\varepsilon_{\Phi,1}$ is updated. A more elegant proportional integral based procedure (from control theory) to control $\varepsilon_{\Phi,1}$ is described in Tucker (2013). Ciardi et al. (2005) attempt to extend the approach of Mary and Sagaut to an unstructured solver

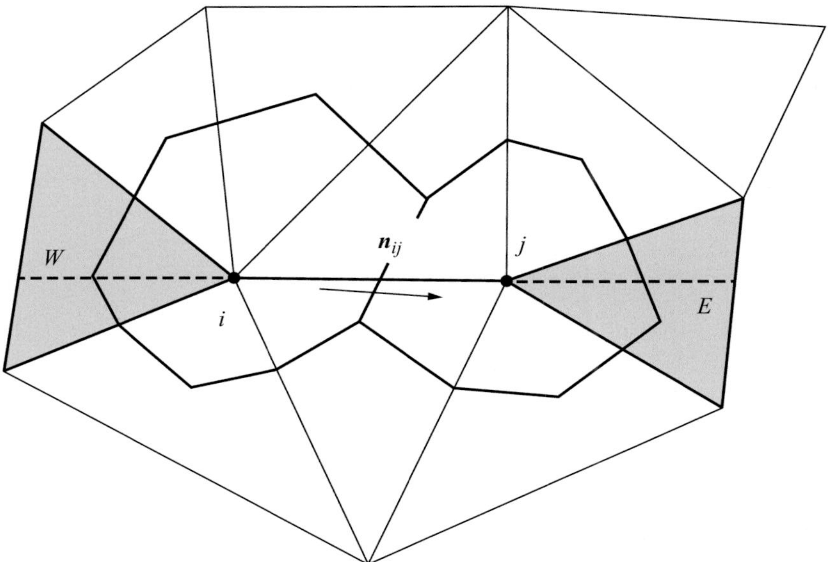

Figure 4.25. Unstructured control volumes and nomenclature for dispersion detector (from Tajallipour et al. (2009).

by selecting grid nodes that form approximate straight lines. However, clearly this has limited applicability. Hence, Tajallipour et al. (2009) try to further extend this approach to unstructured meshes. The following condition is noted for dispersion

$$\begin{aligned}(\phi_i - \phi_{i-1})(\phi_{i+1} - \phi_i) < 0 \\ (\phi_{i+2} - \phi_{i+1})(\phi_{i+1} - \phi_i) < 0\end{aligned} \tag{4.117}$$

where $\phi = \rho, u, v, w$ and p. This can be partly expressed more generally as

$$(\nabla\phi)^P \bullet \mathbf{n}_{ij} = (\phi_j - \phi_i)/|\mathbf{x}_j - \mathbf{x}_i|. \tag{4.118}$$

where the superscript P indicates a central point and $\mathbf{x}_j$ and $\mathbf{x}_i$ are the vector positions of nodes i and j adjacent to P. Hence, observing Figure 4.25, Equalities (4.117) can be more generally expressed as

$$\begin{aligned}\left[(\nabla\phi)_{ij}^W \bullet \mathbf{n}_{ij}\right][(\nabla\phi)^P \bullet \mathbf{n}_{ij}] = \left[(\nabla\phi)_{ij}^W \bullet \mathbf{n}_{ij}\right][(\phi_j - \phi_i)/|\mathbf{x}_j - \mathbf{x}_i|] < \theta \le 0 \\ \left[(\nabla\phi)_{ij}^E \bullet \mathbf{n}_{ij}\right][(\nabla\phi)^P \bullet \mathbf{n}_{ij}] = \left[(\nabla\phi)_{ij}^E \bullet \mathbf{n}_{ij}\right][(\phi_j - \phi_i)/|\mathbf{x}_j - \mathbf{x}_i|] < \theta \le 0\end{aligned} \tag{4.119}$$

The satisfaction of Equalities (4.119), where θ is a pre-set value, means that dispersion is present and that $\varepsilon_{\Phi,1}$ should be increased. This is achieved in a linear fashion. The approach of Ciardi et al. (2005) is a stepwise procedure.

4.4.3.6 Going to High Order with Unstructured Grids Deferred correction is an approach relevant to both structured and unstructured grids. It simplifies going to high order. The approach involves making a stable low order flux calculation, F^L_{CVF}, and correcting this based on data available from the previous iteration but using this in a high order flux F^H_{CVF}. Hence, the following can be written

$$F_{CVF} = F^L_{CVF} + \alpha[F^H_{CVF} - F^L_{CVF}]^{old} \tag{4.120}$$

The bracketed term is evaluated at the previous iteration and $0 \leq \alpha \leq 1$. As can be seen with $\alpha = 1$ at steady state convergence, a solution, F^H_{CVF}, is gained. Typically, the bracketed correction is small relative to the main low order flux. Hence, the approach can give considerable stability benefits.

We noted earlier the quadratic face reconstruction of Barth and Jespersen (1989). The ENO (Essentially Non Oscillatory – see Harten et al. 1987) and WENO (Weighted Essentially Non-Oscillatory – see Liu et al. 1994) schemes can be extended to arbitrary order. Such schemes use a combination of polynomials. These have different levels of upwind bias. An adaptive combination of these stencils is used to control oscillations. These approaches have been mainly used on regular grids but can be extended to unstructured systems (Abgrall 1994, Wolf and Azevedo 2007). Indeed, going to high order on unstructured meshes, unlike structured, is challenging. For example, Wolf and Azevedo require use of an enlarged stencil. There are a range of connected specialist techniques.

For many years, where high accuracy is required, the spectral technique has been popular. This approach typically uses Fourier series as the underlying discretization 'polynomial'. This means the approach has high accuracy with what is described as exponential convergence. This, for smooth solutions, without shocks, is the best possible. Typically, the spectral approach favours flows with multiple periodic boundaries and is an excellent approach for studying the fundamental physics of turbulence. Patera (1984) devised the spectral element method. This element-based technique makes use of Chebyshev and Legendre polynomials. A commercial CFD code (no longer sold), derived from this approach, has found some considerable success.

Another popular class of high order, unstructured methods takes advantage of the fact that many of the approaches discussed are Riemann related and allow for control volume face discontinuities. Thus, just high order functions can be applied inside the element with multiple internal quadrature points – see Figure 4.26. This is the basis of the Discontinuous Galerkin (DG) method. The method is similar to the Galerkin finite element method in that shape functions are used. The discontinuous shape function fields are (weakly) connected at boundaries through boundary conditions. The discontinuous shape function fields are shown in Figure 4.26. The Spectral Difference (SD) method is another discontinuous Galerkin related method. It is intended to be a simpler, more computationally economical alternative to DG. The Spectral Volume (SV – see Liu and Nishimura 2006), and SD methods are similar to DG in making use of piecewise discontinuous polynomials. Also, the DG, SV and SD methods are analogous to Galerkin finite element, finite volume and finite difference (Wang et al. 2007) techniques. The correction procedure via reconstruction (CPR) scheme unifies these approaches (Haga et al. 2011).

Figure 4.27 shows the development of the CPR process in one dimension. Frame (I) shows the points to be collocated and Frame (II) shows a polynomial of degree,

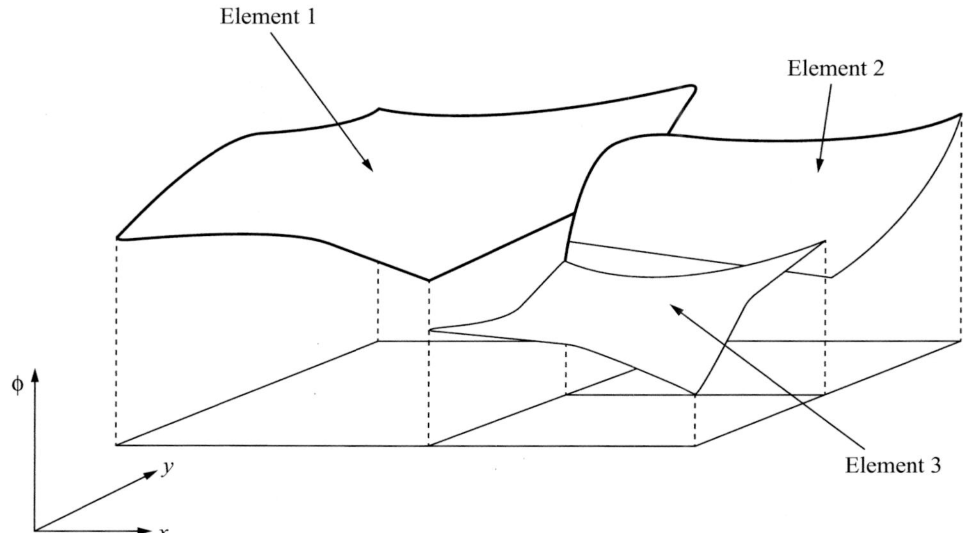

Figure 4.26. Broken functions of high order polynomials used in some high order unstructured meshes (from Tucker (2013) – Published with the kind permission of Springer).

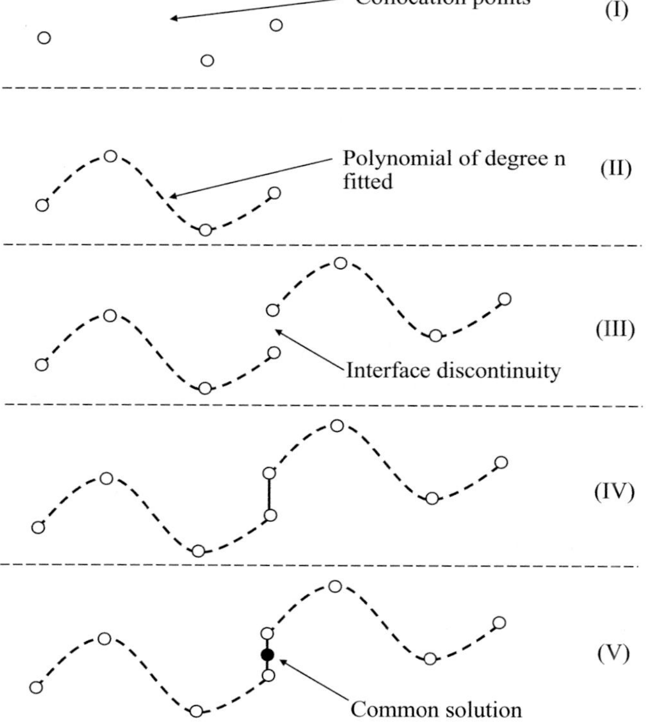

Figure 4.27. The development of the CPR process in one-dimension: (I) points to be collocated, (II) polynomial of degree, n, fitted to points, (III) interface discontinuity arising from just considering local isolated data, (IV) connection of discontinuity and (V) identification of common solution.

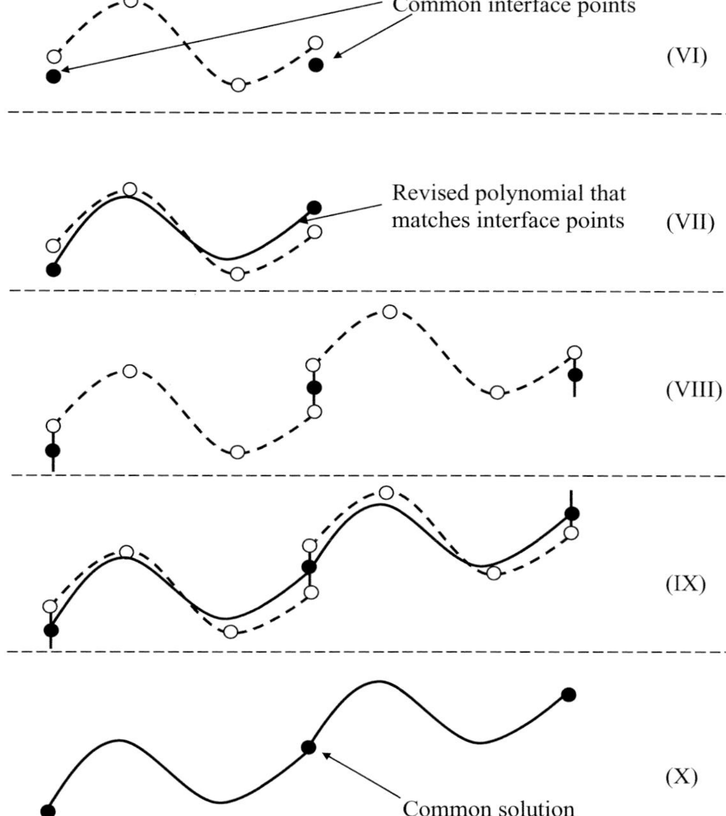

Figure 4.28. The development of continuous solution in the CPR process in one dimension: (VI) common interface points, (VII) development of revised polynomial $(n + 1)$ to match interface points, (VIII) extension of process to include an adjacent cell, (IX) continuous function through two cells and (X) final variable distribution for the solution.

n, fitted to the points. As shown in Frame (III), since the element is considered in isolation, an interface discontinuity arises with the adjacent elements. Frames (IV) and (V) show the connection of the discontinuity and the identification of a common solution. The next stage is shown in Figure 4.28. This involves the development of a continuous solution in the CPR process. This process is achieved by adding a source term to the governing flow equations. This correction field is polynomial based and is essentially zero away from interfaces with values at interfaces that enforce continuity. The common interface flux can involve Roe's approximate Riemann solver. Frame (VII) shows the development of a revised polynomial $(n + 1)$ to match interface points. This is extended to include an adjacent cell in Frame (IX). Finally, Frame (X) gives the final variable distribution for the solution.

Figure 4.29 plots the turbulence energy spectrum predicted by the CPR scheme for different orders but keeping the degrees of freedom fixed. The plot is taken from Vermeire et al. (2014). The degrees of freedom is pinned around 64^3, the simulation

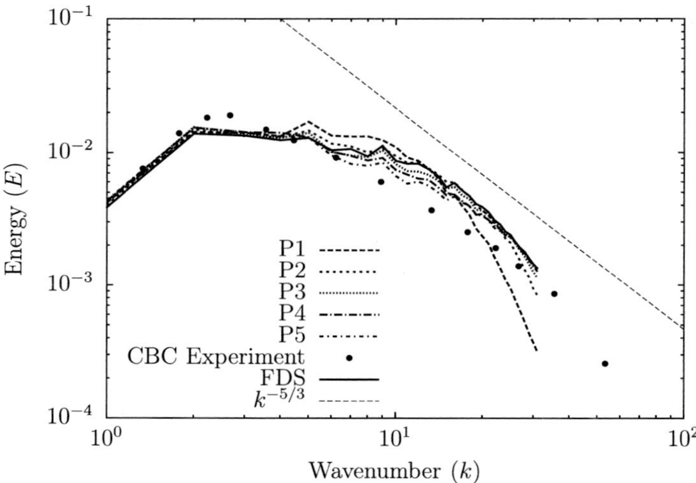

Figure 4.29. Turbulence energy spectrum predicted by high order CPR scheme for different orders but keeping the degrees of freedom fixed (from Vermeire et al. 2014).

being made for homogenous decaying turbulence. Hexahedral elements are used. The improvements from the high order scheme are clear with increasing order (P). Sixth order finite difference results (FDS) are also shown on the plot. None of the CPR results use an LES model, the dissipation at high wave numbers being provided by the scheme.

The key advantage of these methods is that they potentially allow for much lower data exchange overheads on massively parallel simulations through considerably lower grid counts. A key requirement for LES is that as the grid is refined, the numerical contribution should be much lower than any dissipative input from the LES model. Typically for a basic LES model, the dissipative input scales with the square of the grid spacing. This is the same as the rate of decay of the numerical input for a second order scheme. Hence, as pointed out by Ghosal (2002) and later by Chow and Moin (2003), for a low order scheme it is difficult to disentangle the numerical input from that of the LES model. This is shown in Figure 4.30, adapted from Chow and Moin (2003). This figure plots the numerical contribution relative to that from an LES model for a second order numerical scheme. As can be seen, at high wave numbers, the energy contribution from the numerical scheme is much higher than that of the ideal LES model. Note that a log scale is used in the axis for energy. This all gives a strong theoretical justification for using a high order scheme for LES. However, in practical terms, the benefits of high order schemes, for LES, is still unclear. Also, currently there appears to be no modified equation analysis for more advanced high order, unstructured schemes such as CPR. Furthermore, there is some evidence that they can tend to be dissipative at high wave numbers. However, akin to DRP schemes, there could be room for modifications that give better resolution properties at the sacrifice of order. The discontinuous Galerkin approach has seen successful use for acoustic propagation in the commercial ACTRAN (ACoustic

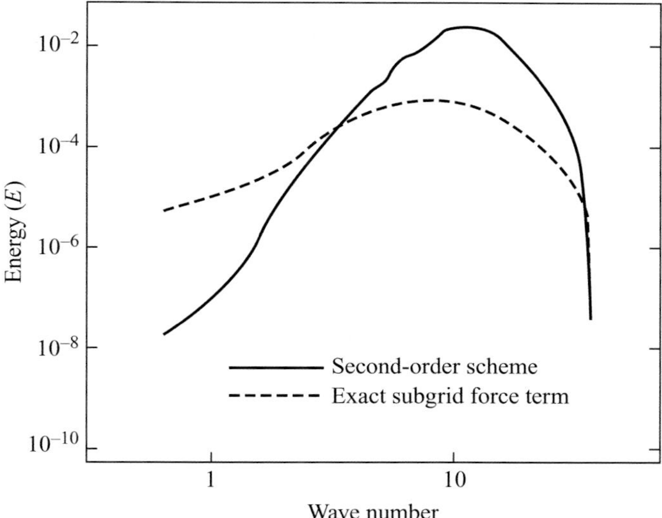

Figure 4.30. Numerical contribution relative to that from an LES model for a second order numerical scheme (from Chow and Moin 2003).

TRANsmission) program. Hence, such methods might have benefits in aeroacoustics problems.

For RANS simulations, the justification for high order unstructured schemes is complicated. Many of the RANS equations, as discussed in Chapter 5, can be more challenging to solve with a high order method. The efficient solution of high order unstructured equation systems is in its infancy and it is unclear how efficient methods such as multigrid convergence acceleration work with high order methods. Also, the core memory requirement scales with the order of such schemes to a high power. Obviously, the use of a higher order scheme only guarantees improved accuracy where grids are sufficiently fine to be within the radius of convergence of the numerical scheme. For an industrial calculation this could correspond to a highly substantial grid.

4.5 Alternative Discretization Methods

4.5.1 *Lattice Boltzmann*

For quite some time there has been a commercial lattice-Boltzmann solver (PowerFLOW). This approach solves the discrete Boltzmann equation (Succi 2001). In a real fluid flow there are obviously particle collisions at a micro scale. The lattice Boltzmann method replicates these. Hence, collision models are needed (see Bhatnagar et al. 1954). A discrete, generally hexahedral mesh is needed and particles around this undergo propagation and collision processes. The rules for the latter must seek to ensure conservation of mass momentum and energy. Using the Chapman-Enskog theory (an expansion, the expansion coefficient being the Knudsen number – see Chapter 2), the Navier-Stokes equations can be derived from the Boltzmann equations. The lattice-Boltzmann approach avoids the need to solve some form of Poisson equation for pressure. As noted previously, the approach still needs a grid and is intrinsically unsteady. Essentially it is an eddy-resolving method and well suited to DNS (Direct Numerical Simulation) and related approaches (see Chapter 5). It does not have the challenging convective term found in classical CFD. As with standard CFD, there is a problem with the range of scales at higher Reynolds numbers and their resolution. Hence, some form of modelling is needed. However, typically this would involve (very) large eddy type simulations where the smaller scales are modelled. The modelled scale, rather than being characterized by a length scale, as in most classical turbulence models, is time scale based. For high Knudsen number flows, promising results are found. Evidently, the approach lends itself naturally to multi-component/phase flows. The lattice-Boltzmann method has the advantage that it is compact to code. Also, it is easily parallelized on many processors. However, it needs substantial storage. In two dimensions, each grid point needs to store nine particle variables. In three dimensions, this increases to nineteen. The method is largely restricted to Cartesian cells and incorporating non-uniform stretched cells is not that easy. However, zones with different square lattice resolutions can be used with a 2:1 refinement ratio. The lattice-Boltzmann method is not ideal for high-speed complex geometry aerodynamic flows where skin friction is important. As with classical LES and DNS, the grid must be sufficiently fine to account for the aerodynamically critical, fine near wall scales and again this is problematic. The Boltzmann equation takes the form

$$\frac{\partial \phi}{\partial t} + c\frac{\partial \phi}{\partial x_i} = \left(\frac{\partial \phi}{\partial t}\right)_{coll} \tag{4.121}$$

where $\phi(\mathbf{x}, \mathbf{c}, t)$ is a distribution function for a single particle. Also, $\boldsymbol{c}$ is a particle velocity at the micro scale. The latter, right-hand side term, called the collision term, and representing particle interactions, can be modelled as

$$\left(\frac{\partial \phi}{\partial t}\right)_{coll} = -\frac{1}{\tau}\left[\phi - \phi^{eq}\right] \tag{4.122}$$

where τ is a relaxation time parameter and ϕ^{eq} is an equilibrium distribution (see Equation 4.130), later). Note that the distribution function and particle velocity are related to the more usual CFD variables through the integral expressions

$$\rho = \int \phi dc \tag{4.123}$$

$$\rho \mathbf{u} = \int \mathbf{c}\phi dc \tag{4.124}$$

Also,

$$\mu = \rho\left(\tau - \frac{1}{2}\right)T \tag{4.125}$$

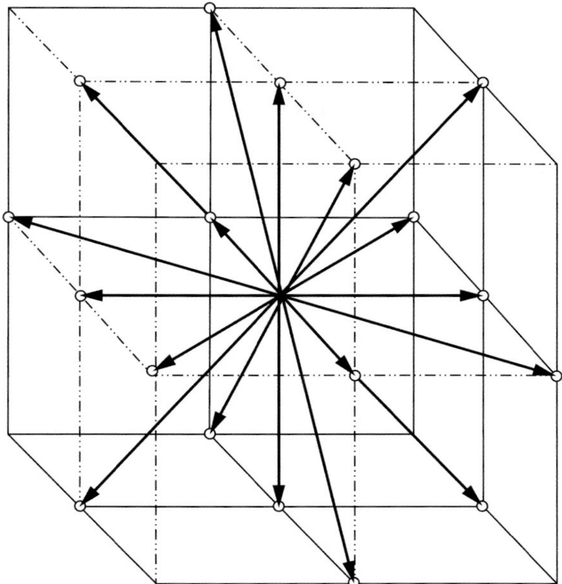

Figure 4.31. Three-dimensional, nineteen state lattice-Boltzmann model.

where T is the lattice temperature which is a constant value for isothermal simulations. Following Marié et al. (2009), the Boltzmann equation for discrete use is given as

$$\frac{\partial \phi_\alpha}{\partial t} + c_{\alpha,i}\frac{\partial \phi_\alpha}{\partial x_i} = -\frac{1}{\tau}\left[\phi_\alpha - \phi_\alpha^{eq}\right] \tag{4.126}$$

The use of nineteen discrete particle velocities, as shown in Figure 4.31, means that the standard velocity-strain relationship becomes modified as

$$\tau_{ij} = \rho v\left(\frac{\partial u_i}{\partial x_j} + \frac{\partial u_j}{\partial x_i}\right) - \tau\frac{\partial \rho u_i u_j u_k}{\partial x_k} \tag{4.127}$$

(see Marié et al. 2009). The final, essentially cubic in Mach number term limits the method to low Mach numbers. For the discrete Boltzmann equation the following can be written

$$\begin{aligned}\rho &= \sum_\alpha \phi_\alpha \\ \rho\mathbf{u} &= \sum_\alpha \mathbf{c}_\alpha \phi_\alpha\end{aligned} \tag{4.128}$$

The ultimate discretized Boltzmann equation is very simple and typically given as

$$\phi_\alpha(\mathbf{x} + \mathbf{c}_\alpha \Delta t, t + \Delta t) - \phi_\alpha(\mathbf{x}, t) = -\frac{\Delta t}{\tau}\left[\phi_\alpha(\mathbf{x}, t) - \phi_\alpha^{eq}(\mathbf{x}, t)\right] \tag{4.129}$$

where the equilibrium distributions, ϕ^{eq}, can be approximated as a series of the form

$$\phi_{\alpha}^{eq}(\mathbf{x},t) = \rho W f_{\alpha}\left[1+\frac{\mathbf{c}_{\alpha}\cdot\mathbf{u}}{T}+\frac{(\mathbf{c}_{\alpha}\cdot\mathbf{u})^2}{2T^2}-\frac{\mathbf{u}^2}{2T^2}+\frac{(\mathbf{c}_{\alpha}\cdot\mathbf{u})^3}{6T^2}-\frac{(\mathbf{c}_{\alpha}\cdot\mathbf{u})}{2T^2}\mathbf{u}^2\dots\dots\dots\dots\right] \tag{4.130}$$

The weighting function values are

$$W f_{\alpha} = \begin{cases} 1/18, & \text{in 6 coordinate directions;} \\ 1/36, & \text{in 12 bi-diagonal directions;} \\ 1/3, & \text{the rest of the particles.} \end{cases}$$

Equations (4.128)–(4.130) with the viscosity Equation (4.125) give the equation set to be solved. For the application of the lattice-Boltzmann approach to a high lift configuration, see Fares and Nölting (2011). Similarly, Armstrong et al. (2013) consider landing gear noise. Bres et al. (2009) study canonical acoustic problems with the lattice-Boltzmann method, finding it to have low dissipative and dispersive properties. Marié et al. (2009) study the dissipative and dispersive properties, again with acoustic propagation problems in mind. They find that the scheme is less dissipative than a classical high order spatial scheme. It is also less dispersive than a second order spatial scheme with a three-stage explicit Runge-Kutta temporal integration. However, it is found to be more dispersive than optimized, classical, high order, finite difference type Navier-Stokes schemes.

4.5.2 *Smooth Particle Hydrodynamics (SPH)*

The SPH approach has been available for several decades – see Gingold and Monaghan (1977). It was originally developed for astrophysics, such as modelling the formation of stars. The method has the attraction of not needing a mesh, this being a key simulation bottleneck (see Chapter 3). The method is Lagrangian. The fluid domain comprises particles. These are spatially separated over what is called the smoothing length, l. The properties at a particular point are established by summing weighted information (using a kernel function) at surrounding points. A popular weighting is Gaussian distribution based. Weighting functions are constructed to give zero weight outside l. SPH is attractive for unsteady free surface flows. For these, volume of fluid and level set methods (see Chapter 7) can be utilized. Also, moving meshes can be used. However, meshes can become of poor quality and tangled. Then complete re-meshing is needed (see Chapter 3). SPH intrinsically conserves mass, since this is just a summation over particles. SPH does not require any complex pressure evaluation procedures. It is just evaluated based on data from weighted sums of surrounding particles. To calculate a quantity $\phi(\mathbf{r}_j)$ of particle j, where $\mathbf{r}$ is the location, the following equation can be used

$$\phi_j = \phi(\mathrm{r}_j) = \sum_i m_j \frac{\phi_i}{\rho_i} Wf(|\mathrm{r}_j - \mathrm{r}_i|, l) \tag{4.131}$$

where Wf is the kernel function and ρ_i and m_i are the density and mass of a particle i. Gradients can be expressed as

$$\nabla\phi(r) = \sum_i m_i \frac{\phi_i}{\rho_i} \nabla Wf(|\mathrm{r} - \mathrm{r}_i|, l) \tag{4.132}$$

The method more naturally lends itself to the concept of adaptivity than standard grid based CFD methods. An increase in resolution is effected by a local decrease in l. A large number of particles are needed to get the levels of accuracy associated with standard grid based CFD methods. SPH also tends to not deal with shock-like discontinuities as well as standard CFD methods. It is intended for highly compressible problems. Hence, SPH is not so easy to apply to incompressible flows, without running into the difficulties associated with preconditioning to be discussed later. Treating boundary conditions with SPH can be challenging. There are a range of approaches. These include the use of dummy, image and mirror particles. The former option is evidently easiest to implement. Currently the SPH approach is of low order, between 1 and 1.5 and no more than 2nd order. This low order is problematic for the use of SPH for eddy-resolving simulations. However, 4th order Gaussian kernels can be used to recover order. The high aspect ratio numerical representations needed for high speed boundary layers will be challenging for SPH and might need specialist flow aligned Gaussian kernels. SPH can also be formulated in a Eulerian fashion. SPH is well suited to things like scouring flows, waves and structural interactions on coasts, fluid-structure interactions, porous media, multiphase flows and lubrication problems with free surfaces. Highly deforming multiphase flows are a particular strength of SPH. For the use of SPH to model the two-phase flow in an aero-engine bearing chamber see Wieth et al. (2015). Marongiu et al. (2007, 2010) use SPH for modelling a Pelton wheel flow. Further examples of the application of SPH are described in Section 6.7 of Chapter 6.

4.5.3 *Vortex Methods*

The basis for the vortex method is the vorticity transport equation

$$\frac{\mathrm{D}\Omega}{\mathrm{Dt}} = \Omega \cdot \nabla\mathbf{u} + \nu\Delta\Omega \tag{4.133}$$

However, this continuous vorticity field equation is represented as a discrete, particle, p, based field. In this the vorticity is defined as

$$\Omega(\mathbf{x}, t) \approx \sum_{p=1}^{N_p} \Gamma_p(t)\delta(\mathbf{x}\text{-}\mathbf{x}_p(t)) \tag{4.134}$$

(see Kirchhart 2013), where δ is the dirac delta function – smooth functions can be used instead. Also, Γ_p is the circulation, $\mathbf{x}_p$ the particle central location and N_p the number of particles. The particles convect with the local fluid velocity, $\mathbf{u}$, as

$$\left[\frac{d\mathrm{x}_p}{dt} = \mathbf{u}(\mathbf{x}_p(t), t), p = 1, \ldots, N_p\right] \tag{4.135}$$

This Lagrangian system derivative is equivalent to the material derivative used in the continuous equation. Hence, the following discrete Lagrangian vorticity representative equation can be written:

$$\frac{d\Gamma_p(t)}{dt} = \Gamma_p(t) \cdot \nabla\mathbf{u}(\mathbf{x}_p(t), t), \; p = 1, \ldots, N_p \tag{4.136}$$

Equation (4.136) is a set of ordinary differential equations and can be integrated through time using approaches to be described later. The key issue is getting the velocity field from the particle vorticity field above. Helmholtz decomposition enables $\mathbf{u}$ to be decomposed into the three components

$$\mathbf{u} = \mathbf{u}_\Omega + \mathbf{u}_\Phi + \mathbf{u}_\infty \tag{4.137}$$

where $\mathbf{u}_\Omega$ is the velocity induced by the vorticity field, $\mathbf{u}_\Phi$ is a potential field and $\mathbf{u}_\infty$ is the velocity at infinity. The velocity associated with the vorticity field can be evaluated using the Biot-Savart law:

$$\mathbf{u}_\Omega(\mathrm{x}, t) = -\frac{1}{4\pi}\int_V \frac{\mathrm{x}-\mathrm{y}}{\|\mathrm{x}-\mathrm{y}\|^3} \times \Omega(\mathrm{y}, t)d\mathrm{y} = -\frac{1}{4\pi}\sum_{p=1}^{N_p} \frac{\mathrm{x}-\mathrm{x}_p(t)}{\|\mathrm{x}-\mathrm{x}_p(t)\|^3} \times \Gamma_p(t) \tag{4.138}$$

The summation over many particles is expensive and, if the computational cost is to be controlled, requires specialist techniques. A boundary element type technique (see Section 4.5.4) can be used to gain the potential field, based around a Laplace equation. Boundary condition specification is not straightforward. The previous discussion is just for inviscid flow and hence extension is needed for viscous flow. Just a surface mesh is essentially required. Also, some interior mesh for a data framework can be used. Attempts have been made to hybridize this approach, using more classical CFD near walls and the vortex method in the core region where perhaps it shows its greatest strength.

4.5.4 *Boundary Element Method*

The boundary element method (BEM), akin to the panel method, has the attraction that no volume mesh is necessary. Instead, just the surface needs to be discretized as a series of elements. This is advantageous for moving geometry problems avoiding the need for volume re-meshing. For simple Laplace and Poisson equations (including harmonic source functions), the method is relatively simple to apply and can be solved using the Fast Multipole Method (FMM) – see in Liu and

Nishimura (2006). However, for more complex differential equations, more specialist techniques such as the Dual Reciprocity Method (DRM – see Partridge et al. 1992) and Multiple Reciprocity Method (MRM – see Nowak and Brebbia 1989) are needed. BEM sees very little use for solution of complex differential equations such as the Navier-Stokes. However, Camacho and Barbosa (2005) successfully apply the approach to some simple flows but find their approach is limited to very low Reynolds numbers.

As to be discussed in Chapter 6, wave propagation, such as acoustic waves, is an important component of numerical analysis of engineering problems. Acoustic wave transmission can be expressed through the Helmholtz equation. This is a Poisson equation with harmonic source terms. This equation can be effectively solved using the BEM method and this procedure is used in the commercial ACTRAN program. The BEM approach is used in Chapter 7 with FMM to solve a Poisson equation to de-feature geometry and also work out nearest wall distances for turbulence models.

The intention of the text in this section based on Coughlin (2010), is to give, for completeness, a very general idea of the nature of the BEM method and the type of mathematics behind it. For simplicity, just the Poisson equation is considered.

4.5.4.1 Fundamental Solution The fundamental solution is a general function that satisfies the partial differential equation being considered. It is general and independent of geometry. Hence, writing the Laplace equation in polar co-ordinates, we have

$$\frac{1}{r}\frac{\partial}{\partial r}\left(r\frac{\partial \Phi}{\partial r}\right)+\frac{1}{r^2}\frac{\partial^2 \Phi}{\partial \theta^2}=0 \tag{4.139}$$

where $x = r\cos\theta$, $y = r\sin\theta$ and $\Phi(r, \theta)$. If Φ is a function of r only, (4.139) reduces to

$$\frac{d}{dr}\left(r\frac{d}{dr}[\Phi(r)]\right)=0 \tag{4.140}$$

Integrating (4.140) twice suggests a fundamental solution of the form

$$\Phi=\frac{1}{2\pi}\ln(r)\rightarrow\frac{1}{2\pi}\ln\sqrt{(x-\xi)^2+(y-\eta)^2} \tag{4.141}$$

4.5.4.2 BEM Equations The Poisson equation

$$\nabla^2\phi = S_\phi \tag{4.142}$$

needs to be expressed in boundary integral form. If the functions ϕ_1 and ϕ_2 satisfy (4.142), then

$$\frac{\partial^2\phi_1}{\partial x^2}+\frac{\partial^2\phi_1}{\partial y^2}=S_1 \tag{4.143}$$

$$\frac{\partial^2\phi_2}{\partial x^2}+\frac{\partial^2\phi_2}{\partial y^2}=S_2 \tag{4.144}$$

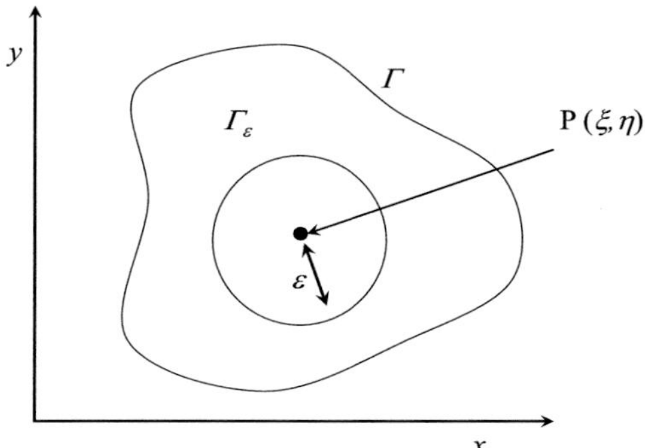

Figure 4.32. BEM schematic for point $P(\xi, \eta)$ lying in the domain V (from Coughlin 2010).

for a domain V bounded by curve Γ, it may be shown that

$$\int_{\Gamma} \left(\phi_2 \frac{\partial \phi_1}{\partial n} - \phi_1 \frac{\partial \phi_2}{\partial n} \right) ds(x, y) = \iint_{V} (\phi_2 S_1 - \phi_1 S_2) dxdy \qquad (4.145)$$

which can be derived from multiplying Equation (4.143) by ϕ_2 and Equation (4.144) by ϕ_1. The difference of the resulting equations, on integration and application of the Divergence (Green's) theorem, gives Equation (4.145), which can be used to obtain a solution to Poisson's equation.

If we take ϕ_1 to be the fundamental solution Φ, and ϕ_2 to be the required solution that satisfies Equation (4.142) and its boundary conditions, it follows that $S_1 = 0$ and $S_2 = S_\phi$, giving

$$\int_{\Gamma} \left[\left(\phi \frac{\partial \Phi}{\partial n} - \Phi \frac{\partial \phi}{\partial n} \right) \right] ds(x, y) = \iint_{V} (\Phi S_\phi) dxdy \qquad (4.146)$$

or

$$\int_{\Gamma} \left[\left(\phi \frac{\partial \Phi}{\partial x} - \Phi \frac{\partial \phi}{\partial x} \right) n_x + \left(\phi \frac{\partial \Phi}{\partial y} - \Phi \frac{\partial \phi}{\partial y} \right) n_y \right] ds(x, y) = \iint_{V} (\Phi S_\phi) dxdy \qquad (4.147)$$

However, our fundamental solution, Φ, is not well defined at $P(\xi, \eta)$. To accommodate point P within the domain, we surround it by a circle, centred on P, with radius ε, as shown Figure 4.32. The fundamental solution, Φ, is well defined in the region between $\Gamma$and Γ_ε. Hence, for $\Gamma$and Γ_ε, in Figure 4.32, the following can be written, based on Equation (4.147):

$$\int_{\Gamma} \left(\Phi \frac{\partial \phi}{\partial n} - \phi \frac{\partial \Phi}{\partial n} \right) ds(x, y) + \int_{\Gamma_\varepsilon} \left(\Phi \frac{\partial \phi}{\partial n} - \phi \frac{\partial \Phi}{\partial n} \right) ds(x, y) = \iint_{V \cup V_\varepsilon} (S_\phi \Phi) dxdy \qquad (4.148)$$

If we let $\varepsilon \to 0^+$, Equation (4.148) may in fact be written as

$$\phi_P - \iint_V (S_\phi \Phi) dxdy = \int_\Gamma \left(\phi \frac{\partial \Phi}{\partial n} - \Phi \frac{\partial \phi}{\partial n} \right) ds(x, y) \qquad (4.149)$$

(see Ang 2007), where ϕ_P is our value at point P. If we then repeat this analysis (see again Ang 2007) for point P lying on the boundary, we obtain

$$\frac{1}{2}\phi_P - \iint_V (S_\phi \Phi) dxdy = \int_\Gamma \left(\phi \frac{\partial \Phi}{\partial n} - \Phi \frac{\partial \phi}{\partial n} \right) ds(x, y) \qquad (4.150)$$

As either ϕ or $\partial\phi/\partial n$ are known from the boundary conditions, Equation (4.150) can be solved (as a series of simultaneous equations based around the Γ surface discretized as a series of elements) to find the unknown value. Interior values may then be found by replacing P by each interior point in turn and solving Equation (4.149). Equations (4.149) and (4.150) may be summarised as

$$\lambda\phi_P - \iint_V (S_\phi \Phi) dxdy = \int_\Gamma \left(\phi \frac{\partial \Phi}{\partial n} - \Phi \frac{\partial \phi}{\partial n} \right) ds(x, y) \qquad (4.151)$$

where $\lambda = 1$ if $\phi_P \in V \cup \Gamma$, $\lambda = 1/2$ if ϕ_P lies on Γ and $\lambda = 0$ if $\phi_P \notin V \cup \Gamma$.

Equation (4.151) not only contains a boundary integral, but also a double domain integral arising from the source term. The source term, which could be the temporal derivative of ϕ, can be treated in a range of ways. For special cases where the source term S_ϕ is a harmonic function such that $\nabla^2 S_\phi = 0$, through the Galerkin vector approach, and making use of a fundamental solution, the volume integral can be reduced to a surface integral. Alternatively, if $S_\phi =$ constant, then ϕ can be redefined to a new function that carries the source information. For example, $\phi = \phi + f(x, y)$ where $f(x, y) = (x^2 + y^2)/4$ gives a the solution for $S_\phi = 1$. This approach, with FMM, is used to solve the Poisson equation used for turbulence length scales and de-featuring as discussed in Chapters 5 and 7. When the source term is the time derivative of ϕ, this derivative can be discretized as a finite difference.

The final BEM computation is split into two parts. First, the system of equations for the boundary element's boundary conditions are evaluated. Once these are solved, it is easy to obtain values for any interior point as a post-processing type operation

Solution of Equation (4.142) using BEM with FMM, to accelerate convergence, along with Generalized Minimum Residual (GMRES) (see Section 4.6) to solve the linear system of equations arising from the discretization gives a method which has a solution time O(N). Note, N is the number of surface elements on the domain boundary Γ. More details of the FMM method can be found in Liu and Nishimura (2006). Notably, the typical problem with more standard BEM techniques is that the cost scales with N^2. This high cost, arising from the approach, is because, unlike the finite volume method, BEM gives densely populated matrices. These are relatively expensive to invert (with the finite volume method the matrices are banded). As can be seen, even for simple partial differential equations, relative to, say, the finite volume method, the BEM approach has substantial mathematical complexity.

PART B – SPECIALIZED SOLUTION PROCEDURES

4.6 Simultaneous Equation Solvers and Preconditioning

Many of the previous equation forms involved a left-hand side term that is a time derivative. For steady flows, it is possible to integrate through time to a steady state using a simple forwards difference. This is akin to a crude iterative solution process. Indeed, due to the non-linear nature of CFD problems, solution procedures that are essentially iterative in nature are preferred. If the steady state is sought, then the time step can be spatially varied to give the quickest convergence to the steady state. However, such methods are slow to propagate low frequency solution content and are best supplemented with multigrid convergence acceleration. This is discussed later, but basically involves assimilating coarser grid data. Such data facilitates the propagation of lower frequency information more quickly. It also allows the use of larger time steps without stability constraints being violated. The stability constraints that must be considered are the Courant number, noted earlier,

$$C = \frac{u\Delta t}{\Delta x} \tag{4.152}$$

For compressible flows, this needs to be re-expressed as $C = (|u| + c)\Delta t/\Delta x$, where c is the speed of sound. Therefore, for compressible flow, the velocity scale is replaced by the maximum acoustic wave speed. The more severe stability constraint is the diffusive. This involves the parameter

$$D = \frac{\Gamma \Delta t}{\rho(\Delta x)^2} \tag{4.153}$$

where Γ is the diffusion coefficient. For a simple, first order, forwards difference in time, $C < 1$ for stability. Hence, as the grid becomes finer, the solution time step, or pseudo-time step, needs to be made smaller. Therefore, the expense of grid refinement is compounded when explicit time marching.

The discretised equations can also be rearranged and treated in some matrix form. Efficient matrix solution algorithms are freely available. Jacobi and Gauss-Siedel are two of the simplest solvers but are relatively inefficient and seldom used on their own. Jacobi simply takes the form

$$\phi_p^n = \frac{\left(\sum_{NB} A_{NB\phi}^{n-1} \phi_{NB}^{n-1} + S_\phi^{n-1}\right)}{A_{P\phi}^{n-1}} \tag{4.154}$$

where n is the iterative level. On the other hand Guass-Siedel exploits the latest ϕ^n values in the left hand side evaluations.

Gaussian elimination is effective in one dimension but for higher dimensions, with sparse, non-linear matrix systems, it is inefficient. A popular hybrid CFD, simultaneous equation solver (see Patankar 1980) is the Tri-diagonal Matrix Algorithm (TDMA, sometimes known as the Thomas algorithm). With this, a simplified Gaussian elimination is used in one coordinate direction. The other directions are treated using Gauss-Siedel. The direction that simplified Gaussian elimination is applied is generally varied with n. Hence, when TDMA is applied, the discretized equations are arranged in the following tri-diagonal form

$$A_{p\phi}\phi_p = A_{I1\phi}\phi_{I1} + A_{I2\phi}\phi_{I2} + S \tag{4.155}$$

where

$$S = A_{I3\phi}\phi_{I3} + A_{I4\phi}\phi_{I4} + A_{I5\phi}\phi_{I5} + A_{I6\phi}\phi_{I6} + S_\phi \tag{4.156}$$

Equation (4.155) is then solved with the subscripts, $I1$–$I6$ arranged as follows:

$$\begin{aligned} &I1 = E, I2 = W, I3 = S, I4 = N, I5 = F, I6 = B;\\ &I1 = N, I2 = S, I3 = E, I4 = W, I5 = F, I6 = B;\\ &I1 = F, I2 = B, I3 = S, I4 = N, I5 = E, I6 = W. \end{aligned} \tag{4.157}$$

Note that the new subscripts F (Front) and B (Back) have been introduced to provide a notation in three dimensions. Effectively, the terms in the source are solved using Gauss-Siedel and the others (the three diagonals in Equation 4.155) with simplified Gaussian elimination. The TDMA algorithm component can be expressed in the tri-diagonal matrix form as

$$\begin{bmatrix} A^1_{P\phi} & -A^1_{I1} & & & 0 \\ -A^2_{I2} & A^2_{P\phi} & -A^2_{I1} & & \\ & -A^3_{I2} & A^3_{P\phi} & -A^3_{I1} & \\ & & \ddots & \ddots & -A^{N-1}_{I1} \\ 0 & & & A^N_{I2} & -A^N_{P\phi} \end{bmatrix} \begin{bmatrix} \phi^1 \\ \phi^2 \\ \phi^3 \\ \phi^4 \\ \phi^5 \end{bmatrix} = \begin{bmatrix} S^1_\phi \\ S^2_\phi \\ S^3_\phi \\ S^4_\phi \\ S^5_\phi \end{bmatrix} \tag{4.158}$$

where $A^1_{I2} = 0$ and $A^N_{I1} = 0$. The first sweep in the process will eliminate the $A^i_{I2} = 0$ elements of the matrix. To do this, the revised coefficients and source term, identified with the $\sim$ overbars, are formulated

$$-\tilde{A}^i_{I1} = \begin{cases} \frac{-A^i_{I1}}{A^i_{P,\phi}} & \rightarrow i = 1 \\ \frac{-A^i_{I1}}{A^i_{P,\phi} - \tilde{A}^{i-1}_{I2}} & \rightarrow i = 2, 3 \cdots N-1 \end{cases} \tag{4.159}$$

$$\tilde{S}^i_\phi = \begin{cases} \frac{S^i_\phi}{A^i_{P,\phi}} & \rightarrow i = 1 \\ \frac{S^i_\phi + A_{I2}\tilde{S}^{i-1}_\phi}{A^i_{P,\phi} - A_{I2}\tilde{A}^{i-1}_{I1}} & \rightarrow i = 2, 3 \cdots N-1 \end{cases} \tag{4.160}$$

Then, the efficient back substitution

$$\begin{aligned} \phi^N &= \tilde{S}_\phi^N \\ \phi^i &= \tilde{S}_\phi^i - \tilde{A}_{I1}^i \phi^{i+1} \quad \rightarrow i = N-1, \ldots 1 \end{aligned} \tag{4.161}$$

gives the solution. An example of a TDMA solver for both periodic and non-periodic flows is given in Appendix C.

For vector and parallel processing, to avoid data dependencies, the red-black procedure is popular. This can be considered as a Jacobi/Gauss-Siedel hybrid. If a structured grid is viewed as a chessboard, with red and black squares, the technique involves applying the Jacobi method to all the red and then finally all the black squares.

For a good initial guess, Newton-Raphson (Newton) methods have especially fast convergence. This property is well suited to the modelling of unsteady flows. If time-steps are sufficiently small, the solution from the previous time-step is a good approximation to the solution for the next. As commonly known, the method solves for the zero of functions and hence, extended to a matrix form, $A(\phi) = 0$. The function, at an iteration, n, can be expressed as the following Taylor series

$$A(\phi) = A(\phi^{n-1}) + A'(\phi^{n-1})(\phi - \phi^{n-1}) \tag{4.162}$$

Setting $A(\phi^n) = 0$ in Equation (4.162), gives the well-known Newton solution equation

$$\phi = \phi^{n-1} - \frac{A(\phi^{n-1})}{A'(\phi^{n-1})} \tag{4.163}$$

For more details on the application of Equation (4.163) in CFD, see Engelman and Sani (1986). As implied in Equation (4.163), ideally the Jacobian $\partial A\ (\phi^{n-1})/\partial\phi$ must be repeatedly re-evaluated. The complexity of $A(\phi)$ makes this expensive. Hence, for unsteady flows Gresho et al. (1980) and Engelman and Sani (1986)) only re-evaluate $A'(\phi)$ periodically with respect to time-steps. It is worth noting here the Secant method. With this, the Jacobian can be approximated as a finite difference. This,

$$\frac{\partial A(\phi^{n-1})}{\partial \phi} \approx \frac{A(\phi^{n-1}) - A(\phi^{n-2})}{\phi^{n-1} - \phi^{n-2}} \tag{4.164}$$

gives the requirement for extra storage of 'n–2' values. This approach has yet to find use in CFD. However, Orkwis and Vanden (1994) compare numerically and analytically evaluated Jacobians. The former are computed using a symbolic mathematics software package and better convergence gained.

More sophisticated solvers include Stone's (1968) method (also called the Strongly Implicit Procedure – SIP), Bi-CGSTAB (biconjugate gradient stabilized method), GMRES (see Campobasso and Giles 2003) and Recursive Projection Methods (RPM – see Campobasso and Giles 2004). Hicken and Zingg (2008) present a parallel Newton-Krylov solver for the Euler equations. These advanced simultaneous equation methods are not outlined here. However, it is important to stress that what are termed strong solvers, such as the Krylov, can force steady state solutions which physically do not exist (see Krakos and Darmofal 2010). The weaker solvers will simply fail to converge and this in a sense is physically helpful.

4.6.1 *Diagonal Dominance*

Simpler solvers need $A_{P,\phi}$ to be of sufficient size. For $S_\phi = 0$, at the steady state, $A_{P,\phi} = \sum A_{NB,\phi}$ is necessary. Otherwise, for a uniform property domain with, say, $\phi = 0$ and $S_\phi = 0$, the wrong ($\phi \neq 0$) solution would be returned. To make $A_{P\phi} = \sum A_{NB\phi}$, following Patankar (1980), in discrete form, the continuity equation multiplied by the solution variable

$$\phi_P^{n+1}\left[\frac{\partial \rho}{\partial t} + \nabla \cdot (\rho \mathbf{u})\right] \tag{4.165}$$

can be subtracted from the governing convection-diffusion transport equations.

Indeed, for convergence of Gauss-Siedel related iterative solvers, this must be extended. The Scarborough criterion requires for at least one equation that

$$\frac{\sum |A_{NB\phi}|}{|A_{P\phi}|} < 1 \tag{4.166}$$

with, for all others,

$$\frac{\sum |A_{NB\phi}|}{|A_{P\phi}|} \leq 1 \tag{4.167}$$

This reflects the need to make, where possible $|A_{P\phi}|$ large – diagonally dominant, giving solution stability by lessening changes in ϕ between iterations. Diagonal dominance can be improved by making under-relaxation an explicit part of the governing equation:

$$\phi_P = \alpha_\phi \phi_P^n + (1 - \alpha_\phi)\phi_P^{n-1} \tag{4.168}$$

where α_ϕ is an under-relaxation parameter. Equation (4.168) is typically amalgamated with the coefficients of Equation (4.155). Care is required. Small α_ϕ can cause excessively slow convergence. In this situation, when convergence is based on changes in variables, convergence can be erroneously assumed. See Chapter 7 for further discussion on this. There are similarities between time-marched and under-relaxed iteratively solved equations. The reduction of $\Delta\tau$ in the former is analogous to reducing α_ϕ. When discretising the governing equations it is important to look carefully at S_ϕ involving ϕ_P terms. If the signs of terms that pre-multiply ϕ_P are consistent with increased diagonal dominance, then these terms can be amalgamated with $A_{P\phi}$. If there is the potential for sign changes in ϕ_P, maximum and minimum value constructs can be used to ensure that when signs are correct, terms are allocated appropriately between $A_{P\phi}$ and S_ϕ.

4.6.2 *The Multigrid Algorithm*

For dense grids, the performance of more basic simultaneous equation solvers can be slow. As noted, they are poor at propagating low frequency solution components from domain boundaries. Through the use of coarser grids, Brandt (1977) greatly speeds up the reduction of low frequency errors. This technique is now outlined. The discretized equations can be expressed in a matrix form as

$$[A_m][\phi_m] = [S_m] \tag{4.169}$$

where ϕ_m is a solution variable, $[A_m]$ the discretized equation coefficients and $[S_m]$ source terms. For an explicit, time marched system, $[A_m]$ will just have diagonals. The subscript m indicates grid levels with $m = 1$ for the finest. An exact solution to Equation (4.169), $[\phi_m]$, is expressed as

$$[A_m][\phi_m^a + \phi_m^c] = [S_m] \tag{4.170}$$

where $[\phi_m^a]$ are approximate values and $[\phi_m^c]$ corrections to the approximations. The following residual equation can be written

$$[R_m] = [S_m] - [A_m][\phi_m^a] \tag{4.171}$$

Equations (4.170) and (4.171) can be combined to eliminate $[S_m]$:

$$[A_m][\phi_m^a + \phi_m^c] = [R_m] + [A_m][\phi_m^a] \tag{4.172}$$

For a coarser, $m + 1$ mesh, Equation (4.172) can be stated as

$$[A_{m+1}][\phi_{m+1}^{a,f} + \phi_{m+1}^c] = [R_m^f] + [A_{m+1}][\phi_{m+1}^{a,f}] \tag{4.173}$$

The superscript f indicates that the value originates from a finer mesh. Equation (4.173) solves for what is termed the *full approximation*, $[\phi_{m+1}^{a,f} + \phi_{m+1}^c]$. This and the right-hand side of the equation are expressed as

$$\begin{aligned} [\phi'_{m+1}] &= [\phi_m^{a,f} + \phi_{m+1}^c] \\ [S'_{m+1}] &= [R_m^f] + [A_{m+1}][\phi_{m+1}^{a,f}] \end{aligned} \tag{4.174}$$

Since S'_{m+1} forces the $m + 1$ solution to have the accuracy of the m grid solution, it is generally called the forcing function. Using Equations (4.174), Equation (4.173) can be restated as

$$[A_{m+1}][\phi'_{m+1}] = [S'_{m+1}] \tag{4.175}$$

giving, for $m > 1$, an equation similar in form to Equation (4.169). Corrections for lower grid levels can then be obtained by subtracting $[\phi_m^{a,f}]$ from $[\phi'_{m+1}]$. For the method to work, ϕ values at m being transferred to level $m + 1$ must be sufficiently smooth. Hence, an adequate simultaneous equation solver is needed. For explicit methods, the residual smoothing technique (see Blazek et al. 1991) can be used. However, generally, for highly stretched grids (which can be most economical when resolving relatively high-speed flow in boundary layers), implicit methods make better smoothers than their explicit counterparts.

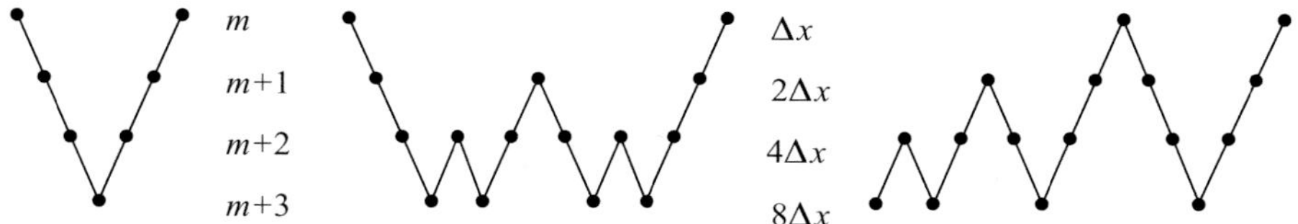

Figure 4.33. Some possible paths that can be used between these different grids: (a) V-cycles, (b) W-cycles and (c) Full Multigrid Cycles.

If the equation system is linear, $[A_{m+1}][\phi^{a,f}_{m+1} \cdots]$ cancels either side of Equation (4.173). Therefore, Equation (4.175) simplifies to the following *Correction Storage* or *Linear multilevel* equation:

$$[A_{m+1}]\left[\phi^{c}_{m+1}\right] = \left[R^{f}_{m+1}\right] \tag{4.176}$$

For a two grid method, ϕ^{c}_{m+1} is interpolated to grid level m where

$$[\phi_m] = \left[\phi^{a,f}_{m}\right] + \alpha\left[\phi^{c}_{m+1}\right] \tag{4.177}$$

and α_ϕ is an under- or over-relaxation parameter. Importantly, a correction equation on grid level $m+2$ could be written for Equations (4.175) or (4.176) – they have the same form as Equation (4.169). This procedure could be repeated for higher m, until no further coarsening can take place. This is the basis of the multilevel method. Brandt (1977) recommends that for optimal performance, for structured grids, coarsening is achieved by missing out every other grid line in the different co-ordinate directions. For unstructured grids, there are different methods for constructing coarsened grids and these are outlined in Chapter 3.

Figure 4.33 illustrates some possible paths that can be used between different grids. Frame (a) shows the popular V-cycle. Frame (c) shows a W-cycle. This spends more time on the cheap coarser grids with the hope that this gives greater efficiency. Frame (c) shows a full multigrid cycle. With this, a solution guess is produced at the coarsest grids and improved by a series of shallow V-cycles. There have been a wide range of studies on the best multigrid cycles and these seem inconclusive; doubtless, to an extent, the best cycle is problem and even solver and grid structure dependent.

As can be seen from Figure 4.33, Equations (4.175) and (4.176) are represented on successively coarse grids to make corrections to solutions on the next lower grid.

To represent variables on, say, level $m+1$ from level m (restriction), a distance weighted averaging is typically used. Fluid properties are generally calculated at and not restricted to the different levels. However, for stability, sometimes the turbulent viscosity values restricted from the finest grid are used at all levels (see Vaughan et al. 1989). Once corrections (ϕ^{c}_{m+1}) have been calculated, they are bilinearly interpolated onto the next finer mesh. This procedure is called prolongation.

Explicit solutions, being time marched to a steady state, can greatly benefit from multilevel convergence acceleration (see Moinier 1999). This is because the method circumvents the Courant number restrictions associated with explicit approaches.

Table 4.4. *Convergence details for* $Gr = 3 \times 10^5$ *heated cavity (from Tucker 2001 – Published with the kind permission of Springer)*

Grid	Iterations (Single level)	CPU(s) (Single level)	Iterations (Multilevel)	CPU(s) (Multilevel)
11^3	65	4.82	9	3.31
17^3	116	42.79	9	13.08
21^3	145	147.70	9	36.77
33^3	269	1379.00	9	184.97
65^3	656	34741.76	10	2234.83

To utilize multilevel algorithms, time-steps are considered equivalent to iterative relaxation steps and coarsened with mesh.

Figure 4.34 plots the residual against the number of iterations for different numbers of grid levels. The simulation is for a heated cavity at a Grashoff number, *Gr*, of 1.44×10^6 with a 65^3 grid. The Figure 4.34 inset shows the mid-plane flow structure for a plane parallel to the heated wall. As can be seen, there are two large scale recirculations. Table 4.4 summarizes the number of iterations for different grid densities. The important aspect of multigrid convergence acceleration is evident in the table. This aspect is that whatever the grid density, the number of iterations is virtually constant.

An alternative to geometric is an algebraic method (Ruge and Stueben 1986 and Raw 1996). With this, instead of using a hierarchy of grids, equation coefficients

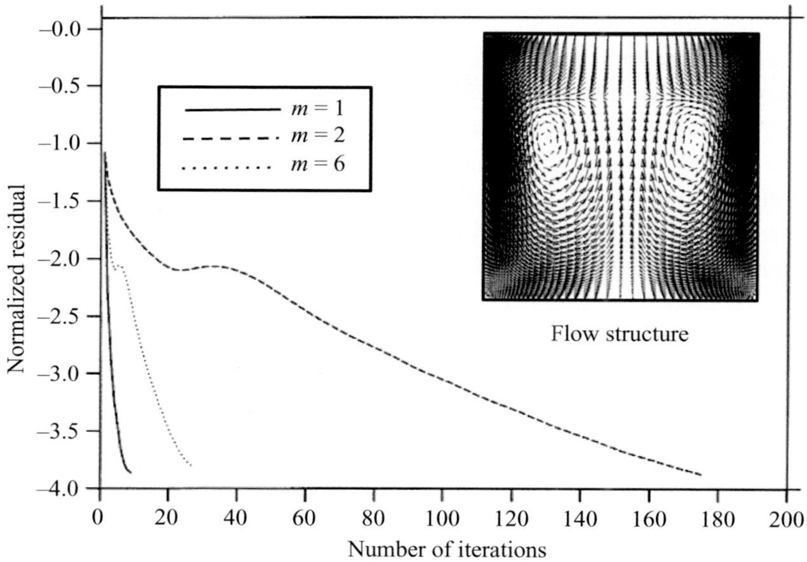

Figure 4.34. Plot of number of iterations against energy residual for $Gr = 1.44 \times 10^6$ and a 65^3 grid (from Tucker (2001) – Published with the kind permission of Springer).

are amalgamated at different levels. This approach is only really efficient for linear equations. A key difficulty with geometric multigrid is that as the grid is successively coarsened, for a complex geometry it is hard for the coarsened mesh to support the geometry. However, even with coarse grids that poorly represent geometries, impressive multigrid performance can sometimes be found. The algebraic multigrid does not suffer from the geometry resolution issue but as noted is limited to linear equations. However, it would be useful for the solution of the Poisson based wall distance equation discussed later in Chapter 7 and other specialist purposes.

4.6.3 *Coupling Enhancements and Coupled Solvers*

The compressible equation systems above could be solved sequentially and looped over until convergence. Similarly, the incompressible pressure based systems, to be considered further later, along with their pressure evaluation, are highly sequential processes. However, for rotating or swirling flows and so forth, there is a strong coupling in the equation systems. For example, if considering the flow in a cylindrical polar coordinate system, the dominant term in the radial momentum equation is the centrifugal force. This involves the square of the tangential momentum. Hence, when solved in a decoupled fashion, without the application of specialist coupling techniques (see Gosman et al. 1976 and Acharya and Jang 1988), convergence can be slow. In the approach of Acharya and Jang (1988), the discretized tangential velocity, v_θ, equation is written as

$$v_\theta = \left(\frac{\left[\sum A_{NB,v_\theta} v_\theta + S_{1,v_\theta} + S_{2,v_\theta} v_\theta \right]}{A_{p,v_\theta}} \right) \tag{4.178}$$

where S_1 and S_2 constitute a decomposed source term. This equation is used to replace v_θ^2 in the source term of the discretized radial velocity, v_r, equation. The extra terms are rearranged so that the radial momentum source term can be written in the linear form, $S_{1,v_r} + S_{2,v_r} v_{r,P}$. The Coriolis term ($2\rho_P v_r \Omega$, where Ω is system angular velocity) in the discretized form of the tangential momentum equation produces terms which augment $S_{2,v_r} v_{r,P}$. Therefore, when these terms are negative, diagonal dominance of the equation set can be improved by adding them to the term A_{P,v_r}. The approach of Gosman et al. (1976) is simpler to implement and generally found, for rotating flows, to be more effective. With this approach, the following group of terms are added to the radial momentum equation source term zone

$$S_{Extra} = \alpha \frac{\rho_P |v_\theta| \left(v_r^{n-1} - v_r^n \right)}{r} \tag{4.179}$$

where α is a tuneable parameter and the superscripts n refer to iterative states. Also, to increase diagonal dominance, v_r^n can be combined with A_{P,v_r}.

For general internal flow systems, the coupling of equations can be useful for iterative convergence and can remove convergence difficulties related to poor grid quality. A coupled algorithm for incompressible flow is described by Mangani et al.

(2013). For compressible flow, a procedure is outlined by Raw (1996). For coupled solutions, solvers that can deal with block matrices are required, (i.e., the array elements normally considered are arrays in themselves). The size of the 'submatrix' depends on the number of differential equations required to model the problem. It would be 5 × 5 for a three-dimensional problem. For a coupled algorithm, the array elements for all the flow governing equations must be stored at the same time. Hence, coupled solvers need more storage than their sequential counterparts. The basic nature of coupled solvers is gained by expressing the flow governing equations as

$$\frac{\overline{\Phi}_P^n - \overline{\Phi}_P^{n-1}}{\Delta t} + \left(\frac{\partial \mathbf{F}}{\partial x}\right)^n = 0 \tag{4.180}$$

which can be considered as a one-dimensional Euler type equation system. The temporal term is discretized using a forwards difference. The over bars or vectors are used to indicate that the terms are arrays. The elements of these would be the usual flux vector and vector of conserved variables. The following can be written

$$\mathbf{F}^n = \mathbf{F}^{n-1} + \frac{\partial \mathbf{F}}{\partial \overline{\Phi}} \Delta \overline{\Phi} + HOT \tag{4.181}$$

and hence

$$\left[\frac{\partial \mathbf{F}}{\partial x}\right]^n = \delta_x \mathbf{F}^{n-1} + \delta_x \frac{\partial \mathbf{F}}{\partial \overline{\Phi}} \Delta \overline{\Phi} + HOT \tag{4.182}$$

where δ_x is some form of, say, finite difference operator, $\partial \mathbf{F}/\partial \overline{\Phi}$ a Jacobian matrix and $\Delta \overline{\Phi} = \overline{\Phi}^n - \overline{\Phi}^{n-1}$. Substituting Equation (4.182) into (4.180) yields the following 'submatrix' Δ form (see later) system

$$\left[\frac{\mathbf{I}}{\Delta t} + \delta_x \left(\frac{\partial \mathbf{F}}{\partial \overline{\Phi}}\right)\right] [\Delta \overline{\Phi}] = -\delta_x \mathbf{F}^{n-1} \tag{4.183}$$

where **I** is a unit matrix. Equation (4.183) can then be assembled into a global solution matrix, each element of which is a submatrix. When using coupled solvers, relative to a sequential, the number of iterations to convergence should decrease. However, this does not necessarily mean that the actual computing time will be lower. This is a significant function of the equation set being solved. For example, with turbulent flows, the momentum and turbulence model equations are generally only weakly linked through diffusion terms. Then it can be better (see Willcox 1998) to keep the momentum and turbulence model equation link sequential (the momentum equations being still solved in a coupled fashion).

4.6.4 *Preconditioning*

At low Mach number as noted earlier (see Equation (4.101)), for compressible solvers, there is a large disparity in eigenvalues. The net result of this is inaccurate solutions and numerical stiffness. Using the Jacobian matrix,

$$\Gamma = \frac{\partial \Phi}{\partial \phi} \tag{4.184}$$

Equation (4.1), written for the compressible flow equations, can be re-expressed as

$$\Gamma \frac{\partial \phi}{\partial t} = A(\Phi) \tag{4.185}$$

where now, for a two-dimensional flow,

$$\phi = \begin{bmatrix} p \\ u \\ v \\ T \end{bmatrix} \tag{4.186}$$

where T is temperature. This new solution vector is instead of $\Phi = [\rho, \rho u, \rho v, \rho E]^T$. Also,

$$\Gamma = \begin{bmatrix} 1/\beta M^2 & 0 & 0 & 0 \\ u/\beta M^2 & \rho & 0 & 0 \\ v/\beta M^2 & 0 & \rho & 0 \\ \frac{e+p}{\rho \beta M^2} - 1 & \rho u & \rho v & -\frac{\gamma p R}{\gamma - 1} \end{bmatrix} \tag{4.187}$$

(see Choi and Merkle 1991), where R is the specific gas constant, $\beta = \gamma R T$ and γ is the ratio of specific heats. The eigenvalues have changed from the following values, noted earlier,

$$\Lambda = \begin{pmatrix} u \\ u \\ u \pm c \end{pmatrix} \tag{4.188}$$

to

$$\Lambda = \begin{pmatrix} u \\ u \\ \dfrac{u\left(1 + \frac{\beta M^2}{\gamma R T}\right) \pm c'}{2} \end{pmatrix} \tag{4.189}$$

where a key step is the introduction of the pseudo-acoustic speed

$$c' = \sqrt{u^2 \left(1 - \frac{\beta M^2}{\gamma R T}\right)^2 + 4\beta M} \tag{4.190}$$

where now $\beta = k \gamma R T$. The parameter k is chosen to ensure that the acoustic wave speed is similar to that of fluid particles. Hence, we have changed the scaling of time derivative in compressible solver through introduction of the Jacobian. The latter,

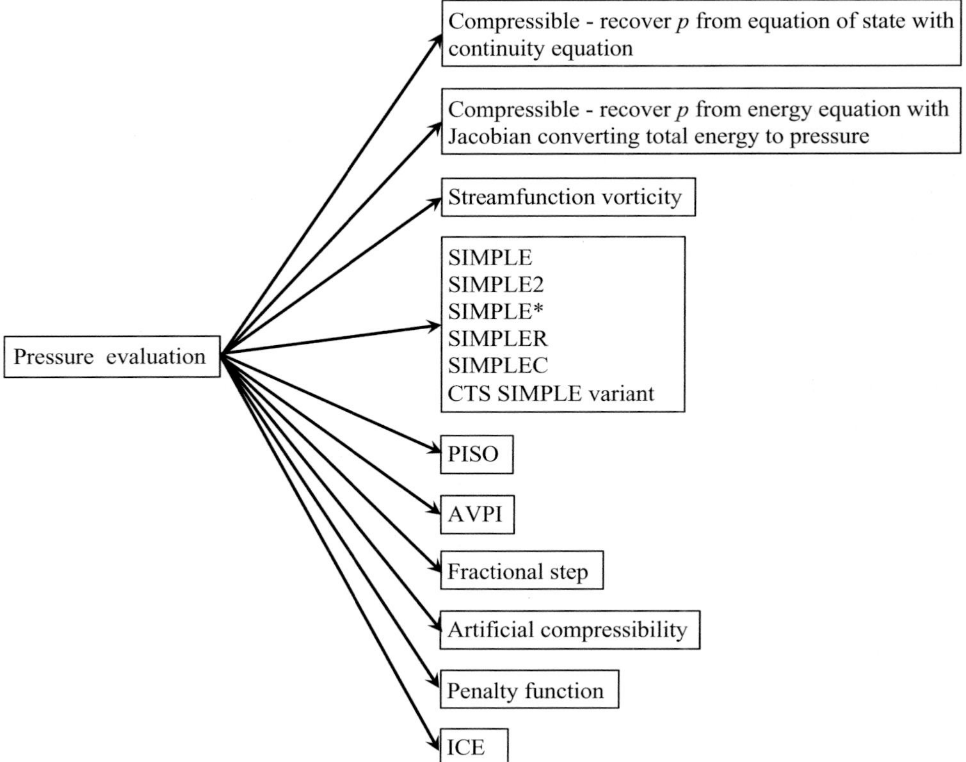

Figure 4.35. Summary of some pressure evaluation methods.

with c', maintains well-conditioned eigenvalues at low speeds. However, the time accuracy has been lost. Hence, Equation (4.185) must be considered as involving a pseudo-time. If time accuracy is needed, then the correct time derivative must be appended and a dual time stepping approach used. The most popular form of preconditioning is that of Weiss and Smith (1995). However, preconditioning approaches, although curing the problem of excessive smoothing, do not deal with the problem of numerical stiffness. Hence, they require a large number of iterations to secure iterative convergence. This is especially problematic for unsteady flow problems. Another issue is that most preconditioning strategies seem to neglect that in a three-dimensional flow, there are three key Mach number components. Amalgamating these into one resolved component neglects the tensorial nature of the discretized equations. Hence, for highly anisotropic Mach number fields, preconditioned results can be inaccurate. Then it seems better to switch to pressure based solution methods. These are discussed next.

4.7 Evaluation of the Pressure Field

There is a wide range of methods for evaluation the pressure field in CFD. Some of these are summarized in Figure 4.35. For compressible flows, the density changes are sufficient (with information provided from the energy equation) for pressure to be recovered reliably from the density change provided in the time derivative of the continuity equation. Alternatively, via a Jacobian it can be recovered from the energy equation in compressible flow (see Moinier 1999). At low Mach numbers, a simple, alternative expedient can be to scale down the geometry and increase velocity to preserve the Reynolds number. The higher velocity ensures a higher Mach number. If this is sufficiently high, then the compressible flow solver will have reasonable computational performance. However, for many flows, this Mach number scaling strategy will give rise to excessive density variations in localized regions. The Mach number required for reasonable compressible solver performance can vary dramatically between different codes. Hence, there is a need to understand the CFD programs traits. Also, the Mach number tends to zero at walls. This can be problematic, especially for eddy-resolving simulations. There are critical fine-scale flow structures near walls that require accurate numerical schemes to capture. For incompressible flow there are the following options:

1) Eliminate the pressure variable through use of the stream function vorticity method (see Chapter 2 for equations) or velocity-vorticity method;
2) Create some pseudo-equation of state that gives sufficient weak density changes to reliably recover pressure and
3) Through combining the Navier-Stokes and continuity equations, create an explicit equation for pressure.

Overall the latter is perhaps the best option. However, if one wishes to model flows over a wide Mach number range the situation is more complex. Wide ranges of Mach number can occur within a particular flow or simply through the range of cases that need to be considered. Examples of the former are the combustor-turbine zone in a gas turbine aeroengine. The combustor flow is low Mach number but the adjacent turbine has choked flow. Similarly, the turbine blade surface is perforated by cooling holes. These have a low Mach number but again the blade flow is choked. This aspect will be discussed later.

4.7.1 *Velocity-Vorticity Approach*

In two dimensions, or for axisymmetric flow, pressure can be eliminated through use of the stream function–vorticity formulation. In three dimensions, solving the vorticity Equation (4.133) can eliminate pressure. This equation is supplemented with the equations for the definition of vorticity and the continuity of flow equation to form a solvable system. Alternatively, the continuity and equation for the definition of vorticity can be combined to give

$$\nabla^2 \mathbf{u} = -\nabla \times \boldsymbol{\Omega} \tag{4.191}$$

Hence, Equations (4.133) and (4.191) can be solved together. As noted by Speziale (1987), although the Navier-Stokes equations need to include centrifugal and Coriolis terms depending on the frame of reference, the vorticity equation remains the same. The boundary conditions just need updating. Hence, the intrinsic numerical nature of the vorticity equation does not change. However, as noted by Davies and Carpenter (2001) the boundary condition for vorticity can be problematic.

4.7.2 *Artificial Compressibility Methods*

If density variations can be completely neglected, the artificial compressibility method of Rogers and Kwak (1990) can be effective. This is based on the approach of Chorin (1967). However, the approach of Chorin was aimed at steady problems. As with Chorin, the pseudo-equation of state used is

$$p = \frac{\rho}{\beta} \tag{4.192}$$

This corresponds to the pseudo-speed of sound:

$$c = \frac{1}{\sqrt{\beta}} \tag{4.193}$$

Using the artificial equation of state, the continuity equation is re-expressed as

$$\beta \frac{\partial p}{\partial \tau} + \left(\frac{\partial u_j}{\partial x_j} \right) + 0 \tag{4.194}$$

The parameter β has a fixed theoretical range (see Rogers et al. 1986). Although Chorin's outlook is incompressible flow, the Rogers and colleagues approach is intended to be used in a compressible solver framework. When used in this framework, like all other equations, the above pressure Equation (4.194) has a smoothing term. If β is too large, in steep pressure gradient areas, this will act as a substantial mass source term. However, with reasonable care the method is effective.

The penalty function approach (see Engelman and Sani 1986) has similarities to the artificial compressibility method. The continuity equation is modified along similar lines to give the following pressure relationship

$$p = -\frac{1}{\beta} \left(\frac{\partial u_j}{\partial x_j} \right) \tag{4.195}$$

where β is a small number, typically $10^{-9} \leq \beta \leq 10^{-4}$. The requirement for small values of β means that, for stability, explicit methods need prohibitively small time-steps.

4.7.3 *SIMPLE Related Methods*

An alternative approach (see Harlow and Welch 1965) is to substitute the momentum equations (i.e., Navier-Stokes equations) into the continuity equation. Then a Poisson equation for pressure can be formed with a rather extensive source term. Making the equation less wieldy, the SIMPLE-Semi-Implicit Method for Pressure-Linked Equations – (Patankar and Spalding 1972) and related methods

SIMPLEC -Semi-Implicit Method for Pressure Linked Equations-Consistent- (Van Doormaal and Raithby (1984)) and SIMPLER (SIMPLE Revised) became popular. These can be viewed as approaches to establish a pressure field that is compatible with mass conservation. Terms that ultimately become zero at iterative convergence are ignored.

The discretized momentum equations can be written as follows, where the pressure gradient term has been separated from source terms

$$A_{P,\phi}\phi_P = \sum A_{NB,\phi}\phi_{NB} + \frac{\Delta p}{\Delta n_\phi} + S_{1,\phi} \tag{4.196}$$

The term $\Delta p/\Delta n_\phi$ is the driving pressure gradient, in discrete form, for the variable ϕ. Therefore, in a Cartesian system when $\phi = u$, $\Delta p/\Delta n_u = \Delta p/\Delta x$, and similarly $\Delta n_v = \Delta y$ and $\Delta n_w = \Delta z$. Assuming approximate values, represented by $*$ superscripts, Equation (4.196) can be written as follows for u, v and w:

$$A_{P,\phi}\phi_P^* = \sum A_{NB,\phi}\phi_{NB}^* + \frac{\Delta p^*}{\Delta n_\phi} + S_{1,\phi} \tag{4.197}$$

The following is then assumed

$$\phi_P' = \phi_P - \phi_P^* \tag{4.198}$$

where ϕ' is a correction that when added to the approximate value ϕ^* will give the correct solution, ϕ. This correction equation can be formed by subtracting Equation (4.197) from Equation (4.196):

$$A_{P,\phi}\phi_P' = \sum A_{NB,\phi}\phi_{NB}' + \frac{\Delta p'}{\Delta n_\phi} \tag{4.199}$$

Note that the source terms are assumed to cancel. Generally, the correction equation is simplified. For the SIMPLE method, drastically, $\sum A_{NB,\phi}\phi_{NB}'$ is neglected. For SIMPLEC, the approximation $\sum A_{NB,\phi}\phi_{NB}' \cong \sum A_{NB,\phi}\phi_P'$ is used. Then, for SIMPLEC, Equation (4.199) can then become

$$A_{P,\phi}\phi_P' = \sum A_{NB,\phi}\phi_p' + \frac{\Delta p'}{\Delta n_\phi} \tag{4.200}$$

which can be re-expressed as

$$\phi_P' = d_{P,\phi}\frac{\Delta p'}{\Delta n_\phi} \tag{4.201}$$

where $\phi = u$, v or w and

$$d_{P,\phi} = \frac{1}{(A_{P,\phi} - \alpha_\phi \sum A_{NB,\phi})} \tag{4.202}$$

Note, an iterative under-relaxation parameter, α_ϕ, has been introduced in the above. This parameter, $0 < \alpha_\phi < 1$, improves the stability of the iterative solution. To obtain the SIMPLE scheme, $\sum A_{NB\phi}$ is just simply ignored. Equations (4.198) and (4.201) allow the following velocity equations to be formulated

$$\phi_P = \phi_P^* + d_{P,\phi} \frac{\Delta p'}{\Delta n_\phi} \tag{4.203}$$

Equations (4.203) are now substituted into the discretized continuity equation $\sum_{\phi=u,v} \Delta\phi_P / \Delta n_\phi = 0$.This gives the *pressure correction* equation

$$A_{P,p} p'_P = \sum A_{NB,\phi} p'_{NB} + S_{1,p'} \tag{4.204}$$

Equation (4.204) coefficients are given in Appendix B when using the simple finite difference-finite volume discretization outlined in Section 4.2.4. Using the pressure correction evaluated from Equation (4.204), the pressure is given as

$$p_P = p_P^* + \alpha_P p'_P \tag{4.205}$$

where α_P is an under-relaxation parameter ($0 < \alpha_P \leq 1$). The velocity components are updated using Equation (4.203). An iteration using the SIMPLE and related methods involves the following stages:

(1) Guess p^*;
(2) Solve momentum equations for the velocities u^*, v^* and w^*;
(3) Solve pressure correction for p';
(4) Apply Equation (4.205) to gain p;
(5) Correct velocities with Equation (4.203) and
(6) Return to (2), until iterative convergence reached.

Assessing convergence is discussed in Chapter 7. Further variants of SIMPLE are the SIMPLE2, SIMPLE* (Barton 1998a) and PISO schemes. SIMPLER (already noted) is similar to SIMPLE but also solves a Poisson equation for pressure rather than just p' (i.e., both p and p' equations must be solved for each iteration). The PISO (Pressure Implicit with Splitting of Operator) scheme of Issa (1986), extends SIMPLE with an additional velocity correction step and Poisson based pressure correction equation. There have been many studies comparing the performances of the above schemes. For steady flows these have been inconclusive. However, PISO is often found superior for unsteady flows – see, for example, Barton (1998a,b) who finds PISO to be more stable for larger time-steps.

For time dependent flows, with sufficiently small time steps, Equations (4.202) can be simplified (see Jones and Marquis 1985 and Henkes 1990), reducing memory requirements to give what is sometimes known as the AVPI scheme. This memory reduction is achieved by avoiding the need to store $d_{P,\phi}$ values. The momentum equation coefficients, for unsteady flow, take the form

$$A_{P,\phi} = \sum A_{NB,\phi} + \frac{\rho_P^{n-1}}{\Delta t} + S_{2,\phi} \tag{4.206}$$

where $S_{2,\phi}$ is a contribution to $A_{P,\phi}$ to improve diagonal dominance. Also, the superscript n represents the time level. Examining Equations (4.202) and (4.206) shows that, when $\alpha_\phi = 1$ and $[\rho_P^{n-1}/\Delta t \gg S_{2\phi}]$, Equation (4.202) can be approximated as

$$d_{P,\phi} \cong \frac{\Delta t}{\rho_P^{n-1}} \tag{4.207}$$

Equation (4.207) can be evaluated on the fly, thus saving storage. For a first order backwards difference (Euler) time scheme (see Section 4.8) and a constant density flow, the AVPI scheme corresponds to the pressure correction equation

$$\nabla^2 p'^n = \frac{\rho}{\Delta t} \nabla \mathbf{u}^* \tag{4.208}$$

Based on the AVPI assumption, Equation (4.203) becomes

$$\phi_P = \phi_P^* + \frac{\Delta t}{\rho_P^{n-1}} \frac{\Delta p'}{\Delta n_\phi} \tag{4.209}$$

which for $\phi = u$ is a finite difference expression for

$$\rho \frac{\partial u}{\partial t} = -\frac{\partial p'}{\partial x} \tag{4.210}$$

Equation (4.210) expresses a temporal change in u directly due to $\partial p'/\partial x$, and thus a time scale in Equation (4.209) and hence SIMPLE. Raithby and Schneider (1979, 1980) suggest that time scale inconsistencies between the discretized momentum equations and Equation (4.209) mean α_P in Equation (4.205) must be small. To remedy this inconsistency, they propose, based on the momentum equation coefficients,

$$\alpha_P = 1 - \frac{\sum A_{NB,\phi}}{A_{P,\phi}} \tag{4.211}$$

Importantly, Equation (4.206) shows that as $\Delta t \to 0$, diagonal dominance of the momentum equations is improved. The reverse is so for the pressure correction equation, their coefficients being inversely proportional to those for the momentum equations. This can lead to instability and the need for special treatments. It was noted that for the AVPI scheme to be applicable there was the condition $[\rho_P^{n-1}/\Delta t \gg S_{2\phi}]$. For rotating flows, for certain momentum equation components, $S_{2\phi}$ can be large. Under these circumstances it can be useful to hybridize the AVPI scheme with one of the SIMPLE related variants described above being used instead of AVPI where $S_{2\phi}$ is large. This will then give a partial storage saving since approximation (4.207) can still be partly applied.

The Implicit Continuous-fluid Eulerian (ICE) scheme of Harlow and Amsden (1971), like SIMPLE, yields a Poisson equation for pressure. However, it is suitable for flows at all speeds (when used for supersonic flows, SIMPLE, ideally, requires modification). Extensions to SIMPLE methods to high-speed flows are described by Demirdžić et al. (1993). As discussed earlier, such extensions are important for flows with disparate Mach number ranges.

4.7.3.1 Fractional Step Methods For eddy-resolving simulations, the computations are unsteady and fractional step methods see a degree of popularity. Strong use of the SIMPLE ingredients is made in a fractional-step, predictor-corrector procedure. For the procedure of Choi and Moin (1994), we rewrite Equation (4.1) in the following finite difference form with the pressure gradient term again isolated

$$\frac{\phi_P^* - \phi_P^{n-1}}{\Delta t} = \frac{1}{2}\left(A(\phi^*) + A(\phi^{n-1})\right) - \frac{1}{\rho}\frac{\Delta p^{n-1}}{\Delta n_\phi} \tag{4.212}$$

Also note that since we are considering incompressible flow, the conserved variables, Φ, are replaced by the velocity components ϕ. Since the approach is intended for a high-fidelity unsteady simulation, the discretization of the function A is expressed as a half and half weighted average of '*' and 'old' ($n - 1$) time levels. This corresponds to the use of a centred difference in time (Crank-Nicolson scheme – see Section 4.8). Equations (4.212) are implicitly solved based on a known $n - 1$ pressure field. The latter aids convergence. Then, half the $n-1$ time level pressure gradient is subtracted from the ϕ_P^* to give ϕ^{**}

$$\frac{\phi_P^{**} - \phi_P^*}{\Delta t} = \frac{1}{2\rho}\frac{\Delta p^{n-1}}{\Delta n_\phi} \tag{4.213}$$

Next, half the n time level pressure gradients $(-\Delta p^n/2\Delta n_\phi)$ are added, to give ϕ_P^n as

$$\frac{\phi_P^n - \phi_P^{**}}{\Delta t} = -\frac{1}{2\rho}\frac{\Delta p^n}{\Delta n_\phi} \tag{4.214}$$

Equation (4.214) (for $\phi = u$ and v) is substituted into the discretized continuity equation to give

$$\frac{\Delta^2 p^n}{\Delta x^2} + \frac{\Delta^2 p^n}{\Delta y^2} = \frac{2\rho}{\Delta t}\left(\frac{\Delta u^{**}}{\Delta x} + \frac{\Delta v^{**}}{\Delta y}\right) \tag{4.215}$$

Once p^n has been evaluated, using the Equation (4.215), Equation (4.214) provides ϕ^n. This procedure corresponds to solving

$$\frac{\phi_P^n - \phi_P^{n-1}}{\Delta t} = \frac{1}{2}\left(A(\phi^*) + A(\phi^{n-1})\right) - \frac{1}{2\rho}\left(\frac{\Delta p^n}{\Delta n_\phi} + \frac{\Delta p^{n-1}}{\Delta n_\phi}\right) \tag{4.216}$$

This does not exactly correspond to the use of the Crank-Nicolson scheme, for which $\phi^* = \phi^n$, but the error is second order.

4.7.4 *Pressure/Velocity Coupling*

The pressure difference across the control volume drives the flow. It is tempting to simply take the average of the two nodes that surround the control volume face to interpolate the pressure to the control volume face such that

$$p_w = \frac{(p_W + p_P)}{2}, \quad p_e = \frac{(p_E + p_P)}{2} \tag{4.217}$$

However, the difference is $p_E - p_W$. This means that the central pressure point is not sensed by the solution. The solution will hence have pressure oscillations. To overcome this problem, for many years the approach of Harlow and Welch (1965) was adopted. This involved the use of staggered control volumes as shown in Figure 4.5b. The momentum control volumes are located such that they are sandwiched between pressure nodes. This also has the highly desirable eddy-resolving simulation property that kinetic energy is conserved (Kim and Moin 1985). More recently, most codes avoid the need for staggered grids, which substantially complicate coding and code maintenance and are highly incompatible with unstructured grids. The interpolation of Rhie and Chow (1983) has become popular with incompressible flow solvers. This and related methods are discussed next.

4.7.4.1 Momentum Interpolation on Collocated Grids If the under-relaxation is made an integral part of the discretized Equation (4.35), with the time derivative included, then the following can be written

$$\phi_P = \frac{\alpha_\phi}{A_P}(A_E\phi_E + A_W\phi_W + A_N\phi_N + A_S\phi_S + B_P) \tag{4.218}$$

where

$$B_P = \frac{(1-\alpha_\phi)}{\alpha_\phi}A_P\phi_P^0 + \frac{\rho\Delta x\Delta y}{\Delta t}\phi_P^{old} + S_\phi \tag{4.219}$$

Please refer to Figure 4.3 to follow the nomenclature in this section. Also, the superscript 'o' refers to an old iteration and the superscript 'old' an old time-step. Note, S_ϕ in the previous contains the pressure gradient term. If this is next taken out, then the following can be written, for the u velocity component and nodes P and E

$$u_P = \frac{\alpha_u\left(\sum_i A_i u_i + B_P\right)_P}{(A_P)_P} - \frac{\alpha_u\Delta y(p_e - p_w)_P}{(A_P)_P} \tag{4.220}$$

$$u_E = \frac{\alpha_u\left(\sum_i A_i u_i + B_P\right)_E}{(A_P)_E} - \frac{\alpha_u\Delta y(p_e - p_w)_E}{(A_P)_E} \tag{4.221}$$

Similarly, patterned on the previous, the necessary velocity component at the 'e' control volume face can be expressed as

$$u_e = \frac{\alpha_u\left(\sum_i A_i u_i + B_P\right)_e}{(A_P)_e} - \frac{\alpha_u\Delta y(p_E - p_P)}{(A_P)_e} \tag{4.222}$$

in which data at the 'e' location must be appropriately interpolated. Similar expressions can obviously be written for the other control volume faces. With the Rhie and Chow interpolation, all the terms in the first right-hand side term of (4.222) and the $(A_P)_e$ term in the last term on the right-hand side of (4.222) are interpolated. To achieve this, a linear interpolation is used, based on the values given in (4.220) and (4.221). Hence, the following can be written

$$\left(\frac{\sum_i A_i u_i + B_P}{A_P}\right)_e = Wf_e\left(\frac{\sum_i A_i u_i + B_P}{A_P}\right)_E + (1 - Wf_e)\left(\frac{\sum_i A_i u_i + B_P}{A_P}\right)_P \tag{4.223}$$

and

$$\left(\frac{1}{A_P}\right)_e = Wf_e\left(\frac{1}{A_P}\right)_E + (1 - Wf_e)\left(\frac{1}{A_P}\right)_P \tag{4.224}$$

The linear weighting factor used in the previous is

$$Wf_e = \frac{\Delta x_P}{2\delta x_e} \tag{4.225}$$

To understand (4.222) better, substitute (4.223) into it, but making use of the components $[(\sum_i A_i + B_P)/A_P]_P$ (4.220) and $[(\sum_i A_i + B_P)/A_P]_E$ (4.221). This yields the following equation (with the temporal terms ignored)

$$u_e = \left[Wf_e u_E + (1 - Wf_e)\,u_P\right] + \left\{\begin{array}{l} -\frac{\alpha_u \Delta y(p_E - p_P)}{(A_P)_e} + Wf_e \frac{\alpha_u \Delta y(p_e - p_w)_E}{(A_P)_E} \\ +(1 - Wf_e)\frac{\alpha_u \Delta y(p_e - p_w)_P}{(A_P)_P} \end{array}\right\}$$

As we can see, this equation involves a simple linear velocity interpolation and then various differences in pressure. The latter gives rise to an enlarged pressure-smoothing stencil. The problem with the above is that the solution is both under-relaxation factor and time-step dependent. Choi (1999) addressed the latter using the following equation, which extends (4.222)

$$u_e = \alpha_u\left(\frac{\sum_i A_i u_i}{A_P}\right)_e - \frac{\alpha_u \Delta y(p_E - p_P)}{(A_P)_e} + (1 - \alpha_u)u_e^0 + \alpha_u \frac{\rho\, \delta x_e \Delta y}{(A_P)_e \Delta t} u_e^{old} \tag{4.226}$$

As shown by Yu et al. (2002), at convergence the following is true: $u_e = u_e^0 = u_e^{old}$, $u_E = u_E^0 = u_E^{old}$, and $u_P = u_P^0 = u_P^{old}$. Then (4.226) simplifies to

$$u_e = \alpha_u\left(\frac{\sum_i A_i u_i}{A_P}\right)_e - \frac{\alpha_u \Delta y(p_E - p_P)}{(A_P)_e} + (1 - \alpha_u)u_e + \alpha_u \frac{\rho\, \delta x_e \Delta y}{(A_P)_e \Delta t} u_e \tag{4.227}$$

which can be further rearranged as

$$u_e = \left(\frac{\sum_i A_i u_i}{A_P}\right)_e - \frac{\Delta y(p_E - p_P)}{(A_P)_e} + \frac{\rho\, \delta x_e \Delta y}{(A_P)_e \Delta t} u_e \tag{4.228}$$

Then application of (4.223) gives

$$\left(\frac{\sum_i A_i u_i}{A_P}\right)_e = \frac{Wf_e(\sum_i A_i u_i)_E (A_P)_P + (1 - Wf_e)(\sum_i A_i u_i)_P (A_P)_E}{(A_P)_P (A_P)_E} \tag{4.229}$$

Similarly, the application of (4.224) gives

$$(A_P)_e = \frac{(A_P)_P (A_P)_E}{Wf_e (A_P)_P + (1 - Wf_e)(A_P)_E} \tag{4.230}$$

Table 4.5. *Some momentum interpolation equations that have found use in CFD and their traits*

Velocity interpolation for east face	
Worker	Comments
Rhie and Chow	Both time-step and under-relaxation dependent solutions
$u_e = [Wf_e u_E + (1 - Wf_e)u_P] + \left\{ \begin{array}{l} -\frac{\alpha_u \Delta y(p_E - p_P)}{(A_P)_e} + Wf_e \frac{\alpha_u \Delta y(p_e - p_w)_E}{(A_P)_E} \\ +(1 - Wf_e)\frac{\alpha_u \Delta y(p_e - p_w)_P}{(A_P)_P} \end{array} \right\}$	
Majumdar	Not under-relaxation factor dependence for steady problems.
$u_e = [Wf_e u_E + (1 - Wf_e)u_P] + \left\{ \begin{array}{l} -\frac{\alpha_u \Delta y(p_E - p_P)}{(A_P)_e} + Wf_e \frac{\alpha_u \Delta y(p_e - p_w)_E}{(A_P)_E} + (1 - Wf_e) \\ \times \frac{\alpha_u \Delta y(p_e - p_w)_P}{(A_P)_P} + (1 - \alpha_u)\left[u_e^0 - Wf_e u_E^0 - (1 - Wf_e)u_P^0\right] \end{array} \right\}$	
Choi	Time-step dependent
$u_e = [Wf_e u_E + (1 - Wf_e)u_P] + \left\{ \begin{array}{l} -\frac{\alpha_u \Delta y(p_E - p_P)}{(A_P)_e} + Wf_e \frac{\alpha_u \Delta y(p_e - p_w)_E}{(A_P)_E} + (1 - Wf_e) \\ \times \frac{\alpha_u \Delta y(p_e - p_w)_P}{(A_P)_P} + (1 - \alpha_u)\left[u_e^0 - Wf_e u_E^0 - \left(1 - Wf_e\right) u_P^0\right] \\ + \left[\frac{\alpha_u A_e^{old}}{(A_P)_e} u_e^{old} - Wf_e \frac{\alpha_u \left(A_P^{old}\right)_E}{(A_P)_E} u_E^{old} - (1 - Wf_e) \frac{\alpha_u \left(A_P^{old}\right)_P}{(A_P)_P} u_P^{old} \right] \end{array} \right\}$	
Issa and Oliveira	Under-relaxation and time-step dependent, used for multiphase flow
$u_e = \frac{\alpha_u Wf_e\left(\sum_i A_i u_i + B_P\right)_E + \alpha_u (1 - Wf_e)\left(\sum_i A_i u_i + B_P\right)_P - \alpha_u \Delta y(p_E - p_P)}{Wf_e (A_P)_E + (1 - Wf_e)(A_P)_P}$	
Yu et al.	Both time-step and under-relaxation factor independent
$u_e = \frac{1}{(A_P)_e} \left\{ \begin{array}{l} [Wf_e (A_P)_E u_E + (1 - Wf_e)(A_P)_P u_P] \\ +\alpha_u [Wf_e (S_C)_E + (1 - Wf_e)(S_C)_P \delta x_e \Delta y \\ -Wf_e (S_C)_E \Delta x_E \Delta y - (1 - Wf_e)(S_C)_P \Delta x_P \Delta y] \\ +\alpha_u [-\Delta y(p_E - p_P) + Wf_e \Delta y(p_e - p_w)_E \\ +(1 - Wf_e)\Delta y(p_e - p_w)_P] \\ +(1 - \alpha_u)\left[u_e^0 - Wf_e u_E^0 - (1 - Wf_e)u_P^0\right] \\ +\alpha_u \left[A_e^{old} u_e^{old} - Wf_e \left(A_P^{old}\right)_E u_E^{old} - (1 - Wf_e)\left(A_P^{old}\right)_P u_P^{old}\right] \end{array} \right\}$	
$(A_P)_e = \begin{array}{l} Wf_e \left(\sum_i A_i\right)_E + (1 - Wf_e)\left(\sum_i A_i\right)_P \\ -[Wf_e (S_P)_E + (1 - Wf_e)(S_P)_P]\delta x_e \Delta y + A_e^{old} \end{array}$	

Then (4.230) and (4.229) can be substituted into (4.228) to give

$$u_e = \frac{[Wf_e(\sum_i A_i u_i)_E (A_P)_P + (1 - Wf_e)(\sum_i A_i u_i)_P (A_P)_E] - \Delta y[Wf_e (A_P)_P + (1 - Wf_e)(A_P)_E](p_E - p_P)}{(A_P)_P (A_P)_E - [Wf_e (A_P)_P + (1 - Wf_e)(A_P)_E]\rho \delta x_e \Delta y/\Delta t} \tag{4.231}$$

It can be seen that the above Equation (4.231) has no under-relaxation dependence. However, the interpolation expression is still a function of time-step. Yu and colleagues revised Equation (4.231) to correct this defect. Table 4.5 summarizes some

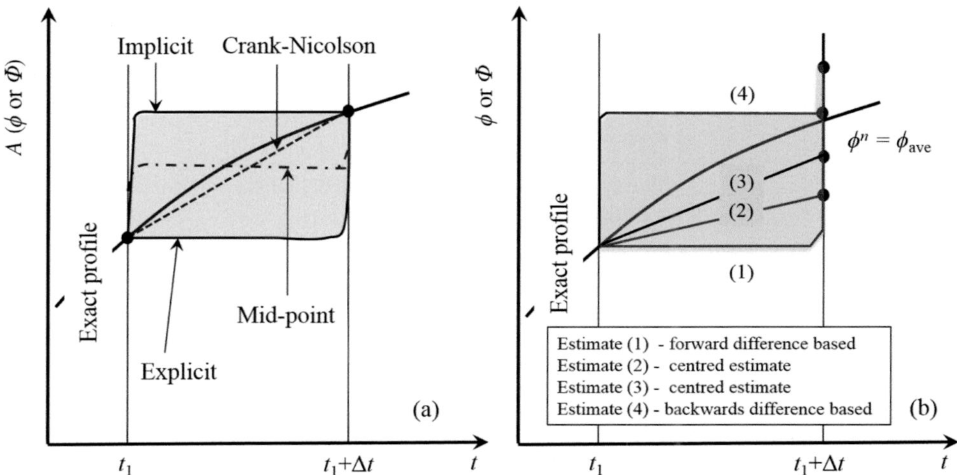

Figure 4.36. Temporal profile assumptions: (a) basic two-point schemes and (b) predictor-corrector procedure (Frame (a) based on Tucker (2013)) – Published with kind permission of Springer).

momentum interpolation equations that have found use in CFD and their traits. Note that in the table, $A_e^{old} = \rho\ \delta x_e \Delta y/\Delta t$ and so forth.

4.8 Temporal Schemes

As with the spatial schemes, the temporal scheme discretization generally involves some form of profile assumption in the right hand side of Equation (4.1), but now in time.

Figure 4.36a gives examples of some profile assumptions. Note that for some multi-stage, predictor-corrector schemes, as shown in Frame (b), the profile assumption can be composite in nature. Such schemes will be discussed later. The basic profiles can be expressed as

$$\phi = Wf^{n+1}\phi^{n+1} + Wf^{n}\phi^{n} + Wf^{n-1}\phi^{n-1} \tag{4.232}$$

where ϕ is some function and

$$Wf^{n} = \lambda, Wf^{n-1} = 1 - \lambda - k, Wf^{n-2} = k \tag{4.233}$$

Typical values of k and λ are considered later. The left hand side temporal derivative in Equation (4.1) can also be broken down and expressed in the weighted sense as

$$(1+\eta)\frac{\Delta[\rho_i\phi_i]^{n}}{\Delta t} - \eta\frac{\Delta[\rho_i\phi_i]^{n-1}}{\Delta t} = A(\phi) \tag{4.234}$$

where

$$\frac{\Delta[\rho_i\phi_i]^{n}}{\Delta t} = \frac{[(\rho\phi_i)^{n} - (\rho_i\phi_i)^{n-1}]}{\Delta t}, \quad \frac{\Delta[\rho_i\phi_i]^{n-1}}{\Delta t} = \frac{[(\rho\phi_i)^{n-1} - (\rho_i\phi_i)^{n-2}]}{\Delta t} \tag{4.235}$$

Table 4.6. *Values of η, λ and k for commonly occurring time schemes (based on Tucker (2013) – published with the kind permission of Springer)*

Scheme	η	λ	k	Stability	Order	Other Features
Explicit (Euler)	0	0	0	$0 \leq C^2 \leq 2D \leq 10 \leq C^2 \leq 2D \leq 1$	1	–
Implicit (Backwards Euler)	0	1	0	Unconditionally Stable	1	–
Crank-Nicolson (one-step trapezoidal)	0	½	0	Unconditionally Stable	2	–
Adams-Moulton	0	5/12	1/12	–	3	–
Exponential	0	$\lambda = \dfrac{1}{1 - e^{A_{i\phi}\Delta t}} - \dfrac{1}{A_{i\phi}\Delta t}$	0	Unconditionally Stable	1	–
Lax-Wendroff	0	0	0	$0 \leq C^2 \leq 2D^* \leq 1$	–	$\Gamma = \Gamma^*$– modified diffusion coefficient

Table 4.6 summarizes some values of η, λ and k for just a few time schemes. Further schemes are tabulated by Beam and Warming (1976). Note that when generating different schemes, Equation (4.232) is applied just to the spatial derivatives – the right-hand-side terms in Equation (4.1). Also, the median dual cell vertex scheme, unlike the cell centred, necessitates the consideration of how the time derivative varies over the control volume. It can simply be assumed constant. This is called a lumped formulation. It can, on the other hand, vary in some fashion consistent with the underlying discretization (see Fletcher 1998). This is achieved through the use of a mass matrix with the intent of giving accuracy benefits. Such matters are not outlined here. Depending on the value of η, Equation (4.234) supports different levels of data in time. Schemes that make use of two levels of data are considered next.

4.8.1 *Two-Level Schemes*

4.8.1.1 Implicit/Semi-Implicit The semi-implicit, Crank-Nicolson scheme, being a central difference in time, is neutrally dissipative. In essence, it is relatively easy to implement, having a weighted average of the right-side data in Equation (4.1). The only question that arises is if, in the function $A(\phi)$, the temporal averaging

$$A(\phi) = \frac{(A(\phi^n) + A(\phi^{n-1}))}{2} \tag{4.236}$$

or

$$A(\phi) = \frac{A(\phi^n + \phi^{n-1})}{2} \tag{4.237}$$

is used. The distinction between Equations (4.236) and (4.237) is drawn by Skelboe (1977), where these variants are called one and two legged schemes. The former has more stability. This has parallels to the Jameson, kinetic energy preserving scheme, discussed earlier in Section 4.4.3.4, where the variables going into the fluxes are centrally averaged, rather than the functions themselves, again giving greater stability. Even though the Crank-Nicolson scheme appears quite simple, a range of further variants can be derived. See, for example: Giles (2004), Hujeirat and Rannacher (1998), Beam and Warming (1976), Briley and McDonald (1975), Lacor (1999) and Jameson (1991). These variants are largely aimed at enhancing stability or computational efficiency in this neutrally dissipative scheme. The neutrally dissipative nature of the Crank-Nicolson scheme makes it well suited to eddy-resolving simulations. The scheme can be hybridized. For example, the explicit Adams-Bashforth scheme (see later in Section 4.8.2) can be used for the convective terms and Crank-Nicolson just for the diffusive. For increasing $\lambda > 0.5$, dissipation can be introduced into the Crank-Nicolson scheme. $\lambda = 0.6$ corresponds to a scheme that can be derived based on the finite element method with linear temporal elements. $\lambda = 1$ corresponds to a first order backwards difference in time. Like the first order upwind in space, this scheme is highly dissipative and hence not suitable for eddy-resolving simulations. For $0.5 \leq \lambda \leq 1$, two-level schemes are unconditionally stable. This does not mean that for $\lambda < 1$, solutions will be physically realistic (as with low dissipation spatial scheme solutions, they can have substantial oscillatory components), but instead that errors will reduce with time.

4.8.1.2 Linearization of Crank-Nicolson Equation Next we will consider how to solve the Crank-Nicolson equations efficiently. We will consider the simple model equation

$$\frac{\partial \phi}{\partial t} + \frac{\partial (\phi\phi)}{\partial x} = 0 \tag{4.238}$$

If we apply averaging to the function and not the individual variables, this can be discretized as

$$\frac{\partial \phi}{\partial t} + \frac{\partial}{\partial x}\left(\frac{(\phi\phi)^{n-1}}{2} + \frac{(\phi\phi)^{n}}{2}\right) = 0 \tag{4.239}$$

Then, discretizing in space, using, for illustrative purposes, simple central differences, gives

$$\frac{\phi_P^n - \phi_P^{n-1}}{\Delta t} + \frac{1}{2}\frac{(\phi_E\phi_E - \phi_W\phi_W)^{n-1}}{2\Delta x_P} + \frac{1}{2}\frac{(\phi_E\phi_E - \phi_W\phi_W)^{n}}{2\Delta x_P} = 0 \tag{4.240}$$

For computational efficiency, ideally we would wish to linearize the $(\phi\phi)$ products. This is done using the Taylor series:

$$(\phi\phi)_P^n = (\phi\phi)_P^{n-1} + \Delta t\left(\frac{\partial \phi\phi}{\partial \phi}\right)_P^{n-1} + \Delta t^2\left(\frac{\partial (\phi\phi)}{\partial t^2}\right)_P^{n-1} \tag{4.241}$$

Using the Jacobian $J = (\partial\,(\phi\phi)/\partial\phi)_P^{n-1}$, Equation (4.241) can be re-expressed as

$$(\phi\phi)_P^n = (\phi\phi)_P^{n-1}\left(\frac{\partial\phi\phi}{\partial\phi}\right)_P^{n-1}(\phi^n - \phi^{n-1}) + O(\Delta t^2) \tag{4.242}$$

Since $J = (\partial\,(\phi\phi)/\partial\phi)_P^{n-1} = 2\phi_P^{n-1}$, this simplifies to

$$(\phi\phi)_P^n = 2\left(\phi^{n-1}\phi^n\right)_P - (\phi\phi)_P^{n-1} \tag{4.243}$$

Equation (4.243) can be substituted into Equation (4.240) to give

$$\frac{\phi_P^n - \phi_P^n}{\Delta t} + \frac{1}{2\Delta x_p}\left[(\phi^{n-1}\phi^n)_E - (\phi^{n-1}\phi^n)_W\right] = 0 \tag{4.244}$$

This equation is linearized and hence cheaper to solve whilst maintaining the second order accuracy of the Crank-Nicolson scheme.

This procedure can also be focused on the differential equation. Firstly, Equation (4.238) is expressed as

$$\frac{\partial\phi}{\partial t} + \lambda\frac{\partial F^n}{\partial x} + (1-\lambda)\frac{\partial F^{n-1}}{\partial x} = 0 \tag{4.245}$$

where $F = \phi\phi$. The equation is then rearranged as

$$\frac{\partial\phi}{\partial t} + \lambda\left(\frac{\partial F^n}{\partial x} - \frac{\partial F^{n-1}}{\partial x}\right) = -\frac{\partial F^{n-1}}{\partial x} \tag{4.246}$$

Based on Equation (4.242), the following can be stated

$$F^n = F^{n-1} + J(\phi)^{n-1}(\phi^n - \phi^{n-1}) \tag{4.247}$$

Then the following can be written based on the previous

$$\frac{\partial F^n}{\partial x} = \frac{\partial F^{n-1}}{\partial x} + \frac{\partial}{\partial x}(J(\phi)^{n-1}\Delta\phi) \tag{4.248}$$

Substituting of this into Equation (4.246) gives

$$\frac{\partial\phi}{\partial t} + \lambda\frac{\partial}{\partial x}(J(\phi)^{n-1}\Delta\phi) = -\frac{\partial F^{n-1}}{\partial x} \tag{4.249}$$

where $\Delta\phi$ is the solution variable and the equation is sometimes called the *Delta*(Δ) *form*.

The Δ form can also be formulated based on the discretized equation. Equation (4.1) can be expressed as the following weighted form for the non-conserved variable

$$\frac{\Delta\phi}{\Delta t} = \lambda A^n + (1-\lambda)A^{n-1} \tag{4.250}$$

Equation (4.242) can again be used to give the following

$$A^n = A^{n-1} + J(\phi)^{n-1}\Delta\phi + O(\Delta t^2) \tag{4.251}$$

where $J(\phi)^{n-1} = \partial A/\partial\phi$. Then substitution of that equation into Equation (4.250) yields

$$[1 - \lambda\Delta t J(\phi)^{n-1}]\Delta\phi = \Delta t A^{n-1} \tag{4.252}$$

This linearization work is related to the work of Beam and Warming (1976). Further variants are described by Briley and McDonald (1975), Jameson (1991) and Lacor (1999).

4.8.1.3 General Explicit Schemes For $\lambda = k = \eta = 0$, the first order explicit forwards difference Euler scheme is gained. The severe stability constraints relating to this scheme are summarized in Table 4.6. The diffusive term restriction $2D \leq 1$ implies that when the mesh spacing is halved, the time-step must be made four times smaller. The scheme thus requires small time-steps and is only naturally suited to problems that need extremely high temporal resolution (i.e., problems where variables do not change slowly with time) with carefully distributed grids that do not have excessively small local cells that exert an excessive global solution impact. High order explicit methods can be formulated, but typically the stability restrictions become ever more severe. The key advantage of explicit methods is that when performing parallel simulations data only needs to be transferred once per time-step. Hence, such methods can be parallelized highly efficiently. The approach is also highly compatible with compact schemes, which when exploited in implicit mode need costly matrix inversions.

The Lax-Wendroff scheme illustrates an important concept used in many flow solvers. It is really a scheme for Euler equation solution. Its basis is the direct application of the Taylor series to model the time derivative as

$$\frac{\partial \phi_P}{\partial t} \approx \frac{\phi_P^n - \phi_P^{n-1}}{\Delta t} - \left(\frac{\Delta t}{2!}\right)\frac{\partial^2 \phi}{\partial t^2} \tag{4.253}$$

Based on dimensional arguments, since $\Delta t = \Delta x/u$, the final term can be approximated as $0.5\,\Delta t\, u^2 \partial^2\phi/\partial x^2$. Hence, the Lax-Wendroff scheme involves adding a diffusive term. The related Lax scheme is used in one of the problem exercises at the end of this chapter. This relates to developing a simple Euler solver. In this exercise a crude Laplacian smoothing term is added to more than cancel the anti-diffusive term arising from the explicit temporal scheme. Many aerospace schemes have an explicit basis and as noted in the spatial integration section, have diffusive terms added. These also counterbalance anti-diffusivity from the explicit temporal integration, derived earlier, thus providing stability. Hence, it can be seen that the temporal and spatial integrations, ideally, need to be considered together – even though the method of lines allows their isolation.

Much like the Lax-Wendroff scheme, Gresho et al. (1984) presented the Balancing Tensor Diffusivity (BTD) scheme. This is based on the work of Dukowicz and Ramshaw (1979) and Crowley (1967). This, scheme, unlike the Lax-Wendroff, looks at convection-diffusion equations of the form $\partial\phi/\partial t + u\partial\phi/\partial x = \Gamma\partial^2\phi/\partial x^2$. BTD corresponds to discretizing this equation with the forward Euler scheme and augmenting the diffusion coefficient to

$$\Gamma^* = \Gamma + u^2\frac{\Delta t}{2} \tag{4.254}$$

For three-dimensional flows, to apply BTD, cross-stream diffusion terms are needed, giving a nineteen-point difference stencil. Hence, the scheme is complicated to implement and hence seldom sees use. For $\Gamma = 0$, (in one-dimension) BTD becomes the Lax-Wendroff method. The scheme is designed to improve both stability and accuracy.

4.8.1.4 Stability and Accuracy A simple stability analysis given by Patankar and Baliga (1978) is helpful in understanding the basic stability traits of some of the schemes given in Table 4.6. It considers numerical solution of the model equation $\partial\phi/\partial t = \Gamma\partial^2\phi/\partial x^2$. This equation just has the most problematic diffusive term. For this simple diffusive model equation, the finite difference discretization is $A_{W\phi} = A_{E\phi} = 2\Gamma/(2\Delta x)^2$, $A_{P\phi} = 2\Gamma/(2\Delta x)^2$. A discretized spatial domain consisting of three grid points can be considered. The following boundary and initial conditions can be applied – for $t \geq 0$, $\phi_W = \phi_E = 0$, and $t = 0$, $\phi_P > 0$. The discretized equation, multiplied by Δt, is

$$\frac{\phi_P^n}{\phi_P^{n-1}} = \frac{1 - A_{P\phi}\Delta t(1 - \lambda_P)}{1 + \lambda_P A_{P\phi}\Delta t} \tag{4.255}$$

For a physically sensible behaviour, ϕ_p^n/ϕ_P^{n-1} should stay positive. It also should tend to zero with increasing $A_{P\phi}\Delta t$. Figure 4.37 shows the variation of ϕ_p^n/ϕ_P^{n-1} with increasing $A_{PQ}\Delta t$ for different values of λ. Clearly, as Δt becomes large, the Crank-Nicolson scheme will give physically unrealistic solutions. Also shown in the figure is the exponential analytical temporal integration scheme given in Table 4.6 (Note, for practical problems it is inadvisable to use the exponential scheme since its formal accuracy is first order and exponentials are expensive to repeatedly compute). It is accepted that for $0.5 \leq \lambda \leq 1$ two-level schemes are unconditionally stable. However, as noted earlier, this does not mean that for excessively large time-steps they give physically realistic solutions. This can be seen from Figure 4.37.

4.8.2 *Higher-Level Schemes*

The Adams-Bashforth is an explicit scheme. It is based around the Taylor series expansion for ϕ^{n+1}:

$$[\phi]_i^{n+1} = [\phi]_i^n + \left.\frac{\partial[\phi]}{\partial t}\right|_i^n \Delta t + \frac{1}{2}\left.\frac{\partial^2[\phi]}{\partial t^2}\right|_i^n \Delta t^2 + O(\Delta t^3) \tag{4.256}$$

The derivative involving the 2nd power is approximated as the finite difference given below

$$\left.\frac{\partial^2[\phi]}{\partial t^2}\right|_i^n = \frac{\partial}{\partial t}\left(\frac{\partial[\phi]}{\partial t}\right)_i^n = \frac{\left.\frac{\partial[\phi]}{\partial t}\right|_i^n - \left.\frac{\partial[\phi]}{\partial t}\right|_i^{n-1}}{\Delta t} + O(\Delta t) \tag{4.257}$$

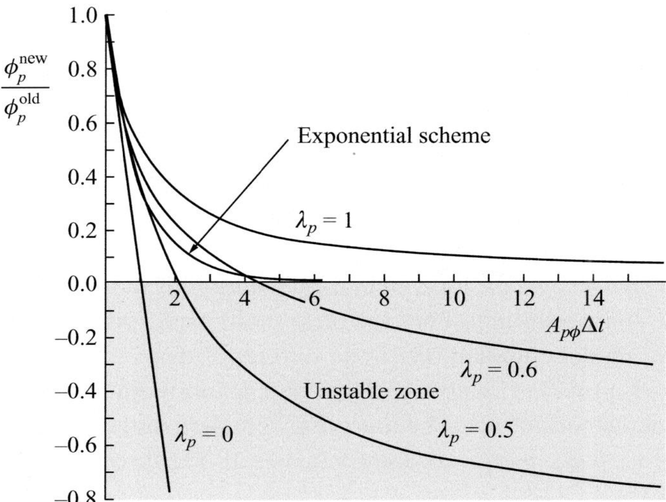

Figure 4.37 Variation of ϕ_p^n/ϕ_P^{n-1} and hence stability with $A_{PQ}\Delta t$ for $\lambda = 0 - 1$ (Adapted from Patankar and Baliga 1978).

Combining Equations (4.256) and (4.257) gives the following

$$[\phi]_i^{n+1} = [\phi]_i^n + \left.\frac{\partial[\phi]}{\partial t}\right|_i^n \Delta t + \frac{1}{2}\left(\frac{\left.\frac{\partial[\phi]}{\partial t}\right|_i^n - \left.\frac{\partial[\phi]}{\partial t}\right|_i^{n-1}}{\Delta t} + O(\Delta t)\right)\Delta t^2 + O(\Delta t^3) \tag{4.258}$$

or alternatively

$$[\phi]_i^{n+1} = [\phi]_i^n + A(\phi)^n + \Delta t + \frac{1}{2}\left(\frac{\left.\frac{\partial[\phi]}{\partial t}\right|_i^n - \left.\frac{\partial[\phi]}{\partial t}\right|_i^{n-1}}{\Delta t} + O(\Delta t)\right)\Delta t^2 + O(\Delta t^3) \tag{4.259}$$

The Adams-Bashforth scheme can be expressed more generally to higher orders as

$$\phi^{n+1} - \phi^n \simeq \Delta t \sum_{t=0}^{p} W f_l \left(\frac{\partial \phi}{\partial t}\right)^{n-1} \tag{4.260}$$

where the parameter, p, determines the number of coefficients involved. Various Adams multipoint methods can be gained by selecting different values of η, λ and k in Equations (4.233) and (4.234). Unlike the Adams-Bashforth, the Adams-Moulton method is implicit. The coefficients for the third order method are given in Table 4.6. Clearly, the more levels used, the more data storage and the more data transfer at parallel interfaces. Hence, in practical simulations, no more than three levels would typically be found. As with high resolution spatial schemes, the coefficients in Equation (4.260) can be modified for specialist purposes (see Tam and Webb 1993).

Table 4.7. *Weighting functions for backwards difference schemes (from Tucker (2013) – published with the kind permission of Springer)*

N_{BD}	Wf^{n+1}	Wf^{n}	Wf^{n-1}	Wf^{n-2}	Wf^{n-3}	Wf^{n-4}
1	1	−1	–	–	–	–
2	3/2	−4/2	1/2	–	–	–
3	11/6	−18/6	9/6	−2/6	–	–
4	25/12	−48/12	36/12	−16/12	3/12	–
5	137/60	−300/60	300/60	−200/60	75/60	−12/60

4.8.2.1 Gear Schemes Another implicit method is the backwards difference Gear scheme. The discrete equation for this is given as

$$\left[\frac{\partial \phi}{\partial t}\right] = \sum_{n-(N-1)}^{n+1} \frac{Wf^n}{\Delta t} = -A^{n+1} \tag{4.261}$$

Table 4.7 gives weighting functions for Equation (4.261). The parameter N_{BD} corresponds to the order.

4.8.3 *Predictor-Corrector Methods*

Predictor-corrector methods are another key class of temporal scheme. Probably the most widely used temporal integration schemes in aerospace engineering, especially for acoustics, are explicit Runge-Kutta schemes. These are used for both unsteady flows and integrating to steady state solutions. Essentially, the Runge-Kutta technique can be considered as using a carefully selected blend of lower order methods (see Figure 4.36b). These are intended to give a desired higher order of accuracy or increased stability. The blend of lower order methods gives rise to multiple stages of solutions with different weighting coefficients, *Wf*. Explicit Runge-Kutta methods are most common. However, there are implicit variants that allow larger Courant numbers. Explicit Runge-Kutta schemes offer the potential for high order with low storage. Most Runge-Kutta schemes are tuned for stability. For a p stage scheme that advances a solution from the nth to the $n+1$ time-step, the following can be written

$$\begin{aligned} &\phi^0 = \phi^n \\ &\phi^l = \phi^n + Wf_l \Delta t A(\phi^{l-1}) \qquad \text{for } l = 1, \ldots\ldots, p \\ &\phi^{n+1} = \phi^p \end{aligned} \tag{4.262}$$

where Wf_l are the coefficients of the algorithm. Table 4.8 shows the coefficients for a standard four-stage scheme. These coefficients can be altered as with the high resolution spatial schemes to give customized numerical properties for different types of numerical problems (see Bogey and Bailly 2004 and Hu et al. 1996).

Equation (4.263) attempts to clarify the practical implementation. The scheme is basically a blend of various temporal differences. All are pivoted around time

Table 4.8. *Standard four-stage Runge-Kutta scheme coefficients*

Wf_1	Wf_2	Wf_3	Wf_4
1	1/2	1/6	1/24

level 'n' (i.e., this remains fixed for all the stages). The gradients, multiplied by the weighting functions, are evaluated based on the latest Runge-Kutta scheme data, from summing the fluxes around the control volume.

$$
\begin{aligned}
[\phi]_1 &= [\phi]^n + \left.\frac{\partial[\phi]}{\partial t}\right|_0 Wf_1\Delta t \\
[\phi]_2 &= [\phi]^n + \left.\frac{\partial[\phi]}{\partial t}\right|_1 Wf_2\Delta t \\
[\phi]_3 &= [\phi]^n + \left.\frac{\partial[\phi]}{\partial t}\right|_2 Wf_3\Delta t \\
[\phi]_4 &= [\phi]^n + \left.\frac{\partial[\phi]}{\partial t}\right|_3 Wf_4\Delta t \\
[\phi]^{n+1} &= [\phi]_4
\end{aligned}
\tag{4.263}
$$

Clearly, each time-step involves more work – four times that needed for a standard first order Euler scheme. However, the idea is that the scheme allows the use of a larger Courant number to thus compensate for the additional effort.

As noted earlier, the Runge-Kutta scheme is often used for time marching to a steady state. For this purpose, good smoothing properties are important when used with multigrid convergence acceleration. To improve stability, the right-hand side of Equation (4.1) can be split into different components. These can be associated with convection, A_c, diffusion, A_d, and numerical smoothing, A_s. A blend of these can be used for different stages of the solution. The following blend is presented by Lacor (1999)

$$A^n = A_c^n + \alpha\left(A_d^n + A_s^n\right) + (1-\alpha)\left(A_d^{n-1} + A_s^{n-1}\right) \tag{4.264}$$

The superscript n indicates the stage of the Runge-Kutta scheme and α is a stability-tuned parameter. A variety of weightings are possible.

Residual smoothing (see Lerat 1979) improves the surrounding support for a grid point and hence allows the use of larger explicit time-steps. Obviously, since the residual is zero at convergence, the averaging process does not impact on the final solution. With this approach, the residual, A_m, is smoothed to $\overline{A}_m$. This is achieved by adding a scaled, by ε, pseudo-Laplacian (exact for a uniform grid) to A_m:

$$\overline{A}_m = A_m + \varepsilon\nabla^2\overline{A}_m \tag{4.265}$$

where

$$\nabla^2 \overline{A}_m = \sum_{NB} (\overline{A}_{NB} - \overline{A}_m) \tag{4.266}$$

The subscript *NB* implies summation over all faces associated with node *m*. Therefore,

$$\overline{A}_m = A_m + \varepsilon \sum_{NB} (\overline{A}_{NB} - \overline{A}_m) \tag{4.267}$$

This can be rearranged and solved using a simple Jacobi iteration (around two being sufficient) as

$$\overline{A}_m = \frac{A_m + \varepsilon \sum_{NB} (\overline{A}_{NB})}{1 + \varepsilon \sum_{NB} (1)} \tag{4.268}$$

As a general rule of thumb, with residual smoothing, the maximum allowable Courant number can be increased by a factor of $\sqrt{1 + 4\varepsilon}$. The typical maximum practical value $\varepsilon = 0.75$ allows a factor of 3 increase in time-step or Courant number. Other variants on this residual smoothing approach exist. For example, when forming the Laplacian, rather than using a centered discretization, upwind based methods can be used (Zhu et al. 1993 and Blazek et al. 1991). To crudely test this approach with the Euler solver that forms part of the problem exercise, it is convenient to smooth $\Delta\Phi$ using a simple averaging of surrounding values based on Equation (4.70). The application of this equation can simply be repeated.

4.8.4 *Splitting Methods*

Splitting methods are a further key class of temporal integration scheme. A disadvantage of implicit methods is the expense of solving large simultaneous equation sets. For explicit methods this is not necessary, but instead there are greater stability restrictions. With splitting methods, implicit and explicit schemes are combined. In certain circumstances, improvements in computational performance are made. The most widely known splitting method is the Alternating Direction Implicit (ADI) scheme of Peaceman and Rachford (1955). With this, in two dimensions, one full time-step Δt essentially consists of two $\Delta t/2$ half steps. For the first half time-step, spatial derivatives in, say, an x coordinate direction are treated implicitly while those in the y direction are treated explicitly. For the second half step, this procedure is reversed. Douglas and Gunn (1964) present a second order splitting scheme which in three dimensions is unconditionally stable.

4.8.5 *Adaptive Time-Steps*

As with spatial steps, time-step sizes or the schemes order can be adapted locally (in time) to optimize accuracy with respect to cost. Broadly the procedures used are akin to those used for adapting the spatial resolution. To assess the truncation errors schemes of different natures (see Gresho et al. 1984), orders or computations with different step lengths can be contrasted. Based on Taylor's series, for a scheme of order $q-1$, the following can be derived

$$\Delta t^n = C_t \Delta t^{n-1} \left| \frac{\varepsilon}{E^{n-1}} \right|^{1/q} \tag{4.269}$$

where the superscripts n and $n-1$ refer to solutions for different step sizes or orders, n being the latest. The parameter, ε, is a pre-set normalised error input value. This enables the maximum temporal solution error to be specified ($\varepsilon = 0.001$ corresponds to about a 0.1% solution error). Also, C_t, is a safety factor. Since the time-step needs to be spatially invariant, the temporal error $E^{n-1} \; \alpha \; \phi_2 - \phi_1$ (the difference in the solution for the differing orders, step lengths or schemes), needs to be averaged or the maximum considered and this difference appropriately normalized to give a meaningful error estimate. In addition to the basic approaches noted here that originate from the classical solution of ordinary differential equations, there are a range of other time-step adaptation approaches. For example Muramatsu and Ninokata (1992) use fuzzy logic and with space-time meshes (see Chapter 3), temporal adaptation can also be utilized (see Mani and Mavriplis 2010)). Notably, akin to using the adjoint method to explore spatial truncation errors, this approach can also be applied temporally (Rumpfkeil and Zingg 2007). However, a key difficulty is the requirement to perform an integration of the adjoint equations backwards in time. This requires storage of the solution through time. The implementation of even the most basic adaptive scheme needs care with regards to the maximum allowable time-step change (swing), correct normalization of errors and the most sensible zone to examine the temporal truncation error. Also, limits need to be placed on the minimum time-step size.

For steady flow solutions, local time-stepping can be used to accelerate convergence. The idea is that in local zones the optimal time-step is used based on the cell size and local velocities. Frink et al. (1991) use the following equation, based on a two-dimensional stability analysis

$$\Delta t_m = \frac{\mu}{\rho} \left(\frac{V_m}{C_{1,m} + C_{2,m} + C_{3,m}} \right) \tag{4.270}$$

where

$$\begin{aligned} C_{1,m} &= [|u_m| + c_m] L_{x.m} \\ C_{2,m} &= [|v_m| + c_m] L_{y.m} \\ C_{3,m} &= [|w_m| + c_m] L_{z.m} \end{aligned} \tag{4.271}$$

The L terms are projected areas. The time-step update can be made periodically.

4.9 Boundary Conditions

The boundary condition specification differs greatly between compressible and incompressible flow solvers. For incompressible solvers, typically at an inflow

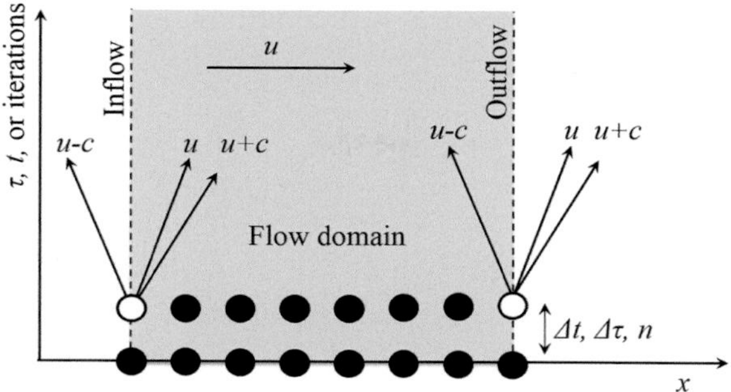

Figure 4.38. Boundary condition specification for compressible, subsonic flow.

boundary the velocity can be specified quite naturally as a Dirichlet boundary condition. At the outflow, differential boundary conditions (of different natures, potentially implemented using finite differences) can be applied to velocities and the pressure specified. The boundary normal velocity should be rescaled to ensure mass conservation. If a pressure correction solver is being used, then no pressure correction need be applied at the boundary where the pressure is specified or known. At solid walls, typically, a differential boundary condition is applied on the pressure correction equation with the normal gradient set to zero. Typically this would be achieved by zeroing the equation coefficient that links the control volume to the wall. The pressure at the inflow can also be specified. Again this would mean that no pressure correction is applied. The wall velocities would then need to be specified in a manner that ensures mass conservation. There are a range of application field specific boundary conditions that can be needed. Some examples of these are given in Chapter 6.

For compressible flow, the boundary condition specification at again the most basic level is quite different. As discussed earlier, for compressible flow there are the three characteristics with eigenvalues u, $u + c$ and $u - c$ and these need to be considered at the boundary. These describe information propagation rates. For subsonic flow, at the inlet, u and $u + c$ are positive, while $u - c$ is negative. Hence, for subsonic flow, as shown in Figure 4.38, we have two 'information' sources going into the domain resulting in the need to specify two flow properties – normally p_0 and T_0, but this aspect is flexible. On the other hand, at the outflow, there is one piece of information coming in relating to the eigenvalue $u - c$. Hence, there is just one thing to specify. This is normally the static pressure. However, if a simple (centred difference) solver is used with a smoother, as outlined previously, this allows more directional information propagation. Hence, there is a need for some other softer boundary conditions. Typically, at inflow

$$\rho_1 = \rho_2 \left(\text{i.e.} \frac{\partial \rho}{\partial x} \sim 0 \right) \tag{4.272}$$

or the same condition can be applied to pressure. At the outflow, extrapolations can be used as

$$\begin{cases} \rho u_I = 2\rho u_{I-1} - \rho u_{I-2} & \left(\text{i.e. } \frac{\partial^2 \rho u}{\partial x^2} \sim 0\right) \\ \rho_I = 2\rho_{I-1} - \rho_{I-2} & \left(\text{i.e. } \frac{\partial^2 \rho}{\partial x^2} \sim 0\right) \end{cases} \tag{4.273}$$

Sometimes, the soft boundary conditions are obvious. For example, at solid surfaces, fluxes can be set to zero. The boundary condition specification for compressible flow is of such complexity that some trial and error can be necessary to ensure compatibility between the solver framework and numerical boundary conditions. For example, using the Euler equation in one dimension, the following Bernoulli type equation follows

$$\rho u \frac{\partial u}{\partial x} = -\frac{\partial p}{\partial x} \rightarrow \rho u du = -dp \left(\text{or since } c = -\sqrt{\frac{dp}{d\rho}}\right) \rightarrow \delta u = -c^2 \frac{d\rho}{du} \tag{4.274}$$

This shows that at low Mach numbers, small density changes cause large velocity changes. This can be problematic when, for example, specifying the inflow primary variables according to Problem (10) in Chapter 2. It will give rise to instability. To avoid this, under-relaxation, via Equation (4.168), of the density at the inflow plane can be used with $\alpha_\rho \approx 0.25$. This crude under-relaxation will destroy time accuracy. If this is needed, more sophisticated boundary condition treatments are needed. Note that use of the equations in Problem (10), Chapter 2, is expected in the compressible solver proposed and in part of the problems section here. Further action can be needed with the density. For example, even with Equation (4.272) it is possible for the inlet density value to rise above that of the inlet stagnation density value ($\rho_0 = p_0/RT_0$). This results in the square root of a negative number being taken in Problem (10). Hence, density clipping can be needed to ensure the inlet density in the iterative process is less than the stagnation value of density at the inlet.

For supersonic flow, at the outflow, no characteristics point into the domain and hence nothing needs to be specified. At the inflow, there are three characteristics pointing into domain. Hence, three things need to be specified. We can define the static pressure at the inflow (note that $p_{0,IN}/p_{IN} = (1 + \gamma - 1/2Ma_{IN}^2)^{\gamma/(\gamma-1)}$). Defining the static pressure at the inflow fixes density i.e., $\rho_{IN} = \rho_{IN,0}(P_{IN}/P_{IN,0})^{1/\gamma}$.

As explored in Chapter 6, for systems with multiple boundaries there is the potential for non-unique solutions. Boundary location is also an important thing to consider. The specification of boundary conditions for unsteady flows have further complications and these are explored next.

For eddy-resolving simulations it can be necessary at the inflow plane to impose time varying vortex distributions with a wide range of scales. This is best done through velocity boundary conditions. On the other hand, at the outflow, a specified static pressure is a fully non-reflecting boundary condition. Figure 4.39 shows the distribution of vertical disturbance velocity for a Tollmien-Schlicting wave in a channel (this was discussed in Chapter 3). For an initial period, the curves have

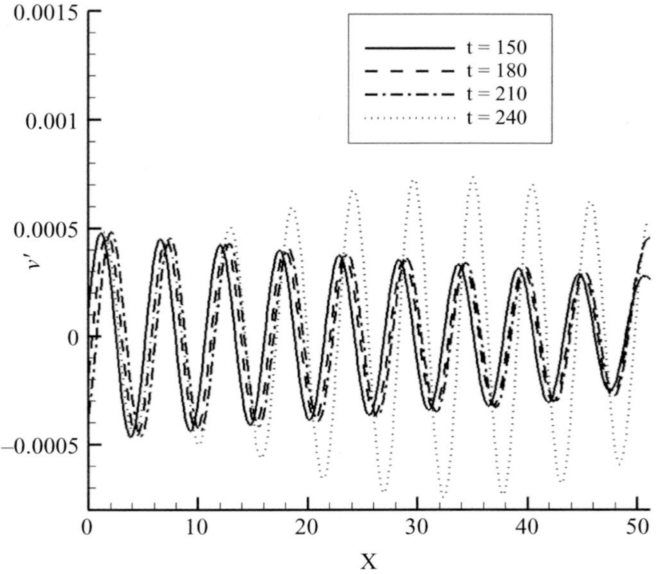

Figure 4.39. Boundary reflection for a Tollmien-Schlichting wave in a channel (wave distributions shown at different points in time).

perfect agreement with the analytical solution but eventually disturbance growth (from reflection) and solution divergence can be seen.

Hence, at the outflow it is important to avoid boundary reflections. Simple strategies include keeping outflow boundaries well away from areas of intense unsteadiness, coarsening grids towards boundaries (so that they will not support unsteady waves) and increasing the numerical dissipation towards boundaries (to damp out unsteadiness). The latter approach is used by Shur et al. (2003). The listed crude strategies have the downside that they are computationally wasteful and this has led to a wide range of research exploring means of allowing compact non-reflecting domains. An effective, one-dimensional, compressible flow, characteristic based, non-reflecting boundary condition is due to Giles (1990). A simpler convective condition is that of Pauley et al. (1990):

$$\frac{\partial \phi}{\partial t} + U_c \frac{\partial \phi}{\partial x} = 0 \tag{4.275}$$

where U_c is taken as the bulk mean velocity. Another simple approach is that of Ashcroft and Zhang (2001). With this, the instantaneous flow, ϕ, is pinned to a target field, ϕ_{target}. If accurate for the mean flow, a RANS predicted target could be used as the target for an eddy-resolving simulation. Alternatively, an initial time averaged, LES field could be used (see Shur et al. 2003). The pinning equation is given as

$$\phi(l) = (1 - s)\phi_{\text{target}}(l) + s\phi(l) \tag{4.276}$$

The instantaneous flow is gradually blended to the target field over a blending zone of extent L. A potential function for this purpose is given as

$$s = |1 - (l/L)^{\beta}| \tag{4.277}$$

where $0 < l < L$, $l = 0$ corresponding to the domain boundaries and $l = L$ inside the domain. Figure 7.22, Frame (a) also shows a potential s distribution.

4.10 Conclusions

In this chapter, just a very few of the potential vast array of numerical schemes have been outlined. Fortunately, as with RANS turbulence models and eddy-resolving models (see Chapter 5), few numerical schemes see main line use. In terms of convective term treatments, for compressible flows the Roe scheme still sees great popularity. This is even though it is problematic for eddy-resolving simulations for which the scheme was never designed. Also, for the compressible solvers, Runge-Kutta based time marching, with multigrid, has some considerable popularity. SIMPLE based solvers have great use with most variants of SIMPLE, including PISO, typically not, for a range of flows, showing dramatic performance differences. High order, unstructured methods are still in their relative infancy. However, these might enable massively parallel, complex geometry simulations to be performed, by enabling lower levels of data transfer across processors. The discontinuous approaches, outlined in the text under the CPR banner, avoid the need for large computational stencils and the data transfer associated with these. However, in implicit mode the mainstream high order unstructured approaches give a substantial memory overhead. Also, there are problems with numerical stiffness. They can be difficult to converge with more advanced RANS models (see Chapter 5). Also, for eddy-resolving simulations they appear sometimes to be too dissipative at high wave numbers. For rotating flows, the use of fully coupled solvers is compelling as with complex internal flows. Even, notionally, the same numerical scheme can be implemented in subtly different ways. This can have a strong solution influence. Most stability enhancements seem to detract from accuracy. This is especially so for eddy-resolving simulations where there is greater sensitivity to both the spatial and temporal schemes and the way these interact. Kinetic energy conservation can restrict the growth of energy associated with divergence. This inherent stability can allow the use of less smoothing which in turn can give highly substantial accuracy gains. Hence, this is an example where improved stability is not traded for accuracy. Cleary the design of numerical algorithms is an extremely important field and has largely stagnated over recent years, but exciting new work is emerging.

Problems

1) Apply the modified equation analysis to the following equation

$$\frac{\partial \phi}{\partial t} + u\frac{\partial \phi}{\partial x} = S \tag{4.278}$$

For a first order offset difference in space, show that the modified equation gives anti-diffusive, diffusive and neutrally diffusive temporal components for the forwards, backwards and centrally differenced Crank-Nicolson schemes. For the former, ensure the modified equation corresponds with that given in the text.

2) In Section 4.2.4 a finite-difference/finite volume discretization for the following equation was given

$$\rho u \frac{\partial \phi}{\partial x} + \rho v \frac{\partial \phi}{\partial y} = -\frac{\partial p}{\partial x} + \mu \left[\frac{\partial^2 \phi}{\partial x^2} + \frac{\partial^2 \phi}{\partial y^2} \right] \tag{4.279}$$

This was for a strictly Cartesian grid. The discretised equation took the following form

$$A_P \phi_P = A_W \phi_W + A_E \phi_E + A_N \phi_N + A_S \phi_S + S_\phi \tag{4.280}$$

The discretization used central differences for the convective terms. Write a computer program in some language such as BASIC, FORTRAN or MATLAB to solve the above equation for the general variable ϕ. Verify that the solver will predict linear Couette flow (see analytical solution in Question 1 of Chapter 2) where $\phi = u$ (x – direction velocity) for a channel where everywhere the pressure gradients ($\partial p/\partial x = 0$) are zero. Initially use a Gauss-Siedel solver. Note, a TDMA solver is given in Appendix C.

3) Adapt the program in Question (2) to investigate the behaviour of computations of the unsteady convection-equation:

$$\frac{\partial \phi}{\partial t} + u \frac{\partial \phi}{\partial x} + v \frac{\partial \phi}{\partial y} = 0 \tag{4.281}$$

where ϕ is a scalar quantity to be convected in time t by a velocity field (u, v) in some two-dimensional domain in x and y. Use a two-dimensional grid of equally spaced points where $-1 <= x <= 1$ and $-1 <= y <= 1$ and specify an initial concentration of ϕ in a cylindrical form. Take the initial concentration at a radius 0.25 centred on (–0.5, – 0.5) to be a maximum concentration of 1.0. Outside of the cone, the initial concentration is zero. Subject the cone to a velocity field which does not vary with time and is given by

$$\begin{aligned} u &= -y \\ v &= x \end{aligned} \tag{4.282}$$

which is a velocity field that rotates anticlockwise. Adapt your computer program to investigate explicit, semi-implicit and fully implicit temporal discretizations. For spatial terms, test second order central differences along with first, second and third order upwind differences. In all cases the boundary condition for ϕ is that the normal derivative across the boundary should be zero. Vary the grid size, time-step and angular rotation.

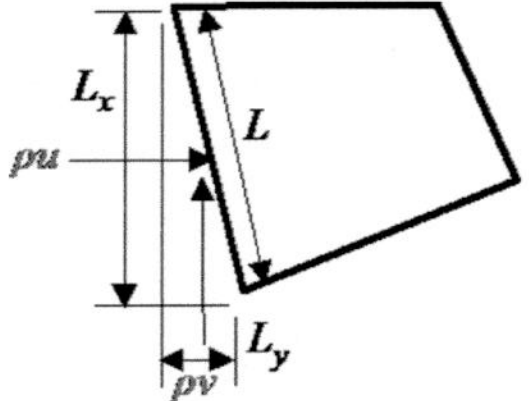

Figure 4.40. Flux evaluation for a west face.

Hints:

- run cases for a time equivalent to moving the cylinder through 90° from its original position to (–0.5, 0.5).
- use the width of the cylinder and its height to help characterise the effect of the numerical diffusion.
- investigate limits where wiggles do not occur.
- investigate the form of the equations and the effect of varying appropriate ratios of the variables.

4) In the Problems section of Chapter 3 there was a mesh generation exercise for supersonic flow over a wedge – Question 3. As part of this, mesh and edge weights where generated (Problems 3–9). Use these to perform control volume flux integration for an Euler solution. Use the initialization process outlined in Problem 9 of Chapter 2. The finite volume discretization can be simply expressed as

$$V\frac{\Delta\rho[?]}{\Delta t} = \sum_{CVS}\delta m(?) = \text{Source} \tag{4.283}$$

where V is the cell volume, in this instance an area. Also, the summation is over the control volume surfaces (CVS). For mass conservation, the bracketed quantities are as follows: [1], (1), Source $= 0$. For momentum, $[u]$ and (u). The source term sums pressures multiplied by the projected face lengths with appropriate signs to express the impact of pressure differences driving the flow. For energy, [E], (h_0). Equation (4.282) can be alternatively expressed as

$$V\frac{\Delta\rho[?]}{\Delta t} = \sum\nolimits_{cvs}\text{Flux} \tag{4.284}$$

and then for a face, L, as shown in Figure 4.40, the following can be written

$$\text{Flux}_{mass} = \dot{m} = \rho\mathbf{u}\cdot\mathbf{L} = \rho u L_x + \rho v L_y \tag{4.285}$$

$$\text{Flux}_{mom} = \dot{m}u = (\rho\mathbf{u}\cdot\mathbf{L})\mathbf{u} \tag{4.286}$$

or

$$\text{Flux}_{x,mom} = (\rho u L_x + \rho v L_y)u + pL_x \tag{4.287}$$

$$\text{Flux}_{y,mom} = (\rho u L_x + \rho v L_y)v + pL_y \tag{4.288}$$

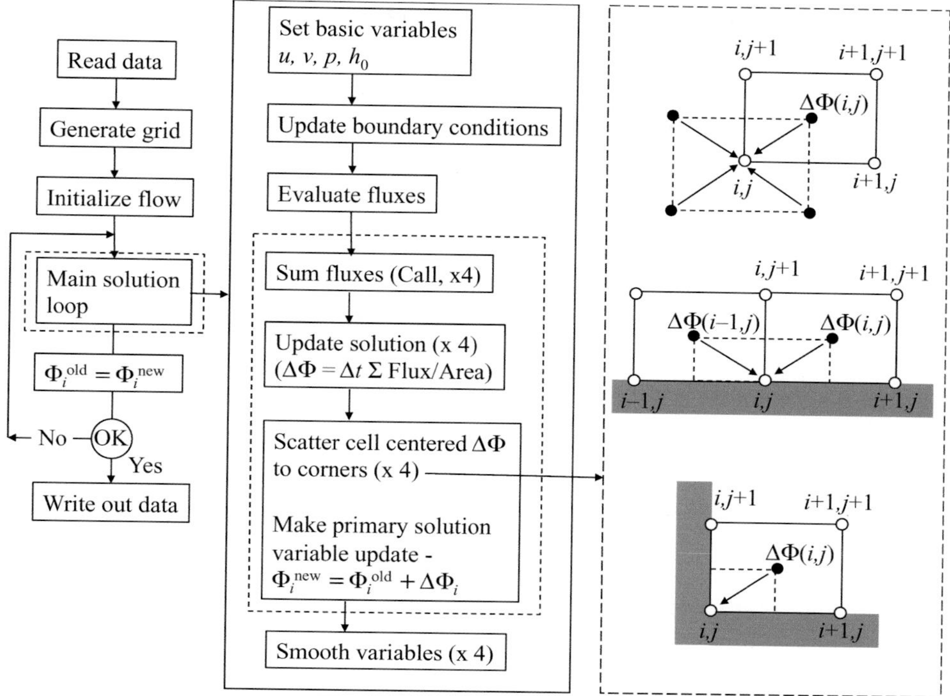

Figure 4.41. A flow chart for the program structure.

Similarly, for the energy equation, the following can be written

$$\text{Flux}_{enthalpy} = \dot{m}h_0 = (\rho\mathbf{U} \cdot \Delta\mathbf{L})h_0 \tag{4.289}$$

or

$$\text{Flux}_{enthalpy} = (\rho u L_x + \rho v L_y)h_0 \tag{4.290}$$

To solve the system of equations, use a crude variant of the Lax-Wendroff method, sometimes called the Lax method. Essentially the following equation is solved

$$\rho^{n+1}[?]^n = \frac{\Delta t}{V}\sum\nolimits_{cvs}\text{Flux} + \rho^n[?]^n + \frac{1}{4}\varepsilon CFL\Big([\rho?]^n_{i,j+1} + [\rho?]^n_{i+1,j} - 4[\rho?]^n_{i,j} + [\rho?]^n_{i-1,j} + [\rho?]^n_{i,j-1}\Big) \tag{4.291}$$

As shown, use a first order explicit difference to advance in time. The velocity scale in the Courant (CFL) number should be the sum of local fluid particle velocity and the local speed of sound. However, to simplify matters and give a conservative time-step value, make the velocity scale as twice the speed of sound. Also, you can make the space step, in the Courant number, the smallest in the whole domain and have a uniform conservative time-step value. To be safe, initially try a Courant number of around 0.5 with ε also around this value. A flow chart for the program structure is given in Figure 4.41. The program is best made modular using subroutines. The left-hand flow chart in Figure 4.41 shows the broad program structure. The central chart gives the main solution loop, which is repeatedly called.

A subroutine to set boundary conditions is needed. However, this should focus on the values at the inflow and outflow. At solid walls it is necessary to set a zero flux condition and this is best done when defining the fluxes. To start with it is best to assume both subsonic inflow and outflow and hence use a modest pressure ratio of say $p_2/p_{01} \approx 0.9$. Hence, in accordance with the main body of the text, it is necessary to specify the total temperature and pressure at the inlet (along with the stagnation density) and also the flow direction. Problem 10 in Chapter 2 outlines equations to specify what is needed at the inflow boundary. Note that at low Mach numbers, the under-relaxation of density and also its clipping can be needed, as discussed in the boundary condition section in the preceding text. At the outlet, it is just necessary to specify the static pressure.

In the main loop, the four key solution fluxes (mass, momentum in the x and y directions and energy) are evaluated. These are then summed. The primary solution variable, Φ, is then updated.

For the overlapping grid, cell vertex method, to set fluxes at faces of constant 'i' value an algebraic expression of the form below is needed

$$\begin{aligned} \mathit{fluxi}\Box(i,j) &= 0.5[\rho u(i,j) + \rho u(i,j+1)]Lix(i,j)\Box(i,j) \\ &\quad + 0.5[\rho v(i,j) + \rho v(i,j+1)]Liy(i,j)\Box(i,j) + S \end{aligned} \tag{4.292}$$

Table 4.9 gives the flux quantities in Equation (4.292). Note that the control volume surface pressure force is combined with the momentum flow, essentially giving a solution to the strong conservation form of the equations as discussed in Chapter 2. For more classical cell centred or cell vertex approaches, the indices in Equation (4.292) would be altered to select nodes that straddle the control volume face rather than sit on its corners.

Similar expressions can be written for j faces. The evaluated fluxes need to be summed, as follows, to give an updated primary variable solution change.

$$\begin{aligned} &\Delta\Box = \Delta t \Sigma \mathit{flux}, \Box \rightarrow \rho, \rho u, \rho v, E \\ &\Delta\Box = \Delta t[\mathit{iflux}(i,j) - \mathit{iflux}(i+1,j) + \mathit{jflux}(i,j) - \mathit{jflux}(i,j+1)]/\mathit{Area}(i,j) \\ &\Box^{new} = \Box^{old} + \Delta\Box \end{aligned} \tag{4.293}$$

For the overlapping grid cell vertex method, although the updated solution variable is at the cell centre, the variables are primarily stored at the cell vertices. Hence, the updated solution changes can be averaged or 'scattered' to the cell corners and then the primary solution variable updated. The scattering process for interior grid points, walls and corners are show in the right-hand group of schematics in Figure 4.41. The top frame shows it for interior points where simply 1/4 values at surrounding grid points are used. At walls half of

Table 4.9. *Flux quantities in Equation (4.292)*

Quantity	□	S
Mass	1	0 – no mass source
x – momentum	$v_{x,\,ave}$	dlix(i,j) p_{ave}
y – momentum	$v_{y,\,ave}$	dliy(i,j) p_{ave}
Energy	h_0	0 – no internal heat generation

the values at the two wall adjacent values are used. For all corners, the nearest interior value is directly injected to the corner node. This consistent averaging is important to maintain time accuracy. The smoothing process involves use of Equation (4.71)[2], with Equations (4.70) and (4.72). The latter is necessary to ensure there is smoothed data at boundaries. Note that this is just second order derivative smoothing and is hence inaccurate but stable. Higher order smoothing can be tried once the basic program is working.

The subroutines can be broken down in various ways. It can be convenient to do the solution update and distribution of the cell centred data to the corners in one subroutine with the overlapping grid cell vertex method.

Initially try the overlapping grid cell vertex method. The inflow/outflow boundary condition is discussed in this chapter. Variables will need representing at walls through taking adjacent first off-wall grid values. This part needs care, especially with the use of overlapping grids and the consequent dual meaning of subscripts.

First try a simple channel flow case, with a bump in the middle as shown in Figure 4.4. Explore for different pressure ratios and levels of smoothing how symmetrical the flow is. Then try supersonic flow as shown in Figure 4.42.

For the supersonic case, shown in Chapter 3 with the wedge angle of 5.7°, check that your solver gives the expected shock angles relative to the analytical solution for such flows and the solution traits (see Chapter 3). Alter the level of explicit smoothing to see how this impacts on the shock strength. Tip: build the solver up slowly, constantly checking that the tasks that you expect the program to perform are actually taking place. Note that for the supersonic flow case, the initialization must be altered to that given in Chapter 2, otherwise a supersonic flow through the domain might not be gained. Instead, use a static pressure defined to give $M = 1.6$ based around the total pressure. Based on isentropic flow and the static pressure, estimate the density and velocity at the outlet and apply these estimates through the complete domain. This will give supersonic flow. Note that for this high-speed flow, the solution is likely to be less stable and so need a smaller Courant number.

2 Note, to ensure values that have already been smoothed are not smoothed again separate arrays must be defined for the smoothed and unsmoothed variables.

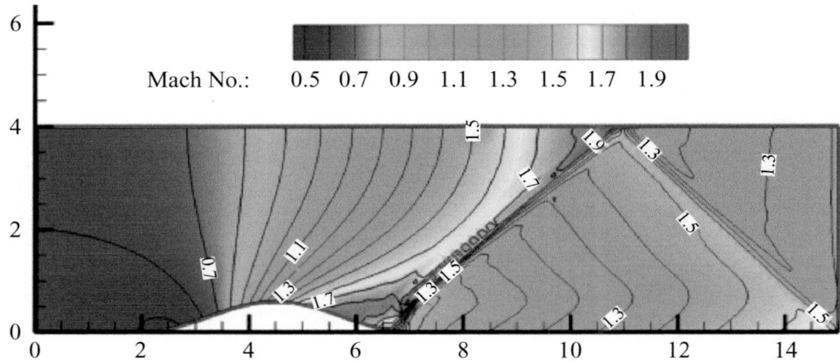

Figure 4.42. Mach number contours for supersonic flow over a bump.

5) When working, the basic Euler code can be extended to make use of a Runge-Kutta 'temporal' integration, fourth order smoothing, deferred correction, residual smoothing and many other techniques described in this text. Explore some of these options. For the fourth order smoothing, a simple way to construct an explicit smoothing expression is to use finite differences (ignoring grid misalignment) or fit a third order polynomial along grid lines, not fitting to the central node point. The polynomial can then be rearranged to make the central grid point its subject. The two polynomials in the vertical and horizontal directions can then be averaged.
6) Try combining the second order Laplacian smoothing with higher order, fourth derivative smoothing. Use a shock switch, as discussed in Section 4.4.2, to just turn on the lower order derivative smoothing around shocks. Demonstrate that this gives rise to improved flow solutions for the flow field shown in Figure 4.41.
7) Using the equation coefficients given in Appendix B, generate a pressure correction flow solver. Define a subroutine to initialize the flow. Then define subroutines to set up the coefficients for two velocity components and also the pressure correction equation. Initially, define $r = 1$ everywhere and solve for simple Cartesian, Couette flow comparing with the analytical solution given in Question 1 of Chapter 2. Then move to cylindrical coordinates and compare with the solution for Poiseuille flow in an annular geometry, again comparing with the analytical solution in Question 4 given in Chapter 2 for verification. Initially use a simple Gauss-Siedel iterative solver, with AVPI or SIMPLE, to converge the iterative equation set. You could also consider TDMA and other pressure correction variants. To start with, for the simple Couette flow, since the pressure is uniform everywhere, the pressure correction at wall can be set to zero and thus the differential boundary conditions at walls for the pressure correction can be initially avoided. (However, these can be readily implemented by setting wall coefficients to zero in the discretized equations.) This is possible because the pressure everywhere is zero (constant). Initially, generate a code where the velocities are collocated. Study the pressure and

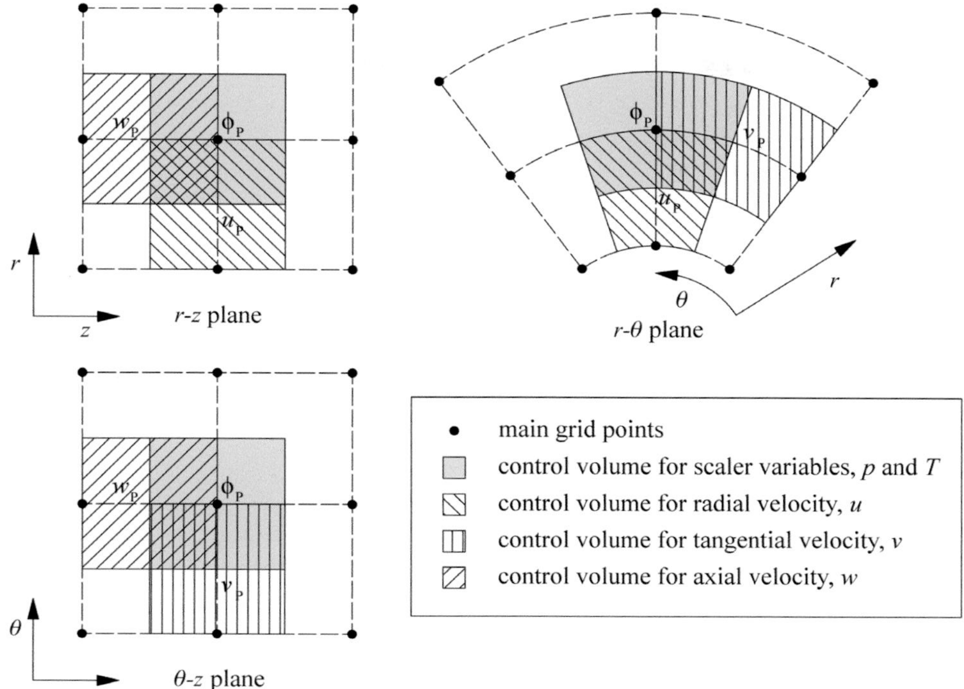

Figure 4.43. Staggered grid arrangement for a cylindrical polar coordinate system (from Tucker (2001) – published with kind permission of Springer).

velocity fields. Next try staggered grids, as shown in Figure 4.43. For initial solutions, use second order central differing. Then you could explore the impact of the use of high order polynomial interpolations to control volume faces and biasing these polynomials in different ways depending on the flow direction.

8) Consider the case where a cylinder wall has rotation. Again contrast with the analytical solution given in Question 3 in Chapter 2. Explore the use of the Gosman et al. (1976) based equation coupling discussed in Section 4.6.3 to couple the radial and tangential momentum equations.

REFERENCES

ABGRALL, R. 1994. On essentially non-oscillatory schemes on unstructured meshes: analysis and implementation. *Journal of Computational Physics*, 114(1), 45–58.

ACHARYA, S. & JANG, D. 1988. Source term decomposition to improve convergence of swirling flow calculations. *AIAA Journal*, 26(3), 372–374.

AGARWAL, A. & MORRIS, P. J. 2000. Direct simulation of acoustic scattering by a rotorcraft fuselage. In Proceedings of Sixth AIAA/CEAS Aeroacoustics Conference, Lahaina, Hawaii, 12–14 June, AIAA Paper No. AIAA-2000-2030.

ANG, W.-T. 2007. *A beginner's course in boundary element methods*, Universal-Publishers.

ARAKAWA, A. 1966. Computational design for long-term numerical integration of the equations of fluid motion: Two-dimensional incompressible flow, Part I. *Journal of Computational Physics*, 1(1), 119–143.

ARMSTRONG, D. B., NAJAFI-YAZDI, A., MONGEAU, L. & RAYMOND, V. 2013. Numerical simulations of flow over a landing gear with noise reduction devices using the lattice-boltzmann method. AIAA Paper No. AIAA-2013-2114.

ASCHER, U. M. & PETZOLD, L. R. 1998. *Computer methods for ordinary differential equations and differential-algebraic equations*, Vol. 61, Society of Industrial and Applied Mathematics (SIAM).

ASHCROFT, G. Z. 2001. A computational investigation of the noise radiated by flow induced cavity oscillations. AIAA 39th Aerospace Sciences Meeting, January 9–11, AIAA Paper No. AIAA- 2001-0512.

ASHCROFT, G. & ZHANG, X. 2001. A computational investigation of the noise radiated by flow-induced cavity oscillations. Proccedings 39th Aerospace Sciences Meeting and Exhibit, January, AIAA Paper 2001-0512.

ATKINS, H. L. & LOCKARD, D. P. 1999. A high-order method using unstructured grids for the aeroacoustic analysis of realistic aircraft configurations. In 5th AIAA/CEAS Aeroacoustics Conference and Exhibit, May, AIAA Paper No. AIAA-1999–1945.

BARTH, T. J. & JESPERSEN, D. 1989. The design and application of upwind schemes on unstructured meshes, 27th Aerospace Sciences Meeting, Reno, Nevada, January 9-12, AIAA Paper No. AIAA-1989-0366.

BARTON, I. 1998a. Comparison of SIMPLE- and PISO-type algorithms for transient flows. *International Journal for Numerical Methods in Fluids*, 26, 459–483.

BARTON, I. 1998b. Improved laminar predictions using a stabilised time-dependent simple scheme. *International Journal for Numerical Methods in Fluids*, 28, 841–857.

BEAM, R. M. & WARMING, R. F. 1976. An implicit finite-difference algorithm for hyperbolic systems in conservation-law form. *Journal of Computational Physics*, 22(1), 87–110.

BEAM, R. M. & WARMING, R. F. 1982. Implicit numerical methods for the compressible Navier-Stokes and Euler equations. *In Von Karman Inst. for Fluid Dyn. Computational Fluid Dyn.*, 99 (SEE N83-19024 09-34), 1.

BELL, B. C. & SURANA, K. S. 1994. A space-time coupled p-version least-squares finite element formulation for unsteady fluid dynamics problems. *International Journal for Numerical Methods in Engineering*, 37(20), 3545–3569.

BHATNAGAR, P. L., GROSS, E. P. & KROOK, M. 1954. A model for collision processes in gases. I. Small amplitude processes in charged and neutral one-component systems. *Physical Review*, 94(3), 511.

BIRKEFELD, A. & MUNZ, C. 2012. Simulations of airfoil noise with the discontinuous Galerkin solver NoisSol. *ERCOFTAC Bull*, 90, 28–33.

BLAISDELL, G., SPYROPOULOS, E. & QIN, J. 1996. The effect of the formulation of nonlinear terms on aliasing errors in spectral methods. *Applied Numerical Mathematics*, 21(3), 207–219.

BLAZEK, J., KROLL, N., RADESPIEL, R. AND ROSSOW, C. C. 1991. Upwind implicit residual smoothing method for multistage schemes, AIAA Tenth Computational Fluid Dynamics Conference, AIAA Paper No. AIAA-91-1533.

BLAZEK, J. 2005. *Computational Fluid Dynamics: Principles and Applications*, Elsevier.

BOGEY, C. & BAILLY, C. 2004. A family of low dispersive and low dissipative explicit schemes for flow and noise computations. *Journal of Computational Physics*, 194(1), 194–214.

BOOK, D. L., BORIS, J. P. & HAIN, K. 1975. Flux-corrected transport II: Generalizations of the method. *Journal of Computational Physics*, 18(3), 248–283.

BORIS, J. P. & BOOK, D. L. 1973. Flux-corrected transport I: SHASTA, A fluid transport algorithm that works. *Journal of Computational Physics*, 11(1), 38–69.

BRANDT, A. 1977. Multi-level adaptive solutions to boundary-value problems. *Mathematics of Computation*, 31, 333–390.

BRANDT, A. 1980. Multilevel adaptive computations in fluid dynamics. *AIAA Journal*, 18(10), 1165–1172.

BRES, G. A., PÉROT, F. & FREED, D. 2009. Properties of the lattice-Boltzmann method for acoustics. Proc. AIAA Aeroacoustics Conference, Miami, Florida, AIAA Paper No. AIAA-2009-3395.

BRILEY, W. & MCDONALD, H. 1975. Solution of the three-dimensional compressible Navier-Stokes equations by an implicit technique. Proceedings of the Fourth International Conference on Numerical Methods in Fluid Dynamics, *Lecture Notes in Physics, Springer-Verlag, Berlin*, *35*, 105–110.

BROECKHOVEN, J. R. & LACOR, C. 2007. *Large-eddy simulation for acoustics*. In C. A. Wagner, T. Huttl & P. Sagaut (eds.), Cambridge University Press.

CAMACHO, R. & BARBOSA, J. 2005. The boundary element method applied to incompressible viscous fluid flow. *Journal of the Brazilian Society of Mechanical Sciences and Engineering*, 27, 456–462.

CAMPOBASSO, M. S. & GILES, M. B. 2003. Effects of flow instabilities on the linear analysis of turbomachinery aeroelasticity. *Journal of Propulsion and Power*, 19(2), 250–259.

CAMPOBASSO, M. S. & GILES, M. B. 2004. Stabilization of a linear flow solver for turbomachinery aeroelasticity using recursive projection method. *AIAA Journal*, 42(9), 1765–1774.

CARTON, D. W., HILLEWAERT, K. & GEUZAINE, P. 2012. DNS of a low pressure turbine blade computed with the discontinuous Galerkin method. ASME Turbo Expo 2012: Turbine Technical Conference and Exposition, ASME Paper No. 2101-2111.

CHAPMAN, M. 1981. FRAM – Nonlinear damping algorithms for the continuity equation. *Journal of Computational Physics*, 44(1), 84–103.

CHEN, C.-J., NASERI-NESHAT, H. & HO, K.-S. 1981. Finite-analytic numerical solution of heat transfer in two-dimensional cavity flow. *Numerical Heat Transfer*, 4, 179–197.

CHESSHIRE, G. & HENSHAW, W. D. 1990. Composite overlapping meshes for the solution of partial differential equations. *Journal of Computational Physics*, 90(1), 1–64.

CHOI, H. & MOIN, P. 1994. Effects of the computational time step on numerical solutions of turbulent flow. *Journal of Computational Physics*, 113, 1–4.

CHOI, S. K. 1999. Note on the use of momentum interpolation method for unsteady flows. *Numerical Heat Transfer, Part A: Applications*, 36, 545–550.

CHOI, Y. & MERKLE, C. L. 1991. Time-derivative preconditioning for viscous flows. AIAA 22nd Fluid Dynamics Conference, Paper No. AIAA-91-1652.

CHORIN, A. J. 1967. A numerical method for solving incompressible viscous flow problems. *Journal of Computational Physics*, 2(1), 12–26.

CHOW, F. K. & MOIN, P. 2003. A further study of numerical errors in large-eddy simulations. *Journal of Computational Physics*, 184(2), 366–380.

CHUNG, Y. M. & TUCKER, P. G. 2003. Accuracy of higher-order finite difference schemes on nonuniform grids. *AIAA Journal*, 41(8), 1609–1611.

CIARDI, M., SAGAUT, P., KLEIN, M. & DAWES, W. 2005. A dynamic finite volume scheme for large-eddy simulation on unstructured grids. *Journal of Computational Physics*, 210(2), 632–655.

COLONIUS, T. & LELE, S. K. 2004. Computational aeroacoustics: progress on nonlinear problems of sound generation. *Progress in Aerospace Sciences*, 40(6), 345–416.

COUGHLIN, G. 2010. *On hexahedral meshing for complex geometry*. MPhil Thesis, University of Cambridge.

CROWLEY, W. 1967. Second-order numerical advection. *Journal of Computational Physics*, 1(4), 471–484.

DAUDE, F., BERLAND, J., EMMERT, T., LAFON, P., CROUZET, F. & BAILLY, C. 2012. A high-order finite-difference algorithm for direct computation of aerodynamic sound. *Computers & Fluids*, 61, 46–63.

DAVIES, C. & CARPENTER, P. W. 2001. A novel velocity-vorticity formulation of the Navier-Stokes equations with applications to boundary layer disturbance evolution. *Journal of Computational Physics*, 172(1), 119–165.

DAVIS, R. & MOORE, E. 1982. A numerical study of vortex shedding from rectangles. *Journal of Fluid Mechanics*, 116(3), 475–506.

DEARDORFF, J. W. 1970. A numerical study of three-dimensional turbulent channel flow at large Reynolds numbers. *Journal of Fluid Mechanics*, 41(2), 453–480.

DELANAYE, M. & ESSERS, J. 1997. Finite volume scheme with quadratic reconstruction on unstructured adaptive meshes applied to turbomachinery flows. *Journal of Turbomachinery*, 119(2), 263–269.

DEMIRDŽIĆ, I., LILEK, Ž. & PERIĆ, M. 1993. A collocated finite volume method for predicting flows at all speeds. *International Journal for Numerical Methods in Fluids*, 16, 1029–1050.

DEMIRDZIC, I. & PERIC, M. 1988. Space conservation law in finite volume calculations of fluid flow. *International Journal for Numerical Methods in Fluids*, 8(9), 1037–1050.

DENTON, J. D. 1992. The calculation of three-dimensional viscous flow through multistage turbomachines. *Journal of Turbomachinery*, 114(1), 18–26.

DOUGLAS, J. & GUNN, J. E. 1964. A general formulation of alternating direction methods. *Numerische Mathematik*, 6, 428–453.

DUCROS, F., FERRAND, V., NICOUD, F., WEBER, C., DARRACQ, D., GACHERIEU, C. & POINSOT, T. 1999. Large-eddy simulation of the shock/turbulence interaction. *Journal of Computational Physics*, 152(2), 517–549.

DUCROS, F., LAPORTE, F., SOULERES, T., GUINOT, V., MOINAT, P. & CARUELLE, B. 2000. High-order fluxes for conservative skew-symmetric-like schemes in structured meshes: application to compressible flows. *Journal of Computational Physics*, 161(1), 114–139.

DUKOWICZ, J. & RAMSHAW, J. 1979. Tensor viscosity method for convection in numerical fluid dynamics. *Journal of Computational Physics*, 32(1), 71–79.

ENGELMAN, M. & SANI, R. 1986. Finite element simulation of incompressible fluid flows with a free/moving surface. *Recent Advances in Numerical Methods in Fluids*, 5, 47–74.

FARES, E. & NOLTING, S. 2011. Unsteady flow simulation of a high-lift configuration using a lattice Boltzmann approach. Proceedings of the forty ninth AIAA Aerospace Sciences Meeting including the New Horizons Forum and Aerospace Exposition, AIAA Paper No. AIAA-2011-869.

FERZIGER, J. H. & PERIC, M. 2002. *Computational methods for fluid dynamics*, Vol. 3, Springer.

FLETCHER, C. A. 1988. *Computational techniques for fluid dynamics. Vol. 1: Fundamental and general techniques*, Springer.

FLETCHER, C. A. 1998. *Computational Techniques for Fluid Dynamics*, Vol. 1, Springer-Verlag.

FRINK, N. T. 1994. Recent progress toward a three-dimensional unstructured Navier-Stokes flow solver. Proceedings AIAA, 32nd Aerospace Sciences Meeting & Exhibit, Reno, Nevada. AIAA Paper No. AIAA-1994-0061

FRINK, N. T., PARIKH, P. & PIRZADEH, S. 1991. A fast upwind solver for the Euler equations on three-dimensional unstructured meshes. Proceedings 29th Aerospace Sciences Meeting. January AIAA paper No. 1991-0102.

FRITSCH, G. & GILES, M. 1992. Second-order effects of unsteadiness on the performance of turbomachines. 37th International Gas Turbine and Aeroengine Congress and Exposition, ASME Paper No. GT-32-389.

GAMET, L., DUCROS, F., NICOUD, F., POINSOT, T., et al. 1999. Compact finite difference schemes on non-uniform meshes: Application to direct numerical simulations of compressible flows. *International Journal for Numerical Methods in Fluids*, 29(2), 159–191.

GHOSAL, S. 1996. An analysis of numerical errors in large-eddy simulations of turbulence. *Journal of Computational Physics*, 125(1), 187–206.

GILES, M. 2004. The Hydra user's guide. Version 6, Rolls-Royce Plc.

GILES, M. B. 1990. Nonreflecting boundary conditions for Euler equation calculations. *AIAA Journal*, 28(12), 2050–2058.

GILES, M. B. 1988. Calculation of unsteady wake/rotor interaction, *Journal of Propulsion and Power*, 4(4), 356–362.

GILES, M. B. 1991. UNSFLO: *A numerical method for the calculation of unsteady flow in turbomachinery*, Gas Turbine Laboratory Report, Massachusetts Institute of Technology, Report No. 205.

GINGOLD, R. A. & MONAGHAN, J. J. 1977. Smoothed particle hydrodynamics: theory and application to non-spherical stars. *Monthly Notices of the Royal Astronomical Society*, 181, 375–389.

GLASS, J. & RODI, W. 1982. A higher order numerical scheme for scalar transport. *Computer Methods in Applied Mechanics and Engineering*, 31(3), 337–358.

GODUNOV, S. K. 1959. A difference method for numerical calculation of discontinuous solutions of the equations of hydrodynamics. *Matematicheskii Sbornik*, 89(3), 271–306.

GOSMAN, A., KOOSINLIN, M., LOCKWOOD, F. & SPALDING, D. 1976. Transfer of heat in rotating systems. Gas Turbine Conference and Products Show, ASME Paper No. 76-GT-25.

GRESHO, P., LEE, R. & SANI, R. 1980. On the time-dependent solution of the incompressible Navier-Stokes equations in two and three dimensions. In C. Taylor & K. Morgan (eds.), *Recent Advances in Numerical Methods in Fluids*, 27–79, Pineridge Press, Ltd.

GRESHO, P. M., CHAN, S. T., LEE, R. L. & UPSON, C. D. 1984. A modified finite element method for solving the time-dependent, incompressible Navier-Stokes equations, Part 1: Theory. *International Journal for Numerical Methods in Fluids*, 4(6), 557–598.

GRINSTEIN, F. F., MARGOLIN, L. G. & RIDER, W. J. 2011. *Implicit Large Eddy Simulation – Computing Turbulent Fluid Dynamics*, Cambridge University Press.

HARLOW, F. H. & AMSDEN, A. A. 1971. A numerical fluid dynamics calculation method for all flow speeds. *Journal of Computational Physics*, 8, 197–213.

HARLOW, F. H. & WELCH, J. E. 1965. Numerical calculation of time-dependent viscous incompressible flow of fluid with free surface. *Physics of Fluids*, 8(12), 2182.

HARTEN, A. 1983. High resolution schemes for hyperbolic conservation laws. *Journal of Computational Physics*, 49(3), 357–393.

HARTEN, A., ENGQUIST, B., OSHER, S. & CHAKRAVARTHY, S. R. 1987. Uniformly high order accurate essentially non-oscillatory schemes, III. *Journal of Computational Physics*, 71(2), 231–303.

HASSAN, Y. A., RICE, J. G. & KIM, J. 1983. A stable mass-flow-weighted two-dimensional skew upwind scheme. *Numerical Heat Transfer*, 6, 395–408.

HE, L. & WANG, D. 2011. Concurrent blade aerodynamic-aero-elastic design optimization using adjoint method. *Journal of Turbomachinery*, 133(1), 011021.

HEDGES, L., TRAVIN, A. & SPALART, P. 2002. Detached-eddy simulations over a simplified landing gear. *Journal of Fluids Engineering*, 124(2), 413–423.

HENKES, R. A. 1990. *Natural-convection boundary layers*. Ph.D. dissertation, Technische University Delft.

HICKEN, J. E. & ZINGG, D. W. 2008. Parallel newton-krylov solver for the euler equations discretized using simultaneous approximation terms. *AIAA Journal*, 46(11), 2773–2786.

HIGNETT, B. P., WHITE, A., CARTER, R., JACKSON, W. & SMALL, R. 1985. A comparison of laboratory measurements and numerical simulations of baroclinic wave flows in a rotating cylindrical annulus. *Quarterly Journal of the Royal Meteorological Society*, 111(467), 131–154.

HIRSCH, C. 2007. *Numerical Computation of Internal and External Flows: The Fundamentals of Computational Fluid Dynamics: The Fundamentals of Computational Fluid Dynamics*, Vol. 1 and 2, Butterworth-Heinemann.

HIRT, C. 1968. Heuristic stability theory for finite-difference equations. *Journal of Computational Physics*, 2(4), 339–355.

HIRT, C., AMSDEN, A. A. & COOK, J. 1974. An arbitrary Lagrangian-Eulerian computing method for all flow speeds. *Journal of Computational Physics*, 14(3), 227–253.

HIXON, R. 2000. Prefactored small-stencil compact schemes. *Journal of Computational Physics*, 165(2), 522–541.

HOLMES, D. & CONNELL, S. 1986. Solution of the 2D Navier-Stokes equations on unstructured adaptive grids. Ninth Computational Fluid Dynamics Conference, AIAA Paper No. 89–1932.

HOLMES, D., CONNELL, S. & ENGINES, G. A. 1989. Solution of the 2D Navier-Stokes equations on unstructured adaptive grids. Proceedings of the 9th Computational Fluid Dynamics Conference. AIAA Paper No. 89-1932-CP.

HORIUTI, K. & ITAMI, T. 1998. Truncation error analysis of the rotational form for the convective terms in the Navier–Stokes equation. *Journal of Computational Physics*, 145(2), 671–692.

HU, F., HUSSAINI, M. Y. & MANTHEY, J. 1996. Low-dissipation and low-dispersion Runge–Kutta schemes for computational acoustics. *Journal of Computational Physics*, 124(1), 177–191.

HU, X., & NICOLAIDES, R. 1992. Covolume techniques for anisotropic media. *Numerische Mathematik*, 61(1), 215–234.

HUJEIRAT, A. & RANNACHER, R. 1998. A method for computing compressible, highly stratified flows in astrophysics based on operator splitting. *International Journal for Numerical Methods in Fluids*, 28(1), 1–22.

HU, X. & NICOLAIDES, R. 1992. Covolume techniques for anisotropic media. *Numerische Mathematik*, 61, 215–234.

ISERLES, A. 1986. Generalized leapfrog methods. *IMA Journal of Numerical Analysis*, 6(4), 381–392.

ISSA, R. I. 1986. Solution of the implicitly discretised fluid flow equations by operator-splitting. *Journal of Computational Physics*, 62(1), 40–65.

ISSA, R. & OLIVEIRA, P. 1994. Numerical prediction of phase separation in two-phase flow through T-junctions. *Computers & Fluids*, 23, 347–372.

JAMES, I., JONAS, P. & FARNELL, L. 1981. A combined laboratory and numerical study of fully developed steady baroclinic waves in a cylindrical annulus. *Quarterly Journal of the Royal Meteorological Society*, 107(451), 51–78.

JAMESON, A. 1991. Time dependent calculations using multigrid, with applications to unsteady flows past airfoils and wings. Proceedings of the tenth Computational Fluid Dynamics Conference, June, AIAA Paper No. AIAA-1991-1596.

JAMESON, A. 2008a. Formulation of kinetic energy preserving conservative schemes for gas dynamics and direct numerical simulation of one-dimensional viscous compressible flow in a shock tube using entropy and kinetic energy preserving schemes. *Journal of Scientific Computing*, 34(2), 188–208.

JAMESON, A. 2008b. The construction of discretely conservative finite volume schemes that also globally conserve energy or entropy. *Journal of Scientific Computing*, 34(2), 152–187.

JAMESON, A., SCHMIDT, W., TURKEL, E., et al. 1981. Numerical solutions of the Euler equations by finite volume methods using Runge-Kutta time-stepping schemes. Proceedings 14th Fluid and Plasma Dynamics Conference, June AIAA Paper No. AIAA-1981-1259.

JEFFERSON-LOVEDAY, R. 2008. *Numerical simulations of unsteady impinging jet flows*. Ph.D. dissertation, Swansea University.

JONES, W. & MARQUIS, A. 1985. Calculation of axisymmetric recirculating flows with a second order turbulence model. *Proceedings of the 5th Symposium on Turbulent Shear Flows*, Cornell University, 20.1–20.11.

JOO, J. & DURBIN, P. 2009. Simulation of turbine blade trailing edge cooling. *Journal of Fluids Engineering*, 131(2), 021102.

NAKAHASHI K., & TOGASHI F. 2000. Unstructured overset grid method for flow simulation of complex multiple body problems. Proceedings of ICAS 2000 Congress, Paper No. ICAS0263.

KARABASOV, S. A. & GOLOVIZNIN, V. M. 2007. New efficient high-resolution method for nonlinear problems in aeroacoustics. *AIAA Journal*, 45(12), 2861–2871.

KARABASOV, S. & GOLOVIZNIN, V. 2009. Compact accurately boundary-adjusting high-resolution technique for fluid dynamics. *Journal of Computational Physics*, 228(19), 7426–7451.

KHATIR, Z. 2000. *Discrete vortex modelling of near-wall flow structure in turbulent boundary layers*. Ph.D. dissertation, The University of Warwick.

KIM, J. & MOIN, P. 1985. Application of a fractional-step method to incompressible Navier-Stokes equations. *Journal of Computational Physics*, 59(2), 308–323.

KIM, J. W. & LEE, D. J. 1996. Optimized compact finite difference schemes with maximum resolution. *AIAA Journal*, 34(5), 887–893.

KIRCHHART, M. 2013. Vortex methods. Handout for the CES-Seminar Talk.

KRAKOS, J. A. & DARMOFAL, D. L. 2010. Effect of small-scale output unsteadiness on adjoint-based sensitivity. *AIAA Journal*, 48(11), 2611–2623.

LACOR, C. 1999. *Industrial computational fluid dynamics*. von Karman Institute for Fluid Dynamics. May 31–June 4 (Eds. J.-M. Buchlin & Ph. Planquart), VKI LS 1999-06.

LAIZET, S. & LAMBALLAIS, E. 2009. High-order compact schemes for incompressible flows: A simple and efficient method with quasi-spectral accuracy. *Journal of Computational Physics*, 228(16), 5989–6015.

LEE, K. R., PARK, J. H. & KIM, K. H. 2011. High-order interpolation method for overset grid based on finite volume method. *AIAA Journal*, 49(7), 1387–1398.

LEITH, C. E. 1965. Numerical simulation of the earth's atmosphere. *Meth. Comp. Phys*, 4, 1–28.

LELE, S. K. 1992. Compact finite difference schemes with spectral-like resolution. *Journal of Computational Physics*, 103(1), 16–42.

LEONARD, B. P. 1979. A stable and accurate convective modelling procedure based on quadratic upstream interpolation. *Computer Methods in Applied Mechanics and Engineering*, 19(1), 59–98.

LERAT, A. 1979. Une classe de schémas aux différences implicites pour les systèmes hyperboliques de lois de conservation. *Comptes Rendus Acad. Sciences Paris*, 288, 1033–1036.

LIEN, F.-S. & LESCHZINER, M. 1994. Upstream monotonic interpolation for scalar transport with application to complex turbulent flows. *International Journal for Numerical Methods in Fluids*, 19(6), 527–548.

LIOU, M.-S. & STEFFEN JR, C. J. 1993. A new flux splitting scheme. *Journal of Computational Physics*, 107(1), 23–39.

LIU, X.-D., OSHER, S. & CHAN, T. 1994. Weighted essentially non-oscillatory schemes. *Journal of Computational Physics*, 115, 200–212.

LIU, Y. & NISHIMURA, N. 2006. The fast multipole boundary element method for potential problems: a tutorial. *Engineering Analysis with Boundary Elements*, 30, 371–381.

LOCKARD, D. P., BRENTNER, K. S. & ATKINS, H. 1995. High-accuracy algorithms for computational aeroacoustics. *AIAA Journal*, 33(2), 246–251.

LU, Y., YUAN, X. & DAWES, W. 2012. Investigation of 3D internal flow using new flux-reconstruction high order method. ASME Turbo Expo 2012: Turbine Technical Conference and Exposition, 2195–2216.

MAJUMDAR, S. 1988. Role of underrelaxation in momentum interpolation for calculation of flow with nonstaggered grids. *Numerical Heat Transfer*, 13, 125–132.

MANGANI, L., DARWISH, M. & MOUKALLED, F. 2013. Development of a Novel Pressure-Based Coupled CFD Solver for Turbulent Compressible Flows in Turbomachinery Applications. American Society of Mechanical Engineers 2013 Fluids Engineering Division Summer Meeting, Paper No. FEDSM2013-16082.

MANI, K. & MAVRIPLIS, D. J. 2010. Spatially non-uniform time-step adaptation for functional outputs in unsteady flow problems. 48th AIAA Aerospace Sciences Meeting Including the New Horizons Forum and Aerospace Exposition, AIAA Paper No. AIAA-2010-121.

MANOHA, E., TROFF, B. & SAGAUT, P. 2000. Trailing-edge noise prediction using large-eddy simulation and acoustic analogy. *AIAA Journal*, 38(4), 575–583.

MARIÉ, S., RICOT, D. & SAGAUT, P. 2009. Comparison between lattice Boltzmann method and Navier-Stokes high order schemes for computational aeroacoustics. *Journal of Computational Physics*, 228, 1056–1070.

MARONGIU, J., LEBOEUF, F. & PARKINSON, E. 2007. Numerical simulation of the flow in a Pelton turbine using the meshless method smoothed particle hydrodynamics: A new simple solid boundary treatment. *Proceedings of the Institution of Mechanical Engineers, Part A: Journal of Power and Energy*, 221(6), 849–856.

MARONGIU, J.-C., LEBOEUF, F., CARO, J. & PARKINSON, E. 2010. Free surface flows simulations in Pelton turbines using an hybrid SPH-ALE method. *Journal of Hydraulic Research*, 48(S1), 40–49.

MARY, I. & SAGAUT, P. 2002. Large eddy simulation of flow around an airfoil near stall. *AIAA Journal*, 40(6), 1139–1145.

MASON, P. J. & CALLEN, N. S. 1986. On the magnitude of the subgrid-scale eddy coefficient in large-eddy simulations of turbulent channel flow. *Journal of Fluid Mechanics*, 162, 439–462.

MOINIER, P. 1999. *Algorithm developments for an unstructured viscous flow solver*. Ph.D. dissertation, Oxford University.

MOSAHEBI, A. & NADARAJAH, S. K. 2011. An implicit adaptive non-linear frequency domain method (pNLFD) for viscous periodic steady state flows on deformable grids. Proceedings of the 49th Aerospace Sciences Meeting, January, Orlando, Florida, Paper No. AIAA-2011-775.

MOULINEC, C., BENHAMADOUCHE, S., LAURENCE, D. & PERIC, M. 2005. LES in a U-bend pipe meshed by polyhedral cells. *Engineering Turbulence Modelling and Experiments*, 6, 237–246.

MURAMATSU, T., & NINOKATA, H., 1992. Thermal striping temperature fluctuation analysis using the algebraic stress turbulence model in water and sodium, *Japan Society of Mechanical Engineers International Journal*, Series 2, 35(4), 486–496.

NASSER, A. & LESCHZINER, M. 1985. Computation of transient recirculating flow using spline approximations and time-space characteristics. Proceedings of the 4th International Conference on Numerical Methods in Laminar and Turbulent Flow, Swansea, 480–491.

NOWAK, A. & BREBBIA, C. 1989. The multiple-reciprocity method: A new approach for transforming BEM domain integrals to the boundary. *Engineering Analysis with Boundary Elements*, 6, 164–167.

ORKWIS, P. D., TURNER, M. G. & BARTER, J. W. 2002. Linear deterministic source terms for hot streak simulations. *Journal of Propulsion and Power*, 18(2), 383–389.

ORKWIS, P. D. & VANDEN, K. J. 1994. On the accuracy of numerical versus analytical Jacobians. Proceedings 32nd AIAA, Aerospace Sciences Meeting & Exhibit, Reno, Nevada, AIAA Paper 94-0176.

ORSZAG, S. A. 1971. Accurate solution of the Orr–Sommerfeld stability equation. *Journal of Fluid Mechanics*, 50(04), 689–703.

OZYORUK, Y. & LONG, L. N. 1997. Multigrid acceleration of a high-resolution. *AIAA Journal*, 35(3), 428–433.

PARTRIDGE, P. W., BREBBIA, C. A. & WROBEL, L. C. 1992. *The dual reciprocity boundary element method*, Computational Mechanics Publications.

PATANKAR, S. 1980. *Numerical heat transfer and fluid flow*, CRC Press.

PATANKAR, S. & BALIGA, B. 1978. A new finite-difference scheme for parabolic differential equations. *Numerical Heat Transfer*, 1(1), 27–37.

PATANKAR, S. V. & SPALDING, D. B. 1972. A calculation procedure for heat, mass and momentum transfer in three-dimensional parabolic flows. *International Journal of Heat and Mass Transfer*, 15(10), 1787–1806.

PATERA, A. T. 1984. A spectral element method for fluid dynamics: laminar flow in a channel expansion. *Journal of Computational Physics*, 54(3), 468–488.

PAULEY, L. L., MOIN, P. & REYNOLDS, W. C. 1990. The structure of two-dimensional separation. *Journal of Fluid Mechanics*, 220, 397–411.

PEACEMAN, D. W. & RACHFORD, J. 1955. The numerical solution of parabolic and elliptic differential equations. *Journal of the Society for Industrial & Applied Mathematics*, 3, 28–41.

PINELLI, A., NAQAVI, I., PIOMELLI, U. & FAVIER, J. 2010. Immersed-boundary methods for general finite-difference and finite-volume Navier–Stokes solvers. *Journal of Computational Physics*, 229(24), 9073–9091.

PITSCH, H. 2006. Large-eddy simulation of turbulent combustion. *Annual Review of Fluid Mechanics*, 38, 453–482.

PREECE, A. 2008. *An Investigation into Methods to aid the Simulation of Turbulent Separation Control*. Ph.D. dissertation, The University of Warwick.

RAITHBY, G. & SCHNEIDER, G. 1979. Numerical solution of problems in incompressible fluid flow: treatment of the velocity-pressure coupling. *Numerical Heat Transfer, Part A: Applications*, 2, 417–440.

RAITHBY, G. & SCHNEIDER, G. 1980. Erratum. *Numerical Heat Transfer*, 3, 513.

RAW, M. 1996. Robustness of coupled algebraic multigrid for the Navier-Stokes equations. Proceedings 34th Aerospace Sciences Meeting and Exhibit, January, AIAA Paper No., AIAA-96-0297.

RAYNER, D. 1993. *A Numerical Study into the Heat Transfer beneath the Stator Blade of an Axial Compressor*. Ph.D. dissertation, University of Sussex.

REINDL, D. T., BECKHAM W. A., MITCHELL, J. W. & RUTLAND, C. 1991. Benchmarking transient natural convection in an enclosure, ASME Paper No. 91-HT-8, pp. 1–7.

RHIE, C. & CHOW, W. 1983. Numerical study of the turbulent flow past an airfoil with trailing edge separation. *AIAA Journal*, 21, 1525–1532.

RIDER, W. & DRIKAKIS, D. 2005. *High-resolution methods for incompressible and low-speed flows*, Springer.

RIZZETTA, D. P., VISBAL, M. R. & MORGAN, P. E. 2008. A high-order compact finite-difference scheme for large-eddy simulation of active flow control. *Progress in Aerospace Sciences*, 44(6), 397–426.

ROACHE, P. J. 1992. A flux-based modified method of characteristics. *International Journal for Numerical Methods in Fluids*, 15(11), 1259–1275.

ROE, P. 1986. Characteristic-based schemes for the Euler equations. *Annual Review of Fluid Mechanics*, 18(1), 337–365.

ROE, P. L. 1981. Approximate Riemann solvers, parameter vectors, and difference schemes. *Journal of Computational Physics*, 43(2), 357–372.

ROGERS, S. E. & KWAK, D. 1990. Upwind differencing scheme for the time-accurate incompressible Navier-Stokes equations. *AIAA Journal*, 28(2), 253–262.

ROGERS, S. E., KWAK, D. & CHANG, J. L. 1986. Numerical solution of the incompressible Navier-Stokes equations in three-dimensional generalized curvilinear coordinates. *NASA STI/Recon Technical Report N*, 87, 11964.

RUGE, J. & STUEBEN, K. 1986. Algebraic multigrid. *Arbeitspapiere der GMD*, 210.

RUMPFKEIL, M. P., ZINGG, D. W. 2007. A general framework for the optimal control of unsteady flows with applications. Proceedings of the 45th AIAA Aerospace Meeting and Exhibit, 8–11 January, Reno, Nevada, Paper No. AIAA 2007–1128.

SANDHAM, N. & YEE, H. 2001. Entropy splitting for high order numerical simulation of compressible turbulence. In *Computational Fluid Dynamics*, 361–366, Springer.

SEGERLIND, L. 1984. *Applied Finite Element Analysis*, John Wiley and Sons.

SEIDL, V., PERIC, M. & SCHMIDT, M. 1995. Space- and time-parallel Navier-Stokes solver for 3d block-adaptive Cartesian grids. *Parallel Computational Fluid Dynamics: Proceedings*, 95, 557–584.

SHUR, M., SPALART, P., STRELETS, M. K. & TRAVIN, A. 2003. Towards the prediction of noise from jet engines. *International Journal of Heat and Fluid Flow*, 24(4), 551–561.

SKELBOE, S. 1977. The control of order and steplength for backward differentiation methods. *BIT Numerical Mathematics*, 17(1), 91–107.

SPALART, P., HEDGES, L., SHUR, M. & TRAVIN, A. 2003. Simulation of active flow control on a stalled airfoil. *Flow, Turbulence and Combustion*, 71(1–4), 361–373.

SPALDING, D. 1972. A novel finite difference formulation for differential expressions involving both first and second derivatives. *International Journal for Numerical Methods in Engineering*, 4, 551–561.

SPEZIALE, C. G. 1987. On the advantages of the vorticity-velocity formulation of the equations of fluid dynamics. *Journal of Computational Physics*, 73(2), 476–480.

SPYROPOULOS, E. T. & BLAISDELL, G. A. 1998. Large-eddy simulation of a spatially evolving supersonic turbulent boundary-layer flow. *AIAA Journal*, 36(11), 1983–1990.

STANIFORTH, A. & COTE, J. 1991. Semi-Lagrangian integration schemes for atmospheric models – a review. *Monthly Weather Review*, 119(9), 2206–2223.

STONE, H. L. 1968. Iterative solution of implicit approximations of multidimensional partial differential equations. *Society for Industrial and Applied Mathematics Journal on Numerical Analysis*, 5, 530–558.

SUCCI, S. 2001. *The Lattice Boltzmann Equation for Fluid Dynamics and Beyond*, Oxford University Press.

SWANSON, R. C. & TURKEL, E. 1992. On central-difference and upwind schemes. *Journal of Computational Physics*, 101(2), 292–306.

TAJALLIPOUR, N., BABAEE OWLAM, B. & PARASCHIVOIU, M. 2009. Self-adaptive upwinding for large eddy simulation of turbulent flows on unstructured elements. *Journal of Aircraft*, 46(3), 915–926.

TALHA, T. 2012. *A numerical investigation of three-dimensional unsteady turbulent channel flow subjected to temporal acceleration*. Ph.D. dissertation, University of Warwick.

TAM, C. K. & SHEN, H. 1993. Direct computation of nonlinear acoustic pulses using high order finite difference schemes. Proceedings 15th Aeroacoustics Conference. October AIAA Paper No., AIAA-93-4325.

TAM, C. K. & WEBB, J. C. 1993. Dispersion-relation-preserving finite difference schemes for computational acoustics. *Journal of Computational Physics*, 107(2), 262–281.

TAM, C. K., WEBB, J. C. & DONG, Z. 1993. A study of the short wave components in computational acoustics. *Journal of Computational Acoustics*, 1(1), 1–30.

TANG, L. & BAEDER, J. D. 1998. Uniformly accurate finite difference schemes for p-refinement. *SIAM Journal on Scientific Computing*, 20(3), 1115–1131.

THOMAS, P. & LOMBARD, C. 1979. Geometric conservation law and its application to flow computations on moving grids. *AIAA Journal*, 17(10), 1030–1037.

TU, C., DEVILLE, M., DHEUR, L. & VANDERSCHUREN, L. 1992. Finite element simulation of pulsatile flow through arterial stenosis. *Journal of Biomechanics*, 25(10), 1141–1152.

TUCKER, P. 1997. Numerical precision and dissipation errors in rotating flows. *International Journal of Numerical Methods for Heat & Fluid Flow*, 7(7), 647–658.

TUCKER, P. G. 2001. *Computation of unsteady internal flows*, Springer.

TUCKER, P. 2002a. Novel multigrid orientated solution adaptive time-step approaches. *International Journal for Numerical Methods in Fluids*, 40(3–4), 507–519.

TUCKER, P. G. 2002b. Temporal behavior of flow in rotating cavities. *Numerical Heat Transfer: Part A: Applications*, 41(6–7), 611–627.

TUCKER, P. G. 2004. Novel MILES computations for jet flows and noise. *International Journal of Heat and Fluid Flow*, 25(4), 625–635.

TUCKER, P. G. 2013. *Unsteady computational fluid dynamics in aeronautics*, Springer.

UZUN, A. & HUSSAINI, M. Y. 2009. Simulation of noise generation in the near-nozzle region of a chevron nozzle jet. *AIAA Journal*, 47(8), 1793–1810.

VAN ALBADA, G., VAN LEER, B. & ROBERTS JR, W. 1982. A comparative study of computational methods in cosmic gas dynamics. *Astronomy and Astrophysics*, 108, 76–84.

VAN DOORMAAL, J. & RAITHBY, G. 1984. Enhancements of the SIMPLE method for predicting incompressible fluid flows. *Numerical Heat Transfer*, 7, 147–157.

VAN LEER, B. 1974. Towards the ultimate conservative difference scheme, II: Monotonicity and conservation combined in a second-order scheme. *Journal of Computational Physics*, 14(4), 361–370.

VAN LEER, B. 1977. Towards the ultimate conservative difference scheme, III: Upstream-centered finite-difference schemes for ideal compressible flow. *Journal of Computational Physics*, 23(3), 263–275.

VAN LEER, B. 1979. Towards the ultimate conservative difference scheme. V. A second-order sequel to Godunov's method. *Journal of Computational Physics*, 32(1), 101–136.

VAUGHAN, C., GILHAM, S. & CHEW, J. 1989. Numerical solutions of rotating disc flows using a non-linear multigrid algorithm. Proceedings of the 6th International Conference on Numerical Methods in Laminar and Turbulent Flow, 63–67.

VERMEIRE, B. C., NADARAJAH, S. & TUCKER, P. G. 2014. Canonical test cases for high-order unstructured implicit large eddy simulation. Proceedings 52nd AIAA Aerospace Sciences Meeting, AIAA Paper No. AIAA-2014-0935.

VISBAL, M. R. & GAITONDE, D. V. 2002. On the use of higher-order finite-difference schemes on curvilinear and deforming meshes. *Journal of Computational Physics*, 181(1), 155–185.

WALLIS, S. G. & MANSON, J. R. 1997. Accurate numerical simulation of advection using large time steps. *International Journal for Numerical Methods in Fluids*, 24(2), 127–139.

WANG Z. J., LIU Y., MAY G., & JAMESON A., 2007. Spectral difference method for unstructured grids II: extension to the Euler equations. *Journal of Scientific Computing*, 32(1), 45–71.

WATERSON, N. P. & DECONINCK, H. 1995. A unified approach to the design and application of bounded higher-order convection schemes. *Numerical Methods in Laminar and Turbulent Flow*, 9, 203–214.

WATSON, R. 2014. *Large eddy simulation of cutback trailing edges for film cooling turbine blades*. Ph.D. dissertation, University of Cambridge.

WEISS, J. M. & SMITH, W. A. 1995. Preconditioning applied to variable and constant density flows. *AIAA Journal*, 33(11), 2050–2057.

WIETH, L., LIEBER, C., KURZ, W., BRAUN, S., KOCH, R., & BAUER, H. J. 2015. Numerical modeling of an aero-engine bearing chamber using the meshless smoothed particle hydrodynamics method. ASME Turbo Expo 2015, Turbine Technical Conference and Exposition, Montreal, Canada, ASME Paper No. Paper No. GT2015-42316.

WILLCOX, D. 1998. *Turbulence modelling for CFD*, DCW Industries Inc.

WOLF, W. & AZEVEDO, J. 2007. High-order ENO and WENO schemes for unstructured grids. *International Journal for Numerical Methods in Fluids*, 55(10), 917–943.

XIA, H. 2005. *Dynamic Grid Detached-Eddy Simulation for Synthetic Jet Flows*. Ph.D. dissertation, The University of Sheffield.

YANG, G., CAUSON, D., INGRAM, D., SAUNDERS, R. & BATTEN, P. 1997. A Cartesian cut cell method for compressible flows, Part B: moving body problems. *Aeronautical Journal*, 101(1002), 57–65.

YAO, Y., SAVILL, A., SANDHAM, N. & DAWES, W. 2000. Simulation of a turbulent trailing-edge flow using unsteady RANS and DNS. In Y. Nagano, K. Hanjalic & T. Tsuji (eds.), *Turbulence, Heat and Mass Transfer*, 463–470, Aichi Shuppan.

YU, B., TAO, W.-Q., WEI, J.-J., KAWAGUCHI, Y., TAGAWA, T. & OZOE, H. 2002. Discussion on momentum interpolation method for collocated grids of incompressible flow. *Numerical Heat Transfer, Part B: Fundamentals*, 42, 141–166.

ZHU, Z. W., LACOR, C. & HIRSCH, C. 1993. A new residual smoothing method for multigrid multi-stage schemes. Proceedings of the 11th AIAA CFD Conference, Paper No. AIAA-93-3356.

ZIENKIEWICZ, O. C. & TAYLOR, R. L. 2005. *The finite element method for solid and structural mechanics*, Butterworth-Heinemann.

ZIENKIEWICZ, O., TAYLOR, R. & NITHIARASU, P. 2005. *The Finite Element Method for Fluid Dynamics, Sixth Edition*, Elsevier.

ZOLTAK, J. & DRIKAKIS, D. 1998. Hybrid upwind methods for the simulation of unsteady shock-wave diffraction over a cylinder. *Computer Methods in Applied Mechanics and Engineering*, 162(1), 165–185.

CHAPTER 5

Turbulence

5.1 Introduction

Turbulence generally plays a key role in drag generation, heat transfer, particle dispersion and scalar mixing along with sound generation. These are all aspects that are vital in aerodynamic design. The modelling of turbulence has strongly defined CFD as a postdictive rather than predictive process. Hence, this chapter has strong importance in relation to the safe and reliable use of CFD. Initially the nature of turbulence is outlined and thus the modelling challenges defined. Then a range of established modelling approaches is outlined, starting with low order and then moving to techniques where CFD becomes predictive with no modelling.

5.1.1 *The Basic Nature of Turbulence*

The formidable task facing turbulence modelling is reflected by Richard Feynman (Physics Nobel Laureate), who wrote, "Turbulence is the last great unsolved problem in classical physics." Indeed we have the situation where we do not even know basic physical constants. For example, for the Karman 'constant', κ, (in $l = \kappa\, d$, where l is the mixing length and d nearest wall distance) the published range is 0.38–0.45) and we do not even know if it is a constant! As noted by Spalart et al. (2006), a 2% decrease in κ gives 1% decrease in predicted aircraft drag. This might not sound like much, but it matters greatly to plane manufacturers. They sell planes before they are made and if the craft do not make the range, serious problems arise.

If a sensitive velocity-measuring device (such as a hot wire anemometer) is placed in a turbulent flow, the trace shown in Figure 5.1 will be produced. Similar graphs can be plotted for the remaining v (y direction) and w (z direction) velocity components. As can be seen, the fluid velocity can be decomposed into mean (Φ) and fluctuating (ϕ') components. Hence, instantaneous velocities can be expressed as

$$\phi = \Phi + \phi' \tag{5.1}$$

Transition to turbulence occurs when inertial forces overwhelm the viscous. The transition process can be observed in the smoke patterns rising from the tip of a cigarette. The initial smoke has a laminar region. This is followed by a turbulent zone, where the fluid flow has a more chaotic appearance. The latter can clearly be seen in the Figure 5.2a schematic adjacent to the flow visualization for a more

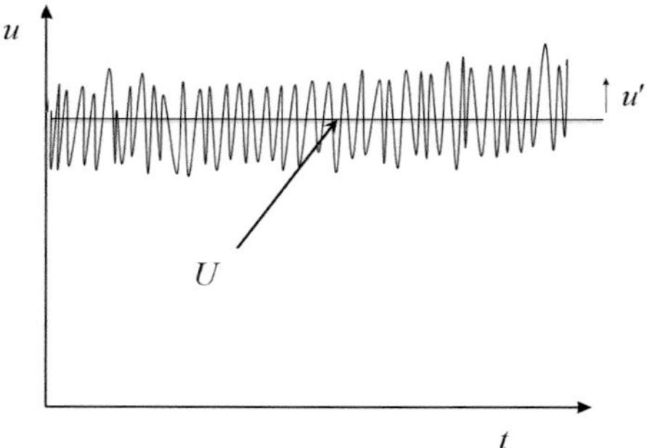

Figure 5.1. Schematic of a trace of velocity against time in a turbulent flow.

fully turbulent jet. Figure 5.2b shows a sketch of da Vinci – for which he describes the "clouds as scattered and torn." This could be regarded as one of the earliest turbulent flow studies. The onset of turbulence is characterized by the Reynolds number (Re) parameter, where $Re = \rho UL/\mu$, and U and L are characteristic length and velocity scales. Turbulent flows can be regarded in a sense as stochastic (they have been modelled with some success using stochastic type models, but as will be seen later, despite the chaotic appearance of the Figure 5.1 trace, some order can frequently be observed). The velocity fluctuations shown in Figure 5.1 are caused by vortices (turbulent eddies) passing over the measuring device. The maximum size of these vortices (which can have highly complex shapes) is of the same order as the flow system size. These larger vortices are often called the *integral* scales. For oceans, the vortices can be several miles in diameter. Atmospheric turbulence is perhaps illustrated in Van Gogh's painting called *La nuit etoilee* (Figure 5.2c). The vortex structures shown in this are similar to those observed in shear flows, where two fluid streams, having different velocities, move parallel to each other. Figure 5.3 shows the situation when $U_1 > U_2$. Frame (a) gives a schematic and (b) gives the experimental observations of Lasheras and Choi (1988). As can be seen, a set of moving vortex structures is produced – Kelvin Helmholtz vortices. When a uniform fluid flow passes through a grid, isotropic and approximately homogenous turbulence

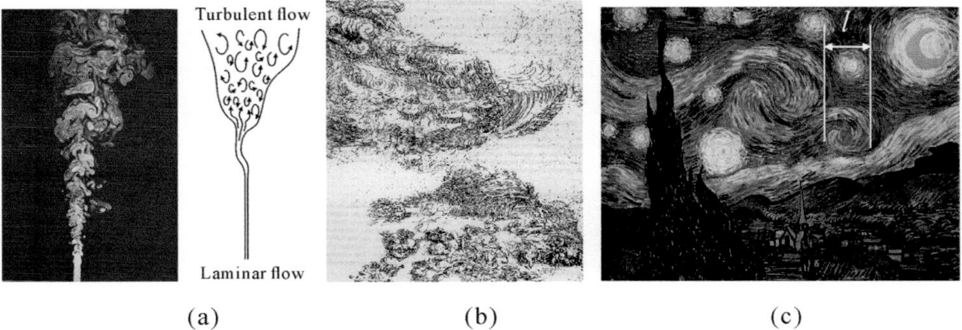

Figure 5.2. Observations of turbulence: (a) Transition in a jet plume, (b) da Vinci and (c) Van Gogh. (The far left-hand, flow visualization image in Frame (a) is from Dimotakis et al. 1983 and used with kind permission of AIP Publishing.)

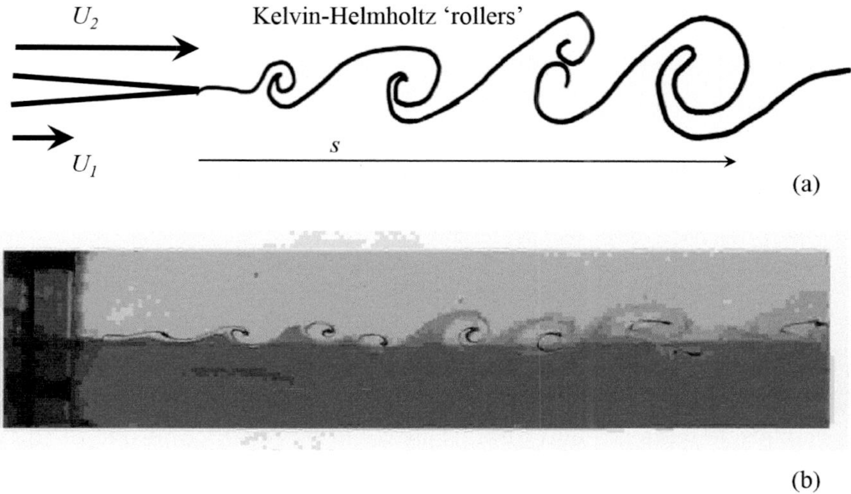

Figure 5.3. Mixing layer: (a) schematic and (b) experimental observation of Lasheras and Choi (1988) – Frame (b) published with the kind permission of Cambridge University Press.

is produced. Homogenous isotropic turbulence is an important canonical flow used for studying the physics of turbulence and developing mathematical models for it. The turbulent fluctuations decay with distance from the grid. This decay is slow and hence the turbulence can be regarded as homogenous.

5.1.1.1 The Energy Cascade Larger vortices, which contain most of the turbulence energy, are formed and maintained by mean flow field shear forces. Without these forces, turbulence would be quickly dissipated, by viscous action, into heat. Progressively smaller vortices are formed from the action of larger vortices. The length scale of the smallest vortices is such that $Re \sim 1$. These smallest scales are called the *Kolmogorov* scales. Because viscous and inertial forces are equal, when $Re = 1$, these smaller vortices quickly become destroyed, dissipated at a rate ε. This draining of turbulent kinetic energy, k, generated by mean flow field shear via progressively smaller vortex structures is called the *turbulence energy cascade.*

Figure 5.4 gives a typical form of the turbulence energy spectrum. Essentially this is a plot of energy, E, against wave number – the latter being proportional to the inverse of eddy size. To get the idea of the energy spectrum, it is perhaps interesting to consider the famous rhyme of Richardson (1922):

Big whorls have little whorls,
which feed on their velocity,
and little whorls have lesser whorls,
and so on to viscosity (in the molecular sense).

The large eddies, passing over a probe, cause the low frequency Figure 5.1 components. The larger vortices produce the greatest differences in u', v' and w' (anisotropy); whilst in smaller vortices, viscous damping tends to make the turbulence more isotropic ($u' = v' = w'$). The smaller scales give the higher frequency

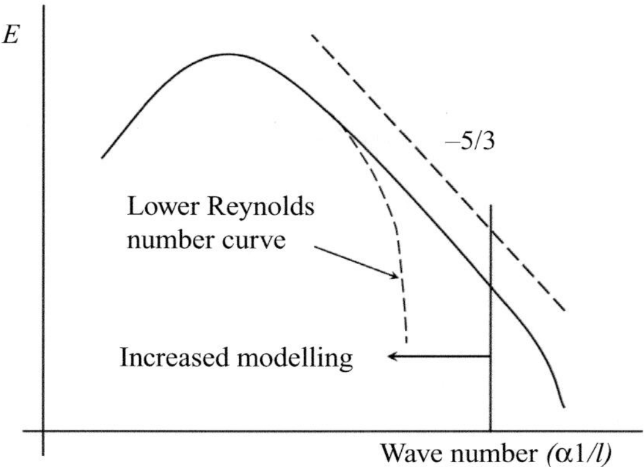

Figure 5.4. Schematic of the form of a typical turbulence energy spectrum.

components. Energy, as in the rhyme of Richardson, transfers from the large scales to the small scales eventually being dissipated by molecular viscosity. As can be seen, consistent with the idea of a turbulence energy cascade, most E is associated the with larger, lower frequency scales. The energy content decreases with smaller scales (higher frequencies). At sufficiently high Reynolds numbers there is a region in the energy spectrum where inertial effects dominate and eddies are about isotropic. Note that in reality even the smallest scales are highly anisotropic elongated structures. The latter means that the eddies have mostly forgot their initial anisotropic state. Through simple analysis, Kolmogorov showed that in this region, called the *inertial sub-range*, the energy spectrum curve should have a –5/3 gradient. This is called the Kolmogorov –5/3 law and represented by the dashed line in Figure 5.4. It should be noted that at high Reynolds numbers the extent of the –5/3 law region is larger. However, at low Reynolds numbers it can be small/negligible and hence difficult to see. Figure 5.5 plots the extent of the –5/3 zone through a

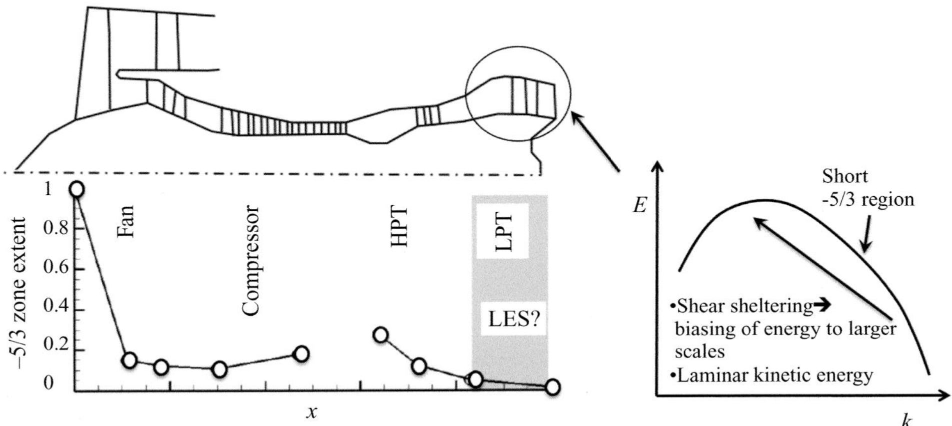

Figure 5.5. Extent of Kolmogorov –5/3 zone through a gas turbine aeroengine (from Tucker 2013a – published with kind permission of Elsevier).

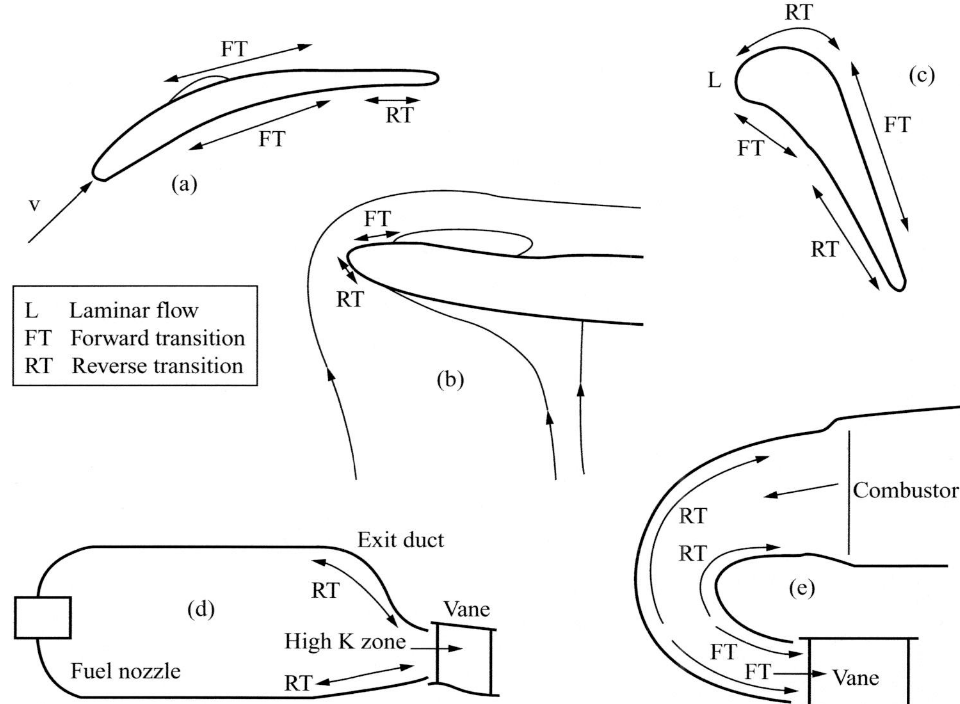

Figure 5.6. Examples of mixed Reverse Transition (RT) and Forward Transition (FT) zones in turbomachinery: (a) compressor blade, (b) intake in crosswind, (c) turbine blade, (d) combustor exit duct and (e) curved combustor exit duct (from Tucker 2013a – published with kind permission of Elsevier).

modern gas turbine aeroengine. The estimates are made, as described in Tucker (2013b). The vertical axis is normalized by the extent of the –5/3 zone for the fan, at the front of the engine. In this zone, the Reynolds number is high and hence the extent of the –5/3 zone large. However, at the other end of the engine, in what is called the low-pressure turbine (LPT) zone, the extent is much lower. These aspects are all of importance for large eddy simulation modelling, as will be discussed later.

5.1.1.2 Forward Transition Transition modelling is a particular challenge. This is especially so since there is an even greater lack of physical understanding in this area and the phenomena has extreme sensitivity. Forward transition (FT), from laminar to turbulent flow, or transition as it is more commonly called, is an especially complex process. In turbomachinery, although there are many regions with considerable Reynolds numbers there can be a substantial amount of laminar flow connected to turbulent zones via transition. Some of these zones are shown on Figure 5.6. The low-pressure turbine is the zone where transition is especially important, this being where around 90% of the engine's thrust comes from. Hence, for this industrial application, the area of transition modelling is of extreme importance. Indeed, this is also, but to a lesser extent, the case on wings, where for accurate CFD modelling it is necessary to define the point along the chord (near the leading edge) where the flow transitions to turbulence. Transition can be influenced by surface roughness, acoustic and external disturbances, pressure gradients or free stream velocity changes, surface curvature, temperature gradients and body forces such as those caused by rotation.

Natural transition is generally associated with the growth of Tollmien-Schlicting (T-S) waves in a low free stream turbulence environment. Hence, with increasing s or Re_s, (i.e., Reynolds number with a length scale based on s) T-S waves (linear two-dimensional waves) will occur. If the waves become amplified, transition takes place. Then the linear T-S wave instability process gives way to a non-linear, three-dimensional instability. This involves the formation of vortex loop structures. This is then followed by the formation of what looks like turbulent spots. These spread downstream to fill the boundary layer. For a pictorial representation of this process, see Schlichting (1979).

Another kind of transition is known as *bypass* transition. With this, the linear T-S wave process is bypassed. Disturbance amplification takes place through a non-linear process. This forms elongated streaks called Klebanoff modes. Bypass transition takes place for free stream intensities greater than 0.5% (see Abid 1993).

In turbomachinery there are multiple blade rows. The wakes from upstream blades will also provide another route to transition. Separation of a laminar boundary layer will also, via Kelvin-Helmholtz vortices and their breakdown, provide a route to transition. Wakes and other transition modes can interact in this process in a complex fashion. The routes to transition are numerous and reflected in the Figure 5.7 schematic by Coull and Hodson (2011). As can be inferred from this figure and the very brief discussion here, the prediction of transition poses a great challenge.

5.1.1.3 Reverse Transition Fluid flows can undergo substantial acceleration. This can give rise to the process of reverse transition (RT) where a turbulent flow becomes laminar. Examples where strong acceleration can be encountered are in the nozzles of rockets and also labyrinth type seals. Further examples for turbomachinery are again given in Figure 5.6. The level of acceleration in a fluid flow is generally characterized by the acceleration parameter K, given as

$$K = \frac{v}{U_\infty^2}\frac{dU_\infty}{ds} \tag{5.2}$$

where, v, is the kinematic viscosity, U_∞ is the velocity at the edge of the boundary layer and s is the streamline direction. However, the previous equation is relatively simplistic. For example, the degree of laminarization is not just a function of the acceleration level. It also depends on the duration to which the fluid has been subjected to the acceleration. For a realistic engineering system, the fluid's acceleration field will be spatially dependent. This will give rise to a lag between the straining induced on the mean flow by the acceleration, and the turbulence response

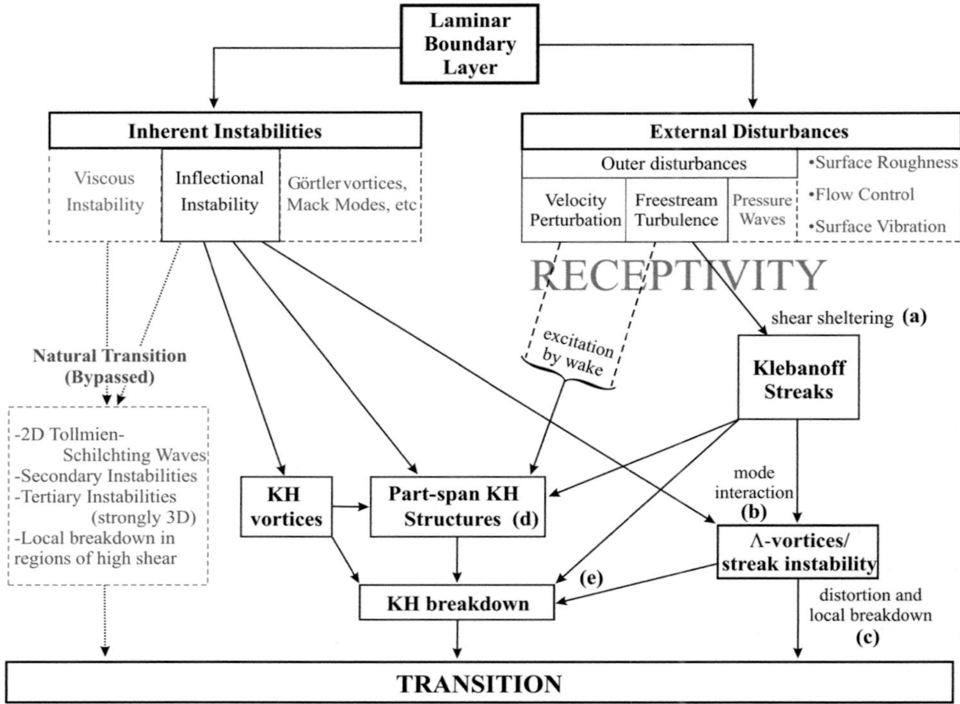

Figure 5.7. Potential routes to transition (from Coull and Hodson 2011 – published with the kind permission of Cambridge University Press).

to this. In many areas of turbulence modelling, it is important and challenging to characterize such response lags. Furthermore, if the acceleration is sufficiently rapid, the turbulence will enter the rapid distortion state. In this, it can be assumed that the state of the turbulence is essentially frozen and consistent with its upstream conditions.

5.1.1.4 Scale of Turbulent Vortices and Their Treatment The vortices found in a turbulent flow act to mix quantities such as thermal energy, momentum, passive scalars (such as dye) and particles. Based on this observation, turbulence is sometimes modelled by augmentation of diffusion coefficients in governing equations. This would replicate the lateral diffusion of a passive scalar and so forth in a uniform flow. Turbulence is three-dimensional. Even if the mean flow is one- or two-dimensional, the turbulent fluctuations will be three-dimensional ($u' \neq 0$, $v' \neq 0$, $w' \neq 0$). Hence, simulations attempting to resolve turbulent eddies (eddy-resolving simulations), to be discussed later, must be performed in three dimensions.

The Navier-Stokes equations account naturally for turbulence, being remarkably exact in doing this. However, numerical predictions for turbulent flows must be made on finite sized grids with finite time-steps Δt. Turbulent kinetic energy destruction by viscous action takes place at around $Re = 1$ (small scales of motion, where viscous and inertial forces balance).

$$Re = \frac{\rho \Delta U L}{\mu} \tag{5.3}$$

For air, $\mu/\rho \approx 1 \times 10^5$, so for $Re = 1$ we can say $L = 1 \times 10^{-5}/\Delta U$. So for just $\Delta U = 100.0$ (say in Figure 5.3), $L \approx 0.0001$ mm. Therefore, the important turbulent destruction scale can be small relative to the size of the engineering system. If the mean flow field velocity U is 100 m/s (in say Figure 5.3), the time scale, Δt, for velocity fluctuations is given by $\Delta t = L/U = 0.1 \times 10^{-3} \ / \ 100 \ = \ 1 \times 10^{-6}$ s. Note that for the shear layer in Figure 5.3, $L \propto s$ and hence as $s \to 0$ (the downstream splitter plate edge), even the large, energy containing scales can be small, with correspondingly small time scales $t \sim s/U$. Furthermore, with increasing Re, the energy in the small scales increases. Therefore, the time scale for turbulent fluctuations is small and generally much less than the time period for the flow to develop. As can be seen, these scales would require the use of impractically fine/small grid and time steps for most type CFD predictions of engineering interest. A simulation that resolves all the scales in a system from the integral down to the Kolmogorov is normally described as a Direct Numerical Simulation. Simulations can instead try to resolve just to scales into the inertial subrange and model the rest. These are called Large Eddy Simulations and will be described (along with related variants) in more detail towards the end of this chapter. Importantly for many engineering calculations, the small-scale unsteady features are not of interest. What is of interest are the averaged velocities (U, V and W) and not the instantaneous (u', v', w') components given by the Navier-Stokes equations (see Chapter 2). Therefore, it is convenient to average fluctuations and then directly solve for averaged variables. Averaged values, Φ, for a general instantaneous flow variable, ϕ, are given by the following integral

$$\Phi = \frac{1}{T} \int_{t_o - T/2}^{t_o + T/2} \phi dt \tag{5.4}$$

(Note that to successfully average the time scales for the mean flow, variations must be much greater than the fluctuating time scale T – see Section 5.4.1). Importantly, despite solving for averaged values, the effect of u', v' and w' on U, V and W is generally significant and so must be taken into account. The Navier-Stokes equations, when time averaged, are described as the Reynolds Averaged Navier-Stokes equations. These will be discussed further later.

5.1.1.5 Turbulent Boundary Layers Near-solid surfaces (roughly inner 1% of a boundary layer) complex flow aligned streak-like structures can be observed – the streaks can be seen in Figure 2.1. There are also bursts and ejections of momentum in the near wall region. A typical turbulent boundary layer can be divided in two regions called the outer and inner. Data in the latter collapses best when the friction velocity, $u_\tau = \sqrt{\tau_w}/\rho$, is defined. Then, the mean velocity is made dimensionless as $U^+ = U/\sqrt{\tau_w}/\rho$. Also, the wall distance is made dimensionless in the form of

Table 5.1. *Summary of mesh densities used to resolve the outer part of a turbulent boundary layer (from Tucker 2013b – used with the kind permission of Springer)*

Reference	δ/Δ
Chapman (1979)	1/13
Piomelli et al. (1989)	1/10
Schumann (1975)	1/5–1/10
Deardorff (1970)	1/7
Tucker and Davidson (2004)	1/11
Average	≈1/10

a Reynolds number such that $y^+ = \rho y u_\tau/\mu$. It is useful (see Table 5.2) to non-dimensionalize grid spacings in this way. The inner layer consists of three sections: (1) *Viscous sub-layer (*where $U^+ = y^+$*)*; (2) *Buffer layer* (approximately logarithmic dependence of U^+ on y^+) and (3) *Fully Turbulent region* where $U^+ = (1/\kappa)\log(Ey^+)$. Region (1) is called the viscous sublayer, where viscous forces dominate. In Region (2) both (molecular) viscous and inertia effects are important. In Region (3), often called the 'logarithmic law layer', the predominant momentum and heat transfer mechanisms will be due to turbulent fluctuations, viscous related effects playing no role. Sometimes, the outer region is called the *law of the wake* zone which is populated by large eddies. The flow dynamics in this region is similar to that found in wakes. These also have large scale flow structures, the scale of which is dependent on the size of the geometrical feature that generated the wake. The eddy size (l) in the outer region scales as $l \propto \delta$, where δ is the boundary layer thickness. Outside the viscous sublayer, in the inner region eddies scale as $l \propto y$. To resolve the outer scales on a CFD grid is not that challenging – even with a second order flow solver. Table 5.1 summarizes the boundary layer thickness, normalized by grid spacing, Δ, used by a range of workers who successfully captured, using eddy-resolving simulations, the flow dynamics in the outer part of turbulent boundary layers. As can be seen, on average, for a cube occupying the outer part of the boundary layer, a mesh of 10^3 grid points is sufficient. The grid requirements in this zone are fairly modest. As noted, the outer part of the boundary layer is often called the 'law of the wake' region. Wake zones have the helpful property that they are Reynolds number independent.

Table 5.2. *Near wall grid requirements for LES (from Tucker 2013b – used with the kind permission of Springer)*

Method	$\Delta x^+/\mathrm{Min}(\Delta y^+)/\Delta z^+$	Number $y^+<10$ points
DNS	(10–15)/1/5	3–5
Wall resolved LES	(50–130)/1/(15–30)	3–5
LES with a wall model	100–600/(30–150)/100–300	–
Hybrid RANS-LES	(100–600)/1/(100–300)	2–5
RANS	1000/1	2–5

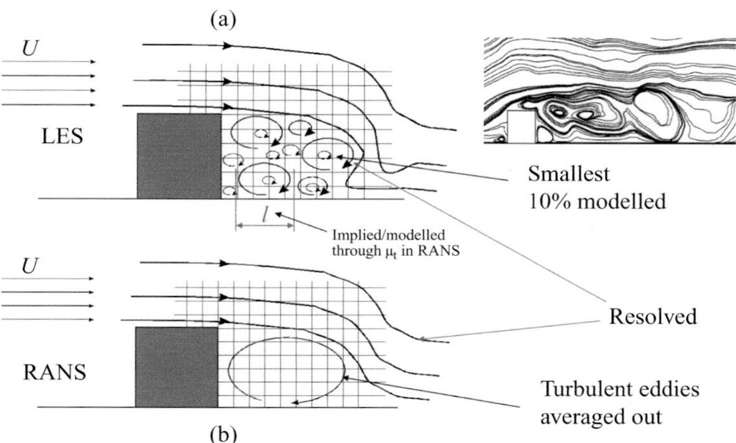

Figure 5.8. RANS and LES of flow over a cube: (a) instantaneous LES flow field and (b) time average of LES flow field or RANS.

Hence, this implies that for the outer part of the boundary, the grid requirements are fairly Reynolds number independent. Therefore, even for high Reynolds number aerospace flows it is not that infeasible to resolve the turbulent eddies on the CFD grid in the outer part of the boundary layer.

The really problematic region of the boundary layer to resolve on a grid is the inner part. This, as shown in Figure 2.1, is populated by the fine elongated streaks. The resolution of these is reflected in Table 5.2. For the inner part of the boundary layer, the most appropriate scalings are generally considered to be based on the so-called wall units, described earlier. In the table, the x coordinate is taken to correspond to the streamwise, y to the wall normal and z to the remaining direction. The inner layer scalings, which involve the wall shear stress, are strongly dependent on the velocity, viscosity and also fluid density and are thus Reynolds number dependent. Indeed, the cost to DNS in this zone scales at around Re^3. The use of some lower order modelling, to be discussed later, reduces this scaling to varying degrees. Notably, to control the computational cost, some hybrid methods attempt to completely model the inner layer, or even the complete boundary layer with low order RANS. This weakens the Reynolds number sensitivity to an exponent of between 0.5 and 1 (rather than Re^3 for DNS). The exponent stated depends on if the boundary layer development from a leading edge needs to be considered. The small scales at the leading edge, where the boundary layer develops from a length scale of zero, places further resolution challenges, thus raising the exponent. The implications of modelling a developing boundary layer with an eddy-resolving simulation method with respect to grid are outlined in Chapter 3.

To further illustrate the nature of eddy-resolving simulations, Figure 5.8 gives an impression of an instantaneous flow field. The image would change substantially with time, the reattachment point varying by several cube heights. The time varying flow field will look very much like the real flow field. With LES, ideally, we

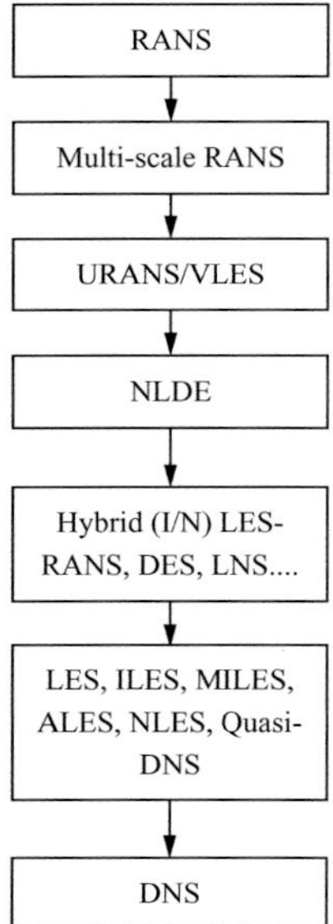

Figure 5.9. Hierarchy of turbulence modelling methods.

try to predict (resolve on the grid) about 90% of the turbulent energy (with the Navier-Stokes equations) leaving the model to account for just roughly 10%. With RANS, the model accounts for 100% of the turbulence.

As indicated by Figure 5.8a, with LES, typically the grid cannot pickup (resolve) eddies smaller than the grid spacings (in fact, several grid spacings are needed to faithfully resolve an eddy). For these, we can use RANS-like models that mostly dissipate energy – referred to as Subgrid Scale (SGS) models. Alternatively, we can dispense with any model, using the numerical scheme to dissipate energy and perform even more complex modelling roles. Many numerical schemes are dissipative, this being a necessary property for CFD code convergence. This approach, which uses the numerical scheme to provide a model, is described by many names, but nowadays Implicit LES seems to be most popular. Figure 5.8b gives a schematic of what the RANS flow field would look like or a time averaged LES. As noted previously, the LES and RANS methods can be hybridized in various ways. Indeed the methods described can be arranged in a hierarchy that are summarized in Figure 5.9.

Having given a brief overview of the essential nature of turbulence and different methods to deal with it, more details of methods will be given in turn and the hierarchy given in Figure 5.9 expanded upon. First the RANS method will be discussed. For this the turbulence is fully modelled.

5.2 Basic RANS Method

5.2.1 *The Reynolds Equations*

The Navier-Stokes equations, as given in Chapter 2, which give instantaneous velocities (u, v, w) can be cast in a form to give the time averaged velocities

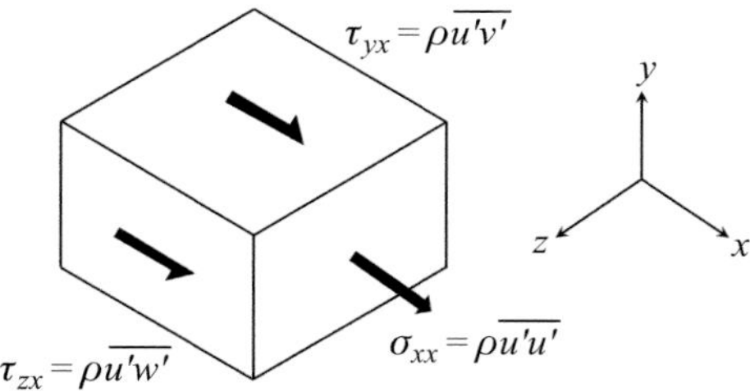

Figure 5.10. Locations where stresses act on fluid element.

(U, V, W). This process is outlined now, but for simplicity we will just consider the right-hand side (the convective terms) of the Navier-Stokes equations. It can be shown, in fact, that the left-hand side of the equation does not change. Also, for brevity, just the 'u' momentum equation will be considered, the process for the others being obvious. The u-momentum equation (see Chapter 2) is stated as

$$\underbrace{\rho\frac{\partial u}{\partial t} + \rho\frac{\partial}{\partial x}(uu) + \rho\frac{\partial}{\partial y}(vu) + \rho\frac{\partial}{\partial z}(wu)}_{\text{LHS(A)}} = \underbrace{-\frac{\partial p}{\partial x} + \frac{\partial}{\partial x}\left(\mu\frac{\partial u}{\partial x}\right) + \frac{\partial}{\partial y}\left(\mu\frac{\partial u}{\partial y}\right) + \frac{\partial}{\partial z}\left(\mu\frac{\partial u}{\partial z}\right)}_{\text{RHS(B)}} \tag{5.5}$$

On substitution of Equations (5.1), the left-hand side of this equation can be written as

$$\rho\frac{\partial U}{\partial t} + \rho\frac{\partial u'}{\partial t} + \rho\frac{\partial}{\partial x}(U + u')(U + u') + \rho\frac{\partial}{\partial y}(U + u')(V + v') + \rho\frac{\partial}{\partial z}(U + u')(W + w') = B \tag{5.6}$$

Averaging over a small time period, where $\overline{u'} = \overline{v'} = \overline{w'} = 0$, gives

$$\underbrace{\rho\frac{\partial U}{\partial t} + \rho\frac{\partial}{\partial x}UU + \rho\frac{\partial}{\partial y}UV + \rho\frac{\partial}{\partial z}UW}_{\text{LHS(C)}} = B - \rho\left[\frac{\partial\overline{u'u'}}{\partial x} + \frac{\partial\overline{u'v'}}{\partial y} + \frac{\partial\overline{u'w'}}{\partial z}\right] \tag{5.7}$$

The $-\rho\overline{u'u'}$, $-\rho\overline{u'v'}$ and $-\rho\overline{u'w'}$ terms are called Reynolds stresses. These terms are momentum flow terms and hence give rise to forces/stresses. The places where these forces act on a fluid control volume are shown in Figure 5.10. The sign convention used here is that for τ_{ij}, j represents the direction the stress acts. The surface on which it acts is normal to the i direction. As previously stated, the left-hand side of Equation (5.5), B, does not yield any extra terms.

Analogous to the work of Stokes, discussed in Chapter 2, who gave for an incompressible fluid $\tau_{ij} = \mu\left[\partial u_i/\partial x_j + \partial u_j/\partial x_i\right]$, Boussinesq (1877) proposed

$$\tau_{ij} = -\rho\overline{u_i'u_j'} = \mu_t\left[\frac{\partial U_i}{\partial x_j} + \frac{\partial U_j}{\partial x_i}\right] \tag{5.8}$$

or rather (from setting $i = j = 1$; $i = 2$, $j = 1$; $i = 3$, $j = 1$)$\sigma_{xx} = -\rho\overline{u'u'} = \mu_t[\partial U/\partial x + \partial U/\partial x]$, $\tau_{yx} = -\rho\overline{v'u'} = \mu_t[\partial U/\partial x + \partial V/\partial y]$, $\sigma_{xz} = -\rho\overline{w'u'} = \mu_t[\partial W/\partial x + \partial U/\partial z]$.

In the previous, μ_t is called the eddy or turbulent viscosity. Substitution of these expressions into Equation (5.7) after rearrangement gives

$$C = B + \frac{\partial}{\partial x}\mu_t\left(\frac{\partial U}{\partial x}\right) + \frac{\partial}{\partial y}\mu_t\left(\frac{\partial U}{\partial y}\right) + \frac{\partial}{\partial z}\mu_t\left(\frac{\partial U}{\partial z}\right) + \left(\frac{\partial \mu_t}{\partial x} + \mu_t\frac{\partial}{\partial x}\right)\left[\frac{\partial U}{\partial x} + \frac{\partial V}{\partial y} + \frac{\partial W}{\partial z}\right] \quad (5.9)$$

But the continuity equation for a constant density flow is $\partial U/\partial x + \partial V/\partial y + \partial W/\partial z = 0$. Hence the latter terms in Equation (5.9) go to zero. Thus, the full 'U' velocity component Reynolds equation can be stated as

$$\rho\frac{\partial U}{\partial t} + \rho\frac{\partial}{\partial x}(UU) + \rho\frac{\partial}{\partial y}(VU) + \rho\frac{\partial}{\partial z}(WU) = -\frac{\partial P}{\partial x}$$
$$\frac{\partial}{\partial x}\left((\mu + \mu_t)\frac{\partial U}{\partial x}\right) + \frac{\partial}{\partial y}\left((\mu + \mu_t)\frac{\partial U}{\partial y}\right) + \frac{\partial}{\partial z}\left((\mu + \mu_t)\frac{\partial U}{\partial z}\right) \quad (5.10)$$

Equations, having a similar form to (5.10) can be written for the U and W velocity components. Equation (5.10) has a convenient form, being virtually identical to Equation (5.5), but the prescription of μ_t is difficult and this is discussed next.

5.2.2 *RANS Model Character*

The key problem faced is that RANS models are to varying extents *phenomenological* – but the phenomena is not fully understood. The most popular aerospace turbulence model is still perhaps the Spalart and Allmaras (1994) SA model. SA is probably the most extreme, widely used, phenomenological model and this is reflected in the first line (p. 5) of their paper: "Abstract – *A transport equation for turbulent viscosity is assembled based on* ***empiricism*** *and* ***arguments of dimensional analysis***." The model is in fact a mix of crude engineering scaling, strong physical insight and mathematical modelling, especially with respect to gaining a model that has robust convergence and has resilience to non-unique solutions. Even so, the model is not free from non-unique solutions (see, for example, Rumsey 2007) arising from erroneous initial/boundary conditions. When looking at the dictionary definition of empiricism, the word quackery occurs. This is not such an inappropriate word, since, go to any turbulence modelling conference and you would be hard pushed to get people to even agree on a definition of turbulence. It is also notable, as pointed out by Spalart (2000), that even a biennial turbulence-modelling symposium can yield around twenty-five new turbulence models. Clearly, the availability of

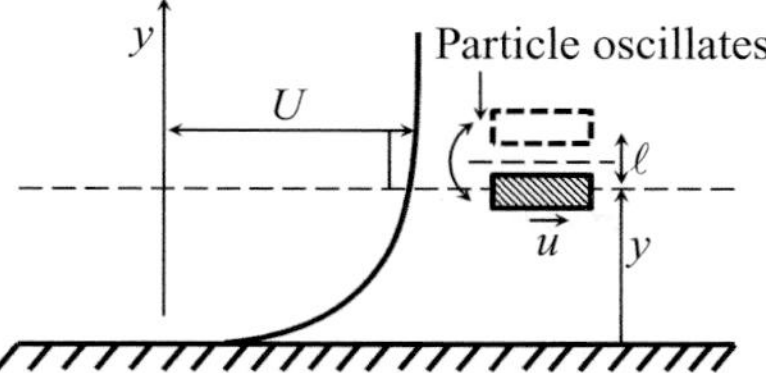

Figure 5.11. Macroscopic momentum transport in a boundary layer.

many hundreds of turbulence models is confusing for an aerodynamacist and also reflects a struggle to remedy defects.

RANS models can range from being highly phenomenological or more theoretically based, but even the most theoretical contain substantial empirical elements (see Spalart 2000). They can also be classified by the number of differential equations they involve. The mixing length model, to be discussed next, has no differential equations – and so it is called a *zero equation model*. The *k-l* model, also to be discussed later, involves one substantial differential equation and hence is called a one-equation model. Two-equation models can also be derived. Also the term *realizability* is used. This refers to whether the model is constrained so that it does not break basic physical principles. For example, a realizable model would ensure all the normal Reynolds stresses remain positive along with turbulence kinetic energy. There are perhaps hundreds of basic RANS models. The natures of some turbulence models are now outlined.

5.2.3 *Zero Equation Models*

5.2.3.1 The Mixing Length Turbulence Model The macroscopic transport of momentum and heat can be considered similar to that at the molecular level. At the molecular level we relate shear stress to fluid velocity in the following way: $\tau_l = \mu\, dU/dy$. This ('laminar') shear force is a result of the momentum transfer as molecules randomly move about (i.e., $u_i' u_j'$ components at the molecular level). Therefore, by analogy (recall Boussinesq), we write $\tau_t = -\rho\, \overline{v'u'} = \mu_t\, dU/dy$. In the analysis of molecular problems, the average distance between collisions (the mean free path) is defined. Prandtl's mixing length theory is again analogous to this. In this theory a length l is defined, which represents the average distance fluid lumps travel normal to the mean flow direction (given by the RANS). Consider the fluid lump shown in Figure 5.11.

The approximate velocity at $y + l$ can be stated as $u(y + l) \approx u(y) + l\partial u/\partial y$ and hence

$$u'(y) \approx u(y + l) - u(y) = l\frac{\partial u}{\partial y} \tag{5.11}$$

The particle amplitude of deviation l is called the *Prandtl mixing length*. Another assumption of Prandtl's (which from continuity of flow considerations does not seem unreasonable) is that v' has the same order of magnitude as u'. Therefore, Equation (5.11) can be written as

$$\tau_t = -\rho\, l^2 \left(\frac{\partial u}{\partial y}\right)^2 = \mu_t \frac{\partial u}{\partial y} \tag{5.12}$$

and so μ_t can be stated as

$$\mu_t = \rho\, l^2 \frac{\partial u}{\partial y} \tag{5.13}$$

or more generally

$$\mu_t = \rho l^2 \left[2 \frac{\partial U_i}{\partial x_j} S_{ij} \right]^{1/2} \tag{5.14}$$

Prandtl further assumed

$$l = \kappa y \tag{5.15}$$

This is the essence of the mixing length turbulence model that furnishes values of μ_t. As to be discussed later, Equation (5.15) is only reasonable outside the viscous sublayer and inside the inner layer.

In the previous, turbulence production is essentially governed by the modulus of straining $S = \sqrt{2 S_{i,j} S_{i,j}}$, that is,

$$S = \sqrt{\frac{1}{2}\left(\frac{\partial U_i}{\partial x_j} + \frac{\partial U_j}{\partial x_i}\right)^2} \tag{5.16}$$

However, as follows, a vorticity based velocity scale is also possible and as will be seen later can give better behaviour in stagnation regions.

$$\Omega = \sqrt{\frac{1}{2}\left(\frac{\partial U_i}{\partial x_j} - \frac{\partial U_j}{\partial x_i}\right)^2} \tag{5.17}$$

The mixing length model could be more simply formulated based on dimensional grounds by stating that $\mu_t \propto$ *Length* $\times$ *Velocity*, with the velocity scale defined as $l\,S$. The model is referred to as incomplete. The equation for l takes different forms for different regions of the flow. In the viscous sublayer ($y^+ < 5$)

$$l = \kappa d Da \tag{5.18}$$

where Da is some form of damping function and $d = y$. The van Driest is the best known, where $Da = 1 - \exp(-y^+/26)$. There are numerous forms of Da (see, for example Launder and Priddin 1973). In the outer part of the boundary layer $l = C\,\delta$, where typically $C = 0.09$. For different free shear flows and wake zones other formulations must be used. For example in a:

mixing layer $l \approx 0.07 \times$ the layer width;
round jet $l \approx 0.075 \times$ the jet half width;
plane jet $l \approx 0.09 \times$ jet half width and
wake $l \approx 0.16 \times$ the wake half width

and so on. As can be seen, the modeller has numerous potential l choices. These, in themselves, can be subjected to further corrections to be discussed later.

A further interesting zero equation model is the LVEL model of Spalding (1994). This makes a Taylor series based fit to the 'law of the wall' (there are three separate equations fitting the inner part of the boundary layer velocity profile). This is differentiated to gain a dimensionless eddy viscosity. The model is useful for geometries that contain many zones supporting channel type flows. The Taylor series based 'law of the wall' can be expressed as

$$y^+ = U^+ + \frac{1}{E}\left[e^{\kappa U^+} - 1 - \frac{\kappa U^+}{1!} - \frac{(\kappa U^+)^2}{2!} - \frac{(\kappa U^+)^3}{3!} - \frac{(\kappa U^+)^4}{4!}\right] \tag{5.19}$$

where $E = 9.8$. This can be differentiated to give the dimensionless effective viscosity. The previous can be differentiated and the result can then be rearranged to yield the following effective viscosity expression:

$$\mu_t = \mu\frac{\kappa}{E}\left[e^{\kappa U^+} - 1 - \frac{\kappa U^+}{1!} - \frac{(\kappa U^+)^2}{2!} - \frac{(\kappa U^+)^3}{3!}\right] \tag{5.20}$$

The dimensionless velocity U^+ in Equation (5.20) can be calculated using a Newton-Raphson procedure needing d.

Zero equation models generally allow more modest grids. Although they might be expected to yield economical solutions, they can in fact result in slow iterative convergence. Although zero equation models can seldom be expected to provide high accuracy, they do impose basic physical turbulence expectations. Hence, their linearity means they are unlikely to yield highly substantial errors. However, this can occur.

5.2.4 *One-Equation Turbulence Models*

Zero-equation models assume the turbulence is in some form of equilibrium state. They assume that the rate of production and dissipation of turbulence at a particular point are in equilibrium. However, this is not always the case. High levels of turbulence can arise at a particular point in space, as a result of being generated in another zone and transported by the fluid motion to that point. Hence the modelling of convective processes can be important. All one-equation models essentially solve a convective-diffusion transport equation for a transported property, say, ϕ, with a source term S_ϕ which can involve a turbulence production term or, for some models, both production and destruction.

5.2.4.1 Eddy Viscosity Transport Models The Secundov et al. (2001), Baldwin and Barth (1990) and SA models solve a convection-diffusion transport equation essentially for μ_t. Perhaps the most basic one-equation model, being the forerunner of those mentioned, is due to Nee and Kovasznay (1969). It is close to the mixing length model described previously. However, to account for non-equilibrium, it solves the convection diffusion equation

$$\frac{\partial \phi}{\partial t} + \frac{\partial U_j \phi}{\partial x_j} = \frac{1}{\rho} \frac{\partial}{\partial x_j} \left(\Gamma_\phi \frac{\partial \phi}{\partial x_j} \right) + S_\phi \tag{5.21}$$

where $\phi = \mu + \mu_t$. Also,

$$\Gamma_\phi = \frac{\mu + \mu_t}{\sigma_\phi} \tag{5.22}$$

where σ_ϕ is a Schmidt number. The source term is

$$S_\phi = P - D \tag{5.23}$$

where P is the production term and D the destruction. For the former

$$P = \mu_t\, S \tag{5.24}$$

The destruction term

$$D = \frac{(\mu + \mu_t)\mu_t\, B}{d^2} \tag{5.25}$$

where this time B is a constant. If it is assumed that the turbulence is in equilibrium, then the turbulence production and destruction terms can be equated as

$$A\mu_t S = \frac{(\mu + \mu_t)\mu_t\, B}{d^2} \tag{5.26}$$

where A is a constant. On rearrangement (note $\mu + \mu_t$), this becomes

$$\mu_t = \left(\frac{A}{B}\right) d^2 S \tag{5.27}$$

which is the Prandtl mixing length model with $l = d(A/B)^{1/2}$. This implies the following constraint between the constants A and B:

$$A = C^2\, B \tag{5.28}$$

Based on the standard boundary layer value of $C = 0.4$, the values of $A = 0.1333$ and $B = 0.8$ follow. As with SA, there are a range of values for A and B that would satisfy the constraint given by C.

5.2.4.2 SA Model Although the SA model is inspired by the Nee and Kovasznay (1969) it is considerably more complex. As implied by Figure 5.12, it could be regarded as modular. Some of the modules of Figure 5.12 will be discussed later. The SA model still retains convection and diffusion terms but even these aspects have differences. For example, the convection term form is non-conservative. The differences in the diffusion term are not outlined here but it is noted that diffusion is also used to provide turbulence dissipation, removing turbulence to a free stream that is intended to be extensive. Essentially the model consists of the following

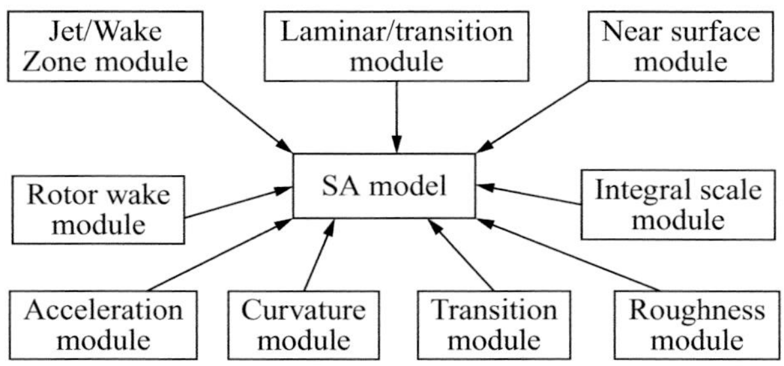

Figure 5.12. Modules available to the SA model (from Tucker 2013a – published with kind permission of Elsevier).

four nested modules: free shear, log-outer layer, buffer and viscous sublayer, and transition. The free shear module is conceptually the easiest, hence best looked at first. SA solves a transport equation for essentially eddy viscosity, much like the Nee and Kovasnay model. There are three constants in the free shear module. These values are found by calibrating to expected peak shear stress values, based on measurements, for a two-dimensional mixing layer $\tau_{\max} = 0.01(\Delta U)^2$ and wake $\tau_{\max} = 0.06(\Delta U)^2$, where ΔU is the velocity difference driving the shear layer. As with the Nee-Kovasnay model, the calibration does not suggest unique constants – a range of values will satisfy the calibration. However, three values are settled on by Spalart and Allmaras. These make the turbulent Schmidt number a more standard value. It is clearly accepted that the model will over-predict the spreading rate by over 38% for a plane jet. Hence, jets are not of interest to Spalart and Allmaras – their key concern being aircraft drag at design conditions. For wall bounded flows, the SA equation has a destruction term that is proportional to $-\mu_t^2/d^2$. Hence, like the Nee-Kovasnay model, in the absence of convection and diffusion, SA reduces to a mixing length model near walls. The much greater complexity of the SA model, relative to the Nee-Kovasnay, makes its calibration much more complex, the SA model having twelve constants. The solution variable for SA is also more complex than used in the Nee-Kovasany model. For SA, this is given s

$$\phi = \tilde{v} = \frac{\mu_t}{\rho f_{v1}} \tag{5.29}$$

where the damping function, $f_{v1}(\tilde{v}/v)$, is defined later. Note that since in high Reynolds number regions (away from walls) f_{v1} tends to unity $\phi = \tilde{v} = \mu_t/\rho$ giving the general character of the key SA solution variable. The production and destruction terms have the following broad forms:

$$P = \gamma \cdot C_{b1}(1 - f_{t2})\hat{S}\tilde{v}, D = f(v, \tilde{v}, d, \kappa, S)\left(\frac{\tilde{v}}{d}\right)^2, \tag{5.30}$$

Note:

$$\hat{S} = \Omega + \frac{\tilde{v}}{(\kappa d)^2} f_{v2} \tag{5.31}$$

where the functions f_{v2} and f_{t2} are as follows

$$f_{v2} = 1 - \frac{x}{1 + x f_{v1}} \tag{5.32}$$

$$f_{t2} = c_{t3} \exp(-c_{t4} X^2) \tag{5.33}$$

Notably, the production is now vorticity based. Hence, it is not that different to when based on strain rate in boundary layer zones. However, helpfully, in stagnation zones (airfoil leading edges) excessive turbulence production is avoided, vorticity being zero in such zones (see later). The key damping function in the SA model is due to Mellor and Herring (1968):

$$f_{v1} = \frac{\tilde{\chi}^3}{\tilde{\chi}^3 + C_{v1}^3} \tag{5.34}$$

This avoids the need to work out the local wall shear stress. For the standard form of the SA model, $\tilde{\chi} = \tilde{v}/v$. Modifications to these terms will be described later. With SA, the near wall destruction term involves a near wall non-viscous destruction element/function, f_w, which increases towards walls. This is because the turbulence destruction term $(\tilde{v}/\tilde{d})^2$ alone decays too slowly in the outer part of the boundary layer. This gives inaccurate boundary layer skin friction values. The f_w term has the following general form:

$$f_w = g\left[\frac{1 + c_{w3}^6}{g^6 + c_{w3}^6}\right]^{1/6}, \quad g = r + c_{w2}(r^6 - r), \quad r = \min\left[\frac{\hat{v}}{\hat{S}\kappa^2 d^2}, 10\right] \tag{5.35}$$

As can be inferred, this function is fairly tortuous in d. The f_w function is a major element that distinguishes SA from the Nee-Kovasany and Secundov (to be described later) models. Also, with SA, near walls essentially (note in the log layer $\mu_t = \rho\kappa d u_\tau$), $\mu_t = \rho\kappa d u_\tau f_{v1}$ implying $\phi = \mu_t/\rho f_{v1} = \kappa d u_\tau$, giving a linear near wall behaviour and hence grid resilience. Outside the near wall region, there is no explicit turbulence length scale. This makes the model less applicable to internal flows and also certain free shear flows. The model is relatively effective for boundary layers with adverse pressure gradients. Hence, SA can be useful for predicting stall, but like most models has a tendency to delay separation. The SA model constants are as follows:

$$\begin{aligned}
&c_{b1} = 0.1355, \ \sigma = \frac{2}{3}, \ c_{b2} = 0.622, \ \kappa = 0.41, \\
&c_{w2} = 0.3, \quad c_{w3} = 2, \quad c_{v1} = 7.1, \quad c_{t3} = 1.2, \quad c_{t4} = 0.5, \\
&c_{w1} = \frac{c_{b1}}{\kappa^2} + \frac{1 + c_{b2}}{\sigma}
\end{aligned} \tag{5.36}$$

5.2.4.3 Secundov v_t-92 model The v_t-92 model (see Secundov et al. 2001) is another one-equation turbulent eddy viscosity model. In it, a transport equation for the kinematic turbulent viscosity, $\phi = v_t = \mu_t / \rho$, is solved. The diffusion term, like SA, is non-standard, having the following addition:

$$Diffusion = \frac{\partial}{\partial x_i}[(C_0 - C_1)\phi - v]\frac{\partial \phi}{\partial x_i} \tag{5.37}$$

For the more normal diffusion term component

$$\Gamma_{vt} = C_0 v_t + v$$

Also, the source term takes the following form

$$\begin{aligned} S_{v_t-92} &= P - D \\ &= C_2 F_2 \left(v_t \Gamma_1 + A_1 v_t^{4/3} \Gamma_2^{2/3} \right) + C_2 F_2 A_2 N_1 \sqrt{(v + v_t)\Gamma_1} + C_3 v_t \left(\frac{\partial^2 v_t}{\partial x_i \partial x_i} + N_2 \right) \\ &\quad - C_4 v_t \left(\frac{\partial U_i}{\partial x_i} + \left| \frac{\partial U_i}{\partial x_i} \right| \right) - \frac{C_5 v_t^2 \Gamma_1^2}{c^2} - \frac{[C_6 v_t N_1 d + C_7 F_1 v v_t]}{d^2}. \end{aligned} \tag{5.38}$$

where c is the speed of sound. The functions F_1 and F_2 are computed, respectively, from the following expressions:

$$F_1 = \frac{N_1 d + 0.4 C_8 v}{v_t + C_8 v}, \quad F_2 = \frac{\chi^2 + 1.3\chi + 0.2}{\chi^2 - 1.3\chi + 1.0}, \tag{5.39}$$

where $\chi = v_t/(7v)$. In Equation (5.38), the parameters Γ_1, Γ_2, N_1 and N_2 are defined as follows:

$$\begin{aligned} \Gamma_1^2 &= \frac{\partial u_i}{\partial x_j}\left(\frac{\partial u_i}{\partial x_j} + \frac{\partial u_j}{\partial x_i} \right), \quad \Gamma_2^2 = \sum_i \left(\frac{\partial^2 u_i}{\partial x_j \partial x_j} \right)^2, \\ N_1^2 &= \sum_i \left(\frac{\partial v_t}{\partial x_i} \right)^2, \quad N_2^2 = \sum_i \left(\frac{\partial N_1}{\partial x_i} \right)^2. \end{aligned} \tag{5.40}$$

The numerous constants in the previous equations are not given here, but can be found in Secundov et al. 2001. Notably, the Secundov model, like the next, has an interesting extension to two-equations. This will be discussed later.

5.2.4.4 Turbulence Kinetic Energy Transport Models Along very different lines, a transport equation for turbulence kinetic energy can be solved. This essentially, once the root is taken, gives the velocity scale needed in the basic mixing length model. The equation for turbulence kinetic energy can be derived as follows:

(1) the Navier-Stokes equation for u is multiplied by u'; the process is repeated for the v and w components and the results are added;
(2) Process (1) is repeated with the RANS equation replacing the Navier-Stokes and
(3) The final equations for processes (1) and (2) are subtracted and some subsequent considerable rearrangements made.

A well-known one-equation model is the *k-l* model (see, for example, Wolfshtein 1969). In this model $\phi = k$ (the turbulence kinetic energy) and l is a mixing length type expression. The standard differential transport Equation (5.21) is used with Equation (5.23) and

$$P = -T_1 \overline{u_i' u_j'} \frac{\partial U_i}{\partial x_j} \tag{5.41}$$

$$D = T_2 \tag{5.42}$$

In the above $\Gamma_k = \mu + \mu_t/\sigma_k$ (σ_k is the diffusion Schmidt number for k), $T_1 = 1$ and $T_2 = \varepsilon$ (the rate of dissipation of turbulence kinetic energy) defined as

$$\varepsilon = \frac{k^{3/2}}{l_\varepsilon} \tag{5.43}$$

where $l_\varepsilon = C_{\varepsilon 0} y(1 - e^{-A_\varepsilon y^+/C_\mu^{1/4}})$ and $y^+ = y\rho k^{1/2} C_\mu^{1/4}/\mu$. Note, the latter equation is only equivalent to the standard y^+ for basic equilibrium boundary layers and hence normally this parameter is expressed as y^*. The term y^* is made use of later where it is explicitly necessary to distinguish more precisely between differing near wall damping behaviours. The turbulent viscosity can be expressed as

$$\mu_t = \rho C_\mu l_\mu k^{1/2} \tag{5.44}$$

where

$$l_\mu = C_{\mu 0} y(1 - e^{-A_\mu y^+/C_\mu^{1/4}}) \tag{5.45}$$

The following standard constants are used in the previous: $\sigma_k = 1$, $A_\varepsilon = 0.263$, $A_\mu = 0.016$, $C_{\varepsilon 0} = 2.4$, $C_{\mu 0} = 2.4$, $C_\mu = 0.09$.

5.2.5 *Two-Equation Turbulence Models*

Broadly, most of the available two-equation models extend the *k-l* described Section 5.2.4.4. In terms of eddy viscosity based transport models, the Secundov model is unusual in that there is also a two-equation extension to this. There are a vast number of two-equation turbulence model variants. Most are based on solving convection-diffusion transport equations for both k and ε, again based around Equation (5.21).

5.2.5.1 k-ε model For the *k-ε* model the eddy viscosity is computed as

$$\mu_t = \rho f_\mu C_\mu \frac{k^2}{\varepsilon} \tag{5.46}$$

For the well-known models of Launder and Sharma (1974) –LS – and Chien (1982) to evaluate k in Equation (5.21), $T_2 = \varepsilon + \tilde{D}$, where $\tilde{D}$ is defined later. Also, Equation (5.21) is used to estimate the rate of dissipation of k, with $\phi = \varepsilon$, $\Gamma_\varepsilon = \mu + \mu_t / \sigma_\varepsilon$ (where σ_ε is the diffusion Schmidt number forε). To do this (evaluate ε) the following alterations are also needed:

$$T_1 = -\frac{C_{\varepsilon l}\varepsilon}{k}$$

Table 5.3. *Constants for the Launder and Sharma and Chien k-ε models*

Model	LS	Chien
$C_{\varepsilon 1}$	1.44	1.35
$C_{\varepsilon 2}$	1.92	1.8
C_{ε}	0.09	0.09
σ_k	1.0	1.0
σ_{ε}	1.3	1.3

and

$$T_2 = C_{\varepsilon 2} f_2 \frac{\varepsilon^2}{k} + E.$$

The constants for these two popular models are summarized in Table 5.3. Further functions are summarized in Table 5.4. Typically the k-ε model involves five to six constants.

5.2.5.2 k-ω models A popular alternative to solving for k and ε is to use k and ω (see Wilcox 1988), where

$$\mu_t = \rho C_\mu \frac{k}{\omega} \tag{5.47}$$

Hence, the turbulence velocity and length scales needed to formulate the eddy viscosity are $u = k^{1/2}$ and $k^{1/2}/\omega$, respectively, ω being a turbulence frequency. The k-ω model shows excessive sensitivity to the specified free stream turbulence intensity level. This has given rise to the zonal Menter (1993) SST model.

With Menter SST, the k-ε model is used away from walls and k-ω near them, the k-ε model being less sensitive to the specified free stream turbulence level. The ε equation is changed into one for ω by the transformation $\varepsilon = k\omega$. Effectively, except for a new production limiter, the k equation remains the same as that used in the basic k-ω model. The two models are blended, to achieve smoothness, using an expression of the form

$$\phi = F_1 \phi_1 + (1 - F_1)\phi_2 \tag{5.48}$$

Table 5.4. *Terms in the Launder and Sharma and Chien k-ε models. $Re_t = k^2/\varepsilon v$ the turbulent Reynolds number (Note, y^+ is defined in its standard from and hence based on local wall shear stress)*

Model	LS	Chien
f_2	$1 - 0.3\exp(-Re_t^2)$	$1 - 0.3\exp[-(Re_t/6)^2]$
f_μ	$\exp[-3.4/(1 + Re_t/50)^2]$	$1 - 0.3\exp(-0.0115y^+)$
$\tilde{D}$	$2v(\partial k^{1/2}/\partial x_j)^2$	$-2vk/y^2$
E	$2vv_t(\partial^2 U_i/\partial x_j \partial x_k)^2$	$2v\varepsilon\exp(-y^+/2)/y^2$

where $F_1(l/d, Re_y)$ (where $Re_y = \rho\omega d^2/\mu, l = \sqrt{k}/\omega$) is a hyperbolic tangent based function involving $1/d^2$ terms. The function F_1 is zero at the walls growing to unity. ϕ relates to a blending for the turbulence constants in the k-ω model to those of the transformed k-ε equation constants. The blending is also applied to a new cross-diffusion term in the ω equation (This term arises directly from the ε diffusion term through the transformation.)

$$\text{Extra Diffusion Term} = 2\rho(1 - F_1)\frac{\sigma_\omega}{\omega}\frac{\partial k}{\partial x_j}\frac{\partial \omega}{\partial x_j} \tag{5.49}$$

Note that $F_1 = 0$, in the transformation, this term relating to the model blending. Importantly, there is also an additional shear stress transport (SST) term in the Menter model. This is intended to add a form of Reynolds stress transport modelling (reminiscent of Reynolds stress models discussed later) to the primary shear stress. This additional element, is expressed as follows (Note, $\tau = -\rho\overline{u'v'}$):

$$\frac{D\tau}{Dt} = \frac{\partial \tau}{\partial t} + u_k\frac{\partial \tau}{\partial x_k} \tag{5.50}$$

Equation (5.50) is essentially active for adverse pressure gradient zones. The additional implied transport equation means that the predicted shear stress no longer responds instantaneously to changes in the strain field. Activation of the SST component is again controlled by a hyperbolic tangent function. The SST component manifests itself as a limiter. Ultimately, it ensures that the turbulent shear stress does not exceed a set fraction of k via the 'Bradshaw assumption' given as

$$\tau = -\rho\overline{u'v'} = \rho a k \tag{5.51}$$

where $a = 0.31$ (the Townsend structure parameter). Reynolds stress related models, more elegantly, replicate this trait. As shown by Georgiadis and Yoder (2013), for shock separated flows $a = 0.35$ is a more suitable value. The limiting is achieved through the eddy viscosity equation given as

$$\mu_t = \frac{\rho\alpha_\kappa}{\max(\alpha_1\omega, \omega F_2)} = \frac{\rho\alpha_1 k}{\max(\omega, \omega F_2/\alpha_1)} \tag{5.52}$$

where a_1 ($= 0.03$) is a constant. Note that F_2 is another blending function. Also, to prevent excessive turbulent kinetic energy in stagnation zones the production term in the k equation is modified as

$$P_k = \min\left(10\beta^*\rho k\omega, 2\mu_t S_{ij}.S_{ij} - \frac{2}{3}\rho k\frac{\partial U_i}{\partial x_j}\delta_{ij}\right) \tag{5.53}$$

where $\beta^* = 0.09$. Whether the k-ε model's insensitivity to specified inflow turbulence levels is entirely a good thing is perhaps debatable.

5.2.5.3 Other two-equation models In terms of other interesting two-equation models, Kim and Chung (2001) solve for μ_t and k. Also, Warner et al. (2005) solve equations for k and kl. The v_t-L extends v_t-92 by including a transport equation for

L. This allows the influence of large free stream eddies approaching an object to be modelled, L having the potential to capture integral scales. The transport equation for L has the usual convection and diffusion terms. A key, L, destruction term is proportional to

$$D \propto \frac{\mu_t}{\rho}\frac{L}{d^2} \tag{5.54}$$

This describes the decrease in L due to wall proximity. The full transport equations for the v_t-L model, respectively, read as follows:

$$\frac{\partial v_t}{\partial t} + \frac{\partial u_i v_t}{\partial x_i} = [C_0(1+\Phi)v_t + v]\frac{\partial^2 v_t}{\partial x_i^2} + P_v - D_v \tag{5.55}$$

$$\frac{\partial L}{\partial t} + u_i\frac{\partial L}{\partial x_i} = [k_0\,(1+\Phi)\,v_t + v]\frac{\partial^2 L}{\partial x_i^2} + \frac{k_1\,(1+\Phi)\,Gv_t}{L} + \frac{k_2 v_t}{L} - \frac{L}{3}\frac{\partial U_i}{\partial x_i}$$
$$- \frac{k_3 L v_t \Gamma_1}{v_t + v} - \frac{k_4\,(1+\Phi)\,Gv_t L}{d^2} \tag{5.56}$$

where

$$\begin{aligned} S_v &= P_v - D_v \\ &= C_2F_2\big(v_t\Gamma_1 + A_1 v_t(v_t+\beta v)(v_t+v)^{-2/3}\Gamma_2^{2/3}\big) + C_2F_2A_2N_1\sqrt{(v+v_t)\Gamma_1} \\ &\quad + C_3 v_t\left(\frac{\partial^2 v_t}{\partial x_i \partial x_i} + N_2\right) + \frac{C_1(1+\Phi)v_t N_1}{L\left(1+\frac{0.1L^2\Gamma_1}{v_t+v}\right)} - C_4 v_t\left(\frac{\partial U_i}{\partial x_i} + \left|\frac{\partial U_i}{\partial x_i}\right|\right) \\ &\quad - \frac{C_5 v_t^2\Gamma_1^2}{c^2} - \frac{[C_6(1+C_9\Phi)v_t(N_1 d) + C_7F_1 v v_t]}{d^2} \end{aligned} \tag{5.57}$$

In the above $G^2 = (\partial L/\partial x_i)^2$, $\Phi = K\,(L^2/v_t)\partial L/\partial x_i\, S_{ik}\, \partial L/\partial x_k$. There are around twenty constants in the previous equation and these are not detailed here. Other quantities not given are the same as in the v_t-92 model.

5.2.6 *Non-linear Eddy Viscosity Models*

The Reynolds stress model, to be discussed later, is costly, solving numerous transport equations for individual Reynolds stresses and other variables. The extreme computational cost of the Reynolds stress model but clear weaknesses of simple linear RANS eddy viscosity models has motivated the quest for some sort of ‘half-way-house’. This has given rise, ultimately, to non-linear eddy viscosity models and also the algebraic Reynolds stress model. The former can all be viewed as supplying extended forms of the Boussinesq approximation with the addition of quadratic (see Speziale 1987 and Gatski and Speziale 1993) and cubic terms (see Craft et al. 1996). The model of Speziale (1987) is based on the constitutive relationship for a non-Newtonian fluid (a Rivlin-Erikson fluid). The Gatski and Speziale equation, called the Explicit Algebraic Stress Model (EASM) is derived from an algebraic Reynolds stress model. The way in which non-linear stress-strain relationships can be derived differs greatly but ultimately involves expansions with strain and vorticity tensors. The quadratic terms allow anisotropy to be modelled and the cubic terms the consequences of streamline curvature. Generally a linear RANS model has a constant value of $C_\mu = 0.09$. The non-linear models can involve variable C_μ formulations based on S and Ω. These will helpfully avoid excessive turbulence prediction at stagnation points.

Despite being often marketed as a means of nearly getting Reynolds stress model performance at linear eddy viscosity model cost, practice shows this can be far from the case. Certain non-linear model solutions can be nearly as expensive as performing a hybrid RANS-LES. Also, results can be worse than for linear models, the added non-linearity giving the potential for increased results deviation. Typical stress-strain relationships for some non-linear models are given as

$$-\overline{u_i'u_j'} = \left[-\frac{2}{3}k\delta_{ij} + 2\frac{\mu_t^*}{\rho}S_{ij}\right]_L + n_1\left[\tilde{A}\frac{\mu_t^*}{\rho}\frac{k}{\varepsilon}\left(S_{ik}S_{kj} - \frac{1}{3}S_{kl}S_{kl}\delta_{ij}\right) + \tilde{E}\right]_{NL} \quad (5.58)$$

where S_{ij} is the usual mean strain rate and $\delta_{i,j}$ is the Kronecker delta which for $i = j$ takes the value of unity otherwise it has the value zero. $n_1 = 0$ gives a standard linear Boussinesq approximation for which $\mu_t^* = \mu_t$. For $n_1 = 1$ the non-linear models are gained. Note that the subscripts L and NL are used to represent linear and non-linear contributions. The forms of $\tilde{A}$, $\tilde{E}$ and μ_t^* for the non-linear model of Speziale (1987), the cubic and also the explicit algebraic stress are outlined next.

5.2.6.1 Speziale Model For the model of Speziale:

$$\tilde{E} = 4C_E C_\mu \frac{\mu_t^*}{\rho}\frac{k}{\varepsilon}\left(\overset{\circ}{S}_{ij} - \frac{1}{3}\overset{\circ}{S}_{ij}\delta_{ij}\right) \quad (5.59)$$

In Equation (5.59), $\overset{\circ}{S}_{ij}$, is the Oldroyd derivative. This is given as:

$$\overset{\circ}{S}_{ij} = \frac{\partial S_{ij}}{\partial t} + U_k\frac{\partial S_{ij}}{\partial x_k} - \frac{\partial U_i}{\partial x_k}S_{kj} - \frac{\partial U_j}{\partial x_k}S_{ki} \quad (5.60)$$

Also,

$$\tilde{A} = 4C_D C_\mu \quad (5.61)$$

and

$$\mu_t^* = \mu_t \quad (5.62)$$

where $C_\mu = 0.09$, $C_D = C_E = 1.68$. This expression extends the Boussinesq approximation such that the Reynolds stress is not an instantaneous function of strain and eddy viscosity. With the Equation (5.60), closer to the Reynolds stress model, where the Reynolds stress is transported, the strain rate is also expressed as a transported quantity. The transport based expression, like the Reynolds stress model, allows a lag between the strain field and Reynolds stress. This is a feature of many more complex turbulent flows and hence allows the potential for greater modelling accuracy.

This lag is discussed further later. Note, also, that the model is quadratic in strain, S_{ij}. For low strains, the extra terms will become negligible and models like the k-ε work well. For large strains the quadratic terms become more important.

5.2.6.2 *Cubic Model* For the cubic model of Craft et al. (1996):

$$\tilde{E} = 4c_2\frac{\mu_t^*}{\rho}\frac{k}{\varepsilon}(\Omega_{ik}S_{kj} + \Omega_{jk}S_{ki}) + 4c_3\frac{\mu_t^*}{\rho}\frac{k}{\varepsilon}\left(\Omega_{ik}\Omega_{jk} - \frac{1}{3}\Omega_{lk}\Omega_{lk}\delta_{ij}\right)$$
$$+ 8c_4\frac{\mu_t^*}{\rho}\frac{k^2}{\varepsilon^2}(S_{ki}\Omega_{lj} + S_{kj}\Omega_{li})S_{kl} + 8c_5\frac{\mu_t^*}{\rho}\frac{k^2}{\varepsilon^2}S_{ij}S_{kl}S_{kl} + 8c_6\frac{\mu_t^*}{\rho}\frac{k^2}{\varepsilon^2}S_{ij}\Omega_{kl}\Omega_{kl} \tag{5.63}$$

$$\tilde{A} = 4c_1, \quad \mu_t^* = \mu_t \tag{5.64}$$

where S_{ij} is again the usual mean strain rate tensor and $\Omega_{ij} = 1/2(\partial U_i/\partial x_j - \partial U_j/\partial x_i)$ is the mean vorticity tensor. The coefficients $c_1, c_2, c_3, c_4, c_5, c_6$ are 0.1, –0.1, –0.26, $10c_\mu^2$, $5c_\mu^2$ and $-5c_\mu^2$, respectively, where

$$c_\mu = \frac{0.3}{1 + 0.35\{\max(\tilde{S}\tilde{\Omega})\}^{1.5}}[1 - \exp(-0.36\exp(0.75\max(\tilde{S}\tilde{\Omega})))] \tag{5.65}$$

with $\tilde{S} = k/\varepsilon\sqrt{2S_{ij}S_{ij}}$ and $\tilde{\Omega} = k/\varepsilon\sqrt{2\Omega_{ij}\Omega_{ij}}$.

5.2.6.3 *Explicit Algebraic Stress Model* For the explicit algebraic stress model,

$$\tilde{E} = 2\mu_t^*\left[-\frac{1}{3}S_{kk}\delta_{ij} + \alpha_3\frac{k}{\varepsilon}(S_{ik}\Omega_{kj} + S_{jk}\Omega_{ki})\right] \tag{5.66}$$

and

$$\tilde{A} = -2\alpha_4 \tag{5.67}$$

with

$$\mu_t^* = \rho\frac{3(1+\eta^2)\alpha_1}{3 + \eta^2 + 6\xi^2\eta^2 + 6\xi^2}\frac{k^2}{\varepsilon} \tag{5.68}$$

where $\eta = 1/2\alpha_2(S_{ij}S_{ij})^{1/2}k/\varepsilon$ and $\xi = 1/2\alpha_3(\Omega_{ij}\Omega_{ij})^{1/2}k/\varepsilon$. The constants $\alpha_1, \alpha_2, \alpha_3, \alpha_4$ are 0.1137, 0.0876, 0.1869, 0.1752, respectively. The cubic model gives the correct sensitization between turbulence production and streamline curvature.

Figure 5.13a, b and c compare predictions and measurements for the variation of normalized $\overline{u'u'}$, $\overline{v'v'}$ and $\overline{w'w'}$ against y/δ, respectively for a fully developed channel flow. The measurements are represented using symbols. The k-l based predictions are given by the dashed and full lines, respectively. As can be seen, the non-linear model gives improved agreement with the measurements. It is well known that when the channel geometry is no longer assumed two-dimensional, the effects of anisotropy on flow structure can also be striking, with the generation of Prandtl motions of the second kind. Non-linear eddy viscosity models are capable of modelling these and

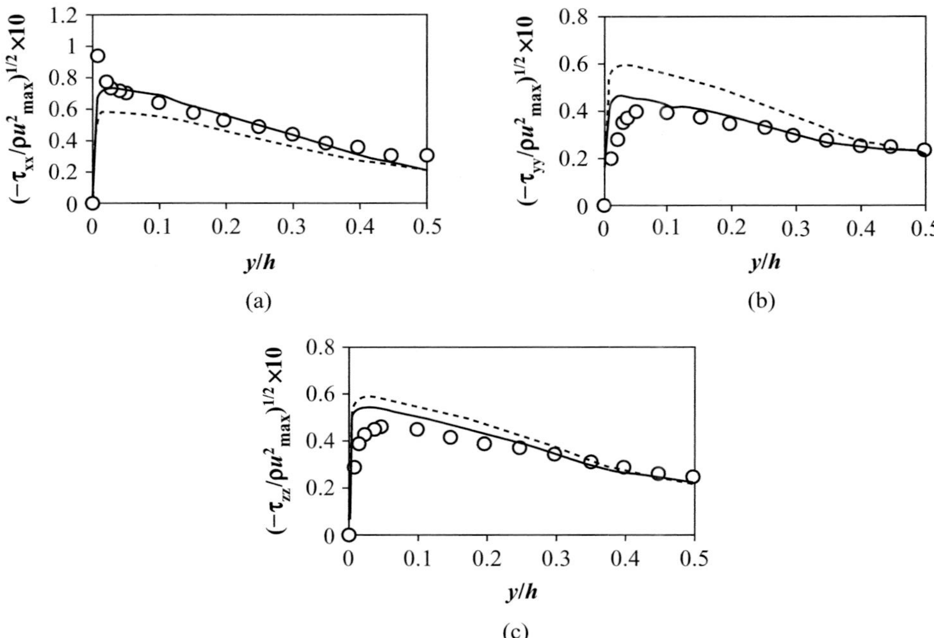

Figure 5.13. Comparison of predictions and measurements for the variation of the normalised Reynolds stresses $\overline{u'u'}$, $\overline{v'v'}$ and $\overline{w'w'}$ against y/h (where $h=\delta$, respectively [o experimental data, —— non-linear model, - - - - - linear model] from Tucker 2003 – published with the kind permission of Wiley).

this can be important for several areas of aerospace application such as the modelling of aeroengine intakes (see Mani et al. 2013). This motion can be seen from Figure 5.14, which shows velocity vectors, weightless particle tracks and the convection of a passive scalar resulting from the secondary motions.

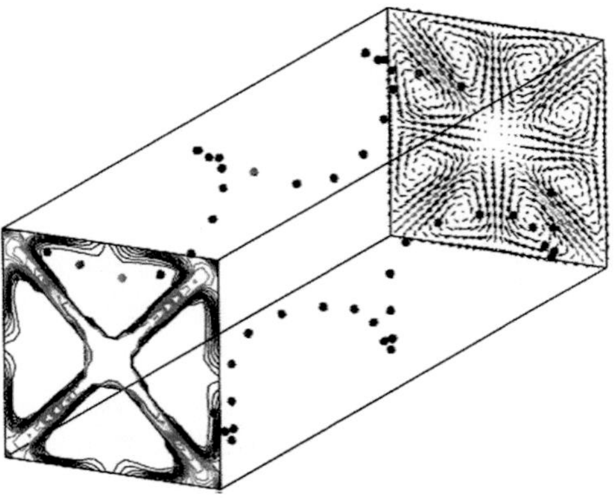

Figure 5.14. Predicted flow structure for a three-dimensional channel flow (Published with the kind permission of Wiley).

5.2.7 Reynolds Stress Model

The Reynolds stress model, which is an 'exact' equation, can be symbolically expressed as

$$\frac{D\overline{u_i'u_j'}}{Dt} = \left[-\overline{u_i'u_k'}\frac{\partial U_j}{\partial x_k} - \overline{u_j'u_k'}\frac{\partial U_i}{\partial x_k}\right] + D + R_{ij} - \varepsilon_{ij} + S_\Omega + S_B \tag{5.69}$$

The right-hand side terms, D (note D no-longer stands for turbulence destruction); S_Ω; S_B; R_{ij} and ε_{ij} correspond to the diffusion, rotation, buoyancy, redistribution and dissipation, respectively. Turbulence production can be exactly expressed by the terms in the square brackets. However, to give a solvable set of equations, D, R_{ij} and dissipation terms need to be modelled. Typically, as with most other turbulence models, following Lien and Leschziner (1994), in commercial codes D is greatly approximated as

$$D = \frac{\partial}{\partial x_j}\left(\Gamma_\phi \frac{\partial \overline{u_i'u_j'}}{\partial x_j}\right), \quad \Gamma_{\overline{u_i'u_j'}} = \frac{\mu + \mu_t}{\sigma_{\overline{u_i'u_j'}}} \tag{5.70}$$

Again, as with many turbulence model formulations μ is ignored when the model is expressed in its high Reynolds number form where the low Reynolds number near wall influences are ignored (see further discussion later). For dissipation the following equation is used

$$\varepsilon_{ij} = \frac{2}{3}\varepsilon\delta_{ij} \tag{5.71}$$

where δ_{ij} is the Kronecker delta defined earlier. Equation (5.71) implies the assumption of isotropy at the small scales – this assumption is problematic near walls. The dissipation in Equation (5.71) is provided by a base RANS eddy viscosity model. The redistribution term acts to redistribute energy amongst the normal Reynolds stresses, such as to make the turbulence more isotropic and also to reduce the Reynolds shear stresses. It characterizes the interaction of pressure and strain. It expresses pressure fluctuations from inter-eddy interactions and also those arising from the interaction of the eddy with the mean velocity gradients. To this end, the pressure-strain modelling has two parts. The so-called slow part enhances isotropy caused by the interaction of eddies. The fast part expresses the interaction between the Reynolds stresses and mean strain and again does not facilitate anisotropy. The simplest model for the slow part takes the following parameter, which expresses the degree of anisotropy of the Reynolds stress tensor:

$$a_{ij} = \frac{\overline{u_i'u_j'}}{k} - \frac{2}{3}k\delta_{ij} \tag{5.72}$$

A characteristic time scale of the turbulence, say k/ε, then divides the previous to give

$$R_{ij} \propto -C_1\varepsilon\left(\frac{\overline{u_i'u_j'}}{k} - \frac{2}{3}k\delta_{ij}/k\right) \tag{5.73}$$

For the fast part,

$$R_{ij} \propto -C_2 \left(P_{ij} - \frac{2}{3} P \delta_{ij} \right) \tag{5.74}$$

5.2.7.1 Deficiencies of the Reynolds Stress Model A key difficulty with the redistribution term is that as the wall is moved towards strong anisotropy can arise. This requirement conflicts with the behaviour of the pressure-strain terms and so near wall corrections are needed. For flows with separation or buoyancy giving rise to large-scale unsteady structures, the Reynolds stress, like all other RANS models, will fail. Its closure is based on eddies at the stochastic level. Even for attached flows, its performance becomes uncertain near walls. A key problem is modelling dissipation. Near walls, the turbulence structures become highly anisotropic and hence so does the dissipation. Also, there are other issues relating to the crude nature of the raw dissipation equation itself which gives rise to many ad hoc potential modifications. The Reynolds stress model involves the resolution of many hundreds of terms. As noted previously, it can also involve solution of a two-equation model to secure closure, even so k can be easily secured by summing the normal Reynolds stresses. There is also the potential need to solve for a further three scalar transport equations (unless turbulence diffusion coefficients are used instead), with added uncertain near wall behaviour terms. Hence, the Reynolds stress model's computational cost can be extreme. Like other advanced models that contain numerous gradient terms, the grid demands (to resolve the gradients) increases.

There are a wide range of fixes and variants for the Reynolds stress model. As one might expect, the spawning of a wide range of options suggests a weakness and unresolved issues. Notably, fourth order moment closures (Poroseva 2013) are possible and these require less ad hoc modelling which does offer potential for improvement. Third order closures are described by Kurbatskij et al. (1995). However, such models, at the time of this writing are at their very early stages. What is more, where closure is needed, the terms will be physically abstract and this makes reliable closure challenging. The previously noted near wall problems have resulted in zonal RANS models, where simpler eddy viscosity models, near walls, replace the Reynolds stress model. The general nature of near wall treatments in RANS is discussed next.

Notably, the Reynolds stress model can be contracted to form an algebraic stress model (see Demuren and Rodi 1984). With this, the convection and diffusion terms from the full Reynolds stress equation system are ignored. This then gives rise to a set of algebraic equations. It can be assumed that the sum of the convection and diffusion terms, in the Reynolds stress model, is proportional to the transport of turbulence kinetic energy. Hence, in this way transport of turbulence can be incorporated through solution of the k-ε equations. Ideally, the model would also need inclusion of the pressure strain terms. The algebraic stress model equation set is not numerically well conditioned. Hence, the explicit algebraic stress model is the more popular surrogate. The algebraic stress model is more of historical and technical process interest.

5.2.8 Scalar Transport

When solving for heat transport or the movement or other scalar equations, velocity-scalar fluctuations arise. These are analogous to the Reynolds stresses having the form $\rho\overline{u_i'\phi'}$. Normally, they are accounted for by augmentation of the diffusion coefficient in the transport equation by $\Gamma_t = \mu_t/\sigma_\phi$. Hence, this is analogous to the use of a Boussinesq type approximation (see Equation 5.8) where this time

$$\rho\overline{u_i'\phi'} = \Gamma_t \frac{\partial \Phi}{\partial x_i} \tag{5.75}$$

where Φ is the time average of ϕ. The Schmidt number, σ_ϕ, expresses the ratio of the turbulent transport of momentum to that of the scalar transport. Indeed, this form of simplified Schmidt number based modelling is used in all (except sometimes the Reynolds stress) the turbulence models discussed. For heat transport, the Schmidt number is called the turbulent Prandtl number and characterizes the ratio of turbulent momentum to thermal diffusivity. There are substantial ranges of reported turbulent Schmidt and Prandtl, Pr_t, numbers in the literature. Indeed, the actual value can vary substantially within the same flow field. For example, with jets there is a turbulence Prandtl number range of $0.4 < Pr_t < 0.9$. For cooling jets used on turbines (film cooling), Konopka et al. (2013) find a variation of between 0.1 and 2.0. Hence, zonalized Prandtl number distributions can be necessary. Zuckerman and Lior (2005) use this procedure for jets in cross flow, again in the context of turbine blade (film) cooling. The turbulent Prandtl number zone values are based on precursor RANS simulations. Notably, eddy-resolving simulations can be used to provide improved turbulent Prandtl/Schmidt number data for RANS models (see Lakehal 2002) and work of this nature is emerging. For this purpose, based on Equation (5.75) and the Boussinesq approximation, in the wall normal direction, y, the following can be derived (see Konopka et al. 2013):

$$Sc_{t,y} = \frac{\overline{\rho u'v'}}{\overline{\rho v'\phi'}} \frac{\partial \Phi/\partial y}{\partial \bar{u}/\partial y + \partial \bar{v}/\partial x} \tag{5.76}$$

An alternative approach used by Konopka et al. (2013) is to use the following estimates

$$\mu_t = \frac{\bar{\rho} k}{\omega}, \quad \omega = \frac{\sqrt{2S_{ij}S_{ij}}}{0.09} \tag{5.77}$$

in conjunction with

$$\Gamma_t = \frac{\sqrt{\overline{\rho u'\phi'}^2 + \overline{\rho v'\phi'}^2 + \overline{\rho w'\phi'}^2}}{\sqrt{\left(\frac{\partial \Phi}{\partial x}\right)^2 + \left(\frac{\partial \Phi}{\partial y}\right)^2 + \left(\frac{\partial \Phi}{\partial z}\right)^2}} \tag{5.78}$$

which forms an isotropic eddy diffusivity based on the Euclidean norm of Equation (5.75).

In a Reynolds stress framework, three extra transport equations for the scalar fluctuation components can be solved. For examples of scalar fluctuation and their dissipation, equations, see Gibson and Launder (1978), Newman et al. (1981) and Karcz and Badur (2005).

5.2.9 *Near Walls*

As partially expressed earlier, the previous models can all be expressed in either low or high Reynolds number forms. With low Reynolds number turbulence models, typically the first off-wall grid nodes need to be located at $y^+ \leq 1$. For low Reynolds number k-ε models, like the Launder and Sharma (1974), the near wall grid nodal location in fact needs to be much less than unity. On the other hand, the SA model's more abstract operating variable is designed, as discussed, to have a benign near wall behaviour. Also, since the model has fewer equations, it is much more forgiving with respect to grid. This aspect is a key attraction. Hence for SA, $y^+ \approx 2$–3 is acceptable. This is also true for the k-l and mixing length models. Also, as noted, zonal models are sometimes formulated. With these, the less grid resilient, generally more complex turbulence model is used away from walls and a lower order model, like the k-l or the mixing length, is used near walls. Notably, the k-ε equations in a low Reynolds number framework can be stiff near walls. This offers another benefit from switching to a simpler model in this zone. Liu et al. (2006) use the k-l model near walls with various other turbulence models used away from them. For the use of zero-equation near wall models coupled to two-equation models, see Iacovides and Chew (1993), and Iacovides and Theofanopoulos (1991)).

There are various ways of interfacing different RANS models. The typical interface location would be around $y^+ = 60$. A potential interfacing strategy for a k-ε/k-l method would involve

$$k^{3/2} = l_\mu \varepsilon \tag{5.79}$$

rearranged for ε. For the k-ε with a mixing length model, the following additional condition could be used:

$$\varepsilon = -\overline{u_i' u_j'} \frac{\partial U_i}{\partial x_j} \tag{5.80}$$

The use of these conditions can give rise to very complex interfaces between the differing RANS models. Figure 5.15 gives an example taken from Tucker (2001) for a complex, system. At the local level, the following expression can be used to smoothly blend the viscosity in the differing model zones:

$$\mu_t = F_\mu \mu_{t,\varepsilon} + (1 - F_\mu)\mu_{t,l} \tag{5.81}$$

where $\mu_{t,\varepsilon}$ is the eddy viscosity from the k-ε model and $\mu_{t,l}$ the μ_t from the one- or zero-equation model. Also, $F_\mu(y^+) = 0$ at the wall and unity above the transition interface.

Low Reynolds number turbulence models are generally sensitized (note that an advantage of the k-ω model is that it does not need near wall damping functions)

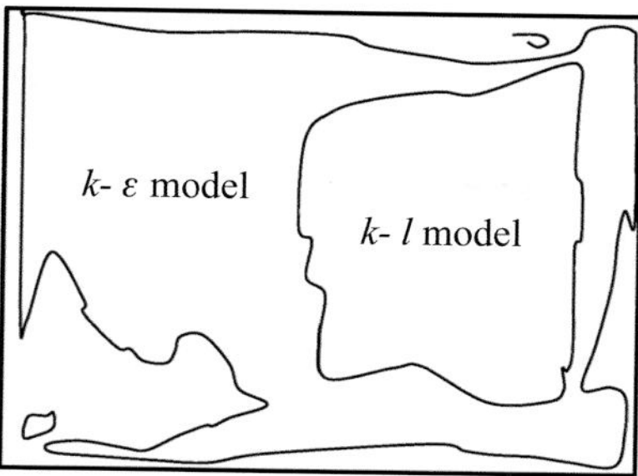

Figure 5.15. Potentially complex interfaces between different RANS models (from Tucker 2001).

to the low Reynolds number zone near walls. Typically, this sensitization is made through the following Reynolds number formulations:

$$Re_T = \frac{\rho k^2}{\varepsilon \mu}, \quad \mathrm{Re}_y = \frac{\rho k^{1/2} d}{\mu}, \quad y^* = \frac{0.55 \rho k^{1/2} d}{\mu}, \quad y^+ = \frac{\rho u_\tau d}{\mu} \tag{5.82}$$

Table 5.5 summarizes for a range of popular turbulence models the different types of Reynolds numbers used in their low Reynolds number forms. The critical difference is whether the Reynolds numbers are wall distance based. Wall distance based turbulence models can be robust and forgiving on coarse grids but are accuracy limited.

Rather than use fine near wall grids with damping functions, computationally economical logarithmic wall functions (see Launder and Spalding 1974) can instead be used. These essentially take the logarithmic velocity profile expression, $U^+ = (1/\kappa)\log(E\,y^+)$, noted earlier, and enforce this expectation at the first off-wall grid cells. When wall functions are implemented, for equations involving a transport

Table 5.5. *Summary of some RANS models, the basis of their damping functions and their number of equations*

Model	Uses d?	Damping basis	No of Eqns
Van Driest damping function based mixing length model	Yes	y^+	0
LVEL model of Spalding	Yes	y^+	0
Wolfshtein k-l	Yes	$y*$	1
Spalart and Allmaras	Yes	–	1
Secundov υ_t-92	Yes	–	1
υ_t-L	Yes	–	2
Launder-Sharma k-ε	No	Re_T	2
Chien k-ε	Yes	y^+, Re_T	2
Lam and Bremhorst (1981) k-ε	No	Re_T	2
Abe et al. (1994) k-ε	Yes	$y*$, Re_T	2
Abid et al. (1996) k-ε	Yes	y^+, Re_y	2
Wilcox k-ω (1988 & 2004 versions)	No	–	2
Menter SST	Yes	–	2
Cubic model of Craft et al.	No	Re_T	2

equation for turbulent kinetic energy, the diffusion of turbulence kinetic energy at walls is set to zero. Also, at first off-wall grid nodes

$$\varepsilon = \frac{Ck^{3/2}}{d} \tag{5.83}$$

where in the standard model

$$C = \frac{C_\mu^{3/4}}{\kappa} \tag{5.84}$$

C is only really applicable for $y^+ > 30$. However, for low Reynolds number flows it can be difficult to ensure that first off-wall grid nodes are positioned in accordance with the y^+ limitation. Therefore, Nishimura et al. (2000) propose the following DNS based C expression:

$$C = \max\left[0.19, 7.90\left(y^+\right)^{-1.89}\right] \tag{5.85}$$

When $y^+ < 11.5$, $\mu_t = 0$, otherwise based on the expected near wall logarithmic velocity profile,

$$\mu_t = \mu(\kappa y^+ / \ln Ey^+ - 1) \tag{5.86}$$

As noted earlier the standard smooth wall value for $E = 9.8$. To account for roughness E can be varied (see Schlichting 1979). Also, if grid nodes tend to be located in the buffer layer, the log-law equation coefficients more appropriate to the buffer layer can be used. Thermal wall functions can also be formulated that take account of the universal temperature behaviour near walls at high Reynolds number (see Launder and Spalding 1974). These are based around

$$T^+ = c_p \rho k^{1/2} C_\mu^{1/4} \frac{(T - T_w)}{q_w} = \sigma_T \left[u^+ + P\frac{Pr}{\sigma_T}\right] \tag{5.87}$$

where q_w is the wall heat flux and T_w the wall temperature. Also, c_p is the specific heat capacity at constant pressure, Pr the Prandtl number of the fluid (see Chapter 2) and σ_T the turbulent Prandtl number. T is the first off-wall value of temperature and P is the pee-function. This is based on the ratio of Pr/σ_T (see Launder and Spalding 1974).

More modern momentum wall functions tend to make use of the Spalding, one-equation, 'law of the wall'. Hence, they are adaptive. As noted previously, a key difficulty with the basic logarithmic wall-function is that the first off-wall grid nodes need to be located in the log law region. Chapter 3 discusses low and high Reynolds numbers near wall grid distributions further. If the near wall grid nodes stray too close to the wall, with a wall function based on Equation (5.86), k grows dramatically. Hence, there can be a severe accuracy impact. For low Reynolds number flows, it can be hard to keep first off-wall grid nodes sufficiently far from walls. Examples would be electronics and room ventilation flows. Under these circumstances the modification of Nishimura et al. (2000) is useful.

Clearly, if use is being made of too coarse a grid (with a Spalding based wall function), it is ideal to ensure that the wall shear stress

$$\tau_w = \mu_{eff} \frac{\Delta U}{\Delta y} \tag{5.88}$$

is plausible. Note that μ_{eff} is the sum of the viscosity and eddy viscosity and ΔU and Δy characterize the near wall velocity gradient. To ensure the correct wall shear stress, a slip velocity can be used at the wall, or the effective eddy viscosity modified. To implement the latter approach, the following near wall Reynolds number can be defined:

$$Re = \frac{\rho \Delta \mathrm{U} \Delta y}{\mu} = u^+ \, y^+ \tag{5.89}$$

Also, the skin friction can be expressed as

$$c_f = \frac{2}{U^{+2}} \tag{5.90}$$

Since $\tau_w = \frac{1}{2}\, \rho C_f \Delta U^2$,

$$\tau_w = \frac{\rho \Delta U^2}{U^{+2}} \tag{5.91}$$

and hence the following expression for effective viscosity can be written

$$\mu_{eff} = \frac{Re}{U^{+2}}\, \mu \tag{5.92}$$

In Equation (5.92) U^+ can be evaluated for Spalding's 'law of the wall'. The U^+ is extracted via Newton-Raphson iteration. Different starting conditions can be chosen depending on the first off-wall grid location – thus speeding up convergence of the iterative procedure. Equation (5.92) will yield a wall value of effective viscosity.

For the Reynolds stress model, commercial solvers can resort to solving the k-ε equations along with the Reynolds stress. In its high Reynolds number form, ε can be evaluated at first off-wall grid nodes using Equation (5.83). Then, considering equilibrium and a log based near wall profile, at first off-wall grid nodes

$$\overline{u'^2_{i,j}} = C\left(\frac{2k}{3}\right)$$

where $C = 1.098; 0.247; 0.655; -0.255$ for $i = j = 1, i = j = 2, i = j = 3$ and $i = 1, j = 2$, respectively. To utilize a Reynolds stress model for heat transfer predictions, whilst controlling computational cost, Craft et al. (2004) make use of an advanced analytic wall function.

5.2.9.1 Boundary Conditions For the low Reynolds number k-ε and k-l models, $k = 0$ is the necessary wall boundary condition for k. For ε, the situation is more complex. Best experimental evidence suggests that near walls ε tends to a large value. Lam and Bremhorst (1981) use the condition $\partial \varepsilon / \partial n = 0$, where n is the normal wall coordinate. Differential boundary conditions are always less efficient to converge relative to Dirichlet boundary conditions. To this end Launder and Sharma (1974) reformulate the ε equation to solve for

$$\tilde{\varepsilon} = \varepsilon - 2v\left(\frac{\partial k^{1/2}}{\partial x_j}\right)^2 \tag{5.93}$$

This gives the more 'solid' $\tilde{\varepsilon} = 0$ wall Dirichlet condition. However, the substitution makes the equation numerically stiff and hence difficult to converge. For SA the boundary condition is again Dirichlet with $\tilde{v} = 0$. For the k-ω model at walls, $\omega \to \infty$, hence a large value can be specified or the formulation of Wilcox (1988) can be used. This uses a hyperbolic ω variation and application of the following expression at the first off-wall grid point:

$$\omega = 6\frac{\mu}{\rho \beta_1 d^2} \tag{5.94}$$

where $\beta_1 = 0.0708$ is a constant.

5.2.9.2 Non-wall boundary conditions For the k-ε model, at inflow boundaries, the following estimates can be used:

$$k = \frac{2}{3}[U_0 T_i]^2, \quad \varepsilon = C_\mu^{3/4}\frac{k^{3/2}}{l}, \quad l = CL \tag{5.95}$$

where U_0 is a reference velocity, say the inflow velocity, and $C \cong 0.1$. Also, L is a system scale that would define the integral eddy size. This could be, for example, a duct hydraulic diameter. More precise l values, relating to a particular problem, could be based on those given for different canonical flows in the discussion on the mixing length (See Section 5.2.3.1). For a boundary layer, rather than use a constant inflow value, it is a nice practice to specify a plausible profile for the turbulence variables. To help understand how the results respond to any boundary condition uncertainties, it is good practice to perform sensitivity checks. However, the k-ε model is typically quite insensitive to the k-ε values specified at the inflow boundary. This is different for the k-ω model, where there can be considerable sensitivity. In a free stream where $k \to 0$ and $\omega \to 0$, as can be seen from Equation (5.47), the evaluation of eddy viscosity becomes problematic and hence a small finite value of ω is needed. Sensible values of ω for internal flows can be found by transforming the Equation (5.95) estimates into the k-ω space.

For external flows the situation is further complicated. For example, if an airfoil in a large domain is being modelled, the decay of the turbulence needs to be accounted for. It needs to be ensured that the turbulence characteristics specified at the potentially distant domain inlet are consistent with that expected in the airfoil leading edge target area. What is more, the atmospheric turbulence contains large scales that would suggest high levels of turbulence intensity. However, these scales will not interact with the relatively small chord scales. Hence, characterizing k and ε or ω on atmospheric data could be highly erroneous. It could result in the eddy viscosity being so high as to render the effective modelled Reynolds number of the wing as many orders of magnitude too low – not far off unity (see Spalart and Rumsey 2007). A further complication is that the decay rates of atmospheric eddies are very different to those for the turbulence relevant to the wing boundary layers.

The complication is extended for turbomachinery, where the presence of upstream blades rows and the turbulence that they generate needs to be accounted for. For a Reynolds stress model, in addition to Equations (5.95), the following can be used

$$\overline{u_1'^2} = k, \overline{u_2'^2} = \overline{u_3'^2} = \frac{1}{2}k, \overline{u_i'u_j'} = 0 \quad (i \neq j) \tag{5.96}$$

Also, the ε distribution will need to be specified. For the SA model, inflow specification is also not completely straightforward. The model has a threshold for it being attracted to a laminar solution through the function f_{t2}:

$$\frac{\tilde{v}}{v} = \sqrt{\log(c_{t3})/c_{t4}} \tag{5.97}$$

This corresponds to $\tilde{v}/v = 0.22 - 0.6$. To avoid this zone and the potential for non-unique solutions, $\tilde{v}/v = 3 - 5$ is recommended (see Spalart and Rumsey (2007)). Note that too high an ambient value will result in the contamination of potential flow regions in close spaces such as those found between a wing and flap.

At outflow boundaries and symmetry planes, the normal gradients of the solved for turbulence variables are set to zero. For free streams with a Reynolds stress model, the Reynolds stresses can be set to zero or their normal gradient. The rate of dissipation or its gradients is also set to zero. For the k-ε model, free stream values of k and ε can be specified or their gradients set to zero. For k-ω simulations, as noted previously, the condition of $\omega = 0$ is problematic in the free stream and instead a small value is needed. For SA, the free stream condition is complicated. If the trip terms are not implemented, then evidently $\tilde{v} = 3 - 5v$ is appropriate. However, if the trip term is used, $0 \leq \tilde{v} < v/10$ seems more appropriate.

For eddy-resolving simulations the inflow and outflow boundary condition specification is considerably more complex and discussed further later and also in Chapter 4.

5.2.9.3 Wall distance As noted previously, some turbulence model damping functions make use of wall proximity. Typically most codes use the nearest wall distance based on search procedures. However, more physics based wall proximity or distance functions can be used. This notion was suggested by Launder et al. (1975), who proposed a solid angle integral based equation. As can be seen from Figure 5.16, particle A is in the concave geometry zone and is surrounded by more surfaces and so will experience extra blocking. On the other hand, B, in the convex geometry zone, will experience less blocking. As shown in Figure 5.17, for a distance-based turbulence model, a fluid particle escaping the trailing edge of an airfoil would experience too much turbulence destruction. The distance contour line will tell the particle it is

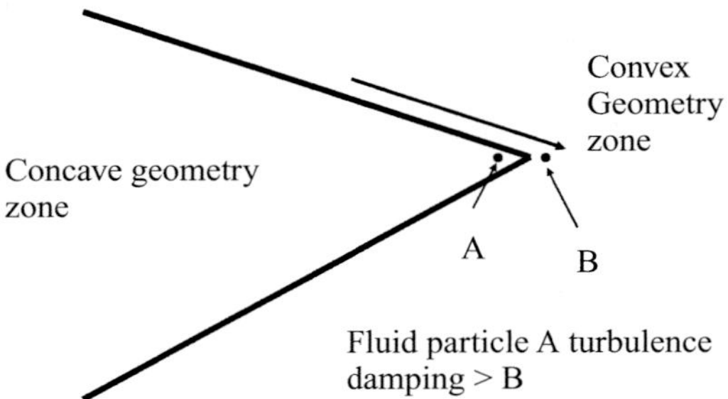

Figure 5.16. Different particle dampings experienced in concave and convex geometry zones.

close to the wall. However, the wall is more like a point and will not provide substantial damping. These issues motivated the solid angle concepts of Launder and colleagues. Differential equations can also provide similar traits to the solid angle based integral equations proposed by them. These are discussed in Chapter 7. As indicated previously, the damping level depends on how much wall the particle can 'see'. The DNS work of Mompean et al. (1996), enforces this idea. It suggests the damping function in sharp convex corners should take the form

$$l \propto (1 - \exp(-x^{+}/17))(1 - \exp(-y^{+}/17)) \tag{5.98}$$

(i.e., the two walls with the normal x and y provide a double damping). Conversely, Figure 5.18a shows eddy viscosity contours around a thin wire. The SA turbulence model is used. As can be seen, the wall distance based damping function erroneously gives a strong wire influence. Frame (b) shows the equivalent information, but this time using a differential wall distance function equation – the Hamilton-Jacobi discussed in Chapter 7. This replicates the features of the integral equation of Launder and colleagues. The equation exaggerates the nearest wall distance around the convex wire but leaves it exact near the channel walls. Note that due to symmetry just half the channel has been modelled. The benefits of using the Hamilton-Jacobi equation are clear. The disrupted eddy viscosity zone (lighter area) around the wire is much smaller.

Figure 5.19 again contrasts results for wall distances computed using a differential equation (this time Poisson, again discussed in Chapter 7) and an exact search. This

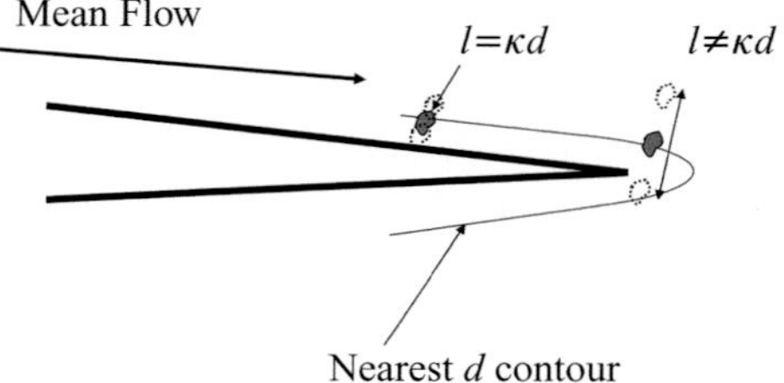

Figure 5.17. Schematic of a fluid particle escaping the trailing edge of an airfoil/blade for a wall distance based turbulence model.

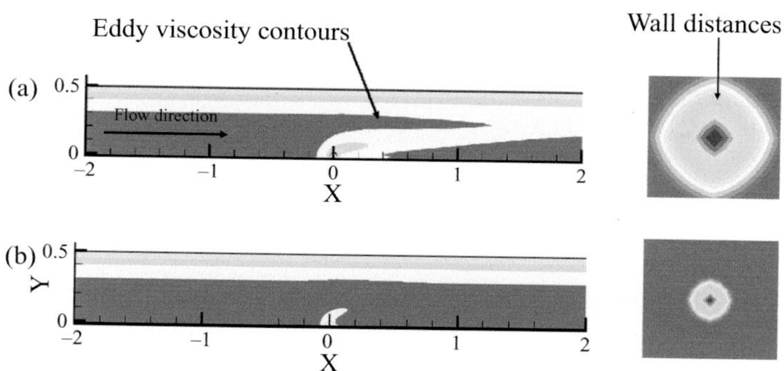

Figure 5.18. Impact of wall distance function on eddy viscosity field around a thin wire: (a) nearest wall distance and (b) differential equation wall distance (from Tucker 2003 – published with the kind permission of Elsevier).

time the trailing edge region of a NACA4412 airfoil is considered. Frames (a) and (b) give SA model, trailing edge region, turbulent viscosity contours for exact and differential distances, respectively. The larger trailing edge differential equation distances reduce the modelled turbulence destruction and so increase μ_t by just under 5%. Frame (c) gives the predicted normal Reynolds stress. Comparison is made with measurements given by the symbols. As can be seen, the differential distance function results in a higher normal stress. Table 5.6 shows the impact of different wall distance function assumptions on predicted lift, C_L, and drag, C_D, coefficients for the NACA4412 airfoil. NSS, in the table, corresponds to a nearest surface search, outlined in Tucker et al. (2005). The eikonal equation result, given in the table, is a differential equation for the exact wall distance. As can be seen, the

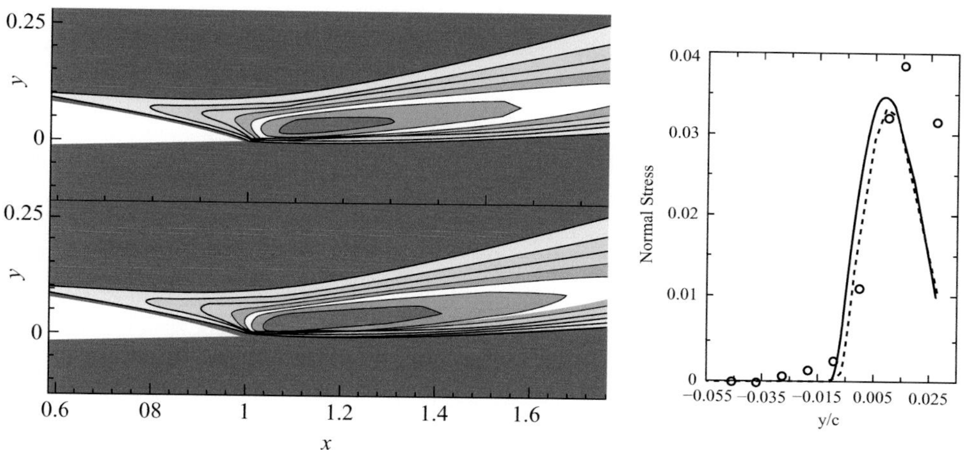

Figure 5.19. Impact of wall distance function field on the flow in the trailing edge region of an airfoil: (a) eddy viscosity field based on nearest wall distance, (b) eddy viscosity field based on a palliative differential equation wall distance and (c) cross-stream variation of turbulence stress downstream of trailing edge (from Tucker and Liu 2007 – published with kind permission of Elsevier).

Table 5.6. *C_L and C_D for different d fields*

Method	C_L	C_D
Eikonal	1.704	0.03466
NSS	1.698	0.03496
Poisson	1.713	0.03543

wall distance formulation has a clear impact and can be used to improve predictive accuracy. Indeed the double damping affect described needed to be accounted for to gain the Prandtl motions of the second kind shown in Figure 5.14.

5.3 RANS Model Performance and Palliatives

5.3.1 *General RANS Performance*

As noted previously, there are a vast range of turbulence models, this being symptomatic of the problem of dealing with something with the tremendous dynamical complexity of turbulence and a phenomenon that is not fully understood. Hence, even without the added dynamic complexity of the interaction with walls, basic shear flows present a challenge. For example, measurements show that a plane jet spreads faster than a round one, but the k-ε model gives the opposite trend. This can be seen from Table 5.7, giving measured and predicted spreading rates for round and plane jets. This issue is called the *round jet/plane jet anomaly*.

For $Ma > 1$, the jet spreading rate decreases, reaching around a factor of 2 at $Ma = 2$. RANS models will not predict this. Rectangular jets, that are horizontally long and thin, will, as they develop, become vertically long and thin, ultimately becoming round. Most eddy viscosity based RANS models will also struggle to capture this. For wall jets, the flow physics becomes considerably more complex with the subtle wall physics coming into play. Then, even the Reynolds Stress model can over-predict the jet-spreading rate by 30%. As shown by Georgiadis and DeBonis (2006), with 'advanced' RANS models we can get a wide range of answers even for basic round jets.

Figure 5.20 further illustrates the problem when considering heat transfer in various zones of a turbine blade. These include impingement cooling in Frame (a), the use of highly separated and hence energized turbulent flow in Frame (b) and shielding through a type of wall jet in Frame (c). The hatched zones show the

Table 5.7. *Measured and predicted spreading rates for round and plane jets*

Flow	k-ε	Measured
Plane jet	0.091	0.1–0.11
Round jet	0.101	0.086–0.096

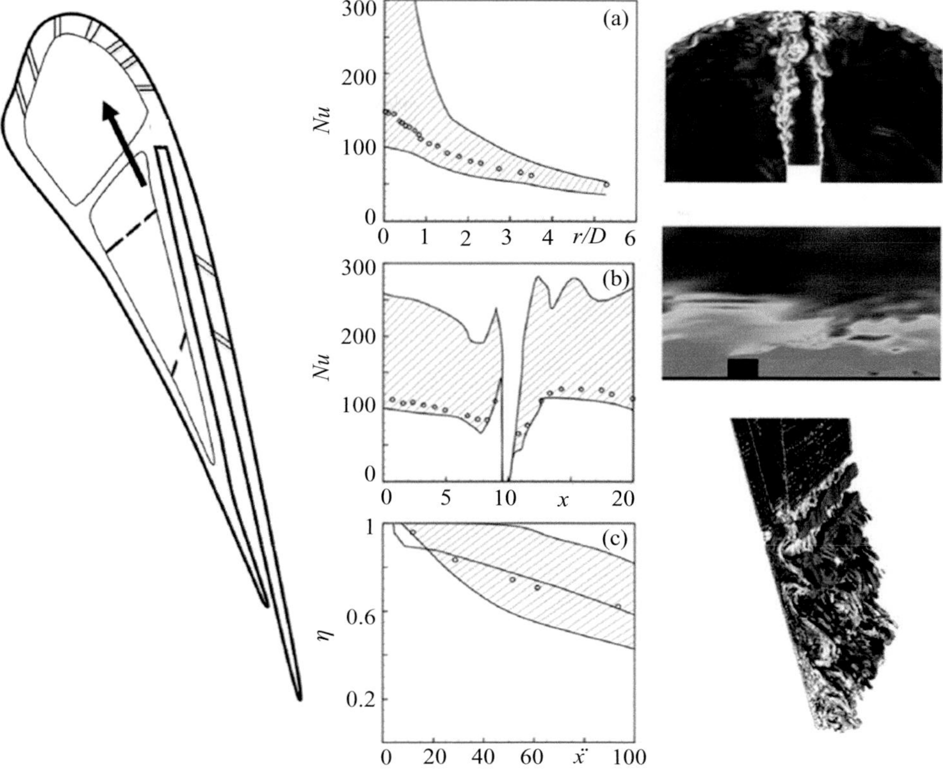

Figure 5.20. Heat transfer in different turbine zones for different turbulence models: (a) convex surface impingement, (b) ribbed passage and (c) cut-back trailing edge.

predicted range in heat transfer for some relatively well-established RANS models. An extensive range of further hatched plots, for the many different zones in a gas turbine aeroengine, are given in Tucker (2013a). Note that for the Figure 5.20 flows, the Reynolds stress model is unlikely to show any great improvement in accuracy. The flows tend to involve large scale unsteady eddies and the Reynolds stress model is not designed to account for these. The next section further explores what causes these noted defects and how they can be dealt with.

5.3.2 *RANS Models Defects and Management*

Models, especially the simpler ones, tend to give:

- erroneous predicted shear for curved shear layers and adverse pressure gradients;
- separation suppression on curved surfaces;
- inability to capture Prandtl's motions of the second kind;
- excessive turbulence in stagnation zones;
- inability to model anisotropy of Reynolds stresses;
- the wrong behaviour for rotating flows;

Table 5.8. *General advantages and disadvantages of different forms of RANS models*

Model	Advantages	Disadvantages	General Comments
Mixing length	Requires more modest grids, easy to understand, useful for near wall LES modelling.	Iterative convergence can be slow since d field potentially non-smooth. Incomplete model needing user insights.	Will not provide high accuracy, but simple. With an expert user, for certain problems, can give reliable results.
One-equation models	Generally conceptually simple and grid resilient, useful for near wall hybrid RANS-LES modelling.	Incomplete models.	Form generally either based on a transport equation for eddy viscosity or turbulence kinetic energy.
Two-equation models	Can model turbulence non-equilibrium effects at modest cost.	Tend, like the above, to be unable to model flows with complex strain fields accurately. Some two-equation models can be numerically stiff.	–
Non-linear models	Can potentially capture anisotropy, streamline curvature and can deal with problems at stagnation points.	Iterative convergence can be difficult to secure. For the wrong flow can give worse accuracy than simpler models.	–
Reynolds stress model	The most mathematically exact description of turbulence and hence has the capability of dealing with complex strain fields and giving anisotropic flow information.	Near walls conflicting modelling requirements arise that can substantially limit accuracy. The numerous non-linear equations that need to be solved can make convergence expensive or difficult to secure. Unforgiving with regards to grid quality.	The model complexity makes verification challenging. Also Reynolds stress models are not as well validated as the more widely used simpler models. The closure assumptions are at the stochastic level. Hence, as with all other RANS models, the Reynolds stress model struggles to cope with flows having large scale unsteadiness.

- insensitivity to density gradients;
- excessive heat transfer at reattachments points (see Leschziner 2000);
- inability to reliably model (forward) transition;
- inability to reliably model reverse transition;
- inaccurate sensitization to surface finish;
- inaccurate sensitization to free stream turbulence and
- problematic behaviour for free shear flows

Table 5.8 summarizes the general advantages and disadvantages of the different turbulence models outlined. More specific advantages and disadvantages for different models are summarized in Table 5.9.

In this section, where possible, we contrast the Reynolds stress model with the Boussinesq approximation. The latter, as noted earlier, is the equation that furnishes Reynolds stresses from eddy viscosity models. This comparison process is helpful in diagnosing why simpler models fail for complex flows yet work for simpler ones. For example, they tend to work well for simple two-dimensional shear flows. However, even for these flows they can be challenged. For simple shear flows, Equation (5.69) reduces to

$$\rho\frac{D\overline{u'v'}}{Dt} = -\rho\overline{v'v'}\frac{\partial U}{\partial y} + \overline{p'\left(\frac{\partial u'}{\partial y} + \frac{\partial v'}{\partial x}\right)} - \rho\frac{\partial}{\partial y}\left\langle\overline{u'v'v'} + \frac{\overline{p'u'}}{\rho}\right\rangle - \rho\varepsilon_{12} \quad (5.99)$$

Taking just the first right-hand side term (involving the gradient of a mean and not fluctuating properties) after multiplying through by a timescale $T = Dt$ and ρ gives

$$-\rho\overline{u'v'} \propto -\rho\overline{v'v'}T\frac{\partial U}{\partial y} \quad (5.100)$$

On dimensional grounds $\mu_t \propto \times$ *Length* $\times$ *Velocity* $\propto \rho\overline{v'v'}T$. Hence, Equation (5.100) can effectively be stated as

$$-\rho\overline{u'v'} = \mu_t\frac{\partial U}{\partial y} \quad (5.101)$$

This is consistent with the Boussinesq approximation. Hence, the general success of eddy viscosity models for simple shear flows, where shear stress is the only important component. However, for more complex free shear flows, as noted with the plane jet/round jet anomaly, issues can arise and these are considered next.

5.3.3 *Free Shear Flow Problem*

For a round jet, the strain field is more complex and gives rise to a vortex stretching process. With regards to the k-ε model there are a range of 'fixes' for free shear flows. Not surprisingly, these are largely focused on the round jet/plane jet anomaly. All these palliatives focus on the ε Equation. This is well known to be the theoretically weak component in the k-ε model. As noted by Pope (1978), simply changing $C_{\varepsilon 1}$ to 1.6 will improve the round jet spreading rate. However, this

Table 5.9. *More specific advantages and disadvantages of different popular turbulence models*

No of equations	Model	Advantages	Disadvantages	General Comments
1	SA	Relatively robust to grid quality, generally does well at predicting separation on curved surfaces but still tends to delay separation.	Not designed for jet flows, not sensitized to streamline curvature, cannot provide/model turbulence anisotropy, insensitive to rotation and buoyancy, struggles for some internal flows partly due to lack of turbulence length scale information. For rapidly changing flows has insufficient transport information.	This is the only one-equation model that sees routine, non-zonalized use. Not suitable as a general purpose CFD turbulence model.
2	k-ε	Widely validated, works well for internal flows where Reynolds shear stresses are important.	In its low Reynolds number form can become numerically stiff. Near the wall the changes in k and ε are rapid, thus giving the need for extremely fine grids. Not popular/well suited for external flows (for example substantially over-predicts the spreading rate for round jets), not sensitized to streamline curvature, cannot provide/model turbulence anisotropy, insensitive to rotation, buoyancy, suppresses separation on curved surfaces, excessive turbulence levels in stagnation zones.	–
2	k-ω	Does not need near wall damping functions and hence avoids any numerical stiffness relating to this.	Free stream sensitivity and non-ideal numerical conditioning in this zone, not sensitized to streamline curvature, cannot provide/model turbulence anisotropy, insensitive to rotation, buoyancy, excessive turbulence levels in stagnation zones.	–

No of equations	Model	Advantages	Disadvantages	General Comments
2	Menter-SST	Does not have the free stream sensitivity of the k-ω model but has this model's superior near wall traits. Generally shows relatively good performance for adverse pressure gradient boundary layer flows with helpful accuracy improvements from turbulence level limiters in stagnation and adverse pressure gradient zones.	Not sensitized to streamline curvature, cannot provide/model turbulence anisotropy, insensitive to rotation, buoyancy.	For shock separated flows $a = 0.35$ is a more suitable value in SST component. Probably the most reliable model for external aerodynamic flows.

gives minimal model generality. Other palliatives are given next. Note ℄ refers to centerline values. Launder (1972) propose the following modification

$$C_{\epsilon 2} = 1.92 - 0.0667\left\{\frac{y_{1/2}}{2U_{℄}}\left(\left|\frac{dU_{℄}}{dx}\right| - \frac{dU_{℄}}{dx}\right)\right\}^{0.2} \tag{5.102}$$

where $y_{1/2}$ is the jet half width, this being the extent at which the jet velocity is half the centerline velocity. Also, x is the stream-wise coordinate. McGuirk and Rodi (1979) propose the following modification, but this time, to the constant $C_{\varepsilon 1}$ takes the functional form

$$C_{\epsilon 1} = 1.14 - 5.31\frac{y_{1/2}}{U_{℄}}\frac{dU_{℄}}{dx} \tag{5.103}$$

Morse (1977) also proposes modifying this constant but this time as

$$C_{\epsilon 1} = 1.4 - 3.4\left(\frac{k}{\epsilon}\frac{dU}{dx}\right)^{3}_{℄} \tag{5.104}$$

Clearly all of these are crude fixes. They can need evaluation of the jet half width (not an easy process in an unstructured flow solver) and imply jet edge behaviour is strongly correlated to the centerline behaviour.

In the modification of Pope (1978), the usual source term in the k-ε model (assuming unit density) $\varepsilon/k\,(C_{\epsilon 1}P - C_{\epsilon 2}\varepsilon)$ changes to

$$S_\varepsilon = \frac{\varepsilon^2}{k}\left(C_{\varepsilon 1}\frac{P}{\varepsilon} - C_{\varepsilon 2} + C\varepsilon 3\chi\right) \tag{5.105}$$

where $C_{\epsilon 3} = 0.79$ is a new constant. Also the following vortex stretching invariant is introduced

$$\chi \equiv \omega_{ij}\omega_{jk}S_{ki} \tag{5.106}$$

where

$$S_{ij} \equiv \frac{1}{2}\frac{k}{\varepsilon}\left(\frac{\partial U_i}{\partial x_j} + \frac{\partial U_j}{\partial x_i}\right) \tag{5.107}$$

and

$$\omega_{ij} \equiv \frac{1}{2}\frac{k}{\varepsilon}\left(\frac{\partial U_i}{\partial x_j} - \frac{\partial U_j}{\partial x_i}\right) \tag{5.108}$$

The idea of this modification is to sensitize the model to the greater vortex stretching experienced in a round jet and the increased dissipation arising from this. For an axisymmetric flow without swirl

$$\chi = \frac{1}{4}\left(\frac{k}{\varepsilon}\right)^3\left(\frac{\partial U}{\partial r} - \frac{\partial V}{\partial x}\right)^2\frac{V}{r} \tag{5.109}$$

Hanjalic and Launder (1980) propose a free shear layer correction where there is the following additional source term in the ε equation:

$$S_\varepsilon, extra = -c_\varepsilon 3(\bar{u}'^2 - \bar{v}'^2)\frac{\partial U}{\partial x}\frac{\varepsilon}{k} \tag{5.110}$$

where

$$(\overline{u'}^2 - \overline{v'}^2) = 0.33k \tag{5.111}$$

Equation (5.110) enhances the influence of normal strains. The diffusion coefficient in the ε equation is also recalibrated. Furthermore, $C_{\epsilon 1}$ is recalibrated such that $C_{\varepsilon 1} = 1.44$, $C_{\varepsilon 3} = 4.44$. The rationale behind the approach is complex. The modification of Pope seems to be the most popular. This is for two reasons. The first is that it has a plausible underlying basis. Also, it has some generality. Notably, the RNG (Renormalization Group) k-ε model (see Yakhot et al. 1992) also involves a modification to $C_{\epsilon 1}$, it again being a function of the strain field. The modification takes the following form

$$C^*_{1\varepsilon} = C_{\varepsilon 1} - \frac{\eta(1-\eta/\eta_0)}{1+\beta\eta^3} \tag{5.112}$$

where

$$\eta = \frac{k}{\varepsilon}S \tag{5.113}$$

and

$$\eta_0 = 4.377, \quad \beta = 0.012 \tag{5.114}$$

This model derives the k-ε equations from the Navier-Stokes equations. The procedure is based on using a statistical mechanics approach utilizing a complex mathematical basis with minimal turbulence assumptions. The only adjustable calibrated constant is β. The RNG term can be helpful for some flows and harmful for others.

To improve the predicted turbulence distributions in free shear layers, Murman (2011) recalibrates the diffusion coefficients in the k and ω equations. Again, such modifications have limited general applicability.

5.3.4 *Compressive/Extensive Strain Problem*

Most eddy viscosity based RANS models are calibrated for flows dominated by shear (strains). If compressive or intensive strains dominate, serious inaccuracies can arise. From the exact production term for turbulence (Equation 5.69) the following kinetic energy production term can be derived for a two-dimensional flow:

$$P_{k,Exact} = -\overline{u'u'}\frac{\partial U}{\partial x} - \overline{v'v'}\frac{\partial V}{\partial y} - \overline{u'v'}S \tag{5.115}$$

where $S = \partial U/\partial y + \partial V/\partial x$. The substitution of the Boussinesq approximation into Equation (5.115) gives

$$P_k = \rho\mu_t S^2 \tag{5.116}$$

In Equation (5.115) the $\partial U/\partial x$ and $\partial V/\partial y$ terms, which arise in the continuity equation, tend to cancel each other. With the former negative and $\overline{u'u'} > \overline{v'v'}$, this allows $P_{k,Exact}$ to be negative. This can easily happen, but Equation (5.116) cannot model this. This can be seen more readily by focusing on the more extreme stagnation zone as shown in Figure 5.21. On line s-s in Region A, the following apply:

$$\left.\begin{array}{c} \dfrac{\partial U}{\partial y} = 0, \ \dfrac{\partial V}{\partial y} \neq 0 \\ \hline \dfrac{\partial U}{\partial x} \ll 0, \ \dfrac{\partial V}{\partial x} = 0 \end{array}\right\} \tag{5.117}$$

The production term, now multiplied with density, expands as follows

$$P = -\rho\overline{u_i'u_{j'}}\frac{\partial U_i}{\partial x_j} = -\rho\overline{u'u'}\frac{\partial U}{\partial x} - \rho\overline{v'v'}\frac{\partial V}{\partial y} - \rho\overline{u'v'}\left(\frac{\partial U}{\partial y} + \frac{\partial V}{\partial x}\right) \tag{5.118}$$

or, using Boussinesq approximation, can be re expressed as

$$P = \left(2\mu_t\frac{\partial U}{\partial x}\right)\frac{\partial U}{\partial x} + \left(2\mu_t\frac{\partial V}{\partial y}\right)\frac{\partial V}{\partial y} + \mu_t\left(\frac{\partial U}{\partial y} + \frac{\partial V}{\partial x}\right)^2 \tag{5.119}$$

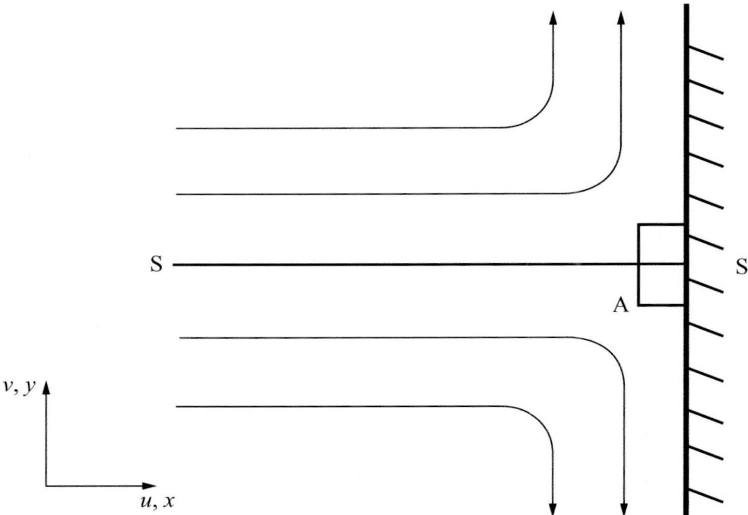

Figure 5.21. Linear RANS model behaviour around stagnation points.

From continuity, $\partial U/\partial x = -\partial V/\partial y$. Combining the continuity equation with Equation (5.58) gives

$$P = \left(2\mu_t \frac{\partial U}{\partial x}\right)\frac{\partial U}{\partial x} - \left(2\mu_t \frac{-\partial U}{\partial x}\right)\left(\frac{\partial U}{\partial x}\right) = 4\mu_t \left(\frac{\partial U}{\partial x}\right)^2 > 0 \qquad (5.120)$$

The exact formula (5.118) reduces, when using the continuity equation for incompressible flow, to

$$P = -\rho\overline{u'u'}\frac{\partial U}{\partial x} + \rho\overline{v'v}'\frac{\partial U}{\partial x} = \left(\rho\overline{v'v'} - \rho\overline{u'u'}\right)\frac{\partial U}{\partial x} \qquad (5.121)$$

Due to presence of the surface $\rho\overline{u'u'} \ll \rho\overline{v'v'}$, and so Equation (5.121) gives

$$P \simeq \rho\overline{v'v'}\frac{\partial U}{\partial x} \qquad (5.122)$$

However, $\partial U/\partial x \ll 0$, so the exact production term, P, will be negative. Unfortunately, the approximate (5.120) is positive and so clearly wrong. The Equations (5.116) and (5.120) quadratic S function means that in stagnation zones, turbulence can be substantially over-estimated. This causes the suppression of leading edge separation on blades/airfoils and excessive predicted heat transfer in this zone.

This problem has given rise to the use of the Kato and Launder (1993) and Yap (1987) corrections (see later). Alternatively, some eddy viscosity models overcome the stagnation problem by modifying C_μ. Note that as indicted previously, the Menter SST model deals with the problem by including a limiter in the turbulence energy production equation. With regards to stagnation zones, in the two-equation model of Secundov and colleagues, an important scaling factor of terms in the differential equation for L is

$$\Phi \propto \frac{\rho L^2}{\mu_t} \frac{\partial L}{\partial x_i} S_{ik} \frac{\partial L}{\partial x_k} \tag{5.123}$$

This phenomenological term becomes maximum in the stagnation region tending to zero in simple shear flows.

5.3.4.1 Yap Correction The correction of Yap (1987), is useful since it also prevents the prediction of excessive heat transfer at re-attachment points. The modification, which is used in the ε transport equation, is given as

$$T_3 = C\frac{\varepsilon^2}{k} \max\left[\left(\frac{k^{3/2}}{\varepsilon l} - 1\right)\left(\frac{k^{3/2}}{\varepsilon l}\right)^2, 0\right] \tag{5.124}$$

where $l = 2.55\ d$ and $C = 0.83$. The Yap correction is usually *most* active just outside the viscous sub-layer, a region that in a high Reynolds number computation is covered with wall-functions. Next, the precise form of the Kato and Launder stagnation correction is outlined.

5.3.4.2 Kato and Launder Stagnation Correction The normal production term given by Equation (5.116) is modified as follows

$$P = \mu_t S\Omega \tag{5.125}$$

where S and Ω are given by Equations (5.16) and (5.17). For the current two-dimensional example, in the stagnation region

$$S = \sqrt{2\left[\left(\frac{\partial U}{\partial x}\right)^2 + \left(\frac{\partial V}{\partial y}\right)^2\right]}, \quad \Omega = \sqrt{2\left[\left(\frac{\partial U}{\partial x}\right)^2 - \left(\frac{\partial V}{\partial y}\right)^2\right]} \tag{5.126}$$

Since $\partial V/\partial y = -\partial U/\partial x\ \Omega = 0$ and so $S\Omega = 0$. Therefore, since in the stagnation zone centre $S\Omega = 0$ there is no turbulence production. In plane shear flows the Kato and Launder modification gives the same turbulence production level as the unmodified equation. There are a range of other ad hoc stagnation corrections – see, for example, Arko and McQuilling (2013). In the work of Lodefier and Dick (2006), the turbulence time scale is limited and the turbulence production disabled in a localized region. Michelassi et al. (2001) also limit the time scale to prevent excessive turbulence production in stagnation zones.

5.3.5 *Curvature Problem*

The exact production term for a two-dimensional shear flow is

$$P_{12,Exact} = -\overline{v'v'}\frac{\partial U}{\partial y} - \overline{u'u'}\frac{\partial V}{\partial x} \tag{5.127}$$

If there is no streamline curvature, the first shear term is greatly dominant. In a flat plate boundary layer the maximum u' is over double that of v'. For a convex curved surface (or concave) as found on an airfoil surface blade, the $\partial V/\partial x$ term comes more strongly into play (i.e., an extra rate of strain becomes active). This, combined with the dominance of $\overline{u'u'}$ and other factors (see Leschziner 2000) makes the last group of terms of some importance. For a convex surface, $\partial V/\partial x$ is negative. Hence, shear stress production and thus shear stress is damped. Indeed experiments show turbulence can be virtually eliminated around convex surfaces (see Sloan et al. 1986). A simple linear eddy viscosity model is unable to model this subtle interplay of terms involving Reynolds stress anisotropy. It must be sensitized in some ad hoc fashion using, for example, the Richardson correction, where essentially

$$Ri = \frac{U_\theta}{R}\frac{\partial U_\theta}{\partial r} \tag{5.128}$$

and R is the streamline radius, and U_θ the stream-wise velocity tangential to the streamline (essentially can sometimes be taken as the surface). Hence, if $l^{new} = l(1 - C\,Ri)$ since $\mu_t \propto Length \times Velocity$, on the suction side of a blade, the correction will reduce l and thus μ_t and the predicted shear.

The 'Richardson number' correction for curvature is derived based on analogy with body forces such as those occurring when there is rotation or density gradients. Fluid particles, when moving along a curved streamline, could be viewed as experiencing a body force. With the Richardson analogy, the extra strains experienced when a shear layer is curved are accounted for by viewing them as a centrifugal body force. The Richardson number can be expressed in gradient, flux, stress and other forms (see Sloan et al. 1986). With the k-ε model, the Richardson correction can be incorporated in the source term of the ε equation. However, it could also be used to scale C_μ. This would mean the flow more immediately responds to curvature.

As noted by Sloan and colleagues, other curvature corrections exist involving, for example, modification of the k equation or 'preferential dissipation' modification of the ε equation. The latter is based on the idea that normal stresses promote dissipation more than shear stresses. Also, C_μ can be modified, based on contractions of more advanced models such as algebraic stress models. Such corrections frequently improve predictions but can also give worse results.

5.3.6 *Body Forces, Rotating and Vortical Flow Problem*

Many flows often involve rotation/swirl. This is especially so for turbomachinery where blade rows rotate. Also, in combustors swirl is used to enhance mixing. The system rotation gives rise to body forces that interact with the turbulence. The nature of the interaction is characterized by the Reynolds stress model source term given as

$$S_\Omega \equiv -2\Omega_k\left\{\overline{u'_j u'_m}\epsilon_{ikm} + \overline{u'_i u'_m}\epsilon_{jkm}\right\} \tag{5.129}$$

where ϵ_{ipq} is the alternating third rank unit tensor. Considering the impact of rotation more simply, the radial pressure gradient in a rotating fluid in the absence of convection and diffusion is governed by

$$\frac{\partial P}{\partial r} = \frac{\rho U_\theta^2}{r} = \frac{\rho(\Omega + U_\theta')^2}{r} \tag{5.130}$$

where U_θ' is the velocity relative to a system rotating at Ω and U_θ is the absolute tangential velocity. If U_θ' radially increases (or more precisely angular momentum $d(rU_\theta)/(rdr) > 0$), pressure increases with radius. Hence, if a fluid particle is displaced radially outwards, it will experience a radially inwards pressure restoring force. The centrifugal force resists inward movements, the centrifugal force of the inward moving particle being greater than that of displaced particle. Now if U_θ' decreases with radius, a radially displaced particle will have a greater centrifugal force than the fluid it is surrounded by and the local pressure will be insufficient to restore its location. To sensitize a turbulence model to rotation, some form of Rossby number is needed of the form $R_o = f(\Omega, S^{-1})$. As noted previously, the effects of local flow curvature can have an analogous influence to rotation. In blade passages and wing tips, vortical structures can be generated. Under these circumstances the rotation corrected version of SA – SARC (SA with Rotation or/and Curvature) (see Spalart and Shur 1997) could be helpful. With the SARC, a portion of the SA model's production term, $C_{b1}(1 - f_{t2})\ \Omega\ \tilde{v}$, is replaced by $C_{b1}(f_{r1} - f_{t2})\ \Omega\tilde{v}$, where f_{r1} is expressed as

$$f_{r1}(r^*, \tilde{r}) = (1 + c_{r1})\frac{2r^*}{1 + r^*}\left[1 - c_{r3}\tan^{-1}(c_{r2}\tilde{r})\right] - c_{r1} \tag{5.131}$$

The dimensionless quantity, r^* is defined as

$$r^* = S/\Omega \tag{5.132}$$

where S and Ω are strain and vorticity related parameters given earlier. Ignoring terms relating to reference frame rotation, the quantity $\tilde{r}$ is defined as

$$\tilde{r} = \frac{2\Omega_{ik}S_{jk}}{D^4}\frac{DS_{ij}}{Dt} \tag{5.133}$$

where DS_{jk}/Dt are the components of the Lagrangian derivative of the strain rate tensor. The usual vorticity tensor takes the form $\Omega_{ik} = 1/2\left[(\partial U_i/\partial x_k - \partial U_k/\partial x_i)\right]$ and

$$D = \sqrt{0.5(S^2 + \Omega^2)} \tag{5.134}$$

Finally, the model constants in Equation (5.131) are as follows: $C_{r1} = 1.0$, $C_{r2} = 12$ and $C_{r3} = 1.0$. However, SARC is rather complex and hence it can sometimes be difficult to secure convergence.

The popular, simplified alternative to SARC is the Approximate SARC or ASARC (see Dacles-Mariani et al. 1995) correction. With this, the SA model production term is formulated as

$$S_{SA} = \Omega + C\min(0, S - \Omega) \tag{5.135}$$

where $C = 2$, and S and Ω are based on equations (5.16) and (5.17). For $C = 0$, the standard SA model is recovered.

Based on eddy-resolving simulations, Bardina et al. (1983) proposed the following additional source, $S_{\varepsilon,extra}$, in the ε equation, to account for the impact of system rotation on turbulence:

$$S_{\varepsilon,extra} = -C_{\varepsilon 3}\Omega\varepsilon \tag{5.136}$$

where Ω is given by Equation (5.17) and $C_{\varepsilon 3}$= 0.15. Inspired by Bardina and colleagues, Wilcox (1988) proposes the following extra source for the ω equation in the k-ω model:

$$S_{\omega,extra} = -C\Omega\omega \tag{5.137}$$

where $C = 3/40$.

If there is rotation with density gradients, the following exact Reynolds stress production term arises:

$$S_B = -\frac{\overline{u_i'\rho'}}{\rho}(\Omega_k x_k \Omega_j - \Omega_j x_k \Omega_j) - \frac{\overline{u_j'\rho'}}{\rho}(\Omega_k x_k \Omega_i - \Omega_k x_k \Omega_i) \tag{5.138}$$

Density gradients within the flow can cause large scale mixing through buoyancy forces akin to the classic Rayleigh-Bernard instability, where cold fluid resting on top of hot, with the gravity vector pointing from cold to hot, creates chaotic mixing. Although the full Reynolds stress model captures the influences of the rotational body forces (hence the physical processes discussed could be elucidated by looking at the terms in this model), it could never account for the consequences of the Rayleigh-Bernard type instabilities noted (see Hanjalic 2002). The closure is at the stochastic level. Following Bo et al. (1995) the k-ε model can be sensitized to buoyancy by extending the production term in the turbulence kinetic energy equation as

$$P_k = \overline{u_i u_j}\frac{\partial U_i}{\partial x_j} - \frac{\overline{\rho' u_i}}{\rho}(\Omega_j x_j \Omega_i - \Omega_j \Omega_j x_i) \tag{5.139}$$

where for a perfect gas at low Mach numbers

$$\frac{\overline{\rho'\mu_i'}}{\rho} = \frac{\overline{u_i'T'}}{T} \tag{5.140}$$

and

$$\overline{u_i'T'} = \frac{\mu_t}{\sigma_T}\frac{\partial T}{\partial x_j} \tag{5.141}$$

5.3.7 *Reverse Transition*

When a fluid flow is subjected to substantial acceleration ($K > 2 \times 10^{-5}$), the flow can laminarize. However, the degree of laminarization of a fluid not just a function of the acceleration level but also depends on the duration for which the fluid has been subjected to this acceleration. For a realistic engineering system, the

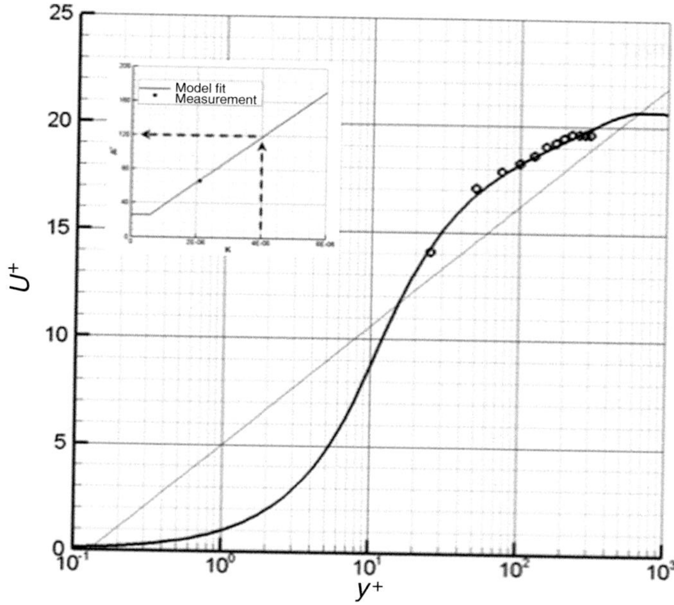

Figure 5.22. Law of the wall for a boundary layer with an acceleration factor of 2.5×10^{-6} (symbols are measurements and full line modified SA model).

fluid's acceleration field will be spatially dependent. Hence to fully characterize the relationship between the acceleration field and the fluid flow would need a lag type equation, as initially proposed by Launder and Jones (1969). Furthermore, if the acceleration is sufficiently rapid, the turbulence will enter the rapid distortion state. In this, it can be assumed that the turbulence is essentially frozen to its upstream state.

Figure 5.22 gives the 'law of the wall' for a boundary layer with an acceleration factor $K = 2.5 \times 10^{-6}$. The symbols represent the measurements. The dashed line shows the standard log-law line. As can be seen from the symbols, the velocity profile shows a substantial rise relative to the standard log-law line. The SA, Menter SST and Chien turbulence models show extreme insensitivity to acceleration. They will give predictions that lay on the standard log-law line even for high K. On the other hand, the low Reynolds number Launder and Sharma k-ε model has the correct sensitization. Oriji et al. (2014a) show that a critical ingredient that allows this in k-ε related turbulence models is the differential form of the 'E' term in the dissipation equation. The use of the turbulence Reynolds number, Re_t, which is non-wall distanced based is also helpful. For example, in the Chein model the Re_y, wall distance based Reynolds number reduces sensitivity to acceleration.

The acceleration thickens the viscous sublayer. Launder and Jones (1969) proposed a provisional linear acceleration parameter against viscous sublayer relationship. This is shown in the inset of Figure 5.22. It is possible to use such viscous sublayer scaling in simpler turbulence models such as the mixing length (see Launder and Jones 1969), k-l (see Oriji et al. 2014a) and SA, to give sensitization to acceleration. For example, for SA, the original damping function given earlier can be revised as follows:

$$f_{v1,new} = \frac{\left(\frac{\chi}{\chi_A}\right)^3}{\left(\frac{\chi}{\chi_A}\right)^3 + C_{v1}^3} \tag{5.142}$$

where

$$\chi_A = \max\left(1.0, \kappa A^+_{f(K)}\right) \tag{5.143}$$

and $A_f^+(K)$ is the function shown in the Figure 5.22 inset. As can be seen from Figure 5.22, use of this term now sensitizes the SA model, there being a strong deviation for the standard log-law line. Note that the function needs further modification for non-equilibrium accelerating flows and this is discussed in Oriji and Tucker (2013). Essentially, the modification requires the inclusion of a lag term expressing the lag in the turbulence response to the strain field.

5.3.8 *Forward Transition Modelling*

The standard, most basic procedure with regards to transition is to estimate the transition point using, for example, experimental correlations. Then, at this location, the turbulence model is forced to trip through, for example, a crude function multiplying eddy viscosity or more elegant procedures, as used with the SA model. An even more refined technique, intended for use in conjunction with the SA model, is proposed by Kožulović and Lapworth (2009). For separated flow transition, the following equations for the intermittency, γ (fraction of the time that the flow is turbulent), distribution are proposed:

$$\gamma = \min\left(C_1 \frac{\Delta S_{up}}{L}, C_2\right) \text{ (Separated)} \tag{5.144}$$

$$\gamma = \max\left(\gamma_{up} - C_3 \frac{\Delta S_{up}}{L}, 1\right) \text{ (Reattached)} \tag{5.145}$$

where L is the length of the separation bubble and ΔS_{up} is the distance from the point of separation. The separation point is identified as where the shear stress goes negative. Hence, $\Delta S_{up}/L = 0$ and unity at separation and reattachment, respectively. Note, γ_{up} is the value of γ at the point of flow reattachment. Also, C_1–C_3 are calibration constants. The intermittency modifies the SA production and destruction terms such that $P = \gamma \cdot C_{b1}(1 - f_{t2})\hat{S}\tilde{v}$ and

$$D = \min\left[\max\left(\gamma, 0.02\right), 1\right] \cdot f\left(v, \tilde{v}, d, \kappa, S\right) \left(\frac{\tilde{v}}{d}\right)^2 \tag{5.146}$$

Oriji and Tucker (2013) use $C_1 = 3.4$, $C_2 = 4$ and $C_3 = 1.1$ in Equations (5.144) and (5.145), but, as ever with transition, careful calibration is needed. The correlation of Abu-Ghannam and Shaw (1980) is a popular for estimating transition onset in

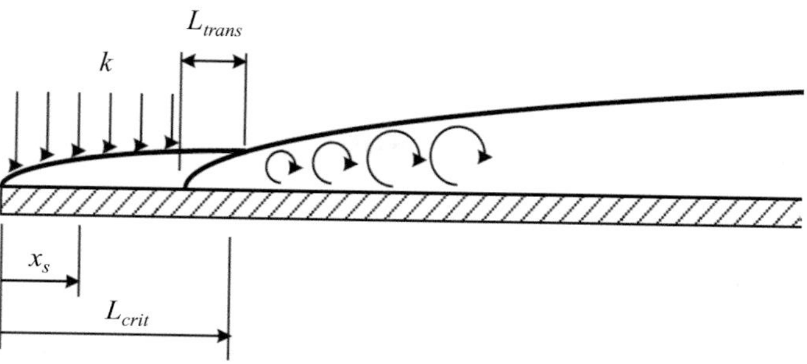

Figure 5.23. Transition in a boundary layer.

turbomachinery, but there are a wide range of other correlations – see, for example, Arnal (1992).

Some turbulence models, under certain circumstances, can model transition processes without recourse to special modifications. However, their accuracy is limited. For example, for a laminar upstream boundary layer feeding into a separation bubble, the transition by contact process as described by Spalart (2000) and Travin et al. (2000) can be used. With transition by contact, the turbulent inflow is set to a low level, allowing generation of a laminar boundary layer and separation. However, the ambient turbulence is initialized to a high level. This allows turbulence generated in a separation bubble to feed into the separated shear layer and for transition to commence. This process has found application with the SA model but, as with bypass transition, it will be highly dependent on the base turbulence model choice.

Abid (1993) explores the performance of several k-ε models at predicting bypass transition over a flat for different free stream turbulence intensities, T_i. To help understand performance of these models it is helpful to consider key elements of the k equation in an averaged form integrated across the boundary layer. Hence we have

$$\frac{d(Uk)_{ave}}{dx} = P_{k,ave} - D_{k,ave} \tag{5.147}$$

where $(Uk)_{ave}$ is the average product of velocity and kinetic energy across the boundary layer. Also, $P_{k,ave}$ and $D_{k,ave}$ are the averaged production and destruction terms for k. As shown in Figure 5.23 the transition process with the k equation takes place in three stages. These are as follows:

(1) Initially k convects and diffuses into the boundary layer but $P_{k,ave}/D_{k,ave} < 1$ so there is insufficient production and the flow stays laminar.
(2) At a critical Reynolds Re_{crit}, $P_{k,ave} = D_{k,ave}$ and k_{ave} starts to grow. This increases μ_t, which in turn increases $P_k = \mu_t\,(du/dy)^2$, and hence there is a substantial k_{ave} growth.
(3) The non-linear growth process continues until transition is fully achieved.

The laminar kinetic energy, k_l, concept of Mayle and Schulz (1997) allows extension of the previous, giving greater physical realism. This is achieved by allowing for the incorporation of the growth of pre-transitional fluctuations. An additional equation for laminar kinetic energy, k_l, is solved. This is again expressed, for illustrative purposes, in the simple averaged form

$$\frac{d(Uk_l)_{ave}}{dx} = P_{k_l,ave} - D_{k_l,ave} - \tilde{T} \tag{5.148}$$

where

$$D_{k_l,ave} = \frac{2\mu\kappa_l}{\rho d^2} \tag{5.149}$$

and $P_{kl,ave}$ is the production term. The formulation of the latter depends on the type of transition being modelled (e.g., either bypass or separated flow transition). The far right-hand side term, $\tilde{T}$ (detailed later), models the transfer of k_l to k. A term of opposite sign is added to the k equation, signifying the transfer of energy from laminar kinetic energy to turbulent kinetic energy.

For illustrative purposes, the laminar kinetic energy model of Pacciani et al. (2013) is given next and discussed in relation to some other models. In line with Equations (5.148) and (5.149), the form of the laminar kinetic energy equation is

$$\frac{Dk_l}{Dt} = P_l - \varepsilon_l + v\nabla^2 k_l - \tilde{T} \tag{5.150}$$

The transfer/transition is triggered when the wall normal Reynolds number Re_y reaches critical amplitude of C_4, the constant values being defined by Pacciani et al. (2013). In the previous,

$$P_l = v_l S^2 \tag{5.151}$$

Also,

$$v_l = C_1 f_1 \sqrt{k_l} \delta_\Omega \tag{5.152}$$

with

$$\delta_\Omega(x) = 0.5U_\infty \left(\frac{\partial U}{\partial y}\right)_{\max}^{-1} \tag{5.153}$$

Furthermore, Equation (5.42), when used in a transport equation for k, becomes modified to $D = T_2 - \tilde{T}$, where $\tilde{T}$ is the energy transfer term such that

$$\tilde{T} = C_2 f_2 \beta^* k_l \omega \tag{5.154}$$

and

$$f_2 = 1 - e^{-\psi/C_3} \tag{5.155}$$

with

$$\psi = \max(0, Re_y - C_4) \tag{5.156}$$

One of the key limitations of the laminar kinetic energy model is the length scale (vorticity thickness, δ_Ω) used in production term of the laminar kinetic energy. The vorticity thickness is a non-local quantity. It is difficult to evaluate in unstructured

Table 5.10. *Summary of different LKE model parameters and calibration constants (From Nagabhushana 2014)*

	Functional form ($\tilde{T}$)	C_2	C_3	C_4
Walters and Leylek (2004), k-ε	$C_2 f_2 k_l \lambda_{eff}/\sqrt{k}$	0.21	8	35
Walters and Leylek (2004), k-ω	$C_2 f_2 k_l \omega(\lambda_T/\lambda_{eff})^{2/3}$	0.08	4	12
Walters and Cokljat (2008), k-ω	$C_2 f_2 k_l \omega/(1 - e^{-0.4(\lambda_{eff}/\lambda_T)^4})$	0.12	0.6	12
Pacciani et al. (2011a, 2011b, 2012, 2013), k-ω	$C_2 f_2 k_l \omega \beta^*$	0.3	8	10

solvers and in complex geometries. The length scales proposed in the laminar kinetic energy models of Walters and Leylek (2004) and Sveningsson (2006), which are based on the local parameters, are more helpful for general transition modelling applications. For example, Walters and Leylek (2004) use

$$\lambda_{eff} = \min(C_\lambda d, \lambda_T) \tag{5.157}$$

and Sveningsson (2006) uses

$$\lambda_{eff} = \min\left(C_\lambda \sqrt{\frac{k}{\omega S}}, \lambda_T\right) \tag{5.158}$$

where λ_T is the standard turbulent length scale given by $\sqrt{k}/\omega$. Turner (2012) proposes a modified λ_{eff} based on Ω instead of S, which is helpful in flows with stagnation regions. This is given as:

$$\lambda_{eff} = \min\left(C_\lambda \sqrt{\frac{k_T}{\omega \Omega}}, \lambda_T\right) \tag{5.159}$$

Table 5.10, from Nagabhushana (2014), gives the variation of the parameters and the corresponding calibration constants for different laminar kinetic energy models. The calibration constant C_4, which dictates the threshold for the transfer of laminar kinetic energy to turbulent kinetic energy is a very sensitive quantity (see Nagabhushana 2014). Hence, the function $\tilde{T}$ and the related constants need further calibration for different test cases.

$$\psi = \max\left(0, \frac{y\sqrt{k}}{v} - C_4\right)$$

or

$$\psi = \max\left(0, \frac{k}{v\Omega} - C_4\right)$$

Importantly, the transition process in a separated shear layer is physically very abrupt and hence in reality is impossible to mimic using a diffusive process. The applicability of the laminar kinetic energy model to wake induced transition is unclear. In the region of the boundary layer bathed by the wake, transition is caused by diffusion of the wake turbulence into the boundary layer. Hence, when this occurs there, should be no generation of k_l.

High levels of intermittent unsteadiness characterize transition. This has given rise to many approaches that solve a transport equation for intermittency. Typically, a standard γ transport equation of the following form would be solved

$$\frac{\partial(\rho\gamma)}{\partial t} + \frac{\partial(\rho u_i \gamma)}{\partial x_i} = P_\gamma + \frac{\partial}{\partial x_i}\left[\left(\mu + \frac{\mu_t}{\sigma_\gamma}\right)\frac{\partial \gamma}{\partial x_i}\right] \tag{5.160}$$

where P_γ is a production term and σ_γ a Schmidt number. The onset of transition location still needs to be specified.

A notable, currently popular, transition model is that of Menter et al. (2006) – the γ-$Re_{\theta t}$ model. The difficulty when using momentum thickness Reynolds number, Re_θ, based correlations, to estimate the transition onset location is that they need integration of properties through the boundary layer. Even for a structured grid, this can give rise to unforeseen complications. However, for an unstructured grid formulation, an efficient, reliable process becomes even more challenging with the need for non-local information. Menter and colleagues elegantly overcome this problem through the solution of a transport equation for the local transition onset momentum thickness Reynolds number $Re_{\theta t}$, along with a transport equation for γ. A range of correlations must be supplied to close the model.

5.3.9.1 Some Future Transition Challenges Current transition models focus on simple airfoil type configurations. However, with aircraft there is the wing body function, and for turbomachinery the blade extremity zones also ideally need some form of transition modelling. These may well be areas of key importance that current transition models are not designed for. Also, both measurements (see Coull and Hodson 2011) and quasi-DNS (Nagabhushana 2014) suggest that Klebanoff-like structures, originating from wakes interacting with *leading edges*, in low-pressure turbines, are of critical importance in transition modelling. The impact of these streaks is currently not accounted for in transition models. Also, the surfaces have roughness and the interaction of this with the transition process may need to be accounted for. A model for this purpose is proposed by Nagabhushana (2014). This is based on extending the laminar kinetic energy concept. Notably, the film cooling holes found in turbines are also likely to influence transition, adding a further level of uncertainty. Mayle and Schulz (1997) note the need not to neglect the impact of surface vibrations on transition. Hence, the ultimate transition model would need to factor this in.

5.3.10 *Free Stream Turbulence*

An important part of turbulence modelling, especially for transition, is to ensure that, as noted earlier, free stream turbulence is accurately transported with the appropriate decay level to a system target zone. Figure 5.24 gives some examples relating to this. Frame (a) relates to external flows (i.e., the flow approaching a wing and nacelle). Frame (b) is for an internal flow – the flow passing through blades rows in axial turbomachinery. For external flow simulations, typically, the domain

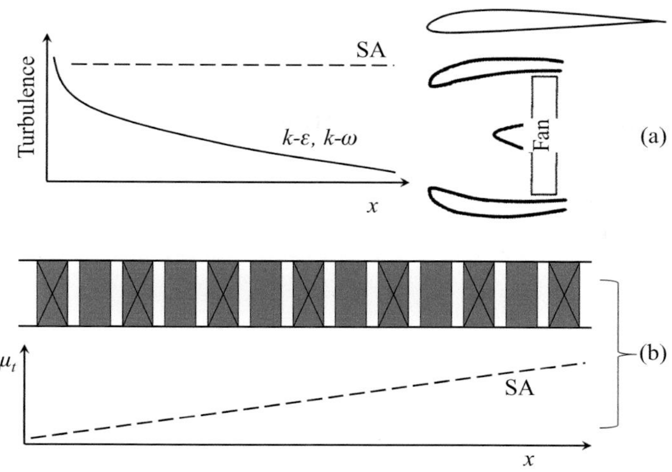

Figure 5.24. Turbulence development in external and internal flows: (a) external flows and (b) internal flow (adapted from Tucker 2013a – published with kind permission of Elsevier).

is large. This allows for a substantial deviation of turbulence from expected levels. The k-ε and k-ω models are designed to give correct decay rates for free stream turbulence. As noted by Spalart and Rumsey (2007), in the uniform flow found in the free stream, the mean velocity gradients vanish and the k-ε equation reduces to

$$U\frac{\partial k}{\partial x} = -\varepsilon \tag{5.161}$$

and

$$U\frac{\partial \varepsilon}{\partial x} = \frac{-C_{\varepsilon 2}\varepsilon^2}{k} \tag{5.162}$$

respectively. Then from analytical solution the following is gained:

$$\frac{k}{\varepsilon} = \left(\frac{k}{\varepsilon}\right)_{FS} + (C_{\varepsilon 2} - 1)\frac{x}{U} \tag{5.163}$$

where subscript FS represents the free stream value and x is the distance from the inflow boundary in the free stream direction. The following can also be written:

$$\begin{aligned} k &= k_{FS}\left[1 + (C_{\varepsilon 2} - 1)\left(\tfrac{\varepsilon}{k}\right)_{FS}\tfrac{x}{U}\right]^{\frac{-1}{C_{\varepsilon 2}-1}} \\ \epsilon &= \epsilon_{FS}\left[1 + (C_{\varepsilon 2} - 1)\left(\tfrac{\varepsilon}{k}\right)_{FS}\tfrac{x}{U}\right]^{\frac{-C_{\varepsilon 2}}{C_{\varepsilon 2}-1}} \\ \upsilon_t &= \upsilon_{tFS}\left[1 + (C_{\varepsilon 2} - 1)\left(\tfrac{\varepsilon}{k}\right)_{FS}\tfrac{x}{U}\right]^{\frac{C_{\varepsilon 2}-2}{C_{\varepsilon 2}-1}} \end{aligned} \tag{5.164}$$

Also, using a similar procedure, the following equation for ω can be generated:

$$\frac{1}{\omega} = \frac{1}{\omega_{FS}} + C_\mu (C_{\varepsilon 2} - 1)\frac{x}{U} \tag{5.165}$$

These above equations can be used to reverse engineer inflow boundary conditions to match requirements to a target zone such as the leading edge of a wing. For a simulation of an external flow, such as that around an aircraft wing, as noted, the domain extent can be extensive and hence the grid spacing at the boundary bigger than one chord. This means that the decay rate of turbulence in the k-ε and k-ω models will be substantially under-estimated. Hence, care is required. Spalart and Rumsey (2007) recommend the use of source terms in the turbulence equations to prevent excessive decay. An alternative, less preferable expedient would be to have floor/limiting values on k and ε or ω. For the k and ω equations in the Menter SST model, Spalart and Rumsey recommend the following additional sources:

$$S_{k,extra} = \beta^* \omega_{amb} k_{amb} \tag{5.166}$$

and

$$S_{\omega,extra} = \beta^* \omega_{amb}^2 \tag{5.167}$$

respectively. The subscripts *amb* indicates target ambient values. If $k = k_{amb}$ and $\omega = \omega_{amb}$, in the free stream, these additional terms will exactly cancel the original destruction terms. Similarly for the k and ε equations, the following sources can be added:

$$S_{k,extra} = \varepsilon_{amb} \tag{5.168}$$

$$S_{\varepsilon,extra} = C_{\varepsilon 2} \frac{\varepsilon_{amb}^2}{k_{amb}} \tag{5.169}$$

The use of source terms is better than floor values. The latter need to be disabled in boundary layers.

Since SA has no explicit destruction term, the inflow value is convected without any decay as shown in the Figure 5.24 schematic. Fares and Schröder (2005) present a transport equation for eddy viscosity that, unlike SA, is sensitized to give correct free stream decay. Figure 5.24b shows a further issue. This relates to multistage turbomachinery, where there are multiple sets of adjacent stationary and rotating blades. With SA, as the flow passes down the machine, there will be turbulence production in the free stream, but no real destruction away from wakes. Hence, there will be a tendency to growth in eddy viscosity as the machine is progressed through. Figure 5.25 shows the axial distribution of the change in mass averaged μ_t distribution, normalized by inlet eddy viscosity, through a 3.5 stage transonic fan/compressor.

The general growth in μ_t is evident. What is more, there is a substantial difference in eddy viscosity levels for the mixing length and SA simulations.

Another related matter is that many turbulence models fail to accurately model skin friction and heat transfer at elevated free stream levels. Turbomachinery can operate with free stream intensities reaching around 30% and the k-ε model can underestimate skin friction by 100% at such levels. Foroutan and Yavuzkurt (2013) review work in this area, proposing a tentative additional term in the k equation for a k-ε model and a revised C_μ function. A key complication is that the response of a boundary layer to free stream turbulence levels is a strong function of the scales

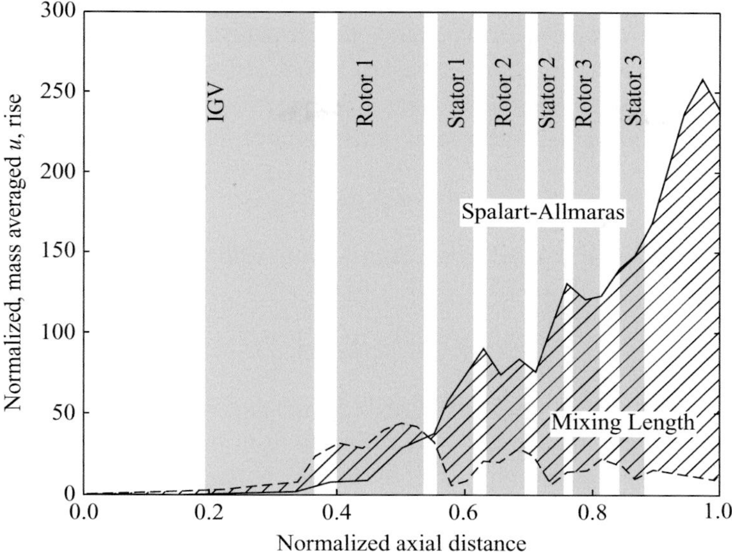

Figure 5.25. Eddy viscosity growth through a 3.5 stage transonic fan/compressor (from Tucker 2013a – published with kind permission of Elsevier).

of the free stream turbulence relative to the boundary layer. Ames et al. (1999) give a simple algebraic, spectral model to deal with elevated free stream turbulence length scales and intensities. For an aircraft located in atmospheric boundary layers, the scales in the boundary layer can be orders of magnitude larger than the scales in the boundary layers of the aircraft. As noted earlier, using the atmospheric boundary layer turbulence kinetic energy levels would give airframe effective Reynolds number tending to unity. Hence, under these circumstances Spalart and Rumsey (2007) recommend using just that part of the turbulence spectrum in the atmospheric boundary layer that has relevance to the airframe turbulence flow physics.

5.3.11 *Surface Roughness/Finish*

The impact on surface finish on gas turbine performance is a subject of much importance (see Fiala and Kügeler 2011, Licari and Christensen 2011, Montomoli et al. 2008 and Suder et al. 1995). It is also a subject of great interest to airframe manufacturers. Roughness can occur through machining processes, manufacturing imperfections and erosion from impinging particles. These processes can give rise to a wide range of surface topologies which can have a profound skin friction impact.

Until relatively recently, the key approach for accounting for roughness was recalibration of the 'law of the wall' constants used in the wall functions of the k-ε model. This involved changing the log-law E value, this time in Equation (5.86), based on sand grain measurements. A more advanced 'wall function' related approach is outlined by Fiala and Kügeler (2011). Inspired by Cebeci and Chang (1978), and work related to mixing length type modelling, Aupoix and Spalart (2003) sensitized the SA model to roughness by incorporating a simple virtual wall distance offset in a low Reynolds number framework. The sand grain roughness based calibration of the offset means that the near wall damping is less active, giving rise to more turbulence. Thus there is sensitivity to roughness. Hence, in the approach of Aupoix and Spalart (2003) the wall distance adjacent to rough surfaces is redefined as

$$\tilde{d} = d + d_R \tag{5.170}$$

where d_R is the shift that is used to account for the wall roughness height. This is defined as

$$d_R = \exp(-H_r\kappa)h_s \approx C_{R2}h_s \tag{5.171}$$

Note that for a fully rough surface, $H_r = 8.5$ and that h_s is the equivalent sand grain roughness height value and $C_{R2} = 0.03$. The wall boundary condition in the convection-diffusion transport equation, of the SA model, is updated to the differential condition

$$\frac{\partial \tilde{v}}{\partial n} = \frac{\tilde{v}}{\tilde{d}} \tag{5.172}$$

where n is the surface normal coordinate. Also,

$$\chi = \frac{\tilde{v}}{v} + C_{R1}\frac{h_s}{\tilde{d}} \tag{5.173}$$

and $C_{R1} = 0.5$.

The application of Equation (5.170) to complex geometries with multiple smooth and rough surfaces is impractical. Step changes in $\tilde{d}$ will occur which are numerically problematic. Hence, the Aupoix and Spalart (2003) modification is most generally implemented using the Hamilton-Jacobi wall distance equation $|\nabla \tilde{d}| = 1 + \varepsilon \tilde{d}\nabla^2 \tilde{d}$, discussed in Chapter 7, subject to the following boundary conditions: $\tilde{d} = 0$ (smooth walls) and $\tilde{d} = d_R$ (rough walls),

The Laplacian ($\nabla^2 \tilde{d}$) ensures a smooth $\tilde{d}$ field. When $\varepsilon \to 0$, the exact wall distance is gained. Notably, the extension just described is for fully rough surfaces. However, much the roughness found in practical problems is transitional. Hence, this is discussed next.

5.3.11.1 Transitional Roughness Boundary Layer Model For a transitionally rough wall, the extension of Aupoix and Spalart (2003) can be modified using the approach of Oriji et al. (2014b) as described next. Note that the approach of Aupoix and Spalart can also model transitional roughness and the approach of Oriji and colleagues is closely related. For sand-grain roughness, based on measurements, the correlation for a transitional roughness function $H_r(h_s^+)$ is given by Ligrani and Moffat (1986).

$$H_r(h_s^+) = \left[\frac{1}{\kappa}\ln(h_s^+) + A\right] \cdot \left[1 - \sin\left(\frac{\pi g}{2}\right)\right] + 8.5\sin\left(\frac{\pi g}{2}\right) \tag{5.174}$$

where, A is the additive constant in the log law. Also, the function g of h_s^+ is

$$g = \begin{cases} 0 & (h_s^+ < h_{smooth}^+) \\ \dfrac{\ln(h_s^+/h_{smooth}^+)}{\ln(h_{rough}^+\big/h_{smooth}^+)} & (h_{smooth}^+ \le h_s^+ \le h_{rough}^+) \\ 1 & (h_{rough}^+ < h_s^+) \end{cases} \tag{5.175}$$

where, $h_{smooth}^+ = 2.25$ and $h_{rough}^+ = 90$. Then a functional relationship,

$$C_{R1} = a_0 + a_1 \cdot f(h_s^+) \tag{5.176}$$

can be used. The composite C_{R1} function proposed by Oriji and colleagues is as follows:

$$C_{R1} = a_0 + a_1 f\left(h_s^+\right) = \begin{cases} \text{if } \left(h_{s,0}^+ < h_s^+ \le h_{s,1}^+\right) \\ a_0 = -0.7675583 \\ a_1 = 0.1947976 \\ \text{if } \left(h_{s,1}^+ < h_s^+ \le h_{s,2}^+\right) \\ f\left(h_s^+\right) = h_s^+ \\ a_0 = 0.7775 \\ a_1 = 0.1275 \\ \text{if } \left(h_{s,2}^+ < h_s^+ \le h_{s,3}^+\right) f\left(h_s^+\right) = \cos\left(\pi.g\right) \\ g = \dfrac{\ln\left(0.55h_s^+/h_{s,1}^+\right)}{\ln\left(0.55h_{s,2}^+/h_{s,1}^+\right)} \end{cases} \tag{5.177}$$

$$\begin{aligned} a_0 &= 0.575 \\ a_1 &= 0.075 \\ f\left(h_s^+\right) &= \cos\left(\pi.g\right) \\ g &= \frac{h_s^+ - h_{s,2}^+}{h_{s,3}^+ - h_{s,2}^+} \\ a_0 &= 0.5 \\ a_1 &= 0 \\ f\left(h_s^+\right) &= 0 \end{aligned}$$

where $h_{s,0}^+ = 4$, $h_{s,1}^+= 8.149783$, $h_{s,2}^+ = 69.826229$, $h_{s,3}^+ = 114.55804$.

Figure 5.26 shows the revised C_{R1} function (see full line) necessary to give plausible behaviour for transitional roughness, based on the measurements of Ligrani and Moffat. Also, the C_{R2} function (see dashed line) is plotted.

Figure 5.27 shows comparison of velocity profiles with sand-grain theory (dashed lines) for smooth, and $h_s^+ = 8$, 16, 31.7 and 62, together with fully rough cases $h_s^+ =$ 128, 252.8 and 494.5. The full line gives the predictions for the model, which show the correct sensitivity to roughness.

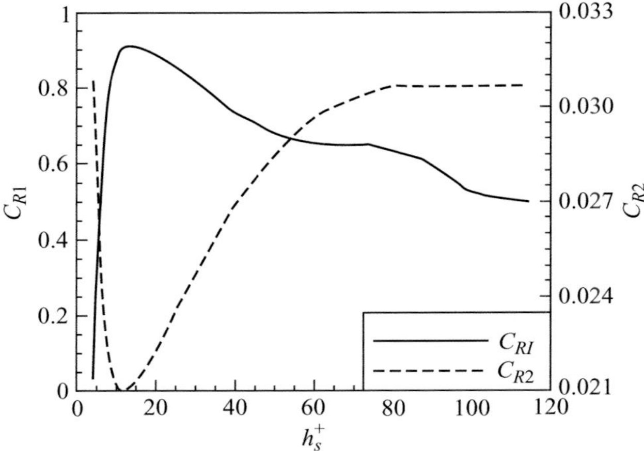

Figure 5.26. Revised C_{R1} function plotted against h_s^+ (from Oriji et al. 2014b – published with the kind permission of the ASME).

Wilcox (1998) proposed sensitizing the k-ω model to wall roughness by just modifying the ω boundary condition as

$$\omega|_w = \frac{u_\tau^2}{\upsilon} S_R, \quad S_R = \begin{cases} \left(\dfrac{50}{h_s^+}\right)^2, & h_s^+ < 25, \\ \dfrac{100}{h_s^+}, & h_s^+ \geq 25 \end{cases} \tag{5.178}$$

Knopp et al. (2009) sensitize both the k and ω boundary conditions. The k wall boundary conditions is as

$$k|_w = \phi_{r1} k_{rough,} \qquad k_{rough} \equiv \frac{u_\tau^2}{\beta_k^{\frac{1}{2}}}, \quad \phi_{r1} = \min\left(1, \frac{h_s^+}{90}\right) \tag{5.179}$$

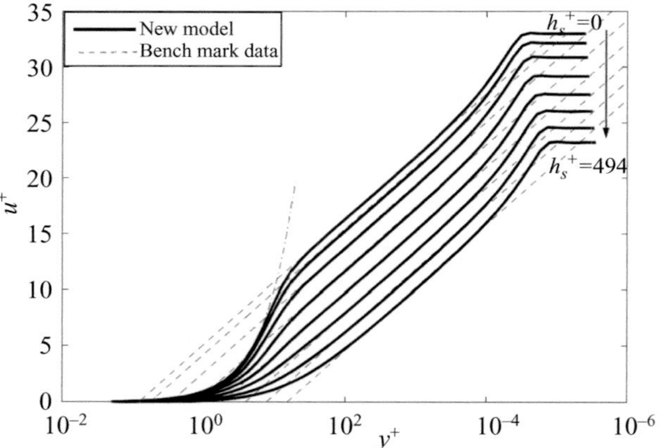

Figure 5.27. Plot of roughness mean velocity profiles (from Oriji et al. 2014b – published with the kind permission of the ASME).

Note that $\beta_k = 0.09$. Also, the following condition for ω is used

$$\omega = \frac{u_\tau}{\beta_k^{\frac{1}{2}} \kappa \widetilde{d}_0}, \quad \widetilde{d}_0 = \phi_{r2} 0.03 h_s \tag{5.180}$$

where

$$\phi_{r2} = \min\left[1, \left(\frac{h_s^+}{30}\right)^{2/3}\right] \min\left[1, \left(\frac{h_s^+}{45}\right)^{1/4}\right] \min\left[1, \left(\frac{h_s^+}{60}\right)^{1/4}\right] \tag{5.181}$$

Note that a limiter is also placed on the wall value for ω (see Knopp et al. 2009).

The approach of Knopp and colleagues is naturally compatible with the Menter SST model. It is also more forgiving with regards to grid, relative to the original roughness modification proposal of Wilcox for the k-ω model. Eça and Hoekstra (2011) explore three further sand-grain roughness sensitization options for the Menter SST model. Durbin et al. (2001) present a zonal k-ε model roughness extension.

However, a key problem is that real roughness is generally very different from sand-grain roughness and can have much greater impact. This adds another layer of modelling complexity (i.e., that it is not really just the roughness height that matters but also the form of the roughness). Many forms of real roughness can be much lower frequency than sand-grain roughness. Thus, they are relatively easy to mesh. The Reynolds numbers in terms of the roughness height can be quite modest. Hence, useful data can be gained from DNS. Therefore, a possibility is to build up such databases to create lookup tables to refine modelling coefficients in RANS models. Stripf et al. (2009a, 2009b) propose a roughness model based on body forces. Relative to more standard approaches that directly modify the turbulence field, this approach gives greater freedom to enforce roughness topology. However, it contains a wide range of ad hoc modelling assumptions.

5.3.12 *Code Implementation and Other Aspects*

Turbulence models can involve numerous additional terms along with complex non-linear switching functions. Hence, as shown by Charbonnier et al. (2008), Tyacke et al. (2012) and Hanimann et al. (2014), very different solutions for the *same* turbulence model, but different numerical solution frameworks can be gained. The more complex turbulence models, with additional gradient terms, like the Reynolds stress model, also place stronger demands on the numerical discretization. The strong sensitivity to code implementation for transition models is outlined by Kelterer et al. (2010). Figure 5.28 shows surface static pressure and passage entropy function contours as stall is approached for a transonic fan. The results are for two different reputable CFD codes with the same reported turbulence model and grid. The solution in Frame (a) has more reversed flow and a reduced pressure ratio. Hence, there is a clear numerical framework and turbulence model implementation impact.

Understanding the model behaviour is further complicated, since multiple solutions in RANS simulations can occur. Solutions can be sensitive to grid, numerical

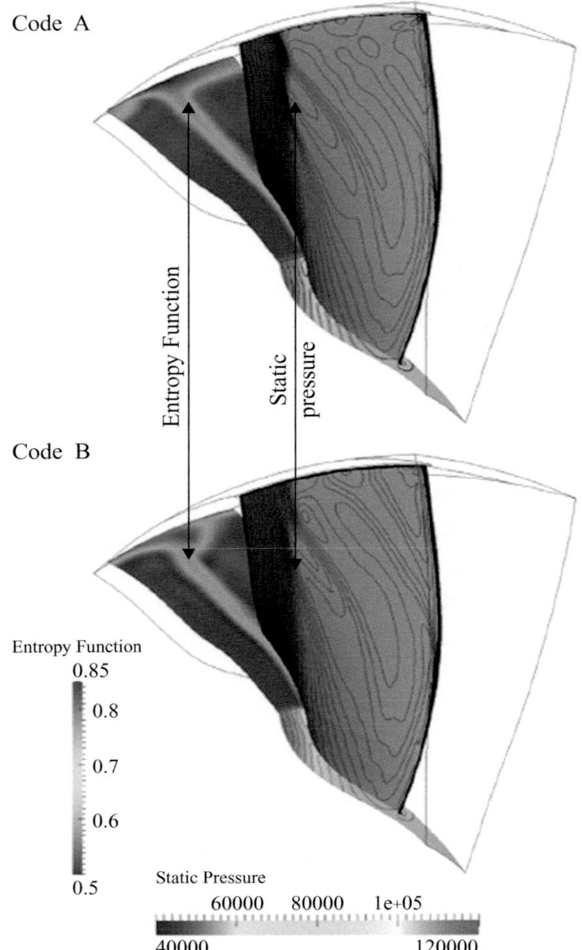

Figure 5.28. Flow simulation for a transonic fan with the same turbulence model but different CFD code (from Tucker 2013a – published with kind permission of Elsevier).

scheme and the initial guess. Using analytical analysis, Rumsey et al. (2006) identify the damping function as the non-uniqueness source in a k-ε model variant. Kamenetskiy et al. (2013) demonstrate non-unique solutions for both the k-ω and SA models when predicting separating flows on high lift devices around stall.

With any turbulence model, care is needed with regards to the numerical implementation. Typically, the turbulence equations can be highly non-linear and unstable. There is, hence, the potential for divisions by zero and parameters violating basic realizability constraints through the iterative process. The SA model has seen such extensive use that such matters are publicized (see, for example, Crivellini et al. 2013) and hence some are noted here. One issue is that the turbulence production function $\hat{S}$ should never become zero or negative. One recommendation is that it should be clipped to always be greater than 0.3*Ω. Another method is reported in Oliver (2008). With this, the following variable is defined:

$$\overline{S} = \frac{\tilde{\upsilon}}{\kappa^2 d^2} f_{\upsilon 2} \tag{5.182}$$

Then

$$\begin{aligned} \hat{S} &= \Omega + \overline{S} \quad \text{when} \quad \overline{S} \geq -c_2\Omega \\ \hat{S} &= \Omega + \frac{\Omega(c_2^2\Omega + c_3\overline{S})}{(c_3 - 2c_2)\Omega - \overline{S}} \text{ when } \overline{S} < c_2\Omega \end{aligned} \tag{5.183}$$

Note that $c_2 = 0.07$ and $c_3 = 0.09$.

When $\tilde{\upsilon} < 0$, Allmaras and Johnson (2012) recommended solution of the following equation instead of the standard SA model equation:

$$\begin{aligned} \frac{\partial \tilde{\upsilon}}{\partial t} + u_j \frac{\partial \tilde{\upsilon}}{\partial x_j} &= c_{b1}(1 - c_{t3})\Omega\tilde{\upsilon} + c_{w1}\left(\frac{\tilde{\upsilon}}{d}\right)^2 \\ &+ \frac{1}{\sigma}\left[\frac{\partial}{\partial x_j}\left((\upsilon + \tilde{\upsilon} f_n)\frac{\partial \tilde{\upsilon}}{\partial x_j}\right) + c_{b2}\frac{\partial \tilde{\upsilon}}{\partial x_i}\frac{\partial \tilde{\upsilon}}{\partial x_i}\right] \end{aligned} \tag{5.184}$$

For ease of comparison, the usual SA model equation is

$$\begin{aligned} \frac{\partial \tilde{\upsilon}}{\partial t} + u_j \frac{\partial \tilde{\upsilon}}{\partial x_j} &= c_{b1}(1 - f_{t2})\hat{S}\tilde{\upsilon} - \left[c_{w1} f_w - \frac{c_{b1}}{\kappa^2} f_{t2}\right]\left(\frac{\tilde{\upsilon}}{d}\right)^2 \\ &+ \frac{1}{\sigma}\left[\frac{\partial}{\partial x_j}\left((\upsilon + \tilde{\upsilon})\frac{\partial \tilde{\upsilon}}{\partial x_j}\right) + c_{b2}\frac{\partial \tilde{\upsilon}}{\partial x_i}\frac{\partial \tilde{\upsilon}}{\partial x_i}\right] \end{aligned} \tag{5.185}$$

In Equation (5.184),

$$f_n = \frac{c_{n1} + \chi^3}{c_{n1} - \chi^3}$$

and $c_{n1} = 16$. Note that the *eddy viscosity*, μ_t, is explicitly set to zero when $\tilde{\upsilon} < 0$. This fix helps deal with negative values of $\tilde{\upsilon}$ that arise through initial solution transients and in zones where the grid quality is poor. It is expected that essentially the original SA model equation will be solved.

In the SA model, there is the function, r, presented earlier. This is essentially $r^2 = l^2/(k^2 d^2)$ and so should not become negative. To avoid this Crivellini et al. (2013) propose the following modification:

$$r^* = \frac{\tilde{\upsilon}}{\hat{S}k^2 d^2}, \qquad r = \begin{cases} r_{max}, & r^* < 0, \\ r^*, & 0 \leq r^* \leq r_{max} \\ r_{max}, & r^* \geq r_{max} \end{cases}$$

where r_{max} is a positive valued limiter.

To restrict the ranges of predicted variables, such as, for example, the eddy viscosity, it is tempting to use hardwired limits in codes. However, care is required when doing this. For example, the dynamic range of the eddy viscosity is large. Hence such practices can be problematic – see Spalart and Rumsey (2007). Modifications to improve iterative convergence of the k-ω model can be found in Bassi et al. (2005).

5.3.13 *More Recent RANS Advancements*

Turbulence model advancements have largely stagnated since the early 1990s. RANS developments for heat transfer lag those for aerodynamics. More recently, specifically for hot jets, Khavaran and Kenzakowski (2007) have proposed revised equations for scalar fluctuations and their dissipation.

Near walls, turbulence is damped by both viscous forces and the wall blocking. The v^2-f RANS model of Durbin (1991) includes an elliptic differential equation to directly account for the latter. In standard eddy viscosity models, the influence of wall blocking on turbulence damping must also be accounted for (wrongly) through viscous modelling. The v^2-f model has shown improvements in predictive accuracy relative to more standard eddy viscosity models – see for example Colban et al. (2007), Hermanson et al. (2003), Luo and Razinsky (2007), Sveningsson and Davidson (2005) and Takahashi et al. (2012)) – for a range of turbomachinery applications. Park and Sung (1997) use a similar equation to Durbin, but this time the motivation is to account for multiple wall proximities and ambiguities regarding the definition of d. Elliptic relaxation equations take the broad form given by Gatski et al. (2007):

$$n\phi - L^2\nabla^2\phi = 1 \tag{5.186}$$

where $\phi = 0$ at walls, $n = 1$ and L is the turbulence integral length scale. For an infinite flat plate, in an extensive domain, the solution to the previous is

$$\phi = 1 - \exp\left(-\frac{d}{L}\right) \tag{5.187}$$

Note that the solution of Equation (5.186) with $n = 0$ (and an auxiliary equation relating ϕ to a wall distance function) links to the d function equation of Spalding (see Chapter 7) used earlier. Hence, the differential wall distance function work, discussed earlier, connects with elliptic relaxation as envisaged by Park and Sung.

For the Reynolds stress model, a key weak link is the numerous closure assumptions and, notably, near walls the pressure-strain term. Thielen et al. (2005) use elliptic blending (Equation (5.186)) to refine the near wall treatment in a Reynolds stress model, finding improved agreement relative to more standard models. High order (fourth order moment) modelling (Poroseva 2013) could resolve outstanding near wall modelling issues related to the pressure strain term. Such modelling evidently potentially yields no unknown coefficients related to modelling turbulent diffusion and can be used to generate two-equation models.

An interesting contraction of the Reynolds stress model is the turbulent potential model reported in Pecnik and Sanz (2007) of Perot (1999). The RANS equations only need the divergence of the Reynolds stress tensor, redefined here as **R**. Using

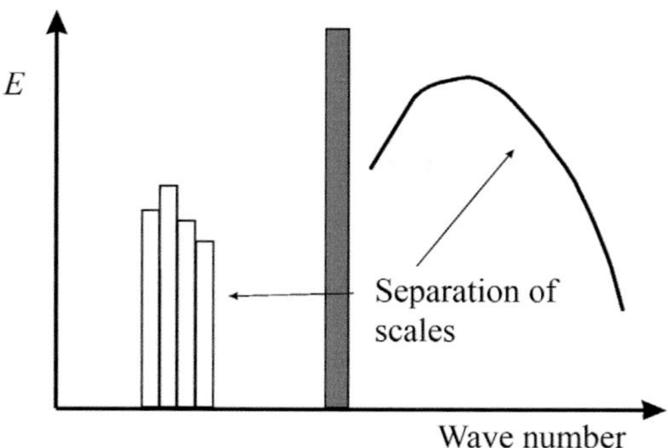

Figure 5.29. Requirement of spectral gap with URANS.

a Helmholtz decomposition this can be expressed as scalar and vector potentials:

$$\nabla \cdot \mathbf{R} = \nabla\phi + \nabla \times \psi \tag{5.188}$$

These, in turn can be directly expressed as simple Poisson equation based functions with Reynolds stress based source terms as

$$\begin{aligned} \nabla^2\phi &= \nabla \cdot (\nabla \cdot \mathbf{R}) \\ -\nabla^2\psi &= \nabla \times (\nabla \cdot \mathbf{R}) \end{aligned} \tag{5.189}$$

These source terms are directly analogous to those for a Reynolds stress model. The scalar and vector potentials are analogous to average pressure drop in turbulent vortices and their vorticity magnitude. In three dimensions, five transport equations need to be solved – the usual k and ε equations along with similar transport equations for ϕ and ψ.

5.4 Unsteady and Eddy Resolving Simulations

5.4.1 *URANS, Very Large Eddy Simulation and Disturbance Equations*

Next we will move through the hierarchy of methods presented in Figure 5.9 to look at unsteady methods. The first to be dealt with is Unsteady RANS (URANS). In practical terms the RANS equations do not change. Just a time derivate is added. Moreover the question is what the theoretical implications of this are. The RANS equation derivation assumed that an averaging took place over a steady base state. If this base state is time varying, say, as the result of the passage of a rotor, then the question arises if the averaging is theoretically valid. Indeed, if the time period of the passage of the rotor becomes close to that of the turbulent time scales, it is not mathematically possible to formulate an average. Hence, for URANS to be valid, some form of spectral gap is necessary. This concept is shown in the Figure 5.29 schematic.

The difficulty is ensuring that this gap exists a priori, and this is not possible. With a k-ε solution, for example, since the time scale is characterized by $C_\mu k/\varepsilon$ (for SA,

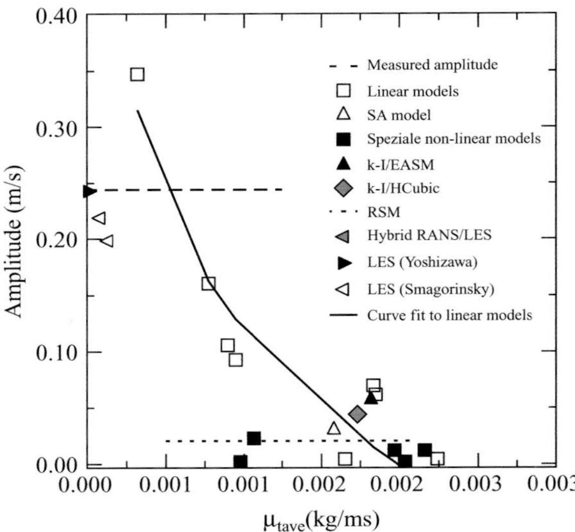

Figure 5.30. Plot of resolved unsteadiness amplitude against average system eddy viscosity (from Liu and Tucker 2007 – published with the kind permission of Wiley).

the inverse of the strain rate can be used), it is possible to make an estimate of the modelled time scales when a simulation has been made. This can then be compared with the resolved time scales. However, if a spectral gap does not exist, then the situation is problematic. Another alternative is to see if there is a separation in a spatial sense – such simulations could then be termed Very Large Eddy Simulations (VLES). Hence, it can be examined if the implied modelled length scale ($\sim k^{3/2}/\varepsilon$, $k^{3/2}/(C_\mu \omega)$) is much smaller than any unsteady structures resolved in the unsteady simulation. However, again this is an a posteriori process.

A critical difficulty with URANS is that the RANS model should be designed to capture all the turbulence. Hence, if a URANS is being made for, say, a bluff body flow, and wake eddies are captured, the solution would be theoretically invalid. A double accounting is taking place with the turbulence being accounted for by both the turbulence model and the eddies being captured in the URANS. Tucker (2013a) explores are range of gas turbine aeroengine flows and using a scaling analysis studies the presence of spectral gaps. This work shows them in many circumstances to be tenuous.

Reverse spectral gaps can also occur. For example, a fast rotating fan blade can be subjected to large scale turbulent eddies. The latter can be generated from separation on an intake duct or atmospheric turbulence. Then, the spectral gap given in Figure 5.29 can be reversed. Under these circumstances, we enter the realm of rapid eddy distortion, where the turbulent time scales are long relative to the external unsteady input that they are being subjected to. Again, this is a regime that is challenging for turbulence models.

Figure 5.30 illustrates another URANS problem that essentially reveals an underlying problem with RANS. The plot gives the resolved unsteadiness amplitude against the average system eddy viscosity for a complex low Reynolds number geometry. The full line is a curve fit to results for a range of RANS models. These include linear and zonal non-linear eddy viscosity models and also a Reynolds stress model simulation. Some eddy-resolving simulation results, using methods to be discussed later, are also included in the figure. The models are detailed in Liu and Tucker (2007). The horizontal long dashed line gives the measured coherent unsteadiness amplitude. The main point to observe from the plot is that the underlying RANS models give very different levels of implied eddy viscosity. The predicted unsteadiness is sensitive to this. Hence, with URANS, a wide range of unsteadiness amplitudes can be predicted depending on which RANS model a user chooses. Notably, the eddy-resolving simulations show much less scatter and are in accord with the measurement.

Eddy viscosity models, unlike the Reynolds stress model, cannot account for some of the effects of stress–strain misalignment in unsteady flows. This can result in them predicting too much turbulent viscosity in separated wake regions. To overcome this, Revell et al. (2011) present an approach that avoids the expense of solving a full Reynolds stress model. Instead, it solves a transport equation for the dot product of S_{ij} and the turbulence anisotropy tensor given by Equation (5.72). This characterizes the lag between the stress and strain field responses. Revell and colleagues found improved agreement with measurements for flow over a NACA0012 airfoil at a high angle of attack.

5.4.1.1 Nonlinear Disturbance Equations (NLDE) Despite these theoretical questions, for quite a few flows there is sound physical evidence for the existence of large-scale turbulent structures that seem distinct from smaller scale structures. This is particularly so for jets. Hence Batten et al. (2007), and others, proposed solving for a non-linear disturbance about a mean RANS solution. Typically, a two-dimensional RANS would be made and then the mesh and solution extruded to three-dimensions. Terracol et al. (2005) apply this approach to a three-element, high-lift wing, the key non-linear disturbance zone being located in the slat cove-wing, leading edge vicinity. The flow in this zone has erratic vortex shedding. Notably, Batten et al. (2002, 2004)) use a stochastic forcing to replicate the input of small-scale turbulence. This produces flows that appear rich in scales and akin to LES. The non-linear disturbance equation can be solved on a much coarser grid than needed for normal eddy-resolving simulations. This is because the approach has a much higher modelled content. A disadvantage of the non-linear disturbance equation approach is that for a truly unsteady flow, where the approach is most attractive, a steady RANS solution is unlikely to be possible to iteratively converge. If a strong solver is used, to secure convergence, then physical realism can be lost. Also, the extrusion from two-dimensional to three-dimensional geometry reduces the class of problems for which the approach can be used.

Bastin et al. (1997) apply a non-linear disturbance equation related decomposition to jet noise. However, in essence a more basic URANS approach is taken. The eddy viscosity coefficient (C_μ – directly scaling eddy viscosity) in a k-ε model

Table 5.11. *Potential LES filter functions expressed in physical and Fourier space (from Tucker 2013b – used with the kind permission of Springer)*

Filter kernel G	Physical space	Spectral space
Box filter	$\frac{1}{\Delta}\text{if } \lvert\mathbf{x}-\mathbf{r}\rvert < \frac{\Delta}{2} \text{ else } 0$	$\frac{\sin\left(\frac{1}{2}w\Delta\right)}{\frac{1}{2}k\Delta}$
Gaussian filter	$\left(\frac{6}{\pi\Delta^2}\right)\exp\left(-\frac{6(\mathbf{x}-\mathbf{r})^2}{\Delta^2}\right)$	$\exp\left(-\frac{w^2\Delta^2}{24}\right)$
Sharp spectral	$\frac{\sin\left(\pi\left(\mathbf{x}-\mathbf{r}\right)/\Delta\right)}{\pi\left(\mathbf{x}-\mathbf{r}\right)}$	$1 \text{ if } \lvert w\rvert \leq w_c \text{ else } 0$

is reduced, allowing the solution to support larger coherent structures. Reau and Tumin (2002) outline a more theoretically rigorous procedure. Disturbance equations for the coherent scales are specifically formulated, closure being achieved through Boussinesq type modelling terms, as used in RANS. Further related techniques are the Coherent Vortex Simulation (CVS – see Farge and Schneider 2001) and semi-deterministic method (SDM) of Ha Minh (reported in Druault et al. 2004). These both isolate coherent structures in differing ways. Farge and Schneider use wavelets to enforce the scale separation considering a mixing layer and other flows.

5.4.2 *Hybrid RANS-LES*

The procedure of deriving the RANS equations from the Navier-Stokes was presented previously and based on a time averaging process. The formal derivation of the LES equations is very different. It is generally based on a spatial averaging or filtering. However, temporal filtering is possible (see Pruett 2000). Also space-time filtering can be formulated (see Dakhoul and Bedford 1986a, 1986b). The standard spatial filtering can be expressed by replacing the averaging Equation (5.4) for RANS with the following:

$$\Phi(\mathbf{x}) = \int_{-\infty}^{\infty} \phi\,(r)\,G\,(\mathbf{x}-\mathbf{r})d\mathbf{r} \tag{5.190}$$

The so-called filter kernel, G, can take various forms. Table 5.11 gives potential filter functions expressed in physical and Fourier space, where w is wave number, w_c its cutoff and Δ is the filter width. Note that discrete/numerical filters can also be formulated (see Brandt 2008). Equation (5.190) can be written as the following convolution operation in a shorthand notation (see Sagaut et al. 2007):

$$\Phi = G * \phi \tag{5.191}$$

Using this notation, it can be seen that the fluctuation field can be recovered as $\phi' = (1-G) * \phi$, and this process can be used to develop actual LES models.

In essence the final LES equation is the same as for RANS. This makes the interfacing of the equations deceptively simple for hybrid RANS-LES approaches.

However, the extreme differences in the physics modelled by these equations gives rise to problems and these will be explored later.

Speziale (1998) proposed an approach where if a grid were fine enough, the RANS model would switch off and a DNS would be performed. On the other hand, for a coarse grid RANS would be recovered. Along similar lines, Batten et al. (2002) devised the LNS (Limited Numerical Scales) method. With this, the eddy viscosity takes the following form

$$\mu_T = \alpha \mu_{RANS} \tag{5.192}$$

where μ_{RANS} is the eddy viscosity that the base RANS model would provide prior to multiplication by α. The variable, α, is called the latency parameter. This is defined as

$$\alpha = \frac{\min[(LV)_{LES,} (LV)_{RANS}]}{(LV)_{RANS}} \tag{5.193}$$

where $(LV)_{LES}$ is the product of the LES velocity and length scales. Also, for RANS, $(LV)_{RANS} = \delta + C_\mu k^2/\varepsilon$. Note that δ is a small number. As can be seen, this is formulated for a k-ε model but this is not a rigid element. The model can either switch to a fully RANS or fully LES mode or some state between these two extremes. This depends on the level of turbulence resolution. An eddy viscosity transport equation based model, of a broadly similar genre, is proposed by Menter et al. (2003). This model makes use of the von Karman turbulence length scale. It is called the Scale Adaptive Simulation (SAS) model. Further related approaches are PANS (Partially Averaged Navier-Stokes – see Girimaji 2006), and the partially integrated transport model (see Chaouat and Schiestel 2005).

Approaches of this nature can yield impressive results for the right cases but need to be used with care since they are not without difficulties. For example, as shown in Figure 5.31, distinct islands of LES and RANS can occur. If RANS islands are upstream of LES, the LES zones will initially have suppressed resolved scales effectively having inappropriate inflow conditions. Similarly, a RANS zone downstream of an LES will initially and erroneously receive resolved turbulence. The interface between RANS and LES is, in hybrid RANS-LES approaches, often called the *grey area*. This is because, in this zone, a range of theoretically vexing questions arise. These have given rise to a multitude of studies that are not outlined here. However, suffice it to say that the global approaches described give rise to a multiplicity of grey zones. Further difficulties with SAS are initial value and mesh sensitivities. However, having said this, for the right flow the method can be highly effective and in certain instances palliatives can be formulated (see Batten et al. 2002).

The obvious attraction with the LNS and related approaches is that they adjust to a given grid attempting to give the most sensible turbulence modelling strategy for a given level of resolution. However, the drawbacks discussed have given rise to approaches with more rigid zonalization between the RANS and LES. Typically, the zonalization is such that just near wall area is RANS modelled.

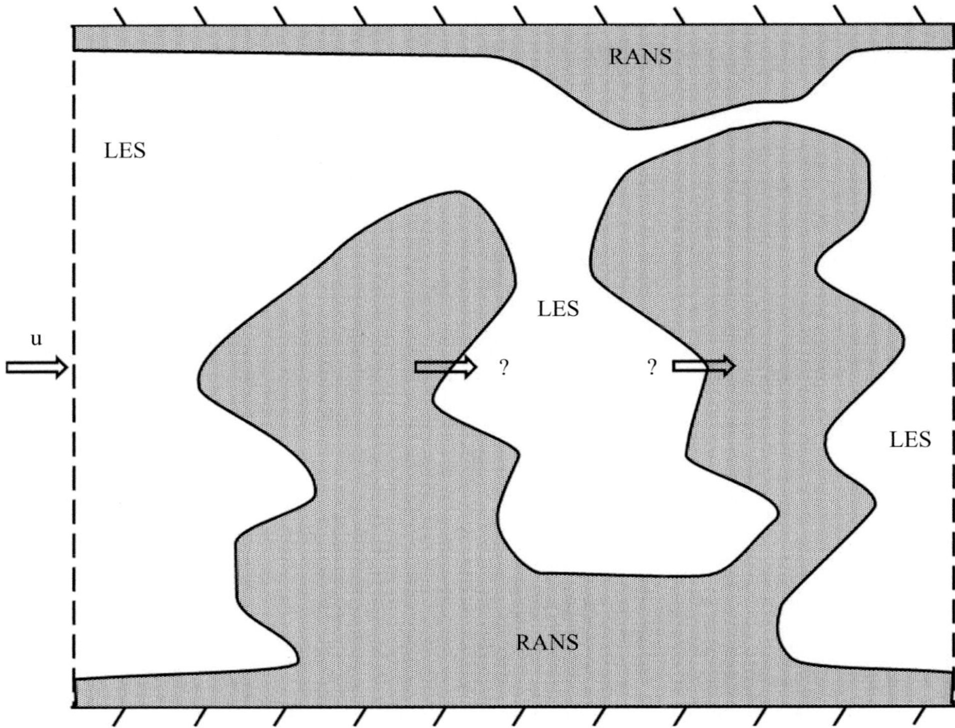

Figure 5.31. Multiplicity of grey zones arising with global hybrid RANS-LES approaches.

5.4.2.1 DES Methods The most widely used approached of this nature is Detached Eddy Simulation. With this, the idea is that the whole boundary layer is treated in RANS mode. Outside this zone LES is used. The initial DES approach, termed here DES97 (see Spalart and Shur 1997), is based around the SA, RANS model. With DES97 the turbulence destruction uses a modified wall distance given by the function

$$\tilde{d} = \min(d, C_{DES}\Delta) \tag{5.194}$$

The constant $C_{DES} = 0.65$ and $\Delta = \max(\Delta x, \Delta y, \Delta z)$. Since it is Δ which controls the RANS-LES interface (RANS is activated $d < C_{DES}\Delta$), the grid needs careful design. Such aspects are dicussed in Chapter 3, where guidance with respect to this is given. Simple scaling analysis shows that for turbulence in equilibrium and away from walls, DES97 reduces $v_t \approx \Delta^2 S$ corresponding to the basic crude Smagorinsky LES model to be discussed later. As noted by Deck et al. (2011), in can be expedient to make the following modifications in the LES zone: $f_{v1} = 1$, $f_{v2} = 0$, $f_w = 1$. These changes prevent the low eddy viscosity found in the LES zone being mistaken for wall proximity. They can also, helpfully, accelerate LES content in grey zones.

If the boundary layer grid has excessive local resolution, the LES zone can make an incursion into the boundary layer. Hence, a further modficiation proposed by

Spalart et al. (2006) is the shielding function

$$r_d = \frac{\upsilon_t + \upsilon}{\kappa^2 d^2 \sqrt{U_{i,j} U_{i,j}}} \tag{5.195}$$

The tensor in the previous may be written more explicitly as

$$\left[\sum_{i,j} \left(\frac{\partial U_i}{\partial x_j} \right)^2 \right]^{1/2} \tag{5.196}$$

The function r_d takes a value of unity in the log layer. It tends to zero towards the boundary layer edge. r_d is used in the further function

$$f_d = 1 - \tanh[(8r_d)^3] \tag{5.197}$$

In the boundary layers, $f_d = 0$ and outside it $f_d = 1$. The DES97 length scale is then re-expressed as the new DDES scale

$$\tilde{d} = d - f_d \max(0, d - C_{DES}\Delta) \tag{5.198}$$

This function is intended to facilitate the boundary layer shielding described previously.

5.4.2.2 *Menter SST Based DES* Most RANS models can be expressed in some form of DES framework. For example, when using a k-ω model, the destruction term in Equation (5.21), formulated for k needs T_2 to be modified as

$$T_2 = \frac{k^{3/2}}{l_{DES}} \tag{5.199}$$

where (see Strelets 2001)

$$l_{DES} = \min(l_{k-\omega}, C_{DES}\Delta) \tag{5.200}$$

This length scale equation ranges between the RANS ($k^{1/2}$ $(\beta^* \omega)$ in the standard k-ω modelling nomenclature) and LES limits. For the DDES equivalent to SA, the following is needed:

$$l_{DDES} = \min\left[\frac{l_{DES}}{(1 - F_{SST})}, C_{DES}\Delta \right] \tag{5.201}$$

The function F_{SST} takes the form of F_2 (see Menter 1993) in the SST model. This function has the opposite traits to f_d taking a value of zero outside the boundary layer, tending to unity inside it.

5.4.2.3 *DES Performance for Separated Flow* A key problem for hybrid RANS-LES methods is that their accuracy can, for certain flows, be strongly controlled by the performance of the RANS model. Figure 5.32 shows flows that are of an increasing challenge to hybrid RANS-LES methods. The first frame (Case I) shows a flow where the separation is well defined. The axial pressure gradient is in fact a

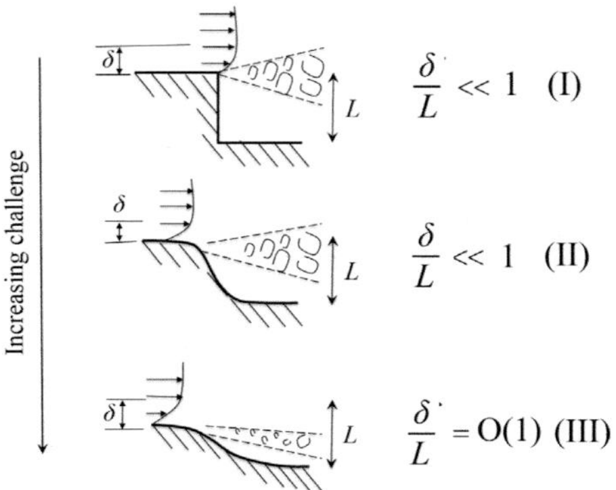

Figure 5.32. Schematics of cases involving separation that are of increasing challenge (from Frames I–III) to hybrid RANS-LES techniques (based on Deck et al. (2011) – published with the kind permission of Springer).

singularity at the corner and tends to an infinite value. Hence, the separation point is fixed by the geometry.

Case II is similar to Case I in that the separation is largely defined by the pressure gradient and so is again benign. However, for Case III, the dynamics of the incoming boundary layer are much more critical to capturing the correct separation point. If this is wrong, it can have a very substantial global solution impact for certain types of flow.

5.4.2.4 Explicitly Zonalized Methods With DES, the interface location is grid driven. There have been other methods that attempt more explicit zonalizations. For example, with a shear layer type flow, Georgiadis et al. (2003) imposed a RANS zone in the largely steady flow area upstream of the shear layer. Then, downstream of the shear layer, LES is applied. Davidson and Peng (2003) defines a RANS near wall zone for $y^+ < 60$ and LES elsewhere. A similar procedure is used by Tucker (2004), but with the LES zone replaced by ILES (see Sections 5.4.3.4 and 4.4.3.3). Keeping a RANS-LES interface at a constant location means that the skin friction more sensibly converges with grid refinement.

Deck et al. (2011) propose a multi-zone hybrid RANS-LES approach. This is based around a structured multi-block solver/mesh framework. Marker zone variables are used to define RANS, DES and LES zones. For example, the RANS-LES zones can be naturally defined as in the work above with an upstream RANS zone or a near wall RANS zone based on y^+. Also, to deal with Case III (see Figure 5.32), a locally embedded wall resolved LES could be defined. This would better ensure the separation point is correctly captured and not subjected to the vagaries of RANS modelling. On top of this zonalization, different filter scales can be defined that are most appropriate to the local flow (see Table 5.13). The strategy needs the user to engage with the flow physics and grid topologies found in different zones but seems sensible. It is consistent with the practical realities of using hybrid RANS-LES methods.

Dual grid methods have been proposed by Piomelli and Balaras (2002) and Leschziner et al. (2009). The grid topologies are designed to be optimal for the differing zones (i.e., high aspect ratio cells in the RANS zone and isotropic in the LES). Hence, the implied grid structure is Chimera-like. Notably, in the RANS zone, in the approach of Piomelli and Balaras, the parabolized RANS equations are solved. A simple mixing length closure, in a boundary layer form is used. A key assumption of this method is that the coupling between the inner and outer parts of the boundary layer is weak but this is questionable (see Leschziner et al. 2009). As noted in Chapter 3, a range of workers have used grid embedding. However, this is really an LES technique with optimal grids.

5.4.2.5 Hybrid RANS-LES Deficiencies With DES type methods, the RANS-LES interface is grid-controlled and hence for complex grid topologies the interface can become a highly non-uniform. This is again so if y^+ is fixed, and what is more, the interface will also be time varying, unless the shear stress is based on initial time invariant RANS estimates. The other alternative is to fix the interface based on spatially averaged wall shear stress values. For the highly complex geometry considered by Liu and Tucker (2007), either approach made little difference to the engineering quantities of interest. As noted previously, fixing the y^+ value of the interface allows sensible wall shear stress convergence with grid refinement.

As also noted previously, the interface between the RANS and LES zones is known as the grey area. This term mostly refers to the wall parallel zone between a near wall RANS layer and the LES region, as one moves in the wall normal direction. However, mildly separated flows present a more complex form of grey area. For these, the upstream local injection of synthetic turbulence is ideally required, as is the careful choice of an LES filter scale and other measures (see Deck et al. 2011) that are conducive to the rapid development of resolved turbulence. The issue with a wall parallel interface in a classical boundary layer grey area is that in the RANS layer, there is a large eddy viscosity. As the wall is moved away from, the RANS layer gives way to the LES zone. In moving between these zones, there is a sudden loss of eddy viscosity. However, in moving from the RANS to LES zone, there is insufficient opportunity for resolved scales to develop (this can also be partly delayed by eddy viscosity 'leaking out' from the RANS zone). This results in a potential dramatic loss of wall shear stress, unless the mean velocity profile makes a strong adjustment. This gives rise to a kink in the velocity profile and a substantial under-prediction of skin friction coefficient (Nikitin et al. 2000). To more rapidly energize resolved scales, Piomelli et al. (2003), Davidson and Billson (2006) and Davidson (2009) inject synthesized turbulence at the RANS-LES interface.

An opposite issue is the excessive energization of the RANS zone. The resolved scales in the LES buffet the RANS layer. Consequently, although the expectation is that the RANS zone is steady, it actually becomes a URANS zone. The result of the buffeting is that when the resolved and modelled turbulence energies are added, the RANS zone shows an excess of turbulence energy. A range of attempts have been made to get around these grey area issues – see Zhong and Tucker (2004), Hamba

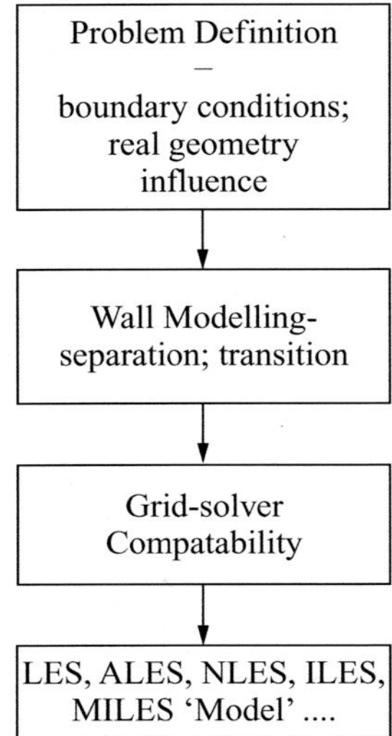

Figure 5.33. LES hierarchy.

(2006), Temmerman and Leschziner (2002) and Gieseking et al. (2011), to name but a few. These have shown some promise but only in a preliminary and ad hoc sense. Next, we move down the Figure 5.9 hierarchy to consider pure LES. As would be expected, all the issues discussed for LES are also pertinent to hybrid RANS-LES.

5.4.3 *Large Eddy Simulation*

Obviously, the aspects of importance for LES are very different from those for RANS. With RANS, solutions have 100% modelled content. Hence the process of model calibration is of extreme importance. The process is postdictive. With LES, the intention is that 90% of the turbulence is resolved on the grid. Hence with LES, on sufficiently fine grids, we move closer to the area of predictive CFD. This means that LES has different modelling priorities than RANS. Figure 5.33 attempts to explore these priorities by means of a hierarchy of elements that are important to successful LES. As would be expected, the modelling is now at the bottom of the hierarchy.

At the top of the Figure 5.33 hierarchy is problem definition. This is referring to the type of boundary conditions used and the extent and level of geometry considered. For example, with LES, small geometrical features such as fillets, welds, surface finish and rivets can have a very clear solution impact. This is quite different from RANS. Also, the insensitivity of RANS models to turbulence properties at inflow boundaries is generally regarded as a positive trait. Whether this is so or not is highly debatable. For many practical flows, especially transitional and a low Reynolds number, the state of the incoming turbulence strongly defines the downstream flow. Hence, since LES is predictive, it can be very important to correctly and accurately characterize the turbulence at the inflow.

The next aspect of the hierarchy is wall modelling. As noted in the discussion at the start of this chapter, for a wall resolved LES (LES where no special wall treatments are needed), the grid requirements at high Reynolds numbers are severe. Hence, even with LES near wall treatments can arise and these will be discussed further later.

Clearly since LES is generally attempting to faithfully resolve approximately 90% of the highly complex turbulence structures, this places much stronger demands on the numerical schemes than for RANS simulations. However, as shown in Chapter 3, the mesh and the numerical scheme interact strongly. Hence care needs to be taken to ensure that the grid topology is compatible with the numerical schemes used. Clearly also the underlying numerical scheme used in the LES must be adequate. Finally, in the hierarchy there is the modelled content supplied. Typically, this is provided by an explicit LES sub-grid scale (SGS) model.

Next we will work through the hierarchy, discussing the aspects noted in more detail.

5.4.3.1 Problem Definition With LES, as noted, the precise form of surfaces can have a strong solution impact. For example, Rao et al. (2014) show the impact of modelling surface roughness over a low-pressure turbine blade. Also, Tucker (2013b) shows the impact of surface vibration on a low-pressure turbine related configuration. Similarly, Raverdy et al. (2003) show the influence of acoustic waves, supported by LES, on a transitional shear layer. Depending on the domain extent and hence problem definition, these acoustic waves could have differing solution impacts. Hence, relative to RANS the problem definition is more of a critical issue. This is complicated by the fact that engineering simulations have an increasingly coupled and global nature, but LES grid requirements push the analysis towards a more restricted domain. Inflow is also a critical factor and this is discussed next – however, challenges with regards to outflow boundary conditions, as discussed in Chapter 4, should not be neglected. The range of inflow boundary conditions is extensive and reviews can be found by Keating et al. (2004) and Tabor and Baba-Ahmadi (2010). Only the briefest discussion is given here.

Fortunately, for jets and many flows involving detached shear layers, there can be strong shearing. Hence, the highly unstable and inflectional velocity profiles will rapidly generate resolved velocity scales. Then, turbulence inflow becomes less critical. There are numerous methods for generating inflow turbulence. A relatively convenient approach is to have an extended upstream domain, with some form of numerical trip (see Mullenix et al. 2011 and Bisek et al. 2013). Obviously, the downside of such approaches is the increased computational cost arising from the need for an extended upstream domain.

5.4.3.1.1 Precursor Simulations Precursor simulations can be made and this data stored in a database. The data can undergo some scaling to make the turbulence more compatible with the desired inflow field. Alternatively, a simulation can be run concurrently.

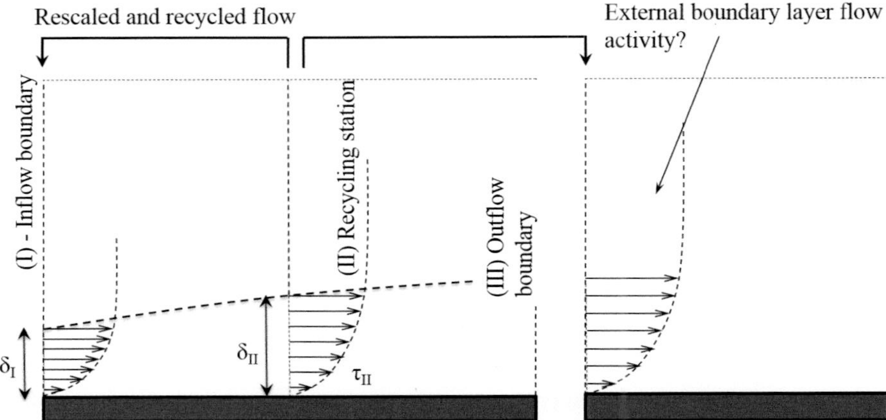

Figure 5.34. Schematic of Lund's recycling procedure (from Tucker 2013b – published with the kind permission of Springer).

A key method for generating turbulence inflow is through use of re-cycling procedures. This can involve simple pipe or channel flow LES/DNS with periodic flow boundary conditions in the streamwise direction, giving rise to fully developed boundary layers. RANS information can be used to set the mean velocity and turbulence levels. The recycled simulation, from a stored database, can then be used to convert this into unsteady inflow information. The following expression is used for this purpose:

$$\overline{u}_{i,LES}(t) = \underbrace{U_{i,RANS}}_{I} + \underbrace{\left[u_{i,DB}(t) - U_{i,DB}\right]}_{II} \cdot \underbrace{\frac{\sqrt{\overline{\overline{u'^2_{i,RANS}}}}}{\sqrt{\overline{\overline{u'^2_{i,DB}}}}}}_{III} \tag{5.202}$$

where the subscript *RANS* identifies RANS data and *DB* data from the database. The double bars represent an ensemble average of the LES. Term I gives the mean RANS velocity profile data. Term II is the difference between the instantaneous and mean velocities at a point taken from the database. Term III, effectively, scales the fluctuations from the database so that they are consistent with the RANS target.

The most popular general procedure, allowing for developing boundary layers, is due to Lund et al. (1998). Figure 5.34 gives a schematic of Lund's recycling procedure. As shown in Figure 5.34, there is an inflow boundary (I) (where a target momentum thickness is specified), a recycling station (II) and an outflow boundary (III). Data is taken from Station II and recycled into the inflow plane. Since the boundary layer is developing, basic boundary layer scaling laws need to be applied. These are based on the expected inner and outer layer boundary layer dynamics. The resulting unsteady data can either be stored or the recycling and main flow simulations carried out together.

For many practical flows, there can be classical boundary layer content and also flow activity external to the boundary layer. This could include wakes from turbine blades along with inter-wake turbulence. When Lund's recycling is used, these two, coupled elements would need to be blended and scaled in some plausible way – see, for example, Wu (2010).

5.4.3.1.2 Synthetic Turbulence Synthetic turbulence can be generated based on RANS information or even measurements. One of the earliest approaches of this type was as developed by Kraichnan (1970). This was for homogenous isotropic turbulence and extended to anisotropic turbulence by Smirnov et al. (2001). Batten et al. (2004) present a simplified version of Smirnov and colleagues' approach. With this, the velocity field is constructed using the following equation:

$$u_i(x_j, t) = a_{ik}\sqrt{\frac{2}{N}}\sum_{n=1}^{N}\left[p_k^n \cos\left(\hat{d}_j^k \hat{x}_j + \omega^n \hat{t}\right) + q_k^n \sin(\hat{d}_j^k \hat{x}_j + \omega^n \hat{t})\right] \quad (5.203)$$

where $\hat{x}_j = 2\pi x_j/L, \hat{t} = 2\pi t/\tau, \hat{d}_j^n = d_j^n V/c^n, V = \frac{L}{\tau}, c^n = \sqrt{\frac{3}{2}\overline{u'_l u'_m}\, d_l^n d_m^n / d_k^n d_k^n}, p_i^n = \varepsilon_{ijk}\eta_j^n d_k^n,\ q_i^n = \varepsilon_{ijk}\xi_j^n d_k^n,\ \eta_i^n, \xi_i^n = N(0,1),\ \omega^n = N(1,1),\ d_i^n = N(0, 1/2)$, and a_{ij} is given by

$$a_{ij} = \begin{pmatrix} \sqrt{\overline{u'_1 u'_1}} & 0 & 0 \\ \overline{u'_1 u'_2}/a_{11} & \sqrt{\overline{u'_2 u'_2} - a_{21}^2} & 0 \\ \overline{u'_1 u'_3}/a_{11} & (\overline{u'_2 u'_3} - a_{21}a_{31})/a_{11} & \sqrt{\overline{u'_3 u'_3} - a_{31}^2 - a_{32}^2} \end{pmatrix} \quad (5.204)$$

The elements are real if the RANS generated Reynolds stresses are realizable. The time and distance components in the Fourier space model are scaled by a time scale, τ, and a velocity scale c^n. In the previous, $N(a,b)$ are normally distributed random numbers with a mean of a and standard distribution of b. Also, ε_{ijk} is the alternating symbol. The problem with synthetically generated turbulence is that it can exhibit an initial decay. This could be partly because of the fidelity of the generated structures but also due to numerical errors, grid resolution and also LES model influences. To overcome this problem, some workers have used downstream control zones where forcing is imposed (see Keating et al. 2004 and Laraufie et al. 2011).

The digital filtering approach of Klein et al. (2003) is also a popular method for generating turbulence inflow. With this, a random data series about a mean of zero is generated. This is then subjected to the digital filter. The filter coefficients can be generated based on a correlation function for the turbulence field that absorbs a modelled turbulence length scale. Proper Orthogonal Decomposition (PODs) can be used, for example, to take time resolved experimental measurements, such as particle image velocimetry, and impose them at turbulence inflow boundaries (see Perret et al. 2008 and Druault et al. 2004) with minimal data storage. In the work of Perret and colleagues, the PODs are just used to capture the low order modes. The smaller scales are imposed using Gaussian random numbers, adjusted to give a correct turbulence energy spectrum.

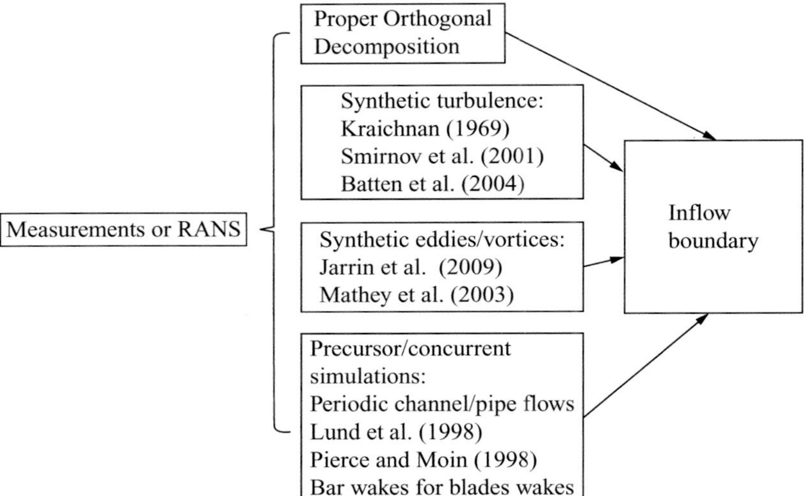

Figure 5.35. Potential inflow options for idealized turbulence (from Tucker 2013b – published with the kind permission of Springer).

5.4.3.1.3 Advantages and Disadvantages The high Reynolds numbers found in real engineering systems place strong limitations on Lund's recycling. With respect to cost and user ease, synthetic turbulence generally seems the best option. However, for simple boundary layers, synthetic turbulence will be less authentic than recycled. Therefore, a longer development zone will generally be needed. Control plane forcing can be used to reduce the extent of the initial turbulence adjustment period. However, this adds another user complication.

Figure 5.35 summarizes the potential inflow options noted previously. Some key elements relating to the relative performance of different inflow techniques are: how long a development length is needed to establish turbulence; how long (for recycling methods) is needed to overcome initial transients and thus establish a developed turbulent state; how practically useable is the method; how Reynolds number limited is the method and how well does the output connect with real engineering systems.

Many of the more idealized inflow approaches, noted previously, in their raw form, have minimal relevance to complex engineering problems. For example, it might be tempting to consider applying Lund's recycling to jet nozzle flows. For the replication laboratory experiments, this has some value. However, with the nozzle used by Bridges and Wernet (2003), for example, the contraction is substantial. This geometry, which is not untypical, will give rise to partial flow relaminarization. Also, as outlined in Eastwood (2010), in a real engine, upstream of the nozzle, there is a compressor, combustor and turbine. There are also numerous other associated geometrical features such as an A-frame, the gearbox shaft, leak flows, temperature and pressure probes, bolts, eccentricities from thermal expansion, ill-fitting components and fastenings. It has been supposed for some time that these may influence the jet noise itself (Moore 1977). Similar issues occur in many other places inside a gas turbine aeroengine, where, as discussed in Chapter 6, recent evidence is emerging

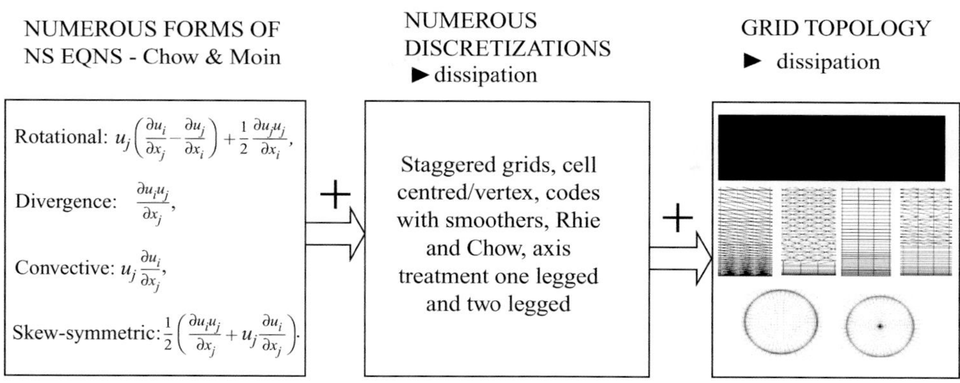

+ turbulence model validity + wall modelling + problem definition + soln uniquence + ...= Do not get too focused on LES model

Figure 5.36. Ultimate numerical influences for LES (from Tucker 2008 – published with kind permission of Elsevier).

of high degrees of coupling between both upstream and downstream flow zones. Hence, there is a strong problem definition aspect relating to turbulence inflow and also outflow boundary conditions.

5.4.3.2 Wall Modelling With a wall resolved LES grid, demands tend to that for a DNS. Hence, this has given rise to a need to alleviate this extreme near wall grid demand. There are essentially two approaches to get around this. Both are essentially RANS based. One approach, which was used in the earliest historical LES (Deardorff 1970 and Schumann 1975), was to use wall functions, in a range of formulations that will not be outlined here. The other is to essentially use a mixing length model near the wall. This is sometimes called the Lilly correction. With this, the usual modelled turbulence length scale, in the Smagorinsky model, $l = C_s\Delta$, is modified to

$$l = \min(kd, C_s\Delta) \tag{5.205}$$

As can be seen, this is in fact close to DES but based around a mixing length model. The key difference is that the application of the previous equation to LES would require a much finer grid. However, an attraction of Equation (5.205) is that for a coarse grid it does tend to a plausible limit that may well still give meaningful CFD results. Clearly Equation (5.205) is not suitable when complex surface topologies of small scale are being considered. Then wall resolved LES tending to quasi-DNS would be needed.

5.4.3.3 Grid-Solver Compatibility Figure 5.36 focuses on the grid-solver compatibility aspect. With regards to numerical contamination, as noted by Chow and Moin (2003), different forms of the Navier-Stokes equations can be chosen (and discretized identically), ultimately giving different numerical traits. A few forms of the

Navier-Stokes equations are given in the left-hand frame of Figure 5.36 and are outlined in Chapter 2.

Once we have decided on the form of the equation we wish to discretize, we then go ahead and discretize it. There are a vast range of different temporal and spatial discretizations. These are discussed in Chapter 4, but to summarize we can have staggered grids; collocated, cell centred or vertex codes; codes with explicit smoothers; Rhie and Chow (1983) interpolation; different mesh singularity treatments; codes that focus on the Lagrangian rather than Eulerian frames or a mix of the two and different inlet/outlet boundary condition treatments. The list is extensive. With jets, for example, far field boundary conditions can play a strong role and there is even emerging evidence of meta-stability of the flow to inflow conditions. Temporal discretizations, like spatial, are numerous and, as with spatial, even nominally the 'same scheme' can be discretized differently (i.e., for example, one-legged and two-legged discretization forms). These terms draw a distinction between whether the interpolation function is applied to the function of the variable or the actual variable being solved for. The discretizations, which tend to migrate to having dissipative elements, must then be represented on a grid, the non-alignment to the flow and skewing in which also tends to produce further dissipation. These aspects, when combined with an LES model, then tend to frequently produce an excess of numerical dissipation.

5.4.3.4 LES Models As with RANS, there are a wide range of LES models emerging with time. They can be broadly classified as functional and structural. The former are quite reminiscent of phenomenological RANS models. They try to capture the turbulence energy spectrum and hence the dissipation of turbulence at high wave numbers through dissipation induced by use of an eddy viscosity μ_{SGS}.

The Smagorinsky (1963) sub-grid scale model is the most popular functional model. It is basically a mixing length model, the von Karman constant being replaced by Smagorinsky's constant, C_s (0.1–0.2). Hence, the model equation is

$$\mu_{SGS} = \rho (C_s \Delta)^2 S \tag{5.206}$$

where S is based on the filtered velocity. The standard constant is derived by assuming an idealized energy spectrum with a sharp cut-off filter to give $C_s \cong 0.18$. However, for a shear flow a lower value is typically needed (see Deardorff 1970). Indeed, the required value will be heavily dependent on the filter choice (see Table 5.13). Hence, there is a large published range for C_s. The Smagorinsky model does not have the correct near wall asymptotic behaviour and this has given rise to the WALE (wall-adapting local eddy viscosity) model of Nicoud and Ducros (1999). The Germano model is sometimes known as the dynamic Smagorinsky model. Essentially it is the Smagorinsky model but with C_s computed based on a double filtering procedure. Note, however, that this double filtering is not restricted to the Smagorinsky model and can be combined with functional, structural and mixed (see later discussion relating to Equation 5.207) models. Cited advantages of the Germano model are that no modifications are needed to the chosen base LES model as the wall is moved towards. Also, if the flow is sufficiently resolved, the scheme automatically reverts to DNS and it can be formulated to naturally compensate for numerical errors. However, the numerous averaging procedures to deal with the potentially extreme values of computed C_s are likely to influence solutions for transitional flow. Procedures to deal with the extreme C_s values and their impact on solution stability, include averaging in homogenous flow directions, local patch averaging, damping (see Breuer and Rodi 1994), averaging along streamlines, averaging in time and clipping.

The Variational Multiscale (VMS) method was first introduced by Hughes et al. (2000) in a variational/finite element type framework. With this approach, three scales are considered. These are the large, small (resolved) and modelled scales, ϕ'. In order to split the total resolved scales into large and small components again, like the Germano model, a test filter is needed. The method is relatively easy to implement and can be based around the Smagorinsky model with a top hat filter in a finite difference or finite volume framework. For wall-bounded flows, the approach has shown considerable improvements over the standard Smagorinsky model.

Amongst other things the Smagorinsky model assumes locality of turbulence production and destruction. The locality aspect can be remedied using the Yoshizawa (1993) model. This solves a transport equation for sub-grid scale turbulent kinetic energy having the form of Equation (5.21), when used in the RANS *k-l* framework but with $\varepsilon = C_\varepsilon {k_{SGS}}^{3/2}/\Delta$, $\mu_t = \rho C_\mu \Delta {k_{SGS}}^{1/2}$.

Whereas the functional models have a more phenomenological outlook, the structural have a more mathematical basis. With them, the sub-grid scale tensor $\tau_{ij} = \overline{u_i u_j} - \overline{u}_i \overline{u}_j$ is evaluated based on approximations (identified by * superscripts) to the unfiltered solution u. Hence, these methods evaluate $\tau_{ij} = \overline{u_i^* u_j^*} - \overline{u_i^*}\, \overline{u_j^*}$. The filtering procedure (see Equations 5.191) is inverted to give $u^* = G^{-1} * U$. The inverse filter G^{-1} can be defined as a series expansion G_n^{-1}. This can be truncated, depending on the level of mathematical exactness required. The most extreme truncation gives the scale similarity model of Bardina et al. (1980). An alternative procedure is to write the resolved velocity using Taylor's series. Then the filter inversion procedure yields the well-known Clark's model (see Clark et al. 1979). Structural models tend to give good correlation (based on DNS evidence) with exact sub-grid stresses. They also appear to model well turbulence anisotropy.

In a sense, some LES models appear to follow a similar historical path to that of RANS. For example, Kosovic (1997), Wong (1992) and Lund and Novikov (1992) produced quadratic and cubic LES extensions to the Boussinesq approximation. Also, the earlier model of Clark et al. (1979) can be viewed as having a non-linear stress-strain relationship. Such models tend to be described in the LES literature as *mixed models*. Hence, they can be expressed as

$$-\overline{u_i' u_j'} = [L] + [\overline{NL}] \tag{5.207}$$

where L and $\overline{NL}$ represent the linear and non-linear terms, respectively. The linear stresses could be characterized using the Smagorinsky model. Some models for $\overline{NL}$ are summarized in Table 5.12.

Table 5.12. *Nonlinear stress expressions for different models using (there is implicit summation over* l*) (from Tucker (2013b) – used with the kind permission of Springer)*

LES Model	$\overline{NL}$ (Non-linear part of τ_{ij})
Kosovic (1997)	$C_k\rho\Delta^2\left(0.5\frac{\partial u_i}{\partial x_l}\frac{\partial u_l}{\partial x_j}+1.5\frac{\partial u_i}{\partial x_l}\frac{\partial u_j}{\partial x_l}-0.5\frac{\partial u_l}{\partial x_i}\frac{\partial u_l}{\partial x_j}\right)$
Leray (see Geurts and Holm (2006))	$C_L\rho\Delta^2\left(\frac{\partial u_i}{\partial x_l}\frac{\partial u_l}{\partial x_j}+\frac{\partial u_i}{\partial x_l}\frac{\partial u_j}{\partial x_l}\right)$
α (see Domaradzki and Holm (2001))	$C_\alpha\rho\Delta^2\left(\frac{\partial u_i}{\partial x_l}\frac{\partial u_l}{\partial x_j}+\frac{\partial u_i}{\partial x_l}\frac{\partial u_j}{\partial x_l}+\frac{\partial u_l}{\partial x_i}\frac{\partial u_l}{\partial x_j}\right)$
Clark et al. (1979)	$C_L\rho\Delta^2\left(\frac{\partial u_i}{\partial x_l}\frac{\partial u_l}{\partial x_j}+\frac{\partial u_i}{\partial x_l}\frac{\partial u_j}{\partial x_l}\right)$

Comparisons with DNS show that mixed models have good modelling traits. The linear components capture dissipation. The non-linear provide backscatter. Hence, they have similar properties to the Germano model but do not need the expensive double filtering and ad hoc averaging that occurs with the Germano model.

An important class of LES is ILES – see Boris et al. (1992) and Margolin et al. (2006). This approach uses the inherent traits of a numerical scheme to naturally provide an LES model. There inevitably will be some form of numerical scheme input for typical engineering problems and hence it seems sensible to make best use of this potentially negative input. The most well-known ILES approach is MILES. With this, schemes that maintain monotonicity (i.e., prevent wiggles in solution variables) are used. Appropriate schemes are discussed in Chapter 4. For flows with shocks the use of MILES becomes especially persuasive (LES models like the Smagorinsky will produce too much viscosity in the vicinity of shocks and need locally switching off). The shock capturing term typically used in numerical schemes is the one that connects best with the process of LES modelling. Hence, there is a natural synergy with respect to meeting the modelling requirements of capturing shocks and draining turbulence at the small scales. Also, some numerical schemes yield elements akin to those found in scale similarity models (Sagaut et al. 2007). Hence, the MILES approach can give, in a sense, mixed type models. An alternative approach that has similarities to MILES/ILES is the LES-RF (LES-Relaxation Filtering) approach. With this, filtering alone is used to replace the role of the LES model.

It is important to note that the LES models do not just have to account for fluid dynamics input but could also be designed to model sub-grid scale acoustic input and also combustion characterization (see Chakraborty and Cant 2009) and general scalar transport (Khan et al. 2010).

All of these models have a range of defects which are extensively outlined in Tucker (2013b). However, in the context of Figure 5.33, their importance is relatively low and hence not detailed here.

Table 5.13. *Some potential filter options (from Tucker 2013b – used with the kind permission of Springer)*

Filter description	Filter length scale in model Δ
Quasi-isotropic Cartesian grids	$(\Delta x \Delta y \Delta z)^{1/3}$
Anisotropic grids	$\sqrt{(\Delta x^2 + \Delta y^2 + \Delta z^2)/3}$
Anisotropic grids	$\mathrm{Max}(\Delta x,\ \Delta y,\ \Delta z)$
Anisotropic grids, Andersson et al. (2005)	$\mathrm{Min}(\Delta x,\ \Delta y,\ \Delta z)$
Anisotropic grids, Zahrai et al. (1995)	$(\Delta x \Delta y \Delta z)^{1/9}(\Delta x_j)^{2/3}$
Anisotropic grids, Scotti et al. (1993)	$(\Delta x \Delta y \Delta z)^{\frac{1}{3}} \cosh(4[(\ln c_1)^2 + (\ln c_2)^2 - \ln c_1 \ln c_2]/27)$
Hybrid LES-RANS, Batten et al. (2011)	$2\mathrm{Max}(\Delta x,\ \Delta y,\ \Delta z,\ \lvert\lambda\rvert \Delta t,\ L)$
Batten et al. (2007)	$2\mathrm{Max}(\Delta x,\ \Delta y,\ \Delta z,\ \lvert u_i\rvert \Delta t,\ L_{vK})$
Hybrid LES-RANS, Mani (2004)	$2\mathrm{Max}(\Delta x,\ \Delta y,\ \Delta z,\ \lvert u_i\rvert \Delta t,\ k\Delta t)$
Hybrid RANS-LES, Batten et al. (2007)	$2\mathrm{Max}(\Delta x,\ \Delta y,\ \Delta z,\ L_{vK})$
Hybrid RANS-LES, Batten et al. (2007)	$2\mathrm{Max}(\Delta x,\ \Delta y,\ \Delta z,\ L_{\min})$
Vorticity aligned with a grid line, Chauvet et al. (2007)	$\sqrt{n_x^2 \Delta y \Delta z + n_y^2 \Delta x \Delta z + n_z^2 \Delta x \Delta y}$
General definition of Chauvet et al.'s – Deck et al. (2011)	$\sqrt{\overline{A}_\omega}$
Curvilinear finite difference	$(J \Delta\xi\, \Delta\eta \Delta\zeta)^{1/3}$
Finite volume	$(Vol)^{1/3}$
Unstructured control volume – Batten et al. (2007)	$4 \max\limits_{k=0\ldots n} [\lvert \mathbf{x}_c - \mathbf{x}_k \rvert]$
Unstructured control volume, Spalart	Maximal circle or sphere that encompasses cell
Farge and Schneider (2001)	Wavelet based

It seems important to stress here that at least as much attention should be paid to the filter as to the LES model. The LES filtering process can take various analytical and numerical forms. The filtering can either be implicit or explicit. Such decisions can have a tangible solution impact. A filter length scale is also used in the LES model equations. Table 5.13 gives a range of filters used for this purpose. For curvilinear, finite differences, J is the usual Jacobian as discussed in Chapter 3. Also, $\Delta\xi$, $\Delta\eta$ and $\Delta\zeta$ are the grid spacings in the computational plane. Furthermore, c_1 and c_2 are grid aspect ratios and Vol is the grid cell volume. The filter of Zahrai et al. (1995) has a tensorial/matrix form. In the filter of Chauvet et al. (2007), $\mathbf{n}$ is the unit vector aligned with vorticity. In the filter of Batten et al. (2007) λ is wave speed and L_{vK} is the von Karman length scale. In Batten and colleagues' unstructured mesh definition, n is the number of faces. Also, $\mathbf{x}_k$ is the midpoint of a face k and $\mathbf{x}_c$ is the cell centroid. The maximal circle or sphere concept of Spalart is unpublished but reported in Batten et al. (2007). Also, in Batten et al. (2007), the actual eddy resolvable, based on the selected time step (and magnitude of local velocity, $\lvert u_i\rvert$) is taken into account. This is consistent with the notion of space-time filtering, discussed earlier. As noted in

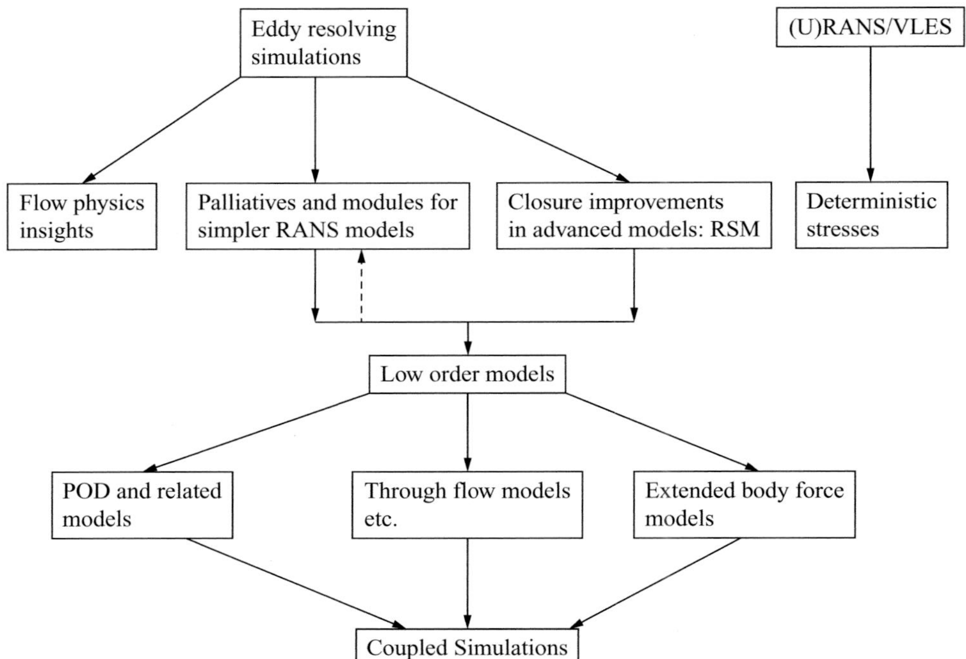

Figure 5.37. Industrial use of eddy-resolving simulations (based on Tucker 2013b – published with kind permission of Elsevier).

Batten et al. (2007), with the definition of Mani (2004), large, transient, spurious, k values can contaminate solutions. In the definition of Deck et al. (2011),$\sqrt{\overline{A}\omega}$ is the average cross-sectional area of a cell. This is normal to the vorticity vector, as utilized by Chauvet and colleagues. Note that $L_{\min}$, in Table 5.13, for hybrid RANS-LES (Batten et al. (2007)) is a user prescribed, minimum cell length scale. This additional parameter can be used to shield a RANS layer zone having dense grid clustering. For high aspect ratio, aerodynamic grids, the cell aspect ratio can be 1000:1. The tabulated filter definitions hence would give a range of this order. Therefore, although there is frequent careful reporting of small variations in, for example, the Smagorinsky constant, it is the filter length scale that has an extremely powerful influence. This can potentially dominate the actual sub-grid scale model choice.

The final level of the hierarchy is DNS. With this approach, the grids must resolve down to the Kolmogorov scales. This typically needs grid spacings of around 10 in wall units. This gives rise to the need for extreme grid densities, making the approach of limited industrial use.

5.5 Future Outlook

Future key improvements in many areas of design will arise through increasingly holistic calculations. As discussed in Chapter 6, automatic design optimization and adjoint-based methods are now becoming popular. Automatic processes need robust turbulence models that can cope with a wide range of grid topologies, thus reducing failed solutions and the need for user intervention. Complex turbulence models are challenging to implement and debug, especially in adjoint design optimization frameworks, and potentially suffer from a lack of robustness. These aspects place limitations on the use of Reynolds stress models and other more advanced turbulence techniques. As shown in Figure 5.12, and noted earlier, the SA model can be viewed as intrinsically modular. Hence, as shown in Figure 5.12 it has a jet/wake module, a laminar/transition zone module and near surface core modules. Auxiliary modules include those for rotation and curvature (Shur et al. 2000), wall roughness (Aupoix and Spalart 2003), compressibility and anisotropy (Spalart and Strelets 2000, Mani et al. 2013). The nature of SA is such that different components/modules are switched on and off through functions sensitized to different flow zones or the user. The idea of activating modules has some attraction and is not new, being essential with incomplete turbulence models, for example. To impact on the aerodynamic design of a system, clearly, the engineer should be considering the flow physics. The other extreme, to the use of, in a sense, an incomplete, fully modular phenomenological strategy would perhaps be an application of a Reynolds stress model. However, although containing much more solidly defined underlying flow physics, such variants are much more mathematically complex and expensive. Also, they cannot naturally account for roughness and are generally insensitive to reverse or forward transition. Hence, the added complexity may well not always make that much of a tangible predictive accuracy impact.

With regards to the notion of modular turbulence modelling, it is instructive to consider Figure 5.7b, which relates to an aeroengine intake in crosswinds. The flow has the following features/aspects:

1) The spatially extensive crosswind contracts to a sudden narrow lip zone, giving rise to a non-equilibrium laminarizing accelerating zone;
2) The flow is subjected to extreme laminarizing curvature;
3) Potentially the flow is subjected to a wide range of roughened lip topologies;
4) The quasi-laminar flow encounters a shock in the proximity of a diffusing zone and there is a shock induced separation;
5) The quasi laminar separation fully transitions and reattaches and
6) A fresh boundary layer redevelopments in the diffusing zone.

The Reynolds stress model would be correctly sensitized to the extreme streamline curvature, and depending on the variant might replicate some of the other non-equilibrium processes. However, tests show for this flow, in its standard form, the performance is disappointing. Oriji and Tucker (2013) use a modular SA approach. The modules deal with the different aspects listed previously, such as acceleration; curvature (using a Richardson number based procedure); transition (Kožulović and Lapworth 2009 correction discussed previously) and roughness (formulated around the Aupoix and Spalart 2003 extension also discussed previously). The transition module is the most critical and based on the above discussion the most challenging.

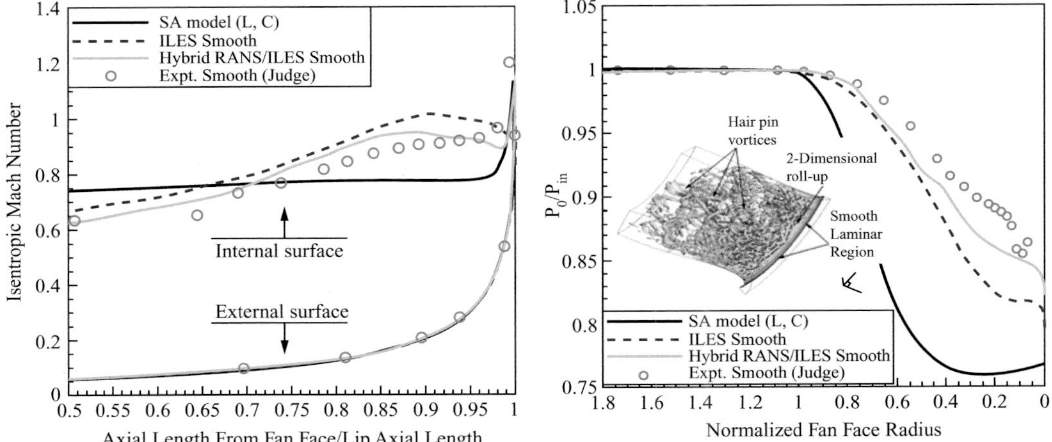

Figure 5.38. Modular/zonal RANS-(I)LES strategy for an idealized aeroengine intake in a crosswind: (a) isentropic Mach number distribution around the lip and (b) total pressure at the downstream fan face (from Oriji et al. (2014b) – insert from Frame (b) published with the kind permission of the ASME).

The modular philosophy discussed connects with the concept of zonalized hybrid RANS-LES of Deck and colleagues, outlined earlier. This is described by these authors as Zonalized DES. The modular approach can be viewed as zonal RANS modelling connecting with the ZDES philosophy. For example, in a ZDES framework, the acceleration, curvature and roughness modules could be used to gain the correct upstream boundary layer state. Then, to avoid the need for a transition module, LES could be used to transition the flow. A downstream near wall RANS zone could then be used to connect the reattached flow to the fan face of the engine. In the work of Oriji et al. (2014b) such simulations have shown promise.

For example, in Figure 5.38, results from using modular RANS in a ZDES type framework are presented. Frame (a) plots isentropic Mach number distribution around the lip. Frame (b) gives the total pressure at the downstream notional fan face. The inset gives the instantaneous flow field. The symbols represent the measurements. The full line is the SA RANS model result without the transition module. Not having this gives rise to the extensive flat horizontal line zone in Frame (a), corresponding to a large separation bubble. Using the SA base model, with the curvature and acceleration modules, and allowing an ILES related approach to resolve the transition (with a downstream RANS zone) gives a substantial improvement in predictive accuracy – see Figure 5.38. This is also true if a pure RANS simulation is made with the transition module (Note, this simulation is not shown). Granted, the transition module needs careful calibration. Interestingly, the grid used of 33×10^6 cells is much too coarse for LES – based on the grid spacing requirements given in Table 5.2. Even so the ILES result shown in Figure 5.38 is not too disappointing (note that it does, however, completely miss the Mach number peak at the right-hand side of Frame (a)). The plausible performance of ILES is probably due to the quasi-laminar incoming boundary layer state arising from the acceleration and curvature. Hence, this flow, even though it has a high system Reynolds number, might not be such a great challenge to LES. Indeed, there are many flows in aerospace systems that are Reynolds number independent. For example, the propulsive jet, once the nozzle has been emerged from, and also the ribbed passages used for cooling purposes in turbine blades. These things make LES relatively accessible in a short time frame.

As noted by Hunt (1973), "the most successful RANS models are those that do the least damage." As also pointed out by Hunt, a key to the success of RANS models is the mean flows insensitivity to μ_t – RANS models can give extreme variations in μ_t. A key difficulty is that transition displays strong sensitivity to a multitude of aspects. Hence, it seems hard to see how RANS based procedures can be reliable for transitional flows. This is especially so for unsteady wake transition in a URANS context. As shown in Figure 5.30, the extreme μ_t variations will give large variations in the flow's temporal content.

Temporal content prediction is also critical for noise. A key noise source is the fourth order space-time correlation of turbulence. Again, RANS based noise procedures are greatly challenged and again there is potential for unreliability in unfamiliar design spaces. In these zones, it might well make sense to push for approaches that have less modelled turbulence content.

LES has clear accuracy gains, however there are substantial potential uncertainties when making hybrid LES-RANS simulations. Also, care is needed with LES and this especially so when such simulations are being performed by engineers heavily engaged in the design process (i.e., non-CFD specialists). Hence, if eddy-resolving simulations are to be used in industry, there is a critical need for new CFD best practices. These could either be generic in nature, like the RANS based ERCOFTAC CFD guidelines, or could be customized to particular field specific zones or even both.

As shown in Figure 5.37, akin to URANS defining deterministic stresses (see right-hand side of Figure 5.37 and Chapter 7), more immediate uses for LES/DNS could be to refine RANS models. This could be at the deeper closure level for Reynolds stress models (see Bentaleb et al. 2012 for an example) or ad hoc fixes, palliatives or modules attached to phenomenological RANS models. As indicated in Figure 5.37, such simulations could be used to refine RANS models that, in turn, could be used, via the production of databases, to refine lower order models. These could then be used in coupled simulations as discussed in Chapter 7.

5.6 Conclusions

The turbulence model choice can have a substantial impact on results making RANS CFD a postdictive process. However, fortunately, generally CFD will predict correct trends even if the turbulence closure precludes capturing precise quantitative levels. If it is desired to predict exact levels of performance then generally strong insights into the turbulence model's nature and many other CFD aspects are necessary. Also, access to reliable validation data to potentially calibrate or find optimal models for a class of flow is necessary. Modular or zonal RANS turbulence modelling can potentially be used. The power of computers continues to steadily rise. Hence, the use of eddy-resolving simulations, in their various forms, will doubtless increase and be used for the refinement of lower order models. The coupled and sometimes multi-physics nature of many problems, to an extent, precludes the wide spread use of pure LES. However, it could be used in a multi-fidelity solution framework, where, for example, LES zones are sandwiched between (U)RANS. The interfaces can be connected by synthetic turbulence reconstruction techniques and carefully formulated time averaging strategies. Hence, there appears useful potential for zonal modelling in both the classical RANS and also eddy-resolving solution frameworks.

Problems

1) Using your incompressible CFD program developed as part of Chapter 4, solve a convection diffusion transport equation for eddy viscosity. Include a turbulence producing source term and an algebraic destruction term. Formulate the algebraic destruction term to capture the correct 'law of the wall' behaviour.
2) Derive the decay rate equations for the k-ω and k-ε equations for flow in a free stream and verify that they agree with the equations given in the text.
3) Based on the grid spacing requirements for different eddy-resolving simulation variants given in Table 5.1 and Table 5.2, estimate the grid requirements for flow over an airfoil with a chord based Reynolds number of 1×10^6. Assume the span is equal to chord. Use basic flat plate boundary layer data, thus ignoring the impact of pressure gradients.
4) For some of the grids shown in Chapter 3, roughly estimate the relative impacts of the different filters given in Table 5.13.

REFERENCES

ABE, K., KONDOH, T. & NAGANO, Y. 1994. A new turbulence model for predicting fluid flow and heat transfer in separating and reattaching flows – I: Flow field calculations. *International Journal of Heat and Mass Transfer*, 37, 139–151.

ABID, R. 1993. Evaluation of two-equation turbulence models for predicting transitional flows. *International Journal of Engineering Science*, 31, 831–840.

ABID, R., MORRISON, J. H., GATSKI, T. B. & SPEZIALE, C. G. 1996. Prediction of aerodynamic flows with a new explicit algebraic stress model. *AIAA Journal*, 34, 2632–2635.

ABU-GHANNAM, B. & SHAW, R. 1980. Natural transition of boundary layers – the effects of turbulence, pressure gradient, and flow history. *Journal of Mechanical Engineering Science*, 22, 213–228.

ALLMARAS, S. R., JOHNSON, F. T. & SPALART. P. R. 2012. Modifications and clarifications for the implementation of the Spalart-Allmaras turbulence model. Seventh International Conference on Computational Fluid Dynamics, Big Island, Hawaii, 9–13 July, Paper No. ICCFD7–1902

AMES, F., KWON, O. & MOFFAT, R. 1999. An algebraic model for high intensity large scale turbulence. ASME, International Gas Turbine and Aeroengine Congress and Exhibition, Indianapolis, Indiana, ASME Paper No. 99-GT-160.

ANDERSSON, N., ERIKSSON, L.-E. & DAVIDSON, L. 2005. LES prediction of flow and acoustic field of a coaxial jet. 11 AIAA/CEAS Aeroacoustics Conference, 23–25 May, Monterey, California, AIAA Paper No AIAA-2884, 2005.

ARKO, B. M. & MCQUILLING, M. 2013. Computational study of high-lift low-pressure turbine cascade aerodynamics at low reynolds number. *Journal of Propulsion and Power*, 29, 446–459.

ARNAL, D. 1992. *Boundary layer transition: prediction, application to drag reduction*. DTIC Document.

AUPOIX, B. & SPALART, P. 2003. Extensions of the Spalart–Allmaras turbulence model to account for wall roughness. *International Journal of Heat and Fluid Flow*, 24, 454–462.

BALDWIN, B. S. & BARTH, T. J. 1990. *A one-equation turbulence transport model for high Reynolds number wall-bounded flows*, National Aeronautics and Space Administration, Ames Research Center. NASA Technical Memorandum No. 102847.

BARDINA, J., FERZIGER, J. & REYNOLDS, W. 1980. Improved subgrid-scale models for large-eddy simulation. Simulation. Proceedings of the 13th AIAA Fluid and Plasma Dynamics Conference, Snowmass, Colo., 14–16 July 1980, AIAA Paper No. AIAA-80-1357.

BARDINA, J., FERZIGER, J. H. & REYNOLDS, W. C. 1983. Improved turbulence models based on large eddy simulation of homogeneous, incompressible turbulent flows. *Stanford University Report*, May.

BASSI, F., CRIVELLINI, A., REBAY, S. & SAVINI, M. 2005. Discontinuous Galerkin solution of the Reynolds-averaged Navier-Stokes and k-ω turbulence model equations. *Computers & Fluids*, 34, 507–540.

BASTIN, F., LAFON, P. & CANDEL, S. 1997. Computation of jet mixing noise due to coherent structures: the plane jet case. *Journal of Fluid Mechanics*, 335, 261–304.

BATTEN, P., GOLDBERG, U. & CHAKRAVARTHY, S. 2002. LNS – an approach towards embedded LES. Fortieth AIAA Aerospace Sciences Meeting & Exhibit. January, AIAA Paper No. AIAA-2002-0427.

BATTEN, P., GOLDBERG, U. & CHAKRAVARTHY, S. 2004. Interfacing statistical turbulence closures with large-eddy simulation. *AIAA Journal*, 42, 485–492.

BATTEN, P., GOLDBERG, U., CHAKRAVARTHY, S. & KANG, E. 2011. Smart subgrid-scale models for les and hybrid RANS/LES. Sixth AIAA Theoretical Fluid Mechanics Conference. Honolulu, HI, 27–30 June, AIAA Paper Number AIAA-2011-3472.

BATTEN, P., SPALART, P. & TERRACOL, M. 2007. Use of hybrid RANS-LES for acoustic source prediction. In T. Huttl, C. W., P. Sagaut (eds.) *Large-Eddy Simulation for Acoustics*, Cambridge University Press.

BENTALEB, Y., LARDEAU, S. & LESCHZINER, M. A. 2012. Large-eddy simulation of turbulent boundary layer separation from a rounded step. *Journal of Turbulence*, 13, 1–28.

BISEK, N. J., RIZZETTA, D. P. & POGGIE, J. 2013. Plasma control of a turbulent shock boundary-layer interaction. *AIAA Journal*, 51, 1789–1804.

BO, T., IACOVIDES, H. & LAUNDER, B. 1995. Developing buoyancy-modified turbulent flow in ducts rotating in orthogonal mode. *Journal of Turbomachinery*, 117, 474–484.

BORIS, J., GRINSTEIN, F., ORAN, E. & KOLBE, R. 1992. New insights into large eddy simulation. *Fluid Dynamics Research*, 10(4-6), 199–228.

BOUSSINESQ, J. 1877. *Theory de lecoulment tourbillant*. Mem Pre. Par .Div. Sav. XXIII, Paris.

BRANDT, T. T. 2008. Usability of explicit filtering in large eddy simulation with a low-order numerical scheme and different subgrid-scale models. *International Journal for Numerical Methods in Fluids*, 57, 905–928.

BREUER, M. & RODI, W. 1994. Large-eddy simulation of turbulent flow through a straight square duct and a 180 bend. *Direct and Large-Eddy Simulation I*, 273–285 Springer.

BRIDGES, J. & WERNET, M. P. 2003. Measurements of the aeroacoustic sound source in hot jets. *AIAA Paper*, 3130, 2003.

CEBECI, T. & CHANG, K. 1978. Calculation of incompressible rough-wall boundary-layer flows. *AIAA Journal*, 16, 730–735.

CHAKRABORTY, N. & CANT, R. 2009. Direct numerical simulation analysis of the flame surface density transport equation in the context of large eddy simulation. *Proceedings of the Combustion Institute*, 32, 1445–1453.

CHAOUAT, B. & SCHIESTEL, R. 2005. A new partially integrated transport model for subgrid-scale stresses and dissipation rate for turbulent developing flows. *Physics of Fluids (1994–present)*, 17, 065106.

CHAPMAN, D. K. 1979. Computational aerodynamics development and outlook. *AIAA Journal*, 17, 1293–1313.

CHARBONNIER, D., OTT, P., JONSSON, M., KOBKE, T. & COTTIER, F. 2008. Comparison of numerical investigations with measured heat transfer performance of a film cooled turbine vane. ASME Turbo Expo 2008: Power for Land, Sea, and Air, 571–582, ASME Paper No. GT2008-50623

CHAUVET, N., DECK, S. & JACQUIN, L. 2007. Zonal detached eddy simulation of a controlled propulsive jet. *AIAA Journal*, 45, 2458–2473.

CHIEN, K.-Y. 1982. Predictions of channel and boundary-layer flows with a low-Reynolds-number turbulence model. *AIAA Journal*, 20, 33–38.

CHOW, F. K. & MOIN, P. 2003. A further study of numerical errors in large-eddy simulations. *Journal of Computational Physics*, 184, 366–380.

CLARK, R. A., FERZIGER, J. H. & REYNOLDS, W. 1979. Evaluation of subgrid-scale models using an accurately simulated turbulent flow. *Journal of Fluid Mechanics*, 91, 1–16.

COLBAN, W., THOLE, K. & HAENDLER, M. 2007. Experimental and computational comparisons of fan-shaped film cooling on a turbine vane surface. *Journal of Turbomachinery*, 129, 23–31.

COULL, J. D. & HODSON, H. P. 2011. Unsteady boundary-layer transition in low-pressure turbines. *Journal of Fluid Mechanics*, 681, 370–410.

CRAFT, T., GERASIMOV, A., IACOVIDES, H., KIDGER, J. & LAUNDER, B. 2004. The negatively buoyant turbulent wall jet: performance of alternative options in RANS modelling. *International Journal of Heat and Fluid Flow*, 25, 809–823.

CRAFT, T., LAUNDER, B. & SUGA, K. 1996. Development and application of a cubic eddy-viscosity model of turbulence. *International Journal of Heat and Fluid Flow*, 17, 108–115.

CRIVELLINI, A., D'ALESSANDRO, V. & BASSI, F. 2013. A Spalart–Allmaras turbulence model implementation in a discontinuous Galerkin solver for incompressible flows. *Journal of Computational Physics*, 241, 388–415.

DACLES-MARIANI, J., ZILLIAC, G. G., CHOW, J. S. & BRADSHAW, P. 1995. Numerical/experimental study of a wingtip vortex in the near field. *AIAA Journal*, 33, 1561–1568.

DAKHOUL, Y. M. & BEDFORD, K. W. 1986a. Improved averaging method for turbulent flow simulation, Part I: Theoretical development and application to Burgers' transport equation. *International Journal for Numerical Methods in Fluids*, 6, 49–64.

DAKHOUL, Y. M. & BEDFORD, K. W. 1986b. Improved averaging method for turbulent flow simulation, Part II: Calculations and verification. *International Journal for Numerical Methods in Fluids*, 6, 65–82.

DAVIDSON, L. 2009. Hybrid LES–RANS: back scatter from a scale-similarity model used as forcing. *Philosophical Transactions of the Royal Society A: Mathematical, Physical and Engineering Sciences*, 367, 2905–2915.

DAVIDSON, L. & BILLSON, M. 2006. Hybrid LES-RANS using synthesized turbulent fluctuations for forcing in the interface region. *International Journal of Heat and Fluid Flow*, 27, 1028–1042.

DAVIDSON, L. & PENG, S. 2003. Hybrid LES-RANS: a one-equation SGS model combined with a k-omega model for predicting recirculating flows. *International Journal of Numerical Methods in Fluids*, 43, 1003–1018.

DEARDORFF, J. W. 1970. A numerical study of three-dimensional turbulent channel flow at large Reynolds numbers. *Journal of Fluid Mechanics*, 41, 453–480.

DECK, S., WEISS, P.-É., PAMIÈS, M. & GARNIER, E. 2011. Zonal detached eddy simulation of a spatially developing flat plate turbulent boundary layer. *Computers & Fluids*, 48, 1–15.

DEMUREN, A. & RODI, W. 1984. Calculation of turbulence-driven secondary motion in non-circular ducts. *Journal of Fluid Mechanics*, 140, 189–222.

DIMOTAKIS, D.E, MIAKE-LYE, R. C. & PAPANTONIOU, D. A. 1983. Structure and dynamics of round turbulent jets. *Physics of Fluids*, 26(11), 3185–3192.

DOMARADZKI, J. & HOLM, D. D. 2001. Navier-Stokes-alpha model: LES equations with nonlinear dispersion in *Modern Simulation Strategies for Turbulent Flow*, ed. by B. J. Geurts. ERCOFTAC Bulletin, 107(48), 2, Edwards Publishing.

DRUAULT, P., LARDEAU, S., BONNET, J.-P., COIFFET, F., DELVILLE, J., LAMBALLAIS, E., LARGEAU, J.-F. & PERRET, L. 2004. Generation of three-dimensional turbulent inlet conditions for large-eddy simulation. *AIAA Journal*, 42, 447–456.

DURBIN, P. A. 1991. Near-wall turbulence closure modeling without "damping functions." *Theoretical and Computational Fluid Dynamics*, 3, 1–13.

DURBIN, P., MEDIC, G., SEO, J.-M., EATON, J. & SONG, S. 2001. Rough Wall Modification of Two-Layer k-ε. *Journal of Fluids Engineering*, 123, 16–21.

EASTWOOD, S. 2010. *Hybrid LES–RANS of complex geometry jets*, PhD Thesis, School of Engineering, University of Cambridge.

EÇA, L. & HOEKSTRA, M. 2011. Numerical aspects of including wall roughness effects in the SST k-ω eddy-viscosity turbulence model. *Computers & Fluids*, 40, 299–314.

FARES, E. & SCHRÖDER, W. 2005. A general one-equation turbulence model for free shear and wall-bounded flows. *Flow, Turbulence and Combustion*, 73, 187–215.

FARGE, M. & SCHNEIDER, K. 2001. Coherent vortex simulation (CVS), a semi-deterministic turbulence model using wavelets. *Flow, Turbulence and Combustion*, 66, 393–426.

FIALA, A. & KUGELER, E. 2011. Roughness modeling for turbomachinery. Proceedings of the ASME Turboexpo, Power for Land, Sea and Air, GT2011, 6–11 June, Vancouver, BC, Canada, ASME Paper No. GT2011-45424.

FOROUTAN, H. & YAVUZKURT, S. 2013. A model for simulation of turbulent flow with high free stream turbulence implemented in OpenFOAM®. *Journal of Turbomachinery*, 135, 031022.

GATSKI, T. & SPEZIALE, C. 1993. On explicit algebraic stress models for complex turbulent flows. *Journal of Fluid Mechanics*, 254, 59–78.

GATSKI, T. B., RUMSEY, C. L. & MANCEAU, R. 2007. Current trends in modelling research for turbulent aerodynamic flows. *Philosophical Transactions of the Royal Society A: Mathematical, Physical and Engineering Sciences*, 365, 2389–2418.

GEORGIADIS, N. J., D. ALEXANDER, J. I. & RESHOTKO, E. 2003. Hybrid Reynolds-averaged Navier-Stokes/large-eddy simulations of supersonic turbulent mixing. *AIAA Journal*, 41, 218–229.

GEORGIADIS, N. J. & DEBONIS, J. R. 2006. Navier-Stokes analysis methods for turbulent jet flows with application to aircraft exhaust nozzles. *Progress in Aerospace Sciences*, 42, 377–418.

GEORGIADIS, N. J. & YODER, D. A. 2013. Recalibration of the shear stress transport model to improve calculation of shock separated flows. 51st AIAA Aerospace Sciences Meeting including the new horizons forum and aerospace exposition, Grapevine(Dallas/Ft.WorthRegion), Texas, AIAA Paper No. AIAA-2013–0685.

GEURTS, B. J. & HOLM, D. D. 2006. Leray and LANS-α modelling of turbulent mixing. *Journal of Turbulence*, 7(10), 1–33.

GIBSON, M. & LAUNDER, B. 1978. Ground effects on pressure fluctuations in the atmospheric boundary layer. *Journal of Fluid Mechanics*, 86, 491–511.

GIESEKING, D. A., CHOI, J.-I., EDWARDS, J. R. & HASSAN, H. A. 2011. Compressible-flow simulations using a new large-eddy simulation/Reynolds-averaged Navier-Stokes model. *AIAA Journal*, 49, 2194–2209.

GIRIMAJI, S. S. 2006. Partially-averaged Navier-Stokes model for turbulence: A Reynolds-averaged Navier-Stokes to direct numerical simulation bridging method. *Journal of Applied Mechanics*, 73, 413–421.

HAMBA, F. 2006. A hybrid RANS/LES simulation of high-Reynolds-number channel flow using additional filtering at the interface. *Theoretical and Computational Fluid Dynamics*, 20, 89–101.

HANIMANN, L., MANGANI, L., CASARTELLI, E., MOKULYS, T. & MAURI, S. 2014. Development of a novel mixing plane interface using a fully implicit averaging for stage analysis. *Journal of Turbomachinery*, 136, 081010.

HANJALIC, K. 2002. One-point closure models for buoyancy-driven turbulent flows. *Annual Review of Fluid Mechanics*, 34, 321–347.

HANJALIC, K. & LAUNDER, B. 1980. Sensitizing the dissipation equation to irrotational strains. *Journal of Fluids Engineering*, 102, 34–40.

HERMANSON, K., KERN, S., PICKER, G. & PARNEIX, S. 2003. Predictions of external heat transfer for turbine vanes and blades with secondary flow fields. *Journal of Turbomachinery*, 125, 107–113.

HUGHES, T. J., MAZZEI, L. & JANSEN, K. E. 2000. Large eddy simulation and the variational multiscale method. *Computing and Visualization in Science*, 3, 47–59.

HUNT, J. 1973. A theory of turbulent flow round two-dimensional bluff bodies. *Journal of Fluid Mechanics*, 61, 625–706.

IACOVIDES, H. & CHEW, J. 1993. The computation of convective heat transfer in rotating cavities. *International Journal of Heat and Fluid Flow*, 14, 146–154.

IACOVIDES, H. & THEOFANOPOULOS, I. 1991. Turbulence modeling of axisymmetric flow inside rotating cavities. *International Journal of Heat and Fluid Flow*, 12, 2–11.

KAMENETSKIY, D., BUSSOLETTI, J., HILMES, C., JOHNSON, F., VENKATAKRISHNAN, V. & WIGTON, L. 2013. Numerical Evidence of Multiple Solutions for the Reynolds-Averaged Navier-Stokes Equations for High-Lift Configurations. 51st AIAA Aerospace Sciences Meeting Including the New Horizons Forum and Aerospace Exposition, Dallas/Ft. Worth, Texas, AIAA Paper No. AIAA-2013–0663.

KARCZ, M. & BADUR, J. 2005. An alternative two-equation turbulent heat diffusivity closure. *International Journal of Heat and Mass Transfer*, 48, 2013–2022.

KATO, M. & LAUNDER, B. 1993. The modeling of turbulent flow around stationary and vibrating square cylinders. Ninth Symposium on Turbulent Shear Flows, pp. 10-4-1–10-4-6.

KEATING, A., PIOMELLI, U., BALARAS, E. & KALTENBACH, H.-J. 2004. A priori and a posteriori tests of inflow conditions for large-eddy simulation. *Physics of Fluids (1994–present)*, 16, 4696–4712.

KELTERER, M., PECNIK, R. & SANZ, W. 2010. Computation of laminar-turbulent transition in turbumachinery using the correlation based γ-Re_θ transition model. ASME Turbo Expo 2010: Power for Land, Sea, and Air, ASME Paper No. GT 2010-22207.

KHAN, M., LUO, X., NICOLLEAU, F., TUCKER, P. & LO IACONO, G. 2010. Effects of LES sub-grid flow structure on particle deposition in a plane channel with a ribbed wall. *International Journal for Numerical Methods in Biomedical Engineering*, 26, 999–1015.

KHAVARAN, A. & KENZAKOWSKI, D. C. 2007. *Progress toward improving jet noise predictions in hot jets.* NASA/CR-2007-214671.

KIM, S. H. & CHUNG, M. K. 2001. New vt-k model for calculation of wall-bounded turbulent flows. *AIAA Journal*, 39, 1803–1805.

KLEIN, M., SADIKI, A. & JANICKA, J. 2003. A digital filter based generation of inflow data for spatially developing direct numerical or large eddy simulations. *Journal of Computational Physics*, 186, 652–665.

KNOPP, T., EISFELD, B. & CALVO, J. B. 2009. A new extension for k-ω turbulence models to account for wall roughness. *International Journal of Heat and Fluid Flow*, 30, 54–65.

KONOPKA, M., JESSEN, W., MEINKE, M. & SCHRÖDER, W. 2013. Large-eddy simulation of film cooling in an adverse pressure gradient flow. *Journal of Turbomachinery*, 135, 031031.

KOSOVIC, B. 1997. Subgrid-scale modelling for the large-eddy simulation of high-Reynolds-number boundary layers. *Journal of Fluid Mechanics*, 336, 151–182.

KOŽULOVIĆ, D. & LAPWORTH, B. L. 2009. An approach for inclusion of a nonlocal transition model in a parallel unstructured computational fluid dynamics code. *Journal of Turbomachinery*, 131, 031008.

KRAICHNAN, R. H. 1970. Diffusion by a random velocity field. *Physics of Fluids (1958–1988)*, 13, 22–31.

KURBATSKIJ, A., POROSEVA, S. & YAKOVENKO, S. 1995. Calculation of statistical characteristics of a turbulent flow in a rotated cylindrical pipe. *High Temperature*, 33, 738–748.

LAKEHAL, D. 2002. Near-wall modeling of turbulent convective heat transport in film cooling of turbine blades with the aid of direct numerical simulation data. *Journal of Turbomachinery*, 124, 485–498.

LAM, C. & BREMHORST, K. 1981. A modified form of the k-ε model for predicting wall turbulence. *Journal of Fluids Engineering*, 103, 456–460.

LARAUFIE, R., DECK, S. & SAGAUT, P. 2011. A dynamic forcing method for unsteady turbulent inflow conditions. *Journal of Computational Physics*, 230, 8647–8663.

LASHERAS, J. & CHOI, H. 1988. Three-dimensional instability of a plane free shear layer: an experimental study of the formation and evolution of streamwise vortices. *Journal of Fluid Mechanics*, 189, 53–86.

LAUNDER, B. E., MORSE, A. P., RODU, W. & SPALDING D. B. 1972. Prediction of free shear flows – A comparison of the performance of six turbulence models. NASASP-311.

LAUNDER, B. E. & JONES, W. P. 1969. *On the prediction of laminarisation*, HM Stationery Office, ARC CP 1036.

LAUNDER, B., REECE, G. J. & RODI, W. 1975. Progress in the development of a Reynolds-stress turbulence closure. *Journal of Fluid Mechanics*, 68, 537–566.

LAUNDER, B. & SHARMA, B. 1974. Application of the energy-dissipation model of turbulence to the calculation of flow near a spinning disc. *Letters in Heat and Mass Transfer*, 1, 131–137.

LAUNDER, B. E. & PRIDDIN, C. 1973. A comparison of some proposals for the mixing length near a wall. *International Journal of Heat and Mass Transfer*, 16, 700–702.

LAUNDER, B. E. & SPALDING, D. 1974. The numerical computation of turbulent flows. *Computer Methods in Applied Mechanics and Engineering*, 3, 269–289.

LESCHZINER, M. A. 2000. Turbulence modelling for separated flows with anisotropy-resolving closures. *Philosophical Transactions of the Royal Society of London Series A: Mathematical, Physical and Engineering Sciences*, 358, 3247–3277.

LESCHZINER, M., LI, N. & TESSICINI, F. 2009. Simulating flow separation from continuous surfaces: routes to overcoming the Reynolds number barrier. *Philosophical Transactions of the Royal Society A: Mathematical, Physical and Engineering Sciences*, 367, 2885–2903.

LICARI, A. & CHRISTENSEN, K. 2011. Modeling cumulative surface damage and assessing its impact on wall turbulence. *AIAA Journal*, 49, 2305–2320.

LIEN, F.-S. & LESCHZINER, M. 1994. Assessment of turbulence-transport models including non-linear RNG eddy-viscosity formulation and second-moment closure for flow over a backward-facing step. *Computers & Fluids*, 23, 983–1004.

LIGRANI, P. M. & MOFFAT, R. J. 1986. Structure of transitionally rough and fully rough turbulent boundary layers. *Journal of Fluid Mechanics*, 162, 69–98.

LIU, Y. & TUCKER, P. 2007. Contrasting zonal LES and non-linear zonal URANS models when predicting a complex electronics system flow. *International Journal for Numerical Methods in Engineering*, 71, 1–24.

LIU, Y., TUCKER, P. G. & LO IACONO, G. 2006. Comparison of zonal RANS and LES for a non-isothermal ribbed channel flow. *International Journal of Heat and Fluid Flow*, 27, 391–401.

LODEFIER, K. & DICK, E. 2006. Modelling of unsteady transition in low-pressure turbine blade flows with two dynamic intermittency equations. *Flow, Turbulence and Combustion*, 76, 103–132.

LUND, T. S. & NOVIKOV, E. 1992. Parameterization of subgrid-scale stress by the velocity gradient tensor. *Annual Research Briefs*, 27–43.

LUND, T. S., WU, X. & SQUIRES, K. D. 1998. Generation of turbulent inflow data for spatially-developing boundary layer simulations. *Journal of Computational Physics*, 140, 233–258.

LUO, J. & RAZINSKY, E. H. 2007. Conjugate heat transfer analysis of a cooled turbine vane using the v2f turbulence model. *Journal of Turbomachinery*, 129, 773–781.

MANI, M. 2004. Hybrid turbulence models for unsteady flow simulation. *Journal of Aircraft*, 41, 110–118.

MANI, M., BABCOCK, D., WINKLER, C. & SPALART, P. 2013. Predictions of a supersonic turbulent flow in a square duct. Fifty-first AIAA Aerospace Sciences Meeting including the New Horizons Forum and Aerospace Exposition, January, AIAA Paper No. AIAA-2013-0860.

MARGOLIN, L., RIDER, W. & GRINSTEIN, F. 2006. Modeling turbulent flow with implicit LES. *Journal of Turbulence,* 7(15), 1–27.

MAYLE, R. & SCHULZ, A. 1997. Heat transfer committee and turbomachinery committee best paper of 1996 award: The path to predicting bypass transition. *Journal of Turbomachinery*, 119, 405–411.

MCGUIRK, J. & RODI, W. 1979. The calculation of three-dimensional turbulent free jets. In *Turbulent Shear Flows I*, Springer, pp. 71–83.

MELLOR, G. L. & HERRING, H. 1968. Two methods of calculating turbulent boundary layer behavior based on numerical solutions of the equations of motion. Proceedings of the Computation of Turbulent Boundary Layers Conference, August, Stanford, California, 18–25.

MENTER, F. R. 1993. Zonal two equation *k*-turbulence models for aerodynamic flows. AIAA Paper *No.* AIAA-1993-2906.

MENTER, F., KUNTZ, M. & BENDER, R. 2003. A Scale-Adaptive Simulation Model for Turbulent Flow Predictions. 41st Aerospace Sciences Meeting and Exhibit, AIAA Paper No. AIAA-2003-767.

MENTER, F., LANGTRY, R., LIKKI, S., SUZEN, Y., HUANG, P. & VÖLKER, S. 2006. A correlation-based transition model using local variables – Part I: model formulation. *Journal of Turbomachinery*, 128, 413–422.

MICHELASSI, V., GIANGIACOMO, P., MARTELLI, F., DÉNOS, R. & PANIAGUA, G. 2001. Steady three-dimensional simulation of a transonic axial turbine stage. ASME Turbo Expo 2001: Power for Land, Sea, and Air, ASME Paper No. 2001-GT-0174.

MOMPEAN, G., GAVRILAKIS, S., MACHIELS, L. & DEVILLE, M. 1996. On predicting the turbulence-induced secondary flows using nonlinear k-ε models. *Physics of Fluids (1994–present)*, 8, 1856–1868.

MONTOMOLI, F., HODSON, H. & HASELBACH, F. 2008. Effect of roughness and unsteadiness on the performance of a new LPT blade at low Reynolds numbers. ASME Turbo Expo 2008: Power for Land, Sea, and Air, ASME Paper No. GT2008-50488.

MOORE, C. 1977. The role of shear-layer instability waves in jet exhaust noise. *Journal of Fluid Mechanics*, 80, 321–367.

MORSE, A. 1977. *Axisymmetric Turbulent Sheer Flows with and Without Swirl.* PhD thesis, University of London.

MULLENIX, N. J., GAITONDE, D. V. & VISBAL, M. R. 2011. A plasma-actuator-based method to generate a supersonic turbulent boundary layer inflow condition for numerical simulations. Proceedings of the 20th AIAA Computational Fluid Dynamics Conference, Honolulu, HI, 27–30 June 2011. AIAA Paper No. AIAA-2011-3556.

MURMAN, S. M. 2011. Evaluating modified diffusion coefficients for the SST turbulence model using benchmark tests, 41st AIAA Fluid Dynamics Conference and Exhibit, June, AIAA Paper No. AIAA-2011-3571.

NAGABHUSHANA, R. V. 2014. *Numerical investigation of separated flows inlow pressure turbines: current status and future outlook*. PhDthesis, University of Cambridge.

NEE, V. W. & KOVASZNAY, L. S. 1969. Simple phenomenological theory of turbulent shear flows. *Physics of Fluids (1958–1988)*, 12, 473–484.

NEWMAN, G., LAUNDER, B. & LUMLEY, J. 1981. Modelling the behaviour of homogeneous scalar turbulence. *Journal of Fluid Mechanics*, 111, 217–232.

NICOUD, F. & DUCROS, F. 1999. Subgrid-scale stress modelling based on the square of the velocity gradient tensor. *Flow, Turbulence and Combustion*, 62, 183–200.

NIKITIN, N., NICOUD, F., WASISTHO, B., SQUIRES, K. & SPALART, P. 2000. An approach to wall modeling in large-eddy simulations. *Physics of Fluids (1994–present)*, 12, 1629–1632.

NISHIMURA, M., TOKUHIRO, A., KIMURA, N. & KAMIDE, H. 2000. Numerical study on mixing of oscillating quasi-planar jets with low Reynolds number turbulent stress and heat flux equation models. *Nuclear Engineering and Design*, 202, 77–95.

OLIVER, T. A. 2008. *A high-order, adaptive, discontinuous Galerkin finite element method for the Reynolds-averaged Navier-Stokes equations*. Ph.D Thesis, Massachusetts Institute of Technology, Dept. of Aeronautics and Astronautics.

ORIJI, U. R. & TUCKER, P. G. 2013. RANS modelling of accelerating boundary layers. ASME 2013 International Mechanical Engineering Congress and Exposition, IMECE2013, November 13-21, 2013, San Diego, CA, Paper No. IMECE2013-63467.

ORIJI, U. R., KARIMISANI, S. & TUCKER, P. G. 2014a. RANS Modeling of accelerating boundary layers. *Journal of Fluids Engineering*, 137, 011202–011202.

ORIJI, U. R., YANG, X. & TUCKER, P. G. 2014b. Hybrid RANS/ILES for aero engine intake. Proceedings of ASME Turbo Expo 2014 GT2014 June 16-20, 2014, Dusseldorf, Germany, Paper No. GT2014-26472.

PACCIANI, R., MARCONCINI, M., ARNONE, A. & BERTINI, F. 2011a. An assessment of the laminar kinetic energy concept for the prediction of high-lift, low-Reynolds number cascade flows. *Proceedings of the Institution of Mechanical Engineers, Part A: Journal of Power and Energy*, 225, 995–1003.

PACCIANI, R., MARCONCINI, M., FADAI-GHOTBI, A., LARDEAU, S. & LESCHZINER, M. A. 2011b. Calculation of high-lift cascades in low pressure turbine conditions using a three-equation model. *Journal of Turbomachinery*, 133, 031016.

PACCIANI, R., MARCONCINI, M., ARNONE, A. & BERTINI, F. 2012. URANS analysis of wake-induced effects in high-lift, low Reynolds number cascade flows. ASME Turbo Expo 2012: Turbine Technical Conference and Exposition, ASME Paper No. GT2012-69479, 1521–1530.

PACCIANI, R., MARCONCINI, M., ARNONE, A. & BERTINI, F. 2013. Predicting high-lift LP turbine cascades flows using transition-sensitive turbulence closures. ASME Turbo Expo 2013: Turbine Technical Conference and Exposition, 3–7 June, San Antonio, TX, ASME Paper No. GT2013-95605.

PARK, T. S. & SUNG, H. J. 1997. A new low-Reynolds-number k-epsilon-$f\mu$ model for predictions involving multiple surfaces. *Fluid Dynamics Research*, 20, 97.

PECNIK, R. & SANZ, W. 2007. Application of the turbulent potential model to heat transfer predictions on a turbine guide vane. *Journal of Turbomachinery*, 129, 628–635.

PEROT, B. 1999. Turbulence modeling using body force potentials. *Physics of Fluids (1994–present)*, 11, 2645–2656.

PERRET, L., DELVILLE, J., MANCEAU, R. & BONNET, J.-P. 2008. Turbulent inflow conditions for large-eddy simulation based on low-order empirical model. *Physics of Fluids (1994–present)*, 20, 075107.

PIOMELLI, U. & BALARAS, E. 2002. Wall-layer models for large-eddy simulations. *Annual Review of Fluid Mechanics*, 34, 349–374.

PIOMELLI, U., FERZIGER, J., MOIN, P. & KIM, J. 1989. New approximate boundary conditions for large eddy simulations of wall-bounded flows. *Physics of Fluids A: Fluid Dynamics (1989–1993)*, 1, 1061–1068.

PIOMELLI, U., BALARAS, E., PASINATO, H., SQUIRES, K. D. & SPALART, P. R. 2003. The inner–outer layer interface in large-eddy simulations with wall-layer models. *International Journal of Heat and Fluid Flow*, 24, 538–550.

POPE, S. 1978. An explanation of the turbulent round-jet/plane-jet anomaly. *AIAA Journal*, 16, 279–281.

POROSEVA, S. 2013. Personal Communication.

PRUETT, C. D. 2000. Eulerian time-domain filtering for spatial large-eddy simulation. *AIAA Journal*, 38, 1634–1642.

RAO, V. N., JEFFERSON-LOVEDAY, R., TUCKER, P. G. & LARDEAU, S. 2014. Large Eddy Simulations in Turbines: Influence of Roughness and Free-Stream Turbulence. *Flow, Turbulence and Combustion*, 92, 543–561.

RAVERDY, B., MARY, I., SAGAUT, P. & LIAMIS, N. 2003. High-resolution large-eddy simulation of flow around low-pressure turbine blade. *AIAA Journal*, 41, 390–397.

REAU, N. & TUMIN, A. 2002. On harmonic perturbations in a turbulent mixing layer. *European Journal of Mechanics-B/Fluids*, 21, 143–155.

REVELL, A. J., CRAFT, T. J. & LAURENCE, D. R. 2011. Turbulence modelling of unsteady turbulent flows using the stress strain lag model. *Flow, Turbulence and Combustion*, 86, 129–151.

RHIE, C. & CHOW, W. 1983. Numerical study of the turbulent flow past an airfoil with trailing edge separation. *AIAA Journal*, 21, 1525–1532.

RICHARDSON, L. 1922. *Weather prediction by numerical process*, Cambridge University Press.

RUMSEY, C. L. 2007. Apparent transition behavior of widely-used turbulence models. *International Journal of Heat and Fluid Flow*, 28, 1460–1471.

RUMSEY, C. L., PETTERSSON REIF, B. A. & GATSKI, T. B. 2006. Arbitrary steady-state solutions with the *k*-epsilon model. *AIAA Journal*, 44, 1586–1592.

SAGAUT, P., HÜTTL, T. & WAGNER, C. 2007. Large-eddy simulation for acoustics. *Large-eddy Simulation for Acoustics*, Cambridge University Press, 89–127.

SCHLICHTING, H. 1979. *Boundarv Layerer Theory*, 7th edition, McGraw-Hill.

SCHUMANN, U. 1975. Subgrid scale model for finite difference simulations of turbulent flows in plane channels and annuli. *Journal of Computational Physics*, 18, 376–404.

SCOTTI, A., MENEVEAU, C. & LILLY, D. K. 1993. Generalized Smagorinsky model for anisotropic grids. *Physics of Fluids A: Fluid Dynamics (1989–1993)*, 5, 2306–2308.

SECUNDOV, A., STRELETS, M. K. & TRAVIN, A. 2001. Generalization of υt-92 turbulence model for shear-free and stagnation point flows. *Journal of Fluids Engineering*, 123, 11–15.

SHUR, M. L., STRELETS, M. K., TRAVIN, A. K. & SPALART, P. R. 2000. Turbulence modeling in rotating and curved channels: Assessing the Spalart-Shur correction. *AIAA Journal*, 38, 784–792.

SLOAN, D. G., SMITH, P. J. & SMOOT, L. D. 1986. Modeling of swirl in turbulent flow systems. *Progress in Energy and Combustion Science*, 12, 163–250.

SMAGORINSKY, J. 1963. General circulation experiments with the primitive equations, I: The basic experiment*. *Monthly Weather Review*, 91, 99–164.

SMIRNOV, A., SHI, S. & CELIK, I. 2001. Random flow generation technique for large eddy simulations and particle-dynamics modeling. *Journal of Fluids Engineering*, 123, 359–371.

SPALART, P. 2000. Strategies for turbulence modelling and simulations. *International Journal of Heat and Fluid Flow*, 21, 252–263.

SPALART, P. & ALLMARAS, S. 1994. A one-equation turbulence model for aerodynamic flows. *Recherche Aerospatiale*, 1, 5–21.

SPALART, P. & SHUR, M. 1997. On the sensitization of turbulence models to rotation and curvature. *Aerospace Science and Technology*, 1, 297–302.

SPALART, P. R. & RUMSEY, C. L. 2007. Effective inflow conditions for turbulence models in aerodynamic calculations. *AIAA Journal*, 45, 2544–2553.

SPALART, P. R. & STRELETS, M. K. 2000. Mechanisms of transition and heat transfer in a separation bubble. *Journal of Fluid Mechanics*, 403, 329–349.

SPALART, P. R., DECK, S., SHUR, M., SQUIRES, K., STRELETS, M. K. & TRAVIN, A. 2006. A new version of detached-eddy simulation, resistant to ambiguous grid densities. *Theoretical and Computational Fluid Dynamics*, 20, 181–195.

SPALDING, D. 1994. Calculation of turbulent heat transfer in cluttered spaces. Proceedings of the tenth International Heat Transfer Conference, Brighton.

SPEZIALE, C. G. 1987. On nonlinear *k-l* and *k-ε*; models of turbulence. *Journal of Fluid Mechanics*, 178, 459–475.

SPEZIALE, C. 1998. Turbulence modeling for time-dependent RANS and VLES: a review. *AIAA Journal*, 36, 173–184.

STRELETS, M. 2001. Detached eddy simulation of massively separated flows. Proceedings of the 39th AIAA Aerospace Sciences Meeting and Exhibit, Reno, NV, USA, 8–11 January 2001. AIAA Paper No. AIAA-2001-0879.

STRIPF, M., SCHULZ, A., BAUER, H.-J. & WITTIG, S. 2009a. Extended models for transitional rough wall boundary layers with heat transfer – Part I: Model Formulations. *Journal of Turbomachinery*, 131, 031016.

STRIPF, M., SCHULZ, A., BAUER, H.-J. & WITTIG, S. 2009b. Extended models for transitional rough wall boundary layers with heat transfer – Part II: Model Validation and Benchmarking. *Journal of Turbomachinery*, 131, 031017.

SUDER, K. L., CHIMA, R. V., STRAZISAR, A. J. & ROBERTS, W. B. 1995. The effect of adding roughness and thickness to a transonic axial compressor rotor. *Journal of Turbomachinery*, 117, 491–505.

SVENINGSSON, A. 2006. *Turbulence transport modelling in gas turbine related applications*, D.Phil. Thesis, Department of Applied Mechanics, Chalmers University of Technology.

SVENINGSSON, A. & DAVIDSON, L. 2005. Computations of flow field and heat transfer in a stator vane passage using the v2− f turbulence model. *Journal of Turbomachinery*, 127, 627–634.

TABOR, G. & BABA-AHMADI, M. 2010. Inlet conditions for large eddy simulation: a review. *Computers & Fluids*, 39, 553–567.

TAKAHASHI, T., FUNAZAKI, K.-I., SALLEH, H. B., SAKAI, E. & WATANABE, K. 2012. Assessment of URANS and DES for prediction of leading edge film cooling. *Journal of Turbomachinery*, 134, 031008.

TEMMERMAN, L. & LESCHZINER, M. 2002. A priori studies of near wall RANS model within a hybrid LES/RANS scheme, Engineering Turbulence Modelling and Experiments, 5, ed. by W. Rodi, N. Fueyo, pp. 326–371.

TERRACOL, M., MANOHA, E., HERRERO, C., LABOURASSE, E., REDONNET, S. & SAGAUT, P. 2005. Hybrid methods for airframe noise numerical prediction. *Theoretical and Computational Fluid Dynamics*, 19, 197–227.

THIELEN, L., HANJALIĆ, K., JONKER, H. & MANCEAU, R. 2005. Predictions of flow and heat transfer in multiple impinging jets with an elliptic-blending second-moment closure. *International Journal of Heat and Mass Transfer*, 48, 1583–1598.

TRAVIN, A., SHUR, M., STRELETS, M. & SPALART, P. 2000. Detached-eddy simulations past a circular cylinder. *Flow, Turbulence and Combustion*, 63, 293–313.

TUCKER, P. 2003. Differential equation-based wall distance computation for DES and RANS. *Journal of Computational Physics*, 190, 229–248.

TUCKER, P. 2008. The LES model's role in jet noise. *Progress in Aerospace Sciences*, 44, 427–436.

TUCKER, P. 2013a. Trends in turbomachinery turbulence treatments. *Progress in Aerospace Sciences*, 63, 1–32.

TUCKER, P. G. 2013b. *Unsteady Computational Fluid Dynamics in Aeronautics*, Springer.

TUCKER, P. G. 2001. *Computation of unsteady internal flows*. Springer.

TUCKER, P. G. 2004. Novel MILES computations for jet flows and noise. *International Journal of Heat and Fluid Flow*, 25, 625–635.

TUCKER, P. G. & DAVIDSON, L. 2004. Zonal *k-l* based large eddy simulations. *Computers & Fluids*, 33, 267–287.

TUCKER, P. G., RUMSEY, C. L., SPALART, P. R., BARTELS, R. B. & BIEDRON, R. T. 2005. Computations of wall distances based on differential equations. *AIAA Journal*, 43, 539–549.

TURNER, C. 2012. *Laminar kinetic energy modelling for improved laminar-turbulent transition prediction*, D.Phil. Thesis, Faculty of Engineering and Physical Sciences, The University of Manchester.

TYACKE, J., JEFFERSON-LOVEDAY, R. & TUCKER, P. 2012. Numerical modelling of seal type geometries. Proc. of ASME Turbo Expo, Copenhagen, Denmark, 11–15 June, ASME Paper No GT2012–68840.

WALTERS, D. K. & COKLJAT, D. 2008. A three-equation eddy-viscosity model for Reynolds-averaged Navier-Stokes simulations of transitional flow. *Journal of Fluids Engineering*, 130, 121401.

WALTERS, D. K. & LEYLEK, J. H. 2004. A new model for boundary layer transition using a single-point rans approach. *Journal of Turbomachinery*, 126(1), 193–202.

WARNER, J. C., SHERWOOD, C. R., ARANGO, H. G. & SIGNELL, R. P. 2005. Performance of four turbulence closure models implemented using a generic length scale method. *Ocean Modelling*, 8, 81–113.

WILCOX, D. 1998. *Turbulence modeling for CFD*, DCW Industries. Inc.

WILCOX, D. C. 1988. Multiscale model for turbulent flows. *AIAA Journal*, 26, 1311–1320.

WOLFSHTEIN, M. 1969. The velocity and temperature distribution in one-dimensional flow with turbulence augmentation and pressure gradient. *International Journal of Heat and Mass Transfer*, 12, 301–318.

WONG, V. 1992. A proposed statistical-dynamic closure method for the linear or nonlinear subgrid-scale stresses. *Physics of Fluids A: Fluid Dynamics (1989–1993)*, 4, 1080–1082.

WU, X. 2010. Establishing the generality of three phenomena using a boundary layer with free-stream passing wakes. *Journal of Fluid Mechanics*, 664, 193–219.

YAKHOT, V., ORSZAG, S., THANGAM, S., GATSKI, T. & SPEZIALE, C. 1992. Development of turbulence models for shear flows by a double expansion technique. *Physics of Fluids A: Fluid Dynamics (1989–1993)*, 4, 1510–1520.

YAP, C. R. 1987. *Turbulent heat and momentum transfer in recirculating and impinging* flows. PhD thesis, University of Manchester, Department of Mechanical Engineering, Faculty of Technology.

YOSHIZAWA, A. 1993. Bridging between eddy-viscosity-type and second-order models using a two-scale DIA. Proceedings of the ninth International Symposium on Turbulent Shear Flow, August, Kyoto, Japan, 16–18.

ZAHRAI, S., BARK, F. & KARLSSON, R. 1995. On anisotropic subgrid modeling. *European Journal of Mechanics B, Fluids*, 14, 459–486.

ZHONG, B. & TUCKER, P. G. 2004. *k–l* based hybrid LES/RANS approach and its application to heat transfer simulation. *International Journal for Numerical Methods in Fluids*, 46, 983–1005.

ZUCKERMAN, N. & LIOR, N. 2005. Impingement heat transfer: correlations and numerical modeling. *Journal of Heat Transfer*, 127, 544–552.

CHAPTER 6

Advanced Simulation

6.1 Introduction

Industrial design tends to be multidisciplinary and multi-physics, with multiple objectives involving coupled systems. Industrial products are complex and there are potentially hundreds of constraints. Treating the system as holistically as possible is of critical importance if invalid or poor designs are to be avoided.

As noted, system coupling can often be important. For example, as noted by Spalart and Bogue (2003) an ultimate aerospace vision would be to model the coupled airframe, engine and pilot interaction. The latter element also would need to include psychological response to events. The analysis might well encompass the need to account for aerodynamics, heat transfer and multi-phase flow. It could include the need to look at structural integrity in terms of peak stress, thermal or fatigue (both high and low cycle) failure and aeroelastic response. Product cost and weight are also key factors, the latter being highly critical for most aerospace applications.

The use of formal design optimization, involving numerous variations of design variables, has now become relatively standard industrial practice. Then, since coupled simulations are necessary low-order models become especially important. Also, there is the likelihood of geometric and boundary condition uncertainties and hence the need for a stochastic analysis level on top of potentially expensive deterministic simulation runs. Hence, this all furthers the need for lower order models. Therefore, with advanced simulations in mind, in this chapter, design optimization, coupled problems and low-order models are considered together. As part of low-order modelling it is necessary to consider the various truncated forms of the Navier-Stokes equations discussed in Chapter 2.

Notably, many industrial simulations involve a wide range of scales (i.e., they are multi-scale). They can also be multi-physics in nature. Hence, the modelling of multi-scale problems is also briefly considered along with more multi-physics aspects such as multi-phase flow, acoustics and fluid solid coupling. First design optimization is discussed. A critical aspect of this is producing low-order surrogate models. Hence, the design optimization discussion is followed by specifically considering low-order models. Then coupled calculations are considered, this again having the need for low-order models. Then, finally, more multi-physics and multi-scale modelling aspects are very briefly considered.

6.2 Design Optimization

6.2.1 *Key Elements and General Process*

Formal design optimization is important, giving an automatic *process* for *verifiably* seeking an improved design. The design optimization can be performed at various fidelities and stages of product design. For example, to get all the components in the right part of the design landscape, an initial, more global, system-level design optimization could be performed at the start of a design campaign. Then, using high fidelity CFD methods, individual components can be optimized. Typically, for complex engineering systems, such as gas turbine aeroengines, for example, the initial, more global, coupled, system-level design optimization is where the greatest potential for substantial improvements for a product can be found. Typically, at the component level, the potential for geometric changes is more minimal and the impact of the design optimization lower. Design optimization has been in use in aerodynamics for many years. For example, the NACA (National Advisory Committee for Aeronautics) four digit airfoil series emerges from attempts to formally study the behaviour of airfoils – granted in wind tunnels. Later Liebeck and Ormsbee (1970) performed computer-based airfoil optimization work. However, this did not involve the use of classical CFD. The design optimization process needs the following elements:

1) The ability to rapidly change geometry (the *design variables*), which is discussed in Chapter 7. Notably, several versions of geometry with different fidelity levels might be involved in the optimization process;
2) The capability to adapt the mesh to the revised geometry (meshing being discussed in Chapter 3);
3) A range of analysis codes at potentially a wide range of fidelities that are also able to model greatly differing physical problems;
4) Bespoke post-processing software (discussed in Chapter 7) than can automatically extract key integrated quantities. These might have the added layer of in-house correlations or data to correct the simulation output;
5) A design optimization tool to enable a search for an improved design to be made and
6) Scripts and tools to automatically link the different process elements listed.

Notably, the programs involved in elements 1–6 will work at different speeds and hence the whole process is challenging in terms of load balancing. The whole design process is likely to take place on a heterogeneous cluster. Hence, as can be seen, performing an automated design optimization is a substantial computational science task. Also, the use of the elements listed requires specialized experience and knowledge. Much of this typically resides with individuals. However, use of knowledge-based engineering is beginning to emerge and this needs to be linked in with these elements. Commercially available tools to perform design optimization

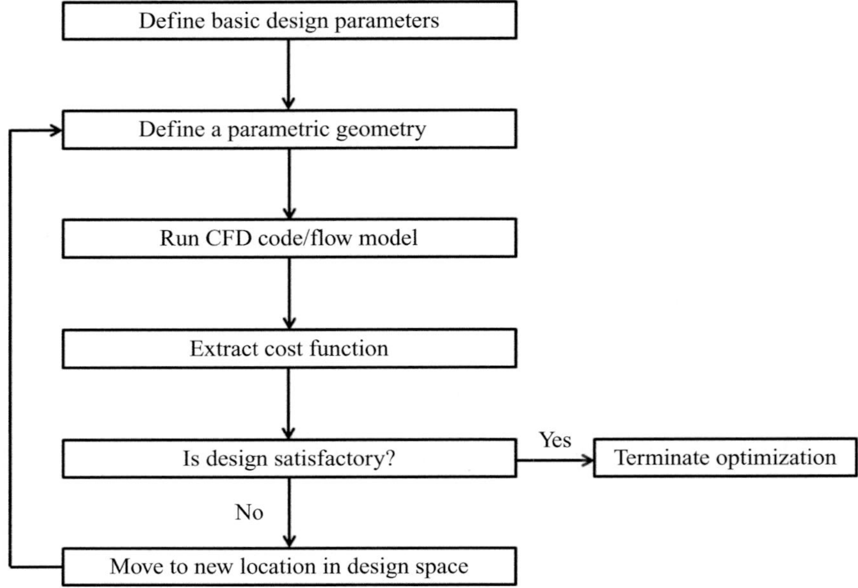

Figure 6.1. Design optimization process in its most basic form.

are DESIGNEXPLORER and ISIGHT. These offer a wide range of optimization algorithms and can take some effort to learn to fully utilize.

With design optimization there are the following key elements:

1) Cost or objective function(s) that it is desired to improve (e.g., drag or lift coefficient, total pressure loss, entropy rise etc.);
2) As already noted, design variables (e.g., geometric changes or changes in mass flow, pressure, etc.) and
3) Constraints (physical parameters that for structural or a multitude of other reasons cannot be violated).

Figure 6.1 shows a design optimization process in its most basic form. As can be seen, the key, first stage to carrying out a design optimization is to parameterize the problem. It can be difficult, without prior experience, to know the key parameters. When designing an airfoil, the lean, sweep, chord, anhedral and dihedral angles and so forth could be part of the parameterization. However, a modern fan blade in an aeroengine is tremendously complex and the above parameterization would be incomplete with regards to creating an industrial design. The nature of the parameterization will vary depending on the type of design optimization approach being used. For example, with adjoint based design optimization, the process is highly local. However, solution of the adjoint equations allows local surface/geometry perturbations to be made at any point on the surface – hence the level of geometrical change for one adjoint simulation can be viewed as unlimited. The downside is that the search is highly local and just one global solution output can be considered, such as a total pressure loss or drag coefficient. For a conventional search, the number of geometric parameters varied needs to be kept low. Otherwise the process cost can get too high. On the other hand, a wide range of solution outputs can be studied. Notably, geometry changes must be propagated through to the mesh. The revised mesh structure must be of good quality. Failed or poorly converged iterative solutions are problematic in an automatic design optimization procedure. Once the mesh has been generated, the CFD solver must be used. Again this presents challenges. Large non-linear equation sets are being solved. This typically needs numerous parameters to be tuned to ensure best convergence (see Chapter 4). Clearly in an automatic design optimization process this aspect is problematic. The other aspect that needs to be considered is that simulations typically nowadays are performed in a parallel framework. Hence it needs to be considered if a multitude of separate design studies will be performed on fewer processors or if more processors will be dedicated to fewer simulation runs. When using commercial CFD programs, the licensing scheme offered by the vendor for parallel processing can heavily define this aspect. Once the solution has been made, post-processing is needed. This typically would need integration of parameters to automatically give design objective functions. This may well also need to be corrected using empirical data to correct for known CFD accuracy defects.

The use of databases is a critical aspect of modern design. The database can comprise both CFD and experimental data. The use of such a database can avoid the running of simulations and also allow the interpolation between simulations in proximity to a desired design location.

6.2.2 *General Overview of Optimization Schemes*

The optimization system suggests geometry (or other design variable) changes to improve the objective functions whilst imposing design constraints. Broadly, there are two main approaches. These are either direct optimization or the use of low-order surrogate or response surface models. The former can be expensive, especially, if the CFD model is costly. Design optimization can be carried out through a range of methods, such as:

1) Steepest descent schemes or gradient based methods/hill climbers;
2) Stochastic evolutionary searches;
3) Surrogate models/response surface methods;
4) Inverse design and
5) Adjoint methods.

Hybrids can also be used and modern commercial design optimization codes facilitate this. Generally, a range of approaches are needed depending on the nature of the problem (i.e., the number of design variables, constraints and objective functions). An example of Approach (1) (steepest descent methods) is the Newton approach. This is normally used to find a value x such that $f'(x) = 0$. The Newton method has a geometric interpretation that involves taking steps towards a minimum and can be extended to multiple dimensions through use of the Hessian matrix. The gradients

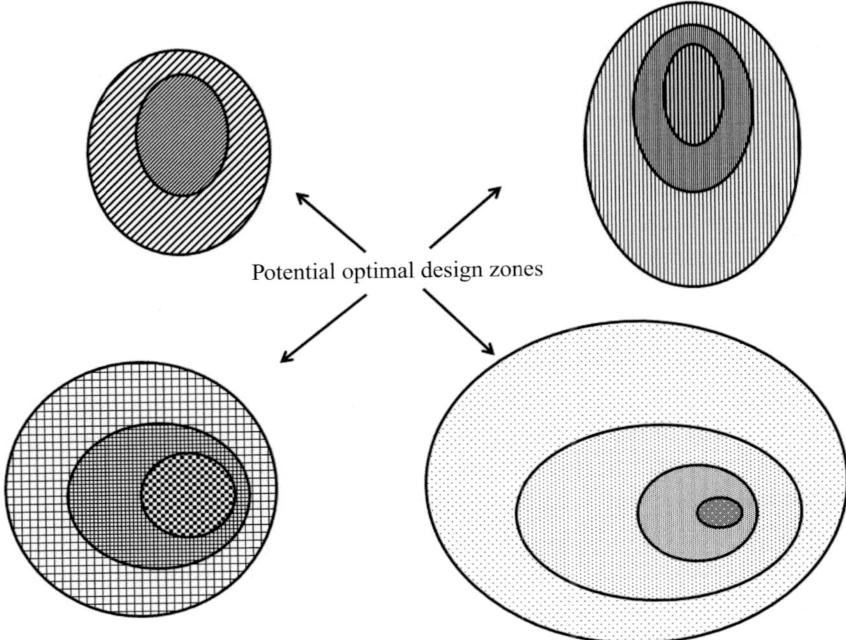

Figure 6.2. Potential for design optimization getting stuck in a local and not global minimum.

in Approach (1) are typically based around finite differences but can come from adjoint solutions. With a basic hill climbing approach, a random point is chosen and the objective function evaluated at this point. Then a seed point in close proximity is randomly chosen. The gradient can easily be evaluated as the difference of the two points over the distance between them. The direction in which a 'positive' gradient is found is selected. A third test point is placed a short distance away in the same direction. This process is repeated and 'hill is climbed' towards a maximum/optimal value. When every possible direction leads to a decrease in the objective function, the optimum has been reached. Obviously if the seed point is in the wrong place it is possible for the search to get stuck in a zone that does not have the best global optimum. Another problem is that the process is intrinsically sequential and this is non-ideal, especially if the objective function evaluations are expensive, say, for example, when making eddy-resolving simulations (see Chapter 5).

For Approach (2), a wide range of designs is simulated and some random/stochastic element is included. It mimics the natural process of evolution, with its natural selection processes, such as random mutation, thus avoiding being stuck in local minima, as shown in Figure 6.2. The most widely known method is the GA (Genetic Algorithm) approach. The alternative simulated annealing method (Kirkpatrick et al. 1983) takes inspiration from the optimization of metallic crystal structure through the annealing (heating and cooling) process. It has a stochastic element, which again avoids the possibility of being locked in a local minimum.

Stochastic evolutionary searches – Approach (2) – are highly parallel, but have relatively slow convergence to an optimum and can need hundreds or even thousands of generations to reach an optimal solution. The former aspect is especially attractive with modern parallel computers. Large companies potentially have a substantial number of free PCs available at certain hours that can be made use of.

Approach (3), at its most basic, for one objective function and design variable is curve fitting. The idea is that a few potentially expensive CFD simulations are made to inform the curve fit which is generally described as a response surface. This expensive data can be supplemented with low-order simulation data and the two fused. This will be discussed further later. Basically, Approach (3) generates a low-order model that can be used, for example, as part of Approaches (1) and (2). Hence, Approach (3) is strictly not an optimization method. To avoid too many computations, it is important that the simulations used to inform the curve fit are judiciously placed. Hence, the initial setup phase will involve the use of a Design of Experiment (DOE) approach – origination from the design of physical experiments. Well-known methods are the Latin hypercube and also the Taguchi. At this point it is important to consider the amount of computational resource available and hence the number of simulations that can be afforded within the design time constraints. The design of experiment should be based around a fraction of the total number of possible runs (unless the only reason for the building of the surrogate models is a qualitative visualization of the design landscape). This is because an important element will be later refinement of the surrogate model. This is called infilling.

Then the response surface will be generated. There are a wide range of methods to construct this and these will be discussed further later. The key point is that the response surface is a low-order fluid/system model – this itself is then placed in a design optimization loop. Hence, in a sense, response surfaces are not really the central element of design optimization but simply a low-order model to replace the CFD. Alternative low-order models will be specifically discussed later in this chapter (and low-order equations have been given in Chapter 2) and again can form part of the design optimization process. Another critical point is that, once the response surface finds an optimum design, this is not the point to stop. At the very least, a full simulation run will be needed to verify that the suggested improved design is in fact an improvement. Also, it is a good idea to perform high fidelity simulations in the vicinity of the current optimum design. It is easy, for example, if the sampling is coarse, to miss the optimal design point. Such issues will be discussed further later. Figure 6.3 shows a response surface optimization strategy as given by Keane and Scalan (2007).

With Approach (4), inverse design, aspects of the flow are predefined and a search made for geometry to match this expected behaviour. In its most standard form, inverse design typically involves the specification of a loading distribution (surface pressure) over an airfoil. The airfoil shape then adapts to match this expected loading distribution. There are wide ranges of inverse design approaches and these are discussed further later. In the work of Page et al. (2013) the loading distribution is specified along with system mass flow rate and a target design pressure ratio for

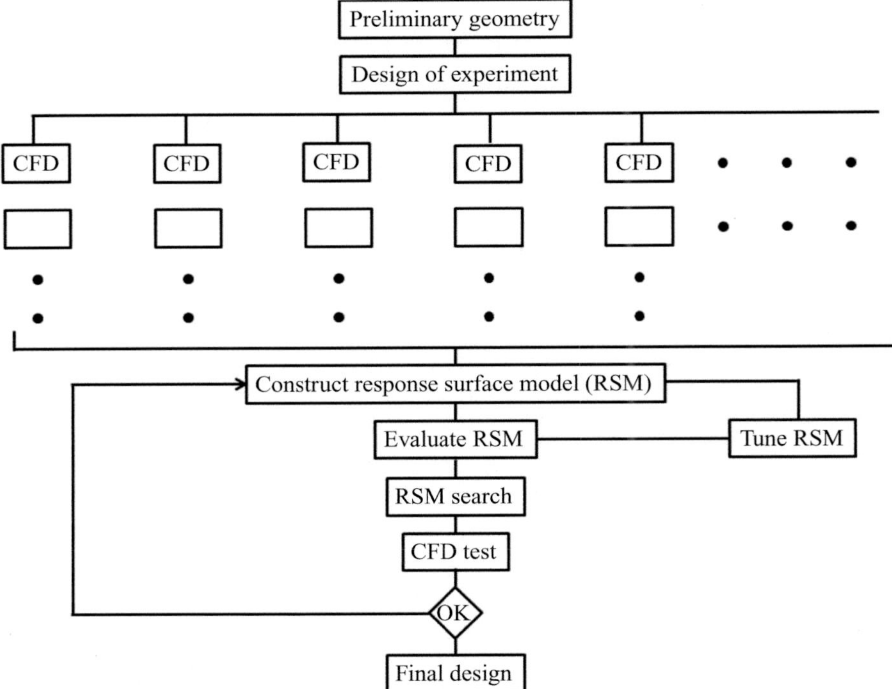

Figure 6.3. Optimization strategy based on an RSM (adapted from Keane and Scanlan 2007)

a transonic fan. Also, Page and colleagues consider a dual operating point strategy intended to design a fan that gives optimum cruise efficiency whilst offering optimum takeoff thrust. Again, this work will be discussed further later. A key issue with inverse design is the specification of the loading distribution for a complex, highly swept blade with, say, transonic flow – this is not easy. However, loading distributions from a successful past design that is close to the new design can also be used. Notably, all of the Approaches (1–4) can be hybridized in various ways.

Approach (5), as stated earlier, allows a multiplicity of geometric changes to be made at very little extra computational cost. The downside is that only one objective function can be considered and the geometric changes must be small perturbations. Also, the adjoint approach is technically challenging to implement. Table 6.1 summarizes advantages and disadvantages of different design optimization approaches. Next these approaches will be outlined in more detail.

6.2.3 *Stochastic Evolutionary Searches*

First, population-based methods are considered. There are a range of approaches, such as firefly, particle swarm, mimetic and genetic algorithms. The latter is focused on here. As noted earlier, the genetic algorithm approach mimics the activity of natural selection to choose creatures which are well suited to survive.

Table 6.1. *Advantages and disadvantages of different design optimization approaches*

	Methods	Advantages	Disadvantages	Other comments
I	Gradient based	Relatively simple	Can get stuck in local minima. Need a continuous function	Restarts can prevent being stuck in a local suboptimum
II	Stochastic evolutionary searches	Not so prone to getting stuck in local minima. Highly parallel	Slow convergence. Need a continuous function	–
III	Surrogate models/ Response surface methods	Cheap low-order models produced. The polynomial coefficients can be used to infer the relative importance of design variables	A wide range of curve fitting methods to choose from. Needs iterative updating – initial response surface will not be accurate. Needs continuous functions	Not strictly an optimizer
IV	Inverse design methods	Very efficient if loading distribution known	Optimal loading distribution can be hard to find. Dealing with shocks can need care	Inverse design can be hybridized with classical design optimization approaches
V	Adjoint	Allows extensive range of geometric changes to be explored at modest cost	Can only deal with small geometric changes. Will not naturally integrate with a CAD geometric parameterization. Can only focus on one objective function	Especially useful for transonic flows where small geometrical changes can have a substantial impact

The value of the objective function is considered to be a measurement of this fitness for survival, and the culling of relatively poorly performing individuals exerts an evolutionary pressure. Each individual point in the design space is encoded in a stream of data, referred to as a genome. The value of each different gene within the genome is known as the genotype. These genotypes and their interplay with others result in an expression of a genotype, called a phenotype. In a genetic algorithm, the performance of each individual – a cluster of phenotypes – is evaluated. The genomes of the best performing individuals (ones whose phenotypes have resulted in superior performance) are then allowed to recombine with each other. The principle is that the genotypes which result in advantageous phenotypes will survive and recombine until the peak value of the test function within the design space is found. In terms of solving the engineering problem, this effectively means that each point in the potential design space must be translated into a gene.

Next, to aid understanding, an example of the application of a genetic algorithm to a turbomachinery flow is considered. The trailing edges of turbine blades can be

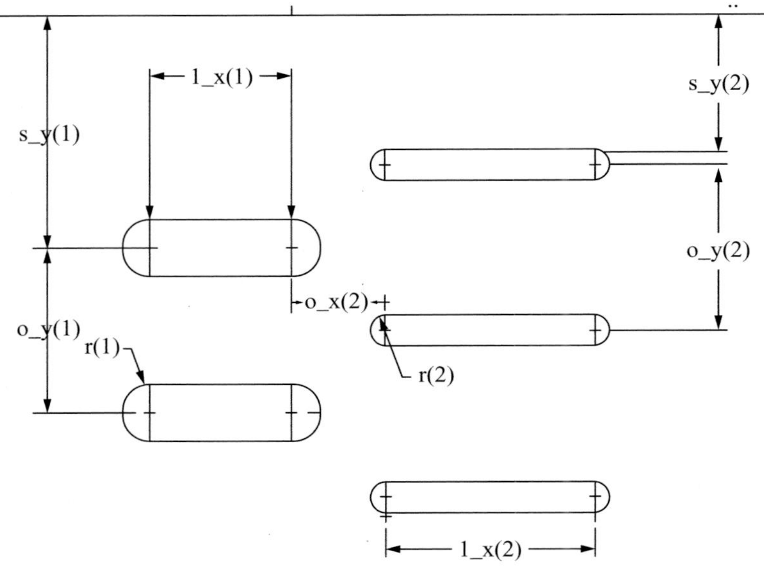

Figure 6.4. Parameterization of turbulator plan form (from Watson and Tucker 2014 – published with kind permission of the American Institute of Aeronautics and Astronautics).

open with air issuing from them. The two halves of the blades are connected by an arrangement of bluff turbulator pin structures. Also, the trailing edge is cut back, leaving an exposed surface that is bathed in the ejected fluid. As shown by Watson et al. (2014), the use of RANS (see Chapter 5) to perform an optimization process would steer the design in the wrong direction. Eddy-resolving methods are needed. In the optimization to be described, initial LES are made on 20 million cell meshes for a range of operation conditions. These show the typical nature of this flow, this being that it is generally dominated by shear layers that produce relatively large-scale structures. This means (as verified through tests) that accurate results can be made using just 3.5 million cells meshes. Hence, the design optimization is based on these coarser grid simulations.

The first step in carrying out a design optimization is to divide the design space into a set of parameters, which, between them, carry the necessary design information. For the purposes of this optimization, the parameters contain the turbulator pin layout data. Figure 6.4 shows an arbitrary pin layout, and its parameterization. The parameters contained in each plan file are listed in Table 6.2. A complete listing of all the values of these parameters for a single design would define the genome. Making use of such a parameterization helps facilitate rapid and automated generation of corresponding CFD meshes from the universal geometry, and provides a quick way to vary the turbulator layout between generations of the optimization.

Geometric constraints must then be placed on the design and these are outlined in Watson and Tucker (2014). The objective is to optimize the adiabatic film cooling

Table 6.2. *Parameterization of turbulator plan forms*

Parameter	Role
n_x	Number of pin rows
l_x(n_x)	Length of each pin row
s_y(n_x)	Spanwise stagger of each pin row
r(n_x)	Width and radius of each pin row
o_y(n_x)	Spanwise separation of each pin row
o_x(n_x)	Streamwise offset of each pin row

effectiveness. A value of unity is ideal, the ejected air being 100% effective in shielding the cut back trailing edge from hot gases that might potentially melt this zone.

Having parameterized the design space, a stochastic initialization is carried out. Effectively, this means that individual members of the first generation are assigned their genes at random, which survive if these genes resulted in a viable phenotype. Figure 6.5 shows a small fraction of turbulator plan forms, which are initially generated. The performance of each plan form is then evaluated by conducting an LES, and calculating the adiabatic film cooling effectiveness. The performance of each individual within the population is then ranked. With this ranking, the evolutionary stage of the optimization proceeds by culling the worst performing 40% of the individuals. The surviving population members are each given a "probability to reproduce" based on their place in the performance hierarchy. The best performing geometries are twice as likely to reproduce as the worst performing which survived the cull, with a linear probability distribution between them. Pairs of parents are then generated based on this probability distribution to produce the next generation of children. The child receives its genome at random from each parent. Each parent's gene has a 50:50 chance of being represented in the child's genome. As well as the transmission of genes, the genetic algorithm allows for some copying errors, as occur in biological DNA copying processes. In terms of the optimization problem, this leads to more exploration of the design space not captured by the genes present

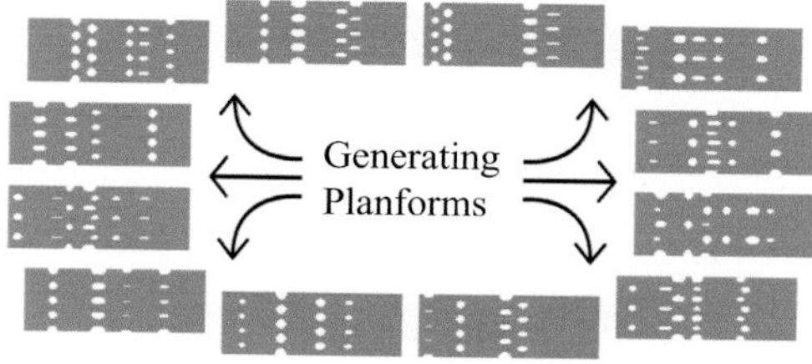

Figure 6.5. Initial random distribution based tabulator plan forms (from Watson and Tucker 2014 – published with kind permission of the American Institute of Aeronautics and Astronautics).

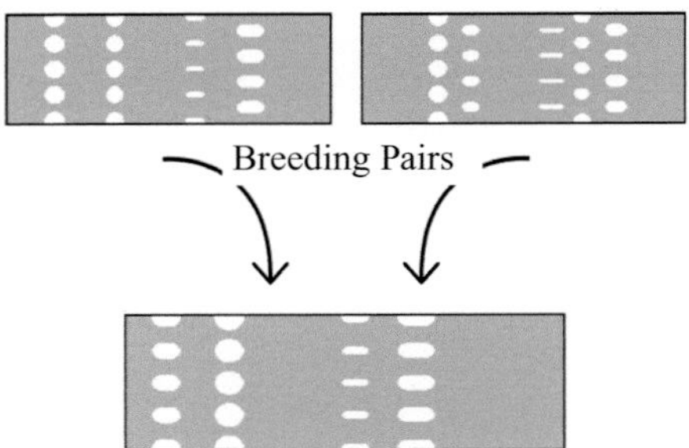

Figure 6.6. Breeding from surviving parents (from Watson and Tucker 2014 – published with kind permission of the American Institute of Aeronautics and Astronautics).

from the first generation. A completed offspring-breeding task is shown in Figure 6.6. If a child is generated which violates a viability constraint, a second child is re-bred from the same parents, and so on, until a viable child is produced.

Figure 6.7 gives a small sample of some of the flows captured during the optimization. Although not so clear from this figure, the turbulator layout can have a dramatic impact on the fluid flow. Figure 6.8 plots the axial variation of adiabatic film cooling effectiveness. The line with the typically higher (on average) values is for the optimized design. Clearly there is a substantial improvement. However, ideally the design should also have considered the discharge coefficient. This is another critical parameter of design interest.

The multitude of data produced in the genetic algorithm based search can be analysed to ascertain the degree of correlation between different variables. This can inform low-order models and subsequent design optimizations.

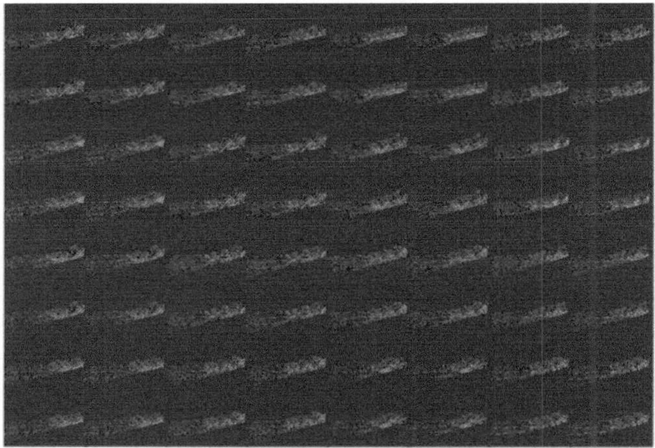

Figure 6.7. Range of flows for different tabulator configurations (from Watson and Tucker 2014 – published with kind permission of the American Institute of Aeronautics and Astronautics).

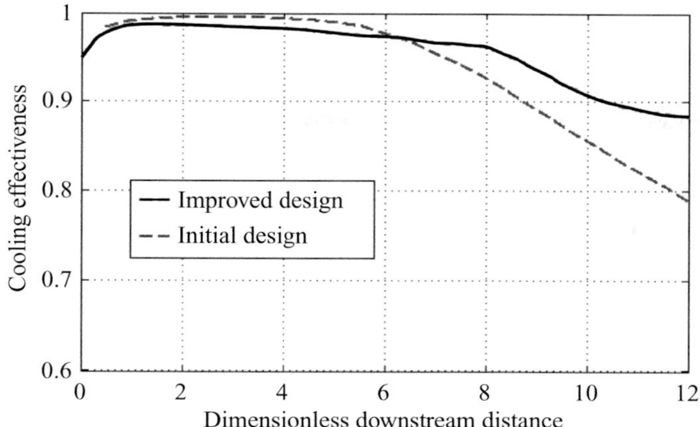

Figure 6.8. Optimized and original axial distribution of film cooling effectiveness (from Watson and Tucker 2014 – published with kind permission of the American Institute of Aeronautics and Astronautics).

6.2.4 *Inverse Design*

Inverse design is mostly used to design airfoils. The idea is that the blade shape will evolve to satisfy a target loading (surface pressure differences between the suction and pressure surfaces). The critical thing is to know the expected loading distribution and this aspect has given rise to criticism of inverse design. For a complex, highly swept wing or fan with shocks, defining the optimum loading is not always easy. However, with transonic fans, for example, the loading distributions are quite rational, the pressure rises being sudden and shock driven, giving rise to rather rectilinear polygon loadings. The shock locations can be readily defined based on physical arguments (see Xu 2012). Also, if there is an existing, successful fan design that is close in scale to a desired new design, this loading distribution can be used. A strategy for manual changes in loading distribution shape was offered by Medd et al. (2003), which concentrated on introducing areas of steep gradient to locate the normal shock over blades. Roidl and Ghaly (2011) tailored the loading distributions to reduce pressure loss coefficients and improve efficiency. Ramamurthy and Ghaly (2010) used a weighted average of two different loading distributions, each with different objectives, to carry out dual point design optimization. As an alternative to manual changes, optimization was used in conjunction with inverse design on centrifugal and axial compressors by Bonaiuti and Zangeneh (2009).

The algorithm of Hield (2008), on the other hand, had an outer performance loop, which automatically ensured the design pressure rise was met for a target mass flow rate. Hence the approach can be readily applied to industrial compressor design. The target mass flow rate and pressure ratio would be supplied to the aerodynamic designer as part of the system specification. Another interesting application, is the use of inverse design in nacelle design. With the approach of Wilhelm (2002) the optimal nacelle shape is designed in isolation from the wing. The pressure distribution from this nacelle is then extracted. When the nacelle is mounted to the wing, the nacelle

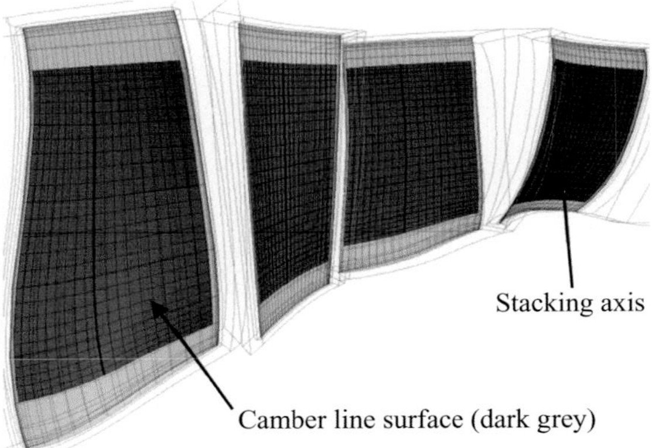

Figure 6.9. Blade geometry, camber line surface (identified by shading in dark grey) and stacking axis for inverse design of a multistage transonic fan (from Page et al. 2013 – published with kind permission of the American Society of Mechanical Engineers).

loading will change through aerodynamics interference. However, if the wing-nacelle geometry is fed into an inverse design tool, the nacelle geometry can, when using inverse design, adjust itself to satisfy the original design intent. Hence, for niche applications inverse design is a powerful tool.

Page et al. (2013) present a dual point inverse design. The approach is applied to a transonic fan, seeking, in essence, to ensure maximum efficiency at cruise and maximum thrust at take-off. This all again is constrained to meet critical system level requirements.

For airfoil sections, typically semi-inverse design is used. With this, an airfoil thickness distribution is specified and the airfoil/blade simply re-cambered. Indeed the thickness distribution must be sensible for structural reasons. Hence, the use of semi-inverse design places a useful natural design constraint. The key-defining feature of inverse design is that the designer is always working with the aerodynamic design intent such as mass flow, pressure ratio, shock location, peak suction surface diffusion and so forth. In normal aerodynamic design, the designer is normally working with geometry and adjusting this to match aerodynamic design intent. Which is easier, is perhaps debatable.

The approach of Page et al. (2013) is discussed next and an example of its application given for a two-stage transonic fan. Figure 6.9 shows the blade geometry with the camber line surface identified by shading in dark grey. The stacking axis is also shown. Along this axis, a set of two-dimensional blade profiles are stacked.

Figure 6.10 shows the process of adjusting the blade geometry to seek the target loading distribution. In Frame (a), the vertical lines identify camber line segments. These are automatically turned until they match the desired loading. The spines, attached normal to the camber line, prescribe the blade thickness distribution. Using the following equation

$$\Delta\alpha = K\frac{(\Delta p_{CFD} - \Delta p_{target})}{p_{SS} - p_{PS}} \tag{6.1}$$

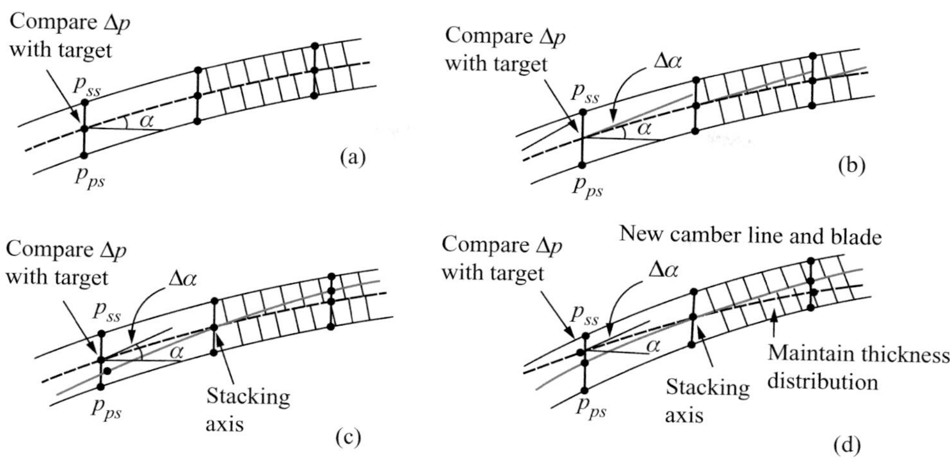

Figure 6.10. Process of adjusting the blade geometry to seek the target loading distribution: (a) initial camber line segments and the thickness distribution, (b) camber line segments updated using Equation (6.1), (c) updated segment alignment and (d) blade surface update based on the thickness distribution (Published with kind permission of the American Society of Mechanical Engineers).

the angle of the segments in Frame (a) are updated to seek the target pressure difference across the blade, Δp_{target}, where

$$\Delta p = p_{SS} - p_{PS} \tag{6.2}$$

is the difference between the pressure surface pressure, p_{PS}, and the suction surface pressure, p_{SS}. The update is based relative to the current pressure difference in the CFD – Δp_{CFD}. An under-relaxation parameter K is needed. However, a wide range of values for this yields satisfactory convergence. The segments that have been updated using Equation (6.1) are shown in Figure 6.10b. As shown in Frame (c), once the segment angles have been updated, they are aligned and relocated to the stacking axis. Then, as shown in Frame (d), the blade geometry is updated based on the thickness distribution. The blade movement is achieved through transfinite interpolation as discussed in Chapter 3. As indicated in Figure 6.9, at the blade extremities, inverse design is not applied. The updated geometry is smoothly blended to the static geometry at the extremities. In essence the process is easy to apply. As noted previously, normally in a design context (for a compressor), the dimensionless mass flow and pressure ratio are prescribed. Then, all that is needed is to define the loading distribution, say based on a past successful design. The fan will then design itself. As in the process of Hield, an outer performance loop is used to increase the blade loading to match the desired pressure ratio.

Figure 6.11 further illustrates inverse design of a multistage transonic fan. Frame (a) shows the final design. Frame (b) shows the mid-span stator geometry before and after design. And Frame (c) shows the initial and final rotor loading distributions. Since the fan is transonic, the geometry changes are quite minor, but Equation (6.1)

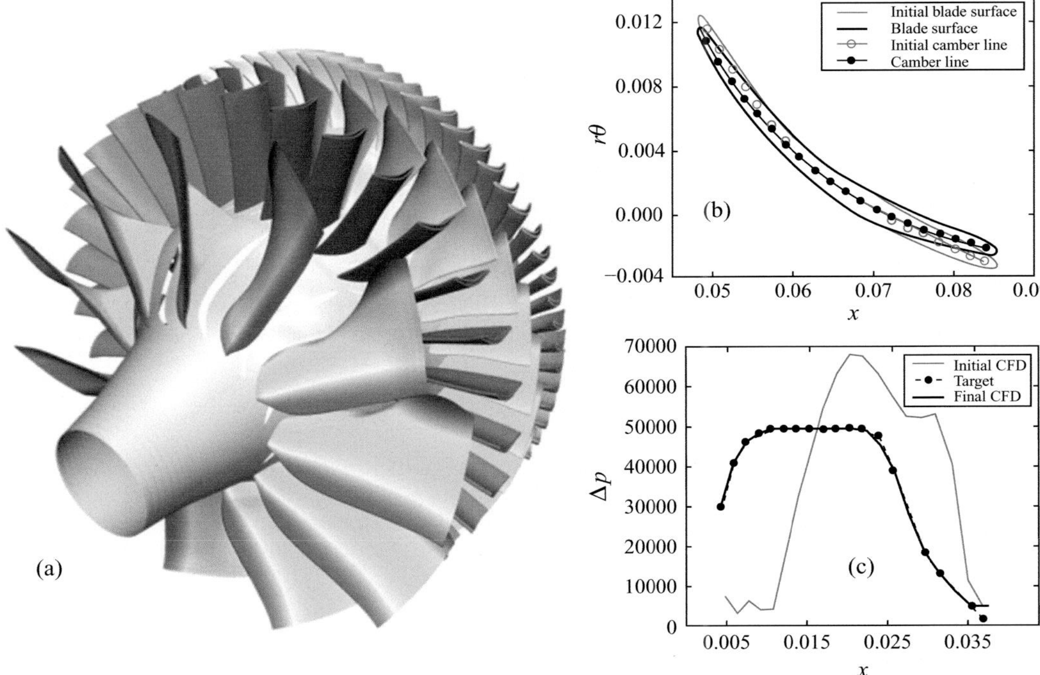

Figure 6.11. Inverse design of a multistage transonic fan: (a) final geometry, (b) mid-span stator geometry before and after design and (c) initial and final rotor loading distribution (adapted from Page et al. 2013 – published with kind permission of the American Society of Mechanical Engineers).

can support substantial geometry changes, yielding successful fan section designs. The simple update Equation (6.1) is highly effective. More complex update expressions can be found in Nill-Ahmadabadi et al. (2010).

6.2.5 *Adjoint Method*

The use of the adjoint method for the 'design' optimization of meshes is outlined in Chapter 3. Here, its use is considered for component optimization. Obviously, the two aspects are closely aligned. However, to be more compatible with design optimization, the Chapter 3 derivation is repeated here, but in a slightly different way that follows Wang and He (2010). The objective function, ϕ, is expressed as

$$\phi = \phi(\boldsymbol{U}, \alpha) \tag{6.3}$$

such that ϕ is a function of the flow field U and also a design variable α. The design and flow variables are related as

$$A(\boldsymbol{U}, \alpha) = 0 \tag{6.4}$$

In classical gradient-based design optimization, such as the steepest descent method, the gradient of the objective function with respect to the design variable is the key information needed. This can be expressed as

$$\frac{d\phi}{d\alpha} = \frac{\partial\phi}{\partial\alpha} + \frac{\partial\phi}{\partial\boldsymbol{U}}\frac{\partial\boldsymbol{U}}{\partial\alpha} \tag{6.5}$$

Looking at the right-hand side of the previous equation, the first term, $\partial\phi/\partial\alpha$, and $\partial\phi/\partial\boldsymbol{U}$ can be evaluated without any expensive iteration. However, the last term is expensive – $\partial\boldsymbol{U}/\partial\alpha$. It involves the solution of the typically non-linear multiple flow equations ($\boldsymbol{U}$ is a vector). The gradient could be evaluated as a finite difference. Alternatively, the following linearized flow equation could be solved

$$\frac{\partial A}{\partial\alpha} + \frac{\partial A}{\partial\boldsymbol{U}}\frac{\partial\boldsymbol{U}}{\partial\alpha} = 0 \tag{6.6}$$

Again, the previous is expensive to solve, needing iteration. What is more, for every design variable the equation would need to be solved. The power of the adjoint method is that with it, Equation (6.6) is recast so that repeated expensive solution of large systems of non-linear equations is not necessary, allowing the exploration of the impact of numerous design variable changes. The exploration of the design space is moved more to a post-processing of existing data. This is achieved by decoupling the objective function sensitivity to the flow field variable sensitivity, $\partial\boldsymbol{U}/\partial\alpha$, evident in Equation (6.5). To do this, Equation (6.6) is multiplied by the adjoint variable (a vector like $\boldsymbol{U}$), $\boldsymbol{\lambda}$, and then subtracted from Equation (6.5) as given:

$$\frac{d\phi}{d\alpha} = \frac{\partial\phi}{\partial\alpha} + \frac{\partial\phi}{\partial U}\frac{\partial U}{\partial\alpha} - \boldsymbol{\lambda}^T\left[\frac{\partial A}{\partial\alpha} + \frac{\partial A}{\partial U}\frac{\partial U}{\partial\alpha}\right] \tag{6.7}$$

which is then rearranged as follows

$$\frac{d\phi}{d\alpha} = \frac{\partial\phi}{\partial\alpha} - \boldsymbol{\lambda}^T\frac{\partial A}{\partial\alpha} + \left[\frac{\partial\phi}{\partial\boldsymbol{U}} - \boldsymbol{\lambda}^T\frac{\partial A}{\partial\boldsymbol{U}}\right]\frac{\partial\boldsymbol{U}}{\partial\alpha} \tag{6.8}$$

It then follows that if the adjoint variable satisfies the following equation

$$\frac{\partial\phi}{\partial\boldsymbol{U}} - \boldsymbol{\lambda}^T\frac{\partial A}{\partial\boldsymbol{U}} = 0 \tag{6.9}$$

that

$$\frac{\partial\phi}{\partial\alpha} = \frac{\partial\phi}{\partial\alpha} - \boldsymbol{\lambda}^T\frac{\partial A}{\partial\alpha} = 0 \tag{6.10}$$

Notably, now unlike Equation (6.5), Equation (6.10) does not depend on the flow variable sensitivity. Also, the ingredient that has replaced this issue, the adjoint variable, given by Equation (6.9), is also not a function of any design variable. The flow variable $\boldsymbol{U}$ is evaluated just once and then the adjoint variable $\boldsymbol{\lambda}$. Next, the gradients can be evaluated using Equation (6.10) for wide ranges of, say, geometric perturbations without any expensive iterations or matrix inversions. The downside is that the design variable changes must be small perturbations. However, for high-speed transonic flows, such perturbations are of critical importance. Typically, the adjoint approach is used to optimize relatively linear problems. Lemke et al. (2014)

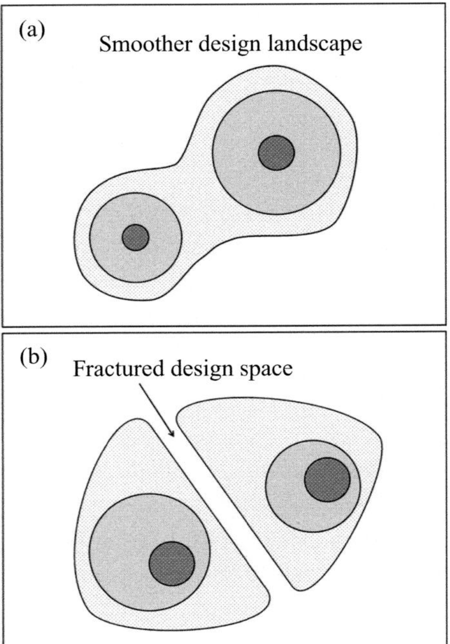

Figure 6.12. Design optimization landscape for low and high fidelity methods: (a) low fidelity and (b) high fidelity.

apply the adjoint method to reactive compressible flow. Even with the non-linearity of the reaction terms, the approach is found to work well.

6.2.6 *Response Surfaces*

The discussion in this section is informed by the work of Keane and Scanlan (2007) and Forrester and Keane (2009). As indicated in Section 6.2.2, response surfaces are a useful design optimization tool. Key assumptions for their use are that the engineering function is smooth and continuous. The latter constraint can be violated by the presence of shocks. If the function is not continuous, then multiple surrogate models can be patched together. With regards to smoothness of the function: it might be intrinsically smooth but the lack of iterative convergence or round of error might give rise to the need for smoothing. Figure 6.12 shows potential design landscapes from both low and high fidelity methods. As Frame (a) indicates, low-order models can tend to have less fractured design landscapes and thus be used to navigate across discontinuities to prevent being locked in a local minimum.

In some physical processes the functional relationship between variables is known. This is useful, since it informs on the ideal function to be used in the surrogate model. However, with non-linear aerodynamic systems, typically the function landscape involved will be complex. Notably, ideally we wish the function landscape to be most accurate around the zone of interest (i.e., the location of the optimal design solution).

6.2.6.1 Polynomial Response Surface The most popular response surface form is the polynomial. In one dimension, this can be expressed as

$$\phi^A = a_0 + a_1 x + a_2 x^2 + \cdots + a_m\, x^m = \sum_{i=0}^{m} a_i x^{(i)} \tag{6.11}$$

where the coefficients $\boldsymbol{a} = \{a_0,\ a_1, \ldots, a_m\}^T$ are approximates, in a least squares sense. This is achieved by solving

$$\mathbf{a} = (\boldsymbol{\Phi}^{\mathrm{T}}\ \mathbf{Wf}\ \boldsymbol{\Phi})^{-1}\ \boldsymbol{\Phi}^{\mathrm{T}}\mathbf{Wf}\ \phi \tag{6.12}$$

where **Wf** has diagonal values of unity and off-diagonals of zero. Also, $\boldsymbol{\Phi}$ is the vector of observed responses and

$$\boldsymbol{\Phi} = \begin{pmatrix} 1 & x_1 & x_1^2 & \cdots & x_1^m \\ 1 & x_2 & x_2^2 & \cdots & x_1^m \\ \vdots & \vdots & \vdots & \ddots & \vdots \\ 1 & x_n & x_n^2 & \cdots & x_n^m \end{pmatrix} \tag{6.13}$$

The exponent m needs to be selected with care. If it is too high, over fitting will occur – the curve for ϕ^A will have substantial oscillations that could lead to inaccuracies. On the other hand, if m is too low, inaccuracies will again occur. As suggested by Forrester and Keane (2009) m can be selected to minimize the cross correlation error (to be discussed later). Note that a weighted least squares procedure can also be used, where

$$\mathbf{Wf} = \begin{pmatrix} wf^{(1)} & \cdots & 0 \\ \vdots & \ddots & \vdots \\ 0 & \cdots & wf^{(n)} \end{pmatrix} \tag{6.14}$$

(i.e., points of the polynomial can be given different weightings, wf). This allows for the fact that different points may not have equal importance in the polynomial fit.

A variant of the weighted least squares approach is the moving least squares method. This is a hybrid approach where, between data points, in some regions, interpolations can be used instead of the polynomial fit. The weighting function is based on the distance of the point considered relative to an observed data point $x^{(i)}$. There are a wide range of potential functions. The Gaussian is popular:

$$wf^{(i)} = \exp\left(-\frac{\sum_{j=1}^{k} \left(x_j^{(i)} - x_j \right)^2}{L^2} \right) \tag{6.15}$$

Note that k represents the spatial dimension and L needs tuning. The latter can be altered to minimize the model error. Altering the weighting function makes the process expensive, since Equation (6.12) must be repeatedly evaluated.

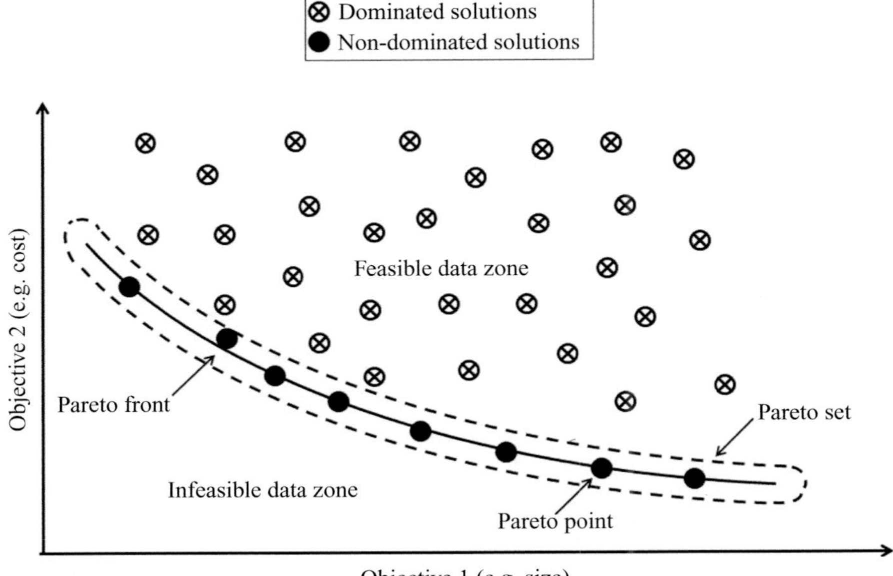

Figure 6.13. The Pareto diagram with the Pareto front.

6.2.6.2 Radial Basis Functions As an alternative to a polynomial fit, radial basis functions can be used. These are essentially weighted combinations of functions to create a best fit. The weighting function can take a range of forms such as linear, cubic, logarithmic and Gaussian. An increasingly popular weighting function, similar to the latter, is the Kriging (based on the work of the South African mining engineer D. Krige who initially developed the method) function. This has additional tuning parameters. These have the downside of making the training period longer but also offer the potential for greater accuracy. Subsets of the Kriging method are universal Kriging and blind Kriging.

The previous approaches have substantially different traits. Hence, it can be beneficial to hybridize them, using a weighted average of different models. Indeed, blind Kriging could be considered to be an approach of this type.

6.2.7 *Multi-Objective Design*

With multi-objective design, different weightings or levels of importance need to be given to different objective functions. The Pareto front is an important element in multi-objective design. If there are two objective functions, it will appear as in Figure 6.13, which could plot, for example, cost against size. The figure indicates that making the component miniature is costly. On the other hand, the larger component is cheaper. The data points in the upper, more right-hand zone are non-optimal. The line is called the Pareto front and solutions on this front are optimal in some sense. However, they involve a trade-off and hence a design decision based on weighted importance.

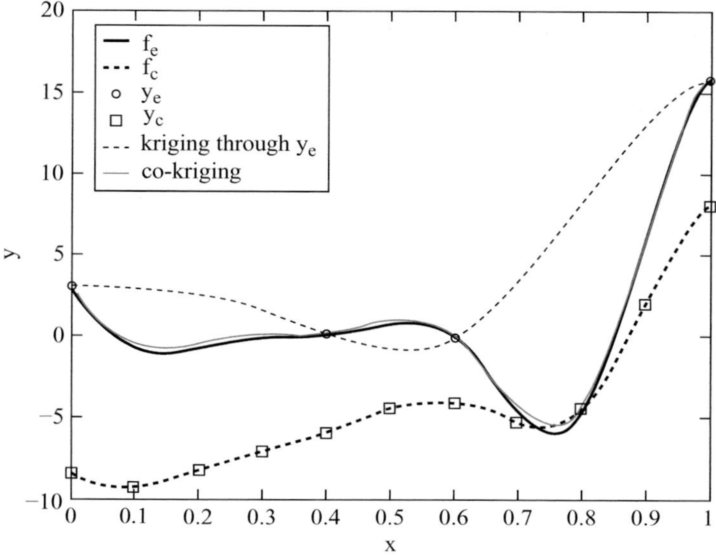

Figure 6.14. One-dimensional example of the multi-fidelity modelling through co-Kriging (from Forrester and Keane 2009 – copyright Elsevier).

The points on the Pareto front form what is termed a Pareto set – all the points being optimum in some sense but the relative weightings to the different objectives yet to be applied. The multiple objective functions can be combined into a single objective via weighting functions, and hence a more simple design optimization procedure applied. The latter functions can be non-linear. It is possible to reverse engineer the Pareto front from such a process. Clearly, construction of a Pareto front is expensive and a surrogate model is useful for this purpose.

6.2.8 *Multi-Fidelity Approaches*

Some CFD approaches such as eddy-resolving simulations or computations involving full Reynolds stress models can be expensive. Then it can be useful to supplement data from high fidelity simulations, ϕ_{acc}, and even expensive rig test data, with lower order modelling data, ϕ_{app}. Cleary, there is a trade-off. The cheap data must be sufficiently cheap that a substantial extra quantity can be generated to benefit the high accuracy model and also the low-order model must be sufficiently meaningful. For example, for flows involving large-scale separation, there is the possibility that cheaper RANS solutions will steer the optimal design in completely the wrong direction – see Watson et al. (2014).

Figure 6.14, taken from Forrester and Keane (2009), shows a simple one-dimensional example of the use of co-Kriging to synthesize low and high-order data. The plot shows that there are four expensive data points. Eleven cheaper data points supplement these. The thinner dashed lines gives the fit, from applying standard Kriging to the high fidelity data. Clearly this misses the minimum in the function at around $x = 0.75$. However, supplementing the high fidelity data with the low

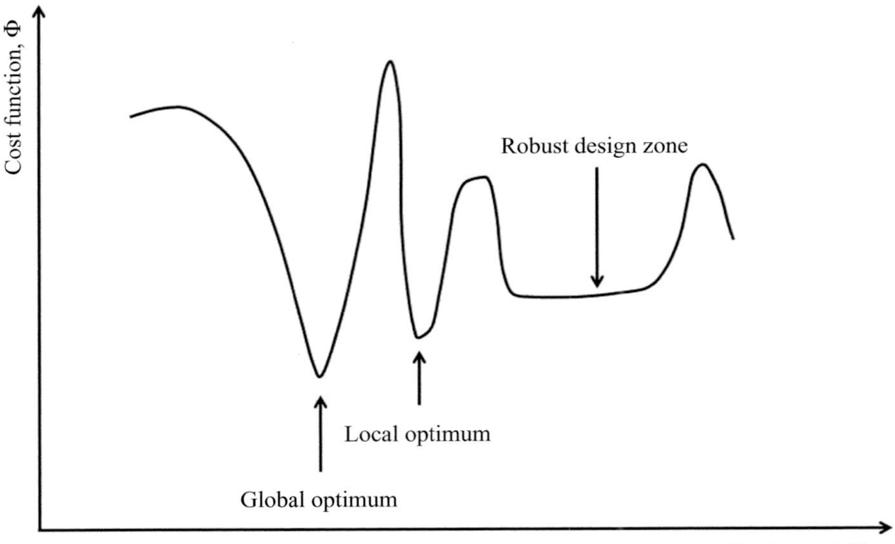

Figure 6.15. Consideration of a robust design optimum rather than a global design optimum.

fidelity through so-called co-Kriging ensures that this feature is present. This can be seen from the thinner full line, which is close the most accurate data (given by the thicker full line).

As show in Figure 6.15, it is not always a good idea to seek the global minimum/optimum. The simulation process can be inaccurate. Hence, with a complex engineering system, when built, if the global optimum is a small zone in the design space, it might actually be missed. Hence, sometimes it is better to aim for a robust minimum, where there is higher confidence that in the manufactured product an improved design will arise.

6.2.9 *Accuracy Checks*

This section is split into two aspects. The first is accuracy checks prior to fully developing the surrogate model. The second is assessing the accuracy of the more mature model. With regards to the former, based on a set of validation data, the accuracy of different surrogates for a particular problem can be assessed. Prior to this, the traits of the different models, as summarized in Table 6.3, can be considered. Using validation, as noted, requires the earmarking of a specific lot of data for this purpose. Hence, if the data is expensive to gain, such a process is not practical. In that case, cross-validation can be used.

6.2.9.1 Cross-Validation The surrogate model/response surface accuracy can be assessed using the cross-validation process. With this, the data is divided into n approximately equal subgroups. The surrogate model is fitted with each individual subgroup removed in turn. Approximates to the removed values, ϕ_i^A, can be

Table 6.3. *Advantages and disadvantages of different response surface models*

Methods	Advantages	Disadvantages	Other comments
Polynomial response surface	Suited to simple design landscapes with low dimensionality where data is cheap to obtain	Care is needed with the selection of the polynomial order. For problems with a high number of dimensions, it is difficult to gain sufficient data. Care is needed with order selection	Order can be selected based on minimization of the cross correlation error
Basic radial basis functions	Good for higher dimensional problems with simple design landscapes	–	Has a wide range of related methods
Kriging	Good control of surface fit in terms of the degree of regression. Sound theoretical basis for judging the level of curvature. Provides an error estimate basis. Informs on the importance of different variables. Suitable for complex design landscapes. Has the potential to give more accurate predictions relative to simpler methods	Expensive to train for higher dimensional problems	Blind Kriging is a jack-of-all-trades approach that is well suited if the design landscape is unknown

evaluated from the surrogate models and errors relative to the full set evaluated. For n approximates, the error is

$$E = \frac{1}{n}\sum_{i=1}^{n}\left[\phi_i^n - \phi_i^A\right]^2 \tag{6.16}$$

Note that ϕ_i^n are values for the full surrogate model without any sets removed.

6.2.9.2 Infilling The surrogate model should be used to gain an initial estimate of the best design. However, once this point has been found, the high fidelity analysis should be used to check how meaningful the surrogate model prediction is. Indeed, to gain the most accurate results, it is necessary to generate what are termed infill points to update the surrogate model in the zone of interest. As can be inferred from Figure 6.14, care is needed with selecting the infill criteria. The actual function has multiple minima and there is a danger of becoming locked in the wrong area. Hence, care is needed with location of the infill and there is an extensive range of methods in this area. Typically, the procedure will stop when the computing resource has run out.

6.2.10 *Constraints*

For engineering problems, there are always regions where it is known a priori that the geometric form will violate one of the typically numerous design constraints.

Hence, the design optimization must be configured so that such constraints are not violated. This can be achieved through the use of penalty functions. This function would penalize the output from the surrogate model so that when the model is explored the constraint zone is kept away from. If a simulation fails or is poorly converged, a penalty could also be applied so that less or no emphasis is applied to this data.

6.2.11 *Hybrid Design Optimization*

Page et al. (2015) wrap a genetic algorithm around inverse design. The genetic algorithm, working together with inverse design seeks to optimize the aerodynamic performance with respect to the objectives of improved aerodynamic efficiency and stall margin. Broadly, the design loading distribution is left the same, in that the shock locations are fixed. However, details of the form of the loading distribution are optimized through the use of Bezier curves offering just more than ten control points over the complete three-dimensional blade surface. Hence, this all allows an experienced designer to place the loading distribution in the correct area of the design landscape through correct shock location. Then, the genetic algorithm makes the subtle refinements to that loading distribution. These can become especially important for transonic flows. The inverse design procedure, by its very nature, ensures that the global design target parameters (e.g., pressure ratio, loading distribution and mass flow for a particular rotational speed) are satisfied. For each design, the stall margin and efficiency data is calculated. The genetic algorithm then ranks the loading distributions based on the two design objectives and mutates them to seek improved generations of loading distributions. This process is summarized in Figure 6.16.

The advantages of using inverse design and the genetic algorithm together are that the former reduces the design space. It does this through fixing the shock position and general performance. Hence, the inverse design places the genetic algorithm in the right part of the design landscape, intelligently reducing the design space to be explored. The genetic algorithm process is also massively parallelizable and avoids becoming trapped in local minima. Figure 6.17 shows the improvement in the fitness (stall margin × efficiency) of the transonic fan design for the first three generations. This is taken from Page et al. (2015).

6.2.12 *Design Optimization with Eddy Resolving Simulations*

As identified towards the end of Chapter 5, eddy-resolving simulations (such as LES and DES) are key emerging technologies for industrial design. Hence, a critical question is how these approaches can be used with design optimization. A key issue with eddy-resolving methods is that long time averages are needed to gain smooth data. To assess the variation of an objective function to a design alteration, typically a finite difference is needed. However, for this to be meaningful, the objective function must be time converged to a high degree. This is costly. The adjoint approach also suffers similar limitations. Figure 6.18, taken from Larsson

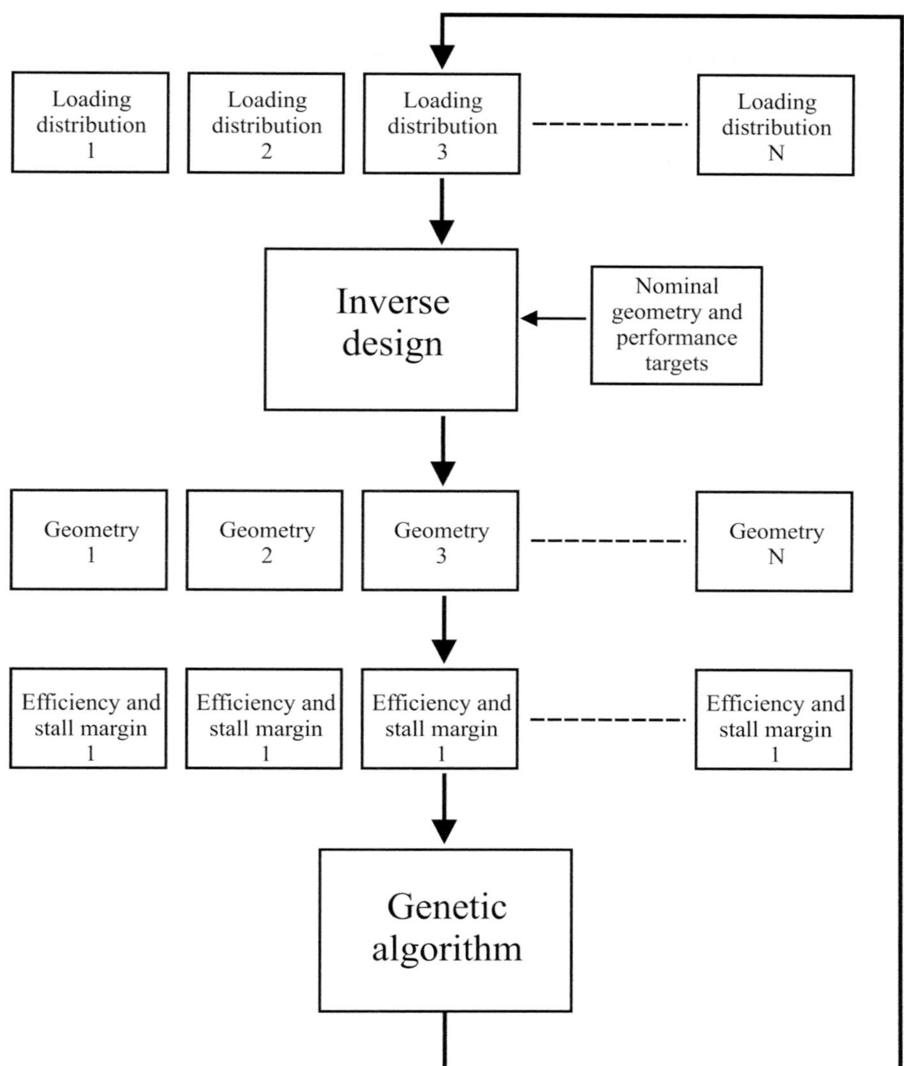

Figure 6.16. Hybrid inverse design procedure involving a genetic algorithm and also inverse design (from Page et al. 2015).

and Wang (2014), analyses the use of design optimization for a surrogate Lorenz (a system of differential equations, first studied by Lorenz, that have chaotic solutions) system. The variable of interest is J and s is a design parameter. The vertical axis is time, t. Frame (a) gives the instantaneous $J(s, t)$ when computed as an initial value problem. Frame (c) gives the time average of this. As can be seen, the field is extremely noisy and unsuitable for extracting gradient information. Frames (b) and (d) repeat the previous when a promising approach called shadowing is used. This involves a linearization of the eddy-resolving equations and is described in Wang (2014). As can be seen, now the gradient information is much smoother.

However, the future of design optimization will involve increasingly multidisciplinary design optimization (MDO) – see Bakker et al. (2012). This is problematic for

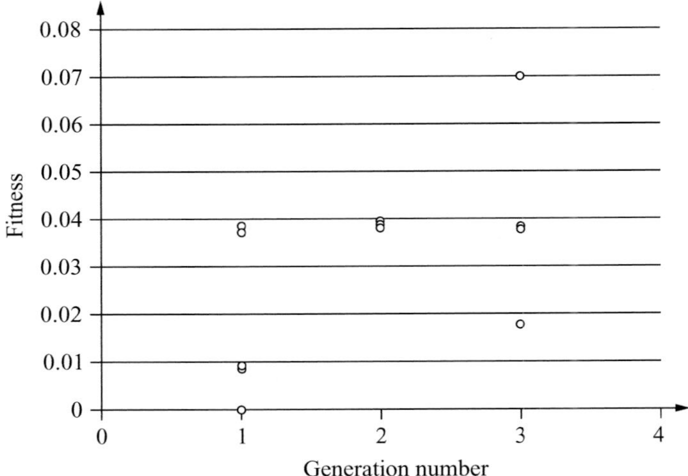

Figure 6.17. Improvement in fitness of design for first three generations (from Page et al. 2015).

expensive eddy-resolving simulations. For some time now, substantial use of MDO has been made use of in aerospace to encompass, for example, economics, propulsion, structural and aerodynamic behaviour. Many other aspects can be potentially included, such as aeroacoustics and electromagnetics linked in with, for example, pilot response to events. All this means that reduced order models, potentially informed by large eddy simulations, are of some importance. Hence, next further forms of reduced order model are considered that could be used as part of a design optimization.

6.3 Reduced Order Models

In Chapter 2, the full Navier-Stokes equations were presented; a range of reduced, related forms of this equation were then given. Notably, these included the following forms:

1) Axisymmetric;
2) Euler;
3) Thin shear layer;
4) Boundary layer;
5) Parabolic;
6) Potential flow;
7) Viscous Reynolds form;
8) Creeping flow/porous media equations and
9) Throughflow equations.

The equations can be considered as low-order models suitable for use in coupled, design optimization problems or both. For some flows, the previous equations, or even the full RANS or Navier-Stokes equations, need to be used in their unsteady

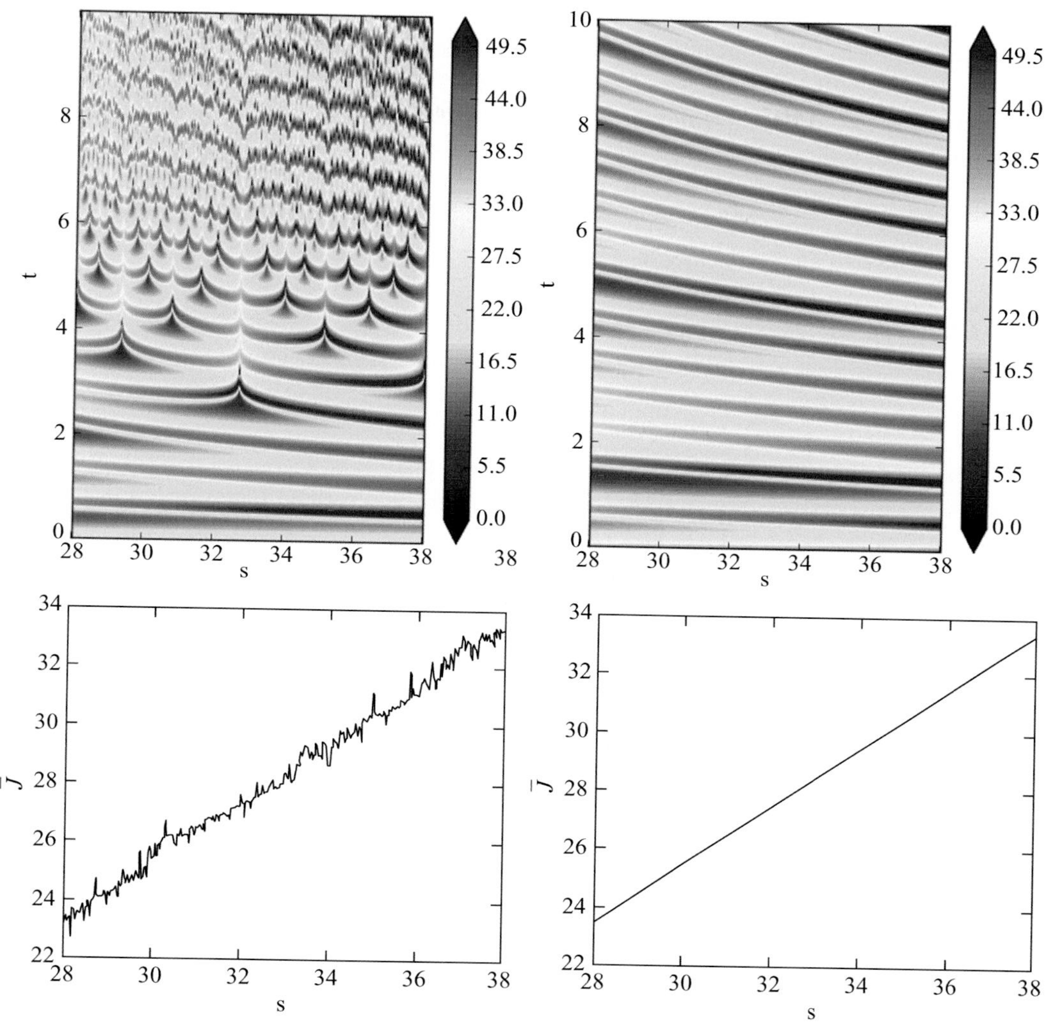

Figure 6.18. Use of design optimization for a surrogate Lorenz system: (a) instantaneous $J(s, t)$, (b) instantaneous $J(s, t)$ when shadowing is used, (c) time average of $J(s, t)$ and (d) time average of $J(s, t)$ when shadowing is used (From Larsson and Wang 2014 – used with the kind permission of the Royal Society).

form. This gives rise to substantially increased computational cost. To reduce this, for some classes of flow it is possible to treat the unsteadiness as a perturbation about a mean. Hence, next, perturbation methods are discussed as a form of lower order model.

6.3.1 *Perturbation Equation Based Methods*

6.3.1.1 Linear Harmonic Method With the linear harmonic approach, essentially we have the following three-component decomposition:

$$\phi = \overline{\Phi} + \phi'' + \phi' \tag{6.17}$$

where $\overline{\Phi}$ is the time mean of the flow. If possible (i.e., if a steady solution can be secured) this can be computed using RANS. If RANS is used to solve for $\overline{\Phi}$, the final ϕ' term will naturally, via, for example, eddy viscosity, be modeled. Then an equation for linear perturbations, ϕ'', about the mean is solved. This can take place either in the time or frequency domains. The frequency(s), ω, of the perturbations are specified. The superposition of these solutions gives the complete solution. The use of linear harmonic methods is especially popular for solving the propagation of acoustic waves. The latter area is discussed later in the chapter.

6.3.1.2 Non-linear Harmonic Methods To understand non-linear harmonic methods, it is helpful to assume that the flow variables, ϕ, vary as

$$\phi(x,t) = \overline{\Phi}(x) + B(x)\cos(\omega t) + C(x)\sin(\omega t) \tag{6.18}$$

Hence, for a specified ω, the unsteady flow time history is described if $\overline{\Phi}(x)$, $B(x)$ and $C(x)$ are known. We assume that the unsteady discretized equation takes the form below

$$\frac{\partial\phi(x,t)}{\partial t} = A(\phi) \tag{6.19}$$

Then combining Equations (6.18) and (6.19) gives

$$\begin{aligned}&\omega(-B(x)\cos(\omega t) + C(x)\sin(\omega t))\\ &\quad = A(\overline{\Phi}(x) + B(x)\cos(\omega t) + C(x)\sin(\omega t))\end{aligned} \tag{6.20}$$

with three unknowns – $B(x)$, $C(x)$ and $\overline{\Phi}(x)$. However, if three t values are specified, for a fixed ω, corresponding to three different points in an unsteadiness cycle, three steady state equations will arise. A more general, non-linear procedure for multiple harmonics (ω values) is given by Hall et al. (2002). Non-linear harmonic methods are popular with turbomachinery and rotor simulations where the unsteadiness is periodic.

The reduction of unsteady problems to steady makes application of the adjoint design optimization approach to unsteady flows much more practical (otherwise the complete time history of the flow solution must be stored and a backwards integration through time performed). For any design variable, α, Equation (6.18) can be re-expressed as

$$\phi(x,t,\alpha) = \overline{\Phi}(x,\alpha) + B(x,\alpha)\cos(\omega t) + C(x,\alpha)\sin(\omega t) \tag{6.21}$$

and the optimization problem reduced to one involving steady equations. The actual adjoint equations(s) can be dealt with in a similar way to the flow equations.

6.3.1.3 Non-linear Disturbance Equations NLDE first requires the mean flow, $\overline{\Phi}$, established from RANS. Then, a disturbance equation is solved. However, with the NLDE approach, there is no specified frequency. The disturbance equation retains non-linear terms. More notably, these disturbances do not interact with the mean flow. As with the linear harmonic method, there is still the potential three-level

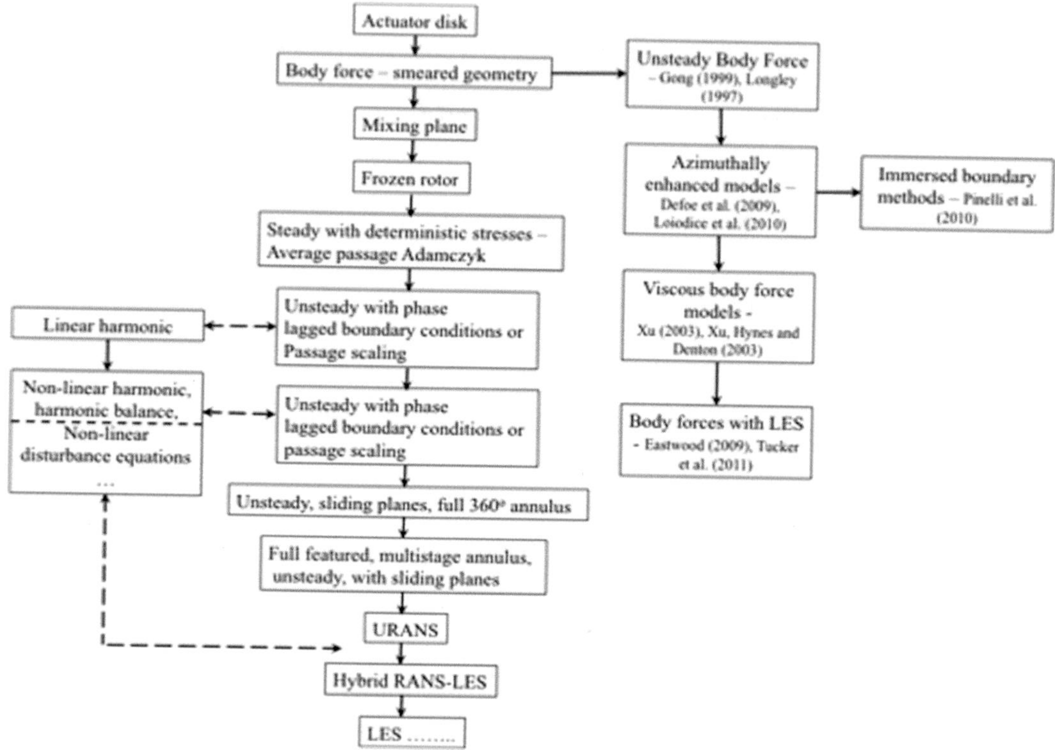

Figure 6.19. Hierarchy of methods associated with the modelling of axial turbomachinery as found in gas turbine aeroengines.

flow decomposition given by Equation (6.17). The fluctuations ϕ'' about $\overline{\Phi}(x)$ are provided by NLDEs – and they no longer need be periodic functions. Distinctly, this time ϕ' are turbulent fluctuations but at a subgrid scale. For more details on the NLDE approach, see Batten et al. (2007).

6.3.2 *Sliding and Mixing Planes and Phase Lagged Boundary Conditions*

Some machines consist of multiple rows of rotating and stationary blades. The unsteady flow arising from this gives rise to a hierarchy of methods. Some of these are shown in Figure 6.19 and will be explored here in different areas of the text. For maximum fidelity, what are termed as sliding planes are needed. With these planes, the rotor mesh(es) slide/rotate relative to the stator(s). They are incremented in the tangential direction every time-step. Clearly an unsteady 360° calculation (i.e., including all blades) is expensive. However, if the number of rotors and stators are the same or integer multiples (see Figure 6.20), their domains can be made to have the same circumferential extent (pitch). Then, if the flow is periodic in the time mean, sliding planes with just single or low numbers of meshed blades can be used. For the single passage calculations, circumferentially periodic boundary conditions are applied at the domain faces coincident with the neglected adjacent blades.

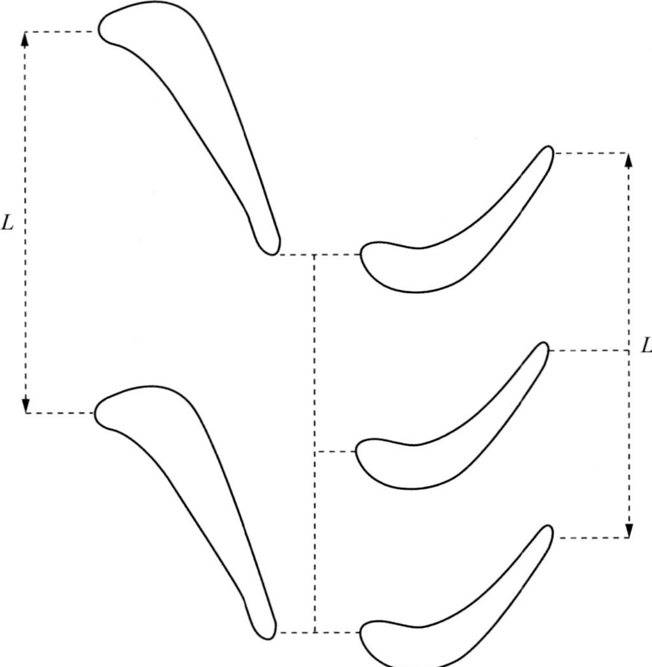

Figure 6.20. Blade configuration that naturally allows use of periodic boundary conditions (Published with kind permission of Elsevier).

Unfortunately, in practical systems the number of rotors and stators are unlikely to allow modelling low blade counts that give natural periodicity. Then, specialized periodicity boundary conditions are needed. Such boundary conditions can involve storing time histories at periodic boundaries and applying an appropriate time shift (see Erdos et al. 1976).

To make the blade row calculations steady, either the frozen rotor technique or so-called mixing planes (see Denton 1992) can be used. The frozen rotor approach keeps the relative positions of the rotor and stator fixed. The governing equations are solved in the most appropriate coordinate system for them (i.e., rotating for the rotor). However, the data transfer is direct, with no temporal information clocking. The solution will depend on the frozen locations of the rotor relative to the stator. Ideally, to get a best idea of the system behaviour, several simulations are needed with the different frozen relative rotor and stator positions.

Mixing planes link the stationary and rotating blade row domains through transferring circumferentially averaged solution variables. They prevent the issue of having solutions that are dependent on the chosen position of the rotor relative to the stator. Figure 6.21, taken from Holmes (2008), shows the data transfer process at a basic mixing plane. Note that the figure shows just one streamwise aligned gridline(s). There will be a series of such lines in the circumferential and radial directions. The process is such that after an initial iteration period, data at plane A is circumferentially averaged and represented at location C. Similarly, a circumferential averaging

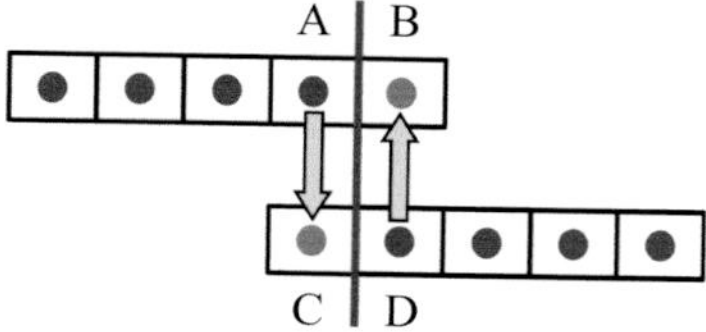

Figure 6.21. Data transfer process for a basic sliding plane (from Holmes 2008 – published with the kind permission of the ASME).

takes place at location D and this data is represented at B and so on. For stability, some under-relaxation is needed. Precisely what variables are averaged is an open question. For mass conservation, density weighting is needed. However, this will not guarantee conservation of other variables. Also, the circumferential averaging implies some mixing out and hence entropy rise. This aspect is empirically modelled. How well this connects with physical reality since wakes gradually mix is uncertain. Holmes (2008) notes the following desirable traits for mixing planes:

1) It should be possible to control the entropy rise at the mixing plane so that the level connects better with physical expectations.
2) More general variable control should be facilitated to enable calibration to rig and high fidelity CFD data. For example, the total pressure could be controlled to allow the matching of stages in multistage machines. This can be especially challenging in compressors. Small deviations become magnified axially inside compressors. Also, these devices tend to have many stages.
3) To speed up convergence in machines with many axial stages, it should be possible to pin the system mass flow at each mixing plane. Otherwise, the system convergence can become slow.
4) The interface should be as non-reflective as possible – this is a challenging issue.
5) The mixing plane should be able to deal with flow reversals and not have directional dependence.
6) The mixing plane should be as conservative as possible

With regards to Point (4), blade rows can be close together in turbomachinery, and in transonic flows there can be a strong interaction of shocks and adjacent blade rows. This can make dealing with reflections especially problematic. In relation to Point (6), Holmes proposes the use of a controller. This approach is shown in Figure 6.22. At each iteration, the difference in fluxes between the different sides of the mixing plane can be evaluated. This difference gives an error signal. The controller seeks to minimize this.

Note that there are a wide range of mixing plane formulations and also issues with regards to their implementation in different CFD solver frameworks. Different implementations can have a clear solution impact (see, for example, Hanimann et al. 2013). As shown by Benichou and Trébinjac (2014), the mixing plane location can also have a substantial impact on results.

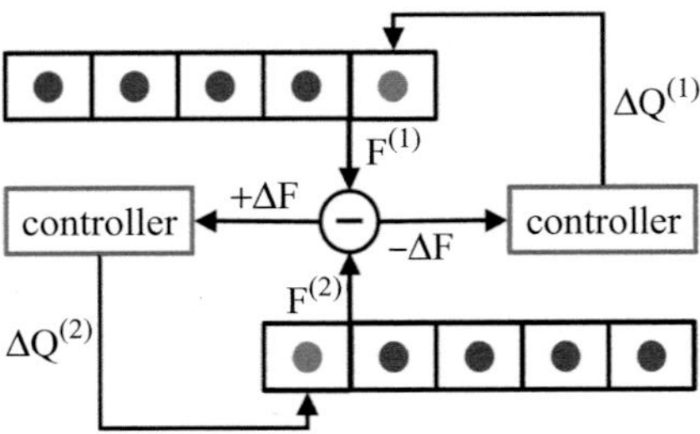

Figure 6.22. Use of a controller system to enforce conservation at a mixing plane (from Holmes 2008 – published with the kind permission of Rolls-Royce).

It is possible to hybridize simulation approaches. For example, Montomoli et al. (2011) model a four-stage (four sets of rotating and stationary blade row units that sequentially compress fluid) compressor. The third stage is sandwiched (through transformation terms) between upstream and downstream mixing plane treated stages. Hence, the first, second and fourth stages are made as steady simulations (RANS with the Spalart and Allmaras turbulence model described in Chapter 5). The third stage is treated as unsteady. Figure 6.23a shows the geometric configuration.

Frame (b) shows a simulation where just upstream of the URANS simulation zone (the third stage) a mixing plane is used. The mid-blade span contours give instantaneous eddy viscosity ratio fields. Note that immediately downstream of this plane the wake vanishes, being instantaneously mixed out. Frame (c) alternatively shows where the Frame (b) mixing plane is replaced by a sliding plane. The greater dynamic complexity of the flow in the stator zone of the third stage is clear from Frame (c), where the modelling fidelity is higher. It is possible to have further combinations of hybridization with a complete axial machine comprising sliding planes, mixing planes and frozen rotors along with lower order body force and actuator disk (see later) stages. These will be discussed next.

6.3.3 *Body Force Modelling*

6.3.3.1 System Scale A further reduction in cost can be achieved by no longer explicitly meshing the blades or rows of them. At the simplest level, and most applicable for, say, a low-speed fan in an electronic system, a planar polynomial momentum source, S_i, could be used. This could take the form

$$S_i = C_0 + C_1 U_j + C_2 U_j^2 \tag{6.22}$$

where U_j is the local normal velocity and C_0, C_1 and C_2 are constants (see Tucker and Pan 2001). On the other hand, losses such as grills can be modelled as quadratic momentum sinks of the form

$$S_i = \frac{1}{2} C_0 \rho U_j^2 \tag{6.23}$$

where S_i is a momentum sink for a fluid with a typically local (not the lower approach velocity) U_j passing through the grill. Appropriate values of C_0 for different grills can be found in design guidelines such as that of Fried and Idelchik (1989).

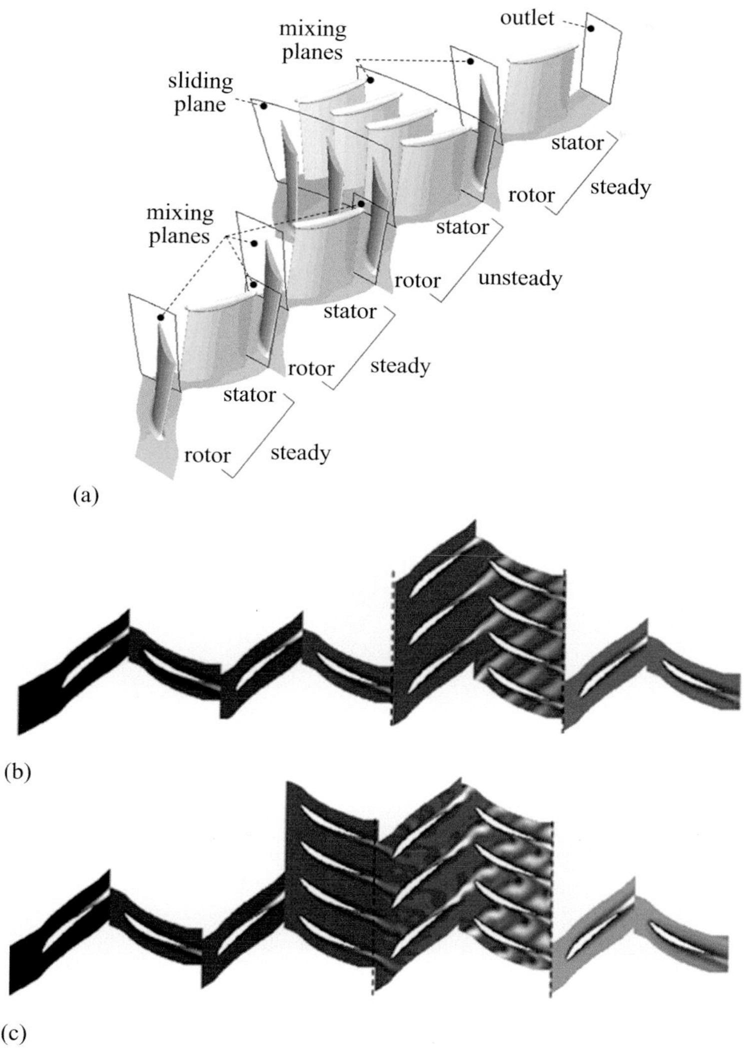

Figure 6.23. Hybridized simulation approach from Montomoli et al. (2011) for modelling a four-stage compressor: (a) geometric configuration, (b) mixing plane is used just upstream of the unsteady RANS simulation zone and (c) mixing plane, in Frame (b) is replaced by a sliding plane. Note that contours give instantaneous eddy viscosity fields. (Figure used with the kind permission of Sage Publishing.)

Figure 6.24 shows the application of these source and sink terms for modelling the flow in an electronic system. Frame (a) gives a schematic of an electronic system with momentum sources and sinks to account for fans and grills. Frame (b) shows the resulting flow field.

As shown on the right-hand branch of Figure 6.19, there are a range of body force models specific to more detailed turbomachinery blade row calculations. They typically solve the Euler equations in an axisymmetric form. Source terms are applied within the blade location zones. Generally the source term field is axisymmetric. They

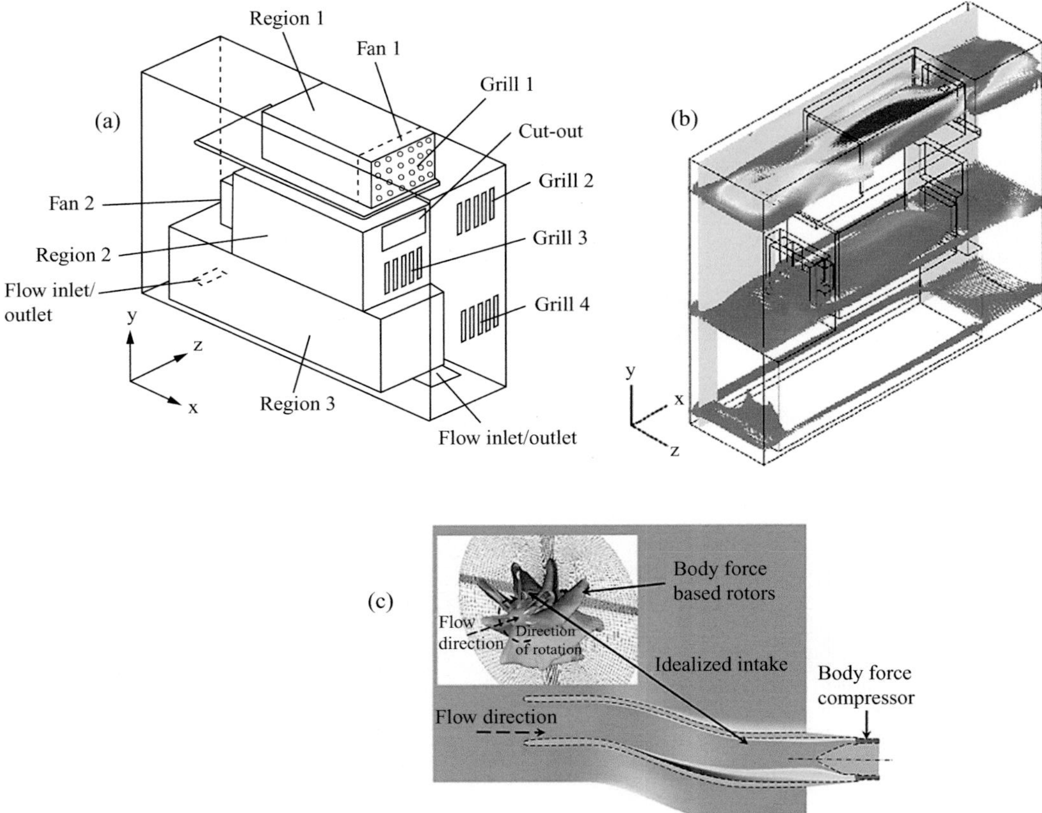

Figure 6.24. Examples of the use of body forces in engineering systems: (a) schematic of an electronic system with momentum sources and sinks to account for fans and grills, (b) the resulting flow field and (c) the use of a body force model to generate open rotor wakes upstream on an idealized aeroengine intake (Frame (c) is published with kind permission of Elsevier).

are designed to introduce appropriate levels of flow turning, essentially forcing the flow to follow local blade metal angles. Calibration is needed to make the model match system characteristics. There are a wide range of body force models – see for example Longley (1997) and Xu (2003). In the model of Gong (1998), the body forces are typically divided into components parallel, F_p, and normal, F_n, to the blade passage. In cylindrical coordinates, these components have the following form:

$$F_{n,x} = \frac{K_n}{L}\frac{V_\theta}{V_{rel}}(V_x \cos\alpha + V_\theta \sin\alpha)(V_\theta \cos\alpha - V_x \sin\alpha) + \frac{1}{\rho}\frac{\partial p}{\partial x}\sin^2\alpha$$

$$F_{n,\theta} = \frac{K_n}{L}\frac{V_x}{V_{rel}}(V_x \cos\alpha + V_\theta \sin\alpha)(V_\theta \cos\alpha - V_x \sin\alpha) + \frac{1}{\rho}\frac{\partial p}{\partial x}\sin\alpha\cos\alpha \tag{6.24}$$

$$F_{n,x} = \frac{K_{vd}}{L}V_x V_{rel}, \quad F_{p,\theta} = -\frac{K_{vd}}{L}V_\theta V_{rel}, \quad F_{p,r} = -\frac{K_{vd}}{L}V_r V_{rel} \tag{6.25}$$

where α now represents the blade metal angle and L is the pitch of the blades. The velocity components V_x, V_r and V_θ correspond to the axial, radial and tangential components. Furthermore, V_{rel} represents the fluid velocity relative to the solid blades. The previous equations have two calibration constants. The constant K_{vd} multiplies the viscous drag terms, and K_n the inertial force related terms. To extend the capability of body force models, it can be useful to generate wake zones and the corresponding momentum deficit and increased turbulence levels associated within these areas (see Loiodice et al. 2010). It is possible to create such zones by locally increasing source terms. An example of a simulation using this approach is shown in Figure 6.24. This shows an idealized open rotor engine intake downstream of a body force modelled rotor. The individual rotor blades are evident. The use of localized source terms, as opposed to a uniform distribution, can be found in immersed boundary methods (see Pinelli et al. 2010). Hence, body force models could also be viewed as low-order immersed boundary methods. For the use of body force models with large eddy simulations, see Eastwood et al. (2009) and Tucker et al. (2012). For an extension of Gong's model to account for azimuthal flow variations, see Defoe et al. (2009).

Xu (2003) and Xu, Hynes and Denton (2003) explore use of a so-called viscous body force modelling approach. With this, essentially just the viscous body force modelling component of Gong is implemented (Equation (6.25)). The inertial component is explicitly resolved through Euler solution of meshed blades.

6.3.3.2 Small Scale Some systems involve massive disparities in geometric scale. An example is a vortex generator on a wing. The difference in scale between this and the wing means that meshing and numerical resolution of such features is challenging. What is termed the BAY model (see Bender et al. 1999) is a body force type model that reproduces the vortex structures generated by vortex generators. Dudek (2010) applies this approach to model vortex generators in a bent idealized engine intake duct, the generators being used to prevent separation in the bend of the duct. Figure 6.25, is taken from Dudek. The mesh used for the S-duct simulation and hence the duct geometry can be seen in the lower right-hand frame of the figure. The approximate location of the vanes, relative to the duct, is also identified in this image. The contours in the figure are for total pressure. The upper right-hand images give measured contours of total pressure in the downstream duct region, where a fan would be located if the duct system was used as an intake component in an aeroengine. The idea is that the vanes reduce the extent of the separated flow zone in the bend, ensuring the flow reaching the fan face is more uniform, hence facilitating better fan performance. The more uniform flow arising from the vanes is clear from the measured total pressures. The Frame (a) predictions are for a clean duct with no vanes. Frame (b) gives the result for a simulation with the vanes resolved by mesh. On the other hand, the Frame (c) simulation is made using the model of Wendt and Biesiadny (2001) and the Frame (e) simulation using the BAY model. The model of Wendt, reminiscent of the actuator disk (see Section 6.3.4), adds vorticity as a step change. As can be seen, both these reduced order models capture the impact of the

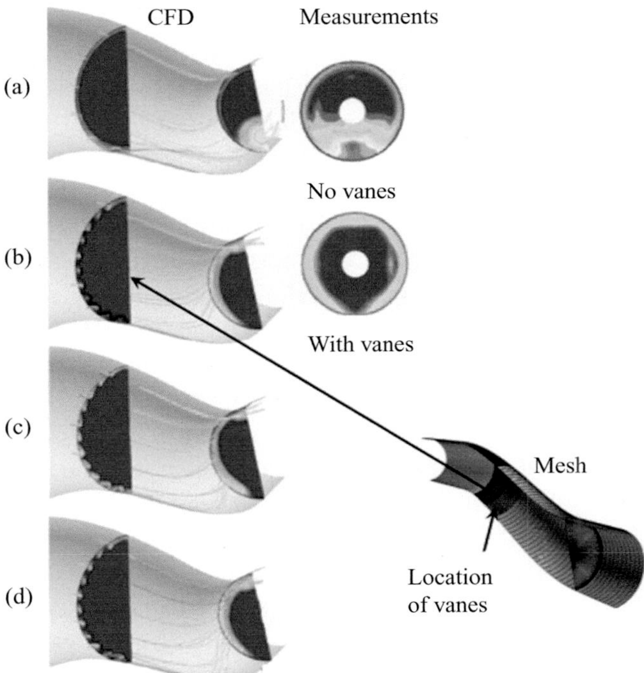

Figure 6.25. The use of low-order models to represent vanes in a duct: (a) predictions for a clean duct with no vanes, (b) simulations with vanes resolved by the mesh, (c) simulation made using the model of Wendt (2001) and (d) simulation made using the BAY model (taken from Dudek 2010 and used with kind permission). (See color plate 6.25.)

vanes without the need to explicitly mesh them and the extra computational cost associated with this. The low-order models also afford the possibility of controlling modelled turbulence length and time scales associated with the vortex generators, thus ensuring even better accuracy of these modelled properties than when using a RANS model with the fully resolved geometry (see Loiodice et al. 2010).

Figure 6.26, from Fujii (2014), relates to a study of plasma actuated flow control. Body forces are used to represent the forces produced by the plasma actuation.

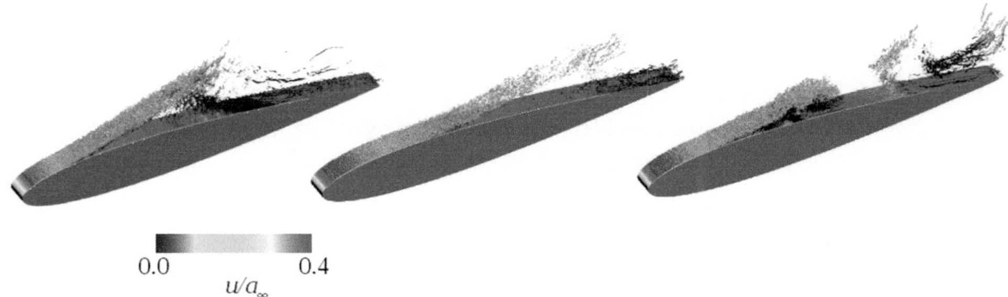

Figure 6.26. Study of plasma actuated flow control through body forces: (a) flow with no control, (b) flow with continuous control and (c) control through input in bursts (from Fujii 2014 – published with the kind permission of the Royal Society). (See color plate 6.26.)

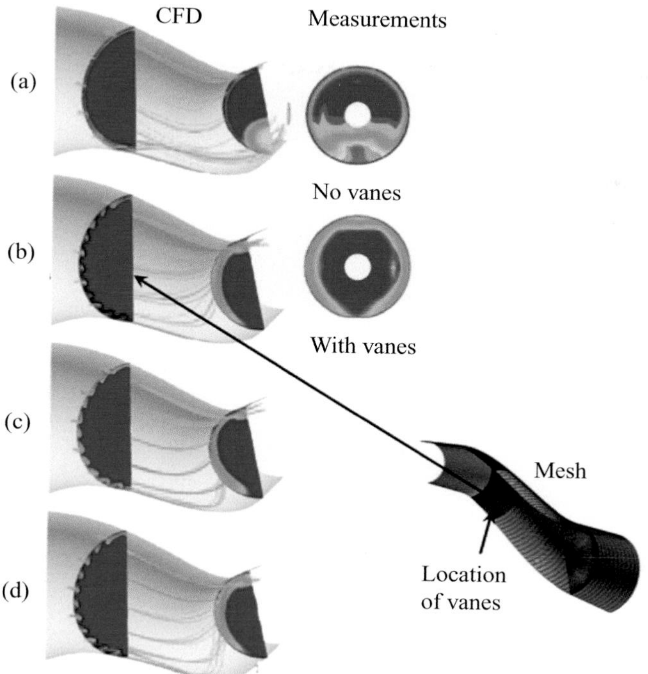

Plate 6.25. The use of low-order models to represent vanes in a duct: (a) predictions for a clean duct with no vanes, (b) simulations with vanes resolved by the mesh, (c) simulation made using the model of Wendt (2001) and (d) simulation made using the BAY model (taken from Dudek 2010 and used with kind permission).

Plate 6.26. Study of plasma actuated flow control through body forces: (a) flow with no control, (b) flow with continuous control and (c) control through input in bursts (from Fujii 2014 – published with the kind permission of the Royal Society).

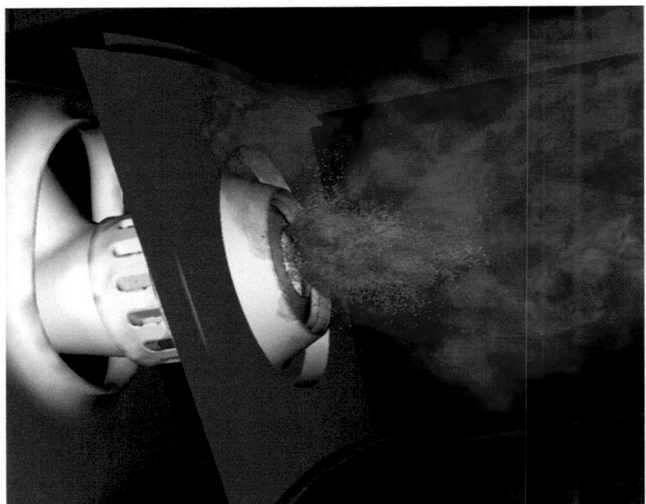

Plate 6.54. Image of a combustion large eddy simulation for a modern gas turbine engine (from Pitsch 2006 – published with the kind permission of the Annual Reviews).

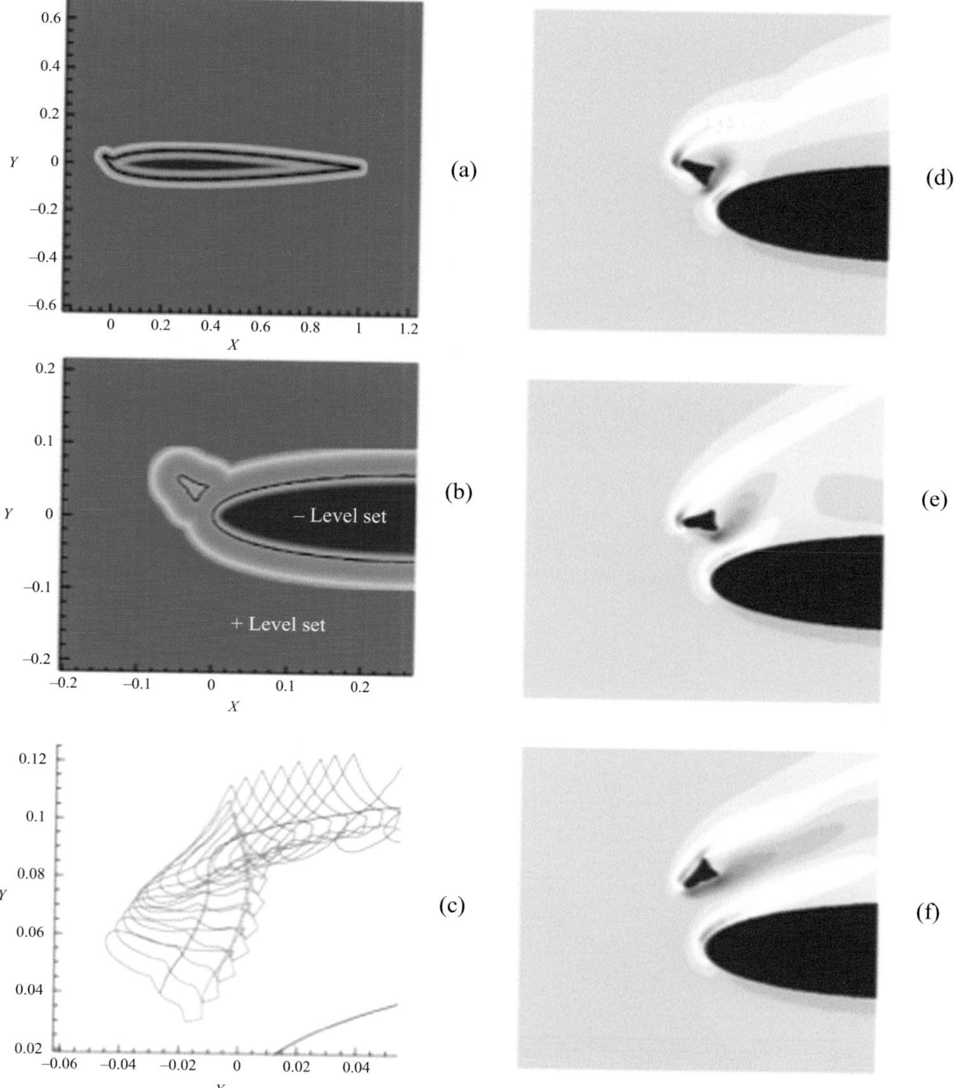

Plate 6.61. Modelling the trajectory of accumulated ice using a level-set based technique: (a) level-set field showing the airfoil and initial ice geometry, (b) zoomed in view of the level-set field, (c) trajectory of ice after mechanical release and (d–f) vorticity field contours and initial ice trajectory at three points in time (from Beaugendre et al. 2011 – Hindawi Publishing Corporation, Open Access).

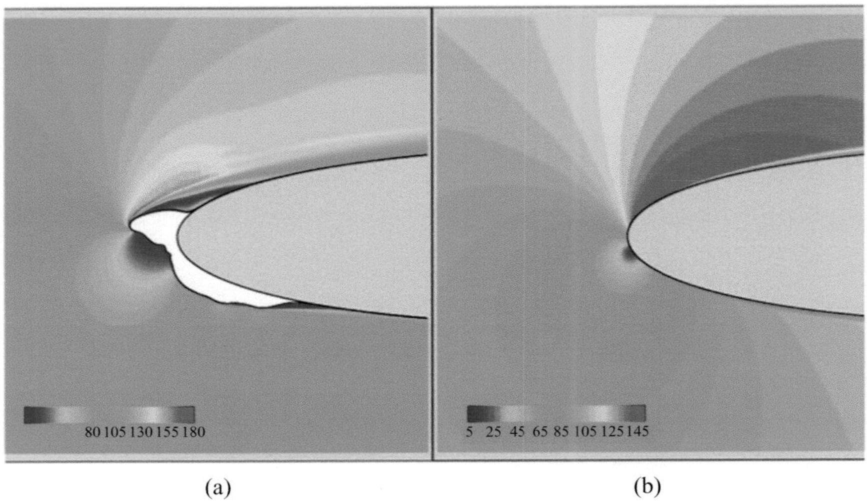

Plate 6.63. Flow field over the ice accreted airfoil and the clean airfoil represented by velocity magnitude contours: (a) airfoil with ice accretion and (b) clean airfoil (from Fossati et al. 2012 – used with the kind permission of John Wiley and Sons Inc.).

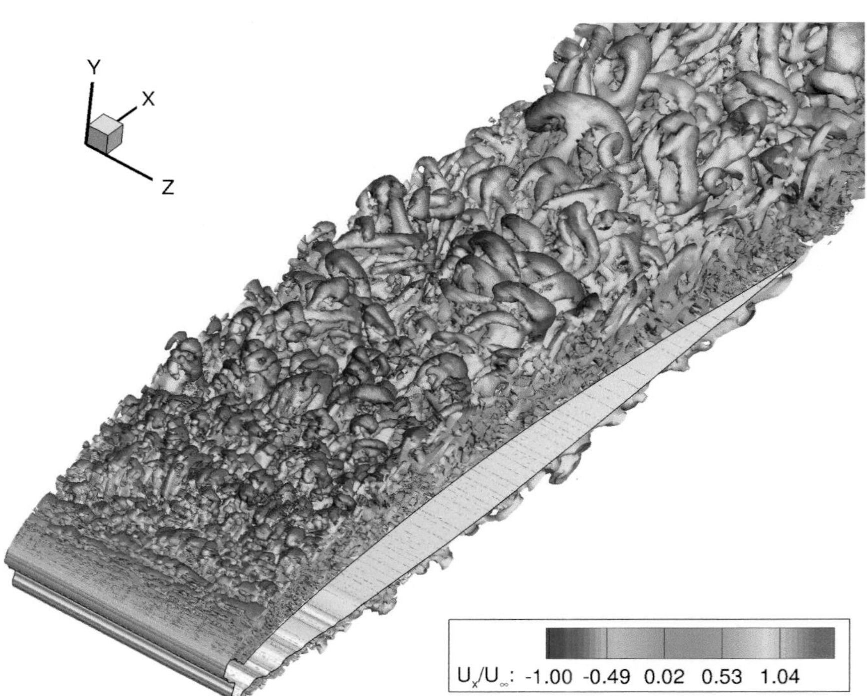

Plate 6.64. Hybrid RANS-LES of flow over a GLC-305 airfoil with a leading-edge ice horn (from Zhang 2015 – used with kind permission).

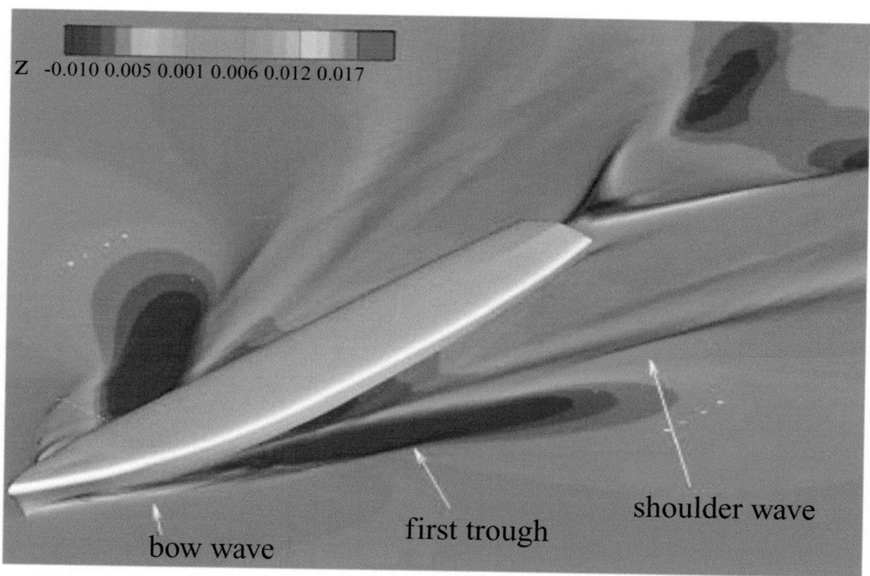

Plate 6.65. Predicted wave elevations through use of level set equations (from Huang et al. 2012 – published with the kind permission of John Wiley and Sons, Ltd.).

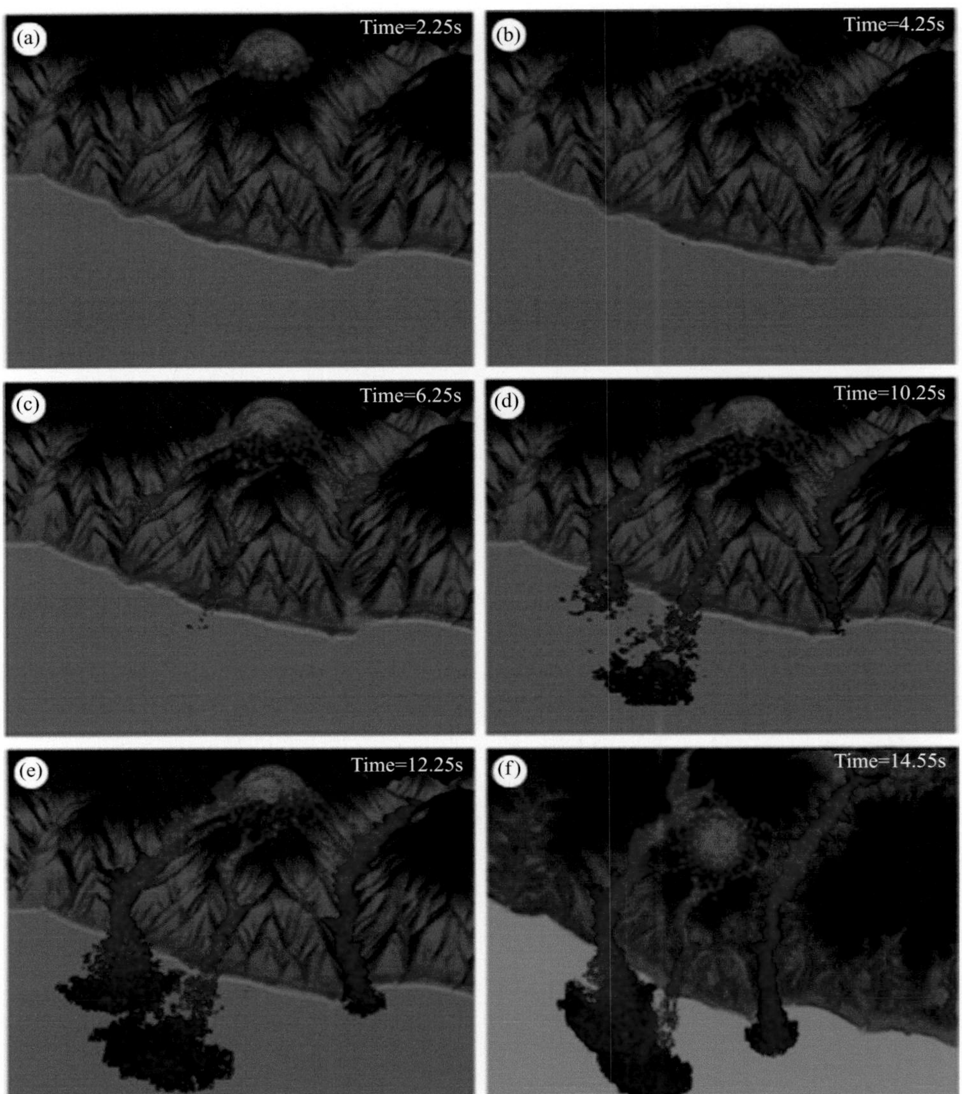

Plate 6.66. Grid free, smooth particle hydrodynamic simulation of a lava flow. The red zones correspond to the lava (from Cleary and Prakash 2004 – used with the kind permission of the Royal Society.)

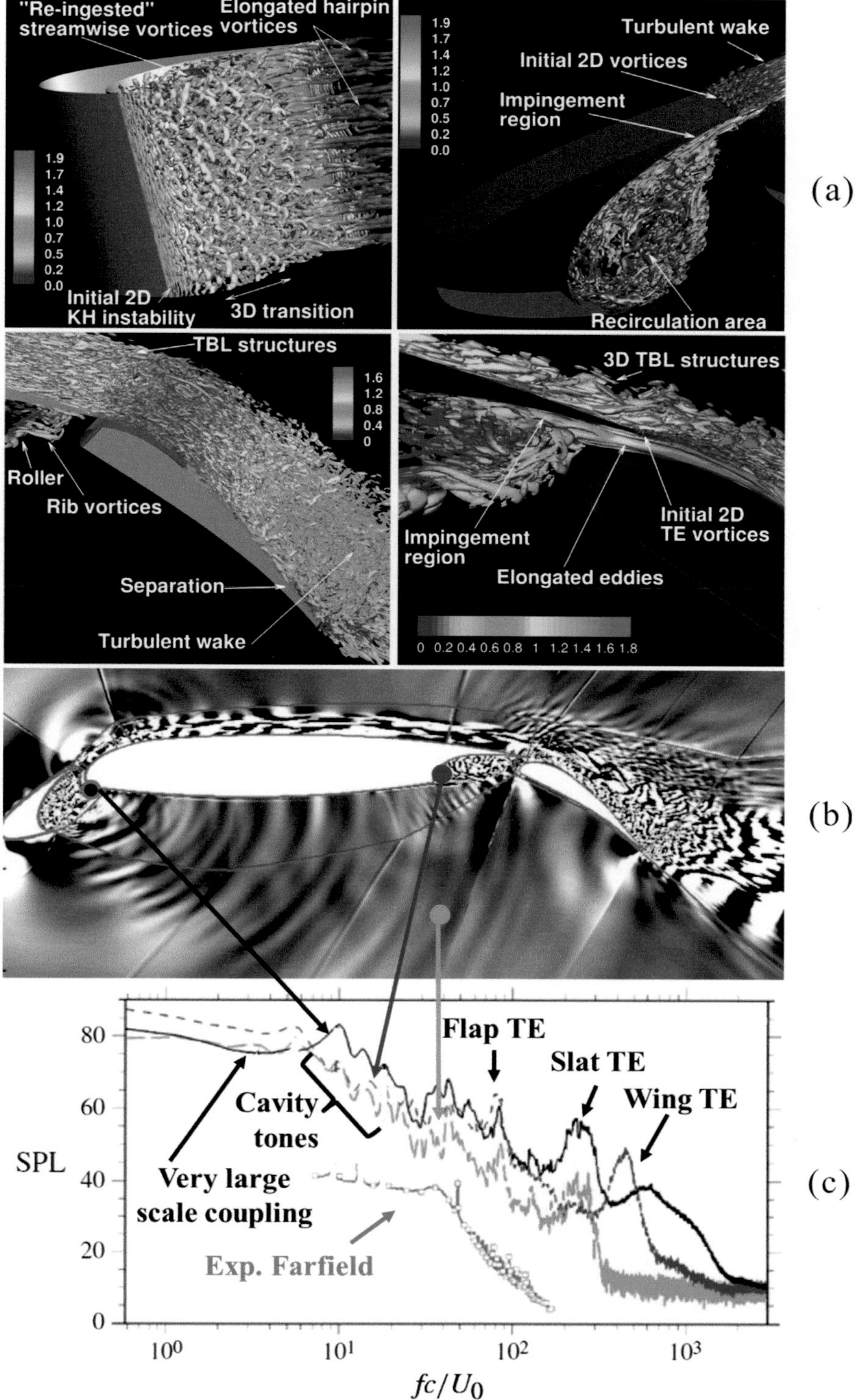

Plate 6.67. Zonal eddy-resolving simulation of aerodynamic noise for a slat-wing-flap configuration at off-design conditions: (a) Q-criterion (see Section 7.10.3.3) isosurface normalized by velocity, (b) near field acoustic waves rendered visible by contours of $(1/\rho)\ \partial\rho/\partial t$ and (c) sound pressure levels (SPL) against dimensionless frequency for different near field zones along with the measured far field sound (from Deck et al. 2014 – published with the kind permission of the Royal Society).

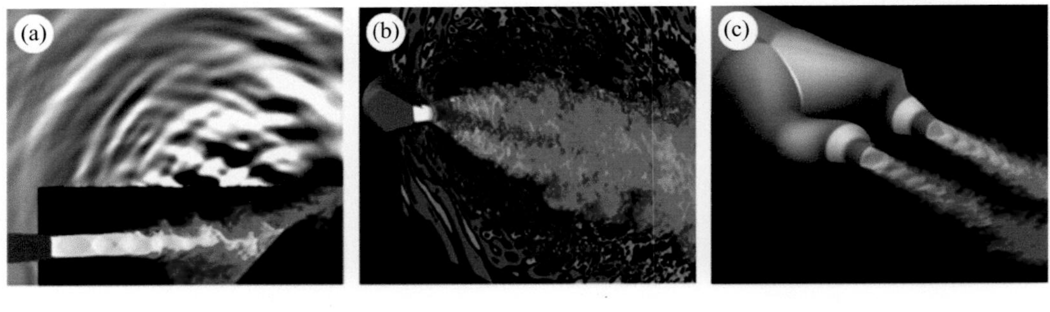

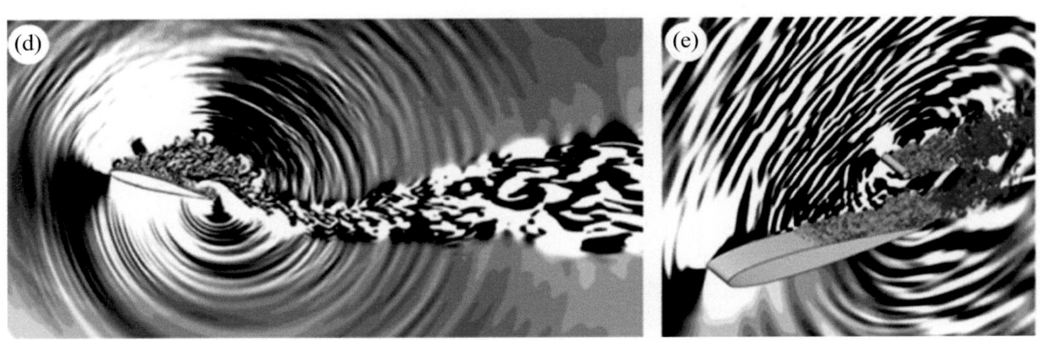

Plate 6.73. Examples of eddy-resolving simulations naturally providing both detailed source information and also near field acoustic propagation data: (a) a supersonic jet impinging on a jet blast deflector, (b) a supersonic jet issuing from a square sectioned chevron nozzle, (c) system providing twin adjacent jets, (d) stalled NACA0012 airfoil and (e) interaction of airfoil and an adjacent rod (from Lele and Nichols 2014 – used with the kind permission of the Royal Society).

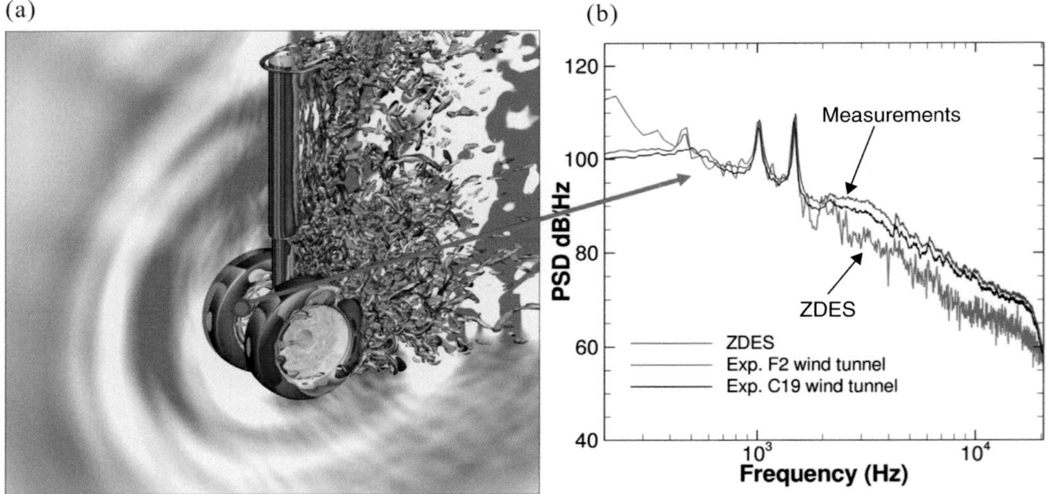

Plate 6.74. Example of an eddy-resolving simulation for a landing gear naturally providing both detailed source information and also near field acoustic propagation data: (a) flow field and acoustic waves and (b) sound pressure spectrum (from Deck et al. 2014 – used with the kind permission of the Royal Society).

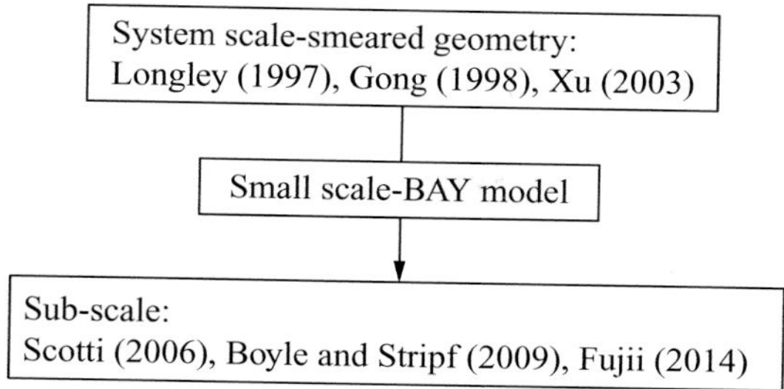

Figure 6.27. Wide range of scales over which body forces can be applied.

Hence, again, the forces are imposed at the small scale. The images give the instantaneous flow for a NACA airfoil profile at an angle of attack around 20°. The actuator is placed at around 15% of chord from the airfoil's leading edge. The Reynolds number of the airfoil, based on chord, is around 1×10^6. The high Reynolds number means that the eddy-resolving simulations need around 1 billion cells. Since a range of cases need to be studied, to optimize flow control parameters, the Japanese Petaflops supercomputer 'K' is used. Frame (a) shows the flow with no control. Frame (b) gives the flow with continuous control and (c) gives control through body force inputs that occur in bursts. Clearly, the latter substantially reduces the separated flow and thus improves aerodynamic performance. Fujii sees such supercomputer simulations as an important part of a new era that moves beyond geometry driven design to 'device design'. Boyle and Stripf (2009), for turbine blades, use body forces to model surface roughness, hence again working at small scale and seeking to interact with turbulence and the transition process. Examples of the use of such body forces can be found in both the RANS and eddy-resolving solution frameworks. For an example of the latter, see Scotti (2006) who studies the impact of roughness on skin friction. Figure 6.27 summarizes the range of flow scales that have been accounted for using body force modelling.

6.3.4 *Unique Incidence and Actuator Disc Models*

Another means of modelling fans and rotating blade rows is through the use of actuators discs. These can come in a range of forms and can also be extended to model unsteadiness (see Joo and Hynes 1997), acoustics (see Cooper and Peake 2000) and also even wakes. In their classical form, they constitute a jump condition in the mesh as shown in Figure 6.28. Hence, the disk is planar. For a compressible flow, the actuator disk equations can be formulated as

$$\begin{aligned} &\text{Continuity: } \rho_1 V_1 = \rho_2 V_2 \\ &\text{Momentum: } p_1 + \rho_1 V_1^2 + F = p_2 + \rho_2 V_2^2 \\ &\text{Energy: } c_p T_1 + \frac{1}{2} V_1^2 + w = c_p T_2 + \frac{1}{2} V_2^2 \end{aligned} \tag{6.26}$$

where V is the velocity magnitude. The first equation expresses mass conservation across the disk, the second relates to momentum and the third energy conservation

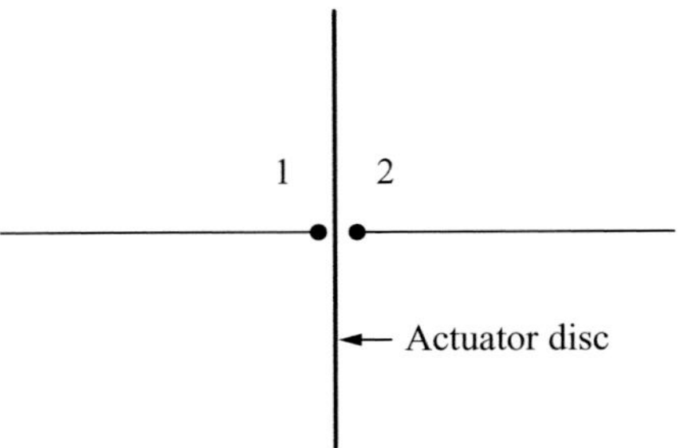

Figure 6.28. Schematic of actuator disk configuration.

across the disk. The subscripts 1 and 2 refer to stations to the left and right of the disk. The term F relates to a highly localized body force arising from the fan. Also, w relates to the work done by this force. This equation set can be expressed in differing ways (see, for example, Taddei and Larocca 2014). All need empirical input based on measurements or high fidelity CFD. Sometimes body force models are described as distributed type actuator disks.

For transonic fans, the unique incidence boundary condition can be used. The shock structure defines the angle at which the fluid approaches the fan (i.e., a unique incidence can be defined to give a low-order fan model). Typically for transonic fans, the flow is subsonic for the inner radial extent of the blade and transonic for the remaining roughly 50%. Hence, it is possible to hybridize the body force and unique incidence models, applying the latter towards the blade outer radius (see Hield 1996). Also, as noted, an actuator disk related model for vortex generators can be used – see Dudek (2005). A key difficulty with actuator disk models is avoiding interface reflectivity.

6.3.5 *Deterministic Stress and Related Approaches*

Deterministic stresses can be used to render steady RANS solutions, hence avoiding the cost of URANS. Some processes for evaluating deterministic stresses are outlined in this section.

6.3.5.1 Mean Source Terms (MST) Following Ning and He (2001), defining the unsteady Navier-Stokes operator, *NS,* we can write the following

$$NS(\phi) = 0 \tag{6.27}$$

which can be integrated in an unsteady fashion using the operator NS for a time varying solution, ϕ. This solution can then be time averaged, assuming the usual decomposition of $\phi = \overline{\Phi} + \phi''$. We could produce a (residual) source term, S, by applying the steady Navier-Stokes operator, NS^S, to the computed mean

$$S = NS^S(\overline{\Phi}) \tag{6.28}$$

The source S accounts for the stresses arising from the fluctuating quantity ϕ''. If we know S, (not an easy task in practice), then we can solve directly for the mean by solving the following equation:

$$NS^S(\overline{\Phi}) - S = 0 \tag{6.29}$$

We could also do this in a URANS to steady RANS process using the following symbolic operators, *URANS* and *RANS*, respectively. Hence, we solve

$$URANS(\Phi) = 0 \tag{6.30}$$

with the decomposition $\langle\Phi\rangle = \overline{\Phi} + \phi''$ and $\phi = \overline{\Phi} + \phi'' + \phi'$, thus $\phi = \langle\Phi\rangle + \phi'$. We then have $S = RANS(\overline{\Phi})$, where *RANS* is a symbolic steady RANS operator. For RANS/URANS, ϕ' is accounted for by the RANS model. The source accounts for ϕ''. $RANS(\overline{\Phi}) - S = 0$ is then the final steady equation system.

6.3.5.2 Deterministic Stress Modelling (DSM) With URANS, or even unsteady Navier-Stokes solutions, we could produce steady state solutions using deterministic stress modelling. For example, the URANS system could be expressed as

$$URANS(\Phi) = \frac{\partial\Phi}{\partial t} + RANS(\Phi) = 0 \tag{6.31}$$

Full closure of this system will have terms of the following form: $\overline{\phi_i'\phi_j'}$, $\overline{\phi_i''\phi_j''}$ and other non-linear terms. With deterministic stress modelling, the key task is to devise a model for the deterministic stress terms $\overline{\phi_i''\phi_j''}$. These are analogous to the Reynolds stresses. Like them, they enter the time-averaged equations in a differentiated form, $\partial\overline{\phi_i''\phi_j''}/\partial x_j$. The $\partial\overline{\phi_i'\phi_j'}/\partial x_j$ is accounted for by the usual RANS model. For the Navier-Stokes equations these terms will be zero. Both the MST and DSM approaches are effective in yielding steady solutions. However, alone, they are of no practical use. The ultimate source terms cannot be determined a priori. Hence, some form of model or database is needed.

For turbomachinery blade rows it is possible to devise simple models. The relative rotation between blade rows in turbomachinery can mean that the wake relative tangential movements can give rise to deterministic stresses (see Adamczyk 1985). This is shown in Figure 6.29. To model the arising stresses, Hall (1997) essentially takes standard wake correlations and applies minor modifications to them. The wake correlations, describing a momentum deficit, are swept around at an appropriate relative speed. This, when referenced to a mean state, gives a velocity fluctuation estimate that can be converted into a deterministic stress. Hence, the approach of Hall is based around phenomenological modelling for the deterministic stresses. Adamczyk (1985) presents an alternative, more complex approach that needs multiple overlapping meshes. This approach needs less empiricism.

Both the Adamczyk and Hall models assume that the deterministic stress distributions are axisymmetric. However, they can have extreme azimuthal variations. To account for these, Meneveau and Katz (2002) make multiple RANS simulations.

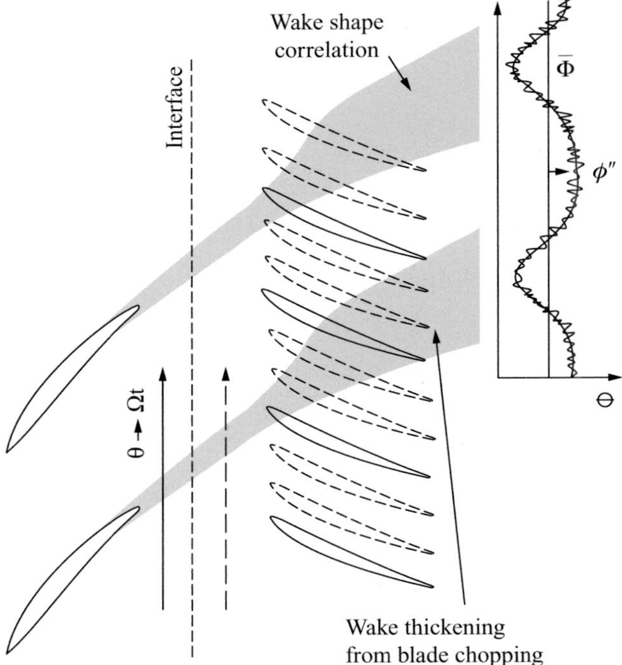

Figure 6.29. Schematic of deterministic model of Hall (from Tucker 2011a – published with kind permission of Elsevier).

The deterministic stresses are formulated by averaging these simulations with appropriate weightings to reflect the time duration that the steady state is most representative of.

Busby et al. (2000) use coarser grid inviscid simulations to evaluate the deterministic stress terms. These stresses are used to secure steady viscous simulations. Orkwis et al. (2002) extend the work of Busby and colleagues by evaluating the deterministic stress terms based on linearized unsteady inviscid solutions. Stollenwerk and Kugeler (2011) extract deterministic stresses at one operating condition from URANS. These stresses are fixed and used in calculations for other operating points.

For separated flows, Benning et al. (2001), Gangwar (2001) and Lukovic (2002) use URANS simulations as training data for a neural network deterministic source models. Hence, the neural network is used as a form of interpolator in the design space. As can be seen from Figure 6.30, there is a wide range of potential input sources for supplying deterministic stresses. The user must appreciate which is the most appropriate. Hence, it needs to born in mind if, for example, viscous processes are important or not. If not, then inviscid solutions on coarser meshes can be used. The work of Bardoux and Lebouf (2001) is indicative of the challenging task that an engineer faces in grappling with complex non-linear flow physics and the most appropriate model for it. LES type simulations can used to validate a deterministic or other low-order modes.

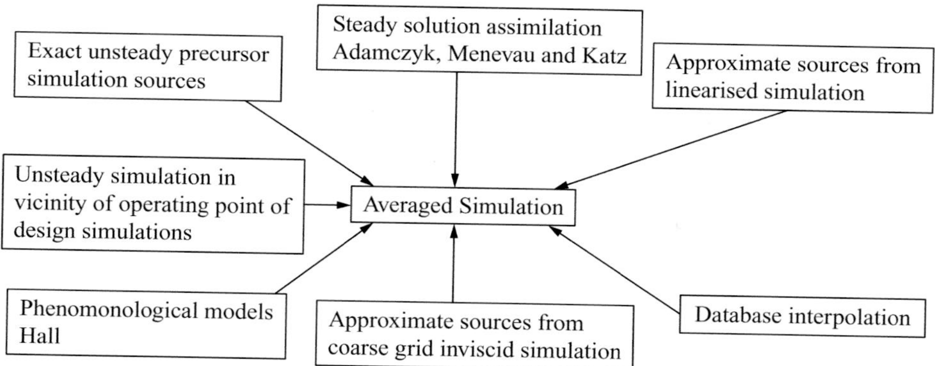

Figure 6.30. Source inputs for deterministic stress simulations (from Tucker 2011a – published with kind permission of Elsevier).

6.3.6 *Plenum Type Models*

Many engineering systems have substantial geometrical complexity. This is often neglected. For example, as explored in Chapter 7, turbine blades in axial gas turbines are often considered as blades located within a clean annular passage. However, there is substantial geometrical complexity on the radial annulus walls of these passages. For example, on the inner surfaces, there will be what are called rim seals (similar types of features can also be observed on the outer walls – see Jefferson-Loveday et al. 2012). These occur between the rows of stationary and rotating blades. In-between these blades there are gaps, through which cooler air is ejected into the main blade passage. This is intended to prevent the ingestion of hot gasses. Resolving the rim seal geometry gives added simulation cost. As discussed in Chapter 3, it also gives a numerical stiffness problem. This relates to a substantial disparity in Mach numbers in the flow. In the main passage, the flow is choked. In the rim seal zone there are considerably lower Mach numbers. However, the Mach number distribution is tensorial. Therefore, explicitly modelling the rim seal is challenging. This is especially so since RANS models will struggle to deal with the rim seal flow, which has separated, large-scale eddies. These, as noted in Chapter 5, are problematic for RANS models. Under these circumstances, it can be better to set turbulence levels based on plausible empirical evidence.

Differing in nature from rim seals, labyrinth type seals can be found at blade extremities. For these, the reduced order k2k model of Wellborn et al. (1999) is often employed. Labyrinth seals also have substantial Mach number disparities, so their explicit resolution is again challenging. Hence, with the k2k model, no geometry is explicitly resolved. Instead, loss coefficients are used to characterize the rim seal zone total pressure loss. The two ends of the missing labyrinth seal zone are connected via an analytical low-order type of model.

Similarly, in the rim seal zone, Jefferson-Loveday et al. (2012) created a low-order model. Basically, the rim seal zone is represented as plenum. Then the expected CFD boundary conditions are applied to the periphery of the plenum. The plenum could be

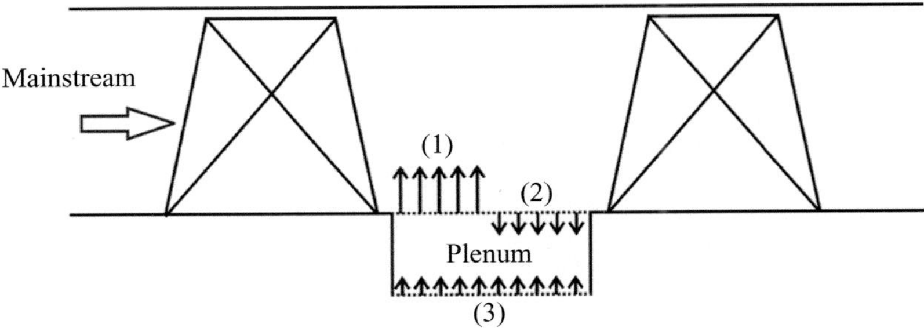

Figure 6.31. Macro-control volume used to resolve a rim seal in a gas turbine (from Jefferson-Loveday et al. 2012 – used with the kind permission of Rolls-Royce plc).

viewed as a macro control volume. Hence, the whole rim seal is represented as a single control volume. This volume is shown in Figure 6.31 with the associated boundary conditions (1–3). In the work of Jefferson-Loveday and colleagues, the boundary conditions are characteristic-based, the simulations taking place in a compressible flow solver framework. At Boundary (3), an inflow conditions is applied. This sets turbulence level expectations based on a mixing length hypothesis which accounts for eddy scales potentially being of the rim seal's scale.

Figure 6.32 shows the radial variation of total pressure loss and exit flow angle downstream of a rotor blade in a 1.5 stage turbine. The dashed line is for a fully resolved (i.e., the rim seal zone is explicitly meshed) unsteady flow simulation. The chain dashed line is a simulation using the plenum boundary condition. As can be seen, there is a substantial improvement in predictions, relative to simulations with no rim seal accounted for given by the full line. The reason for this improvement

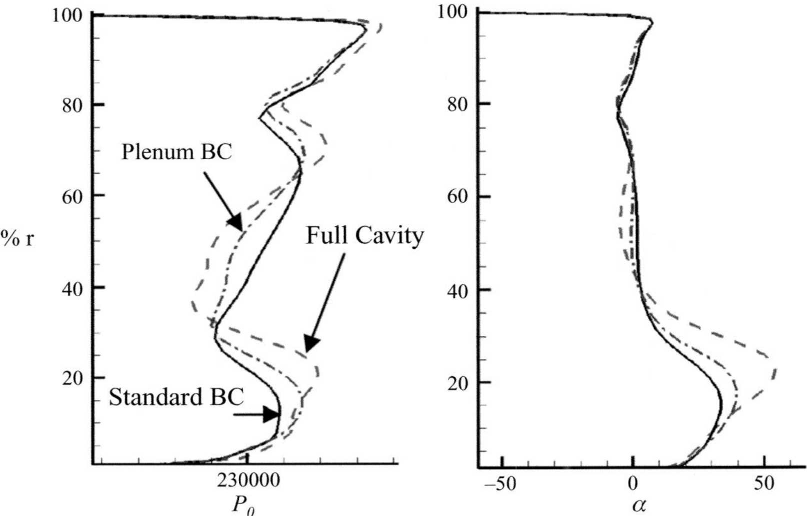

Figure 6.32. Radial profiles of total pressure and flow angle downstream of the rotor blade in a 1.5 stage turbine (from Jefferson-Loveday et al. 2012 – used with the kind permission of Rolls-Royce plc).

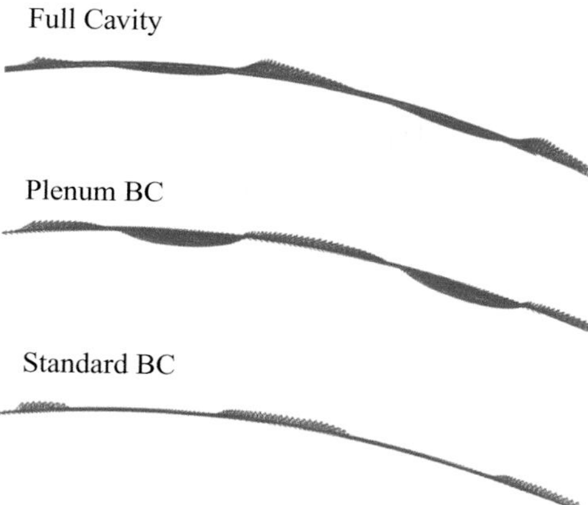

Figure 6.33. Flow vectors for a slice in an r-θ plane passing through the rim seal of a 1.5 stage turbine (from Jefferson-Loveday et al. 2012 – used with the kind permission of Rolls-Royce plc).

is shown in Figure 6.33. This figure gives flow vectors for a slice in an r-θ plane passing through the rim seal. As can be seen from this figure, in the rim seal zone there are both substantial zones of radial inflow and out flow. Hence, if in a standard CFD simulation just an inflow is prescribed on the annulus surface, the imposed flow physics will be erroneous.

6.3.7 *Proper Orthogonal Decomposition*

POD can be used to form low-order models. POD is discussed in Chapter 7 for post-processing purposes. For this, the approach can characterize the flow as a series of modes and then components of the different modes can be plotted, the lower order modes having the most energy. Alternatively if a sufficient number of modes are considered together, akin to Fourier series, with the low-order modes having most energy, a model of the flow can be made. The flow model is 'trained' through a matrix manipulation process on high fidelity data that could include eddy-resolving simulations. For more advanced forms of POD, such as Balanced POD (BPOD), see Ilak and Rowley (2008). Carlson and Feng (2005) show the application of POD to a coupled fluid-structure problem.

6.4 Multi-Scale Problems

The vortex generators discussed earlier give rise to what can be viewed as multi-scale problems, where the disparity in scale presents a substantial modelling challenge. With regard to turbulent flow, the critical structures associated with turbulence production are of a scale of a just a few wall units and hence their deep control suggests the use of small devices relative to the scale of, say, a wing. A wide range of small-scale flow control devices have been explored by various researchers including surface perforations. As shown in Figure 6.34, for electronic and avionics systems,

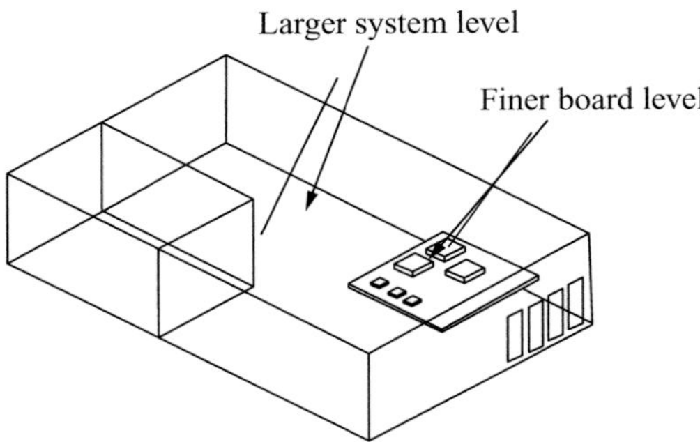

Figure 6.34. Disparity of scales in an electronic system (from Tucker 1997 – published with permission from the IEEE).

there can also be a large disparity in scales between the system and component level. For example, circuit boards are often populated by countless small-scale components and these are sometimes treated as roughness elements. Electronic systems can also have heat sinks comprising small geometric elements – relative to the system scale. Then it has been known to treat the heat sink geometry as a porous media – thus avoiding resolution of the potentially complex heat sink geometry.

A niche challenge is when there is a disparity in scale and a multiplicity of small-scale repeated features. The block spectral method of He (2012, 2013) seeks to address this type of problem. An example of where such systems can occur is the film cooling used over turbine blades. With film cooling, the turbine blade surface is covered with numerous fine scale holes. Fluid issues from these with the intent of protecting the blade surface from hot gases. Another area where there are small-scale holes on a large-scale system are the holes associated with the acoustic liners in aeroengine intakes. With the block spectral method of He, a local spectral representation of the flow fields is 'trained' at a few key points in the flow domain. These 'trained' models are then used at a multitude of locations. Figure 6.35 shows a schematic of the approach. Frame (a) shows a surface with numerous small-scale features, such as film cooling holes, that could be directly solved with blocks mapped at some spectral level. On the other hand, Frame (b) shows the reduced order block spectral method. With this, the un-shaded blocks are solved for and then the reduced order model is mapped to the remaining shaded blocks.

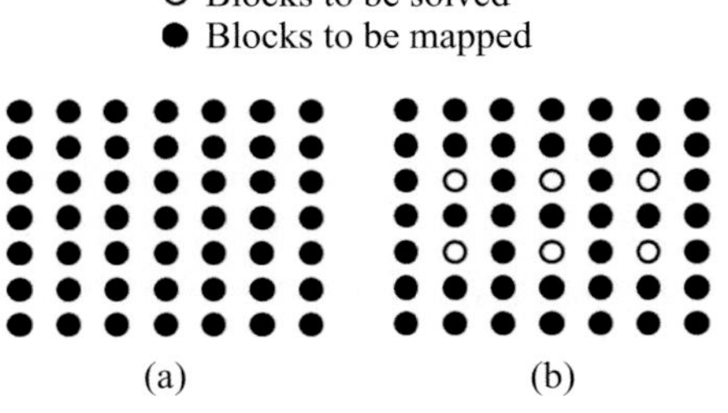

Figure 6.35. Schematic showing the block spectral method of He (2013).

6.4.1 *Particle/Droplet Transport Equations*

Another form of multi-scale problem is where a fluid is populated with fine particles or droplets. If the particles/droplets do not influence the flow, then due to the lack of fluid particle coupling, the task is relatively simple. The solution of particle/droplet equations is important in many fields of engineering and in particular combustion, where it can be necessary to model droplets of fluid. For spherical particles both trajectory

$$\frac{d\boldsymbol{x}}{dt} = \boldsymbol{V}_p \tag{6.32}$$

and momentum

$$m\frac{d\boldsymbol{V}_p}{dt} = \sum \boldsymbol{F} \tag{6.33}$$

equations must be solved, where m is the particle mass, $\boldsymbol{x}$ the particle position vector, $\boldsymbol{V}_p$ the particle velocity and t time. $\sum \boldsymbol{F}$ represents the sum of external forces acting on a particle. The forces on the particle can be numerous and even include electrostatic and thermal gradient related force (thermophoresis). The most notable force is the classical drag

$$\boldsymbol{F}_D = \frac{1}{2}\rho C_D \pi \frac{d^2}{4}|\Delta \boldsymbol{V}|\Delta \boldsymbol{V} \tag{6.34}$$

where $\Delta \boldsymbol{V} = \boldsymbol{V} - \boldsymbol{V}_p$ and $\boldsymbol{V}$ is the fluid velocity, d the particle diameter and C_D a drag coefficient. Also, $\boldsymbol{F}_g = (\rho_p - \rho)\pi d^3 \mathbf{g}/6$-the gravitational force where, here, in Equation (6.33) $\sum \boldsymbol{F} = \boldsymbol{F}_D + \boldsymbol{F}_g$. The particle Reynolds number is defined as $Re_p = \rho d|\Delta \boldsymbol{V}|/\mu$. As a particle drag estimate, for $Re_p < 1$, $C_D = 24/Re_p$. If $1000 \geq Re_p > 1$, the following can be used:

$$C_D = \frac{24}{Re_p}(1 + 0.15 Re_p^{0.687}) \tag{6.35}$$

and when $Re_p > 1000$, $C_D = 0.44$. Equation (6.32) can generally be expressed with sufficient accuracy in the following straightforward discretized form:

$$\boldsymbol{x}^{new} = \boldsymbol{x}^{old} + \left[(1-\lambda)\boldsymbol{V}_p^{old} + \lambda \boldsymbol{V}_p^{new}\right]\Delta t \tag{6.36}$$

where, for example with $\lambda = 1/2$, a time-centred approximation is gained. Incorporating Stokes drag, Equation (6.33) can be re-expressed as, allowing an exponentially based temporal integration,

$$\frac{d\boldsymbol{V}_p}{dt} = \frac{\boldsymbol{V} - \boldsymbol{V}_p}{\tau} + \frac{\sum \boldsymbol{F}}{m} \tag{6.37}$$

If $\rho_p \gg \rho$ (ρ_p is the particle density), the particle relaxation time (the time required for a suddenly released stationary particle to reach around 2/3 the fluid's velocity) can be expressed as

$$\tau = \frac{\rho_p d^2}{18\mu} \tag{6.38}$$

The fourth order Runge-Kutta technique is popular for solving Equation (6.37) but has the restriction $\Delta t < 2\tau$ (see Barton 1996). If it is assumed that the fluid velocity $\boldsymbol{V}$ remains constant through a time-step Δt, Equation (6.37) can be integrated to give the following (see Crowe et al. 1997) stable discretized particle momentum equation:

$$\boldsymbol{V}_P^{new} = \boldsymbol{V}_P^{old} e^{-\Delta t/\tau} + \left(\boldsymbol{V}^{old} - \frac{\sum \boldsymbol{F}}{m} \right) \left(1 - e^{-\Delta t/\tau}\right) \tag{6.39}$$

However, Equation (6.39), although popular, has the disadvantage that as τ tends to zero (with $\sum \boldsymbol{F} = 0$), it gives $\boldsymbol{V}_P^{new} = \boldsymbol{V}^{old}$, erroneously implying the particle velocity lags the fluid's by one time increment. This results from assuming $\boldsymbol{V}$ is constant through the increment. To remedy this, Barton (1996) proposes, prior to the analytical integration of Equation (6.37), that the following Taylor series based fluid velocity variation assumption is made:

$$\boldsymbol{V}(t) = \boldsymbol{V}^{old} + \boldsymbol{C}_0 t + \boldsymbol{C}_1 t^2 \tag{6.40}$$

where

$$\boldsymbol{C}_0 = \frac{\boldsymbol{V}^{new} - \boldsymbol{V}^{older}}{2\Delta t} \tag{6.41}$$

and

$$\boldsymbol{C}_1 = \frac{\boldsymbol{V}^{new} - 2\boldsymbol{V}^{old} - \boldsymbol{V}^{older}}{2\Delta t^2} \tag{6.42}$$

If desired, it can be assumed that $\boldsymbol{C}_1 = 0$, giving a less accurate linear fluid velocity variation assumption. When using Equations (6.41) and (6.42), velocities need to be stored at three time levels. Also, $\boldsymbol{V}^{new}$ is itself an unknown. To deal with this, Barton proposes a predictor corrector approach. With this, based on latest particle positions, $\boldsymbol{V}^{new}$ is repeatedly updated. Evidently, generally only one correction/update is sufficient. When using Equation (6.40), as τ tends to zero, the sensible limit $\boldsymbol{V}_{\boldsymbol{p}}^{new} = \boldsymbol{V}^{new}$ is now gained. For most practical engineering applications, with small enough time-steps Equation (6.39) is sufficient.

Turbulence can substantially impact on particle trajectories. A useful validation test for the previous equations is to explore the lateral dispersion of particles in (homogenous, isotropic) turbulence. Figure 6.36 shows the mean square displacements of particles with time. The symbols are measurements and the lines the simulation. This result is taken for Lo Iacono and Tucker (2004). The flow field is based on a large eddy simulation. The strong dispersion arising just from the turbulence is clear from the figure, there being no lateral mean flow. This test case can also be performed with RANS. However, the modelling techniques described next are then required.

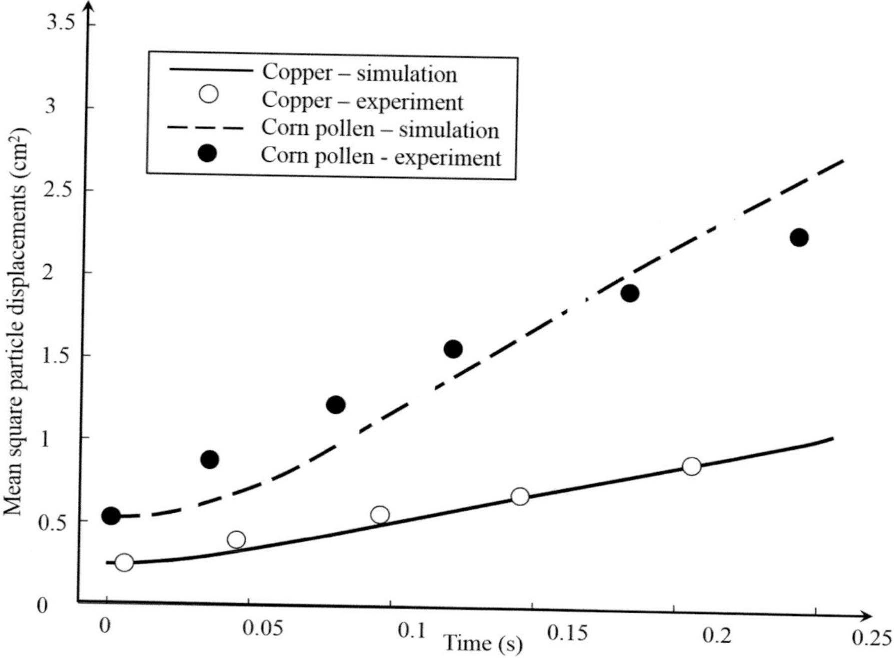

Figure 6.36. Computation of particle dispersion in homogenous turbulence for copper and corn pollen.

6.4.2 *RANS Modelling Particle Trajectories*

When making RANS based simulations, the velocity field with which the particle interacts is expressed as the sum of the usual RANS averaged velocity and a turbulence fluctuation magnitude about this. Therefore, we have

$$u = U + N\sigma_1,\ v = V + N\sigma_2,\ w = W + N\sigma_3 \tag{6.43}$$

For isotropic turbulence predictions,

$$\sigma_1 = \sigma_2 = \sigma_3 \approx \sqrt{2k/3} \tag{6.44}$$

where, k, is the turbulence kinetic energy as defined in Chapter 5. For anisotropic turbulence $\sigma_1 = \sqrt{\overline{u_1'u_1'}}$, $\sigma_2 = \sqrt{\overline{u_2'u_2'}}$, $\sigma_3 = \sqrt{\overline{u_3'u_3'}}$, where the individual Reynolds stress components can be gained from a Reynolds stress turbulence model or a non-linear eddy viscosity model (see Chapter 5). N is a Gaussian random number with a standard deviation of unity. It is updated either when an eddy life (t_l) is exceeded or the particle traverses an eddy. The latter takes a time t_t, the eddy transit time. To evaluate the previous two time scales, an assumption about the eddy size, l_e, is required. For this, the following expression can be used:

$$l_e = C_\mu^a \frac{k^{3/2}}{\varepsilon} \tag{6.45}$$

A range of values for $2/3 < a < 1$ can be used (see Milojević 1990) to fit data. The eddy life time can be approximated as

$$t_l = \frac{l_e}{\sigma} \tag{6.46}$$

The transit time t_t can be expressed as

$$t_t = \frac{l_e}{v_d} \tag{6.47}$$

where v_d is the drift velocity $|\boldsymbol{V} - \boldsymbol{V}_p|$. However, within an eddy v_d is not constant. Therefore, small time-steps are needed with v_d computed each time-step. Values of v_d are used to calculate relative displacements within an eddy. When the total of these is greater than l_e, it is known the particle has left the eddy and N is re-computed.

For unsteady flows, the stochastic particle trajectory algorithm described can be solved sequentially with the flow field – that is, once the flow field has been advanced one time-step, the particle trajectory equation is also advanced this period. This procedure is then repeated. In certain circumstances, for stability, accuracy or computational economy reasons, the time-steps used in the fluid simulation may not be the same as those used in the particle integration. This is fairly common in coupled problems. When this is the case, differing numbers of time-steps (preferably integer multiples to avoid interpolation) might be used for the particle and fluid equation systems to reach the same point in time. Alternatively, the order of one of the schemes can be altered to give greater time-step compatibility or a combination (different order and numbers of time-steps) of these approaches used.

Many particles, in real life, are non-spherical, and the particle shape can have a strong impact on particle deposition. Particles can, for example, be elongated into fibre type structures. Modelling these is considered next.

6.4.3 *Non-Spherical Particles*

The procedure below is based on that discussed in Lo Iacono et al. (2005). Figure 6.37a shows an idealized cylindrical particle. The particle coordinate system is $\boldsymbol{x} = [x, y, z]$, with its origin at the mass center. The orientation of the cylindrical particle is determined by the angle $\theta = [\theta_x, \theta_y, \theta_z]$. The subscripts x, y, z refer to axis of rotation. The axis of symmetry of the cylinder, represented by the vector $\boldsymbol{d}$, is referred to as the principal axis. The radius, r, of the section of the cylinder is assumed to be much smaller than its length, L. The main aspects of the model are explained in Figure 6.37a and b. Here, for simplicity, this is assumed to be in the x-z plane. The total force $\boldsymbol{F}$ acting on this portion of cylinder (near point P) by the surrounding fluid is decomposed into components perpendicular, and parallel, to the principal axis of the cylinder, called respectively D and S.

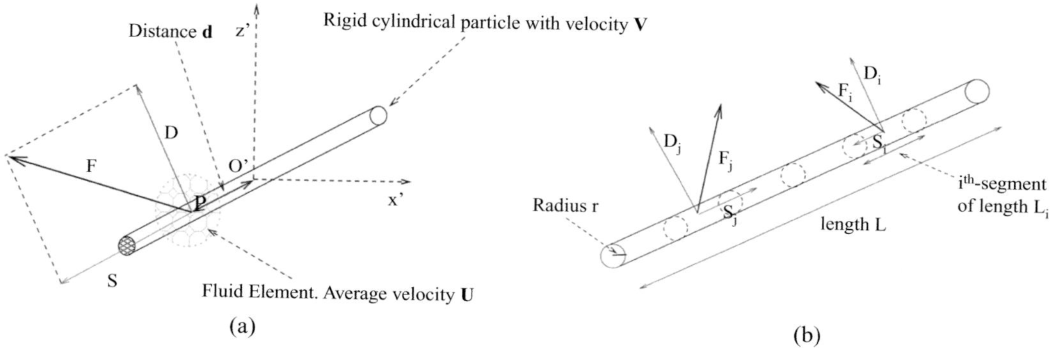

Figure 6.37. Nomenclature for cylinder and integrated segments: (a) cylinder and (b) integrated segments (from Lo Iacono et al. 2005 – published with kind permission of Elsevier).

The normal component D is modelled according the classical drag expression for a cylinder of infinite length (see Schlichting and Gersten 2000):

$$D = \frac{2C_D \upsilon Re_d \rho}{\pi (2r)^2 \rho_p} (\boldsymbol{V} - \boldsymbol{V}_p) \cdot \left\{ \left[\frac{(\boldsymbol{V} - \boldsymbol{V}_p)}{|(\boldsymbol{V} - \boldsymbol{V}_p)|} \times \frac{\boldsymbol{d}}{|\boldsymbol{d}|} \right] \times \boldsymbol{d} \right\} \tag{6.48}$$

The parallel component, S, is relatively easy to estimate and can be modelled as the friction force over a flat plate of equivalent surface area. Note that cylinder end influences are ignored. Only the component normal to the cylinder axis (i.e., the D) generates a torque $\boldsymbol{T} = \boldsymbol{d} \times \boldsymbol{F}$ with module $|\boldsymbol{T}| = D|\boldsymbol{d}|$ while S, being parallel to the axis, causes no rotation. The fibre can be decomposed in a series of i-segments as shown in Figure 6.37b. Then, contributions D, S and $\boldsymbol{T}$ of all i elements along the fibre are calculated. The total forces D and S cause translation of the cylinder. The torque, $\boldsymbol{T}$, causes rotation around the mass centre. The equations of motion (translation and rotation) for a cylinder are therefore given by

$$\begin{aligned} m\frac{du_1}{dt} &= D \\ m\frac{du_2}{dt} &= S \\ I\frac{d\omega_p}{dt} &= \boldsymbol{T} \end{aligned} \tag{6.49}$$

and

$$\begin{aligned} u_1 &= \frac{dr_1}{dt} \\ u_2 &= \frac{dr_2}{dt} \\ \boldsymbol{\omega}_p &= \frac{d\boldsymbol{\theta}_p}{dt} \end{aligned} \tag{6.50}$$

where u_1, u_2 and r_1, r_2 are the translational velocities and displacements, respectively. These are perpendicular and parallel to $\boldsymbol{d}$. Also, I is the moment of inertia. While

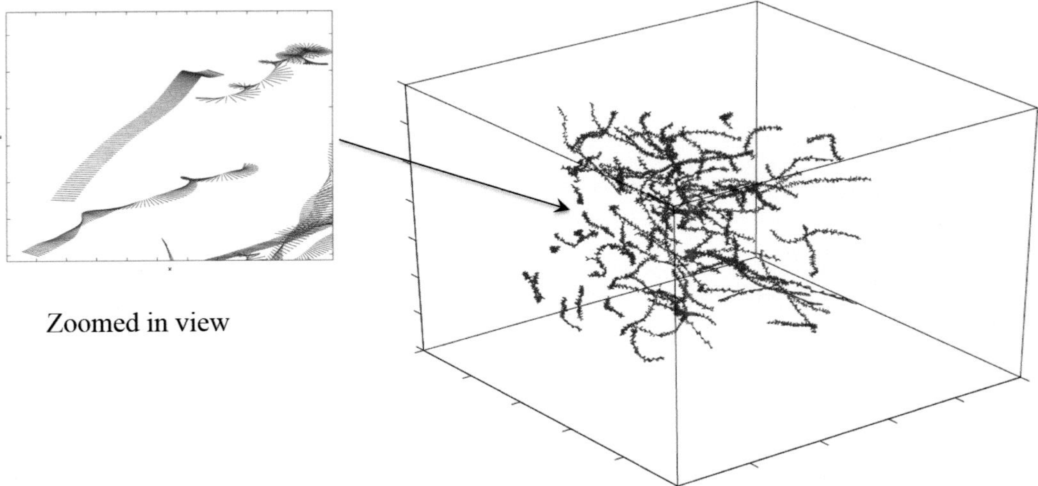

Figure 6.38. Snapshot of particle trajectories of fibers in turbulent flow (from Lo Iacono et al. 2005 – published with kind permission of Elsevier).

$\boldsymbol{\omega}_p$ and $\boldsymbol{\theta}_p$ are the angular velocities and displacements. The same field as used for the earlier particle dispersion is considered (i.e., homogenous, isotropic turbulence from a large eddy simulation). Figure 6.38 shows the integration of these equations through a turbulent field computed using large eddy simulation.

The complex fibre trajectories can be observed and the 'particle' shape can have a strong impact on their deposition (see Lo Iacono et al. 2005). Hence, this aspect is important, the assumption that particles can be treated as spherical being inadequate.

6.5 Coupled Simulations

Many engineering problems need the analysis to take into account coupling between different components and physical processes. This coupling can also be needed at dramatically differing scales. Hence, this aspect is discussed next.

6.5.1 *Multi-scale Coupled Problems*

With electronics/avionics, the critical information is often the junction temperature within the chip. Hence, there is a massive disparity in scale between the desired quantity and the quantities that control it – the system airflow. To gain the junction temperature, the airflow can be modelled using CFD and the junction temperature and heat transfer within the chip using a full finite element (FE) model – the chip structure is highly complex. Alternatively, a calibrated low-order model can be used. Typically this will be a resistor network. Figure 6.39 shows potential resistor network configurations. In Frame (b) both conduction of heat above the chip and also towards the circuit board is accounted for. In Frame (c), more degrees of freedom are added to the model, including conduction through the leads of the chip. This reduced order model can be fully integrated into the multi-scale CFD model. Such

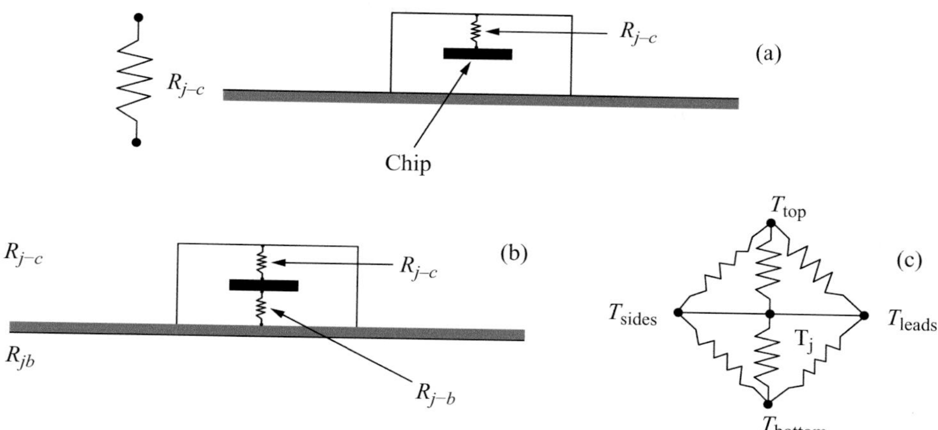

Figure 6.39. Resistor networks of increasing complexity to connect junction temperature of an integrated circuit to a core CFD model.

resistor networks can also find application in turbomachinery heat transfer modelling. They avoid the need for coupling the CFD code to a full structural model. Coupling CFD codes with heat conduction codes is discussed further later.

As noted, for dilute particulate flows there is just a one-way coupling (i.e., only the fluid flow influences the trajectories of the particles). However, for non-dilute flows, such as found in the atomizer region of a combustor, the fluid and particle fields are coupled. Then the particle-source-in-cell (PSI-CELL) – see Crowe et al. (1977) – can be used. With this approach, the particles in the mesh cells are integrated (their flows across cell boundaries being recorded) and their impact on the fluid, mass, momentum and energy balance accounted for through source terms. Figure 6.40, adapted from Crowe et al. (1977) shows the potential coupling phenomena.

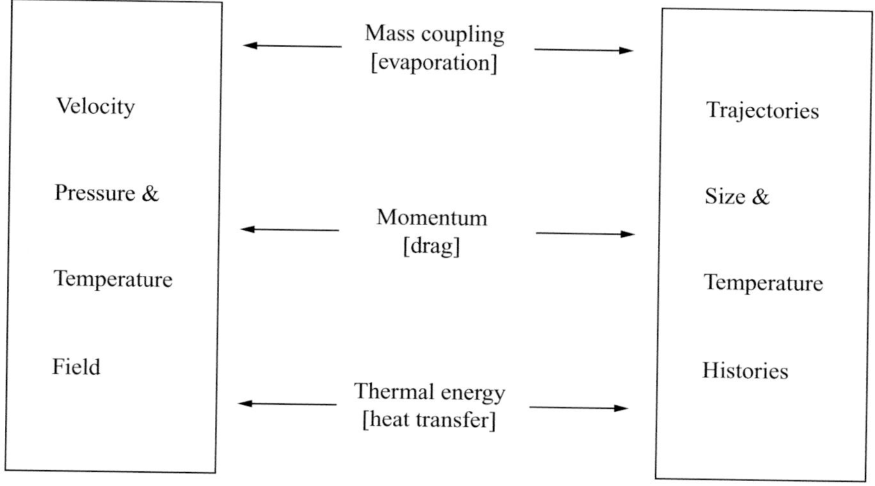

Figure 6.40. Potential coupling phenomena in gas-droplet flows (based on Crowe et al. 1977).

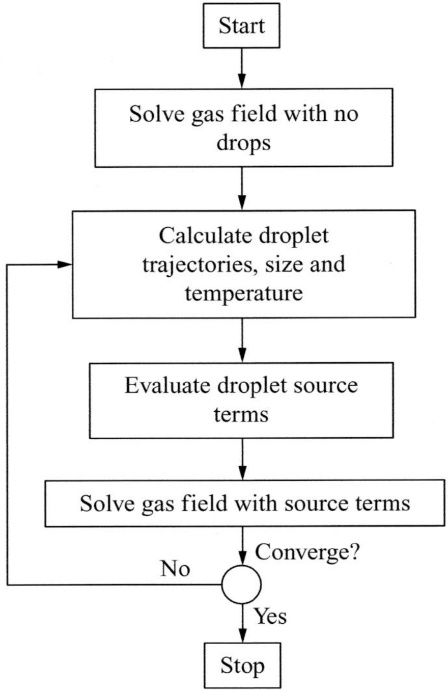

Figure 6.41. Coupling process between Eulerian fluid phase and Lagrangian droplet phase in gas-droplet flows (based on Crowe et al. 1977).

This coupling process is outlined in Figure 6.41 through a flow chart. Notably, for simple, non-spherical particles, corrections can be applied to the Lagrangian transport equation system. However, for highly non-spherical particles there are substantial modelling challenges. This aspect was partially considered earlier. Notably, as to be discussed later, mass transport of a second phase can also be approximated in an Eulerian sense. Another important application for particle tracking is exploring the trajectories of ice particles ingested in aeroengines at high altitudes and also volcanic ash ingestion. The latter is an area of increasing importance.

6.5.2 *Compatible Length Scale Coupled Problems*

Moving on from multi-scale coupled problems, Figure 6.42 gives an example of a coupled problem with less disparity in the component scales. The figure shows a time sequence of the release of a generic store from an F16 aircraft, for simulations performed by Hassan at al. (2007). The accurate prediction of the stores release requires fully coupled solutions that solve the full momentum equations for the stores ensuring conservation of both linear and angular momentum. The equations solved are akin to those considered for the transport of fibres. However, now there is two-way coupling. Also, moving meshes are used to resolve the moving surfaces.

Figure 6.42. Time sequence of F16 generic stores release (from Hassan et al. (2007) – Published with kind permission of the Royal Society).

6.5.2.1 Aeroelasticity Next, simulations that explore the interaction of the inertial and elastic forces in solids with the aerodynamic flow field will be discussed – called the field of aeroelasticity. This is a critical element in aerospace design analysis. Such coupled simulations are important for both the manufacturers of airframes and also the gas turbine engines that propel them. CFD codes can either be coupled to a reduced order model for the structure (modal equations) or a full FE model. For the design of an aerodynamic system, numerous aeroelasticity calculations are necessary. Hence, as noted, reduced order models are also needed in the design process (see Silva and Bartels 2004).

Figure 6.43, from Muir and Friedmann (2014), shows the first five vibration mode shapes for an aeroengine fan. These would represent the vibration modes found in a vacuum environment.

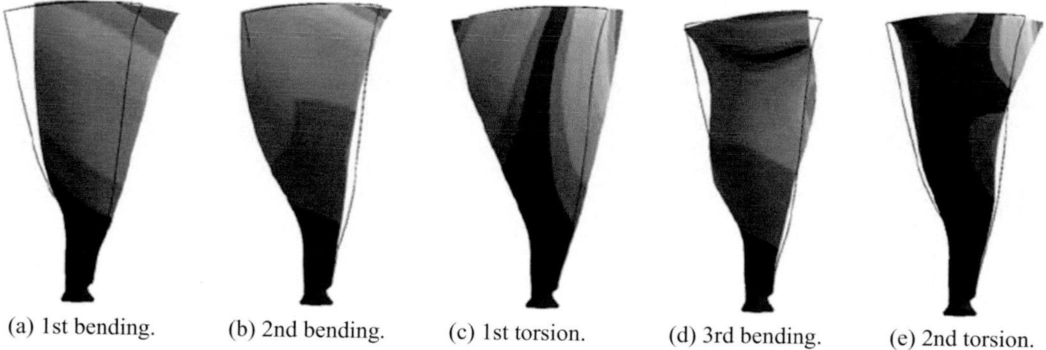

(a) 1st bending. (b) 2nd bending. (c) 1st torsion. (d) 3rd bending. (e) 2nd torsion.

Figure 6.43. First five mode shapes for an aeroengine fan (from Muir and Friedmann 2014 – used with the permission of P. Friedmann).

The equation for a rotating structure can, following Muir and Friedmann (2014), be defined as

$$[m_S]\{\ddot{d}\} + [K_S(d)]\{d\} = \{F_\Omega\} + \{F_{aero}\} \tag{6.51}$$

where d is a displacement, $[K_S]$ the stiffness matrix and $[m_S]$ is the mass matrix for the structure. Importantly, F_Ω is the assembled centrifugal load force vector and F_{aero} the assembled aerodynamic loads vector on the structure. For a non-rotating system, $F_\Omega = 0$. The aerodynamic force vector requires the surface static pressure and wall shear stresses to be surface integrated for elements as:

$$F_{aero,i} = \int_A [p(t) + \tau(t)] \cdot \mathbf{n} dA \tag{6.52}$$

where i, relates to a node, τ is the surface wall shear stress and A is the surface area of a CFD element. It is best to ensure that the force interpolations are conservative (see Doi 2002). The finite element shape function naturally provides a framework to interpolate to the generally finer CFD mesh.

As the first stage in an aeroelasticity study, as with a conventional CFD study, it is needed to ascertain the geometry of the system when running. This is generally known as the 'hot geometry'. Hence an FE model could be run, at the rotational speed of, say, a fan, to ascertain its untwisted geometry. If it was desired to study bird strike, as in the work of Muir and Friedmann (2014), then the non-linear FE code, LS-DYNA, could be used to ascertain the revised geometry arising from the bird impact. This can then be represented in the FE and CFD models. Basically, in aeroelasticity, there are two strategies. The first is to assume a one-way coupling between the fluid and structure. This is called the *forced response* approach. Then, there is the more realistic, two-way coupling. First we will discuss the one-way coupling process. Figure 6.44 shows a forced response procedure for aeroelasticity from Muir and Friedmann (2014).

With this procedure, the hot geometry is established from an FE model. Then, the interface zone between the differing FE and CFD meshes is defined. A steady state CFD calculation is then performed. This will provide initial flow conditions. Then a full unsteady CFD calculation is made. This unsteady load data is imposed on the FE model. The critical point is that the unsteady movement of the solid and the consequences of this are not imposed on the CFD solution. The structural equation system, with forced response, can be expressed as

$$[m_S]\{\ddot{d}\} + [K_S(d)]\{d\} = \{F_\Omega\} + \{F_{aero}(t)\} \tag{6.53}$$

where $F_{aero}(t)$ is the time varying force field provided by the unsteady CFD simulation.

Next a two-way coupled procedure is outlined. Now, as indicted in the following, F_{aero} is a function of the structural displacement vector and its time variation.

$$[m_S]\{\ddot{d}\} + [K_S(d)]\{d\} = \{F_\Omega\} + \{F_{aero}(t, d, \dot{d}, \ddot{d})\} \tag{6.54}$$

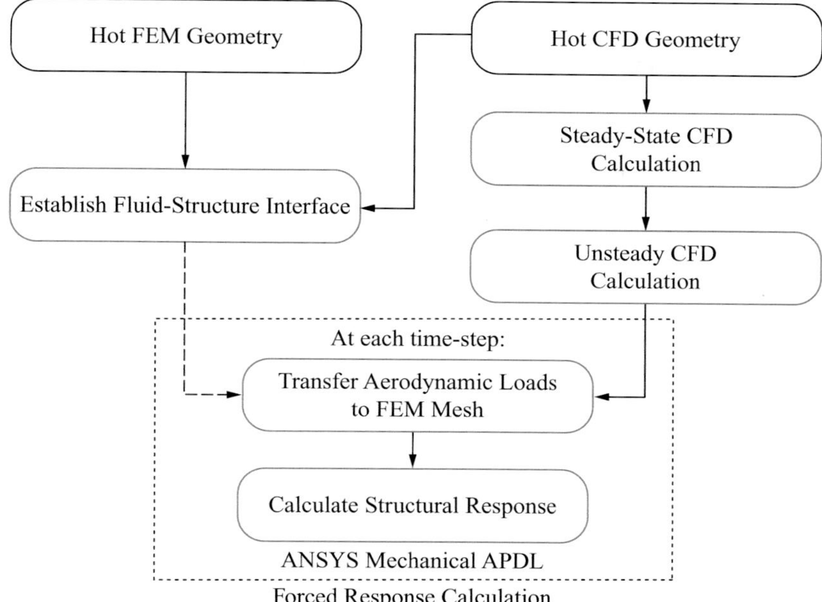

Figure 6.44. A forced response procedure for aeroelasticity (from Muir and Friedmann 2014 – used with permission of P. Friedmann).

Rather than sequential solution between the fluid and structure equations, this can, if desired, be expressed as a coupled system (see Dhopade et al. 2012) combined directly with the fluid flow equations. Figure 6.45 shows a two-way coupled process chain, again taken from Muir and Friedmann (2014). As before, the hot geometry must be established along with some form of mapping between CFD surface mesh and the FE model. Then again, a steady state CFD simulation is performed. The steady state CFD loads are applied to the FE model. Then, the FE simulation is run. The displacements from this are transferred to the CFD surface mesh. The surface displacements are propagated throughout CFD mesh using, for example, the spring analogy mesh movement strategy – see Chapter 3.

As with any moving mesh CFD problem, the space conservation law must be imposed (see Demirdžić and Perić 1988). This ensures that if the CFD mesh moves through a static flow, the fluid would remain static and not be affected by any numerical errors. Once the displaced mesh has been established, the CFD simulation is run. Then the surface loads are transferred to the FE model. If this process is carried out implicitly, then a number of iterations will need to be performed around the finite element and CFD models for a single particular time-step. It will need to be judged when the structural displacements and aerodynamic loads have converged. Once this has occurred, the next time-step is moved to.

When exploring aeroelastic behaviour, the aerodynamic work is an important parameter. It can be expressed as

$$W(t) = \int_{0}^{t} \mathbf{F}_{aero}(t) \cdot \dot{\mathbf{d}}(t) dt \tag{6.55}$$

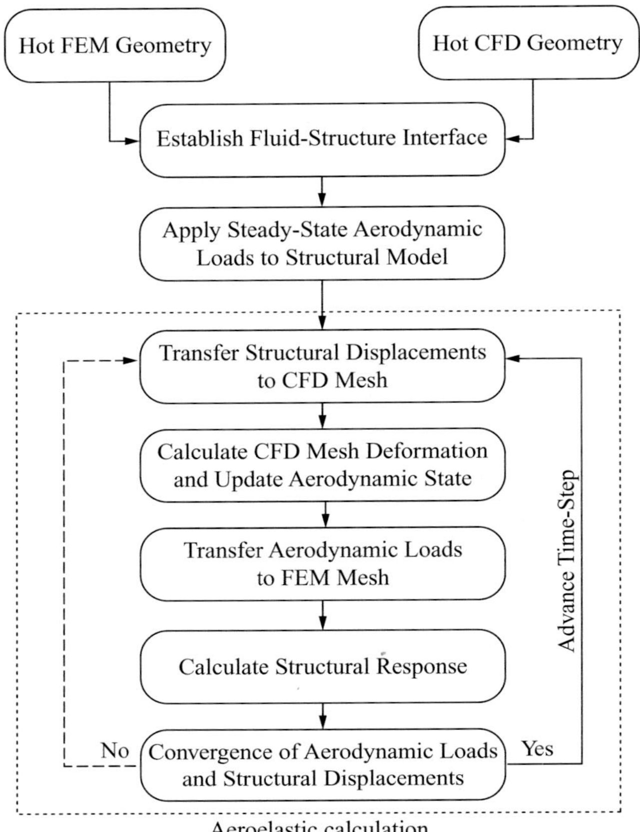

Figure 6.45. Coupled aeroelastic process chain (from Muir and Friedmann 2014 – published with permission of P. Friedmann).

Negative aerodynamic work, W, means that work is being transferred from the structure to the fluid. This is stable. On the other hand, a positive W means that work is transferred from the flow to the structure, implying instability.

Figure 6.46 shows an alternative aeroelasticity process chain as used by Kersken et al. (2012) for flutter prediction in turbomachinery. Looking at responses to perturbation, a three-dimensional FE model captures mode shapes of the structure and frequencies. The surface deformations gained from the FE analysis are fed into modules that allow deformations of the CFD mesh. The non-linear CFD solver is used to gain a steady base field around the static solid geometry. This is fed into the linearized CFD solver. This solver gives the flow field variation resulting from the blade vibration. The mode shape and blade surface pressure perturbation allows the blade surface displacements to be calculated along with other critical aeroelasticity related parameters (e.g., the system damping coefficient). Doi (2002) couple an FE program to an unsteady Navier-Stokes solver to explore flutter prediction for transonic fans.

When performing simulations where the structure deforms, aligning the domains for the fluid and structure is not trivial. Voigt et al. (2010) align the centres of area

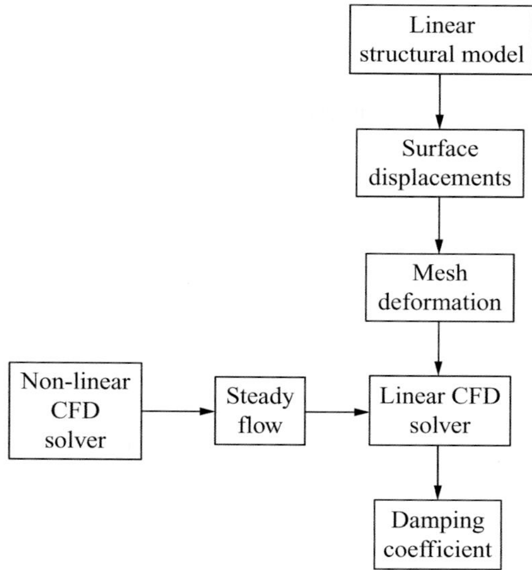

Figure 6.46. Potential aeroelasticity process chain (based on Kersken et al. 2012 – published with kind permission of Springer).

and principal axes of the two surfaces that connect the domains. Even this relatively complex approach is too simplistic for many practical problems.

In addition to the need for geometry alignment, surface data needs to be interpolated between the fluid and structure so as to ensure conservation of energy, momentum and forces. For a discussion on such aspects and the general field of elasticity, see Bartels and Sayma (2007).

Most aeroelasticity calculations are linear, however, non-linear aeroelasticity can be needed for a range of problems such as the high aspect ratio wings used on long endurance aircraft. In 2004, NASA received a mandate to return astronauts to the moon, using this as a base camp for the exploration of Mars. As part of this, heavy payloads need to be landed on Mars and also brought back to the earth. To assist with this (see Bartels and Sayma 2007), NASA has been considering the use of flexible inflatable decelerators. The flexibility of the inflatable structures means that the structural equations are highly non-linear. Also, the structure will get extremely hot, making thermal influences important. For such calculations, the CFD solver will generally be coupled with a full, non-linear FE structural solver. Aerothermoelastic modelling under these conditions, using such a modelling strategy, is described by Bartels and Sayma (2007). Linear material properties are assumed in the non-linear structural model. Figure 6.47 shows the calculations of Bartels and Sayma (2007). Frame (a) gives the mesh used for the CFD. Frame (b) shows the mesh used in the structural FE model. On the other hand, Frame (c) gives surface deflections from wind tunnel measurements and Frame (d) the predicted surface deflections. Impressively, there is just around a 5% discrepancy between the simulated and measured deflections.

Mishra et al. (2013) perform adjoint based design optimization aeroelasticity analysis for a three-dimensional helicopter rotor. Normally the adjoint design

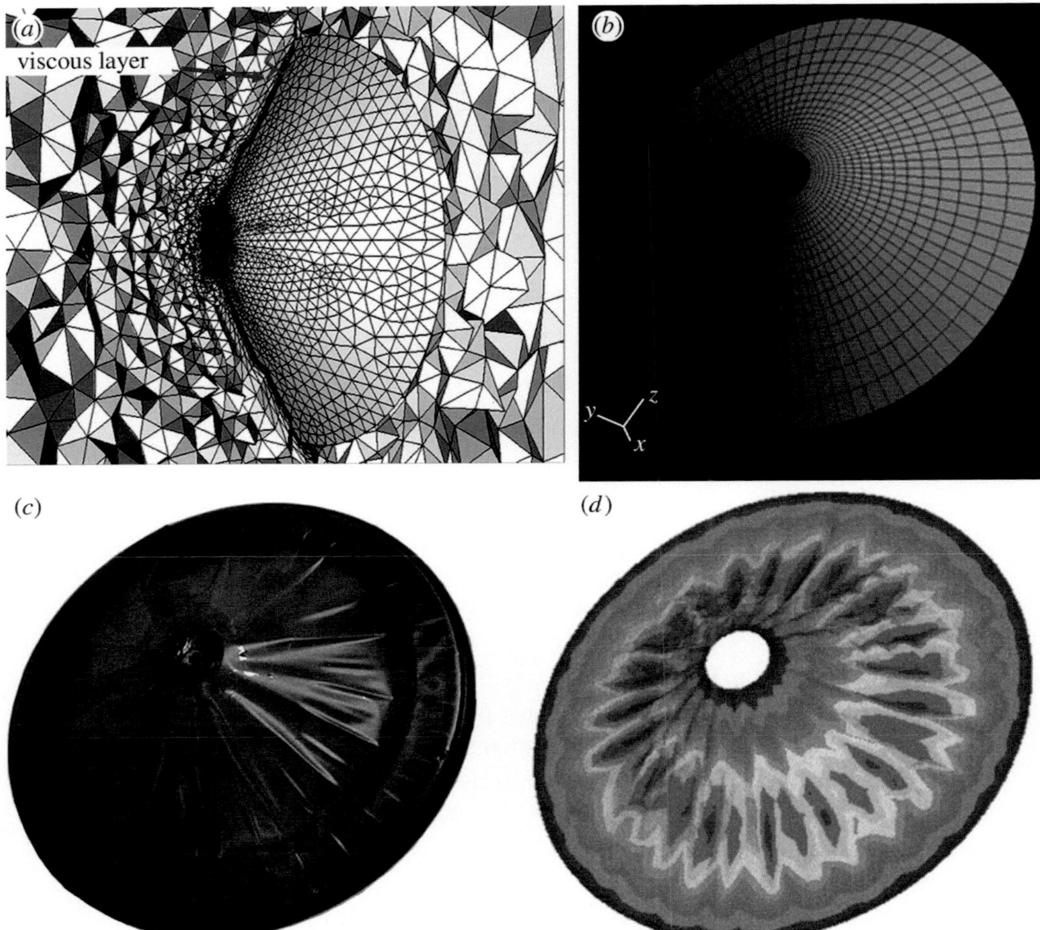

Figure 6.47. Non-linear aerothermoelasticity calculations for a decelerator intended for an atmospheric re-entry vehicle: (a) mesh used for CFD, (b) mesh used in finite element model, (c) surface deflections from measurements and (d) predicted surface deflections (from Bartels and Sayma 2007 – published with the kind permission of the Royal Society).

optimization approach is applied to steady flows and in an uncoupled manner. Hence, this extension to an unsteady coupled problem is advanced. A beam type finite element solver is used for the structure. The optimized rotor achieves a 2% torque reduction.

6.5.2.2 Conjugate Modelling Conjugate modelling is important in many fields of application including electronics (as discussed previously), high-speed flight vehicles and gas turbine aeroengines where the turbine components can be subjected to temperatures above the melting point of normal metals. The conjugate fluid–solid problem can either be solved in one code as one large system of equations (see

Rahman et al. 2005) or using coupled codes (see Duchaine et al. 2009). The latter allows highly optimized bespoke codes to be used. However, in a parallel environment, the efficient connection of the codes and parallel load balancing becomes a challenge. The basic code interfacing requirements are in essence simple. They just need to impose compatibility of temperature, T, at the fluid–solid boundaries and continuity of the heat flux, q. These conditions can be expressed as

$$\begin{aligned} T_{f,i} &= T_{s,i} \\ q_{f,i} &= q_{s,i} \end{aligned} \tag{6.56}$$

However, stability, physical realism requirements and numerical stiffness (the solid time scales are much longer than the fluid) make direct use of the basic conditions given by Equation (6.56) complex (see Giles 1997, Duchaine et al. 2009, Ganine et al. 2013). This is especially so in a parallel environment. To remove numerical stiffness, He (2010) uses a Fourier based temporal discretization for the solid zone of a turbine blade. A time-domain discretization is used for the fluid. The use of the former allows a small set of steady conduction related equations to be solved for in the solid domain. Conjugate sequential and parallel code coupling strategies are proposed by Duchaine et al. (2009).

Rather than exchange temperatures and heat fluxes, Verdicchio et al. (2001) replace the latter exchange with a heat transfer coefficient and ambient temperature. Based on two CFD solutions (one perturbed by δ, hence $T_2 = T_1 + \delta$), the heat transfer coefficient, h, can be estimated as

$$h = \frac{\partial q}{\partial T} = \frac{q_1 - q_2}{T_1 - T_2} \tag{6.57}$$

in which the subscripts identify the two CFD solutions. The local nodal ambient temperatures are evaluated based on continuity of heat flux considerations.

Again care is needed to ensure conservation at the solid–fluid interfaces. This time, conservation of thermal energy is needed. For fluid–solid domains, treated within the same solver, harmonic means for the thermal conductance can be needed at the control volume faces between the two domains. For a face exactly half way between solid and fluid nodes straddling it, the harmonic mean takes the form

$$\phi = \frac{2\phi_i \phi_{i+1}}{\phi_i + \phi_{i+1}} \tag{6.58}$$

where ϕ is the relevant diffusion coefficient in the governing equation. When different solvers are used for the fluid and solid, again conservative spatial interpolations can become important. This is especially so when there are substantial disparities in mesh resolution.

6.5.2.3 Radiative Heat Transfer The thermal radiation and the flow fields are indirectly coupled. For high temperature systems, the radiative heat transfer can be a substantial component of the convective. Most commercial CFD programs have a radiative heat transfer capability. Volume mesh surfaces on boundaries can be used as a basis for the thermal radiation analysis. For a non-participating medium, such as found on heat rejection systems on spacecraft and some turbomachinery applications, the method used to calculate heat flux described by Long (1986) can be used. With this, the system is considered to be an enclosure comprising N isothermal grey surfaces, each of constant emissivity ε_i. The radiative heat flux at the surface i, $q_{R,i}$, is obtained by considering that from the other ($j = 1, 2, \ldots N$) surfaces of the enclosure. The resulting set of N simultaneous equations can be written as the summation:

$$\left\{\sum_{j=1}^{N} F_{ij}(J_i - J_j)\right\} = E_{b,i} - \frac{J_i}{\{(1-\varepsilon_j)/\varepsilon_j\}} = q_{R,i} \tag{6.59}$$

where the unknown radiosity, J, is expressed in terms of the black body emission, $E_{b,i}$ ($E_{b,i} = \sigma T_i^4$), the local emissivity ε_i and view factor F_{ij} (the ratio of radiation received by a surface j to that emitted by a surface i). The variation of emissivity with temperature can be evaluated using fits to tabulated data (see, for example, Touloukian and DeWitt 1970). View factors can be gained from solving the double integral

$$F_{ij} = \left(\frac{1}{\pi A_i}\right) \int_{A_j}\int_{A_i} \frac{\cos\theta_i \cos\theta_j dA_i dA_j}{L^2} \tag{6.60}$$

where A_i and A_j are the areas of surfaces i and j respectively. The distance between the two surfaces is equal to L. The terms θ_i and θ_j are the angles that the line L makes with the normal to the surfaces A_i and A_j. Equation (6.60) can be integrated analytically for simple geometries. The results for analytical integrations can easily be found as tabulated expressions, for example see Howell (1982). However, for complex geometries, the double integral is best integrated numerically. For example, on the face of each surface element, a vector can be defined. The vector passes through the centre of area of the element. The scalar product is used to define an orthogonal vector. The vectors cross at the centre of area of the element considered. The vector product is then used to calculate a unit normal vector to the two lines and therefore the element. A vector **L** is defined between arbitrary elements i and j. A dot product can then be used to find the cosine of the angle between **L** and the unit normal vector to each of the elements i and j. Integral (6.60) can then be solved in the form

$$F_{ij} = \left(\frac{1}{\pi A_i}\right) \sum_{A_j}\sum_{A_i} \frac{\cos\theta_i \cos\theta_j dA_i dA_j}{L^2} \tag{6.61}$$

If the areas of elements i and j are sufficiently small, (6.61) can be simplified to

$$F_{ij} = \left(\frac{1}{\pi}\right) \frac{\cos\theta_i \cos\theta_j A_j}{L^2} \tag{6.62}$$

A check must be made to see if the 'view' (line of **L**) between the emitting element i and the receiving element j is obscured by internal geometry and so forth. If the

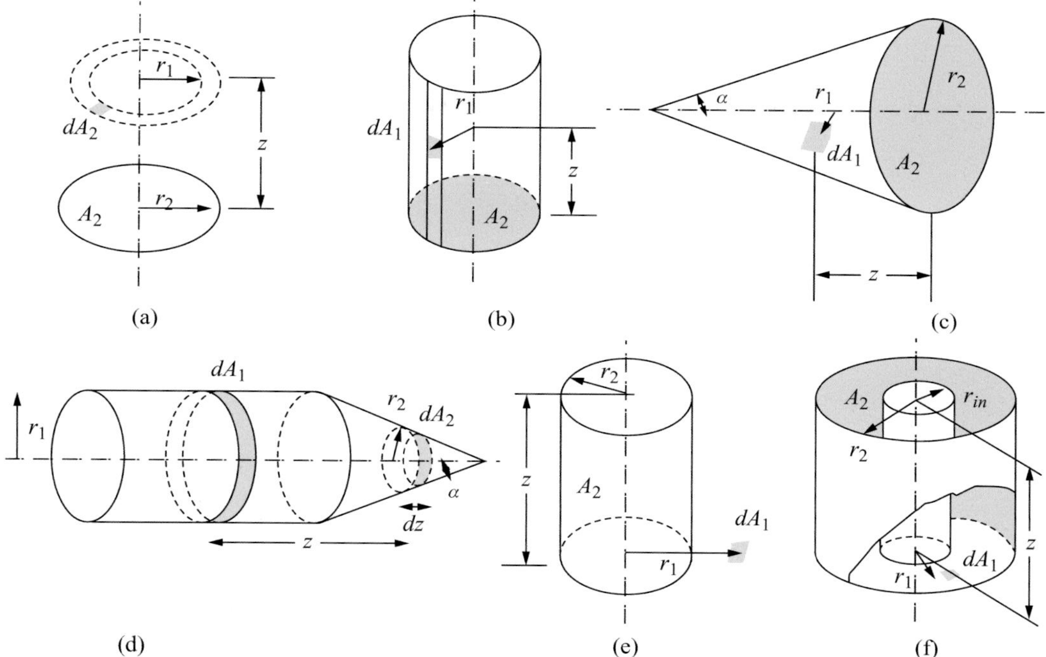

Figure 6.48. Geometries for evaluating views factors: (a) a differential element emitting thermal radiation to a coaxial parallel disc, (b) an element emitting thermal radiation on the inside of a right circular cylinder to the base of the cylinder, (c) an element on the internal surface of a right circular cone emitting thermal radiation to the base of the cone, (d) a ring element on the interior of a right circular cylinder emitting thermal radiation to a ring element on the interior of a coaxial right circular cone, (e) a coaxial ring element emitting thermal radiation to a circular cylinder and (f) a coaxial ring element on an annulus between coaxial cylinders emitting thermal radiation to the outer cylinder (based on Howell 1982 and Leuenberger and Person 1956).

L is obscured, the value F_{ij} is set to zero. The element size needs refinement when calculating view factors between surfaces close to acute angles. To test if the mesh is sufficiently fine in these regions, it should be successively refined until changes in the computed view factor are negligible. Alternatively, the computed view factors can be tested for grid independence using the so-called summation rule. This states that, for an emitting surface i, contained in an enclosure with N surfaces, the sum of all the view factors F_{ij} will equal unity:

$$\sum_{j=1}^{N} F_{ij} = 1 \tag{6.63}$$

To test the performance of such an approach, the six geometries shown in Figure 6.48 are considered. These are: (1) a differential element emitting thermal radiation to a coaxial parallel disc; (2) an element emitting thermal radiation on the inside of a right circular cylinder to the base of the cylinder; (3) an element on the internal surface

Table 6.4. *Computed view factors for different geometries and different numbers of circumferential elements*

	Computed view factors, F					
Geometry	$N = 8$	$N = 16$	$N = 32$	$N = 64$	$N = 128$	Analytical solution
1	–	0.331	0.328	0.327	0.327	0.327
2	–	0.437	0.409	0.407	0.407	0.407
3	–	0.706	0.699	0.697	0.697	0.696
4	0.032	0.031	0.030	0.030	0.030	0.029
5	0.259	0.215	0.213	0.212	0.212	0.212
6	–	0.582	0.572	0.574	0.572	0.572

of a right circular cone emitting thermal radiation to the base of the cone; (4) a ring element on the interior of a right circular cylinder emitting thermal radiation to a ring element on the interior of a coaxial right circular cone; (5) a coaxial ring element emitting thermal radiation to a circular cylinder and (6) a coaxial ring element on an annulus between coaxial cylinders emitting thermal radiation to the outer cylinder. More details of the characteristic dimensions for these geometries are given in Tucker (1993). For Geometry (1), comparison is made with an analytical solution by Leuenberger and Person (1956), and for Geometries (2–6), comparisons are made with analytical solutions given by Howell (1982). Schematics of the geometries can be found in these texts. For Geometries (1), (2) and (3), the performance of the numerical method when the disc-like surfaces are divided into sixty radial sectors is considered. The computed view factors for the different geometries and different numbers of circumferential elements, N, are given in Table 6.4. Results from the analytical solutions are also shown in the table. As can be seen, using this simple method, view factors can be computed accurately with relatively few ($N = 64$) circumferential elements.

For combustors found in turbomachinery, the fluid medium participates in the radiative heat transfer and more complex modelling is needed than that described here. Such models include: a Monte Carlo method, discrete transfer method, discrete ordinates method and finite volume based methods. For a brief overview of these, see Versteeg and Malalasekera (2007).

6.5.2.4 Coupled Aerodynamic Simulations The use of coupled aerodynamic multi-fidelity simulations is increasing. For example, Chima et al. (2010a) perform coupled intake-fan simulations. The simulation involves a range of fidelities with axisymmetric equations for the intake being coupled to a full three-dimensional CFD model with body force terms to approximate some geometrical elements. Coupled intake duct-fan simulations have also been performed by Chima et al. (2010b) and Loiodice et al. (2010), the fans being modelled using body force models as discussed previously. Notably, Medic et al. (2007) perform a CFD simulation through a complete aeroengine. The combustor zone is treated using LES and the upstream compressor

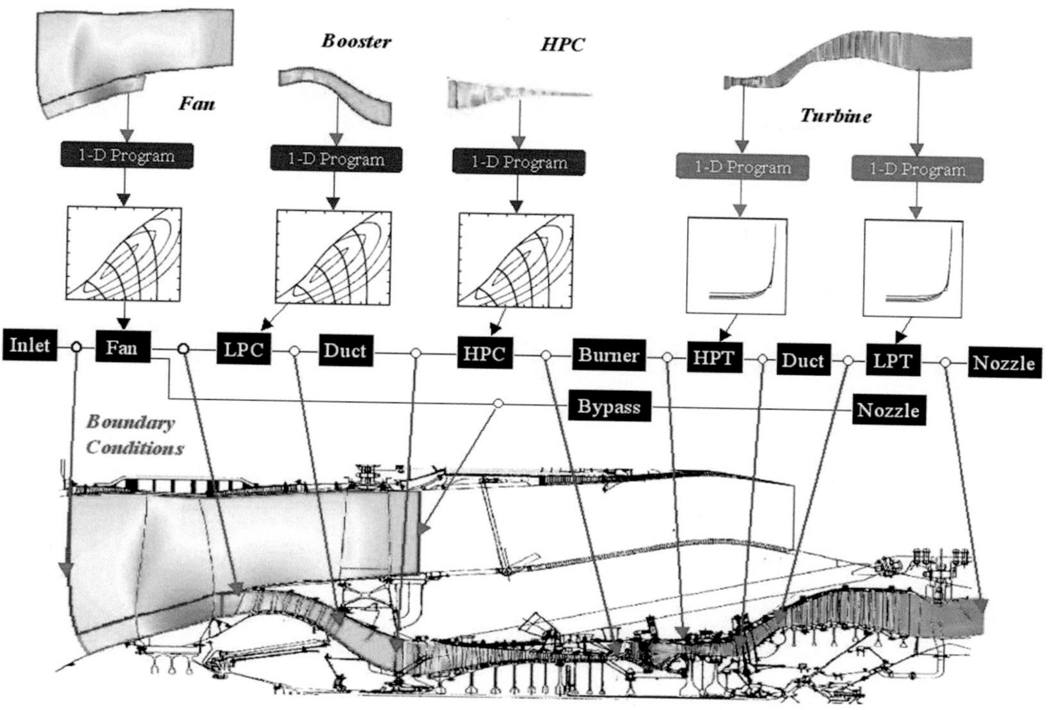

Figure 6.49. Example of the coupling of three-dimensional CFD models with zero dimensional models for a modern gas turbine aeroengine (from Turner et al. 2004 – used with kind permission of the ASME).

and downstream turbine stages using (U)RANS. A key gas turbine aeroengine zone where code coupling is generally needed is the combustor turbine interface. In the combustor, the Mach numbers are low, and specialized (with a range of combustion models) pressure correction (see Section 4.7.3) solvers are generally regarded as most ideal. For the turbine zone, at realistic engine conditions, density based solvers are most efficient, the turbine blade passages having transonic or high subsonic flow. Salvadori et al. (2012) perform coupled combustor turbine simulations with various levels of coupling (i.e., strong and loose coupling). Dhopade et al. (2012) look at high cycle fatigue failure of fan blades, coupling separate structural and CFD codes. Hence, as can be seen, a large range of types of coupled simulations are beginning to emerge. In terms of aerospace, as noted by Spalart and Bogue (2003), the ultimate ideal would be coupled airframe–engine–pilot interaction simulations. Such simulations would need to involve an extreme range of model fidelities. Turner et al. (2004) present simulations relating to the analysis of a complete modern gas turbine aeroengine. Figure 6.49 indicates the wide range of components involved. The analysis involves the fusion of both three-dimensional, full system CFD models with a zero-dimensional thermodynamic cycle type of analysis. Characteristic maps from the three-dimensional models are formulated via models of intermediate dimensionality. These graphical maps are then used to refine the zero-dimensional

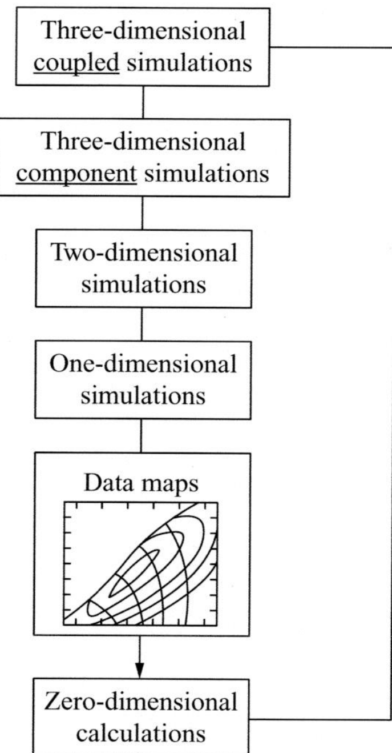

Figure 6.50. The use and interactions in multi-fidelity coupled simulations.

thermodynamic model. In turn, this model is used to provide data such as shaft rotational speeds that enable the fully (one-way) coupled CFD simulation to be run. Hence, the analysis involves both zooming in and out at different fidelity levels.

Figure 6.50 indicates the fidelities involved in the previous analysis and the way that there is a two-way communication path between the fidelity extremes.

Next an example of a recent multi-fidelity-coupled simulation is discussed which relates to an aeroengine mounted to an idealized wing with a deployed flap. The simulation is hybrid RANS-ILES based. The complete, outer bypass duct of the engine is modelled. The fan and downstream vanes are treated using the body force model discussed in Section 6.3.3 extended to replicate wakes. The latter component is calibrated to established wake data. The A-frame (that connects the inner engine to the nacelle) and gearbox shaft in the bypass duct of the engine are modelled using a crude immersed boundary method. The flow then emerges from the jet nozzle. It interacts on its way out of the engine with an engine-mounting pylon (see Figure 6.51). The pylon connects the engine to the wing. The flow then interacts with the wing and a deployed flap. As the diameter of engine continues to increase (to improve efficiency) the interaction of the propulsive jet with the wing and flap is of ever increasing importance.

Figure 6.51 shows an instantaneous isosurface of axial velocity, making visible the internal geometry modelled using the body force model and crude immersed boundary method. The outlet guide vanes, A-frame and gearbox geometry and

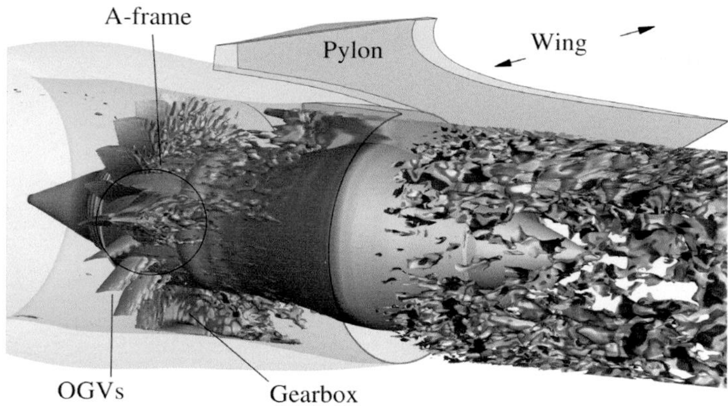

Figure 6.51. Isosurface of instantaneous axial velocity coloured by radial velocity (adapted from Tyacke et al. 2015 – published with kind permission of Springer).

wakes are indicated along with the pylon and wing (not shown). This all clearly creates a complex bypass flow.

Figure 6.52 shows axial velocity contours at different axial planes. These are just downstream of the internal geometry, at the bypass exit plane and halfway along the pylon.

Figure 6.53 shows a short time mean of the axial Reynolds stresses both without and with internal geometry. Without internal geometry, there is no upstream disturbance in the bypass duct (a). In contrast, Frame (b) shows relatively high turbulence levels. This produces a more rapid thickening of the outer and inner shear layers, which may in turn affect the noise produced.

6.5.2.5 Code Coupling Software The CHIMPS code coupling software is used in the pioneering Stanford whole engine simulation work (see Alonso et al. 2006, Medic et al. 2007). It has the ability to interpolate between differing interface meshes.

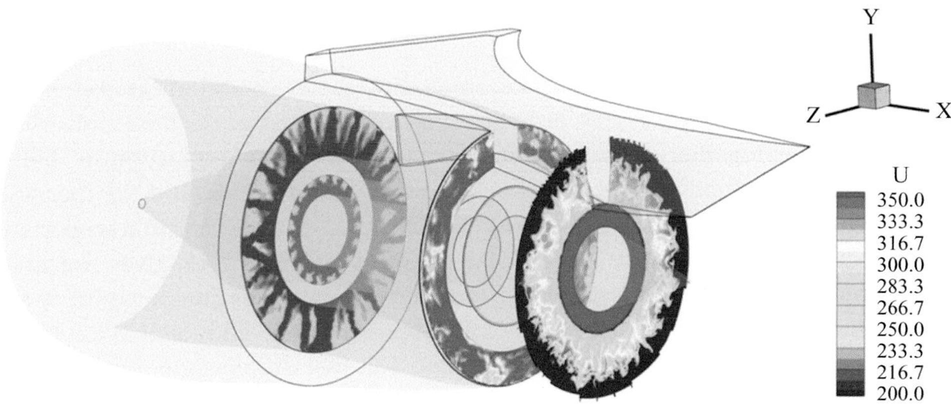

Figure 6.52. Instantaneous axial velocity contours at different axial locations.

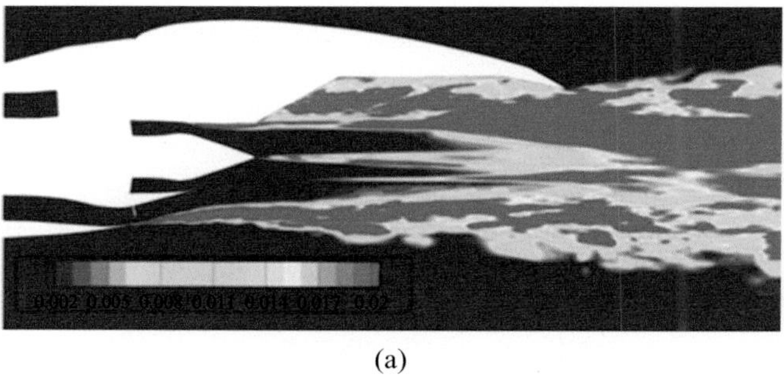

(a)

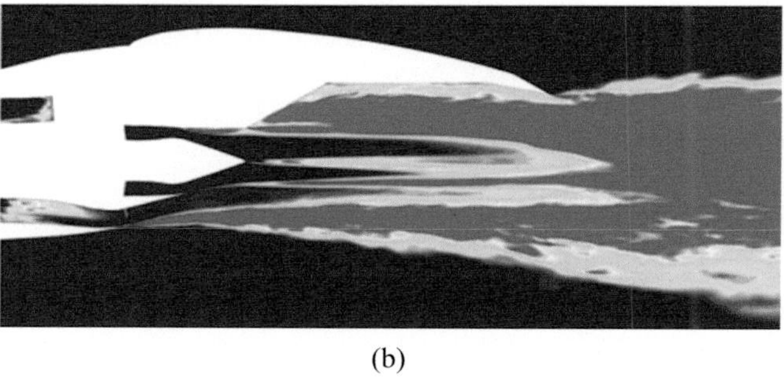

(b)

Figure 6.53. Axial normal Reynolds stress contours: (a) no internal geometry and (b) with internal geometry.

MpCCI (commercial software), PALM (see Buis et al. 2005) and MDICE (Multidisciplinary Computing Environment – see Kingsley et al. 1998), are all relatively accessible code coupling software. The ANSYS Multifield (MFX) is commercial coupling software. When coupling different codes, load balancing is complex. The FE/conduction solution is generally likely to be much faster than that of the fluid. Care is required with communication frequencies between different codes if aliasing errors and physically unrealistic solutions are not to arise.

6.6 Multi-Physics Problems

Many problems are multi-physics. Some of these have been indicated already. For example, it can be necessary, as in aeroelasticity, to look at the fluid and structural behaviour together. It can be necessary to look at the conduction in solids and also radiative heat transfer between fluid and solid systems, and the thermal stresses arising within a solid from this heat transfer. Droplets can be transported within a flow and this process can involve chemical reactions. This gives rise to a multi-scale and multi-physics problem. The disparity in scales can partly be seen from Figure 6.54. This shows an image of a combustion large eddy simulation from Pitsch (2006) for a modern gas turbine engine. The fuel droplets are given by the smaller scale more spherical isosurfaces. The remaining isosurfaces relate to temperature, the elevated temperatures arising from the combustion involving complex

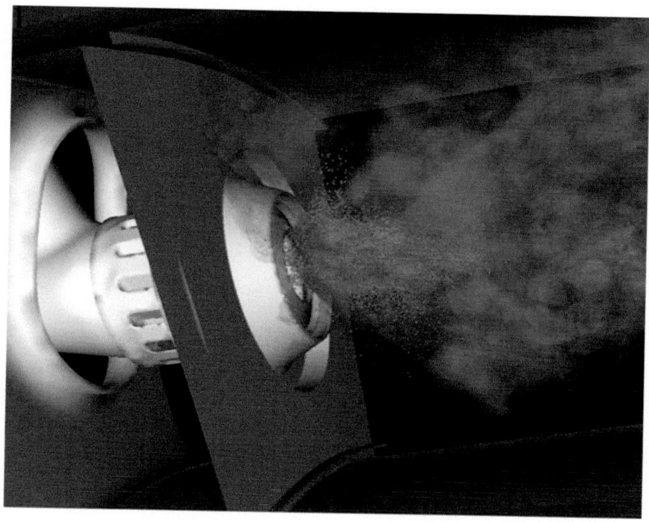

Figure 6.54. Image of a combustion large eddy simulation for a modern gas turbine engine (from Pitsch 2006 – published with the kind permission of the Annual Reviews). (See color plate 6.54.)

chemical process. The simulations involve Lagrangian particle tracking along with models for breakup and evaporation of liquid sprays. As will be discussed further later, and is very much part of combustion modelling, there can be moving interfaces with complex physical processes and forces acting at these interfaces. The engineering problem can involve the need to study wave propagation. Fluid flow can both generate sound and then propagate it. For aerospace systems, this field of computational aeroacoustics (to be discussed later) can be of some importance. With regards to aircraft, it can also be necessary to study their radar signature through solving the equations for electromagnetics. The Navier-Stokes equations can be adapted to study problems of this type. Inclusion of electromagnetics moves the complete analysis to a decoupled but multi-physics and multi-scale endeavour. Figure 6.55, from El Hachemi et al. (2004), shows the scattering of a planar wave on an aircraft in relation to electromagnetics. Such problems are well suited to high-order solvers, it being challenging to capture the high frequencies of interest. There are a range of commercial multi-physics programs, for example, ANSYS Multiphysics, Altair, Abaqus, ADINA, CFD-FASTRAN, COMSOL, CFD-ACE+ Multiphysics, Flex-PDE, FEATool, LS-DYNA, OOFELIE and PHYSICA.

6.7 Interface Tracking for Multi-Phase and Free Surface Flow

6.7.1 *Free Surface Modelling*

Essentially there are the following key free surface flow modelling techniques:

(1) Use a grid that extends beyond the free surface. The fluid boundary, within the extended domain, is then tracked (Harlow and Welch 1965).

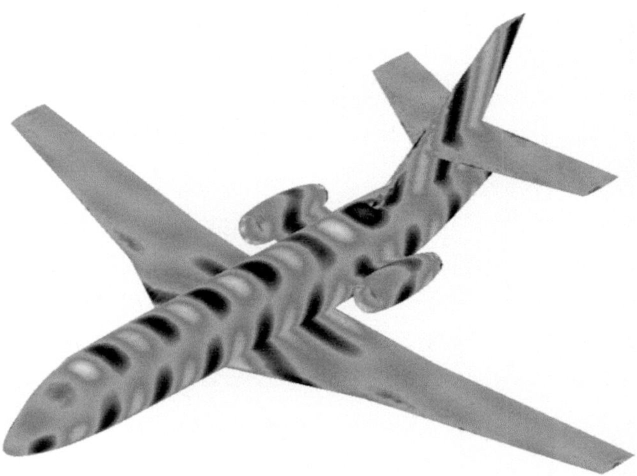

Figure 6.55. Scattering of a planar wave on an aircraft in relation to electromagnetics (from El Hachemi et al. 2004 – published with the kind permission of the Royal Society).

(2) Have a deforming mesh that conforms to the free surface region and moves with it (see Tanner et al. 1975 and Engelman and Sani 1986).
(3) Use a mesh free method, such as smooth particle hydrodynamics (see Chapter 4).

These ideas can also be applied to multi-phase flows and also ice accretion modelling on aeroengine intakes. However, for Approach (2) the mesh needs to encompass both phases, with the mesh adapting to keep the interface zone between the phases defined. For both free surface and multi-phase flows, some sort of interface tracking algorithm is needed. These are considered next. First techniques specific to free surfaces are considered.

6.7.2 *Free Surface Tracking Techniques*

As indicated in Figure 6.56a, the simplest free surface tracking technique (see Miyata and Nishimura 1985) is to define a free surface height, $\phi = h\,(x, z, t)$, relative to a datum and solve for this as

$$\frac{\partial \phi}{\partial t} + u\frac{\partial \phi}{\partial x} + w\,\frac{\partial \phi}{\partial z} = v \tag{6.64}$$

where v is the wall normal, y, velocity component. The remaining components u and w are the wall parallel in the x and z directions, respectively. This approach has the attraction that it needs relatively little storage and computational effort. However, unfortunately, $h\,(x, z, t)$ can become multivalued at the same x and t. Figure 6.56b shows an ad hoc technique resorted to by Saito and Scriven (1981) to overcome this problem by making use of a polar coordinate description of h. Figure 6.57 shows a more flexible variant, used by Engelman and Sani (1986). This uses a more complex

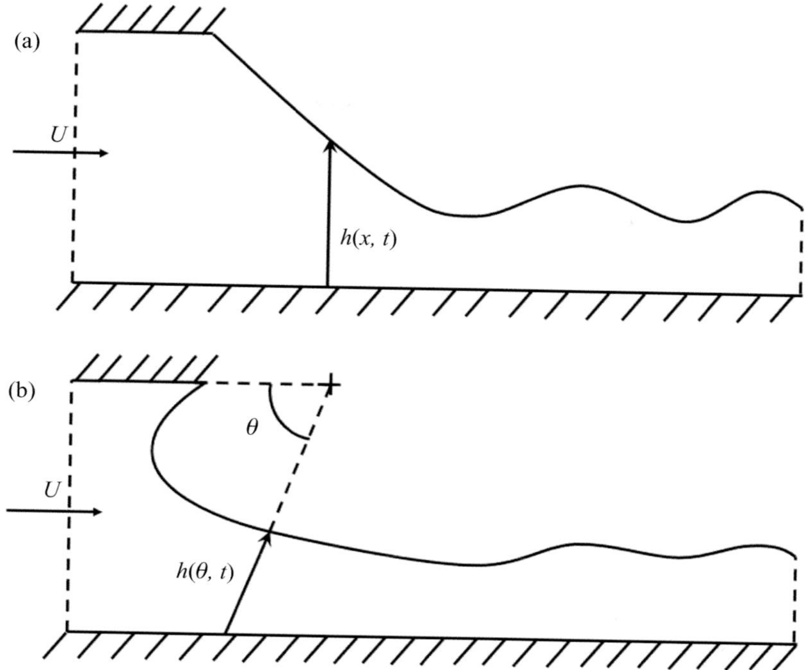

Figure 6.56. Description of free surface profile using a height variable: (a) profile where $h\,(x, t)$ is not multivalued and (b) profile where h can be multivalued (adapted from Tucker 2001).

reference frame for h given by the thicker line in the figure. This line can be defined through use of a parametric equation.

An alternative to the use of Equation (6.64) is to seed the surface with weightless particles and solve the particle Equation (6.32). (Note, however, that Miyata and Nishimura also seed their surface with particles, iteratively locating these so that at the end of time-steps they conveniently align with cell centres.) Alternatively, not just the surface, but all the fluid region can be seeded with particles. The left-hand frames of Figure 6.58 shows the use of this process for a buoyant plume of fluid rising

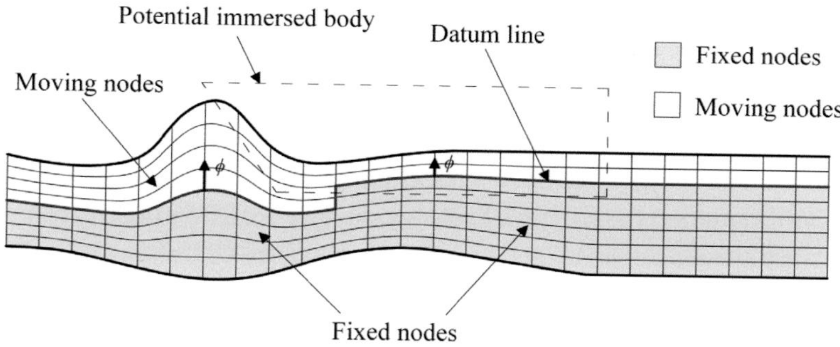

Figure 6.57. More flexible 'height' function variant (adapted from Engelman and Sani 1986).

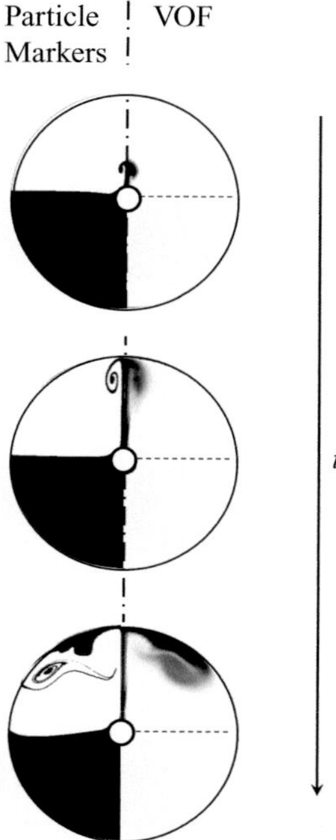

Figure 6.58. Development of interface with time for a suddenly heated cylinder using particle tracking (the left-hand frame shows results of MAC-related method and the right-hand frame of the VOF method).

around a heated cylinder. The figure shows the plume development at different points in time. The seeding of the entire fluid region with weightless particles is used in the popular marker and cell (MAC) approach of Harlow and Welch noted earlier. Unlike Equation (6.64), recovery of the free surface from Equation (6.32) is considerably more complex. However, clearly Equation (6.64) is less generic.

A simple, related, general interface tracking technique is to solve a transport equation for ϕ, the fractional volume of cell occupied by the fluid (see Jun 1986).

$$\frac{\partial \phi}{\partial t} + u\frac{\partial \phi}{\partial x} + v\frac{\partial \phi}{\partial y} + w\,\frac{\partial \phi}{\partial z} = 0 \tag{6.65}$$

Notionally, the value, say, $\phi = 1$ corresponds to a cell completely full of fluid and $\phi = 0$ the reverse. Equation (6.65) is simple to implement, but specialist convection schemes are needed if the interface is not to appear substantially smeared. The ϕ distribution allows recovery of the interface. A key disadvantage is that, unlike the MAC method, Equation (6.65) cannot resolve sub-grid scale features. It will also not capture surface breakup. The right-hand frames of Figure 6.58 show the application of this approach.

Fuel tank sloshing is an important free surface and multi-physics problem where it can be important to extract both structural deflection and acoustic information.

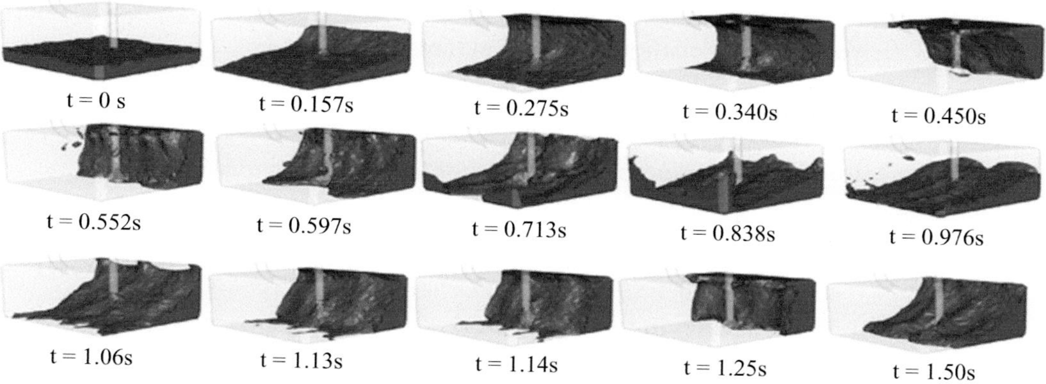

Figure 6.59. The use of the volume of fluid approach to study fuel tank sloshing. Isosurfaces show the air–fuel interface at different points in time (from Singal et al. 2014 – Open Access Publication).

Sances et al. (2010) study fuel sloshing in spacecraft propellant tanks with a view to controlling the craft's stability. A flexible control diaphragm is used and hence the problem also involves fluid-structure interactions. The volume of fluid method is used to track the fee surface interface. Singal et al. (2014) also used the volume of fluid approach to study fuel tank sloshing with kerosene. Figure 6.59 from this work shows isosurfaces of the air-fuel interface at different points in time. The tank is given a sudden lateral acceleration of 9.81 m/s for a duration of 1.5 s. Hence, extreme surface movements can be seen. In a design analysis, this CFD framework could be used to optimize the location of baffles within the fuel tank.

The level-set equation, also discussed in Chapter 3 and Chapter 7, is useful for tracking interfaces. It is used in several forms for this purpose – see for example Sussman et al. (1994) and Wang and Wang (2005). The thin, time-dependent flame front in turbulent combustion is a popular application for level sets. As shown in Figure 6.60, the front is an interface between the burnt and unburned fuel and can move both forwards and backwards in time. Hence, a time-dependent level-set equation formulation is necessary. If the level-set function is taken as a scalar field

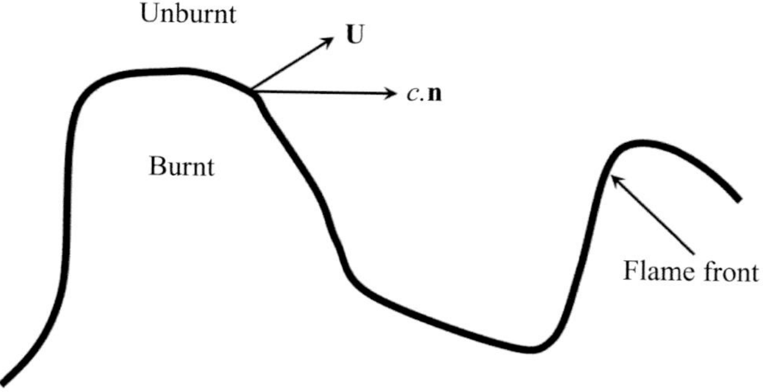

Figure 6.60. Schematic of front position.

ϕ, then ϕ_0 could define the level-set corresponding to the flame front. The equation can be derived by differentiation of $\phi\ (\boldsymbol{x}, t)$ in time as

$$\frac{\partial \phi}{\partial t} + \frac{\partial \boldsymbol{x}_f}{\partial t} \cdot \nabla \phi = 0 \tag{6.66}$$

where $\boldsymbol{x}_f$ is the local position of the flame front. This can be evaluated from the following:

$$\frac{\partial \boldsymbol{x}_f}{\partial t} = \mathbf{V} + s_l \boldsymbol{n} \tag{6.67}$$

The velocity $\mathbf{V}$ corresponds to the flow velocity. Also, s_l corresponds to the laminar flame burning velocity with $\boldsymbol{n}$ being the unit normal to the flame front. The latter can be expressed as

$$\boldsymbol{n} = -\frac{\nabla \phi}{|\nabla \phi|} \tag{6.68}$$

Rearranging equations (6.66) to (6.68) gives the instantaneous flame front level-set equation

$$\frac{\partial \phi}{\partial t} + (-s_l)|\nabla \phi| = -\mathbf{V} \cdot \nabla \phi \tag{6.69}$$

Notably, Equation (6.69) can clearly be used for other interface tracking applications and has strong connections to the eikonal equation used for acoustics (discussed later in this chapter). If $s_l = 0$ (ie., there is no burning velocity), then a level-set equation becomes like Equation (6.65) that can track a fluid interface as needed for multi-phase flows.

Ice build-up on wings and other aerodynamic sections, such as fan blades (see Hayashi and Yamamoto 2013) will increase drag and decrease stall margin. The ice can also break away from aerodynamic surfaces and cause downstream damage. This damage can occur in the core of gas turbine aeroengines or on airframes. There are several instances where ice build-up has resulted in crashes and associated fatalities. The modelling of the ice build-up is very much a multi-physics problem involving supercooled water droplets accreting to a surface. There are two types of ice build-up. For temperatures below –10°C, 'rime ice' is generated, where the ice droplets freeze instantaneously. For temperatures between –10°C to 0°C, 'glaze ice' is generated. With this process, the freezing is not instantaneous and the ice structures run along the body. The arising ice topologies can be extremely geometrically complex and the physics of their formation is also complex. Beaugendre et al. (2011) study the trajectory of ice once it has been mechanically forced normally away from an airfoil surface. The initial ice topology is prescribed based on other work. To parameterize the complex ice topology and its trajectory, a level-set approach is again used. Results from the work of Beaugendre et al. (2011) are shown in Figure 6.61. Frame (a) gives the level-set field showing the airfoil and initial ice geometry. The darker contour colours inside the airfoil correspond to negative level-set values in solid zones. The shades outside the airfoil are for the positive level-set values in the fluid zone. Frame (b) shows a zoomed in view of the level-set field. Frame (c) shows the trajectory of

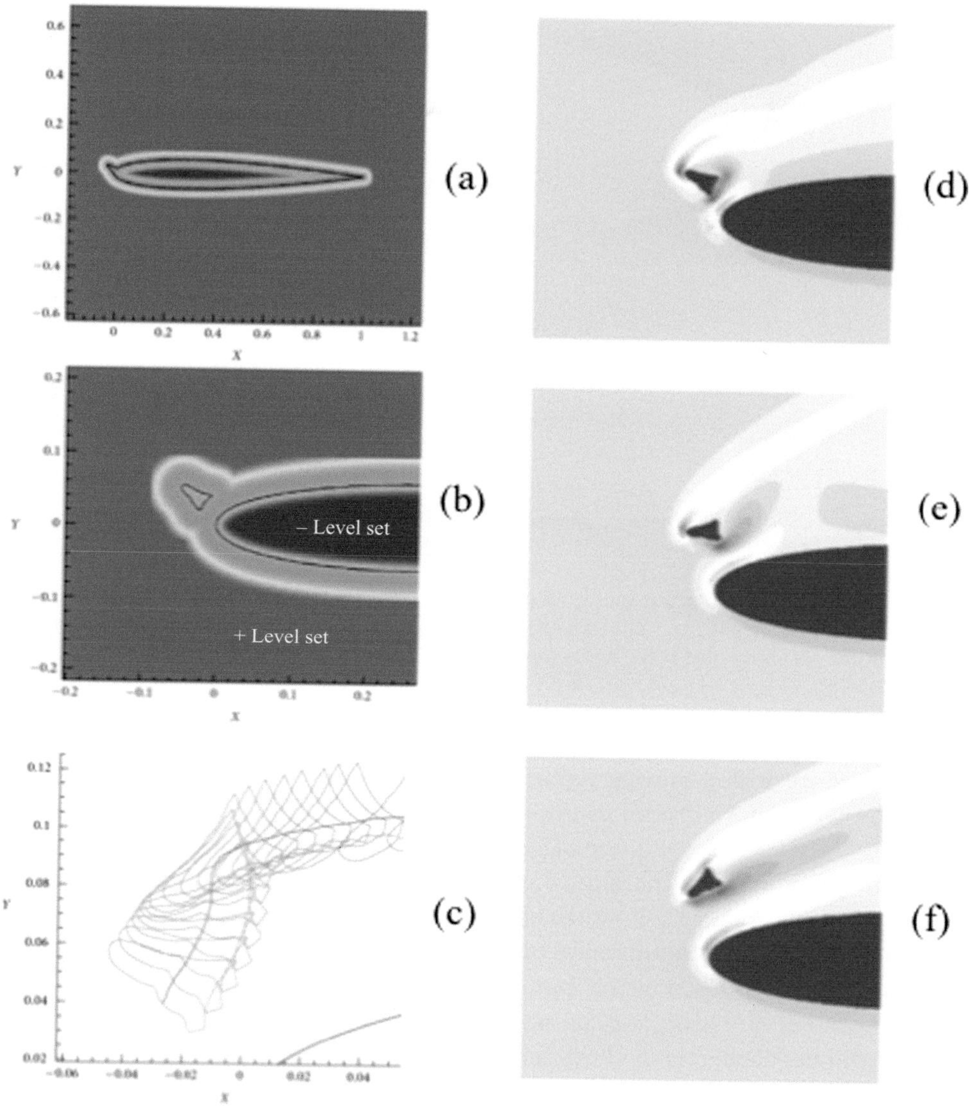

Figure 6.61. Modelling the trajectory of accumulated ice using a level-set based technique: (a) level-set field showing the airfoil and initial ice geometry, (b) zoomed in view of the level-set field, (c) trajectory of ice after mechanical release and (d–f) vorticity field contours and initial ice trajectory at three points in time (from Beaugendre et al. 2011 – Hindawi Publishing Corporation, Open Access). (See color plate 6.61.)

ice after mechanical release, near the airfoil leading edge. Finally frames (d–f) show contours of the vorticity field and initial ice trajectory at three early points in time after the ice is released. The flow field is computed on a Cartesian background mesh. The solid geometry moves through this. The convective terms are solved using a vortex method and the diffusive terms are made using a grid-based solution process. Hence, the solution strategy for the flow field is hybrid.

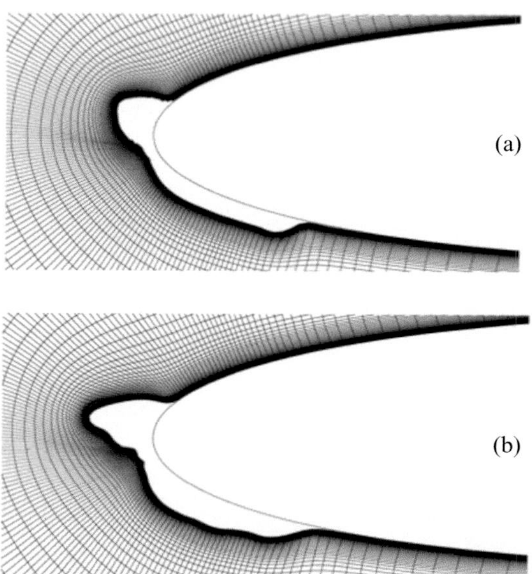

Figure 6.62. Evolution of ice with time when mapping ice build-up with an ALE formulation: (a) $t = 150$ s and (b) $t = 225$ s (from Fossati et al. 2012 – used with the kind permission of John Wiley and Sons Inc.).

If there is no need to model the ice breakaway process, then it is possible to map the surface ice with a moving mesh. For this purpose, the Arbitrary Lagrangian-Eulerian (ALE) formulation can be used. For example, Figure 6.62, from Fossati et al. (2012), shows the evolution of ice with time when mapping ice build-up with an ALE formulation. The two frames correspond to the build-up of ice at two different points in time. Figure 6.63 shows the flow field over the ice accreted airfoil and also the clean airfoil. The contours are of the velocity magnitude. Frame (a) gives the flow for the ice accreted airfoil and (b) the clean airfoil. Again, these results are from Fossati et al. (2012). As can be seen from contrasting the two frames, there is a substantial change in the flow field. Indeed the ice-accreted geometry presents substantial challenges for the turbulence model. The high levels of ice build-up shown will produce substantial levels of separated flow with large-scale eddies. As discussed in Chapter 5, such flows are a great challenge to RANS models. Indeed, as the ice front evolves it will penetrate different boundary layer zones, presenting very different flow physics modelling challenges. The complexity of this turbulence can be partly appreciated from Figure 6.64. This gives an instantaneous field predicted from the used of hybrid RANS-LES for a GLC-305 airfoil with a leading edge ice-horn. Further examples of this type of work can be found in Pan and Loth (2005) and Zhang et al. (2013).

This turbulence modelling complexity is all coupled with the modelling challenges of accurately capturing the multi-physics ice build-up process. Details of a preliminary procedure for capturing this can be found in Blake et al. (2015). In this work, nucleation is assumed to take place instantaneously upon water droplet

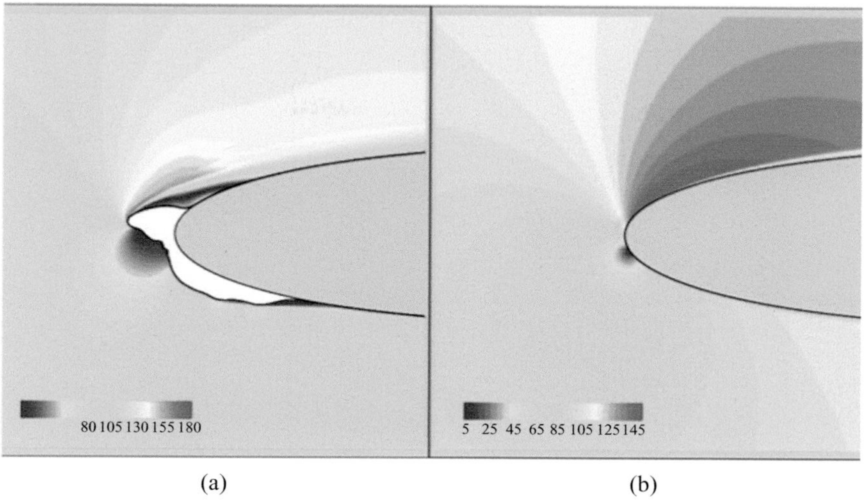

Figure 6.63. Flow field over the ice accreted airfoil and the clean airfoil represented by velocity magnitude contours: (a) airfoil with ice accretion and (b) clean airfoil (from Fossati et al. 2012 – used with the kind permission of John Wiley and Sons Inc.). (See color plate 6.63.)

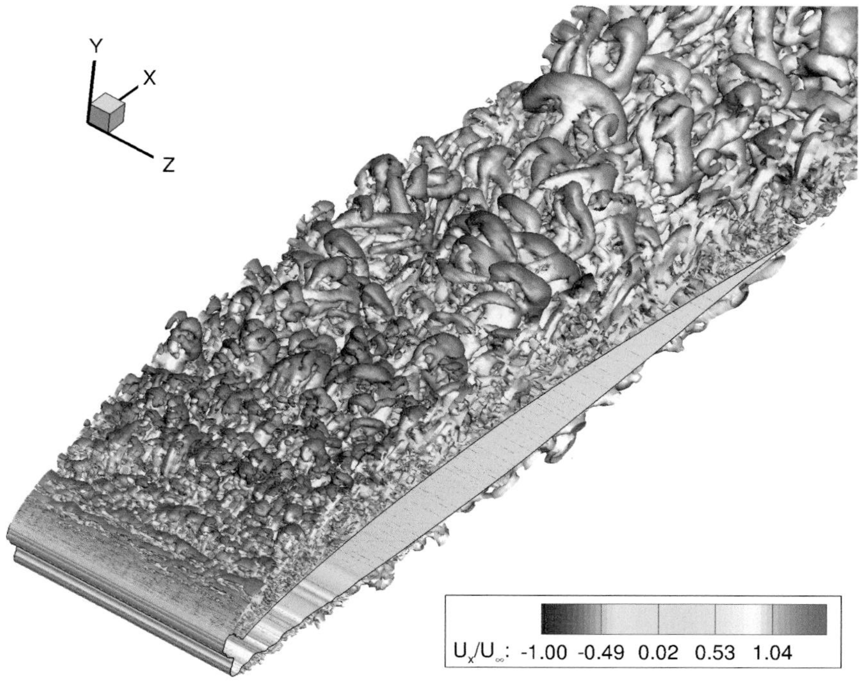

Figure 6.64. Hybrid RANS-LES of flow over a GLC-305 airfoil with a leading-edge ice horn (from Zhang 2015 – used with kind permission). (See color plate 6.64.)

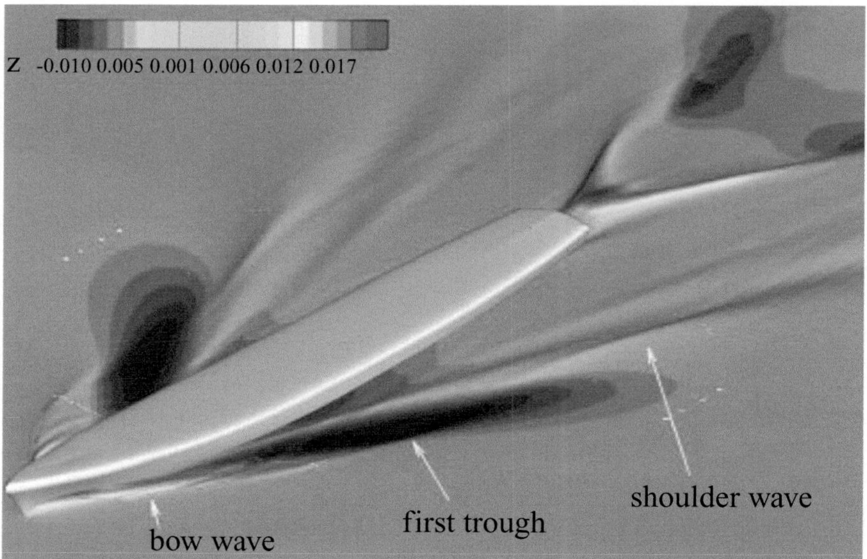

Figure 6.65. Predicted wave elevations through use of level set equations (from Huang et al. 2012 – published with the kind permission of John Wiley and Sons, Ltd.). (See color plate 6.65.)

impact. The approach involves both volume of fluid and level-set methods. If it is desired to extend the concept of fully resolving the ice features in the mesh when the ice breaks away, an overset grid modelling strategy would be needed with adaptive hole cutting – see Chapter 3.

Huang et al. (2012) use overset meshes with the level-set method to model ship hydrodynamics. With ships, a major part of the drag component is associated with the surface waves generated by the passage of the ship. Figure 6.65 shows the level-set evaluated free surface elevations from simulations. As can be seen, the free surface flow is dynamically complex. The level-set is configured so that it is positive in water and negative in air, with the interface zone defined with a level-set of zero. The equation takes the form

$$\frac{\partial \phi}{\partial t} + \mathbf{V} \cdot \nabla \phi = 0 \tag{6.70}$$

This is solved with the re-initialization equation

$$\mathbf{n} \cdot \nabla \phi = \text{sign}(\phi_0) + \varepsilon \nabla^2 \phi \tag{6.71}$$

where ϕ_0 is the level-set function prior to initialization and $\mathbf{n}$ is as defined previously. The re-initialization stage is evidently necessary to ensure the level-set function maintains precision. The Laplacian is for stability where ε is a small number. The above, non-linear equation is solved from the interface. Note, for points adjacent to the interface, the level-set field is algebraically updated. The points are then used as Dirichlet boundary conditions.

An alternative to using interfacing tracking methods on grids is to use grid free methods for flows involving free surfaces. For example, as discussed in Chapter 4,

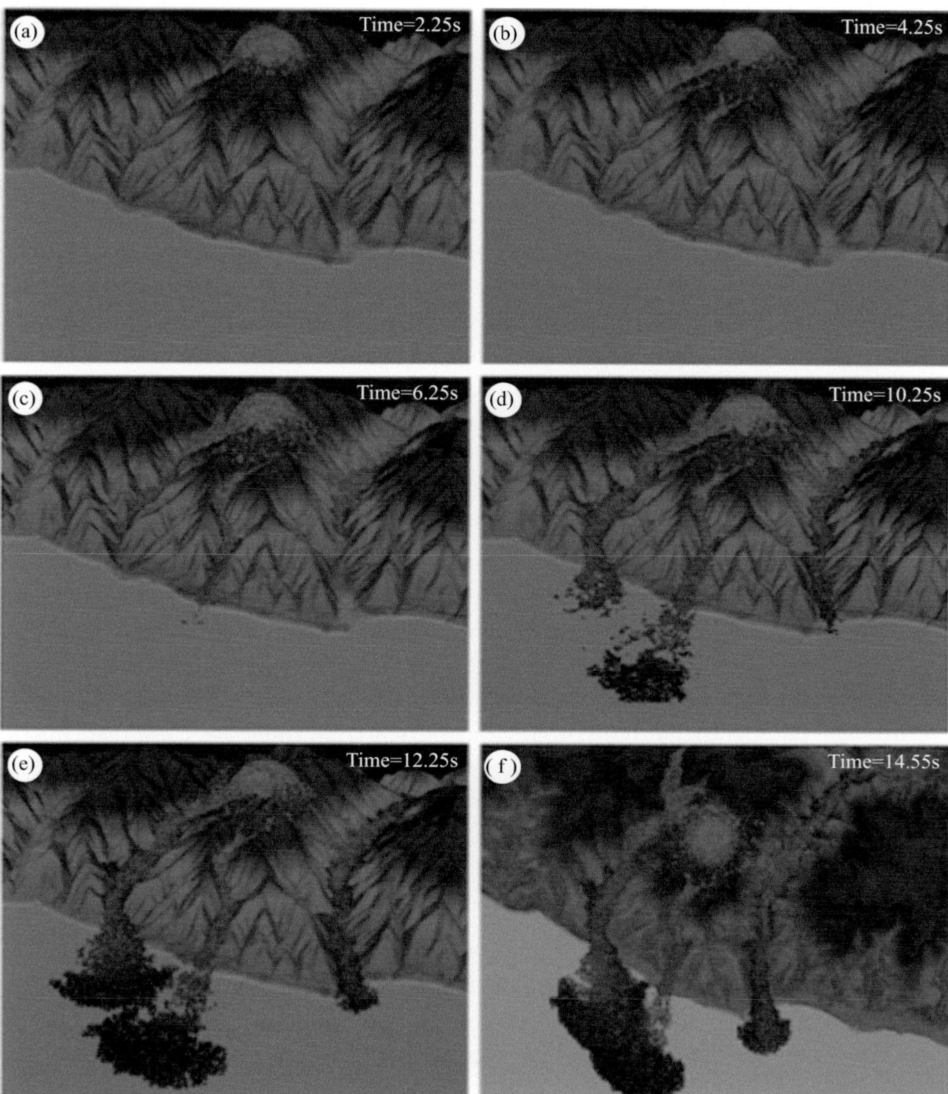

Figure 6.66. Grid free, smooth particle hydrodynamic simulation of a lava flow. The red zones correspond to the lava (from Cleary and Prakash 2004 – used with the kind permission of the Royal Society.) (See color plate 6.66.)

Marongiu et al. (2007, 2010) used smooth particle hydrodynamics for Pelton wheel design. Banim et al. (2006) apply it to aircraft wing fuel tank sloshing. They find that the approach is faster than grid based methods and also captures the free surface breakup. Volcanic activity can be highly problematic to aircraft. The 'ash' consists of fragments of pulverized rock and other materials. The ash material can cause damage to aircraft engines. Ash can also be formed when magma interacts with water. Figure 6.66 shows a smooth particle hydrodynamic simulation of a lava flow from Cleary and Prakash (2004). The lava can be seen to interact with the surrounding water. CFD simulations of ash transport, potentially real time, could be of tremendous future importance to air travel.

6.8 Computational Aeroacoustics

6.8.1 *Introduction*

Noise analysis can be split into two key subsets – tonal and broadband. For the former, URANS can be used and the computational challenge is much less. An example of a tonal noise study would be exploring the noise from a fan in an aeroengine and its interactions with adjacent components. However, this interaction also has a broadband component, coming from the turbulence, hence the noise spectra has a much broader character. Figure 6.67 explores acoustics for a slat-wing-flap configuration at off-design. The simulations are of a zonal hybrid RANS-LES type with a compressible flow solver. Since the hybrid RANS-LES largely resolves most of the unsteady turbulent structures, the turbulent source information is automatically captured. This is shown in Frames (a) through Q-criterion isosurfaces (see Section 7.10.3.3) normalized by velocity. The compressible flow solver then captures the near field acoustic waves. These can be seen in Frame (b) through contours of $(1/\rho)\,\partial\rho/\partial t$. Frame (c) plots predicted sound pressure levels (SPL) against dimensionless frequency for different near field zones, along with the measured far field sound. As can be seen, the spectra has both broadband and tonal components. Capturing the far field is a critical aspect for simulations. Hence, considerable attention is paid here to methods to economically extract this from simulations. Most far field sound prediction approaches have an economic to computer sound wave propagation operator

$$L(\boldsymbol{U}, \phi') = S_{\phi'} \tag{6.72}$$

In this, $\boldsymbol{U}$ is the mean flow velocity field vector. This field can refract (bend) the acoustic perturbation component ϕ', which could be, for example, fluctuating density, ρ', or pressure, p'. The source term $S_{\phi'}$ is unsteady. Notably (Ewert and Schroder 2003), the source/turbulence length scales are typically an order of magnitude smaller than the acoustic the disparity scaling as $1/M$, where M is Mach number. In contrast, the acoustic amplitudes are orders of magnitude lower than the fluctuating amplitudes of the turbulent sources. These things mean that for acoustic wave propagation, specialist numerical schemes are generally needed (see Chapter 4) and care taken with rounding error. The key aspect of far field noise prediction is that $L(\boldsymbol{U}, \phi')$ should be cheap to compute.

6.8.1.1 Noise Prediction Hierarchy and Methods Figure 6.68 gives a hierarchy of wave propagation operators, $L(\boldsymbol{U}, \phi')$.

These will be discussed further later. Obviously, the critical question for the designer is at what stage in the design should the different fidelity levels be used – if at all. Figure 6.69 summarizes some of the potential paths from different flow field evaluation procedures and Figure 6.68 wave propagation techniques to far field

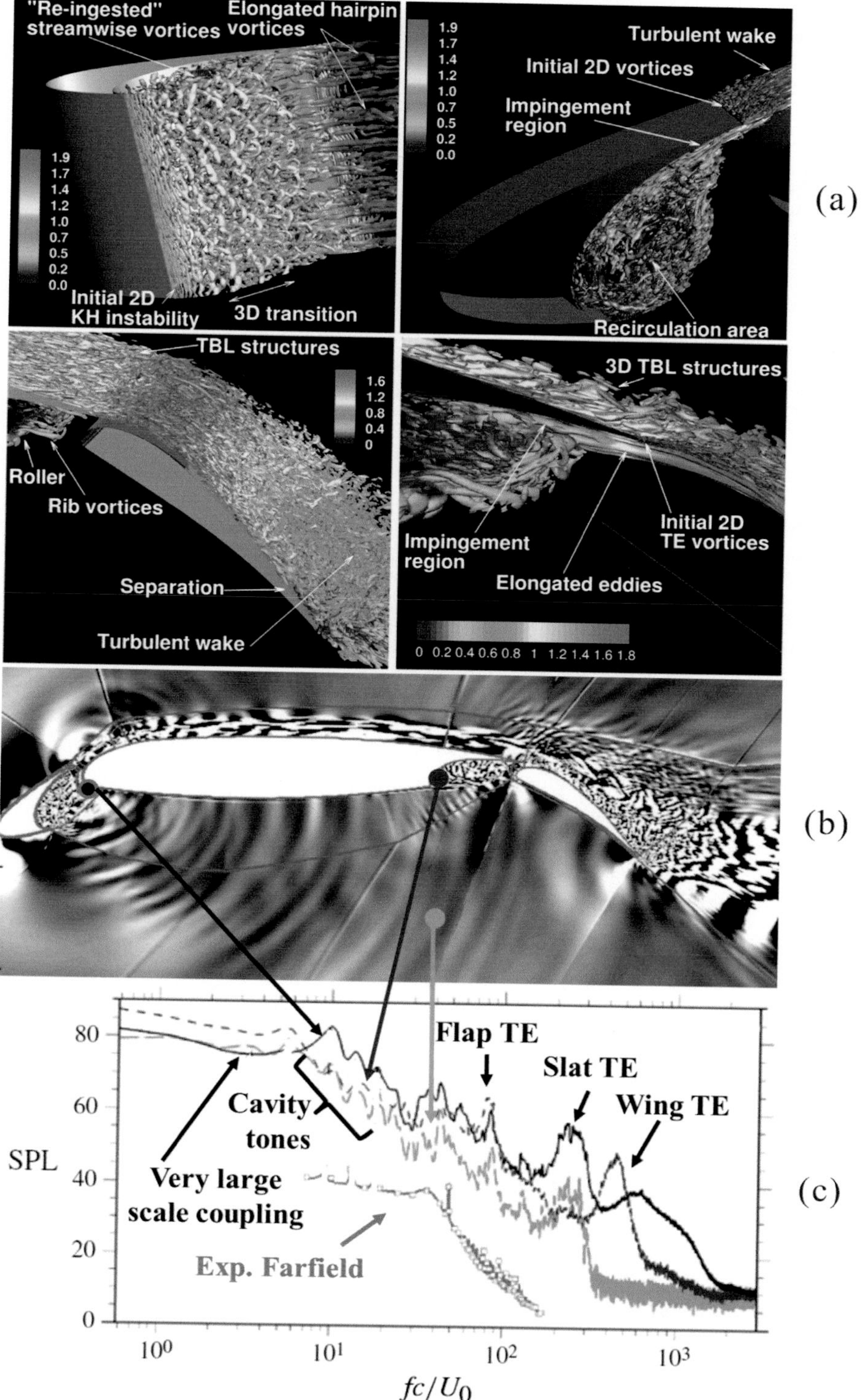

Figure 6.67. Zonal eddy-resolving simulation of aerodynamic noise for a slat-wing-flap configuration at off-design conditions: (a) Q-criterion (see Section 7.10.3.3) isosurface normalized by velocity, (b) near field acoustic waves rendered visible by contours of $(1/\rho)\,\partial\rho/\partial t$ and (c) sound pressure levels (SPL) against dimensionless frequency for different near field zones along with the measured far field sound (from Deck et al. 2014 – published with the kind permission of the Royal Society). (See color plate 6.67.)

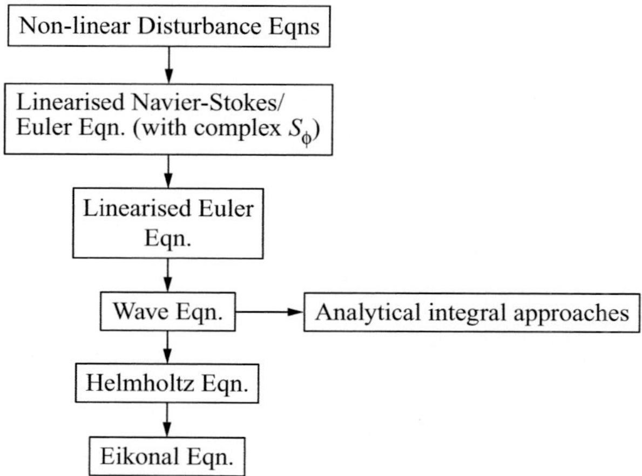

Figure 6.68. Wave propagation hierarchy (from Tucker 2013 – published with kind permission of Springer).

sound predictions. These will be considered in this chapter. Next, forms of $L(\boldsymbol{U}, \phi')$ are discussed in approximately decreasing order of fidelity.

Notably, there are two distinct paths. One is where the unsteady source information is resolved – eddy-resolving branch. The other is the RANS branch. With this, the modelled turbulence is synthesized or models are used to convert the RANS suggested length and time scales into source data in a statistical way. Notably, there are quite a few options for both of these branches. Each has a considerable level of modelling and mathematical complexity. All involve some form of wave operator,

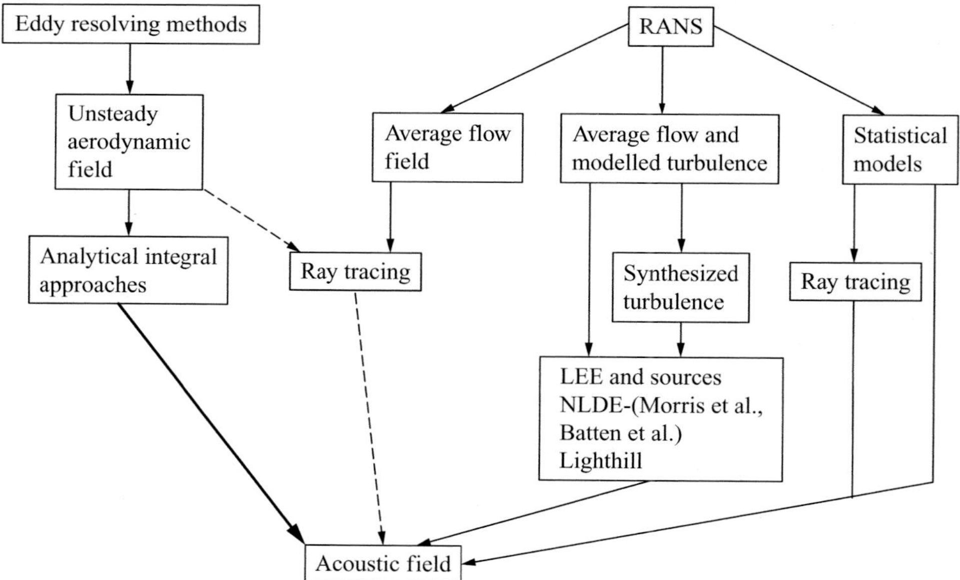

Figure 6.69. Some potential combinations of CFD and CAA for far-field noise prediction (adapted from Tucker 2013 – published with kind permission of Springer).

$L(\boldsymbol{U}, \phi')$, or hybrids of it to propagate sound to the far field. This aspect is considered next. The potentially highest fidelity approaches are considered first.

6.8.2 *Differential Wave Propagation Operators*

6.8.2.1 Non-linear Disturbance Equation First the NLDEs, sometimes called non-linear acoustic solver (NLAS), noted earlier, are considered. Note, the nomenclature used here is consistent with that used in Chapter 2. The usual momentum flux equations can be expressed as

$$\frac{\partial \phi}{\partial t} + \frac{\partial E}{\partial x} + \frac{\partial F}{\partial y} = \text{Viscous terms} \tag{6.73}$$

where $\phi = [\rho, \rho u, \rho v, e]^T$, $E = [\rho, \rho uu + p, \rho uv, eu + pu]^T$ and $F = [\rho v, \rho uv, \rho vv + p, ev + pv]^T$. To give the general idea, the NLDE is presented here in two dimensions. For the full equations, see Morris et al. (1997). The vector ϕ is decomposed into time mean, $\overline{\phi}$, and fluctuating, ϕ', components as

$$\phi = \overline{\phi} + \phi' \tag{6.74}$$

Substituting (6.74) into (6.73) ultimately gives, after ignoring the viscous perturbation terms,

$$\frac{\partial \phi'}{\partial t} + \frac{\partial E'}{\partial x} + \frac{\partial F'}{\partial y} + \frac{\partial \tilde{E}'}{\partial x} + \frac{\partial \tilde{F}'}{\partial y} = S_\phi \tag{6.75}$$

where the linear disturbances are

$$\phi' = \begin{bmatrix} \rho' \\ \overline{\rho}u' + \rho'\overline{u} + \rho' u' \\ \overline{\rho}v' + \rho'\overline{v} + \rho' v' \\ e' \end{bmatrix} \tag{6.76}$$

The terms with tildes are non-linear disturbances. The source term takes the form

$$S_\phi = -\left(\frac{\partial \tilde{E}}{\partial x} + \frac{\partial \tilde{F}}{\partial y}\right) + \text{Mean viscous stresses} \tag{6.77}$$

The relevant flux terms are given Table 6.5.

The form of the mean viscous fluxes is not outlined here. Typically, a RANS solution provides the mean flow field. The perturbation equation provides non-linear perturbations around this. For an application of the NLDE approach to a three-element, high-lift wing, see Terracol et al. (2005).

The NLDE approach displays less sensitivity to numerical errors. There are a wide range of other NLDE related decompositions – see Tucker (2013).

6.8.2.2 Linearized Navier-Stokes Equations Goldstein's (2003) Linearized Navier-Stokes Equations (LNSE) method takes the Navier-Stokes equations and makes the usual following decomposition of variables: $\phi = \overline{\phi} + \phi'$. This is substituted into the Navier-Stokes equations. Again, as with the NLAS/NLDE approach, $\overline{\phi}$ is supplied

Table 6.5. *Flux-like terms for non-linear disturbance equation (from Tucker 2013 – used with the kind permission of Springer)*

	Φ'	$\tilde{\Phi}'$	$\tilde{\Phi}$
E	$\overline{\rho}u' + \rho'\overline{u}$	$\rho' u'$	$\overline{\rho u}$
	$\rho'\overline{u}^2 + 2\overline{\rho u}u' + p'$	$2\rho' u'\overline{u} + \overline{\rho}u'^2 + \rho' u'^2$	$\overline{\rho u}^2 + \overline{p}$
	$\overline{\rho u}v' + \overline{\rho v}u' + \rho'\overline{uv}$	$\overline{\rho}u'v' + \rho' v'\overline{u} + \rho' u'\overline{v} + \rho' u'v'$	$\overline{\rho uv}$
	$u'(\overline{e} + \overline{p}) + \overline{u}(e' + p')$	$u'(e' + p')$	$\overline{u}(\overline{e} + \overline{p})$
F	$\overline{\rho}v' + \rho'\overline{v}$	$\rho' v'$	$\overline{\rho v}$
	$\rho'\overline{v}^2 + 2\overline{\rho v}v' + p'$	$2\rho' v'\overline{v} + \overline{\rho}v'^2 + \rho' v'^2$	$\overline{\rho v}^2 + \overline{p}$
	$\overline{\rho u}v' + \overline{\rho v}u' + \rho'\overline{uv}$	$\overline{\rho}u'v' + \rho' v'\overline{u} + \rho' u'\overline{v} + \rho' u'v'$	$\overline{\rho uv}$
	$v'(\overline{e} + \overline{p}) + \overline{v}(e' + p')$	$v'(e' + p')$	$\overline{v}(\overline{e} + \overline{p})$

by RANS equations. These are subtracted from the equation resulting from the combination of the Navier-Stokes equations and the decomposition $\phi = \overline{\phi} + \phi'$. This subtracted equation set is then rearranged to yield a linearized, left-hand side Euler operator with an exact source term (involving fourth order space-time correlations of turbulence). The resulting equations are an exact consequence of the Navier-Stokes equations.

6.8.2.3 Linearized Euler Equations (LEE) When considering the linearized Euler equation, Equation (6.76) truncates to

$$\phi' = \begin{bmatrix} \rho' \\ \overline{\rho}u' \\ \overline{\rho}v' \\ p' \end{bmatrix} \tag{6.78}$$

Also, Equation (6.75) can be expressed as (see Bailly and Juve 1999)

$$\frac{\partial \phi'}{\partial t} + \frac{\partial E'}{\partial x} + \frac{\partial F'}{\partial y} = S_\phi \tag{6.79}$$

Here the primed terms again involve mean quantities and disturbances – but this time the latter are linear. This full source term is given by Bailly and Juve (1999). As well as characterizing the sound source, it has terms modelling refraction.

The solution of the linearized equations can be more challenging than would be expected. One issue is that if the mean flow profile about which the linearized equations are being solved is highly inflectional, the inflection can give rise to instability that contaminates the linearized Euler equation solution. The next level of simplification for the operator L is the wave equation.

6.8.2.4 Wave Equations The wave equation

$$\nabla^2\phi' - \frac{1}{c^2}\frac{\partial^2 \phi'}{\partial t^2} = 0 \tag{6.80}$$

is the obvious choice of operator for propagating source information but does not account for refraction. In the previous, c, is the speed of sound.

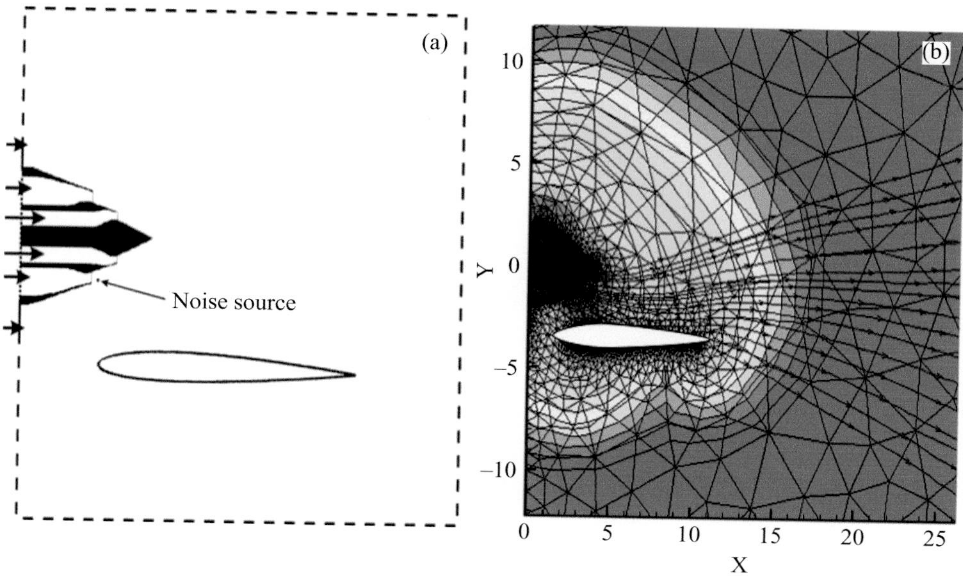

Figure 6.70. Wing-jet schematic and eikonal solutions: (a) schematic of geometry and (b) eikonal solution (Copyright Multiscience Publishing, used with permission).

The Helmholtz is a simplified wave propagation equation. Assuming $\phi' = e^{i\omega t}$ where ω is the sound wave frequency, in radians, the wave equation can be directly re-expressed as $-(\omega^2/c^2)\phi' = \nabla^2\phi'$. Defining the wave number, w, of the sound field as $w = \omega/c$ allows the following Helmholtz equation to be written

$$\nabla^2\phi' = -w^2\phi' \tag{6.81}$$

The eikonal equation is the high frequency limit to the wave equation.

$$1 - U_n\frac{\partial\phi^2}{\partial x_n} - c^2\frac{\partial\phi^2}{\partial x_n} = 0 \tag{6.82}$$

In the previous, ϕ is called the eikonal. This is proportional to the propagation time of the sound ray. The high frequencies can be most annoying to the human ear. Hence, this makes the eikonal equation solutions potentially useful. Nonetheless, the equation is physically crude. Essentially it traces streamlines based on a velocity field which is the summation of the mean fluid velocity and the local speed of sound. The eikonal equation can be solved in either Lagrangian (see Secundov et al. 2007) or Eulerian frameworks. The latter is perhaps its most natural form. Figure 6.70 gives a schematic of an idealized wing-'jet' geometry along with an eikonal solution as discussed in Tucker (2013). Frame (a) gives the geometry. Frame (b) gives eikonal equation computed acoustic wave fronts and acoustic rays emanating from a sound source positioned inside the nozzle. Note, the darker contours correspond to higher eikonal values. The idea is that the wing is positioned to shield a ground observer from noise.

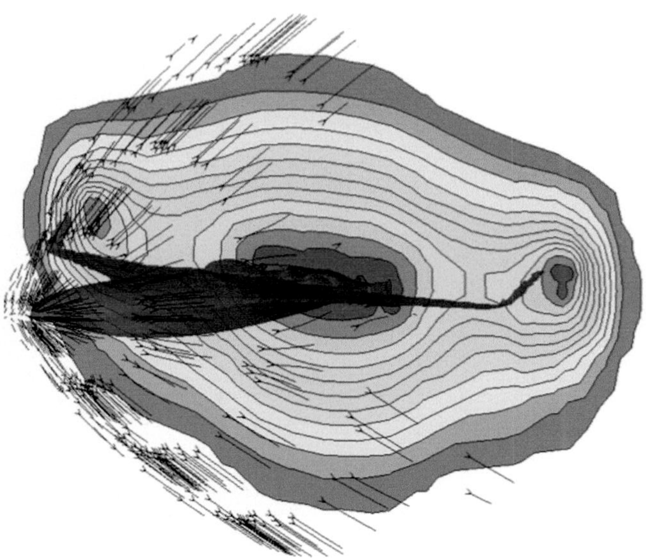

Figure 6.71. Acoustic rays predicted from solution of the eikonal equation and emanating from a sound source placed at the nose of a blended wing body aircraft.

Figure 6.71 again shows use of the eikonal equation to study acoustic ray trajectories. This time a sound source is placed at the nose of a blended wing body aircraft. The expected reward directivity of the rays can be observed. The contours are actual wave fronts. These are directly provided by the eikonal equation. The contours further away from the surface have higher eikonal values. The normals to these fronts enable the acoustic rays to be generated.

Table 6.6 summarizes some of the advantages and disadvantages of the different wave propagation operators noted previously.

6.8.3 *Integral Equation Based Approaches*

Analytical methods can be used to extrapolate near field source data to the far field. Such methods are often called integral methods. They essentially take differential operators that can resolve acoustic waves and also potentially source information and analytically solve them.

6.8.3.1 The Lighthill Equation For example, Lighthill (1952) rearranged the Navier-Stokes equations to give the wave equation operator, $L(\rho')$ and source term

$$L(\rho') = \frac{\partial^2 \rho'}{\partial t^2} - c^2 \nabla^2 \rho' \quad \text{and} \quad S_{\rho'} = \frac{\partial^2 T_{ij}}{\partial x_i \partial x_j} \tag{6.83}$$

If c and ρ can be considered constant,

$$T_{i,j} = \rho u_i' u_j' \tag{6.84}$$

Table 6.6. *Advantages and disadvantages of different acoustic wave operators*

Approach	Advantages	Disadvantages	Other comments
NLDE/NLAS	Low sensitivity to numerical errors. Well suited to jet noise problems	Relatively computationally expensive	There are a wide range of related decompositions
LNSE	Exact equation(s) that gives good physical insights into the sound source mechanisms	Relatively computationally expensive. Can be numerically challenging to solve for some mean flows	–
LEE	Good accuracy relative to computational cost	Can only capture first order influences. The scattering of sound by turbulence is ignored. Can be numerically challenging to solve for some mean flows	–
Wave equation	Cheap to solve	Does not account for refraction	–
Helmholtz equation	Can be solved using the volume mesh free BEM method	Solved for individual specified frequencies	–
Eikonal equation	Cheap, fast means of looking at sound propagation in long-range problems	Limited to high frequencies	Valid for $\lambda \ll L$ (where, here, L is a characteristic length scale) or rather $\omega L/c \gg 1$. Note: λ is the sound wavelength

showing that turbulence is a key sound source. If the unsteady turbulence field is known, the Equation (6.83) source term could be directly specified. Equation (6.83) is exact. For the calculation of far field sound, Lighthill derived the following analytical solution to Equation (6.83):

$$4\pi p'(\boldsymbol{x}, t) = \frac{x_i x_j}{|\boldsymbol{x}|^3} \int \frac{\partial \left[T_{ij}(\boldsymbol{y}, t_r) \right]}{c^2 \partial t^2} dV \tag{6.85}$$

where in the volume (V) integral $\boldsymbol{y}$ is the location of the turbulent sound source. The observer location is $\boldsymbol{x}$. The sound generated at $\boldsymbol{y}$ arrives at $\boldsymbol{x}$ at the later time $|\boldsymbol{x} - \boldsymbol{y}|/c$. Hence, the quantities in square brackets are computed at a retarded time, t_r, which is the moment when sound was generated and where $t_r = t - |\boldsymbol{x} - \boldsymbol{y}|/c$. The integral Lighthill equation (unlike the exact differential equation) cannot model refraction influences.

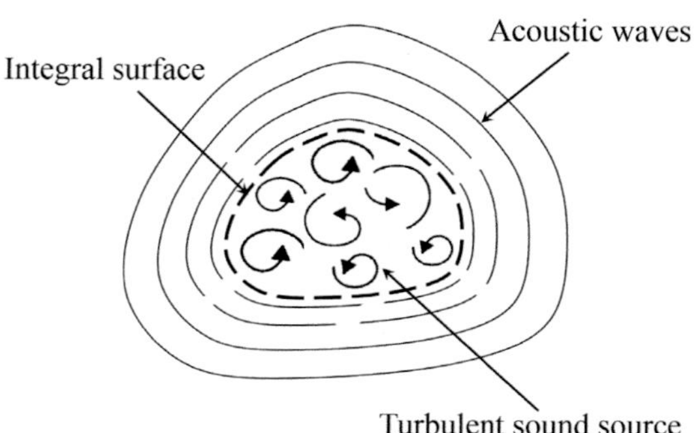

Figure 6.72. Location of integral surface relative to a turbulent sound source.

6.8.3.2 Surface Integral Equations The Ffowcs Williams and Hawkings (FWH) theory generalizes Lighthill's acoustic analogy. The wave propagation operator is identical. However, now

$$S_{\rho'} = \underbrace{\frac{\partial^2 T_{ij}}{\partial x_i \partial x_j}}_{\text{Volume integral}} \underbrace{- \frac{\partial}{\partial x_i}[p'_{ij} n_j \delta(\phi)] + \frac{\partial}{\partial t}[\rho_o u_n \delta(\phi)]}_{\text{Surface integrals}} \tag{6.86}$$

where $p'_{ij} = p_{ij} - p_0 \delta_{ij}$ is a perturbation stress tensor and δ_{ij} is the Kronecker delta. Also, ϕ is a level-set variable. It takes a value of zero on a surface, is negative inside s and positive outside it. On analytical solution of the FWH equation, the first source term gives rise to a volume integral. The latter two source terms give surface integrals. The Equation (6.86) solution yields the following relation for the far field acoustic pressure:

$$\begin{aligned} 4\pi p'(\boldsymbol{x}, t) = & \frac{x_i x_j}{|\boldsymbol{x}|^3 c^2} \frac{\partial}{\partial t^2} \left[\int_V T_{ij} dV \right] \\ & + \frac{x_j}{|\boldsymbol{x}|^2 c} \frac{\partial}{\partial t} \left[\int_s p' n_j + \rho u_j u_n ds \right] + \frac{1}{|\boldsymbol{x}| c} \frac{\partial}{\partial t} \left[\int_s \rho u_n ds \right] \end{aligned} \tag{6.87}$$

where V is the volume outside s, n_j are the projections of the vector of outer normal to s, u_j are velocity components and u_n the velocity component normal to s.

There are various FWH formulations intended for different purposes. For a wing or landing gear, the surfaces could define s. There could also be a volumetric turbulent source zone outside of this. Ideally, for many applications (see Figure 6.72), the surface would be placed around the turbulent source and then the volume integral in the FWH equation ignored. This is useful, since when performing eddy-resolving simulations, storing the time history of the data over a large volume can need considerable storage. A principle difficulty with the FWH approach is enclosing all the noise sources within s. For example, with jets, eddies convect downstream and pass through s. Indeed, results can be sensitive to the location of the surface and

Table 6.7. *Advantages of disadvantages of different integral methods for far field sound extrapolation*

Method	Advantages	Disadvantages
Lighthill	For eddy-resolving simulations, the grid in the source region tends to be well resolved, hence yielding far field sound at higher frequencies than FWH approach	Cannot model refraction influences. Storing noise source as a volume integral can require substantial storage when making eddy-resolving simulations
FWH	If all noise sources are inside s, then the first term on the right-hand side of Equation (6.87) can be omitted. This has a substantial storage savings when making eddy-resolving simulations	Care is needed in choosing the most appropriate FWH formulation for the problem at hand. Closing the surface so that no turbulent eddies pass through it is impossible for many practical problems
Kirchoff	Just pressure needs to be stored at integral surface, unlike FWH where pressure, velocity and density fluctuations need to be specified. In rotating systems, a stationary Kirchoff surface can be used	Surface location sensitivity is greater than for FWH approach. The surface normal pressure gradient needs to be stored at the surface integral location. Still has same surface closing issue as the FWH approach

hence this aspect needs to be carefully considered. There are a range of specialized approaches that attempt to deal with the fact that it is often impossible to entirely contain the turbulent source zone within an enclosed surface (see, for example, Shur et al. 2005).

The Kirchoff approach is applicable to any phenomena governed by the linear wave equation. Unlike with the FWH approach, just pressure fluctuation needs to be stored at the integral surface (i.e., there is no volume source information supported).

The advantages and disadvantages of the integral equation approaches discussed in this section are summarized in Table 6.7. Notably, the approaches can be combined or hybridized. For example, the linearized Euler or even Euler operators can be used in a near body non-uniform flow zone, then where the flow becomes uniform this can be connected to one of the above integral operators. See, for example, Ozyoruk and Long (1997) and Agarwal and Morris (2000) for engines and airframe problems. Mincu and Manoha (2012) use a complex superposition of methods to explore engine airframe interaction, with information being first propagated from the engine to the fuselage, then the reflected noise from the fuselage propagated to an observer.

6.8.4 *Turbulence Source Descriptions*

When performing eddy-resolving simulations, the turbulence source description is relatively easy – the necessary unsteady flow data and the frequency and length scale are naturally available. This can be seen from Figure 6.73. This figure gives examples of eddy-resolving simulations that show the natural provision of both detailed source information and also near field acoustic propagation data. Frame (a)

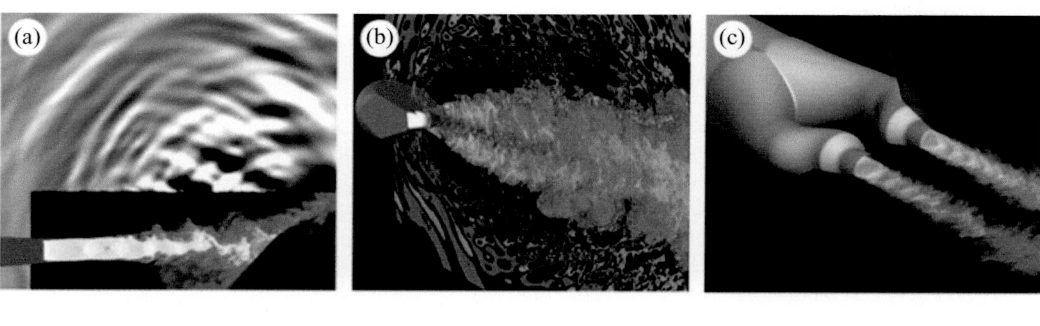

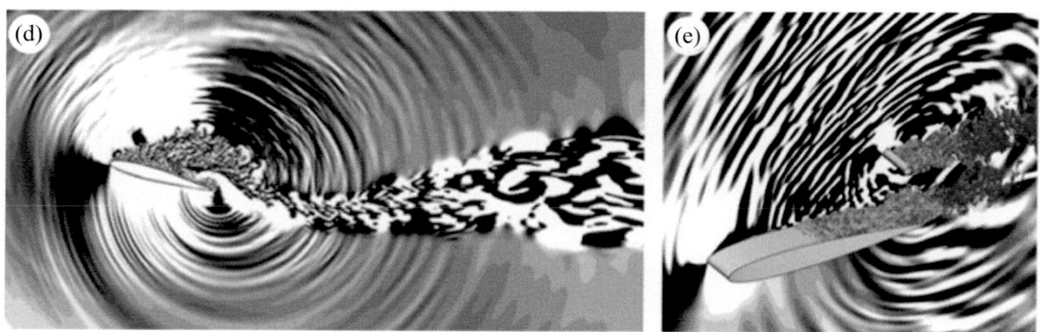

Figure 6.73. Examples of eddy-resolving simulations naturally providing both detailed source information and also near field acoustic propagation data: (a) a supersonic jet impinging on a jet blast deflector, (b) a supersonic jet issuing from a square sectioned chevron nozzle, (c) system providing twin adjacent jets, (d) stalled NACA0012 airfoil and (e) interaction of airfoil and an adjacent rod (from Lele and Nichols 2014 – used with the kind permission of the Royal Society). (See color plate 6.73.)

is for a supersonic jet impinging on a jet blast deflector. Frame (b) represents a supersonic jet issuing from a square sectioned chevron nozzle. Frame (c) shows a system that provides twin adjacent jets. Frame (d) is for a stalled NACA0012 airfoil and (e) gives interaction of an airfoil and an adjacent rod. The airfoil simulations are performed with a compact, high-order scheme, structured flow solver. The jet simulations involve use of an unstructured flow solver. In all simulations, except that for the twin-jet, both the turbulent source field and arising acoustic waves can be seen. Another example, this time for landing gear, can be seen in Figure 6.74. Frame (a) shows the flow field and acoustic waves and Frame (b) the sound pressure spectrum. The spectrum is taken on the surface of the wheel and some tonal peaks can be seen mixed in with the broad band spectrum. Note that these are hybrid RANS-LES simulations and the presence of the surface RANS layer has not resulted in any great loss of accuracy in the surface pressure spectrum predictions.

For RANS, defining the noise sources is challenging. The RANS equations just supply the *mean* flow field and incomplete turbulence source description information. For RANS, typically, the $L(\boldsymbol{U}, \phi') = S_{\phi'}$ system, which has a time varying source to model turbulence, is cast into the frequency domain to yield a steady

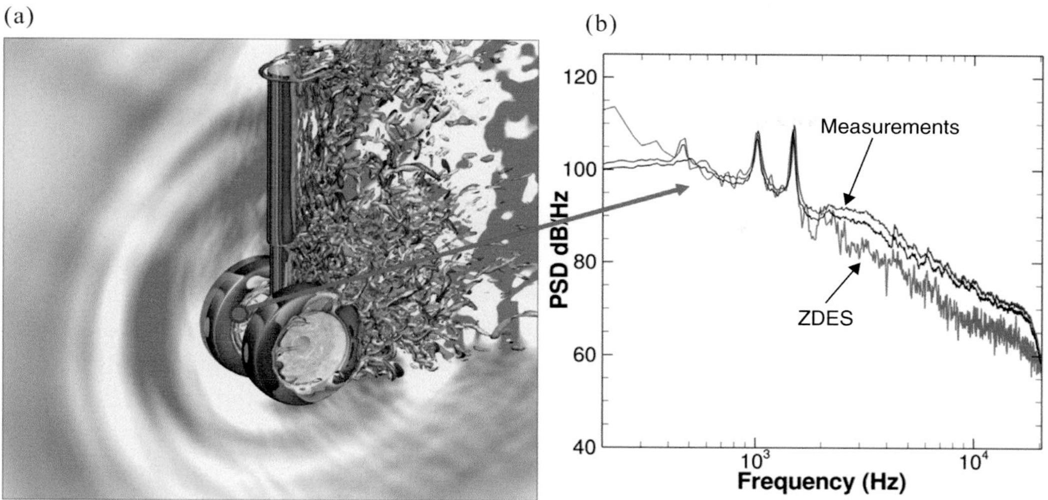

Figure 6.74. Example of an eddy-resolving simulation for a landing gear naturally providing both detailed source information and also near field acoustic propagation data: (a) flow field and acoustic waves and (b) sound pressure spectrum (from Deck et al. 2014 – used with the kind permission of the Royal Society). (See color plate 6.74.)

equation. Using the Green's function (with assumptions such as axisymmetric flow and just consideration of higher frequencies), this equation can be converted into one that directly yields acoustic fluctuations ϕ'. Then ϕ' can be converted into, for example, a noise spectrum. The key aspect, where there is considerable variation in techniques, is the characterization of $S_{\phi'}$, the turbulent source. Most critically, the RANS procedure will typically be highly inaccurate in capturing the length and time scale associated with turbulence and cannot capture the critical high-order correlations that characterize the turbulent noise sources. Hence, very careful calibration is required, making the RANS based noise 'prediction' process highly postdictive. Another option is that the turbulence energy and length scales in the RANS are fed into a synthetic turbulence generation procedures. This synthetic turbulence can go into the noise source terms discussed previously, associated with a particular $L(\boldsymbol{U}, \phi')$ operator.

6.9 Conclusions

In this chapter a wide range of approaches have been considered. These include a brief overview of some methods relevant to the study of multi-phase flows, acoustics, coupled fluid structure interactions and multi-scale problems. Also, some attention has been given to design optimization and the generation of low-order models. Importantly CFD analysis in industry can involve a wide range of these aspects. For example, in a gas turbine aeroengine, there is a combustor. This involves multi-phase, chemically reacting flow and also a very disparate range of length scales from the Kolmogorov up to the very large scale of the combustor geometry. Then, the dynamics of the combustor is strongly coupled to that of the geometry immediately upstream of it. Also, there is an important interaction between the combustor and the downstream turbine. Then, there is the critical matter of exploring how the noise from this process finally becomes emitted from the engine, through the subsequent downstream stages of the turbine. Hence, it is clear that coupled simulations are of critical importance, as is the process of formal design optimization. It can also be seen, from this chapter, that many of the mathematical operators and processes have strong similarities between different fields of application. This is helpful. However, their numerical simulation can often be problematic, with apparently similar equations needing quite different numerical treatments depending on the area of application. Importantly, just within each field, there is a tremendous range of approaches that need to be selected and understood. For example, even within the area of design optimization the correct choice of surrogate model is clearly a specialist skill. The harnessing of all this knowledge to perform reliable CFD simulations presents a great challenge. There could perhaps be a future role for expert systems in this area.

Problems

1) Write a module to integrate weightless and then finite weight particles inside your Euler solver.
2) Generate a cylindrical grid and particle track in the cylindrical polar coordinate system. Use the following equations

$$d\theta_P/dt = \Omega_P \tag{6.88}$$

where Ω_P is the particle's angular velocity and

$$dr_P/dt = u_P \tag{6.89}$$

where u_P is radial velocity. These may be written in finite difference form using the first order forward Euler approach as

$$\theta_P = \theta_P^{old} + \Omega_P \Delta t \tag{6.90}$$

$$r_P = r_P^{old} + u_P \Delta t \tag{6.91}$$

where θ_P and r_P are new positions of the particle at $t + \Delta t$, and Ω_P and u_P are the current velocities evaluated at the old particle locations θ_p^{old} and r_p^{old}. To validate your program use a uniform vertical velocity field and see if the particle trajectories are vertical.
3) Using finite differences, express a convection operator in a form to track a passive phase in an $r - \theta$ coordinate system.
4) Assuming a range of non-uniform imposed vertical velocity distributions, contrast distributions based on particle tracking and passive scalar equation solutions in the $r - \theta$ coordinate system.

REFERENCES

ADAMCZYK, J. 1985. Model equation for simulating flows in multistage turbomachinery. ASME, International Gas Turbine Conference and Exhibit, 30th, Houston, TX, Mar. 18-21 ASME Paper No. 85-GT-226.

AGARWAL, A. & MORRIS, P. J. 2000. Direct simulation of acoustic scattering on a rotorcraft fuselage. Sixth Aeroacoustics Conference and Exhibit. June AIAA Paper No. AIAA-2000–2030.

ALONSO, J., HAHN, S., HAM, F., HERRMANN, M. & IACCARINO, G. 2006. Chimps: A high performance scalable module for multi-physics simulations. 42nd AIAA/ASME/SAE/ASEE Joint Propulsion Conference & Exhibit. July, AIAA Paper No. AIAA-2006-5274.

BAILLY, C. & JUVE, D. 1999. A stochastic approach to compute subsonic noise using linearized Euler equations. Fith AIAA/CEAS Aeroacoustics Conference and Exhibit, May, AIAA Paper No. AIAA-99-1872.

BAKKER, C., PARKS, G. T. & JARRETT, J. P. 2012. On the application of differential geometry to MDO. Twelth Aviation Technology, Integration and Operations (ATIO) Conference and 14th AIAA/ISSMO Multidisciplinary Analysis and Optimization Conference, AIAA Paper No. 2012-5556.

BANIM, R., LAMB, R. & BERGEON, M. 2006. Smoothed particle hydrodynamics simulation of fuel tank sloshing. Twenty fith International Congress of the Aeronautical Sciences, 3–8 September 2006, Hamburg, Germany, Paper No. ICAS 2006-2.7.4

BARDOUX, F. & LEBOUEF, F. 2001. Impact of deterministic correlations on the steady flow field. *Proceedings of the Institution of Mechanical Engineers, Part A: Journal of Power and Energy*, 215, 687–698.

BARTELS, R. & SAYMA, A. 2007. Computational aeroelastic modelling of airframes and turbomachinery: progress and challenges. *Philosophical Transactions of the Royal Society, A: Mathematical, Physical and Engineering Sciences*, 365(1859), 2469–2499.

BARTON, I. E. 1996. Exponential-Lagrangian tracking schemes applied to Stokes law. *ASME Journal of Fluids Engineering*, 118, 85–89.

BATTEN, P., SPALART, P. R. & TERRACOL, M. 2007. Use of hybrid RANS-LES for acoustic source prediction. In T. Huttl, C. Wagner, P. Sagaut (eds.), *Large-Eddy Simulation for Acoustics*, Cambridge University Press.

BEAUGENDRE, H., MORENCY, F., GALLIZIO, F. & LAURENS, S. 2011. Computation of ice shedding trajectories using cartesian grids, penalization, and level sets. *Modelling and Simulation in Engineering*, 2011, 15.

BENDER, E. E., ANDERSON, B. H. & YAGLE, P. G. 1999. Vortex generator modelling for Navier-Stokes codes. ASME Paper No. FEDSM99-6929.

BENICHOU, E. & TRÉBINJAC, I. 2014. Application of an analytical method to locate a mixing plane in a supersonic compressor. ASME Turbo Expo 2014: Turbine Technical Conference and Exposition. ASME Paper No. GT2014-25103.

BENNING, R. M., BECKER, T. M, & DELGADO, A., 2001. Initial studies of predicting flow fields with an ANN hybrid. *Advances in Engineering Software*, 32, 895–901.

BLAKE, J., THOMPSON, D., RAPS, D. & STROBL, T. 2015. Simulating the freezing of supercooled water droplets impacting a cooled substrate. *AIAA Journal*, 53(7), 1725–1739.

BONAIUTI, D. & ZANGENEH, M. 2009. On the coupling of inverse design and optimization techniques for the multiobjective, multipoint design of turbomachinery blades. *ASME Journal of Turbomachinery*, 133(1), 021014.

BOYLE, R. J. & STRIPF, M. 2009. Simplified approach to predicting rough surface transition. *Journal of Turbomachinery*, 131, 041020.

BUIS, S., PIACENTINI, A. & DéCLAT, D. 2005. PALM: a computational framework for assembling high performance computing applications. *Concurrent Computing,*18(2), 231–245.

BUSBY, J., SONDAK, D., STAUBACH, R. & DAVIS, R. 2000. Deterministic stress modeling of hot gas segregation in a turbine. *Transactions of the ASME Journal of Turbomachinery,* 62, 62–67.

CARLSON, H. A. & FENG, J. Q. 2005. Computational models for nonlinear aeroelasticity. 43rd AIAA Aerospace Sciences Meeting and Exhibit, AIAA Paper No. 2005-1085.

CHIMA, R. V., CONNERS, T. R. & WAYMAN, T. R. 2010a. *Coupled Analysis of an Inlet and Fan for a Quiet Supersonic Jet*, NASA/TM-2010-216350.

CHIMA, R. V., AREND, D. J., CASTNER, R. S., SLATER, J. W. & TRUAX, P. P. 2010b. *CFD models of a serpentine inlet, fan, and nozzle*, NASA/TM-2010-216349.

CLEARY, P. W. & PRAKASH, M. 2004. Discrete-element modelling and smoothed particle hydrodynamics: potential in the environmental sciences. *Philosophical Transactions of the Royal Society A: Mathematical, Physical and Engineering Sciences*, 362, 2003–2030.

COOPER, A. J. & PEAKE, N. 2000. Trapped acoustic modes in aeroengine intakes with swirling flow. *Journal of Fluid Mechanics*, 419, 151–175.

CROWE, C. T., SHARMA, M. P. & STOCK, D. E. 1977. The Particle-Source-In Cell (PSI-CELL) model for gas-droplet flows. *Journal of Fluid Engineering*, 99(2), 325–332.

DECK, S., GAND, F., BRUNET, V. & KHELIL, S. B. 2014. High-fidelity simulations of unsteady civil aircraft aerodynamics: stakes and perspectives. Application of zonal detached eddy simulation. *Philosophical Transactions of the Royal Society A: Mathematical, Physical and Engineering Sciences*, 372, 2013-0325.

DEFOE, J., NARKAJ, A. & SPAKOVSZKY, Z. 2009. A novel MPT noise prediction methodology for highly-integrated propulsion systems with inlet flow distortion. 15th AIAA/CEAS Aeroacoustics Conference (30th AIAA Aeroacoustics Conference). AIAA Paper No. AIAA-2009-3366.

DEMIRDŽIĆ, I. & PERIĆ, M. 1988. Space conservation law in finite volume calculations of fluid flow. *International Journal for Numerical Methods in Fluids*, 8, 1037–1050.

DENTON, J. 1992. The calculation of three-dimensional viscous flow through multistage turbomachines. *ASME Journal of Turbomachinery*, 114, 18.

DHOPADE, P., NEELY, A. J., YOUNG, J. & SHANKAR, K. 2012. High-cycle fatigue of fan blades accounting for fluid-structure interaction. Proceedings of ASME Turbo Expo 2012, ASME Paper No. GT2012-68102.

DOI, H. 2002. *Fluid/structure coupled aeroelastic computations for transonic flows in turbomachinery*. DPhil Thesis, Stanford University, Department of Aeronautics and Astronautics.

DUCHAINE, F., CORPRON, A., PONS, L., MOUREAU, V., NICOUD, F. & POINSOT, T. 2009. Development and assessment of a coupled strategy for conjugate heat transfer with Large Eddy Simulation: Application to a cooled turbine blade. *International Journal of Heat and Fluid Flow*, 30, 1129–1141.

DUDEK, J. C. 2005. An empirical model for vane-type generators in a Navier-Stokes code. AIAA Paper No. AIAA-2005-1003.

DUDEK, J. C. 2010. *Modelling vortex generators in the wind – US code*, NASA/TM-2010-216744.

EASTWOOD, S. J., TUCKER, P. G., XIA, H. & KLOSTERMEIER, C. 2009. Developing large eddy simulation for turbomachinery applications. *Philosophical Transactions of the Royal Society A: Mathematical, Physical and Engineering Sciences*, 367, 2999–3013.

EL HACHEMI, M., HASSAN, O., MORGAN, K., ROWSE, D. & WEATHERILL, N. 2004. A low-order unstructured-mesh approach for computational electromagnetics in the time domain. *Philosophical Transactions of the Royal Society A: Mathematical, Physical and Engineering Sciences*, 362, 445–469.

ENGELMAN, M. S. & SANI R. L. 1986. Finite element simulation of incompressible fluid flows with a free/moving surface. *Computational Techniques for Fluid Flow*, 5, 47–74.

ERDOS, J. I., ALZNER, E. & McNALLY, W. 1976. Numerical solution of periodic transonic flow through a fanstage. American Institute of Aeronautics and Astronautics, 9th Fluid and Plasma Dynamics Conference, San Diego, California. AIAA Paper No. AIAA-76-369.

EWERT, R., SCHRODER, W., 2003. Acoustic perturbation equations based on flow decomposition via source filtering. *Journal of Computational Physics*, 188, 365–398.

FORRESTER, I. J. & KEANE, A. J. 2009. Recent advances in surrogate-based optimization. *Progress in Aerospace Sciences*, 45, 50–79.

FOSSATI, M., KHURRAM, R. & HABASHI, W. 2012. An ALE mesh movement scheme for long-term in-flight ice accretion. *International Journal for Numerical Methods in Fluids*, 68, 958–976.

FRIED, E. & IDELCHIK, I. E. 1989. *Flow Resistance: A Design Guide for Engineers*, Hemisphere.

FUJII, K. 2014. Impact of cutting-edge CFD over the finding of separation flow control mechanism with micro devices. *Submitted to the Proceedings of the Royal Society, Series A.*, 372, 20130326. (doi:10.1098/rsta.2013.0326)

GANGWAR, A. 2001. *Source term modeling of rectangular flow cavities*. MSc Thesis, University of Cincinnati, Department of Aerospace Engineering.

GANINE, V., HILLS, N. J. & LAPWORTH, B. L. 2013. Nonlinear acceleration of coupled fluid–structure transient thermal problems by Anderson mixing. *International Journal of Numerical Methods in Fluids*, 71(8), 939–959 .

GILES, M. B. 1997. Stability analysis of numerical interface conditions in fluid–structure thermal analysis. *International Journal for Numerical Methods in Fluids*, 25(4), 421–436.

GOLDSTEIN, M. E. 2003. A generalized acoustic analogy. *Journal of Fluid Mechanics*, 488, 315–333.

GONG, Y. 1998. *A computational model for rotating stall inception and inlet distortions in multistage compressors*. Ph.D. thesis, Massachusetts Institute of Technology.

HALL, E. 1997. Aerodynamic modelling of multistage compressor flow fields – Part 2: Modelling deterministic stresses. *Proceedings of Institution of Mechanical Engineers, Journal of Aerospace Engineering, Part G*, 212, 91–107.

HALL, K. C., THOMAS, J. P. & CLARK, W. S. 2002. Computation of unsteady nonlinear flows in cascades using a harmonic balance technique. *AIAA Journal,* 40, 879–886.

HANIMANN, L., MANGANI, L., CASARTELLI, E., MOKULYS, T. & MAURI, S. 2013. Development of a novel mixing plane interface using a fully implicit averaging for stage analysis. ASME Turbo Expo 2013: Turbine Technical Conference and Exposition. Paper No. GT2013-94390.

HARLOW, F. & WELCH, J. 1965. Numerical calculation of time-dependent viscous incompressible flow of fluid with free surface. *Physics of Fluids*, 8(12), 2182.

HASSAN, O., MORGAN, K. & WEATHERILL, N. 2007. Unstructured mesh methods for the solution of the unsteady compressible flow equations with moving boundary components. *Philosophical Transactions of Royal Society – A*, 15, 2531–2552.

HAYASHI, R. & YAMAMOTO, M. 2013. Numerical simulation on ice accretion phenomena in rotor-stator interaction field. ASME Turbo Expo 2013: Turbine Technical Conference and Exposition. Paper No. GT2013-95448.

HE, L. 2010. Fourier methods for turbomachinery applications. *Progress in Aerospace Sciences*, 46(8), 329–341.

HE, L. 2012. Block-spectral approach to film-cooling modelling. *Journal of Turbomachinery*, 134(2).

HE, L. 2013. Fourier spectral modelling for multi-scale aero-thermal analysis. *International Journal of Computational Fluid Dynamics*, 27(2).

HIELD, P. 1996. A study of the flow capacity of a low hub-tip ratio high speed fan for Pegasus uprate applications using a time-marching axisymmetric throughflow. *Rolls-Royce-Royce plc,* Technical Report No. DNS26773.

HIELD, P. 2008. Semi-inverse design applied to an eight stage transonic axial flow compressor. ASME Turbo Expo 2008: Power for Land, Sea, and Air, ASME Paper No. GT2008-50430.

HOLMES, D. G. 2008. Mixing planes revisited: A steady mixing plane approach designed to combine high levels of conservation and robustness. ASME Turbo Expo 2008: Power for Land, Sea, and Air. ASME Paper No. GT2008-51296.

HOWELL, J. R. 1982. *A catalog of radiation configuration factors*, McGraw-Hill.

HUANG, J., CARRICA, P. M. & STERN, F. 2012. A geometry-based level set method for curvilinear overset grids with application to ship hydrodynamics. *International Journal for Numerical Methods in Fluids*, 68, 494–521.

ILAK, M. & ROWLEY, C. W. 2008. Modelling of transitional channel flow using balanced proper orthogonal decomposition. *Physics of Fluids*, 20(3), 034103-034103-17.

JEFFERSON-LOVEDAY, R., NORTHALL, J. & TUCKER, P. 2012. Implementation of an advanced plenum boundary condition for turbine design calculations. ASME Turbo Expo 2012. Paper No. GT2012-68090.

JOO, W. G. & HYNES, T. P. 1997. The application of actuator disks to calculations of the flow in turbofan installations. *Journal of Turbomachinery*, 199, 723–732.

JUN L. 1986. *Computer modelling of flows with a free surface.* PhD Thesis, University of London.

KEANE, A. J. & SCANLAN, J. P. 2007. Design search and optimization in aerospace engineering. *Proceedings of the Royal Society A: Mathematical, Physical and Engineering Sciences*, 365, (1859), 2501–2559.

KERSKEN, H.-P., ASHCROFT, G., FREY, C., PUTZ, O., STUER, H. & SCHMITT, S. 2012. Validation of a linearized Navier-Stokes based flutter prediction tool, Part 1: numerical methods. Proceedings of ASME Turbo Expo 2012, ASME Paper No. GT2012-68018.

KINGSLEY, G. M., SIEGEL, J. M., HARRAND, V. J., LAWRENCE, C. & LUKER, J. 1998. Development of a multidisciplinary computing environment (MDICE). Seventh AIAA/USAF/NASA/ISSMO Symposium on Multidisciplinary Analysis and Optimization, AIAA Paper No. 98-4738.

KIRKPATRICK, S., GELATT, Jr, C. D. & VECCHI, M. P. 1983. Optimization by simulated annealing. Science,220(4598), 671–680.

LARSSON, J. & WANG, Q. 2014. The prospect of using large eddy and detached eddy simulations in engineering design, and the research required to get there. *Philosophical Transactions of the Royal Society A: Mathematical, Physical and Engineering Sciences*, 372, 20130329.

LELE, S. K. & NICHOLS, J. W. 2014. A second golden age of aeroacoustics? *Philosophical Transactions of the Royal Society A: Mathematical, Physical and Engineering Sciences*, 372, 20130321.

LEMKE, M., REISS, J. & SESTERHENN, J. 2014. Adjoint based optimisation of reactive compressible flows. *Combustion and Flame*, 161, 2552–2564.

LEUENBERGER, H. & PERSON, R. 1956. Compilation of radiation shape factors for cylindrical assemblies. ASME Paper No. 56-A-144.

LIEBECK, R. R. & ORMSBEE, A. I. 1970. Optimization of airfoils for maximum lift. *Journal of Aircraft*, 7(5), 409–416.

LIGHTHILL, M. J. 1952. On sound generated aerodynamically I – General theory. *Proceedings of the Royal Society, Series A*, 211, 564–587.

LO IACONO, G. & TUCKER P. G. 2004. LES computation of particle transport in rib roughened passages. Proceedings of CHT-04, Advances in Computational Heat Transfer III, (eds G. de Vahl Davis and E. Leonardi), Paper No. CHT-04-103.

LO IACONO, G., TUCKER, P. G. & REYNOLDS, A.M. 2005. Predictions for particle deposition from LES of ribbed channel flow. *International Journal of Heat and Fluid Flow*, 26, 558–568.

LOIODICE, S., TUCKER, P. G. & WATSON, J. 2010. Coupled open rotor engine intake simulations. Forty eigth AIAA Aerospace Sciences Meeting and Exhibit, January, Orlando, Florida, Paper No. AIAA-2010-840.

LONG, C. 1986. *Transient analysis of temperature measurements on discs of the RB 199 H.P. vented rotor compressor.* Report no. 86/TFMRC/87. School of Engineering and Applied Sciences, University of Sussex.

LONGLEY, J. 1997. Calculating the flow field behaviour of high-speed multi-stage compressors. ASME Paper No. 97-GT-468.

LUKOVIC, B. 2002. *Modeling unsteadiness in steady simulations with neural network generated lumped determinstic source terms.* D.Phil Thesis, University of Cincinnati, Department of Aerospace Engineering.

MARONGIU, J.-C., LEBOEUF, F., CARO, J. & PARKINSON, E. 2010. Free surface flows simulations in Pelton turbines using an hybrid SPH-ALE method. *Journal of Hydraulic Research*, 48, 40–49.

MARONGIU, J., LEBOEUF, F. & PARKINSON, E. 2007. Numerical simulation of the flow in a Pelton turbine using the meshless method smoothed particle hydrodynamics: a new simple solid boundary treatment. *Proceedings of the Institution of Mechanical Engineers, Part A: Journal of Power and Energy*, 221, 849–856.

MEDD, A. J., DANG, T. Q. & LAROSILIERE, L. M. 2003. 3D inverse design loading strategy for transonic axial compressor blading. ASME Turbo Expo 2003, collocated with the 2003 International Joint Power Generation Conference, ASME Paper No. GT2003-38501.

MEDIC, G., KALITZIN, G., YOU, D., VAN DER WEIDGE, E., ALONSO, J. J. & PITSCH, H. 2007. Integrated RANS/LES Computations of an Entire Gas Turbine Jet Engine. Forty fith Aerospace Sciences Meeting & Exhibit, Reno, Nevada. AIAA Paper No. 2007-1117.

MENEVEAU, C. & KATZ, J. 2002. A deterministic stress model for rotor-stator interactions in simulations of average-passage flow. *Transactions of the ASME Journal of Fluids Engineering*, 124, 550–554.

MILOJEVIÉ, D. 1990. Lagrangian stochastic-deterministic (LSD) predictions of particle dispersion in turbulence. *Particle & Particle Systems Characterization*, 7, 181–190.

MINCU, D.-C. & MANOHA, E. 2012. Numerical and experimental characterization of fan noise installation effects. *ERCOFTAC Bulletin*, 90, 21–27.

MISHRA, A., MANI, K., MAVRIPLIS, D. & SITARAMAN, J. 2013. Time dependent adjoint-based optimization for coupled aeroelastic problems. Thirty first AIAA Applied Aerodynamics Conference. AIAA Paper No. 2013–2906.

MIYATA, H. & NISHIMURA, S. 1985. Finite difference simulation of non-linear ship waves. *Journal of Fluid Mechanics*, 157, 327–357.

MONTOMOLI, F., HODSON, H. & LAPWORTH, L. 2011. RANS–URANS in axial compressor, a design methodology. *Proceedings of the Institution of Mechanical Engineers, Part A: Journal of Power and Energy*, 225, 363–374.

MORRIS, P., LONG, L., BANGALORE, A. & WANG, Q. 1997. A parallel three-dimensional computational aeroacoustics method using nonlinear disturbance equations. *Journal of Computational Physics*, 133(1), 56–74.

MUIR, E. R. & FRIEDMANN, P. P. 2014. Aeroelastic response of bird-damaged fan blades using a coupled CFD/CSD framework. Fifty fith AIAA/ASMe/ASCE/AHS/SC Structures, Structural Dynamics, and Materials Conference. Paper No. AIAA 2014-0334.

NILL-AHMADABADI, M., HAJILOUY-BENISI, A., GHADAK, F. & DURALI, M. 2010. A novel 2D incompressible viscous inverse design method for internal flows using flexible string algorithm. *Journal of Fluids Engineering*, 132(3), 031401.

NING, W. & HE, L. 2001. Some modeling issues on trailing-edge vortex shedding. *AIAA Journal*, 39(5), 787–793.

ORKWIS, P., TURNER, M. & BARTER, J. 2002. Linear deterministic source terms for hot streak simulations. *Journal of Propulsion and Power*, 18(2), 383–389.

OZYORUK, Y. & LONG, L. N. 1997. Multigrid acceleration of a high-resolution. *AIAA Journal*, 35(3), 428–433.

PAGE, J. H., HIELD, P. & TUCKER, P. G. 2013. Inverse design of 3D multi-stage transonic fans at dual operating points. *ASME Journal of Turbomachinery*, 136(4), 041008.

PAGE, J., WATSON, R., ALI, Z., TUCKER, P. & HIELD, P. 2015. Advances of turbomachinery design optimization. AIAA SciTech Conference, 5–9 January, Kissimmee, Florida. AIAA Paper No. AIAA-2015-0629.

PAN, J. & LOTH, E. 2005. Detached eddy simulations for iced airfoils. *Journal of Aircraft*, 42, 1452–1461.

PINELLI, A., NAQAVI, I. Z., PIOMELLI, U. & FAVIER, J. 2010. Immersed-boundary methods for general finite-difference and finite-volume Navier–Stokes solvers. *Journal of Computational Physics*, 229, 9073–9091.

PITSCH, H. 2006. Large-eddy simulation of turbulent combustion. *Annual Review of Fluid Mechanics*, 38, 453–482.

POTSDAM, M. A. & GURUSWAMY, G. P. 2001. A parallel multiblock mesh movement scheme for complex aeroelastic applications. Thirty ninth Aerospace Sciences Meeting and Exhibit, January, AIAA Paper No. AIAA-2001-0716.

RAHMAN, F., VISSER, J. A. & MORRIS, R. M. 2005. Capturing sudden increase in heat transfer on the suction side of a turbine blade using a Navier-Stokes solver. *Journal of Turbomachienery*, 127(3), 552–556.

RAMAMURTHY, R. & GHALY, W. 2010. Dual point redesign of an axial compressor airfoil using a viscous inverse design method. ASME 2014 Twelth Biennial Conference on Engineering Systems Design and Analysis, ASME Paper No. GT2010–23400.

ROIDL, B. & GHALY, G. 2011. Redesign of a low speed turbine stage using a new viscous inverse design method. *ASME Journal of Turbomachienery*, 133(1), 011009.

SAITO, H. & SCRIVEN, L. E. 1981. Study of coating flow by the finite element method. *Journal of Computational Physics*, 42, 53–76.

SALVADORI, S., RICCIO, G., INSINNA, M., & MARTELLI, F. 2012. Analysis of combustor/vane interaction with decoupled and loosely coupled approaches. Proceedings of ASME Turbo Expo GT2012, Copenhagen, Denmark, ASME Paper No. GT2012-69038.

SANCES, D. J., GANGADHARAN, S. N., SUDERMANN, J. E. & MARSELL, B. 2010. CFD fuel slosh modeling of fluid-structure interaction in spacecraft propellant tanks with diaphragms. Proceedings of Fifty first AIAA/ASME/ASCE/AHS/ASC Structures, Structural Dynamics, and Materials Conference, Orlando, Florida, AIAA Paper No. AIAA-2010–2955.

SCHLICHTING, H. & GERSTEN, K. 2000. *Boundary-layer Theory*, Springer.

SCOTTI, A. 2006. Direct numerical simulation of turbulent channel flows with boundary roughened with virtual sandpaper. *Physics of Fluids (1994–present)*, 18, 031701.

SECUNDOV, A., BIRCH, S. & TUCKER, P. 2007. Propulsive jets and their acoustics. *Philosophical Transactions of the Royal Society A: Mathematical, Physical and Engineering Sciences* 365(1859), 2443–2467.

SHEVTSOV, M., SOUPIKOV, A. & KAPUSTIN, E. 2007. Highly parallel fast kd-tree construction for interactive ray tracing of dynamic scenes. *EUROGRAPHICS*, 26(3).

SHUR, M. L., SPALART, P. R. & STRELETS, M. K. 2005. Noise prediction for increasingly complex jets, Part I: Methods and tests. *International Journal of Aeroacoustics*, 4(34), 247–266.

SILVA, W. A. & BARTELS, R. E. 2004. Development of reduced-order models for aeroelastic analysis and flutter prediction using the CFL3Dv6.0 code. *Journal of Fluids and Structures*, 19, 729–745.

SINGAL, V., BAJAJB, J., AWALGAONKARA, N. & TIBDEWALC, S. 2014. CFD Analysis of a Kerosene Fuel Tank to Reduce Liquid Sloshing. Twenty-Fourth DAAAM International Symposium on Intelligent Manufacturing and Automation, 1365–1371, Elsevier.

SPALART, P. R. & BOGUE, D. R. 2003. The role of CFD in aerodynamics, off-design. *Aeronautical Journal*, 107(1072), 323–329.

STOLLENWERK, S. & KUGELER, E. 2011. Unsteady flow modelling using deterministic flux terms. Ninth European Turbomachinery Conference, Paper No. 161.

SUSSMAN, M., SMEREKA, P. & OSHER, S. 1994. A level set approach for computing solutions to incompressible two-phase flow. *Journal of Computational Physics*, 114, 146–159.

TADDEI, S. R. & LAROCCA, F. 2014. An actuator disk model of incidence and deviation for RANS-based throughflow analysis. *ASME Journal of Turbomachinery*, 136, 021001-1.

TANNER, R. I., NICKELL, R. E. & BILGER, R. W. 1975. Finite element methods for the solution of some incompressible non-Newtonian fluid mechanics problems with free surfaces. *Computer Methods in Applied Mechanics and Engineering*, 6, 155–174.

TERRACOL, M., MANOHA E., HERRERO, C., LABOURASSE, E., REDONNET, S. & SAGAUT, P. 2005. Hybrid methods for airframe noise numerical prediction. *Theor. Comput. Fluid Dyn.*, 19, 197–227.

TOULOUKIAN, Y. S. & DEWITT, D. P. 1970. *Thermophysical Properties of Matter – The TPRC Data Series, Vol. 7: Thermal Radiative Properties*. IFI / Plenum.

TUCKER, P. 1993. *Numerical and experimental investigation of flow structure and heat transfer in a rotating cavity with an axial throughflow of cooling air*. D.Phil. Thesis, The University of Sussex, School of Engineering.

TUCKER, P. G. 1997a. CFD Applied to Electronic Systems: A Review. *Trans. IEEE (CPMTA)* 20(4), 518–529.

TUCKER, P. G. 1997b. Numerical precision and dissipation errors in rotating flows. *International Journal of Numerical Methods for Heat & Fluid Flow*, 7(7), 647–658.

TUCKER, P. G. 2001a. *Computation of unsteady internal flows*, Springer.

TUCKER, P. G. 2011b. Computation of unsteady turbomachinery flows, Part 1 – Progress and challenges. *Progress in Aerospace Sciences*, 47(7), October, 522–545 .

TUCKER, P. G. 2013. *Unsteady computational fluid dynamics in aeronautics*, Springer.

TUCKER, P. G., EASTWOOD, S., KLOSTERMEIER, C., XIA, H., RAY, P., TYACKE, J. & DAWES, W. 2012. Hybrid LES approach for practical turbomachinery flows – Part 2: Further applications. *Journal of Turbomachinery*, 134, 021024.

TUCKER, P. G. & PAN, Z. 2001. URANS Computations for a complex internal isothermal flow. *Computer Methods in Applied Mechanics and Engineering*, 190, 2893–2907.

TURNER, M. G., REED, J. A., RYDER, R. & VERES, J. P. 2004. Multi-fidelity simulation of a turbofan engine with results zoomed into mini-maps for a zero-d cycle simulation. ASME Turbo Expo 2004: Power for Land, Sea, and Air, Paper No. GT2004-53956, 219–230.

TYACKE, J. C., NAQAVI, I. Z., & TUCKER, P. G. 2015. Body force modelling of internal geometry for jet noise prediction: Advances in simulation of wing and nacelle stall (Radespiel, R., Niehuis, R., Kroll, N., Behrends, K. (Eds.)), *Notes on Numerical Fluid Mechanics and Multidisciplinary Design*, Springer, 131, 97–109.

VERDICCHIO, J. A., CHEW, J. W. & HILLS, N. J. 2001. Coupled fluid/solid heat transfer computation for turbine discs. Proceedings of ASME Turbomachinery Exposition 2001, ASME Paper No. GT2001-0205.

VERSTEEG, H. K. & MALALASEKERA, W. 2007. *An introduction to computational fluid dynamics: the finite volume method*, Pearson, Prentice-Hall.

VOIGT, C., FREY, C. & KERSKEN, H.-P. 2010. Development of a generic surface mapping algorithm for fluid structure-interaction simulations in turbomachinery. 5th European Conference on Computational Fluid Dynamics ECCOMAS CFD, (eds. J. C. F. Pereira, A. Sequeira, and J. M. C. Pereira), Lisbon, Portugal, 14-17 June.

WANG, D. X. & HE, L. 2010. Adjoint aerodynamic design optimization for blades in multistage turbomachines – Part 1: methodology and verification. *Journal of Turbomachinery*, 132, 021011-1.

WANG, Q. 2014. Convergence of the least squares shadowing method for computing derivative of ergodic averages. *SIAM Journal on Numerical Analysis*, 52, 156–170.

WANG, Z. & WANG, Z. J. 2005. Multi-phase flow computation with semi-Lagrangian level set method on adaptive Cartesian grids. AIAA Paper No. 2005-1390.

WATSON, R. & TUCKER, P. G. 2014. Evolving solutions: distributed computing and cutback trailing edges. 52nd AIAA Aerospace Sciences Meeting, 13–17 January, Maryland, Paper No. AIAA-2014-0227.

WATSON, R., TUCKER, P. G. & WANG, Z. 2014. Unsteady simulation paradigms for trailing edge. 52nd AIAA Aerospace Sciences Meeting, Paper No. AIAA-2014-1425.

WELLBORN, S. R., TOLCHINSKY, I. & OKIISHI, T. H. 1999. Modeling shrouded stator cavity flows in axial-flow compressors. *Journal of Turbomachinery*, 122, 55–61.

WENDT, B. J. & BIESIADNY, T. 2001. *Initial circulation and peak vorticity behavior of vortices shed from airfoil vortex generators*, NASA/CR-2001-211144.

WILHELM, R. 2002. Inverse design method for designing isolated and wing-mounted engine nacelles. *Journal of Aircraft*, 39(6), 989-995.

XU, L. 2003. Assessing viscous body forces for unsteady calculations. *Journal of Turbomachinery*, 125, 425–432.

XU, L. 2012. Transonic axial flow fans, cambridge turbomachinery course. *The Møller Centre,* 2, 475–513.

XU, L., HYNES, T. & DENTON, J. 2003. Towards long length scale unsteady modelling in turbomachines. *Proceedings of the Institution of Mechanical Engineers, Part A: Journal of Power and Energy*, 217(1), 75.

ZHANG, Y. 2015. *Turbulent unsteady separated flows in multidisciplinary computational fluid dynamics applications*. PhD Thesis, McGill University.

ZHANG, Y., HABASHI, W. G. & KHURRAM, R. A. 2013. A hybrid RANS/LES method for turbulent flow over iced wings. Twenty first Annual Conference of the CFD Society of Canada, 6–9 May, Quebec, Canada.

CHAPTER 7

Pre- and Post-Processing

7.1 Introduction

In this chapter, pre- and post-processing is discussed along with related aspects and the CFD process chain. Examples of the latter are given in Figure 7.1. Much of the contents link, in different ways, to those in other chapters. Grid generation is a key aspect of pre-processing but this was discussed in Chapter 3. Prior to grid generation, geometry is required. Hence in this chapter, geometry handling is first discussed. Once the geometry has been defined, the mesh can be generated. Then boundary conditions must be defined. However, during the mesh generation process, a clear idea of boundary conditions is needed and hence some understanding of the flow physics. For example, boundaries must be set sufficiently far away so that any uncertainties/deficiencies arising from them do not contaminate the solution. They ideally should be located where most information is available and hence the uncertainties relating to this aspect are minimal. The mesh may need to be deliberately constructed to dissipate waves through coarsening. Hence, as with most other aspects of CFD, there are intrinsic links. Once the boundary conditions have been defined, then numerous other simulation parameters need to be correctly specified. Hence, this aspect is also considered.

Generally, the governing equations are solved in an iterative fashion and hence there is the need to judge *iterative convergence*. This is also discussed in this chapter. However, the process of judging convergence also relates strongly to the numerical schemes used and hence this area could also have been addressed in Chapter 4. Another aspect is simulating unsteady flows. These can have transients, periodicity, a more stochastic nature or combinations of these things. For such flows it is necessary to detect the end of initial transients and also judge when the flow has reached a statistical steady state. It also needs to be further judged how long an averaging period is needed to gain statistically stationary data. This aspect is also discussed here. The wide range of post-processing techniques available is considered. Both flow visualization and approaches for gaining more quantative data are discussed. In this chapter, the assessment of the solution quality is also considered. However, the starting point for this chapter is the aspect of initial planning and also reporting on the simulation process to give consistency.

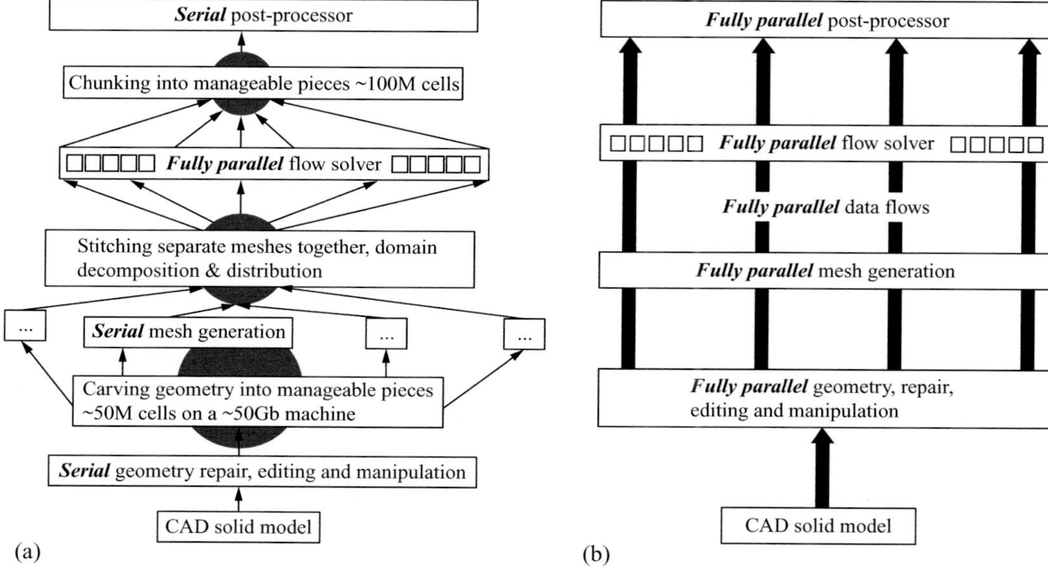

Figure 7.1. CFD process chains: (a) typical approach and (b) ideal future approach (from Tucker 2013 – published with kind permission of Springer).

7.2 Initial Planning and Reporting and Consistency

7.2.1 *Initial Planning*

Prior to making any simulations, some initial planning is vital. There must be clarity on the objective of the simulation. It needs to be considered what level of accuracy is needed and the time and resources available to make the simulation. Perhaps the simulation is simply needed to look at the flow patterns or to gain a rough estimate of pressure loss or a component temperature. Alternatively, it might be necessary to predict skin friction to a high degree of accuracy.

Early on in the design process, typically it is desired to explore a wide range of design scenarios relatively quickly. Hence, lower order methods can be most appropriate. For example, in turbomachinery design there will initially be simulations looking at the basic engine cycle analysis. Later, throughflow equations will be solved along streamlines (see Chapter 2 and Section 7.10.1). Next two-dimensional simulations might be made to improve specific blade designs and then three-dimensional calculations. Finally, three-dimensional, multi-component, potentially coupled simulations can be made. Indeed, even eddy-resolving simulations might be used. Then it is of importance that estimates are made to ensure that the computational resource is available, along with the substantial time frame to perform the simulation in compressed design time periods. Figure 7.2 shows the hierarchy of methods discussed.

The next level is the inclusion of real geometry (annulus) features. This level is indicated in Figure 7.3. Fillets and leakage flows are known to play a surprisingly

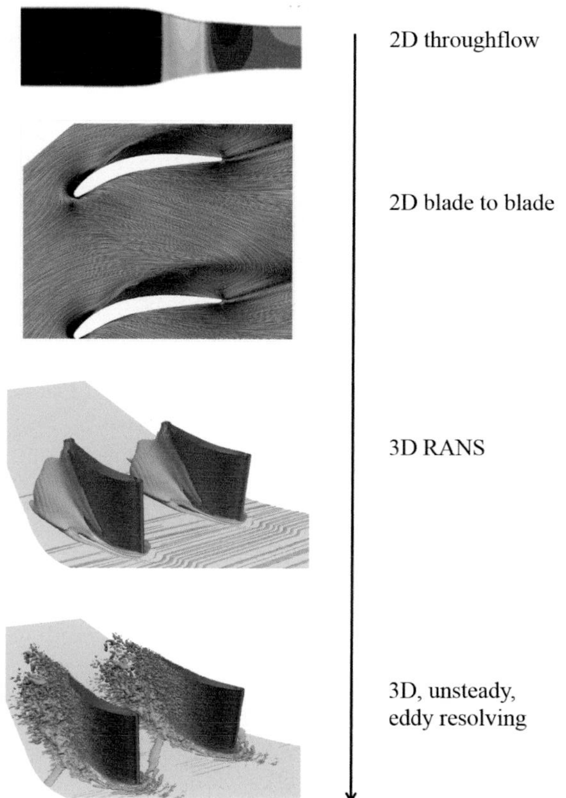

Figure 7.2. Hierarchy of methods for turbomachinery CFD.

strong aerodynamic role (Riera 2014). There are critical features at both blade radial extremities. Turbine blade surfaces are populated by fine cooling holes. Real surface roughness can play a critical role (see Yang 2014). Once all the real annulus features have been included, then multistage calculations can be made. Then coupled simulations that consider the structural and thermal behaviour of the blade can be considered. Complicating decisions relate to if the real annulus features, which

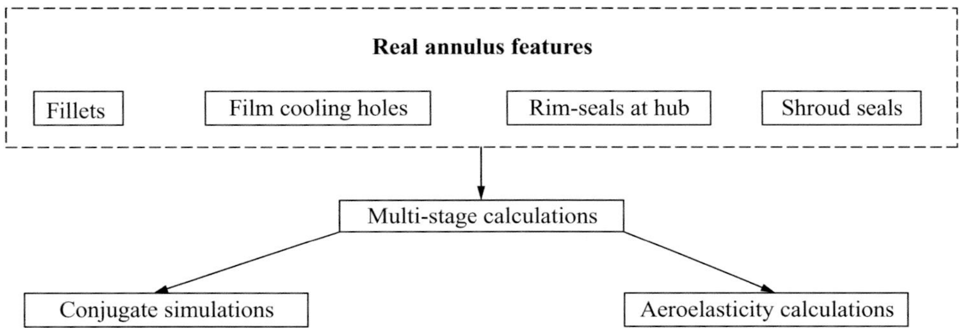

Figure 7.3. Modelling of real geometry (annulus) features.

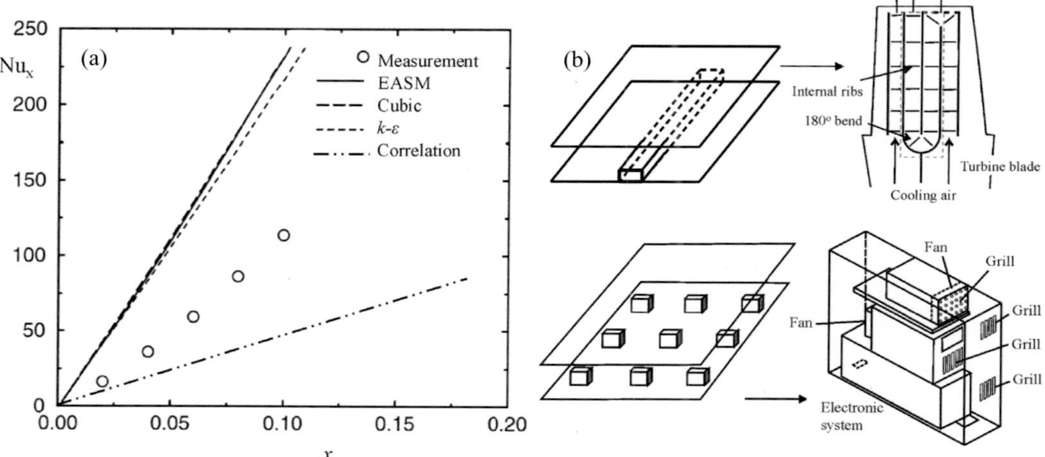

Figure 7.4. Predictions of Nusselt number for different advanced turbulence models and their performance relative to flat plate data: (a) predicted heat transfer and value from correlations and (b) examples of systems where correlations can be relatively useful (Frame (a) from Liu and Tucker 2007 – published with kind permission of Wiley).

involve leakage flows, are to be explicitly meshed or treated using reduced order models.

RANS CFD is sometimes described as postdictive, with LES being viewed as more a predictive approach. Hence, with RANS, for new problems, some thought will be required on the availability of validation data and even the availability of work from other people, using CFD. The latter should give some indication of the accuracy that might be expected from the simulation. However, other people's experiences on the accuracy of CFD approaches for different applications can vary considerably. It depends on the objectives of the reported work, the experience of the analyst performing the work and also the scope of the study. A careful point for consideration, early on, is if CFD is a sensible option. For example, for some complex turbine cooling systems and also electronics, the use of empirical correlations can give better accuracy than CFD – unless eddy-resolving simulations are resorted to. For example, Figure 7.4a shows predictions of Nusselt number in a complex system (the far right-hand, Frame (b) system) for different advanced turbulence models. Their performance is contrasted with that for flat plate correlation data. The turbulence models considered are the Explicit Algebraic Stress (EASM), cubic and a low Reynolds number k-ε model. These models are outlined in Chapter 5. The measurements are given by the symbols. The key thing to observe is that all three turbulence models show considerable error and less accuracy than use of the simple correlation for convective heat transfer over a flat plate. This is given by the chain double-dashed line. The aspect of considering what detail is necessary in an analysis is discussed further in Section 7.3.

7.2.2 *Reporting and Consistency*

Particularly for industrial design, it is important to use a consistent CFD process. If nothing else, this gives a basis for assessment and improvement of the process. Hence, documentation of the process is important with some justification for the choice of the different models and approaches used. How-to guides can be useful and also 'CFD Best Practices'. Examples of such documents are available through organizations like NAFEMS (National Finite Element Methods and Standards) and ERCOFTAC. (European Research Community On Flow Turbulence And Combustion) If a new user attempts to perform CFD simulations in a design environment in a specialized application area, it is quite likely that poor results will arise. This can be avoided through use of a documented process for the user to follow that has been provided by an experienced practitioner. Although such aspects sound dull and obvious, they are of considerable importance. Detailed documentation on CFD codes and the algorithms that they use are also necessary. Even the most basic scheme, for example the Crank-Nicolson time integration, can have different practical implementations, which can have a clear solution impact – see for example Tucker (2013). Turbulence model palliatives can emerge as unknown features in a particular code's implementation of a 'standard' turbulence model. This can make it hard, even for the more experienced user, to be fully aware of the features of the turbulence model being used and hence the likely traits of predictions. Since CFD is dealing with non-linear equations and highly complex physical phenomena, and seeking to generate meaningful solutions for important practical problems, continual attention should be paid to personnel development. Hence, attending conferences, reading journals and becoming involved in working and special interest groups is important.

7.3 Problem Definition

Considerable industrial sector knowledge is required to obtain meaningful CFD data. For example, the modelling of flow and heat transfer in avionics or electronics systems, on the face of it, does not sound that difficult. The flow is low speed and does not involve phase changes, chemical reactions or moving surfaces and so forth. However, even for electronics, there is a lot of knowledge needed to perform reliable simulations. For example, geometries can be tremendously complex. Therefore, the first question is whether a less detailed, more system-level calculation is needed, or a calculation looking at local details around a particular component. For system-level calculations, for example, heat sinks can be treated as porous media and fans and grills as planar momentum sources and sinks, respectively. Coming up with sensible modelling coefficients for these is not a trivial task and will involve access to experimental data and experience. The surfaces of circuit boards, populated by integrated circuits and other low profile components, are sometimes treated as idealized roughness. However, ascertaining an appropriate roughness height is far from easy. The wire harnesses found in such systems can offer tremendous geometric complexity. Therefore, it needs to be considered if it is necessary to include them. A critical interest in integrated circuits is their junction temperature. To get this, ideally a conjugate solution is needed, with a highly detailed, but even so quite idealized conduction model to estimate the junction temperature. Alternatively, the chip can be characterized as a resistor network and this integrated into the CFD model. At the crudest level, a surface heat transfer coefficient, from the CFD, can be used. Then, with a local ambient temperature and a simple, idealized, integrated circuit model, formulated as a one-dimensional resistance, an estimate of the chip junction temperature can be made. Hence, there are a wide range of assumptions and strategies, some outlined in Chapter 6 under multi-scale modelling. Expert knowledge is needed to be able to understand the reliability of the output and generate it. Another problem is boundary conditions. It has been observed that the surrounding environment in which an avionics or electronics system is put in can have a strong impact on the level of heat transfer inside the system. Hence, to be safe, it would be best to model the system and the environment (perhaps a room) that contains it. The turbulence model will have a strong impact on the results and importantly the type of near wall modelling. The user needs to decide on the turbulence modelling fidelity. As noted, there is a reasonable body of evidence that suggests that for avionics/electronics, sometimes it can be more accurate to get the heat transfer by simply using flat plate correlations. Hence, it needs to be considered very carefully if a CFD simulation is really sensible/warranted.

Similarly, for turbomachinery, a vast amount of user knowledge is needed to setup and understand the reliability of a CFD simulation. The user needs to consider if mixing or sliding planes are needed between the rotating and stationary blade rows. Perhaps the blade rows could be modelled using actuator disks or body forces. The relative movement of wakes between rotating and stationary stages can be modelled using a range of approaches that typically treat the wakes as deterministic stresses. The user needs to understand, if unsteady simulations are needed, the most apt turbulence model – whether, for example, to model the fine cooling holes over the surface of the turbine blades and the coolant that is ejected from their trailing edges. Also, accounting for roughness can be important. The turbine has numerous fine scale features such as seals at the extremities of blades. Decisions need to be made about whether these should be treated using specialized reduced order models or fully meshed. The level of knowledge needed in fact seems daunting (see Tucker 2011a, 2011b). Notably, for the internal cooling inside turbines blades, unless the expense of making eddy-resolving simulations is gone to, greater accuracy can be gained by using empirical heat transfer correlations. Hence, again, a careful decision needs to be made.

For combustion modelling, the complexity for the user to consider seems even worse. The CFD practitioner needs to deal with highly complex geometries and a range of chemical reactions and explore the formation of products such as soot. Multiphase flow has to be handled with droplet transport and their breakup. Hence, even when basic CFD methods are well understood, substantial, field specific, specialized knowledge is required and this all feeds into the aspect of problem definition. Therefore, this stage is of critical importance. Consequently, ahead of setting up a simulation it is first necessary to carefully consider what level of accuracy or detail is actually necessary and if this is actually feasible. Based on knowing the flow physics and the level of accuracy required, the appropriate governing equations, outlined in Chapter 2, can be selected. At the preliminary design stages, as noted earlier, a large number of lower fidelity calculations is generally most appropriate, with the fidelity level increasing as the final design is approached. The lower fidelity level calculations enable automatic design optimization procedures to rapidly move through the design landscape. They can readily pass through any discontinuities in this.

Hence, a critical stage, prior to starting the pre-processing, is the problem definition stage. By this it is meant that the user must decide what level of detail and boundary conditions are needed to gain the desired level of engineering insight necessary for the task at hand. For an aeroengine, it needs to be decided if there is substantial system coupling and hence if large scale coupled simulations are needed. Other questions are: If there is coupling, is it one or two way? Also, how close does the coupling need to be?

Many engineering systems have a vast degree of complexity, but considering all of this can sometimes be unnecessary in relation to the degree of solution fidelity desired. For example, Figure 7.5a shows the local geometric detail in an electromechanical component. If it is desired to study the system level cooling and airflow, the level of detail shown is unnecessary. Hence, a key first stage is to de-feature the geometry. Another key initial stage can be CAD clean up. Then it needs to be decided if it's necessary to model the whole system or just a local area.

At this stage, these decisions might be strongly influenced by the level of unknown boundary conditions arising from the selected system extent. Such assumptions may well need to be checked through multiple simulations exploring the sensitivity of solutions to different types of boundary conditions. If solutions involve exploring thermal fields, it will need to be considered if adiabatic boundary conditions are sufficient or if spatially varying isothermal boundary conditions need to be defined. Alternatively, full conjugate simulations might be necessary to gain the desired fidelity.

The geometric complexity to be considered and also the nature of the flow will both strongly control the choice of CFD solver. The idea is to have a range of solvers optimized for different tasks. However, this comes at considerable cost in terms of software licenses or code development effort and some compromise is generally required.

Figure 7.1a shows a typical current CFD process chain. Notably, there are serious serial solution process bottlenecks. These currently, often, restrict the problem scales that can be readily be dealt with and give the necessity for specialized serial computers with large amounts of RAM. Figure 7.1b shows the more ideal fully parallel solution process chain. The process chain outlined in Figure 7.1 is now discussed in more detail.

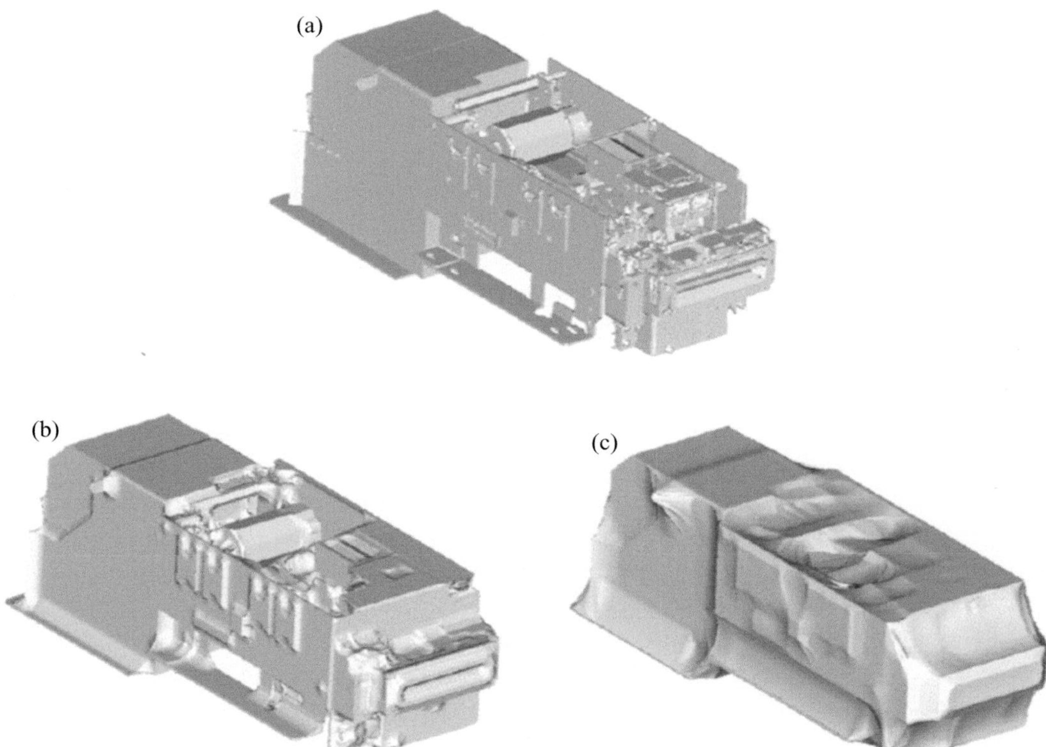

Figure 7.5. Geometric de-featuring: (a) original geometry, (b) slightly de-featured geometry and (c) more heavily de-featured geometry (from Xia et al. 2012 – published with kind permission of Springer).

7.4 Geometry Handling

The level of focus on geometry in CFD largely depends on whether CFD is being used for design or analysis. With the former, the emphasis is more on handling a relatively unknown geometry and, based on CFD insights, producing a design. Analysis involves taking an existing design, meshing and running the CFD, and thus confirming performance or giving some new performance insights. Since geometry is so central to design, CFD codes are emerging which are totally geometry centred. For, example, with FloEFD, the user totally interacts with commercial CAD software. Similarly, the BOXER commercial software is highly geometry focused. Both of these programs use octree type meshes (Chapter 3). Notably, for a typical complex engineering problem, data must be fused from a variety of sources. There might be some CAD (in various forms), hand drawings, point strings and so forth.

For analysis of simple geometries, such as airfoils, point strings can be used to define surface geometry. These can be stacked to extend the geometry to three dimensions. For defining more complex geometries for use with mesh generators, a popular file format is the STL (STereoLithography) also known as Standard Tessellation Language. This format works in a Cartesian coordinate system and

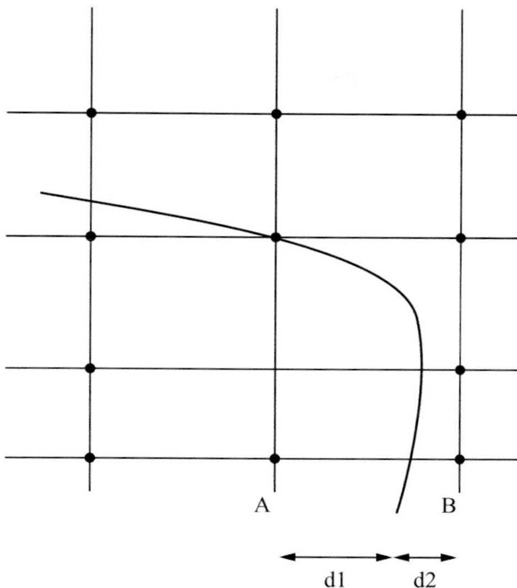

Figure 7.6. The use of level sets to define geometry.

defines the surface as unstructured triangles storing vertices and surface normals. Another popular tessellated surface format is VRML (Virtual Reality Modelling Language).

Two ways of traditionally defining geometry are constructive solid geometry (CSG) and BREP (Boundary REPresentation). CSG is based on the definition and manipulation of simple analytical bodies. These are bodies like cylinders, spheres, cones, cuboids, pyramids and prisms. Also, there are so-called allowable operations which are Boolean on sets such as unions, differences and intersections. The approach is simple but has difficulty in representing free-form shapes often encountered in engineering.

BREP, on the other hand, fits spline-like panels to surfaces, attempting to enforce edge connectivity. The approach is very compatible with curvilinear grid solutions. However, it is more problematic for highly complex geometries.

A simple and effective way to define geometry can be simply to use a sufficiently dense point cloud (see Löhner 1996).

For extremely complex geometries, level-sets can be used. Basically, a surface distance field is stored on a Cartesian grid. This can be of an octree type. The isosurface of zero wall distance, d, is used to define the surface geometry. The process can be better appreciated by considering Figure 7.6 and line a-b in it. Notably, the approach also allows sculpting of the geometry through editing the distance field. CSG can also be applied. For example, Boolean sums can be made

$$P = Q \cup R \tag{7.1}$$

This forms the union of A and B. With distance fields, this is simply achieved using the cheap to evaluate equation

$$d_p = \min\left(d_Q, d_R\right) \tag{7.2}$$

Table 7.1. *Advantages and disadvantages of different geometry definition techniques*

General method	Method	Advantages	Disadvantages
Implicit geometry	CSG	Simple, compact storage. Can deal with relatively complex geometry.	Does not naturally handle more free form shapes used in aerodynamics.
	BREP	Compact storage.	Difficult to maintain watertight geometry.
Explicit geometry	Spatial occupancy	Simple, fast, readily supports complex geometries and geometry changes. Easy to parallelize.	Higher storage.

Notably, explicit geometry is used in medical imaging where it is needed to work rapidly with highly complex geometries. As noted by Dawes (2007), explicit geometry offers the possibility of sculpting geometry real time. This can be done in a virtual reality (VR) environment, modifying geometry with a VR glove. Tools can be defined to cut material and also spray surfaces. Notably, explicit geometry procedures are easy to parallelize. This explicit geometry approach will be discussed further later. Table 7.1 gives advantages and disadvantages of the different geometry definition techniques discussed in this section.

Most well-known commercial CFD systems follow a BREP type procedure. This is illustrated in Figure 7.7 from Dawes (2007) for a compressor volute. As can be seen, even though the geometry is relatively trivial, a large number of curves and patches are required and boundaries are created that require compatibility. The surfaces can then be triangulated. These surfaces can then be volume meshed. Note that for design optimization, the geometry definition needs to be automatically edited.

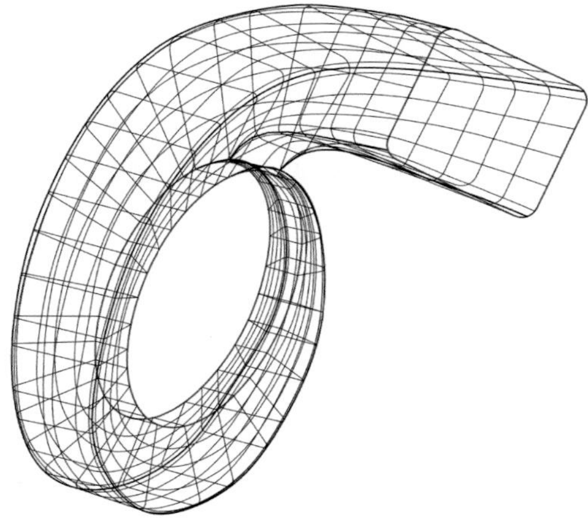

Figure 7.7. The use of BREPs to define geometry (from Dawes et al. 2000 – published with the kind permission of the ASME).

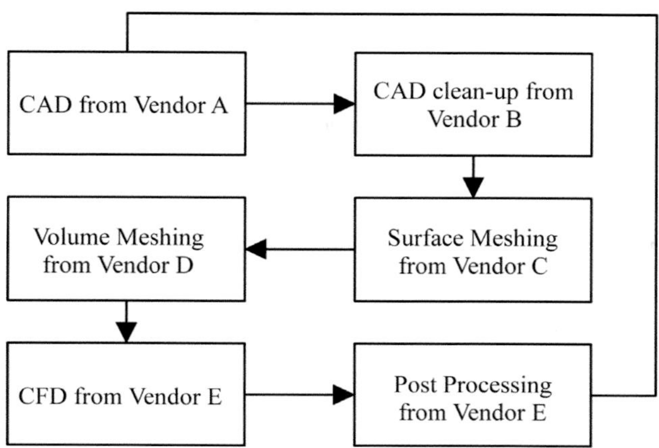

Figure 7.8. Typical CFD process chain (from Dawes 2007 – published with the kind permission of the Royal Society).

A range of techniques can be used, such as free form deformation and surrogate equations. Approaches are outlined in Dawes (2007).

Figure 7.8, from Dawes (2007), shows a typical CFD process chain. For the CAD import, normally the STL, IGES or STEP file formats are used. As can be seen, the process is quite cumbersome. Critically, it depends on reliable and robust data transfer. Data transfer, even after many years, is still problematic with incompatibility issues. For design optimization of large problems, ideally this process needs to be done in parallel. As noted by Dawes (2007), a trend in design optimization is to directly interact with the CAD package and this again really needs to be performed in parallel processing mode.

Figure 7.9 shows the explicit geometry procedure for capturing a turbine geometry used in the commercial BOXER code. Frame (a) gives the tessellated geometry read into BOXER. Frame (b) shows the octree mesh. The signed distance field which now defines geometry is stored on this. Based on the distance field sign, the mesh inside the geometry can be identified. Frame (c) shows the flow mesh. Frame (d) gives the distance field, the dark inner shade giving negative values. This is clipped, enabling just the surface geometry zone to be focused on.

Figure 7.10 shows the sculpting of an extra, new hole through the distance field using a virtual cutting tool given in black. Figure 7.11 shows the updated distance field and revised mesh. The integrated, explicit geometry procedure is much faster and tidier than that of the BREP based method illustrated in Figure 7.8. However, for certain areas of industrial design, the solution accuracy is paramount and the geometry changes relatively subtle. Then more bespoke processes are appropriate.

CFD codes are emerging that can work in real time, allowing users, through hand sketching on touch-type screens, to handle simple geometries. Clearly such approaches are not that practical in three dimensions and have limited physical accuracy. Nonetheless, clearly for the inexperienced fluid dynamicist, quick but crude simulations could give some helpful flow insights.

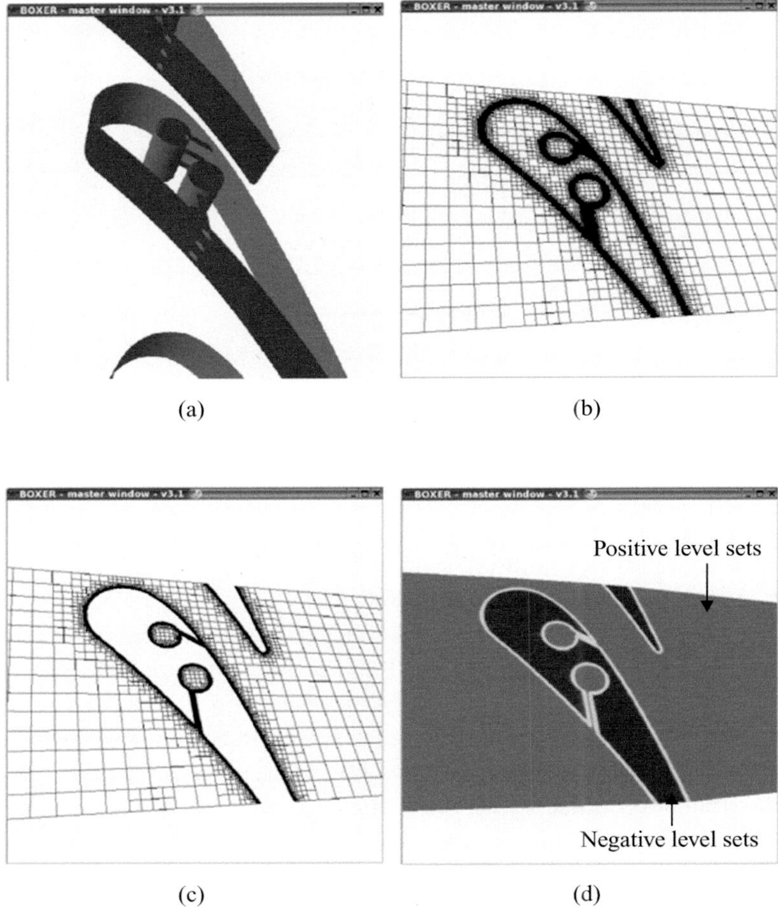

Figure 7.9. Explicit geometry procedure for capturing a turbine geometry: (a) tesselated geometry, (b) octree mesh, (c) flow mesh and (d) clipped distance field (from Dawes 2007 – published with the kind permission of the Royal Society).

7.4.1 *Feature Control*

Due to numerical tolerance and other matters, complex CAD exported geometry generally needs cleaning (especially if exported using the IGES – International Graphics Exchange Standard), and surface gaps will need healing. This can take a substantial amount of time. For a complex engineering system, the analyst can be faced with potentially thousands of CAD surface patches that need to be merged or trimmed. Also, sometimes, de-featuring can be needed as shown in Figure 7.5. Much of the detail shown in this figure is unnecessary for a system level design calculation. There is a range of procedures for de-featuring. It can be done using, for example, a shrink-wrap procedure (see Xia et al. 2012). With this, a distance field isosurface is taken at a fixed offset distance from the surface. This is then projected back to the surface. Figure 7.12 shows an octree-mesh and flow solution on the de-featured

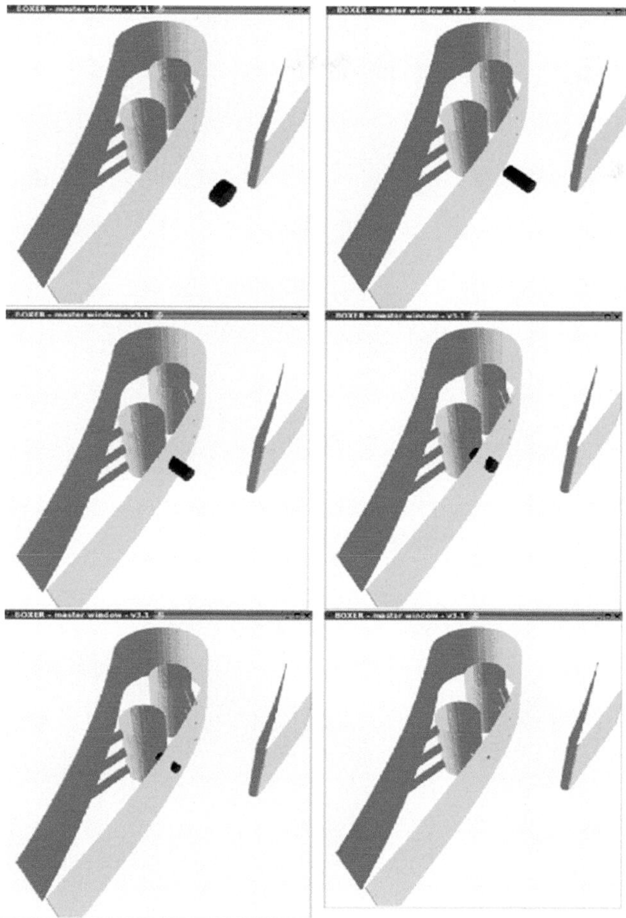

Figure 7.10. The sculpting of a new hole through the distance field using a virtual cutting tool – black – (from Dawes 2007 – published with the kind permission of the Royal Society).

Figure 7.5 geometry. Note that the meshing process is such that the geometry surface has a castellated nature, without any of the cell trimming that better resolves the geometry. The mesh generated for the de-featured model contains only 75,000 cells, whereas the full model has 2.5 million cells. The flow solution is shown in Frame (b), which gives velocity contours and flow streamlines (see Section 7.10.1).

The de-featuring process is shown more simply in Figure 7.13, for a two-dimensional geometry. The two lines are d contours for two distances from the wall. The one corresponding to the furthest d shows the greater de-featuring. As noted, these contours can be projected back to the wall to give a de-featured geometry. The normal, to project back along, can be expressed as

$$\mathbf{n} = \frac{\nabla d}{||\nabla d||} \tag{7.3}$$

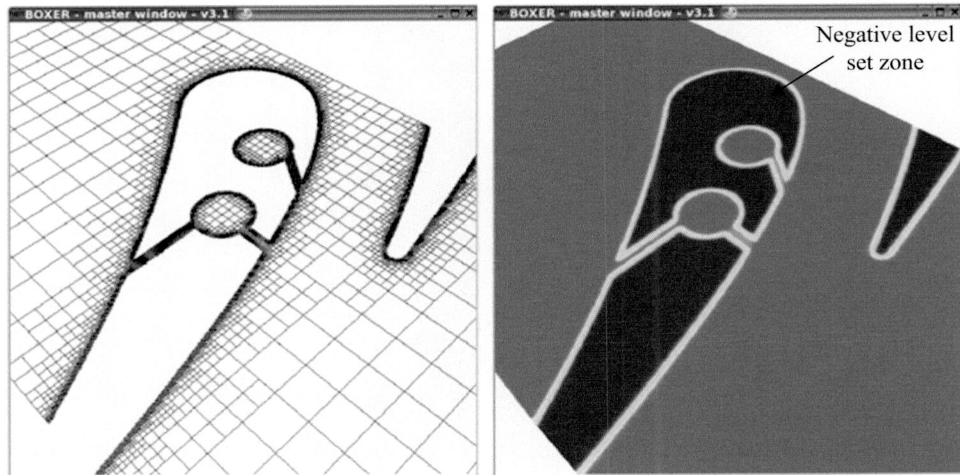

Figure 7.11. Newly sculpted distance field and mesh (from Dawes 2007 – published with the kind permission of the Royal Society).

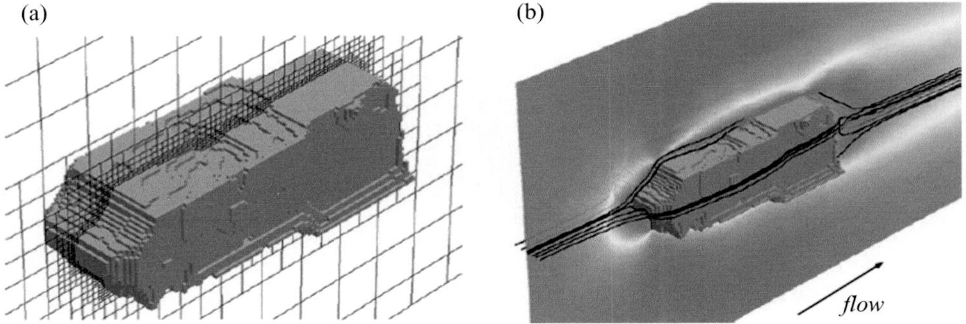

Figure 7.12. Mesh and flow solution for a de-featured geometry: (a) octree, castellated mesh and (b) flow field (from Xia et al. 2012 – published with the kind permission of Elsevier).

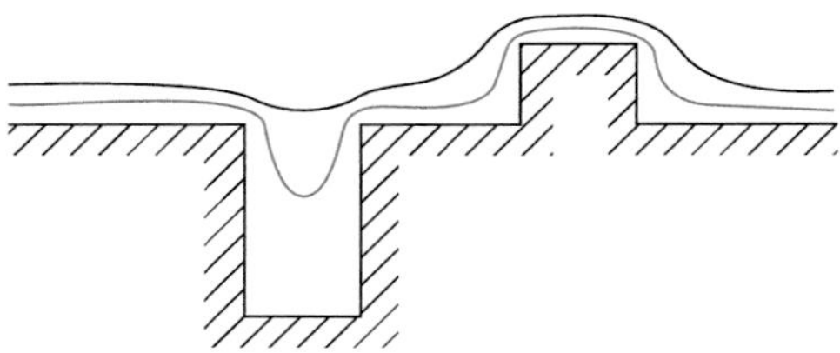

Figure 7.13. A simple two-dimensional example of de-featuring (Published with the kind permission of Elsevier).

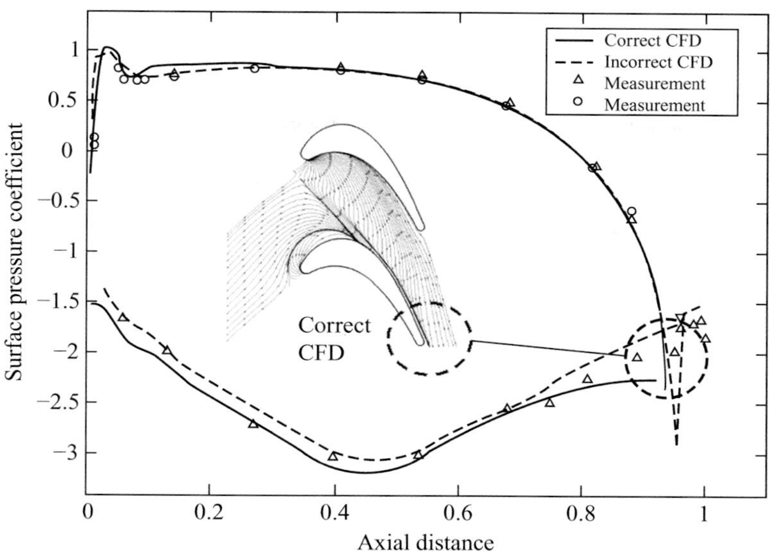

Figure 7.14. Surface pressure distribution on a turbine blade (from Holley et al. 2006 – published with the kind permission of the American Society of Mechanical Engineers).

This can then be projected back to the new surface, at x', using the following equation:

$$\mathbf{x}' = \mathbf{x} - d\mathbf{n} \tag{7.4}$$

A more minor aspect of feature control relates to airfoil trailing edges. A phenomena called ‘fish tailing’ can occur. With this, the surface pressure plot, as shown in Figure 7.14, has what looks like a fish tail near the trailing edge zone. This is a result of the flow staying attached too long around the pressure surface near the trailing edge. The problem is especially severe for blades with thick trailing edges, which is the case for the geometry considered by Holley et al. (2006). Indeed the level of fish tailing in the plot could be a lot worse. The defect can be dealt with either through changing the trailing edge geometry by blunting or the addition of a cusp. Alternatively, localized mesh coarsening or body forces can be used to force separation (Denton 2010). Although fish tailing can cause substantial errors in predicted turning and loss, care must also be taken when modifying geometry to compensate for it. This, in itself, can cause significant errors if not carried out with care.

7.5 Mesh Generation

Once the geometry is available, the mesh can be generated. Generation of an optimal mesh topology requires considerable understanding of the flow physics. The extent of the domain and location of boundaries is also heavily defined by knowing a priori the flow physics and critical locations that have boundary condition sensitivity. Specific mesh generation strategies are discussed in Chapter 3. It is sufficient to say here that this can be by far the most time consuming aspect of running a CFD simulation. Frequently, based on a current solution, some re-meshing is often required, even if this is just to perform a basic grid independence study. Care can be required with the numerical precision at which meshes are exported. For, example, if there is insufficient accuracy, any periodic boundaries might fail to be detected when making searches with unstructured grids. For turbomachinery problems, where there are sliding planes, mesh extrusion can be needed to produce an overlapping grid zone. Also, care is needed to ensure that when the two grids rotate relative to each other, cell faceting does not result in neighbouring domain nodes not being found.

7.6 Setting Parameter Fields

7.6.1 *Boundary Conditions*

Next boundary conditions must be defined. Boundary condition labelling can sometimes take place within the grid generator. As noted previously, for problems with periodicity, periodic neighbouring cells will need to be detected and marked. If measurements are available, then the profiles from the measurements must be imposed in some way on the mesh distribution through, for example, bilinear interpolation. For steady RANS (and related) simulations, with compressible flow solvers, typically the total pressure and total temperatures are specified at inflow and static quantities at outflow. To gain directionality, flow angels at inflow can be specified. For incompressible pressure correction flow solvers, the velocity components present natural boundary conditions. As noted in the turbulence modelling chapter, turbulent quantities must be specified at all boundaries. Some care can be necessary to ensure that these are of a plausible scale.

For eddy-resolving simulations, where time varying inflow quantities are required, the specification process can be complex. Some approaches are summarized in Figure 7.15. These approaches include the use of proper orthogonal decomposition (see Section 7.12.4). This is generally best for capturing large-scale unsteady features. Smaller scales can be infilled using stochastic modelling. Turbulence can be generated using upstream elements such as trips or even idealized components. These elements can all be modelled using body forces and this saves explicit meshing. There are numerous forms of mathematically synthesized turbulence. Most of these are based around Fourier series characterizations. However, some (synthetic eddy methods) are based around considering physical turbulent structures and converting these into unsteady signals at grid points. Precursor simulations can be made. These can be based around cylinders; for example, to replicate, say, blade wakes from upstream blade rows in turbomachinery. These wakes can then be swept across an inflow plane. An example of the use of this process for turbine blades is shown in Figure 7.16. The pre-computed wakes, from a DNS, are swept across the left-hand inflow plane. Periodic pipe, channel or transformed boundary layer flows can be run in an eddy-resolving mode to gain unsteady inflow data. This unsteady data can, if necessary, be pivoted around RANS or measured profile data

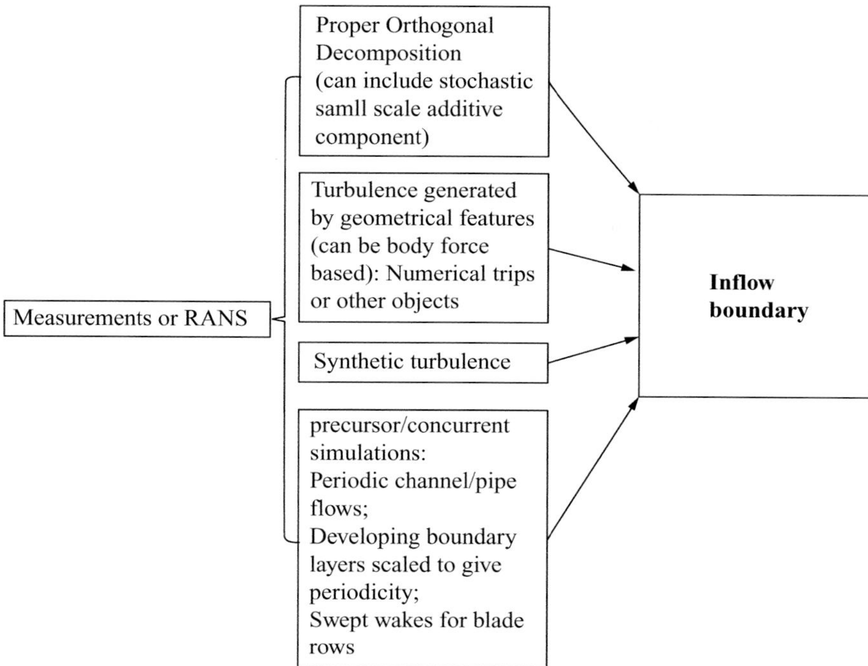

Figure 7.15. Summary of some potential LES inflow methods.

for some time mean profile. For all approaches, measurements or RANS are needed to provide various levels of input information, such as the boundary layer thickness, turbulence length scale, or turbulence intensities to be represented on some form of synthetic turbulence technique. Further details on LES inflow conditions are given in Chapter 5.

Specialized non-reflecting inflow and outflow conditions can be required and target zones (sometimes based on specified RANS flow fields) for sponges specified

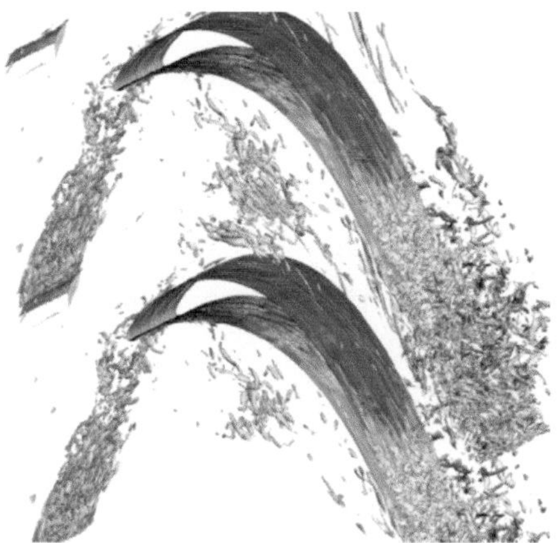

Figure 7.16. Sweeping of DNS pre-computed wakes over the inflow plane of a turbine LES simulation.

(see Shur et al. 2003). For jets, special entrainment conditions can be needed (see Andersson et al. 2005). Hence, again this stage can be quite involved. Non-reflecting boundary conditions are briefly discussed in Chapter 4.

7.6.2 *Initial Flow Guess*

Ideally a good initial guess to the flow solution is desirable. This could be based on lower fidelity solutions. For example, in turbomachinery, data from quasi-two-dimensional throughflow solutions could be used. Clearly, a previous, perhaps semi-converged or closely related CFD solution can provide an alternative initial solution guess.

7.6.2.1 Interpolation Particularly when dealing with large meshes, fast interpolation tools can be helpful. These can be used to interpolate initial flow guesses to finer meshes. Also, for very large simulations it can sometimes be helpful to thin meshes out for qualitative post-processing operations. Interpolators, like the k-dimensional (k-d) tree search procedures (based on space partitioning) can be remarkably efficient for interpolating between meshes. Octree based searches are also relatively efficient. CFD codes can internally make use of multigrid interpolation operators to automatically interpolate a coarse grid solution to finer grids. However, often for RANS, little benefit is observed from this. For eddy-resolving simulations, substantial computational savings can be observed. For such simulations, evolving a statistically stationary flow solution can take a considerable time. Also, the small scales, that are not resolved by the grid, naturally have fast response times. These aspects could be why such interpolations from coarse to fine grids are more useful for eddy-resolving simulations. However, for compressible flow solutions, the interpolation can give rise to strong acoustic wave components. These in turn can energize upstream shear layers. The settling down process can then take a long time. Indeed, it can then be quicker to start a fresh solution.

7.6.3 *Nearest Wall Distance*

The most popular aerospace turbulence models need d. For problems with stationary geometry this would be regarded as a post-processing exercise. However, the evaluation of d can take place outside or within the flow solver. For time varying geometry, it would need to be within the solver along with some degree of re-meshing capability. The wall distance evaluation is generally carried out using search procedures. However, such approaches can be slow and difficult to parallelize. Typically their cost scaling is $O(N_v\, N_s)$ operations where N_s and N_v correspond to the number of surface and internal node points. Wigton (1998) and Boger (2001) present more efficient search procedures needing $O\left(N_v\sqrt{N_s}\right)$ and $O\left(N_v \log\left(N_s\right)\right)$ operations, respectively. The cost of searches can result in codes using the following dangerous approximations (see Spalart 2000):

1) Computing distances down gridlines and not allowing for grid non-orthogonality;

2) Computing d as the distance between a field point and the nearest wall grid point and
3) In multiblock solutions, ascertaining d on a purely block-wise basis, ignoring the possibility that the nearest wall distance might be associated with another block.

The latter can create large inaccuracies and also non-smooth, unhelpful to convergence d distributions. In relation to (3), for overset grids the situation can arise where the same point in space has different equations.

7.6.3.1 Poisson Equation The wall distance can be evaluated from differential equations. For example, Spalding (see Tucker 2003) proposed the use of the Poisson equation

$$\nabla^2\phi = -1 \tag{7.5}$$

The right-hand side value is an arbitrary source. The solution variable ϕ is set to zero at solid boundaries and its normal derivative of ϕ zeroed at inflow–outflow boundaries. The previous Poisson equation gives a ϕ distribution such that close to walls its gradient is high. Away from walls it decreases. An auxiliary equation is used to relate ϕ gradients to d. The equation is analytically derived. For turbulence modelling purposes, accurate wall distances are only needed close to walls. In this region, non-wall-normal gradients of ϕ are small. Hence, it is possible to ignore the wall parallel derivatives in Equation (7.5), reducing it to the one-dimensional equation

$$\frac{\partial\phi^2}{\partial n^2} = -1 \tag{7.6}$$

where $n = x$, y or z, depending on which surface is extensive. Hence, the previous can be re-expressed as

$$\frac{\partial}{\partial n}\left(\frac{\partial\phi}{\partial n}\right) = -1 \quad \text{or} \quad d\left(\frac{\partial\phi}{\partial n}\right) = -1.dn \tag{7.7}$$

and integrated to give

$$\frac{\partial\phi}{\partial n} = -n + C_o \tag{7.8}$$

where

$$C_o = \frac{\partial\phi}{\partial n} + n \tag{7.9}$$

Equation (7.9) can be further integrated as

$$\int \partial\phi = \int n\partial n + C_o \int dn \tag{7.10}$$

to give

$$\phi = \frac{-n}{2} + C_o n + C_1 \tag{7.11}$$

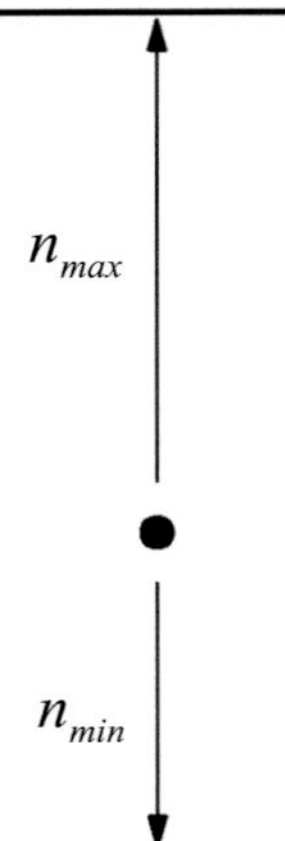

Figure 7.17. Near and far wall distances provided by the Poisson equation.

However, at $n = 0$, $\phi = 0$, hence $C_1 = 0$, and so

$$\phi = \frac{-n}{2} + C_o n \tag{7.12}$$

From Equations (7.9) and (7.12), the following can be written

$$\begin{array}{ccccc} \frac{1}{2}n^2 & + & \frac{\partial\phi}{\partial n}n & - & \phi = 0 \\ \uparrow & & \uparrow & & \uparrow \\ a & & b & & c \end{array} \tag{7.13}$$

which can be solved as

$$n = \frac{-b \pm \sqrt{b^2 - 4ac}}{2a} \tag{7.14}$$

where $a = 1/2$, $b = \partial\phi/\partial n$, $c = -\phi$, or alternatively, where n is now the wall distance, we can write

$$n = -\frac{\partial\phi}{\partial n} \pm \sqrt{\left(\frac{\partial\phi}{\partial n}\right)^2 + 2\phi} \tag{7.15}$$

where in three dimensions

$$\frac{\partial\phi}{\partial n} = \sqrt{\left(\frac{\partial\phi}{\partial x}\right)^2 + \left(\frac{\partial\phi}{\partial y}\right)^2 + \left(\frac{\partial\phi}{\partial z}\right)^2} \tag{7.16}$$

Expression (7.15) gives two wall distances. These are near and far, as shown in Figure 7.17.

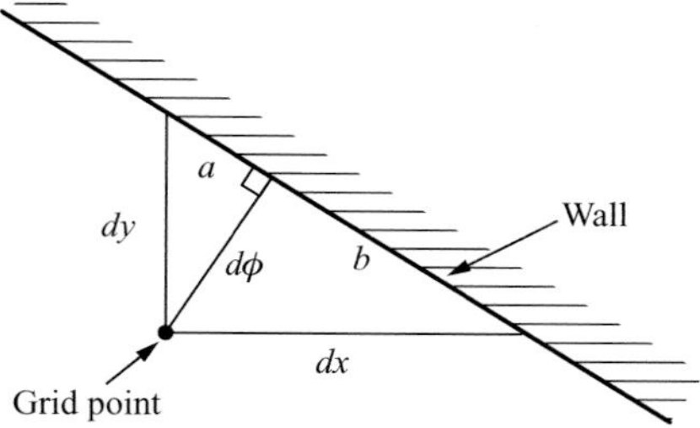

Figure 7.18. Schematic for simple geometric derivation of the eikonal equation.

Equations (7.15) and (7.16), combined with $\tilde{d} = n$, the tilde being used to identify that an approximate function to wall distance is being used:

$$\tilde{d} = \pm\sqrt{\sum_{j=1,3}\left(\frac{\partial\phi}{\partial x_j}\right)^2} + \sqrt{\sum_{j=1,3}\left(\frac{\partial\phi}{\partial x_j}\right)^2 + 2\phi} \tag{7.17}$$

Notably, the two roots of Equation (7.17) are useful. Their sum gives the local system extent and hence a rough estimate to the largest scale of a turbulent eddy that could be contained in the system. This information is potentially useful for incomplete turbulence models such as the mixing length model (see Chapter 5). This is because it crudely allows the mixing length to be plausibly capped to some rational level.

7.6.3.2 Eikonal Equation The eikonal equation gives an exact equation for wall distance. It can very simply be derived, in two dimensions, from the theorem of Pythagoras with reference to Figure 7.18. Using the nomenclature in this figure, the following can be written:

$$(d\phi)^2 = (dy)^2 - a^2 \tag{7.18}$$

$$(d\phi)^2 = (dx)^2 - b^2 \tag{7.19}$$

$$(a + b)^2 = dx^2 + dy^2 \tag{7.20}$$

From Equations (7.18) and (7.19), we have

$$a = \sqrt{(dy)^2 - (d\phi)^2} \tag{7.21}$$

$$b = \sqrt{(dx)^2 - (d\phi)^2} \tag{7.22}$$

Substitution of Equations (7.21) and (7.22) into Equation (7.20), to eliminate a and b, and rearranging, gives

$$\left(\frac{d\phi}{dx}\right)^2 + \left(\frac{d\phi}{dy}\right)^2 = 1 \tag{7.23}$$

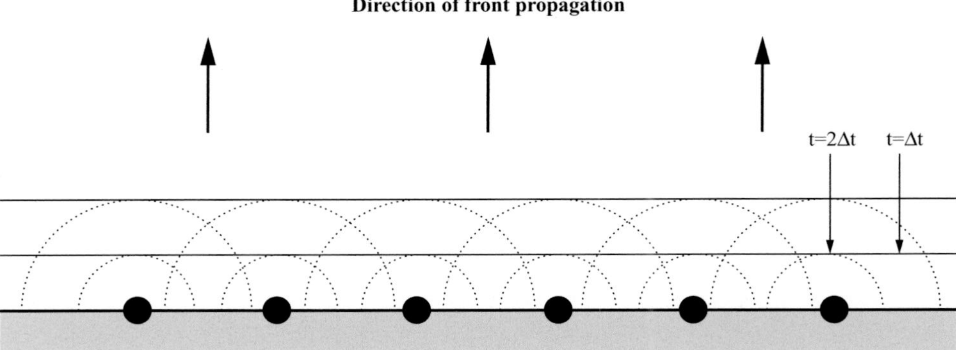

Figure 7.19. Front propagation implied in the eikonal equation.

or more generally, and hence extendable to three dimensions and other coordinate systems,

$$|\nabla \phi| = 1 \tag{7.24}$$

The eikonal equation is in fact an equation for the first arrival time of a high frequency wave front – see Figure 7.19. However, the right-hand side of unity implies a front propagation of unity and hence the first arrival time is equal to the wall distance where $d = \phi$. A viscosity term (where $\Gamma \to 0$, giving viscosity solutions) needs to be added, as follows, to allow the hyperbolic equations numerical solution

$$|\nabla \phi| = 1 + \Gamma \nabla^2 \phi \tag{7.25}$$

The diffusion coefficent can be a function of the wall distance itself with a finite constant, ε, that is,

$$\Gamma = \varepsilon \phi \tag{7.26}$$

Then a Hamilton-Jacobi (HJ) equation is gained. This has similar wall distance traits to the Poisson equation and hence $\phi = \tilde{d}$ (i.e., the exact wall distance is no longer solved for but some function of it). These traits are: in convex geometry zones the wall distance is overestimated and in concave undersestimated. However, as indicated by Launder et al. (2006), these traits are largely desirable and consistent with turbulence modelling physics.

7.6.3.3 Integral Wall Distance Equations Integral equations for wall distance can be formulated. They are based on elemental solid angles $d\omega$. The equation of Launder et al. (2006) is given as

$$\frac{1}{\tilde{d}} = \frac{1}{\pi} \int \frac{d\omega}{d'} \tag{7.27}$$

where d' is a continuous wall distance with no restrictions on surface orthogonality. The angles essentially, as indicated in Figure 7.20, physically represent how much solid surface can be seen by a fluid particle. For example, for a fluid particle that

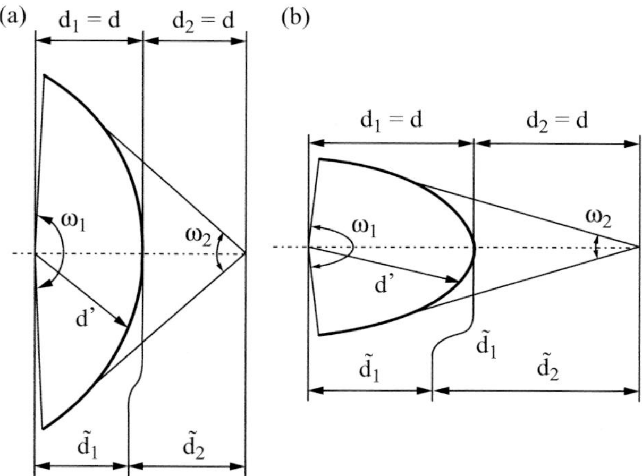

Figure 7.20. Schematic showing nature of solid angle based wall distance equation of Launder and colleague: (a) mildly convex/concave geometry and (b) more heavily convex/concave geometry (from Tucker and Liu 2007 – published with the kind permission of Elsevier).

has passed an airfoil trailing edge, not much solid surface is in view. This equates to the integral equation exaggerating the nearest wall distance, implying the surface will impose little impact on a particle in terms of turbulence damping. Although the integral equations can represent turbulence physics well, they are numerically challenging to solve. It should be pointed out that solid angles are a three-dimensional concept. However, to make Figure 7.20 easier to visualize, usual angles have been considered. An alternative integral wall distance equation to that of Launder et al. (2006) is Spalart's (1992) POEM equation. The latter offers greater wall distance control to the user.

7.6.3.4 Inverse Eikonal Equation For $\phi = 1/d$, the eikonal equation, following (Fares & Schroder 2002), can be re-expressed as

$$\left(\frac{\partial \phi}{\partial x}\right)^2 + \left(\frac{\partial \phi}{\partial y}\right)^2 + \left(\frac{\partial \phi}{\partial z}\right)^2 = \phi^4 \tag{7.28}$$

Since, from the product rule of differentiation, the following is true

$$\frac{\partial}{\partial x}\left(\phi \frac{\partial \phi}{\partial x}\right) = \phi \frac{\partial^2 \phi}{\partial x^2} + \left(\frac{\partial \phi}{\partial x}\right)^2 \tag{7.29}$$

Equation (7.28) can also be re-expressed exactly as

$$\frac{\partial}{\partial x}\left(\phi \frac{\partial \phi}{\partial x}\right) + \frac{\partial}{\partial y}\left(\phi \frac{\partial \phi}{\partial y}\right) + \frac{\partial}{\partial z}\left(\phi \frac{\partial \phi}{\partial z}\right) - \phi\left(\frac{\partial^2 \phi}{\partial x^2} + \frac{\partial^2 \phi}{\partial y^2} + \frac{\partial^2 \phi}{\partial z^2}\right) - \phi^4 = 0 \tag{7.30}$$

Fares and Schroder then add the elliptic term, $\Gamma \nabla^2 \phi$, to Equation (7.30). This then gives the traits of the HJ and Poisson equations, described previously, around convex and concave features. Hence, Equation (7.30) becomes

$$\frac{\partial}{\partial x}\left(\phi\frac{\partial\phi}{\partial x}\right)+\frac{\partial}{\partial y}\left(\phi\frac{\partial\phi}{\partial y}\right)+\frac{\partial}{\partial z}\left(\phi\frac{\partial\phi}{\partial z}\right)-(\Gamma-1)\,\phi\left(\frac{\partial^2\phi}{\partial x^2}+\frac{\partial^2\phi}{\partial y^2}+\frac{\partial^2\phi}{\partial z^2}\right)-\gamma\phi^4=0 \tag{7.31}$$

Note, also, the introduction of the term γ. This is to compensate for the additional Laplacian. The γ term ensures an exact wall distance computation for a plane. A plane in three dimensions can be defined by the angles α and β. Then the following exact equation can be written

$$\phi=\frac{1}{x\cos(\alpha)\cos(\beta)+y\cos(\alpha)\sin(\beta)+z\sin(\beta)+d_0} \tag{7.32}$$

where d_0 is a wall distance offset. This is needed in the boundary condition to ensure that ϕ is finite at solid walls. With Equation (7.31), Equation (7.32) implies

$$\gamma=(1+2\Gamma) \tag{7.33}$$

The linear relationship that ensures for a plane the wall distance computation is exact, gives an error for a sphere that is $O(\Gamma/(r^2+\sqrt{r}(d_0+r_0)^3))$ where r is radius and r_0 the radius of the sphere. The noted error traits are generally compatible with turbulence modelling physics. For a hybridization of the Poisson and eikonal wall distance approaches, to provide greater numerical robustness – see Tucker (2011c).

7.6.3.5 Solution Speeds The advantage of the Poisson equation is that it can be solved relatively easily and robustly. Rapid convergence can be gained with specialized simultaneous equations solution approaches such as multigrid convergence acceleration (see Chapter 3). The eikonal equation can be solved efficiently using specialized fast marching (see Sethian 1999) and sweeping (Zhao 2005) techniques. The former is relatively fast, having a complexity $O(N_v\log(N_v))$, and the latter with Gauss-Siedel iteration is $O(N_v)$. Notably, the eikonal equation can be written in an analogues form to a convective transport equation. Hence, with care, it can be solved using more standard CFD solution techniques (see Tucker et al. 2005). Differential wall distance function equations can be especially helpful for moving grid problems. However, then the procedure needs to be incorporated into the main CFD solver and no longer becomes a pre-processing activity beyond gaining an initial distance field. Notably, sometimes to account for roughness, a wall distance offset is used (see Aupoix & Spalart 2003 and Chapter 5). Hence, the condition $d=0$ at walls is no longer used. Instead, the wall distance is set to a finite wall value. This means that the turbulence model destruction terms are less active, giving rise to more turbulence. In this way (with other measures), the impact of roughness is modelled. Incorporating such wall-offset approaches, for complex geometries with local roughness patches is complicated when using search procedures. However, differential equations such as the Hamilton-Jacobi can naturally implement such approaches. The wall distance offset is easily set through Dirichlet boundary conditions. Also, the Laplacian, in the Hamilton-Jacobi equation, ensures that the computed distance field is smooth with

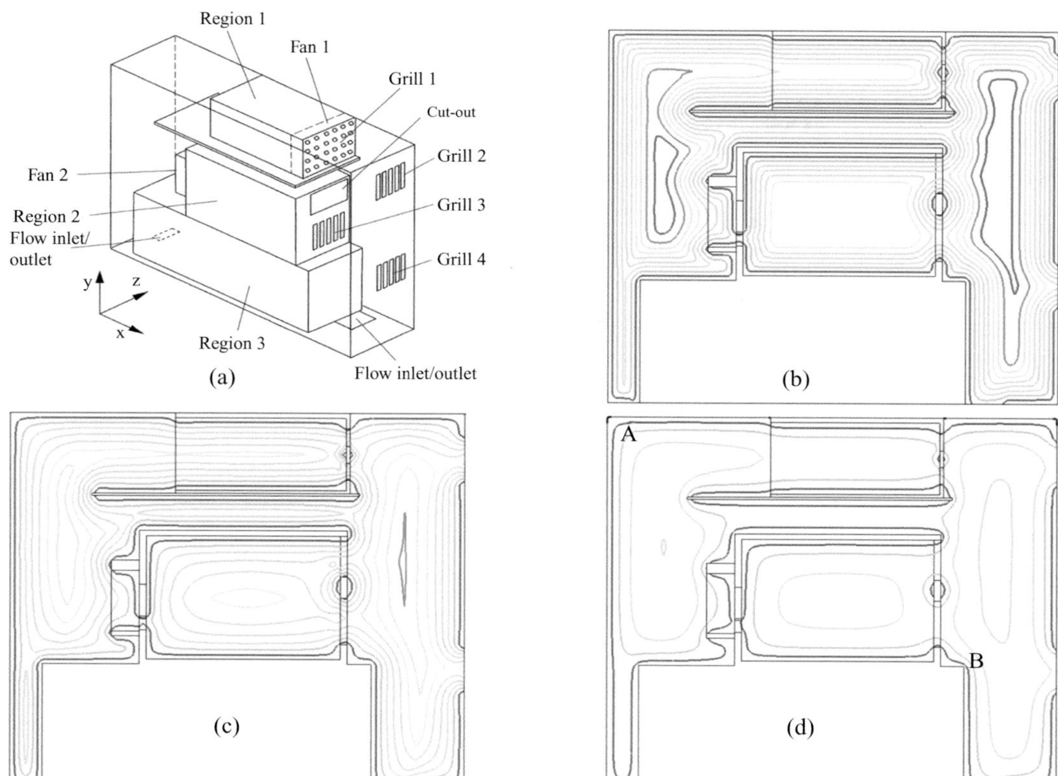

Figure 7.21. Wall distance function contours for a complex electronic geometry: (a) schematic of geometry, (b) eikonal equation solution and (c) Poisson equation solution and Hamilton-Jacobi equation solution.

no sudden jumps at the interfaces between the smooth and rough surface segments (see Oriji & Tucker 2013).

7.6.3.6 Examples of Physical Traits Figure 7.21 gives examples of wall distance function contours from some of the previous equations for a complex electronic geometry. Frame (a) gives a schematic of the geometry. Frame (b) gives the eikonal equation solution, and Frame (c) the Poisson equation. The final Frame (d) gives the Hamilton-Jacobi equation solution with a relatively large diffusion coefficient. The large diffusion coefficient means that in the concave geometry zone (A), the wall distance is underestimated and in the convex zone (B) it is overestimated. As noted previously, these traits are consistent with turbulence modelling physics. For example, for a flow into the page, in zone (A) fluid particles would experience more damping due to the presence of two walls in close proximity. Hence, the underestimated wall distance would give the increased damping arising from this. Conversely, in zone (B), the fluid particles would see less solid surface.

Table 7.2 summarizes the typical advantages and disadvantages of the wall distance evaluation approaches.

Table 7.2. *Advantages and disadvantages of different wall distance estimation*

Method	Advantages	Disadvantages
Search	Resilient to grid quality and no solution robustness issues. Largely accuracy is grid independent.	For massive problems can be slow and hence approximations made. No control of distance field.
Integral methods	Can control distance field.	Hard to solve integral equations for high aspect ratio cells.
Eikonal	Can be solved with efficient methods such as fast sweeping and marching. Can naturally be solved in parallel.	Efficient solution processes hard to implement on unstructured meshes. Iterative convergence and accuracy will depend on grid quality.
Poisson	Simple equation to solve and amenable to solution using the meshless boundary element method. Can naturally be solved in parallel. Easy to implement in a complex industrial solver.	No control of wall distance field to improve modelling of turbulence physics. Makes aggressive wall distance alternations around sharp convex and concave features. Iterative convergence and accuracy will depend on grid quality.
HJ/Inverse eikonal	Good control of wall distance traits. Can naturally be solved in parallel. Relatively easy to implement in a complex industrial solver.	Iterative convergence and accuracy will depend on grid quality. Will need advanced numerical techniques for efficient solution.

7.6.4 *Other Field Control Data*

Also, for hybrid RANS-LES with hard interfaces, the extent of the RANS zone will need to be defined, in some way, to help define the modelled turbulence length scale field. The use of level-sets for this, and other purposes, in jet noise simulations, is shown in Figure 7.22. For example, Frames (a–d) give variable fields that define the level of blending between dissipative upwind and centred convective schemes, the level of blending with a steady target field, the reduction in numerical upwind scheme order towards boundaries (to dissipative waves) and the defined turbulence length scale field. The Frame (b) field would be used in Equation (4.276). As noted previously, in the jet noise work of Andersson et al. (2005), a precursor RANS simulation is used to define entrainment boundaries in a smaller LES domain. Also, for noise problems, integral surfaces can need to be defined to extrapolate near field sound information to give far field sound data. Such aspects were discussed in Chapter 6. Figure 7.23, shows a partly level-set generated surface for a jet wing configuration. Notably, for some approaches it can be necessary to define multiple integral

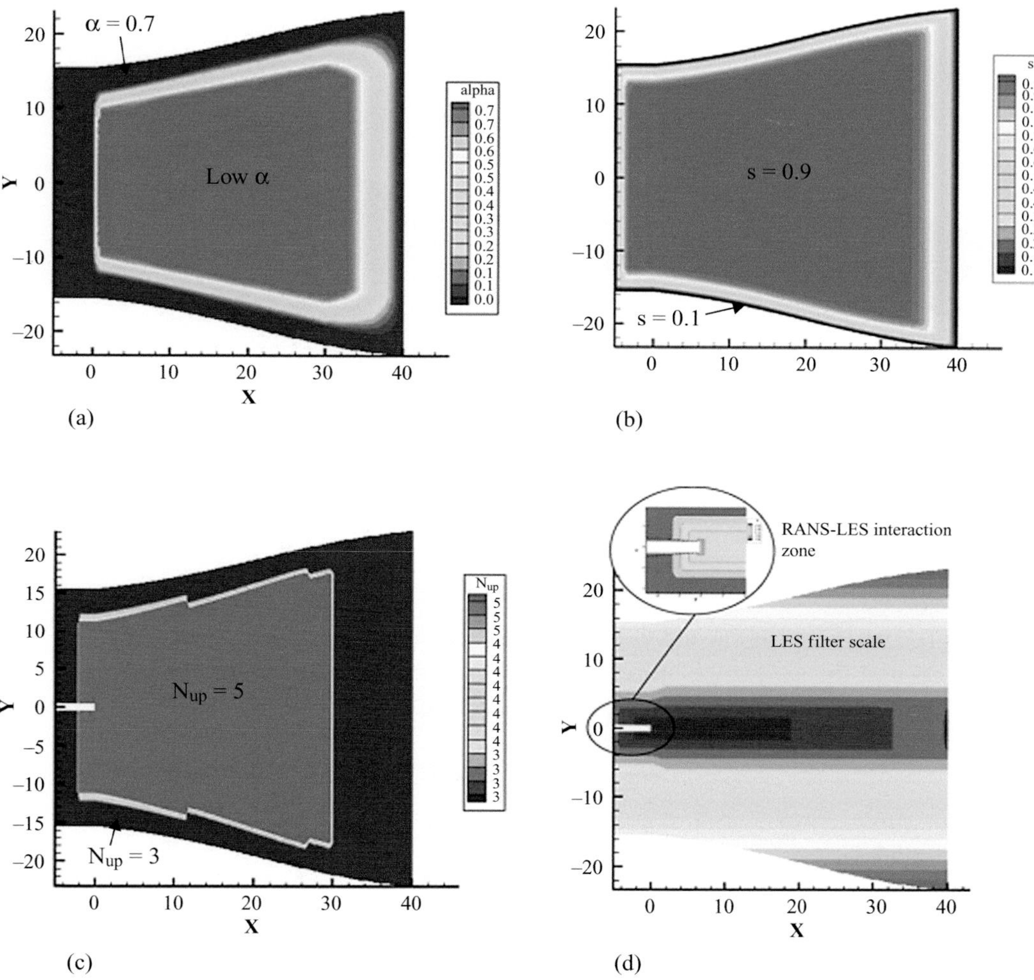

Figure 7.22. Level-set based weighting function distributions used for jet acoustic simulations: (a) upwind/centred scheme blending function, (b) blending function to steady target field, (c) distribution to reduce order of upwinding at boundaries and (d) modelled turbulence scales (from Tucker 2013).

surfaces and then integrate the final solution, once the flow is known, to ensure that the integral surface in the correct location is used. This aspect is discussed later.

For large-scale coupled simulations and design optimization problems it can be necessary to use scripting to ensure that different codes are connected correctly. This scripting will need to call separate interfacing codes to convert file formats to those suitable for the differing codes. Hence, all these elements need to be setup in some way.

For inverse design, target distributions of aerodynamic loadings and also global design constraints or targets must be specified (see Chapter 6). Hence, pre-processing can be complex and involve the specification of a wide range of parameters. Ideally,

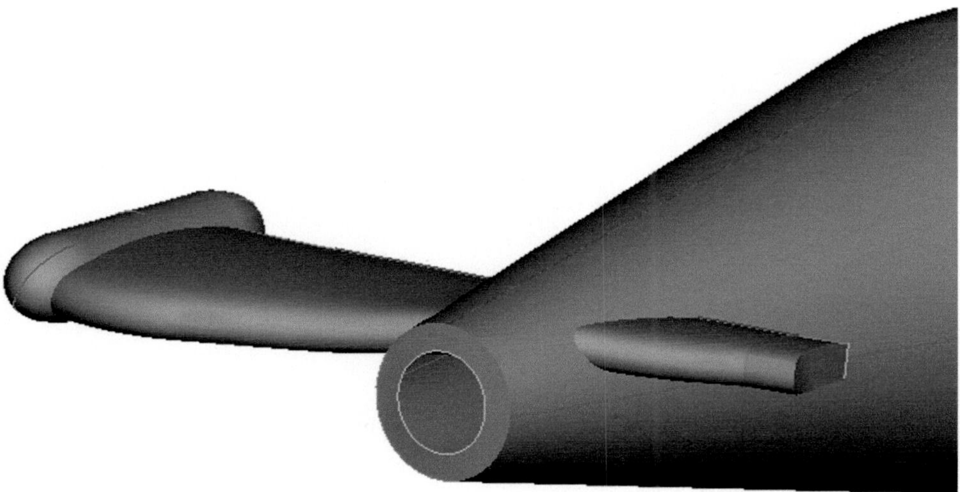

Figure 7.23. Partly level-set generated, integral acoustic surface for a jet wing configuration.

the process should be as automated as possible, thus ensuring consistency of the process and the avoidance of user error. If the run is to be made just a few times, then some written protocol could be produced to ensure consistency of the setup process.

7.6.5 *Summary of Typical Parameters That Need To Be Defined when Setting up and Running a New Case*

Partly, in summary, the parameters/identifiers that will need to be specified when a case is being setup are:

Equations to be Solved Linearized, non-linear harmonic/disturbance, Euler, compressible or incompressible flow, real gas effects; equations for density, viscosity and other fluid properties; frame of reference, rotation, gravitational or body force terms; conjugate, isothermal or non-isothermal, including compressive or frictional heating

Choice of Solution Variables Gravitationally or rotationally reduced pressure; solve energy equation for temperature, enthalpy or constant stagnation enthalpy

Variable Initialization Initial flow field, wall distance computation, sponge zone variables such as increases in numerical dissipation, target variables (for sponge zones or say inverse design) and design constraints

Additional Physical Models Deterministic stresses, body forces, radiation, multiphase, combustion, particles and degree of coupling

Numerical Parameter Settings Spatial numerical discretization: order and type (centred, upwind); Temporal numerical discretization: order and nature of scheme; Simultaneous equation solver: type of multigrid, number of multigrid levels, type of multigrid cycle, number of Krylov vectors, sweep numbers, sweep directions, stability parameters such as under-relaxation and clipping/limiting parameters, convergence criteria, pseudo-time-steps, coupling parameters (e.g., frequency of data exchange, conservation options with coupling, time-steps in the differing domains and other zone coupling parameters)

Turbulence Select RANS, LES or hybrid RANS-LES model; define near wall treatment, palliative fixes, zonal turbulence model interface zones

Boundary Conditions Free stream, mass flow, pressure, weak differential conditions, thermal boundary conditions, RANS model variables or synthetic turbulence inflow specification (from database or on the fly – see also coupling), non-reflecting boundary conditions, and so forth

Specialist Interfaces Define free surface, solid fluid interfaces, mixing planes, sliding planes, frozen rotor assumption and so forth

Coupled Problems Specify nature of problem (aeroelasticity, conjugate, coupling different CFD codes) and codes to be coupled with; if one-way or two-way coupling, equation coupling method (loose or tightly coupled), if recycling in LES define type of coupling and if there are forcing plane targets (see Chapter 5)

Post-Processing Key post-processing monitoring points or integral surfaces, assembly of time averaged data (LES Reynolds stresses and in particular what range of correlations of fluctuating properties need to be stored to minimize data storage, etc.), data write out frequency, data write out of planes for animations and time series analysis

Design Optimization Specify parameters to be used in any design optimization including design constraints, aerodynamic targets, and types of search algorithms.

As can be seen, potentially there can be a lot of parameter fields to set. Hence the user interface needs to be designed with some care. Ideally, the input can be layered with sensible default values to enable the user to safely make a basic CFD run and then increase the complexity. Some codes allow more specialist activities to be defined through user-defined functions. Once the case has been setup it needs to be submitted. This aspect can also have some complexity.

Job Submission When running parallel simulations, the number of processors will need to be given. For hybrid parallel processing codes, which make use of both

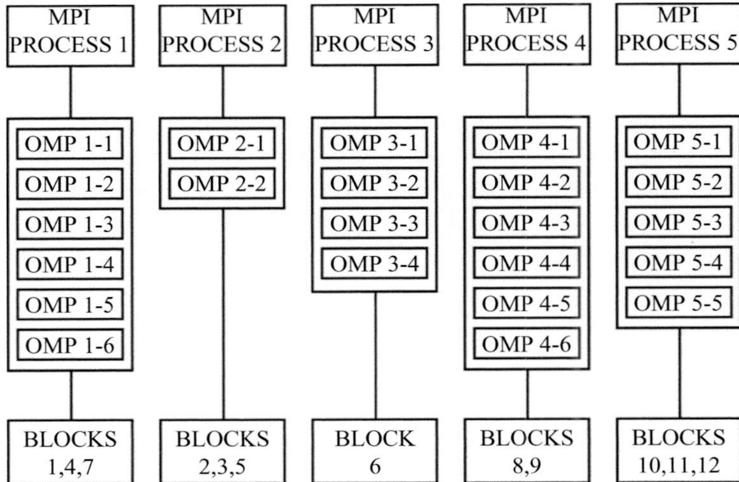

Figure 7.24. Mapping mesh zones to computer processors.

distributed and shared memory directives, an optimal specification is needed for mapping the grid zones to the computer architecture to ensure optimal performance. See Figure 7.24, which indicates how different grid blocks are mapped onto the computer.

Load balancing can be carried out manually, say, for example, on structured grids where grid blocks are subdivided or even merged. On unstructured meshes, load balancing can be automatically achieved using software such as ParMETIS (Parallel Graph Partitioning and Fill-reducing Matrix Ordering). For heterogeneous systems (grid-based computing that connects a range of computers on a wider network), mesh partitioning can be much more challenging, needing specialist approaches (see Mesri et al. 2005).

Scripting When making more practical CFD calculations, as part of the setup process, it can be necessary to loosely connect different pieces of software. Then some knowledge of programming languages that allow scripting is helpful. Such languages are extensive and include Python and TCL (Tool Command Language). Clearly, when connecting programs using scripting, an understanding of file formats is needed so that files can be read in and necessary data transferred to another program.

7.7 Solution File Formats

The simplest file format is plot3d, developed at NASA by Walatka et al. (1990). This format is for structured, multiblock grids. The format has the flexibility to blank some cells out so that they are rendered inactive. The grid and solution files are stored separately. Typically plot3d grid and solution file formats are given as follows:

```
Grid file format
read() nblocks ! Number of grid blocks
read() (ni(n), nj(n), nk(n), n = 1, nblocks)
    ! Number of nodes in each block
do n = 1, nblocks
read()
1 (((x(i,j,k,n), i = 1,ni(n)), j = 1,nj(n)), k = 1,nk(n)),
1 (((y(i,j,k,n), i = 1,ni(n)), j = 1,nj(n)), k = 1,nk(n)),
    ! Read block grid coordinates
1 (((z(i,j,k,n), i = 1,ni(n)), j = 1,nj(n)), k = 1,nk(n))
enddo
Solution file format
read() nblocks
read() (ni(n), nj(n), nk(n), n = 1, nblocks)
do n = 1, nblocks
read() mach, alpha, re, time
    ! General variables that define case
read()
1 ((((q(i,j,k,n,nv), i = 1,ni(n)), j = 1,nj(n)), k = 1,nk(n)),
    nv = 1,5)
! Read five key solution variables
Enddo
```

Generally, the expectation is that the ‘q’ variables are density, the product of the three velocity components with density and total energy per unit volume. With plot3d, a function file can also be defined. This can be for data sets of different dimensions (even one dimension) and also planar data. Basically, the function file contains a range of extra field variables that a user might wish to contour. Blanked out cells are stored in an ‘iblank’ array located at the end of the grid file. The plot3d format is a good choice of format if you write any of the codes associated with the problems in this text.

The AIAA recommend file format is CGNS (CFD General Notation System). This format was initialized in a joint project between Boeing and NASA in 1994. The format is platform independent. The use of a consistent format is clearly useful when handling data sets and looking back at older data that has been retrieved for analysis. However, as noted by Dawes (2007), different interpretations of this ‘standard’ format are possible. The CGNS format is a compact binary and compatible with most standard computer languages. It is also highly flexible. Hence, it works with structured, unstructured and hybrid grids. It is also applicable to the other extensive ranges of scenarios that can occur in CFD, including even the use of reduced order equations such as the potential flow. Notably, CGNS files are able to describe themselves. Therefore, they do not need any additional information to describe their structure or content. At a lower level, CGNS can make use of either

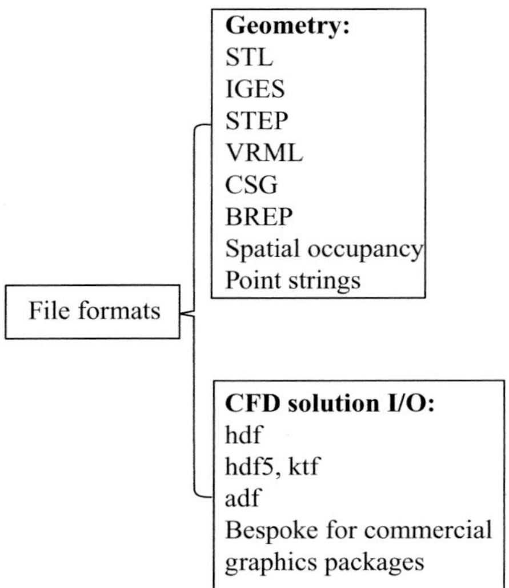

Figure 7.25. Potential file formats for CFD.

the adf of hdf5 file formats. The latter was developed by the National Center for Supercomputing Applications. It has a hierarchical data format, hence the name hdf. Notably, the latest variant of hdf, hdf5, allows parallel input and output of data. The hdf5 format can interact with a range of commercial software ktf is another parallel file format. As noted previously, parallel input and output is especially critical when performing massive simulations. For large meshes, with serial input and output the read in and write out can take a highly substantial fraction of the total run time. A crude parallel strategy can be to write the solution out to multiple files. This has the advantage that the final data set sizes are manageable but is rather messy and untenable for simulations of massive scale.

Figure 7.25 summarizes potential CFD file formats for both geometry and solution input/output.

7.8 Assessing Convergence

7.8.1 *Iterative Convergence*

A popular way of judging convergence is to look at the changes in variables with iterations or pseudo-time-step. However, care is needed. If heavy under-relaxation is used, the changes will be small even though convergence is poor. Similarly for small physical or pseudo-time-steps changes will again be small even though convergence is far off. Also, if the matrix system being solved has a large spectral radius/eigenvalues (λ), convergence can be slow. For example, as shown by Ferziger & Perić (1996) the number of iterations, N, scales as

$$N \propto \frac{\lambda_{\max}}{\ln \lambda_{\max}} \tag{7.34}$$

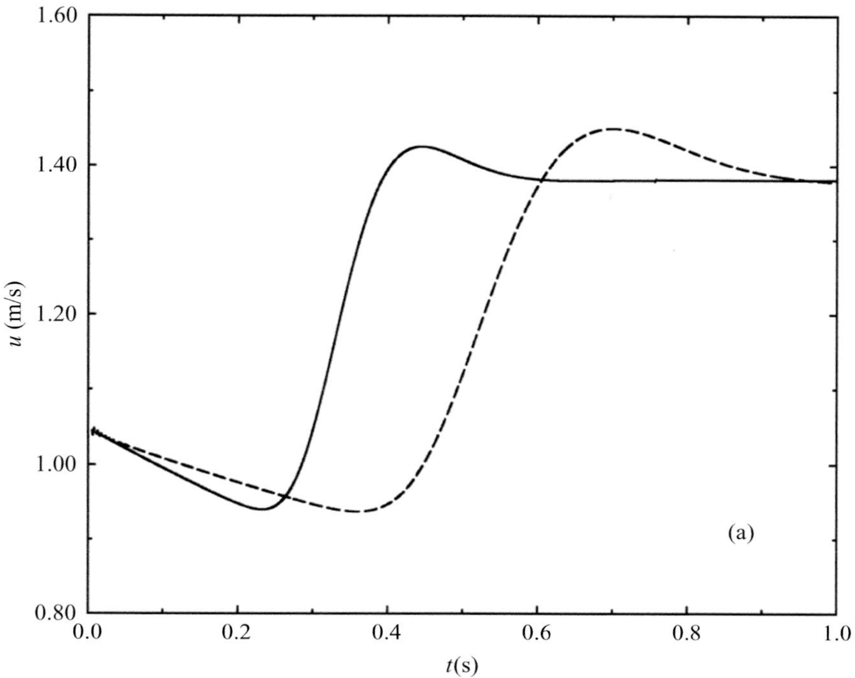

Figure 7.26. Transient flow, velocity-time traces illustrating the effects of poor convergence (from Tucker 2001 – published with the kind permission of Springer).

Hence, if $\lambda_{max} \to 1$, convergence will be slow. Following Ferziger & Perić (1996), for a linear system of equations, which is a reasonable assumption near convergence, the convergence error E can be approximated as

$$E \approx \frac{\|\Delta\phi\|}{\lambda_{max} - 1} = \frac{\sqrt{\sum (\Delta\phi)^2}}{\lambda_{max} - 1} \tag{7.35}$$

where $\Delta\phi$ is the change in the solution variable between iterations. However, some iterative systems have complex eigenvalues. Then, the simple Equation (7.35) estimate becomes less reasonable. More complex mathematical analysis is then needed to relate convergence to changes in solution variables.

The normalized root mean square (*rms*) changes of the solution variables, expressed as follows, can be a reasonable initial tool for assessing convergence:

$$rms_\phi = \frac{\|\Delta\phi\|}{\|\phi\|} = \sqrt{\frac{\sum (\Delta\phi)^2}{\sum (\phi)^2}} \tag{7.36}$$

As noted previously, for unsteady flows with strong under-relaxation, the judgment of convergence using *rms* changes can be dangerous. The impact of poor convergence is illustrated in the Figure 7.26 velocity time traces for a transient low prediction. Convergence is based on the normalized *rms* change given above. The full line is the prediction assuming convergence for about $rms_{u,v} < 1 \times 10^{-5}$ and the dashed 1×10^{-4}. As can be seen from the velocity agreement at $t = 1$ second, the latter looser condition (which is generally accurate enough for steady state solutions) is insufficient. The error is mostly in phase.

Based on the limitations discussed here, the more reliable way to judge convergence, rather than looking for changes in solution variables, is to look at residual values, summed over the solution domain. Assuming we have a discretized matrix equation of the form

$$[A][\phi] = [S_\phi] \tag{7.37}$$

where ϕ is the variable being solved for and A and S_ϕ are matrices of equation coefficients and sources. The residual matrix $[R]$ of the previous Equation (7.37) can be defined as

$$[R] = [A][\phi] - [S_\phi] \tag{7.38}$$

At convergence, ideally the components of R and hence the sum of them would be zero. However, with discrete computing this will not be so and the question arises as to what is a small enough residual. A good approximate guide is that, say, the residual should drop by three to four orders of magnitude. Each order of magnitude drop corresponds to roughly 10%, 1% and 0.1% and so forth convergence errors. However, this all depends rather on the quality of the initial guess. These ideas assume that the initial solution guess $\phi = 0$ corresponds to a 100% error. If the initial guess is good, then a three orders of magnitude residual drop could be hard to achieve. Notably, a poorly formulated matrix system can also give rise to low residuals when convergence is far off. If the under-relaxation forms an explicit part of the discretized equation system, this will impact on the residual magnitude as will the time-step.

Although it should be possible for the solver to drive the residual to machine round off, this is wasteful and sometimes not possible. For example, there might be a minor level of flow unsteadiness that prevents full convergence. Hence, ideally, if residuals are dimensional it is best to sensibly normalize them. For example, the mass, momentum and energy equation residuals should be normalized by sensible scales such as the system mass flow, momentum and enthalpy flows. Care is needed that the form of the discretized equations is known, so that they can be rendered truly dimensionless and hence percentage errors evaluated. Also, for complex systems, care is required so that the velocity and other scales in the non-dimensionalization correctly reflect the scales that are of interest for the analysis. If the length of a low-speed recirculation is of interest, a large velocity scale external to this might not fully reflect the ideal velocity scale to use in the non-dimensionalization. Figure 7.27 gives a potential example where care is needed with the residual normalization. The flow in Zone (2) might be the one of key interest. However, if this has relatively low momentum and enthalpy flows relative to Zone (1), which drives the Zone (2) flow, care is required. Convergence would look deceptively good if based on the mass, momentum and enthalpy flows through Zone (1).

Contouring the residual field can be useful for understanding locations where a solution is proving difficult to converge.

It can also be useful to perform integrations of mass, momentum and energy around system boundaries. (Note that for a conservative scheme the internal fluxes

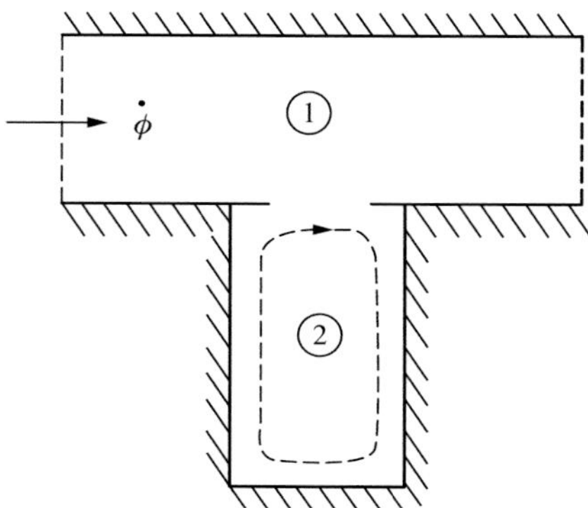

Figure 7.27. System where care is needed in residual normalization.

should cancel and hence this process has strong connections to testing the global residual.) For rotating systems it is also a good idea to check that angular momentum is conserved. However, again some thought is required. Some forms of equations and their discretizations are chosen to ensure momentum conservation. Others (see Jameson 2008) are designed to conserve kinetic energy or total energy. For a compressible flow it is not possible to conserve both momentum and kinetic energy. However, for eddy-resolving simulations the latter can be of more importance; but if just kinetic energy is conserved, a global integration of momentum will reflect some error. It must be considered carefully if this is acceptable.

Notably, for unsteady simulations, convergence becomes more complex to assess. When using explicit time integration, there is no iterative convergence to consider. With implicit temporal integration, the convergence needs to be assessed at each time-step. Care needs to be taken relating convergence to *rms* values, as noted. Residuals will also vary strongly with time-step size. Confusingly, for a pressure correction solver in its natural form, the momentum and pressure correction equation residuals will have opposite growth trends with time-step size. The time-step enters the discretized equations in different ways (see Chapter 4). Also, as a solution evolves, the residual level will generally naturally become lower. For a cyclic flow the residual pattern will ultimately evolve to a cyclic state. However, clearly, since the residual level will partly depend on the maturity of the evolved flow, it becomes harder to judge what an acceptable level is. As indicated previously, the convergence error at each time-step, although relatively small, can have a large additive impact over long time periods. Hence, as noted previously, greater care is needed with unsteady flows when judging convergence.

7.8.2 *Grid/Time-Step Convergence*

7.8.2.1 Taylor Series Based For any CFD solution it is of course necessary to attempt to assess if the grid used is sufficiently fine to make the impact of any numerical disctretization error insignificant relative to the engineering parameter of interest. The ideal procedure for this is to generate solutions on fine and coarse grids, the finer having double the number of grid nodes in all coordinate directions.

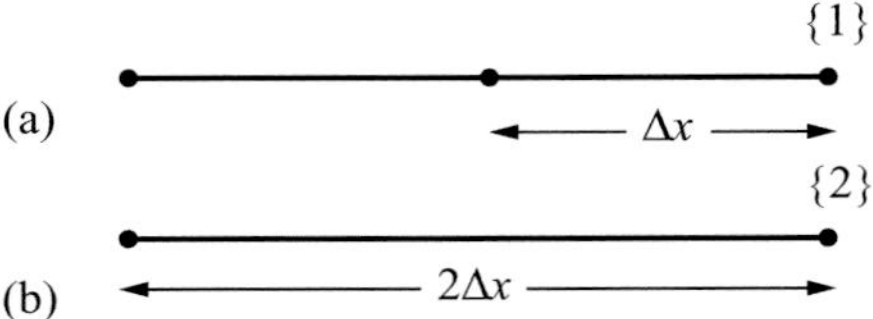

Figure 7.28. Estimating solution error based on two grids.

Unfortunately, for large simulations, this is not always possible. Then some form of less rigorous assessment needs to be performed. This could involve doubling the number of grid nodes in independent coordinate directions, or even coarsening the grid. Alternatively, altering the order of the discretization could be helpful whilst keeping the grid count the same. Essentially the same test will need to be made with time-step, which is akin to defining a grid in time in a parabolic sense. Generally, if solutions are available on two grids, it will be examined how the engineering parameter of interest varies on these two grids. If the variation is relatively small, in relation to the desired accuracy, then the solution is considered to be sensibly grid independent. It is possible to be more rigorous than this (even though such rigorous approaches are seldom used) and formally define the error on the finer grid. For example, Taylor series based analysis can be used. For an extrapolation from a grid point, W, over an interval Δx, to a grid point E using Taylor's series, the following can be written:

$$\phi_E = \phi_W + \frac{\Delta x}{1!}\frac{d\phi}{dx}\bigg|_W + \frac{(\Delta x)^2}{2!}\frac{d^2\phi}{dx^2}\bigg|_W + HOT \tag{7.39}$$

From Equation (7.39), we can express the error, ε, in the approximation as equal to the neglected leading truncation term

$$\varepsilon = \frac{(\Delta x)^n}{n!}\frac{d^n\phi}{dx^n} \approx \phi - \phi_E \tag{7.40}$$

where the far right-hand side ϕ and ϕ_E terms are the exact answer and numerical solution. Now we consider solving a problem on grids ofΔx and $2\Delta x$ as shown in Figure 7.28 when using, say, upwinding.

Using Equation (7.40) we can write the errors for the two solutions as

$$\varepsilon_1 = \phi - \phi_1 = \frac{2(\Delta x)^n}{n!}\frac{d^n\phi}{dx^n} \tag{7.41}$$

$$\varepsilon_2 = \phi - \phi_2 = \frac{(2\Delta x)^n}{n!}\frac{d^n\phi}{dx^n} \tag{7.42}$$

where ϕ_1 and ϕ_2 are solutions (at point E) for the fine and coarse grids, respectively. Combining Equations (7.41) and (7.42) gives the error for the *finer* grid (ε_2) as

$$\varepsilon = \frac{2(\Delta x)^n}{n!}\frac{d^n\phi}{dx^n} = \frac{\phi_2 - \phi_1}{2^{n-1} - 1} \tag{7.43}$$

where the 2 subscript has been deliberately dropped. Therefore, ε from Equation (7.43) can be used to estimate the solution error. Note that in Equation (7.43), *n-1* is, in fact, the order of the numerical scheme (there is a division through by a space-step when making the operator apply to a differential, reducing the order by a factor of one). Different, closely related finite difference error equations, of similar form to (7.43), can be derived for evaluating errors for different orders or natures of finite difference schemes on the same grid. Hence, an assessment of error can be made by simply keeping the order of the scheme the same, but changing its nature (say, contrast forwards and backwards differences). When the refinement ratio between grids, λ, is not a ratio of 2, as assumed in Equation (7.43) the following more general expression can be written

$$\varepsilon = \frac{\phi_2 - \phi_1}{\lambda^{n-1} - 1} \tag{7.44}$$

or the fractional error expressed as

$$\tilde{\varepsilon} = \frac{\hat{\varepsilon}}{\lambda^{n-1} - 1} \tag{7.45}$$

where

$$\hat{\varepsilon} = \frac{\phi_2 - \phi_1}{\phi_1} \tag{7.46}$$

Roache (1994) defines the grid convergence index (GCI) on the fine grid

$$GCI_1 = C\frac{|\hat{\varepsilon}|}{\lambda^{n-1} - 1} \tag{7.47}$$

C is a safety factor where $C = 3$ for two grids. If performing a design optimization, it can be important to run the numerous simulations on the coarser grid. Hence, it is important to attempt to quantify the accuracy on this grid. For the coarse grid accuracy, the following can be written:

$$GCI_2 = C\frac{|\hat{\varepsilon}|\,\lambda}{\lambda^{n-1} - 1} \tag{7.48}$$

Sometimes, the grid is refined independently in different directions; then Roache suggests the following summation

$$GCI = GCI_x + GCI_y + GCI_z \tag{7.49}$$

where the subscripts indicate the different coordinate directions. As noted, the processes discussed are equally applicable to unsteady flows and then (7.49) becomes modified to

$$GCI = GCI_x + GCI_y + GCI_z + GCI_t \tag{7.50}$$

where GCI_t represents the temporal GCI component. For more complex grids, it can be hard to define λ, and then the following estimate can be used:

$$\lambda_{eff} = \left(\frac{N_1}{N_2}\right)^{1/N_d} \tag{7.51}$$

where N_1 and N_2 are the number of grid nodes on the fine and coarse grids and N_d the dimension of the domain.

If the order of the scheme is unknown, to estimate it, following Davis (1983), we then re-write Equation (7.40) as:

$$\phi = \phi_1 + \frac{(\Delta x)^n}{n!} \frac{d^n \phi}{dx^n} \tag{7.52}$$

Defining the following:

$$c = \frac{1}{n!} \frac{d^n \phi}{dx^n} \tag{7.53}$$

allows Equation (7.52) to be written as

$$\phi = \phi_1 + c(\Delta x)^n \tag{7.54}$$

where c is assumed independent of Δx. Then, following Davis (1983), n comes from solving Equation (7.54) with $\Delta x = \Delta x_1$, Δx_2 and Δx_3 to give

$$\frac{\phi_1 - \phi_2}{\phi_2 - \phi_3} = \frac{\Delta x_1^n - \Delta x_2^n}{\Delta x_2^n - \Delta x_3^n} \tag{7.55}$$

If $\Delta x_1 / \Delta x_2 = \Delta x_2 / \Delta x_3 = \lambda$, we can write

$$n = \ln \frac{\left[\frac{\phi_1 - \phi_2}{\phi_2 - \phi_3}\right]}{\ln \lambda} \tag{7.56}$$

In practice, the Taylor series approaches outlined here are seldom reliable. Equations (7.55) and (7.56) can give n far from the formal order of the scheme used. Also, n can vary greatly from grid point to grid point. Similarly, the error suggested by Equation (7.43) can be very unreliable. It is tempting to make use of the error suggested by Equation (7.43) to correct the fine resolution solution. However, this will generally be unreliable. The lack of reliability noted for assessing the order of a scheme and also the truncation error relates to the small Taylor series radius of convergence. For an example of the application of the above extrapolation approaches to engineering predictions, see Prakash & Lounsbury (1986). The procedure for exploring the error on two grids and using it to improve (extrapolate) the fine grid solution is called Richardson extrapolation, after Richardson (1911). For unsteady flows, it is also of course necessary to assess that the solution is independent of the time-step. The Richardson based expressions described can also be used in this context. Automatic time-step adaptation can be exploited during the solution. This seems a more practical option relative to automatically adapting meshes. The latter typically results in meshes that are way too large.

7.8.2.2 Adjoint Based As indicated in Chapter 3, adjoint analysis can also be used to assess zones where the grid (see Venditti & Darmofal 2002)) or time-step (Mani & Mavriplis 2007)) size needs to be improved. However, this approach can add substantial code complexity and for unsteady flows massive storage. This is because for unsteady flows, the solution needs to be stored through time to be backwards

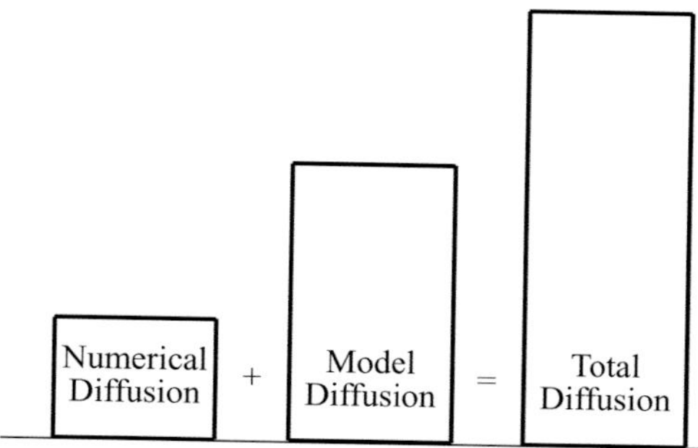

Figure 7.29. Excessive diffusion due to overactive numerical scheme.

integrated. Adjoint methods have the attraction that essentially the grid is checked and refined for a specific output functional like, for example, drag. Their rigorous mathematical basis makes them very revealing with regards to locations that need improved grid resolution, giving sometimes counter intuitive information. For example, errors will tend to convect with the flow. Hence, a poorly resolved upstream grid zone can give errors in downstream zones that are in fact highly resolved. The Taylor series based error analysis, described Section 7.8.2.1, is not well suited to analysing errors of this nature. The backwards propagation of information used in adjoint analysis ensures the zone of grid refinement is reliably detected.

7.8.2.3 Assessing Numerical Input In LES and LES Quality Another way to explore the impact of a numerical scheme is presented by Batten et al. (2011). In the original work, the approach is presented as a smart subgrid scale model. However, much of the method is centred on exploring the numerical impact of the key numerical diffusive gradients associated with shearing in the flow and contrasting these with any turbulence modelling. As noted, the context of the work of Batten and colleagues was LES. However, the approach could also be applied to RANS simulations and hence of course hybrid LES-RANS. For both LES and RANS, for a simple shear flow, the primary modelled viscous flux input can be expressed as

$$F_v = \mu_t \left(\frac{\partial u}{\partial y} \right) \tag{7.57}$$

where μ_t is the eddy viscosity. This can be evaluated through a range of means for both LES and RANS and these are outlined in Chapter 5. The critical area of interest is to ensure that the numerical viscosity is much smaller than the eddy viscosity. Otherwise, the impact of the numerical scheme is likely to be excessive. The situation in an LES context is shown in Figure 7.29.

As outlined in Chapter 4, the popular, compressible flow Roe scheme can be formulated such that the convective flux, F_c, at a control volume face can be expressed as a centred average and an additional smoothing term. These terms are given as

$$F_{c,CVF} = \frac{1}{2}\left(F_N + F_S\right) - \frac{\mathbf{T}\,|\Lambda|\,\mathbf{T}^{-1}}{2}\left(U_S - U_N\right) \tag{7.58}$$

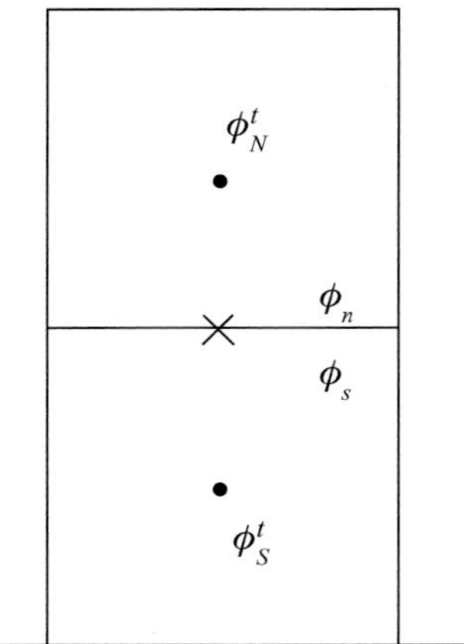

Figure 7.30. Control volumes and labelling relating to the approach of Batten et al. (2011) for estimating the numerical scheme error impact on turbulent flow simulations.

The nomenclature for the previous equation can be found in Figure 7.30. Note that the 't' superscripts are used to imply tangential components are used, such as U. The final right-hand side group of smoothing terms relate to the multi-component diffusive term, D, given as

$$D = \frac{\mathbf{T}\,|\Lambda|\,\mathbf{T}^{-1}}{2}\left(U_S - U_N\right) \tag{7.59}$$

For a general variable ϕ, and considering the primary component, in a simple shear, the previous can be expressed as

$$D\left(\phi\right) = \frac{\rho\,|\lambda|}{2}\left(\phi_S - \phi_N\right) = \frac{\rho\,|\lambda|\,\delta y}{2}\left(\frac{\partial \phi}{\partial y}\right) \tag{7.60}$$

where λ is a velocity scale (eigenvalue wave speeds) and the density is assumed constant. The eddy viscosity that goes into Equation (7.57), is on dimensional grounds, based on the product of length, velocity and density scales. Hence, a direct analogy with Equation (7.60) is

$$F_v = \mu_{\text{num}}\left(\frac{\partial \phi}{\partial y}\right) \tag{7.61}$$

where

$$\mu_{\text{num}} = \frac{\rho\,|\lambda|\,\delta y}{2} \tag{7.62}$$

or more generally, following Batten et al. (2011),

$$\mu_{\text{num}} = \frac{\rho\,|\lambda|\,\overline{V}}{2\,|A_i|} \tag{7.63}$$

The $\overline{V}$ is the average volume/area of two adjacent cells and A is a face area/length. Looking specifically at velocity, Equation (7.60) can be re-written as

$$D\left(u^t\right) = \frac{\rho\left|\lambda\right|}{2}\left(u_s^t - u_n^t\right) \tag{7.64}$$

where u^t indicates the tangential velocity component, giving the primary shear stress and thus connecting with Equation (7.57). The previous can be re-written as

$$D\left(u^t\right) = \frac{\rho\left|\lambda\right|\delta x^t}{2}\left[\frac{u_s^t - u_n^t}{u_S^t - u_N^t}\right]\left(\frac{\partial u^t}{\partial x^t}\right) \tag{7.65}$$

It then follows that

$$\mu_{\text{num}} = \frac{\rho\left|\lambda\right|\overline{V}}{2\left|A_i\right|}\left[\frac{u_s^t - u_n^t}{u_S^t - u_N^t}\right] = \frac{\rho\overline{V}D\left(u^t\right)}{\left|A_i\right|\left(u_S^t - u_N^t\right)} \tag{7.66}$$

Equation (7.66) can be written more generally, in terms of a scheme using a centred difference with an additional (upwind) smoother, F_U, as

$$\mu_{\text{num}} = \frac{V\left[\frac{1}{2}\left(F_N^t - F_S^t\right) - F_U^t\right]}{\left|A_i\right|\left(u_S^t - u_N^t\right)} \tag{7.67}$$

Following Batten et al. (2011), the velocity components and fluxes corresponding to maximum shear are defined as

$$\begin{aligned} F_t &= F_i t_i / \sqrt{t_k t_k} \\ u_t &= u_i t_i / \sqrt{t_k t_k} \\ t_t &= u_i - u_k n_k n_i \end{aligned} \tag{7.68}$$

where the n terms are the face normal vectors. The t_i vector is the component of the interpolated face-velocity vector, which is perpendicular to the face normal vector, n_i. Note that t_i is only ever used after being normalized (i.e., $t_i/\sqrt{t_k t_k}$ is a unit vector tangent to the face). For additional scalar fluxes and turbulent heat flux and so forth, it is clearly best to include a separate μ_{num} component, normalized by an appropriate Schmidt number. This would be computed from the local jumps in the scalar (rather than tangential velocity component). Note that care is needed around shocks. This analysis assumes that the primary gradients considered are due to shearing and not the velocity jumps around shocks. Also, it is necessary to prevent divisions by zero that occur in benign flow regions. In such zones it can be assumed that μ_{num} is approximately equal to zero.

Celik et al. (2009) present a wide range of parameters to assess the sufficiency of grids for LES. Some involve the use of two grids but this, due to computational cost, is impractical for most eddy-resolving simulations. Celik and colleagues define the activity parameter

$$\boldsymbol{s}^* = \frac{\boldsymbol{v}_t + \boldsymbol{v}_{\text{num}}}{\boldsymbol{v}_t + \boldsymbol{v}_{\text{num}} + \boldsymbol{v}} \tag{7.69}$$

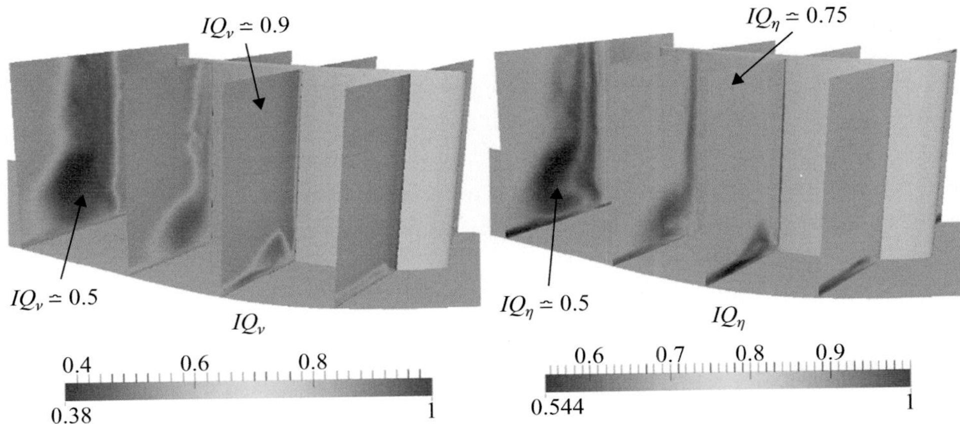

Figure 7.31. Application of LES quality indices to hybrid RANS-LES simulation of flow over a compressor blade: (a) use of IQ_v and (b) use of IQ_η.

where the < > indicates a time averaged property. The difficulty with this parameter is that s^* is always close to unity. Hence, Celik and colleagues propose the following quality index function to give a more sensible operating range of the assessment parameter:

$$IQ_v = \frac{1}{1 + \alpha_v \left(\frac{s^*}{(1-s^*)}\right)^n} \tag{7.70}$$

which is equivalent to

$$IQ_v = \frac{1}{1 + \alpha_v \left(\frac{v_{t,eff}}{v}\right)^n} \tag{7.71}$$

where $v_{t,eff} = v_t + v_{\text{num}}$. In the Equation (7.71) the parameters $\alpha_v = 0.05$ and $n = 0.53$ are proposed. These values mean that IQ_v varies between 0 and 1, with 0.8 considered consistent with a good LES and a value of 0.95 consistent with DNS. The following parameter is also considered by Celik and colleagues:

$$IQ_\eta = \frac{1}{1 + \alpha_\eta \left(\frac{\Delta}{L_\eta}\right)^m} \tag{7.72}$$

where $L_\eta = \left(v^3/\varepsilon\right)^{1/4}$ is the Kolmogorov length scale. However, the evaluation of this scale is adapted to include the numerical dissipation such that $\varepsilon = \varepsilon_{\text{res}} + \varepsilon_{\text{num}}$ is used where ε_{res} is the resolved dissipation and ε_{num} the numerical. In the above $\alpha_\eta = 0.05$ and $m = 0.5$. If IQ_η is low, the grid spacing is substantially larger than the Kolmogorov scales. According to Celik and colleagues, grid spacing values of around 25 L_η is sensible for LES.

Figure 7.31 shows the application of the LES quality indices of Celik and colleagues to hybrid RANS-LES simulation of flow over a compressor blade. Frame (a) shows use of IQ_v and Frame (b) use of IQ_η. Both parameters suggest that the

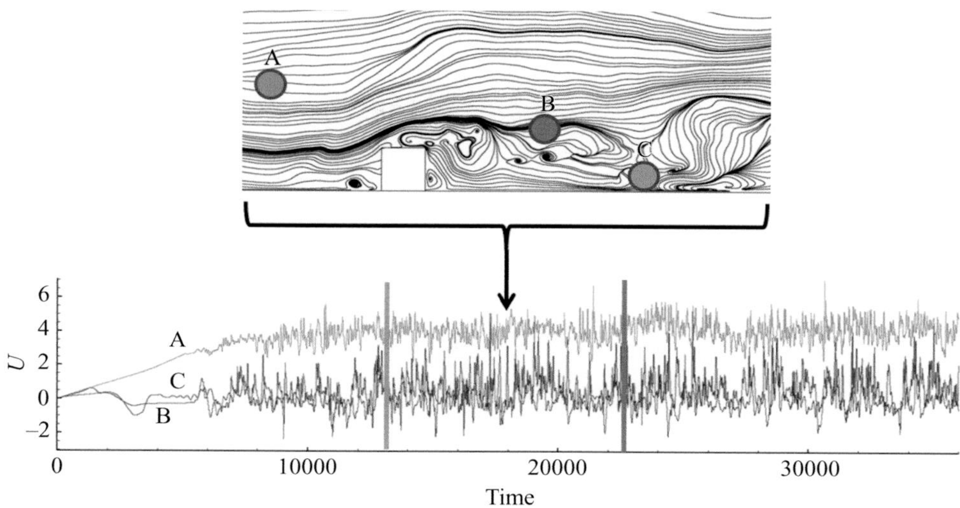

Figure 7.32. Development of flow in a ribbed passage to a statistically stationary state independent of the initial transient for an eddy-resolving simulation.

grid is too coarse in the corner separation regions. The parameter is not applicable in the near wall RANS zones.

7.8.3 *Convergence to a Statistically Stationary or Cyclic State*

For unsteady simulations, it can be necessary to detect when flow has reached a cyclic or statistically stationary state, where all initial transients have gone. See for example, Figure 7.32. This shows the development of flow to a statistically stationary state, independent of the initial transient, for an eddy-resolving simulation. The simulation involves flow in a ribbed passage. The flow evolution at different points is shown as a trace of streamwise velocity against time.

7.8.3.1 Standard Approaches Normal approaches for judging the above are based on the visual inspection of variable traces with respect to time. Alternatively, for LES, past experiences based on the number of large eddy throughflow times can be used. For LES, substantial periods can be required to evolve a flow and then to average it. For example, with jets, Shur et al. (2003) report complete LES needs time periods around $1000L/U$, where L is the jet diameter (giving a large eddy scale) and U the jet velocity. For swirling jets, Tucker (2004) uses a total simulation time of $2000L/U$. With turbine blade simulations with no wakes typically a time period of around $15L/U$ is needed, where L is the blade chord and U the inlet velocity. Alternatively, for flows with upstream, moving blade wakes, fifteen wake passing periods can typically be needed for complete simulations. Hence, looking at the experiences of other people can be useful. However, this is no certain guide since estimates can vary substantially and might well have limited relevance to the flow being considered.

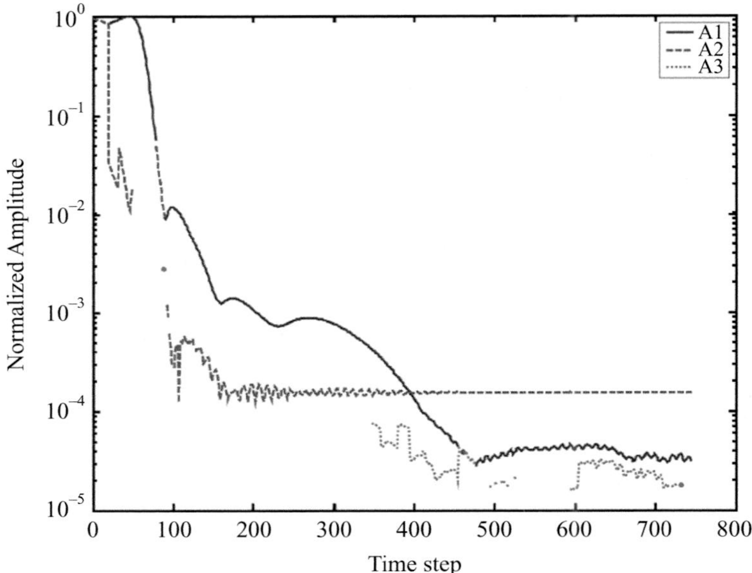

Figure 7.33. Variation of amplitudes, A, of the three most dominant Fourier coefficients with time for a jet in cross flow (from Ahmed & Barber 2005 – Published with the kind permission of the AIAA).

7.8.3.2 Use of FFTs for Periodic Flows For simulations that are periodic, with no stochastic components, a Fast Fourier Transform (FFT) procedure can be used to assess periodicity devised by Ahmed & Barber (2005). With this, different variables at various locations are monitored and analysed using an FFT with padding. The amplitudes of the dominant frequencies are monitored and their amplitudes used to judge convergence to a cyclic state. For example, Figure 7.33 shows the variation of amplitudes, A, of the three most dominant Fourier coefficients with time for a jet in cross flow (from Ahmed & Barber 2005). The second most dominant amplitude, A_2, appears to converge to the cyclic state first. Generally, to build accurate spectra substantial amounts of data are needed and hence the practical value of this FFT based approach is possibly case dependent. Note, also, that this procedure is impractical for LES where there is not the clear periodicity element. For an assessment of this FFT based approach, see Tyacke (2009).

7.8.3.3 Statistically Stationary LES Data An alternative, more automatable technique is to store data in chunks. Then, through making comparisons, going backwards from the final chunk of data it should be possible to gain some insight into the level of data maturity. Mockett et al. (2010), for statistically stationary data, attempt to:

1) Detect the period, t_t, of initial transients and
2) Once the transient has been removed, statistically define the level the accuracy of the arising results for a given time sample.

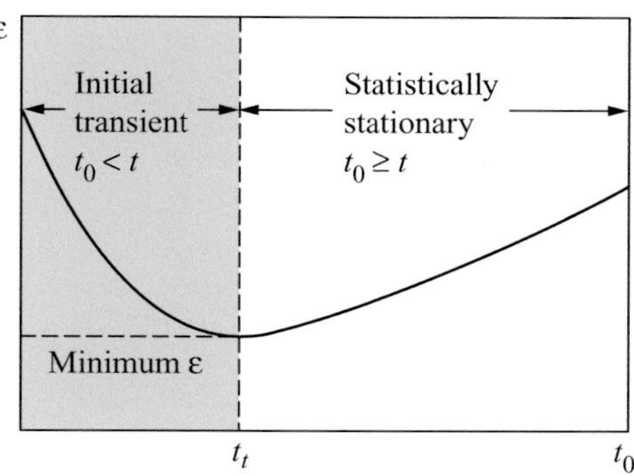

Figure 7.34. Initial transient detection based on statistical error (based on schematic of Mockett et al. 2010).

Aspect (2) has two stages. The first is estimating statistical error. This is achieved by dividing a time series of length, T, into shorter windows of length T_w. The basis of the statistical error, $\varepsilon\left(\tilde{\phi}\right)$, on the data average, $\varepsilon\left(\tilde{\phi}\right)$, is

$$\varepsilon\left(\tilde{\phi}\right) = \frac{\sqrt{\langle(\tilde{\phi} - \phi)\rangle}}{\phi} \tag{7.73}$$

where $\tilde{\phi}$ relates to data in the window of length T_w. On the other hand, ϕ is based on data over the full available time period T. Also, $< >$ represents averaging over the available window data. The error given by Equation (7.73) will vary depending on T and T_w. To gain an idea of error for a fixed T, T_w can be varied but for a finite T such as process is rather ad hoc[1]. Hence, a critical second stage is the ability to rationally and automatically estimate the error trend for differing sample lengths. Mockett et al. (2010) achieve this through the use of standard analytical expressions for Gaussian white noise. Usefully these statistical error estimates can also be used to forecast the amount of additional time steps that would be needed to meet a desired statistical accuracy.

To estimate t_t, the start of the considered data is defined as t_o. This is moved forwards through the data set and the statistical error, ε, (based on the ideas discussed) observed. As shown in Figure 7.34, initially, as t_o increases ε will decrease as the initial transient is removed. Then as t_o increases, ε will increase as the extent of the statistical sample is lost. The point where the minimum error is observed gives t_t. Note that the statistical error is based on the statistical theory informed by white noise. The utilities described are available in the commercial software called MEANCALC.

[1] To further clarify, the Section 7.8.3.3 procedure uses both 'lots of chunks of T_w' and a variable range in T_w. The latter (T_w) is varied to measure error for different values of T_w. For each T_w, lots of windows are needed to be able to calculate the error. For example, to calculate the error on mean for one value of T_w, the following procedure is followed: Divide signal up into windows of size T_w; Compute mean *inside* each window (called 'sample mean'); Subtract the global mean from each sample mean; Square this and average over all windows. Then T_w is varied and the procedure repeated.

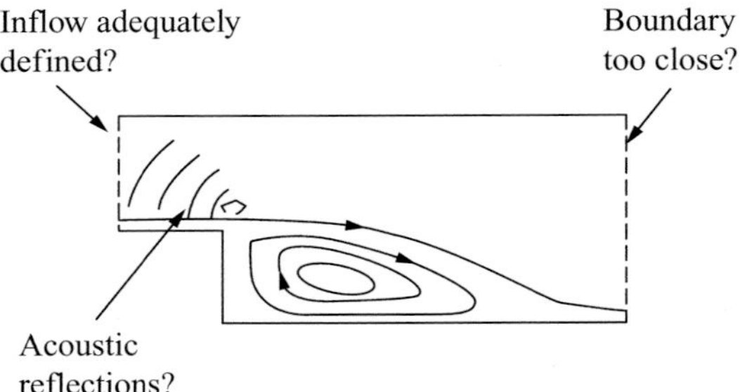

Figure 7.35. Correct location of outflow boundary conditions.

7.9 Assessing Boundary Condition Uncertainty

At some stage it is important to assess the uncertainty arising from boundary conditions. The initial inspection of results can suggest locations where there might be issues. For example, zones where the outflow boundary condition is too close to a recirculation might be observed as shown in Figure 7.35. Also, inflow boundaries can be too close to an area of unsteadiness activity that generates wave modes. These could create solution contamination through reflections at the inflow boundary.

Notably, for multiple inflow/outflow boundaries defined with, say, pressure based boundary conditions (as shown in Figure 7.36) and differential conditions on velocity, non-unique solutions can arise. Hence, such systems can be sensitive to boundary conditions and their formulation. For LES, at inflow, the nature of the synthetic turbulence defined can strongly impact on results. Hence, some form of assessment is ideally needed with boundary conditions being perturbed or varied to determine sensitivity. A formal sensitivity analysis could be performed with some adjoint based procedure. Many problems are coupled. Hence, based on uncoupled CFD results, it is worth revisiting how valid the assumption of neglecting coupling to other components was and if it needs to be reconsidered.

It should be noted here the emerging and important field of formal uncertainty analysis in CFD. With the probabilistic approach, a distribution of potential input boundary condition uncertainties is formulated along with, potentially, many other uncertainties, such as the fluid and geometrical properties, along with physical models and so forth. Then, a probabilistic response of a design output to these is generated. This can be achieved using, for example, the expensive Monte Carlo approach, or

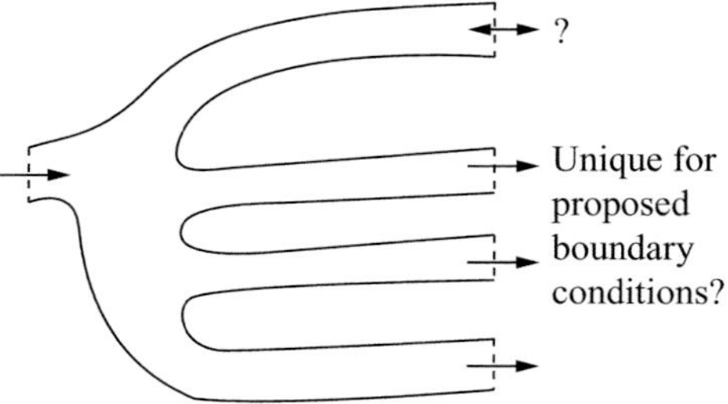

Figure 7.36. Potential for non-unique solutions for flows with multiple boundaries

more sophisticated approaches like the spectral Galerkin. See Iaccarino (2008) for a discussion of this subject.

7.10 Visualizing Flows

A key, intial stage, of post proecssing would be to look at the flow and see if it appears physically sensible. The most basic vizualization techniques include using contour, vector plots and isosurfaces. However, multicomponent plots can be used that include all of these elements. When dealing with three-dimensional data sets, it can be useful to just visualize data on two dimensional slices. Hence, for large three-dimensional data sets, there is the challenge of developing software that can perform planer slices at a sensible rate and hence this is an area of research. Line extraction is also needed for plotting graphs. This can be needed over relatively complex surfaces such as aerodynamic surfaces.

7.10.1 *Streamlines, Streaklines and Related Methods*

The trajectores of weightless particles in steady flow that yield, for example, streamlines are useful for understanding fluid flows (streaklines, pathlines and timelines can also be considered). These can sometimes be in the form of a ribbon. Streamlines can be coloured by a scalar quantity. The upper frame in Figure 7.32 shows streamlines seeded in an instantaneous flow snapshot. Streamlines can be generated through the solution of Lagrangian weightless particle equations (see Chapter 6), the particle trajectories being connected by a line. For a procedure for the generation of streak surfaces, to help better understand flow physics, see McLoughlin et al. (2010).

For visualizing vector fields, an alternative to streamline plots is Line Integral Convolution (LIC) plots. This is an Eulerian approach. It is a sub member of a group of methods called texture advection methods. With LIC, unlike with streamlines and related approaches, the specification of seed points is not needed. Physically, the approach is like different areas of the flow being coloured with different coloured dyes. The dye movement reveals streamlines but there is also a colour mixing process. In practice, the approach is pixel based. Seed pixles are convected both forwards and backwards and a weighted integral of values is used to change the local pixel shading. The weighting functions (convolution kernels) can be time varying, giving an animated effect. Unsteady variants of LIC – ULIC – exists (see Shen & Kao 1998, Grabner and Laramee 2005) suitable for eddy-resolving related simulations. Figure 7.37 gives an LIC plot, based on wall shear stress, for an axial compressor blade flow. Texture based methods have GPU compatability.

7.10.2 *Feature Detection*

When handling large data sets, automatic feature detection is important. Features that can be sought are shock surfaces, separation lines and vortices. The former are looked at next.

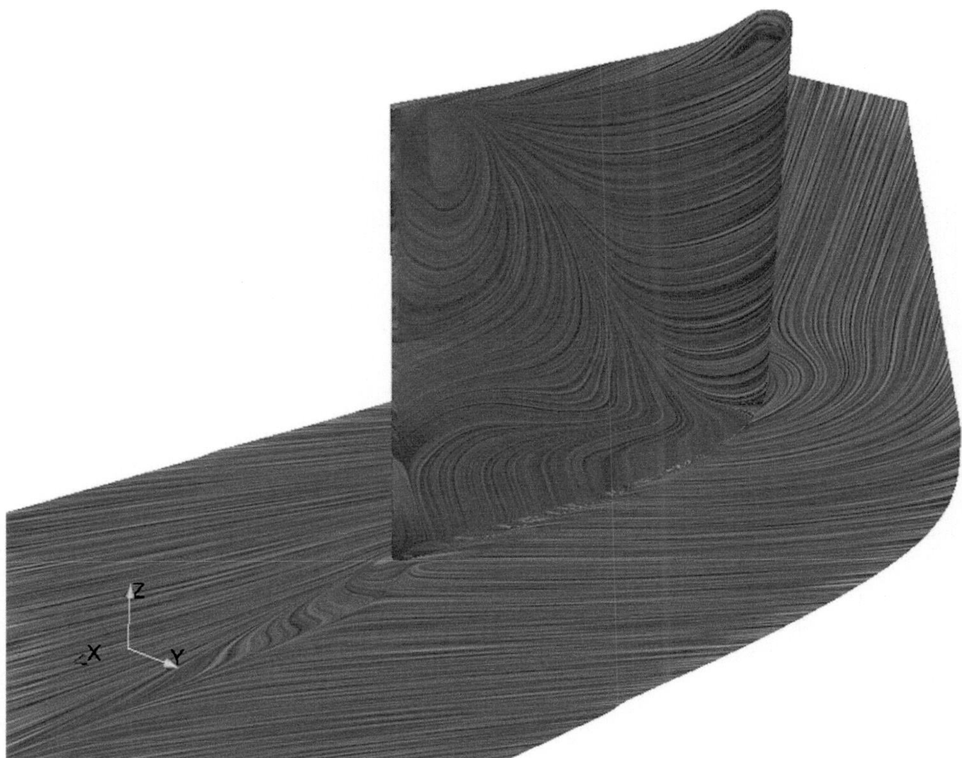

Figure 7.37. Linear integral convolution plot, based on wall shear stress, for an axial compressor blade flow.

7.10.2.1 Shock Identification Walatka et al. (1990) consider the following approaches for the detection of shock features:

1) Look at the Mach number distribution aligned with the pressure gradient. If the Mach number goes through unity and its distribution is decreasing, this is marked as a shock. This process consists of contouring the following shock function given for the unity value

$$\text{Shock Function} = \frac{\mathbf{U}}{c} \cdot \frac{\nabla p}{|\nabla p|} \tag{7.74}$$

Note that $\mathbf{U}$ is the fluid velocity vector and c the speed of sound. For $|\nabla p| < 0.1$, the shock function is set to zero.

2) The divergence of velocity can be used. This is sometimes called the total dilation or the normal strain rate. The conservation of mass equation, using the product rule, can be re-expressed as

$$\frac{D\rho}{Dt} + \rho\nabla.\mathbf{U} = 0 \tag{7.75}$$

Hence,

$$\nabla.\mathbf{U} = -\frac{1}{\rho}\frac{D\rho}{Dt} \tag{7.76}$$

For steady flow, this reduces to

$$\nabla.\mathbf{U} = -\frac{1}{\rho}\mathbf{U}\cdot\nabla\rho \tag{7.77}$$

Note that the shock switches discussed in the numerical methods sections could also be adapted to a post-processing context.

7.10.3 *Vortex Identification*

To better understand the nature of many flows, especially turbulent, when making eddy-resolving simulations, it it important to visualize vortices. There are two key approaches to doing this:

1) Visualize the full extent of the vortex zone through examining some scalar quantity and
2) Visualize the location of the vortex core alone.

Potential quantities that can be examined for the first listed aspect are discussed next.

7.10.3.1 Pressure At a vortex core, low pressure is expected and hence pressure can, in principal, be used to observe vortices. The signature from the pressure field can be succesively amplified through differentation by, for example, the use of the Laplacian of pressure. However, in flows with strong acceleration, streamline curvature and rotation, the background pressure field can swamp the pressure variation associated with the vortex. The use of reduced pressure (see Chapter 2) could be used to help to an extent, however, in general terms the use of pressure is limited.

7.10.3.2 Vorticity, Enstrophy and Helicity An obvious choice for vortex identification would be the magnitude of vorticity $\boldsymbol{\Omega}$, where

$$\boldsymbol{\Omega} = \frac{1}{2}\begin{pmatrix} 0 & -\omega_3 & \omega_2 \\ \omega_3 & 0 & -\omega_1 \\ -\omega_2 & \omega_1 & 0 \end{pmatrix} \tag{7.78}$$

and $\omega = (\omega_1, \omega_2, \omega_3) = \nabla \times \mathbf{U}$. However, even for a simple laminar boundary layer which is free from vortices, the vorticity magnitude will be finite. Hence, this approach is too simplistic. Enstrophy, related to the square of vorticity, can also be used.

If the vorticity is projected using the dot product onto the flow vector, helicity is gained. This overcomes the problem relating to vorticity noted previously, but only for straight streamlines.

7.10.3.3 Q-criterion Most more advanced methods for vortex identification are based on expressions using a mix of the strain rate vector **S** and vorticity $\boldsymbol{\Omega}$, which together define velocity gradients in the flow

$$\nabla \mathbf{U} = \mathbf{S} + \boldsymbol{\Omega} \tag{7.79}$$

They also, potentially, involve the pressure gradient in some sense. As indicated previously, in vorticity there is an obvious connection to vortex identification. However, the strain is more associated with the production of vorticity and straining of vortices.

The Q-criterion for vortex detection uses matrix norms

$$Q = \frac{1}{2}\left(\|\boldsymbol{\Omega}\|^2 - \|\mathbf{S}\|^2\right) \tag{7.80}$$

Positive Q, where vorticity dominates strain, will indentify vortex regions. Equation (7.80) is Galilean invariant and can quickly computed (see Sahner 2009). Jeong and Hussain (1995) note that Q can be regarded as a source term in the Poisson equation for pressure, that is,

$$\nabla^2 p = 2\rho Q \tag{7.81}$$

Hence, physically there is a correspondence with detecting pressure mimima and maxima. Figure 7.16 gives an example of an instantaneous flow being visualized using the Q-criterion.

7.10.3.4 Lambda 2 With the λ_2 approach, of Jeong and Hussain (1995), the three eigenvalues ($\lambda_1 \leq \lambda_2 \leq \lambda_3$) of the matrix $\mathbf{S}^2 + \boldsymbol{\Omega}^2$ are considered. For a minima in pressure, corresponding to a vortex core, there must be two negative eigenvalues. With the λ_2 approach, vortex cores are identified when $\lambda_2 < 0$. Positive values evidently have no physical meaning and the method is most effective for flows with high strain. However, since $Q = -1/2(\lambda_1 + \lambda_2 + \lambda_3)$, the approach is quite closely related to the the Q-criterion. Hence, for the Figure 7.16 Q-criterion plot, the λ_2 plot looks very similar. As shown by Roth & Peikert (1996), for a turbomachinery flow, λ_2 can be everywhere positive, rendering the approach ineffective.

When performing experimental flow visualization experiments, great effort is needed to reliably render the flow field in a way that is clear. This is also true for vortex identication, where it can be necessary to try a range of methods with a wide range of thresholds to render the vortex structures clearly visible. The need for care with thresholds is a signicant drawback. A further vortex identification criteria is the M_z criteria of Haller (2005). It is mathematically complex and based on Lagrangian trajectory information (see Sahner 2009). Note that the M is the strain acceleration tensor. Also, there is the Δ criteria, as discussed in Sahner (2009). This seeks to identify spiralling vortex lines. Table 7.3 summarizes advantages and disadvantages of different vortex indentification approaches.

Rather then try to understand the extent and structural detail of a vortex, just the locus of the vortex core can be extracted. This vortex line extraction has the

Table 7.3. *Advantages and disadvantages of different vortex identification approaches*

Method	Advantages	Disadvantages
Vorticity	Simple	Strongly finite in non-vortical flows; threshold dependence
Enstrophy	Simple	Strongly finite in non-vortical flows; threshold dependence
Helicity	Relatively conceptually simple	Strongly finite in non-vortical flows with complex strain fields; threshold dependence
Pressure and functions of it	Simple	Inflows with strong acceleration; streamline curvature and rotation; the background pressure field can swamp the vortex pressure variation; threshold dependence
Q	Physical meaning for both positive and negative Q values, Galilean invariant	Threshold dependence
Lambda 2	For large S magnitude detects vortices more reliably; Galilean invariant	For $\lambda_2 > 0$, lacks a physical interpretation; threshold dependence
M_Z	–	Complex theory and implementation

attraction of being relatively automatible. There are a wide range of vortex core identification techniques based on eigenvalue/vector analysis. Miura & Kida (1997) attempt to use loci of pressure minima but this results in disconnected lines.

7.10.4 *Acoustic Vizualization*

Contouring instantaneous pressure fields can enable acoustic waves to be seen in flows. However, contouring $\partial \rho/\partial t$ is more effective. Rather, than saving the density fields as two time-steps and differentiating, it is more convenient to use the continuity equation to provide the time derivative

$$\frac{\partial \rho}{\partial t} + \nabla \cdot \rho \mathbf{U} = 0 \tag{7.82}$$

and thus contour $\nabla \cdot \rho \mathbf{U}$.

Figure 7.38 shows a multicomponent plot. The inner core turbulent field can be seen. This shows the sound source. The acoustic wave fronts emantating from the source are also evident.

Through solution of the high frequency wave equation (the eikonal equation) or specialist ray tracing (where the acoustic velocity is added to the fluid velocity and used in a weightless particle tracking procedure), sound rays can be tracked. Figure 7.39 shows sounds rays through a time average of an unsteady coaxial jet flow field (a) and also through the instantaneous flow field (b). Note that both plots are based

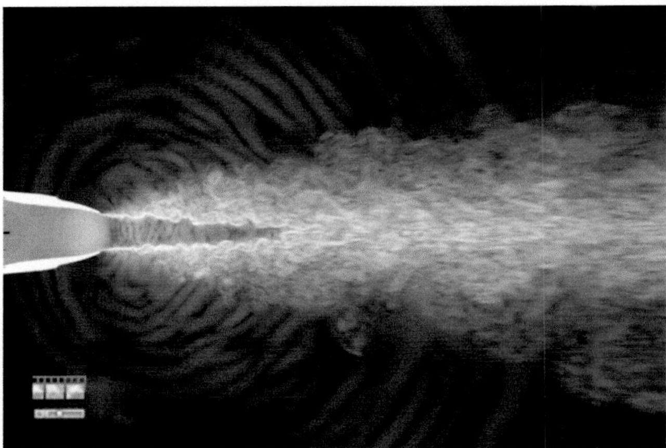

Figure 7.38. Visualization of sound waves through contours of $\partial\rho/\partial t$.

on eddy-resolving simulation data sets. Such acoustic ray plots can be considered as the acoustic equivalents of streamline plots. Note that Frame (b) is traced through a temporally frozen flow.

Tecplot, Fieldview and EnSight are popular commerical post-processing software packages. There are quite a few other free packages and these include: Paraview, Vislt, OpenDX and VIGIE. Notably, Paraview is free and allows parallel post-processing and hence the handling of larger data sets. It is important, for the future development of CFD, that parallel processing can be used to allow the post-processing of large data sets. The first parallel post-processing package was pV3, developed by Haimes (1994). Most of these packages will directly supply many of the flow visualization parameters discussed in this Section 7.10 or enable them to be evaluated through user-defined functions.

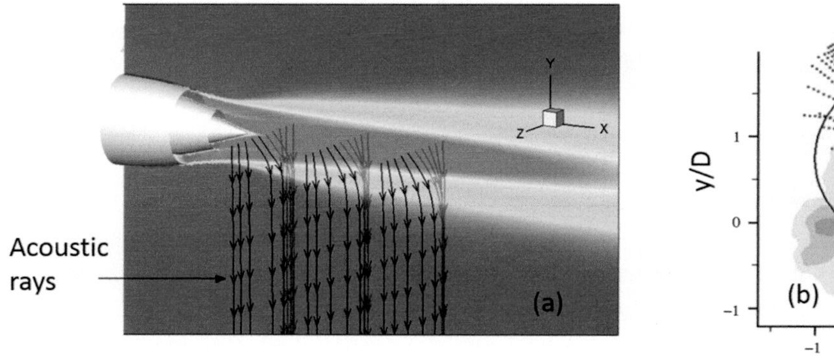

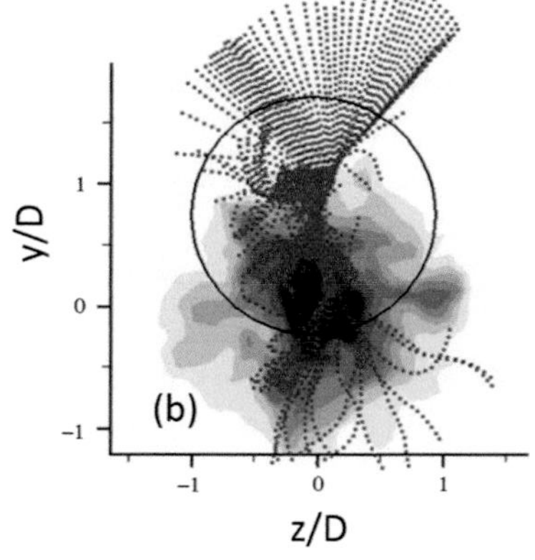

Figure 7.39. Acoustic streamlines in a coaxial propulsive jet: (a) seeded in time mean flow and (b) seeded in an instantaneous flow (Adapted from Tucker 2013 – Frame (b) used with permission from the Royal Society).

7.11 Auxiliary Post-Processing Tools

Many auxiliary tools can be needed for pre- and post-processing. For example, it can be useful to extract data on global lift and drag, total pressure loss, displacement, momentum and basic boundary layer thickness (based perhaps on a vorticity threshold). Boundary layer edge velocity or isentropic Mach number distributions can be needed. It can be useful to break drag down into its constituent parts such as skin friction, wave and wake drag. It can be useful to automatically plot first off-wall y^+ values and hence quickly assess the quality of near wall grids used and areas that need refinement. The development of these capabilities can be time consuming, especially so for unstructured grids.

Examining entropy fields can also be informative. As shown by Denton (1993), entropy generation along a streamline can be expressed as

$$UT\frac{ds}{dx} = f\left(\tau_P, \nabla T_{\perp}\right) \tag{7.83}$$

where x is the streamline coordinate and U is the fluid velocity along the streamline. The fluid temperature is T. On the right-hand side, τ_P is essentially shearing aligned with the flow direction and $\nabla T_{\perp}$ are temperature gradients orthogonal to the streamline. Note that entropy can be generated due to shocks or sudden expansions. Entropy is also generated when there is heat transfer with finite temperature differences and also perhaps most importantly when there is shearing in, for example, boundary layers, or free shear flows. Entropy, given by Equation (2.161) in Chapter 2, characterizes loss. Hence its contouring is useful when attempting to understand the performance of aerodynamic systems and reducing loss in them.

Essentially, as suggested by Equation (7.83), the entropy can be imagined to be a passive scalar convected through the system. Local sources arise from shearing, temperature gradients and also shocks (not identified in Equation (7.83)). Note:

$$s - s_{ref} = c_p \ln\left(\frac{T}{T_{ref}}\right) - R\ln\left(\frac{P}{P_{ref}}\right) \tag{7.84}$$

where all the quantities can be static or total. Since in a stationary adiabatic system total enthalpy is conserved, the following is also true

$$\Delta s = -R\ln\left(\frac{P_{02}}{P_{01}}\right) \tag{7.85}$$

where the subscripts indicate total quantities. Hence, total pressure loss can be used as a measure of loss. This can be usefully evaluated as a mass weighted integral of total pressure loss, allowing designs to be ranked more easily by a single parameter – the total pressure loss coefficient. The entropy equation(s) can be formulated in different ways to give different functions to monitor entropy (see for example Walatka et al. 1990). Often, entropy is plotted using what is called the entropy function, $\exp(-s/R)$. This gives rise to a useful range of values to contour.

Figure 7.40 shows entropy contours for unsteady flow in a turbine stage, taken from Denton (2010). The increased generation in the blade wake zone is evident

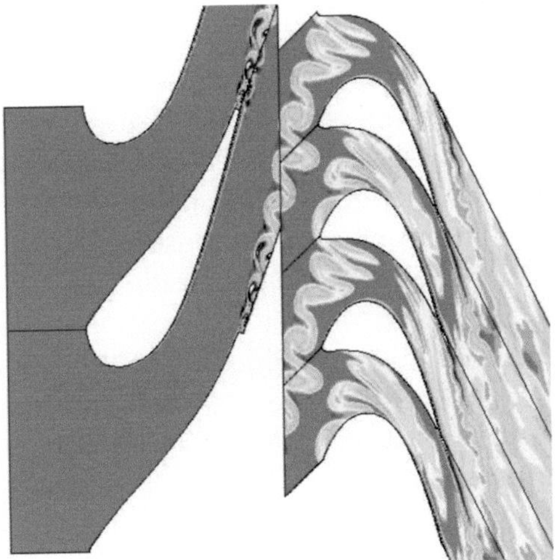

Figure 7.40. Entropy contours in a turbine stage (from Denton 2010 – published with the kind permission of the ASME).

and its subsequent convection downstream. Spurious entropy can also be numerically generated and have a downstream contamination impact.

Since entropy has a convective nature, a more convenient way of exploring local loss behaviour can be through looking at the local entropy generation rate. This process is outlined by Pullan et al. (2006). The entropy generation rate per unit volume can be expressed as

$$\dot{s}_{gen} = \frac{1}{T}\tau_{ik}\frac{\partial U_i}{\partial x_k} + \frac{k}{T^2}\frac{\partial T}{\partial x_i}\frac{\partial T}{\partial x_i} \tag{7.86}$$

where k is thermal conductivity and τ is stress. The equation can then be broken down as follows

$$\dot{s}_{gen} = \frac{1}{T}\left(\frac{\partial \tau_{ik} U_i}{\partial x_k} - U_i\frac{\partial \tau_{ik}}{\partial x_k}\right) + \frac{k}{T^2}\frac{\partial T}{\partial x_i}\frac{\partial T}{\partial x_i} \tag{7.87}$$

This can be integrated over a cell volume to give

$$\dot{s}_{gen} = \frac{1}{TV}\left(\underbrace{\sum \tau_{ik} U_i A_k}_{\dot{W}_{visc}} - \underbrace{U_i \sum \tau_{ik} A_k}_{\dot{W}_{mech}}\right) + \frac{k}{T^2 V}\left(\sum T A_i\right)^2 \tag{7.88}$$

where the A terms are cell face areas and V is the cell volume. Notably, two work terms are evident. These relate to viscous work and potentially useful mechanical work. Note that the stress term involves the product of viscosity and velocity gradients. Hence, a further potential break down is to consider entropy generation from numerical viscosity, based on the numerical viscosity estimates due to

Batten et al. (2011) discussed in Section 7.8.2.3, and entropy generation due to eddy viscosity. Basha et al. (2009) give the above entropy generation equations in an expanded Cartesian form, again using them for CFD post-processing.

7.12 Eddy-Resolving Simulations

The great detail provided by eddy-resolving simulations (time histories of data at potentially billions of spatial grid points) allows for a deeper interrogation of flow physics than is possible for RANS based CFD. This also gives rise to new challenges. LES can be used to provide spectral data for both aerodynamic and aeroacoustic data. At a basic level, it is necessary to establish time averages of solution variables and Reynolds stresses to yield the data akin to that provided by RANS. The Reynolds stresses are formed by making mean flow variable averages, which are squared and also forming running averages of the square of instantaneous flow variables. The differences give the Reynolds stress components. Clearly, as given next in Equation (7.89) there are many fluctuating products of the variables, say, ϕ_i' and ϕ_j', that give turbulent fluxes

$$-\overline{\phi'_i\phi_j} = \overline{\phi_i\phi_j} - \overline{\phi}_i\overline{\phi}_j \tag{7.89}$$

To save storage, prior to making the simulation it is worth considering which of the many fluctuating products is of interest. Hence, the user needs to be selective to ensure that there is not too much data to handle. For far field noise prediction techniques such as, for example, the FWH equation (see Chapter 7), integral surface time history data needs to be stored – unless the code has been developed to evaluate data on the fly. Writing out this time history data can be expensive, especially since, as will be seen later, the data can need to be written out at multiple surfaces. The multiple surfaces are then later interrogated to ensure use of a surface in the irrotational region of the flow where the time history data is purely acoustic.

7.12.1 *Correlation Coefficients*

Correlations, of for example, velocity, can be important to evaluate. They, for example, give physical insight into the structure of turbulence and also if numerical domains, in any periodic directions, are of sufficient extent to contain the turbulence. There are a range of correlation forms, but all of them essentially need time series data, stored at a range of aligned adjacent spatial points. These can then be multiplied and summed in various simple ways to form correlation coefficients. Figure 7.41 gives an example of point arrays needed for generating correlation coefficients for a chevron nozzle. The flow structure for this type of nozzle varies azimuthally, reflecting the azimuthal variations in geometry. Hence, Frame (a) shows the radial location of cluster centres on a cut plane that connects chevron tips. Frame (b) shows the radial location of cluster centres on cut plane that radially connects the chevron roots. Finally, Frame (c) shows the form of the clusters of points focused around the subpoint points in Frames (a) and (b). Note that there are also extensive

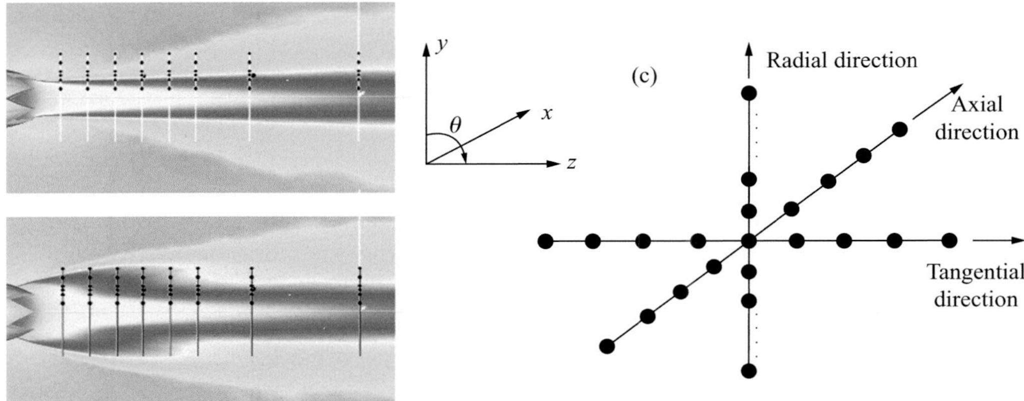

Figure 7.41. Identification of point arrays needed for generating correlation coefficients for a chevron nozzle: (a) radial location of cluster centres on tip cut plane, (b) radial location of cluster centres on root cut plane and (c) cluster of points focused around points in frames (a) and (b).

arrays of points in the azimuthal direction but these are not shown. The second order space-time correlation of turbulence can be expressed as

$$R(x, \Delta, \tau) = \frac{\overline{\phi'(x,t).\phi'(x+\Delta, t+\tau)}}{\overline{\phi'(x,t).\phi'(x,t)}} \tag{7.90}$$

where ϕ' is now a fluctuating turbulence quantity. Also, Δ represents a spatial shift of a 'probe' (relating to Frame (c) and being the distance of points from the central point) and τ a time shift. The correlation Equation (7.90) is normalized by the auto-correlation of the signal for zero time delay.

Figure 7.42 takes results from Xia and Tucker (2011) for essentially LES predicted flow for a serrated nozzle. The reference probe point is fixed at $x/D = 2.0$, where D is the jet diameter and a radius of $0.5D$. This location is identified in the Figure 7.42 insets. The results for the R_{11} correlation (corresponding to an axial probe shift involving the fluctuations in axial velocity) are plotted in the upper frames of Figure 7.42. Each point on the space correlation, representing a spatial separation, becomes a curve in the R_{11}-τ space. The envelopes (shown as dashed lines) of these curves (for increasing separation as the inner curves move from left to right) are the auto-correlation in a frame moving with the energy containing eddies at the convection speed U_c. The integral of the envelope is the moving-axis time-scale of the turbulence. Figure 7.42 shows correlations on both tip and root planes in Columns (a) and (b), respectively. The envelope curve for the tip plane is replicated to the upper frame of Column (b). The difference between the envelopes indicates a time-scale disparity. The correlations can also be presented as iso-correlation contours in the Δ-τ space. This is shown in the lower frames of Figure 7.42. At each contour level, max(τ) can be located in the Δ-τ space. The dashed lines are linear fits of these locations. The slope of the dashed lines gives the convection speed U_c. The difference in the slopes indicates that there is nearly a 20% difference in the

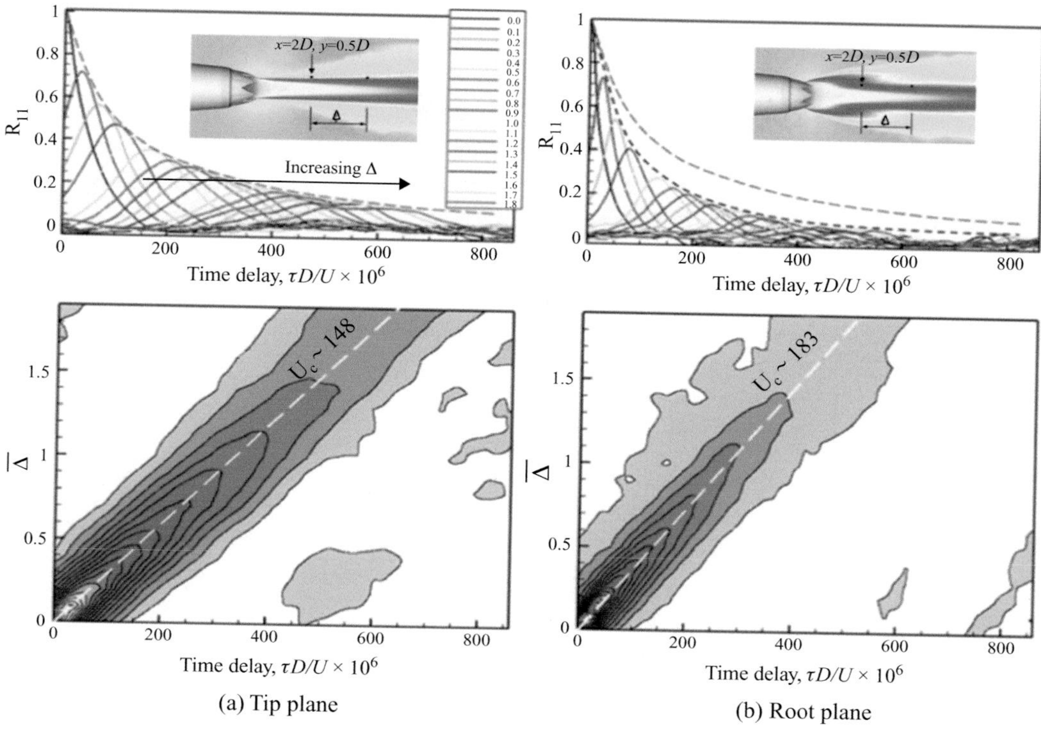

Figure 7.42. Cross correlation of axial velocity fluctuations with downstream separation (upper frames) and iso-correlation contours of cross-correlations (lower frames). Different $\bar{\Delta}$ values are listed in a top-down order in the upper frames (from Xia & Tucker 2011 – published with the kind permission of Springer).

convection velocity between the tip and root zones. Such information is useful for RANS based acoustic models and to understand flow physics at a basic level.

As shown by Goldstein (2003), it is the fourth-order space-time correlations that are key to characterizing turbulent flow noise sources and LES can be compared with measurements of this for validation purposes (see Eastwood 2009). It is important to extract the fourth-order space-time correlations of the turbulence to feed into low-order acoustic models.

The decorrelation for increasing spatial shifts enables the size of the large turbulent scales to be estimated and hence, from such evaluations, as noted previously, it can be determined if domain extents in periodic flow directions are sufficient to contain the turbulence. Multiple large eddy scales area ideally needed (say, five) in periodic directions to ensure too small a domain is not influencing the solution.

7.12.2 *Spectral Information*

Spectral information is useful for understanding physics, providing a deeper level of validation against aerodynamic and aeroacoustics measurements, and exploring numerical scheme fidelity. For example, for eddy-resolving simulations,

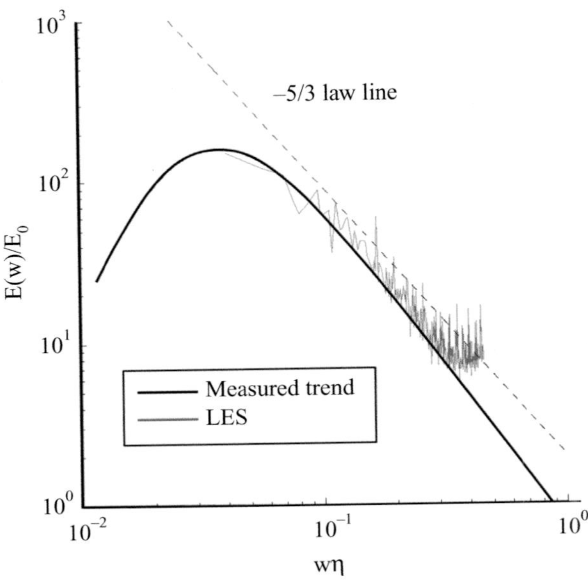

Figure 7.43. Turbulence energy spectrum from an LES.

assembling turbulence energy spectra will allow the user to see if the grid resolves the intertial subrange (assuming the Reynolds number is high enough for this to be present) or Kolmogorov scales. It will also allow the user to explore the performance of the numerical scheme or subgrid scale model. For example, the spectrum shown in Figure 7.43 indicates a lack of dissipation at the small scales, with the turbulence energy not decaying fast enough. Most LES models assume that the extent of the Kolmogorov –5/3 law zone, shown by the dashed line in Figure 7.43, is long. If this is not the case, care is needed (see Razafindralandy et al. 2007), and spectra allows this aspect to be explored.

When performing URANS simulations, it is important to explore the presence of a spectral gap. By this it is meant that there needs to be a clear separation of the modelled and resolved scales of turbulence (see Chapter 5). This can be difficult to estimate a priori, being a strong function of the chosen turbulence model. The modelled time scales can be estimated from the turbulence model equations given in Chapter 5. The resolved turbulence scales can be estimated by assembling turbulence spectra. However, normally the examination of dominant time periods in simple time trace plots is adequate, there generally being few harmonics in URANS simulations.

Figure 7.44c shows acoustic spectra from Xia et al. (2009) for a chevron jet nozzle. The limit on frequency by the imposed grid is clear. In this case, the frequency limit is not due to the resolution of the turbulent scales that define the sound source. The sound is propagated to the far field via the analytical equation of Ffowcs Williams and Hawkings (see Chapter 6, Section 6.8.3.2). This process is based on surface integral data stored at the irrotational zone region of the flow. The first challenge is ensuring that this data is extracted in the irrotational region of the flow. This is not possible to

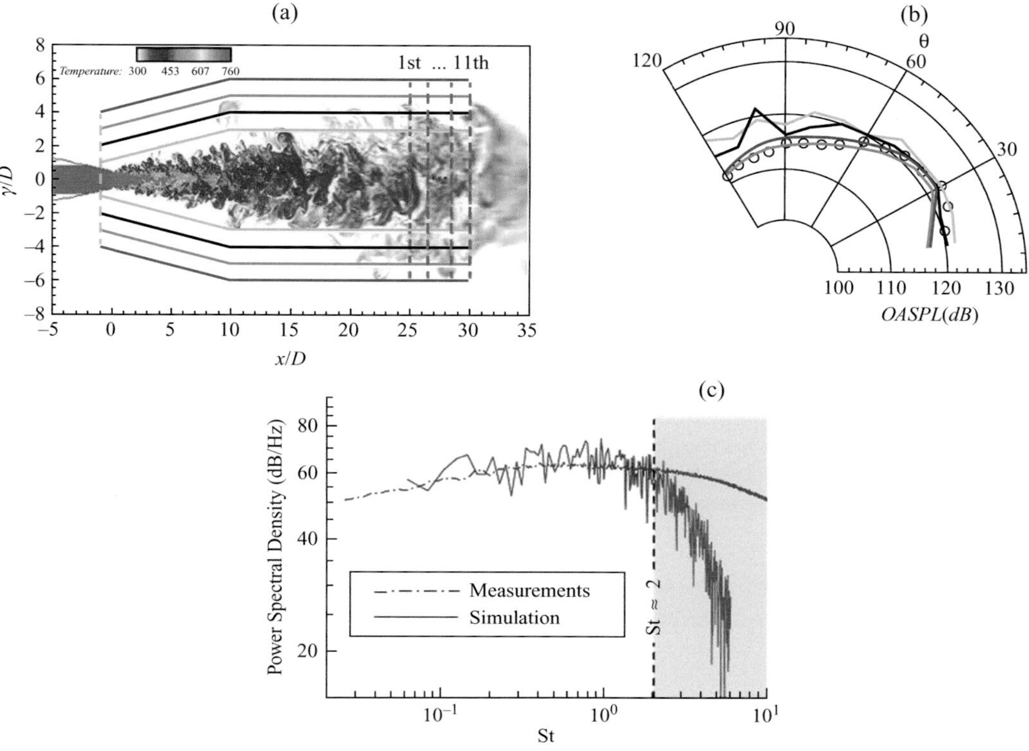

Figure 7.44. Post-processing far field sound from jet noise simulations: (a) surfaces where unsteady data stored, (b) plots of far field sound against observer angle and (c) spectrum of far field sound (Frame (c) adapted from Xia et al. 2009).

know, with any certainty, ahead of making the simulation. Hence, it is best to store data at multiple surfaces. This avoids expensive mutiple reruns. An example of this process for jets is shown in Figure 7.44a. The assembled far field sound directivity plot is shown in Frame (b). This is based on integration of spectral data. Clearly the surfaces in the rotational flow zone have poor accord with the measurements. Returning to the spectral plot (this time for a different jet nozzle) the vertical line gives an estimate of the *St* limit for the grid, based on assuming four grid points are needed to resolve an acoustic wave length for an ambient speed of sound of c_∞. Hence, $St_{max} = c_\infty \ / \ 4\Delta$, where Δ is the grid spacing at the integral surface.

As can be seem, this simple Strouhal number estimate is in reasonable accord with the observed simulation traits. The other aspect to note is that to improve the shape of the spectrum, at the low frequency end a larger data time sample would be helpful. However, with the small time-steps necessary to resolve the high frequencies, this makes the run times extremely long, presenting a substantial challenge. On the other hand, to resolve the higher frequencies, by say a factor of 2, would give rise, for a three-dimensional problem, to an increase of nearly a factor of 10 in grid.

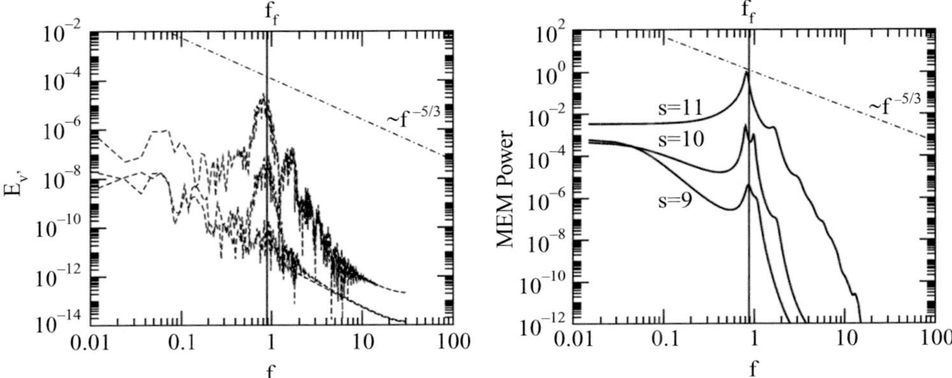

Figure 7.45. Energy spectra evaluted from Discrete Fourier Transform – DFT – (left) and Maximum Entropy Method – MEM – (right) for three axial locations in a shear layer (chain dashed line indicates –5/3 law) (from Balzer & Fasel 2013 – published with the kind permission of the ASME).

To form a spectrum, a Discrete Fourier Transform (DFT) can be used, for the fluctuating quantity of inteterest, ϕ', such that

$$\phi'_\omega = \frac{1}{N}\sum_{n=0}^{N-1}\phi'_n e^{i\omega t_n} \tag{7.91}$$

where N is the number of time samples and $\omega = 2\pi f$ is the circular frequency. Then the spectral energy can be expressed as

$$E_{\phi'} = \frac{1}{2}\phi'_\omega \phi'^{*}_\omega \tag{7.92}$$

Note that the * represents the complex conjugate. Ghil (2002) gives a range of alternative approaches for producing energy spectra applicable to different types of data sets. The Maximum Entropy Method (MEM) is intended for use with broadband noisy signals. Figure 7.45 compares energy spectra evaluted from DFT (left) and MEM (right) for three axial locations in a shear layer. These graphs are taken from Balzer & Fasel (2013). The chain dashed line indicates the –5/3 law gradient. The results for the two sets of results are broadly similar and indicate tonal elements to the flow. These could be associated with Kelvin-Helmholtz rollers, for example, in a transitioing shear layer. For the use of MEM with DNS for a turbine related flow see Balzer & Fasel (2013).

7.12.3 *Stability Analysis*

Performing a stability analysis based on mean CFD velocity profiles can be useful. For two-dimensional parallel viscous flow, the Orr-Sommerfeld equation can be used to determine if the Navier-Stokes equations are unstable and the frequency

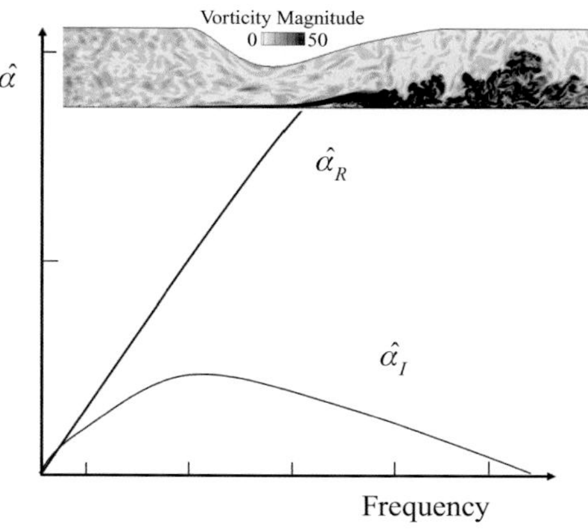

Figure 7.46. Linear stability analysis and the corresponding dispersion plot form.

of the instability. The Orr-Sommerfeld equation for a vertical velocity perturbation v', perpendicular to a local parallel flow velocity U_0, is given as

$$\left[\left(U_0 - \frac{\omega_R}{\hat{\alpha}}\right)\left(\frac{d^2}{dy^2} - \hat{\alpha}^2\right) - \frac{d^2}{dy^2}U_0 + \frac{i}{\hat{\alpha}Re}\left(\frac{d^2}{dy^2} - \hat{\alpha}^2\right)^2\right]v' = 0 \tag{7.93}$$

The disturbance takes the form $v' \propto e^{\omega t - i(\hat{\alpha}_R + i\hat{\alpha}_i)x}$. Also, $\hat{\alpha}$ is the complex (streamwise) wave number $\hat{\alpha} = \hat{\alpha}_R + i\hat{\alpha}_I$ and ω_R the real frequency. Note that $\hat{\alpha}_R$ gives the wave number (an inverse function of wavelength). On the other hand, $\hat{\alpha}_I$ defines stability. The flow is stable for $\hat{\alpha}_I > 0$ with an exponential decay. Conversely it is unstable $\hat{\alpha}_I < 0$ with an exponential growth. For invsicid flow, letting Re tend to infinity in Equation (7.93) gives the simpler Rayleigh equation that can be solved instead of the Orr-Sommerfeld. Figure 7.46 shows the traits of a dispersion plot. The process of generating this is to form a mean velocity profile from the eddy-resolving simulation and then feed this into the stability equation. At the maximum growth rate $(\hat{\alpha}_I)$, the dominant cyclic frequency $(\omega = 2\pi f)$ and wave number $\left(\hat{\alpha}_R = 2\pi/\lambda\right)$ can be estimated. This will generally give frequencies and wavelengths consistent with the LES data from the inset above Figure 7.46. For the use of the Rayleigh equation to sudy LES based flow field stability, see Nagabhushana (2014).

7.12.4 *Proper Orthogonal Decompostion*

Eddy-resolving simulation data can be analysed using POD, thus allowing a low-order system model to be produced. This model could be used as part of a large-scale coupled simulation or as part of a flow control model. Large-scale coherent structures embedded in a turbulent flow can be identified using POD. The input to the POD analysis is a number, N, of time snapshots of the flow field that are absorbed into a spectral type representation of the flow. The velocity components of the snapshots can be formed into the matrix

$$\mathbf{U} = \left[u^1....u^N, v^1....v^N, w^1....w^N\right] \tag{7.94}$$

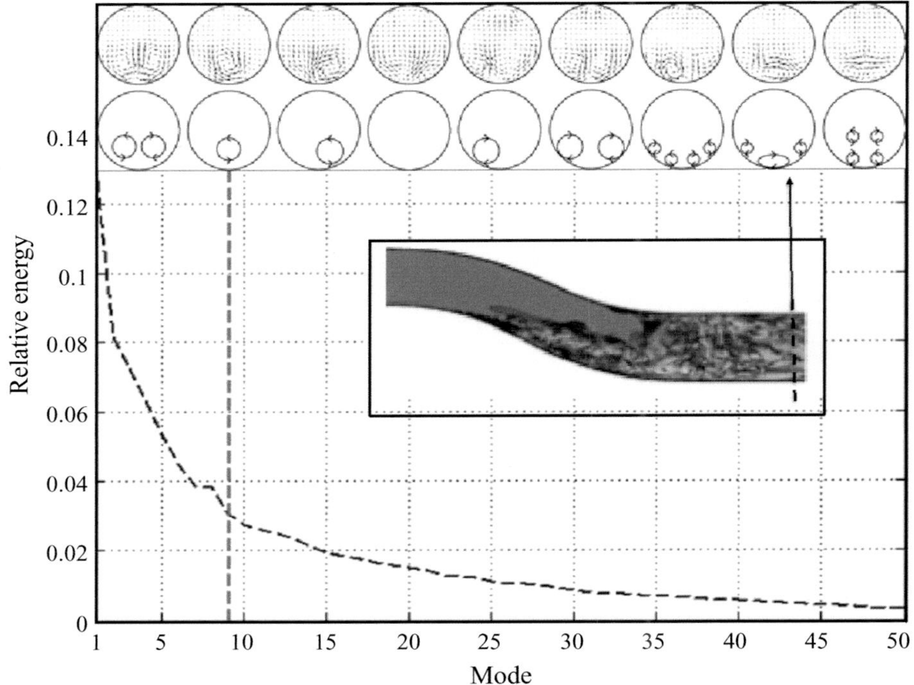

Figure 7.47. Relative energy or eigenvalue spectrum flow in a duct (upper images give the shapes of the POD modes).

A time correlation tensor, C, is then formed as a product of all the snapshots (see Nagabhushana 2014). Through solving the eigenvalue problem

$$CA_i = \lambda_i A_i \tag{7.95}$$

the eigenvalues, λ_i, of the matrix C are found. Note that A_i is the eigenvector. The eigenvalues are arranged in ascending order to form a spectrum. An example of the use of POD is given in Figure 7.47. The plot gives relative energy against mode number for the flow in an idealized aeroengine intake duct. The POD analysis is performed at the right-hand side of the duct shown in the inset of the figure. The idea is to study the nature of the unsteady flow experienced by a fan dowsntream of the duct. The flow in the inset is rendered visible through isosurfaces of the Q-criterion. The upper sequence of images in the figure identify the various POD modes in the duct. The energy in these modes decreases from left to right and this is reflected in the energy plot. To the left of the vertical dashed line, 90% of the flow energy resides. This corresponds to the first nine POD modes.

The flow complexity modelled or observed (see insets) can be increased by considering more modes. Typically, for flows that best suit POD representation, the lower order modes have distinctly more energy. A more modern POD variant is the BPOD approach of Ilak & Rowley (2008). Standard POD is good at characterizing the generally most important, most energetic modes, in the flow. However, for certain flows the modes with low energy levels can be of importance. For the use of POD to analyse DNS for a turbine representative flow, see Balzer & Fasel (2013).

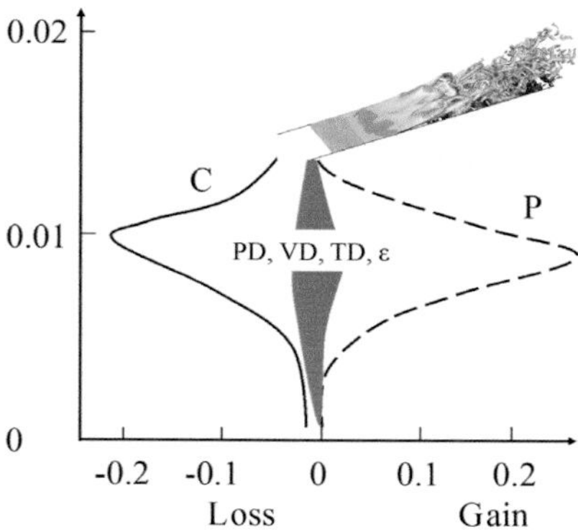

Figure 7.48. Pretransitional turbulence energy budget form for a separated shear layer.

7.12.5 *Turbulence Energy Budgets*

For eddy-resolving simulations, for assessing and refining turbulence models it can be helpful to extract turbulence energy budgets. These can be expressed as

$$\frac{\partial k}{\partial t} + C = P - \varepsilon - \frac{\partial}{\partial x_i}(TD + PD + VD) \tag{7.96}$$

where

$$\begin{aligned} P &= -\left\langle u'_i u'_j \right\rangle \frac{\partial \langle u_i \rangle}{\partial x_j}, \varepsilon = \upsilon \left\langle \frac{\partial u'_i}{\partial x_j}\left(\frac{\partial u'_i}{\partial x_j} + \frac{\partial u'_j}{\partial x_i}\right)\right\rangle, C = \langle u_i \rangle \frac{\partial k}{\partial x_i}, \\ TD &= 0.5\left\langle u'_i u'_j u'_k \right\rangle, PD = \langle p' u'_i \rangle, VD = -\upsilon \frac{\partial k}{\partial x_j} \end{aligned} \tag{7.97}$$

These terms correspond to the turbulence production, dissipation and convection, respectively. Furthermore, *TD*, *PD* and *VD* correspond to the diffusion budget, representing turbulent, pressure and viscous diffusion, respectively.

Figure 7.48 gives representations of pretransitional turbulence energy budgets for a separated shear layer. From the energy budgets it is evident, for example, that the production and convection terms are dominant in the separated shear layer. Note the *PD*, *VD*, *TD* and ε components are similar and simplify the plot represented by the shaded envelope. Using energy budgets for both rough and smooth surfaces, Nagabhushana (2014) devise a transition model for roughened surfaces. Notably, it is possible to freeze mean flow fields from the LES and, based on the energy budgets, devise precise model terms to match them. However, there is no guarantee that when the velocity and turbulence model equations are run together that the target eddy-resolving result will be replicated.

7.12.6 *Anisotropy Invariant Map and Space-Time Plots*

The Lumley triangle or anisotropy invariant map (see Pope 2000) can be useful for understanding turbulence physics and thus potentially refining turbulence

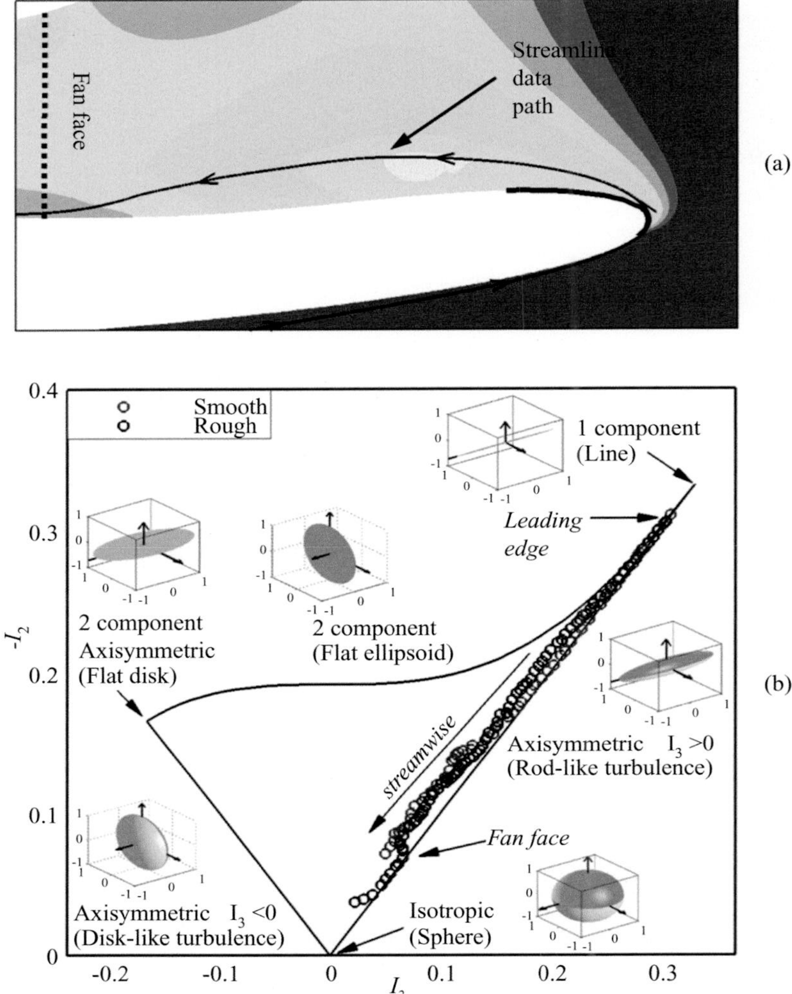

Figure 7.49. Anisotropy invariant maps for the Reynolds stress profiles extracted in a separated flow region (from Oriji et al. 2014 – published with permission of the ASME).

models. Figure 7.49, takes Reynolds stress profiles through the separation bubble from an eddy-resolving simulation in an aeroengine intake configuration. The precise location of the line along which the data is analysed is given in the image above the invariant map. The map shows the nature of the turbulence in the different zones. It indicated that with increasing distance from the intake lip (right-hand side), the turbulence initially has a quasi one-component state (zone indicated in the diagram). This ultimately leads, after passage through the shear layer, to relatively isotropic turbulence. For certain flows, the map can suggest the importance of accounting for anisotropy when modelling turbulence.

Table 7.4. *Output parameters for eddy-resolving simulations*

Quantity	Uses	Disadvantgaes
Reynolds stresses	Understanding flow physics, validation	The run time period increases dramatically with the descending location in this table
Correlation coefficents	Understanding flow physics and simulation fidelity, deeper level of validation, explore if extent of periodic domain boundaries is sufficient	Increasing computational cost and storage
Turbulence energy budgets	Useful for understanding flow physics and also refining RANS models	Quite involved to evaluate
Spectra	Understanding flow physics and simulation fidelity, deeper level of validation	Increasing computational cost. Long run time is needed to generate a clean spectra

Contours of flow variables in a space-time plane can be useful for understanding flow physics, the nature of turbulence structures and their convection velocity. They are quite often used for this purpose in turbomachinery blade flows (see, for example, Nagabhushana 2014).

Table 7.4 summarizes some of the potentitial outputs from eddy-resolving simulations and some uses and disadvantages.

7.13 Post Processing Massive Data Sets

7.13.1 *Data Mining and Analytics*

Data mining and analytic techniques are useful for spotting trends/patterns in massive data sets and hence see some use in CFD – Gosnell et al. (2012). Such techniques involve the following fields: statistics, operations research, artificial intelligence, machine learning (systems that can learn from data) and database management systems. A necessary prerequisite is the ability to automatically detect flow features.

Although the speed of computers in the raw processing of numbers has dramatically increased, the ability to read in and write out data has not. Hence, there is an increasing disparity between what can be computed and the handling of data. As noted by Gourdain et al. (2014), for an eddy-resolving simulation of a compressor, to write out the data for a full revolution would need about 100Tb of disk space. The large amount of data involved in eddy-resolving simulations and the time in writing it out pushes CFD practitioners to the following practices:

1) Writing out data for selected planes.
2) Evaluating quantities of interest on the fly and thus avoiding having to write out large amounts of data to do the associated post-processing later.

As noted by Nichols (2013), the issue of quickly interrogating large data sets to gain needed information can be assisted by using technology such as used by Google in their search engines. Clearly the technology used by Google has the ability to search for data across disparate, many thousands of disks, at high speed. What is more, the technology is resilient to broken (searching on backup disks) or disks being replaced ('hot swapping').

They key is the use of the 'MapReduce' query. With this, the code (post-processing task) is sent to the data (the map phase) on multiple disks. Then the key result is returned – reduce phase. Hence, this is opposite to the usual process, where the data is sent from the numerous disks to the code. The process results in dramatically less data being transferred. The use of MapReduce to extract far field sound for an eddy-resolving simulation is shown by Nichols. Difficulties with the approach are that the Google based procedure is text based and converting floating point numbers to text, for text-based searches is not that efficient. Stored binary numbers are very compact and computers are not efficiently configured to make such transformations. They process numbers more efficiently. However, it seems that these are not insurmountable problems. Note that open source implementations are available such as Hadoop (see Gourdain et al. 2014)).

7.14 Verification and Validation

7.14.1 *Basic Core Solver Tests*

Continuous tests need to be made to ensure that the basic core solver functionality has not been contaminated by newer code developments. For a large industrial organization, an automated test suite is important. For RANS/steady flows, quite penetrating tests can be made relatively quickly. Nonetheless, for an outline of the complex challenges involved in the verification (and validation) of CFD with RANS turbulence models, see Rumsey (2014). A critical part of this, prior to validation, is using the method of manufactured solutions (MMS) to check the numerical implementation. Such manufactured solutions can be formulated with turbulent flow in mind. For eddy-resolving simulations, the task is more challenging. Tollmien-Schlichting waves and other more canonical unsteady flow cases that have analytical solutions look suitable. However, they generally do not have the wide range of scales found in turbulent flows and hence are not as challenging and revealing.

Homogenous decaying turbulence can be run relatively quickly. However, triple periodic boundary conditions are needed. Hence, for a large and complex industrial, perhaps unstructured solver the case can be quite challenging to initialize, run and post process. Ambiguities can arise regarding the treatment of domain corner cells. If the flow solver is compressible then careful Mach number scaling is needed or the complexity of using preconditioning introduced. For academic solvers, clearly tests are needed but generally these are less automated.

The Taylor-Green vortex has a simpler initialization than homogenous decaying turbulence (However, it is probably not as effective a test case). It has the form

$$\begin{aligned} u &= +U_o sin\,(x)\, cos\,(y)\, cos\,(z)\,, \\ v &= -U_o cos\,(x)\, sin\,(y)\, cos\,(z)\,, \\ w &= 0, \\ p &= P_o + \tfrac{\rho_o U_o^2}{16}\,(cos\,(2x) + cos\,(2y))\,(cos\,(2z) + 2) \end{aligned} \tag{7.98}$$

where the subscript 'o' quantities are reference values. Figure 7.50 shows the development of the flow with time. Initially, there are large-scale inviscid structures and these rapidly break down, resulting in a range of scales. The dissipation is directly related to enstrophy and hence easily evaluated. Similarly, the turbulent kinetic energy is easily evaluated through time and its decay rate exponents ($\partial k/\partial t \propto t^{-n}$) compared with established n values and other more detailed DNS data. A bi-periodic channel flow case is perhaps less challenging to set up than homogeneous decaying turbulence. However, the run time is considerably longer. Validation data is readily available in terms of the 'law of the wall' and associated turbulence quantities.

It needs to be remembered that RANS based simulations are broadly regarded as postdictive. Hence, when it has been verified that the correct flow governing equations are being solved, then some form of validation against experimental data is necessary. This is especially so when applying CFD to a new area. Generally, there is the situation where the data closest to the practical design problem being solved is sparse and of limited accuracy. Typically, accurate detailed data is only available for more canonical cases. Hence, validation would typically involve canonical cases that are as close as possible to the intended industrial application, supplemented by more complex geometry cases. For complex geometry codes that have more advanced turbulence models, it can be hard to gain consistent results for essentially the same parameters and grids. Figure 5.28 shows surface static pressure and passage entropy function contours as stall is approached for a transonic fan. The results are for two different reputable CFD codes but the same reported turbulence model and grid. The solution in Frame (a) has more reversed flow and a reduced pressure ratio. Hence, there is a clear numerical framework impact or some other aspect that is hard to define.

When considering validation, it can be helpful to define approximate validation levels. Table 7.5 extends that given by Sagaut & Deck (2009), defining levels of validation largely aimed at eddy-resolving simulations in aerodynamics. Based on an extensive review of literature, this grading system is used by Tucker (2013), to explore the level of validation currently used for various areas of propulsion unit and airframe design.

For acoustics and coupled problems, this table can be adapted and extended. For example, Table 7.6 gives additional acoustic validation levels that can be used in conjunction with those in Table 7.5. Table 7.6 indicates if an integral quantity for far field sound is validated against or comparison extended to the spectral level.

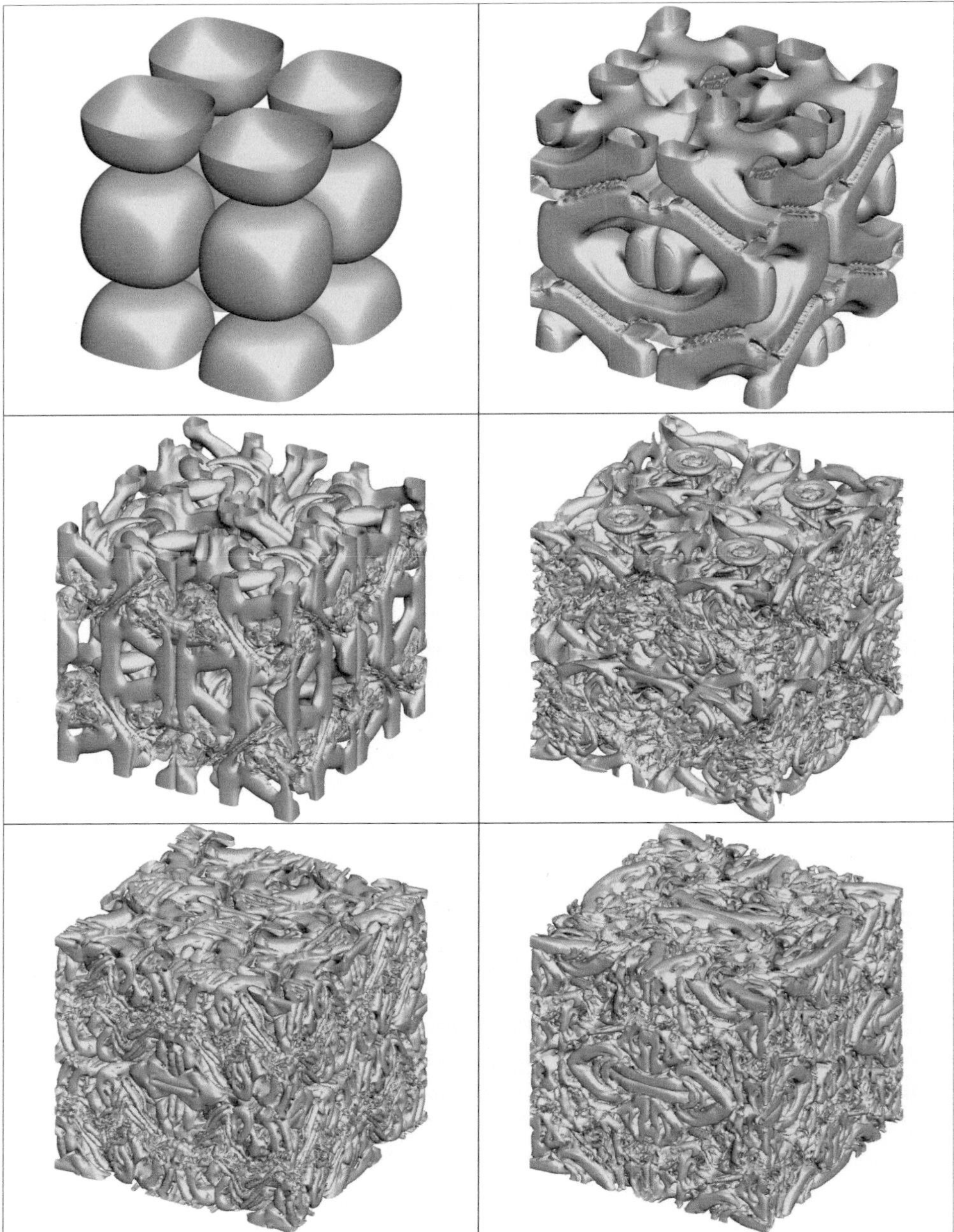

Figure 7.50. Pressure isosurfaces coloured by Mach number for a Taylor-Green vortex at different points in time (from Vermeire et al. 2014).

Note that Grade 3 includes flow field measurements, explicitly tailored to explore sound source information. Tucker (2013) gives an equivalent, tentative validation level table for coupled problems. Clearly, for combustion, multi-phase problems and other more specialized niche industrial applications, substantially altered levels of validation could be formulated.

Table 7.5. *Level of simulation validation (extended version of that given by Sagaut & Deck 2009) for aerodynamics (from Tucker 2013 – used with the kind permission of Springer)*

Grade	Validation Level
1	Integrated forces on structure – lift, drag and moments, etc.
2	Characteristic curve points or flow coefficient, sealing effectiveness, etc.
3	Time averaged velocity, pressure, temperature, surface adiabatic film cooling effectiveness and Nusselt number profiles. Also, could include exit flow angle, total pressure loss profiles and surface shear stress
4	Second order statistics (*rms* quantities) typically for turbulence
5	One-point spectral analysis
6	Two-point space/time correlations
7	High-order space/time correlations

Table 7.7 (again from Tucker 2013) summarizes the typical validation level found in the propulsion and airframe zone aerodynamics excluding acoustics and coupled problems. The data is based on an extensive literature survey. A key point to observe is that for airframes, there is more availability of higher-level data and hence more in-depth eddy-resolving simulation validation. The lower level, for propulsion validation data, perhaps partly relates to the fact that measurements in rotating machinery are generally more challenging, particularly if combustion is involved.

Many potential validation data sets, for commercial confidence reasons, do not define geometry and other potentially key parameters. This restricts the practical extent of their use to a very limited number of stakeholders. The crucial need for validation data, in combustion modelling, is discussed by Swaminathan and Bray (2011). The lack of basic validation data for combustion models is also noted by Pitsch (2006). With regards to validation data, the need to accurately measure inflow conditions is often critical. However, frequently data sets do not even give mean inflow quantities. Typically, relative to pure aerodynamic and aeroacoustic problems, for coupled problems, the general level of validation is lower. This is perhaps not surprising. The general complexity of an experiment for a coupled system is higher. Where this involves moving components, making aerodynamic measurements is more complex.

Table 7.6. *Validation level definitions for aeroacoustics to be used in conjunction with those for aerodynamics (from Tucker 2013 – used with the kind permission of Springer)*

Grade	Validation data
1	Integral far field sound quantity
2	Far field sound spectra
3	Space-time source correlations or source spectral information linked to far field sound

Table 7.7. *Summary of validation levels in the propulsion and airframe zones for aerodynamics*

	Available validation data		LES and hybrids validation	
Level	Propulsion (%)	Airframe (%)	Propulsion (%)	Airframe (%)
1	62	81	11	59
2	100	100	75	71
3	17	81	14	46
4	7	28	0	24
4+	1	0	0	0

Problems

1) Derive the three eigenvalues $\lambda_1 \leq \lambda_2 \leq \lambda_3$ of the symmetric matrix $S^2 + \Omega^2$
2) Show that $Q = -(\lambda_1 + \lambda_2 + \lambda_3)/2$
3) Show that the eikonal equation can be written as a convective transport equation for wall distance with a right-hand source term of unity.
4) Considering three solutions ϕ_1, ϕ_2 and ϕ_3 on grids of Δx_1, Δx_2 and Δx_3, show using Taylor's series that the following can be written

$$\frac{\phi_1 - \phi_2}{\phi_2 - \phi_3} = \frac{\Delta x_1^n - \Delta x_2^n}{\Delta x_2^n - \Delta x_3^n} \tag{7.99}$$

5) Show, based on numerical solutions on three different grids, that the order, n, of a numerical scheme can be evaluated from

$$n = \ln \frac{\left[\frac{\phi_1 - \phi_2}{\phi_2 - \phi_3}\right]}{\ln \lambda} \tag{7.100}$$

Note λ is a grid spacing ratio. Comment on the limitations of the use of this equation.

6) Implement the use of the plot3d file format for the input and output used in your Euler or Navier-Stokes solvers.
7) For your Euler solutions, plot contours of total pressure, stagnation enthalpy and entropy function. Explore these contour fields for different levels of numerical smoothing and different numerical schemes. These aspects are outlined in Chapter 4.
8) Use the Taylor series expression (7.100) to estimate the order of your Navier-Stokes or Euler solver. (Note that this will require surprisingly fine grids to be inside the radius of convergence of Taylor's series.)

REFERENCES

AHMED, M. H. & BARBER, T. J. 2005. Fast fourier transform convergence criterion for numerical simulations of periodic fluid flows. *AIAA Journal*, 43(5), 1042–1052.

ANDERSSON, N., ERIKSSON, L.-E. & DAVIDSON, L. 2005. Large-eddy simulation of subsonic turbulent jets and their radiated sound. *AIAA Journal*, 43(9), 1899–1912.

AUPOIX, B. & SPALART, P. 2003. Extensions of the Spalart–Allmaras turbulence model to account for wall roughness. *International Journal of Heat and Fluid Flow*, 24(4), 454–462.

BALZER, W. & FASEL, H. F. 2013. Direct numerical simulations of laminar separation bubbles on a curved plate: part 1 – simulation setup and uncontrolled flow. *Proceedings of ASME Turbo Expo 2013, 3–7 June, San Antonio*, Texas, ASME Paper No. GT2013–95277, 2013.

BASHA, M., AL-QAHTANI, M. & YILBAS, B. S. 2009. Entropy generation in a channel resembling gas turbine cooling passage: Effect of rotation number and density ratio on entropy generation. *Sadhana*, 34(3), 439–454.

BATTEN, P. et al., 2011. Smart sub-grid scale models for LES and hybrid RANS/LES. 6th AIAA Theoretical Fluid Mechanics Conference, 27–30 June, Honolulu, Hawaii, AIAA Paper No. AIAA-2011–3472 (2011).

BOGER, D. A. 2001. Efficient method for calculating wall proximity. *AIAA Journal*, 39(12), 2404–2406.

CELIK, I., KLEIN, M. & JANICKA, J. 2009. Assessment measures for engineering LES applications. *Journal of Fluids Engineering*, 131, 031102.

DAWES, W. N. 2007. Turbomachinery computational fluid dynamics: asymptotes and paradigm shifts. *Philosophical Transactions Series A: Mathematical, Physical, and Engineering sciences*, 365(1859), 2553–85.

DAWES, W. N., DHANASEKARAN, P. C., DEMARGNE, A. A. J., KELLAR, W. P., & SAVILL, A. M. 2000. Reducing bottlenecks in the cad-to-mesh-to-solution cycle time to allow CFD to participate in design, *ASME Journal of Turbomachinery*, 123(3), 552–557.

DENTON, J. D. 1993. The 1993 IGTI scholar lecture: loss mechanisms in turbomachines. *Journal of Turbomachinery*, 115(4), 621.

DENTON, J. D. 2010. Some limitations of turbomachinery CFD. ASME Turbo Expo 2010: Power for Land, Sea, and Air, ASME Paper No. GT2010-22540, (7), 735–745.

DE VAHL DAVIS, G. 1983. Natural convection of air in a square cavity: A bench mark numerical solution. *International Journal for Numerical Methods in Fluids*, 3(3), 249–264.

EASTWOOD, S. J. 2009. *Hybrid LES – RANS of complex geometry jets*. PhD Thesis, University of Cambridge.

FARES, E. & SCHRODER, W. 2002. A differential equation for approximate wall distance. *International Journal for Numerical Methods in Fluids*, 39(8), 743–762.

FERZIGER, J. & PERIĆ, M. 1996. *Computational methods for fluid dynamics*, Springer.

GHIL, M. 2002. Advanced spectral methods for climatic time series. *Reviews of Geophysics*, 40(1), 1003.

GOLDSTEIN, M. E. 2003. A generalized acoustic analogy. *Journal of Fluid Mechanics*, 488, 315–333.

GOSNELL, M., WOODLEY, R. & GORRELL, S. 2012. Results of mining data features during computational fluid dynamics simulations. International Conference on Data Mining, DMIN'12, pp. 3–9.

GOURDAIN, N., SICOT, F., DUCHAINE, F. & GICQUEL, L. 2014. Large eddy simulation of flows in industrial compressors: a path from 2015 to 2035. *Philosophical Transactions of the Royal Society A: Mathematical, Physical and Engineering Sciences*, 372, 20130323.

GRABNER, M. AND LARAMEE, R. S. 2005. Image space advection on graphics hardware. *Proceedings of the Twenty first Spring Conference on Computer Graphics 2005* (SCCG 2005), May 12–14, 2005, Budmerice, Slovakia, 75–82.

HAIMES, R. 1994. pV3: A distributed system for large-scale unsteady CFD visualization. 32nd Aerospace Sciences Meeting and Exhibit, January, AIAA Paper No. AIAA-94-0321.

HALLER, G. 2005. An objective definition of a vortex. *Journal of Fluid Mechanics*, 525, 1–26.

HOLLEY, B. M., BECZ, S. & LANGSTON, L. S. 2006. Measurement and Calculation of Turbine Cascade Endwall Pressure and Shear Stress. *Journal of Turbomachinery*, 128(2), 232.

IACCARINO, G. 2008. *Quantification of uncertainty in flow simulations using probabilistic methods*, VKI Lecture Series, Non-equilibrium gas dynamics from physical models to hypersonic flights, 8–12 September.

ILAK, M. & ROWLEY, C. W. 2008. Modeling of transitional channel flow using balanced proper orthogonal decomposition. *Physics of Fluids*, 20(3), 034103.

JAMESON, A., 2008. The Construction of Discretely Conservative Finite Volume Schemes That Also Globally Conserve Energy or Entropy. *Journal of Scientific Computing*, 34(2), 152–187.

LAUNDER, B. E., REECE, G. J. & RODI, W. 2006. Progress in the development of a Reynolds-stress turbulence closure. *Journal of Fluid Mechanics*, 68(3), 537.

LIU, Y. & TUCKER, P. G. 2007. Contrasting zonal LES and non-linear zonal URANS models when predicting a complex electronics system flow. *International Journal for Numerical Methods in Engineering*, 71(1), 1–24.

LÖHNER, R. 1996. Progress in grid generation via the advancing front technique. *Engineering with Computers*, 12, 186–210.

MANI, K. & MAVRIPLIS, D. 2007. Discrete adjoint based time-step adaptation and error reduction in unsteady flow problems. 18th AIAA Computational Fluid Dynamics Conference, Reston, Virigina. AIAA Paper No. AIAA-2007-3944.

MCLOUGHLIN, T., LARAMEE, R. S., AND ZHANG, E. 2010. Constructing streak surfaces in 3d unsteady vector fields, Proceedings of the 26th Spring Conference on Computer Graphics 2010 (SCCG 2010), 13–15 May, Budmerice, Slovakia, 25–32.

MESRI, Y., DIGONNET, H. & GUILLARD, H. 2005. Mesh partitioning for parallel computational fluid dynamics applications on a grid. *Finite Volumes for Complex Applications IV*. Hermes Science Publisher, 631–642.

MIURA, H. & KIDA, S. 1997. Identification of tubular vortices in turbulence. *Journal of the Physics Society Japan*, 66(5), 1331–1334.

MOCKETT, C., KNACKE, T. & THIELE, F. 2010. Detection of initial transient and estimation of statistical error in time-resolved turbulent flow data. Proceedings of ETMM8, Marseille, France, June 9–11.

NAGABHUSHANA, R. V. 2014. *Numerical investigation of separated flows in low pressure turbines: current status and future outlook*, PhD Thesis, School of Engineering, University of Cambridge.

NICHOLS, J. 2013. Aeroacoustic post-processing with MapReduce. Annual Research Briefs, Centre for Turbulence Research, Stanford University.

ORIJI, U. R. & TUCKER, P. G. 2013. Modular turbulence modeling applied to an engine intake. *Journal of Turbomachinery*, 136(5), 051004.

ORIJI, U. R., YANG, X. & TUCKER, P. G. 2014. Hybrid RANS/ILES for aero engine intake. *ASME Turbo Expo 2014: Turbine Technical Conference and Exposition*, American Society of Mechanical Engineers, ASME Paper No. GT2014-26472.

PITSCH, H. 2006. Large eddy simulation of turbulent combustion. *Annual Review of Fluid Mechanics*, 38(1), 453–482.

POPE, S. 2000. *Turbulent flows*, Cambridge University Press.

PRAKASH, C. & LOUNSBURY, R. 1986. Analysis of laminar fully developed flow in plate-fin passages: effect of fin shape. *Journal of Heat Transfer*, 108(3), 693.

PULLAN, G., DENTON, J. & CURTIS, E. 2006. Improving the performance of a turbine with low aspect ratio stators by aft-loading. *Journal of Turbomachinery*, 128(3), 492.

RAZAFINDRALANDY, D., HAMDOUNI, A. & BÉGHEIN, C. 2007. A class of subgrid-scale models preserving the symmetry group of Navier–Stokes equations. *Communications in Nonlinear Science and Numerical Simulation*, 12(3), 243–253.

RICHARDSON, L. F. 1911. The Approximate Arithmetical Solution by Finite Differences of Physical Problems Involving Differential Equations, with an Application to the Stresses in a Masonry Dam. *Philosophical Transactions of the Royal Society A: Mathematical, Physical and Engineering Sciences*, 210(459–470), 307–357.

RIERA, W. 2014. *Evaluation of the ZDES method on an axial compressor: analysis of the effects of upstream wake and throttle on the tip-leakage flow*. PhD Thesis, Onera Meudon.

ROACHE, P. J. 1994. Perspective: a method for uniform reporting of grid refinement studies. *Journal of Fluids Engineering*, 116, 405–413.

ROTH, M. & PEIKERT, R. 1996. Flow visualization for turbomachinery design. Proceedings of the 7th Conference on Visualization, 381–384.

RUMSEY, C. L. 2014. Turbulence modeling verification and validation. AIAA SCiTech. AIAA Paper No. AIAA-2014–0201.

SAGAUT, P. & DECK, S. 2009. Large eddy simulation for aerodynamics: status and perspectives. *Philosophical Transactions Series A: Mathematical, Physical, and Engineering Sciences*, 367(1899), 2849–2860.

SAHNER, J. 2009. *Extraction of vortex structures in 3D flow fields*. PhD Thesis, der Otto-von-Guericke-Universität Magdeburg.

SETHIAN, J. 1999. *Level set methods and fast marching methods: evolving interfaces in computational geometry, fluid mechanics, computer vision, and materials science*, Cambridge University Press.

SHEN, H. & KAO, D. 1998. A new line integral convolution algorithm for visualizing time-varying flow fields. *Visualization and Computer Graphics, IEEE Transactions*, 4(2), 98–108.

SHUR, M. et al., 2003. Towards the prediction of noise from jet engines. *International Journal of Heat and Fluid Flow*, 24(4), 551–561.

SPALART, P., 2000. Trends in turbulence treatments. American Institute of Aeronautics and Astronautics Fluids Conference, 19–22 June, Denver, CO, AIAA Paper No. AIAA-2000-33784.

SPALART, P. R. 1992. Sophisticated Formulas for d in POEM, Boeing Internal Document, October.

SWAMINATHAN, N. & BRAY, K. N. C., 2011. *Turbulent premixed flames*, Cambridge University Press.

TUCKER, P. G. 2003. Differential equation-based wall distance computation for DES and RANS. *Journal of Computational Physics*, 190(1), 229–248.

TUCKER, P. G. 2004. Novel MILES computations for jet flows and noise. *International Journal of Heat and Fluid Flow*, 25(4), 625–635.

TUCKER, P. G. 2011a. Computation of unsteady turbomachinery flows: Part 1 – Progress and challenges. *Progress in Aerospace Sciences*, 47(7), 522–545.

TUCKER, P. G. 2011b. Computation of unsteady turbomachinery flows: Part 2 – LES and hybrids. *Progress in Aerospace Sciences*, 47(7), 546–569.

TUCKER, P. G. 2011c. Hybrid Hamilton–Jacobi–Poisson wall distance function model. *Computers & Fluids*, 44(1), 130–142.

TUCKER, P. G. 2013. Trends in turbomachinery turbulence treatments. *Progress in Aerospace Sciences*, 63, 1–32.

TUCKER, P. G. et al., 2005. Computations of Wall Distances Based on Differential Equations. *AIAA Journal*, 43(3), 539–549.

VENDITTI, D. A. & DARMOFAL, D. L. 2002. Grid adaptation for functional outputs: application to two-Dimensional inviscid flows. *Journal of Computational Physics*, 176(1), 40–69.

TUCKER, P. G. & LIU, Y. 2007. Turbulence modeling for flows around convex features giving rapid eddy distortion. *International Journal of Heat and Fluid Flow*, 28, 1073–1091.

TYACKE, J. 2009. *Low Reynolds number heat transfer prediction employing Large Eddy Simulation for electronics geometries*, Ph.D Thesis, *Civil and Computational Engineering Centre, Swansea University*.

VERMEIRE, B. C., NADARAJAH, S. & TUCKER, P. G. 2014. Canonical test cases for high-order unstructured implicit large eddy simulation. Fifty second Aerospace Sciences Meeting, Reston, Virginia, AIAA Paper No. AIAA-2014-0935.

WALATKA, P. P. et al., 1990. PLOT3D User's Manual. *NASA Ames Research Center*, NASA-TM-101067.

WIGTON, L. 1998. Optimizing CFD codes and algorithms for use on Cray computer. *Frontiers of Computational Fluid Dynamics* (Eds. Caughey A. & M. Hafez M. M.), World Scientific Publishing Co., 1–15.

XIA, H. & TUCKER, P. G. 2011. Numerical simulation of single-stream jets from a serrated nozzle. *Flow, Turbulence and Combustion*, 88(1–2), 3–18.

XIA, H., TUCKER, P. G. & COUGHLIN, G. 2012. Novel applications of BEM based Poisson level set approach. *Engineering Analysis with Boundary Elements*, 36(5), 907–912.

XIA, H., TUCKER, P. G. & EASTWOOD, S. 2009. Large-eddy simulations of chevron jet flows with noise predictions. *International Journal of Heat and Fluid Flow*, 30, 1067–1079.

YANG, X. 2014. *Numerical investigation of turbulent channel flow subject to surface roughness, acceleration, and streamline curvature*. PhD Thesis, The University of Cambridge.

ZHAO, H., 2005. A fast sweeping method for eikonal equations. *Mathematics of Computation*, 74(250), 603–627.

CHAPTER 8

Simulation in the Future

As for the future, your task is not to foresee it, but to enable it
– Antoine de Saint-Exupery, French pilot and writer

8.1 Introduction

Slotnick et al. (2014) note that in 2025 air transport will contribute 1.5 billion tonnes of CO_2 emissions annually. The aerospace industry has largely given rise to the inception of computational fluid dynamics and much of its subsequent development. It seems likely, that with modern CFD developments, it will play a paramount role in continuing to help alleviate the environmental impact of aircraft. In aerospace, CFD originally would focus on studying basic airfoils. However, nowadays it has gone well beyond this. To meet the modern multi-objective, multi-scale, coupled industrial requirements, CFD has become a highly multi-disciplinary subject area. It is hoped that this aspect has come across in the preceding text.

Figure 8.1 shows the growth of computers. The dotted line shows the growth with year of the number one fastest computer. The full line gives the 500th fastest. The dashed line gives the growth in the use of LES in an applied, industrial turbomachinery journal – the ASME *Journal of Turbomachinery*. There is a clear trend in the growth in the use of eddy-resolving simulations with increased computing power. Clearly the industrial use of eddy-resolving simulations will lag this more academic use. However, as noted by Lele and Nichols (2014), for non-wall bounded flows, eddy-resolving simulations could be used in industry in the next five to ten years. Indeed there is some current use for specific circumstances. Also, notably, Morton et al. (2007) performed eddy-resolving simulations for F/A–18 fighter configuration. Tail buffet compared well with real flight spectral data.

In this chapter, it is explored how CFD will look in the next decade and beyond. Much of these ideas are based on the contributions to the Royal Society Theme issue of Tucker and DeBonis (2014).

8.2 Computer Science and Computers

As can be seen from Figure 8.1, there is healthy growth in computing power. As noted by Jameson (2008), in the past twenty-five years, computers have become a million times faster. However, algorithmic improvements are needed that are customized for high-performance computers. It is quite power intensive to move data around in memory. Hence, to bring down power consumption, less data should be moved around. Giles and Reguly (2014) estimate 80% of CPU cycles are wasted (just 20% of peak performance due to this) because of clock cycles wasted waiting

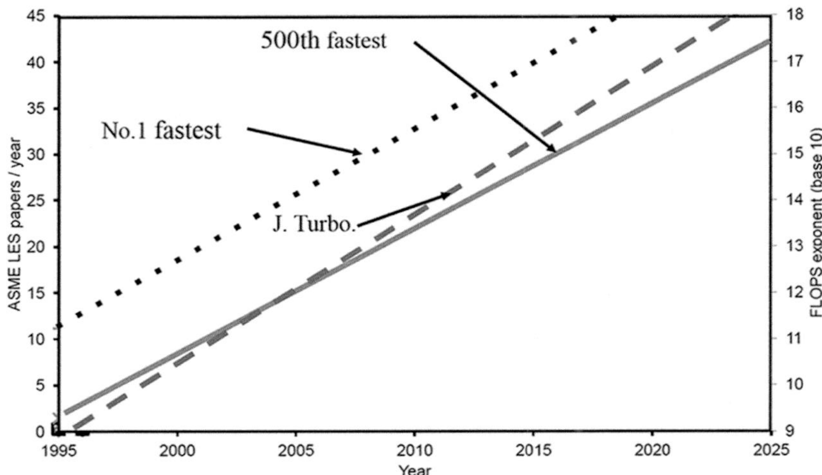

Figure 8.1. Growth of computers and eddy-resolving simulations.

for data. Also, Löhner and Baum (2013) estimate 85–90% of compute is wasted due to computations waiting for memory transfer. This is only going to increase as a problem as we move to future billion core machines. The current state of the art is a million cores. Photonics based technology (exploiting data transfer using light and its high speed) might possibly offer a way forwards (see Giles and Reguly 2014) to improve data transfer rates.

Fortunately, in the past ten years the number of floating point operations normalized by the power consumed to perform the computations has increased by around a factor of 1000. However, even so, projections forwards suggest the growth of simulations to desired scales is currently unsustainable in terms of energy resource. Notably, Graphical Processor Units (GPUs) present good energy efficiency. Improvement in load balancing could be made, going beyond just considering grid partitioning. The balancing could be adapted (using self-monitoring/learning) real time, based on iterative convergence rates. Even now, the problem sizes tackled substantially lags that of available computational resources (see Gourdain et al. 2014). A key difficulty is that CFD code time scales are much longer than computer architecture time scales. Hence, it is hard to adapt the CFD program numerics to map to the architectures. Indeed, as pointed out by Slotnick et al. (2014), exascale (10^{18}) computing is due 2018, but it is projected that CFD codes will not harness them until 2020. However, future machines are likely to be mixed architectures with CPUs and GPUs, and so forth. Intermediary software is beginning to emerge that can interface between the CFD code and computers, allowing the CFD source to be left untouched. See, for example, SBLOCK (Brandvik and Pullan 2010) and OPS for structured codes and OP2 for unstructured (Giles et al. 2011).

Fault tolerant computing will be needed. As the number of processors tends to billions, the chance of a failure substantially increases and this needs to be dealt with. Check pointing involves periodically saving data. However, this again presents a problem. Although computer raw processing has increased substantially, the input/output of data has lagged and the pre- and post-processing of the large data sets presents a great problem. Indeed, just periodically writing out data can be time consuming. This makes check pointing challenging. The volume of data is substantial. As pointed out by Gourdain et al. (2014), one revolution of a rig scale aeroengine compressor simulation (needing around a billion cells) can require 100 Tb of data. Also, the energy consumption for one revolution of the simulation will exceed that of a rig. However, just six revolutions will typically allow the cyclic flow to be reached and the rig test will involve typically millions of revolutions. With regards to reducing data output volume, a simple strategy is to just write data out at points or planes. However, it is difficult to know ahead of time which are the most important locations to write data. Google/Facebook technology can quickly search massive data sets and interrogate them (see Gourdain et al. 2014, Lele and Nichols 2014). In essence, rather than let a computer code deal with the whole data set, code is sent to the data and the processed result sent back. This type of Google technology allows disks to be removed while the machine is still running (hot swapping), without the job being interrupted or any loss of data. This is really important when the computer might have thousands of disks.

As noted by Lele and Nichols (2014), it is important, where possible, that as well as algorithm developments seeking to gain maximum compatibility with computer architectures, developments in computers should try to consider core operation and algorithm needs in the area of computer science.

8.3 Geometry Handling and Grid Generation

Even now, as discussed in Chapter 7, geometry handling for highly complex geometries is still a challenging and time consuming process. The use of explicit geometry could help with this. Also, for deforming geometries, the process of aligning the geometries for a fluid and structure robustly is not always trivial. See for example Voigt et al. (2010) for a state of the art approach that still has limitations for highly complex geometries.

There is also substantial room for improvements in mesh generation. Current eddy-resolving grids are far from optimal, with many wasted cells. Indeed the optimal eddy-resolving simulation mesh topology, even for a simple boundary layer is highly complex as shown in Chapter 3. Work in the area is beginning to emerge – see Addad et al. (2008). It is tempting when considering moving to eddy resolving simulations to reduce cost through exploiting the greater speed of structured grid CFD solvers. However, this neglects consideration of the flow physics driven, complex mesh topologies, needed for optimal eddy resolving simulation grids.

If high-order methods take root, then there will be the critical need for high-order mesh generators. This is an area currently in its infancy. There is the critical need to mesh highly complex geometries. As shown in Figure 8.2, Frame (a) octree meshes can be applied for highly complex geometries but they can be wasteful. Also, they are suboptimal for high speed boundary layers and for the case shown are not-ideal for capturing the detached shear layers. The quasi-axisymmetric shear layer zones will be supported by differing grids. Hence, it will be hard to consistently

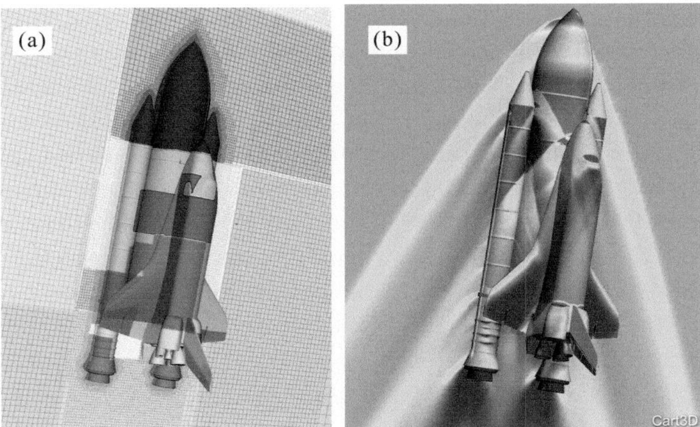

Figure 8.2. Use of octree meshing for the complex geometry space shuttle launch vehicle: (a) mesh and (b) flow field (Kindly supplied by M. Aftosmis).

resolve the shear layers. Note, the shades in Frame (a) indicate the parallel domain decomposition and Frame (b) gives the flow field for a flight Mach number of 2.6. It is likely that a range of mesh generators will be needed that are optimal for different applications. Certainly, octree meshes are a powerful tool for massively complex geometries. On the other hand, RANS based high-speed boundary layer computations need extremely high aspect ratio near wall cells.

8.4 Algorithms

Figure 8.3 illustrates some interesting potentially high-performance computing compatible algorithms/approaches. Frame (a) contrasts high-order unstructured discontinuous schemes such as the discontinuous Galerkin (DG) and standard, high-order finite differences (FD). Frame (b) relates to parallel processing in time and Frame (c) relates to the tiling approach (see Giles and Reguly 2014), to be discussed later.

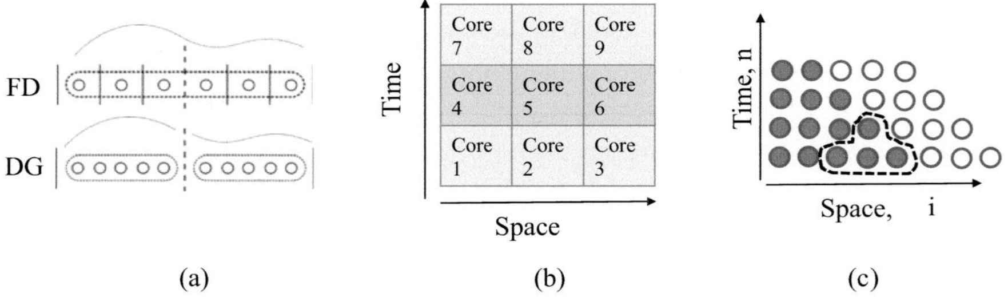

Figure 8.3. Interesting potentially high performance computing compatible algorithms/approaches: (a) high order – discontinuous (like DG) v. finite difference (FD), parallel processing in time and (c) tiling (used with the kind permission of the ASME – adapted from Tyacke and Tucker 2015).

As noted by Slotnick et al. (2014), algorithm development has stagnated. Hence, more work is needed in a wide range of areas. The gas lattice-Boltzmann method could be explored further and this approach can be zonalized with classical CFD. The smooth particle hydrodynamic approach is useful for niche applications such as multiphase flows with high interface deformations. However, this approach is currently limited to low numerical orders (certainly no greater than 2). High-order unstructured methods offer potential but also substantial numerical challenges to alleviate stiffness, reduce implicit mode memory requirements (memory going around the scheme order to the power six), enhance stability and better understand their dissipative and dispersive traits. Techniques like the boundary element method (see Chapter 4) offer potentially useful peripheral uses in CFD. For example, acoustics (see Chapter 6), the evaluation of nearest wall distances (required for some turbulence models) and defeaturing geometries (see Chapter 7). Dealing with flows spanning a wide range of Mach numbers, as ever, presents challenges.

For eddy-resolving simulations, whether explicit or implicit time stepping can be used is not always as clear as with RANS (see Larsson and Wang 2014). There is the potential for zonalization of these schemes in an overall domain.

For many years the use of parallelization in time (i.e., having blocks in time – see Figure 8.3b) has been considered. Recent work in this area for turbulent flows (parareal) is discussed by Reynolds-Barredo et al. (2012). For certain periodic flows, this approach is quite possible. There are also other interesting works exploring the exploitation of domains of independence in the space-time plane that can alleviate communication expenses. This approach is outlined in Giles and Reguly (2014) and a schematic is given in Figure 8.3c. Many approaches seem to be in their infancy and have serious technological barriers but still need to be fully explored.

As ever, there is the need for improved inflow boundary conditions for eddy-resolving simulations that connect better with real engineering systems. Also, to reduce domain sizes and hence computational costs, more robust, non-reflecting outflow boundary conditions are needed.

8.5 Physical Modelling

There is the need for improved physical models and eddy-resolving simulations offer a great opportunity to advance this. For example, see Bisetti et al. (2014) and the use of DNS to improve combustion modelling. New rational couplings and zonalization between different fidelity levels need to be established with eddy-resolving simulations in mind. Such work is being pioneered by Deck et al. (2014), largely for airframes. For a wing with a deployed flap, an eddy-resolving simulation could just be used for the flap zone when it is at a high angle to the flow. For aeroengines, a potential option is coupling an eddy-resolving simulation for the combustor to upstream and downstream RANS zones for the compressor and turbine. Figure 8.4 shows some potential aeroengine zonalization. The left-hand frame expresses system level zonalization and the right-hand frame component level zonalizations.

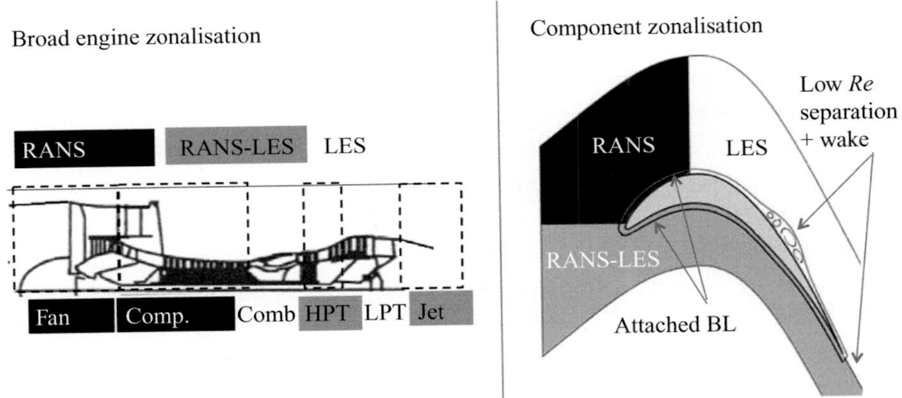

Figure 8.4. Potential aeroengine zonalizations for hybrid RANS-LES (used with the kind permission of the ASME – from Tyacke and Tucker 2015).

There are currently many subgrid scale models for eddy-resolving simulations. However, most practical simulations use Smagorinsky based models.

More fruitful exploration for subgrid scale models would be perhaps to examine those that are customized for use with high-order discontinuous unstructured methods. Evidence suggests that these can be dissipative at the high wave numbers. Hence, a useful exercise may be to alter the dissipation and dispersion properties of these schemes, potentially sacrificing order and adapting subgrid scale models so that the numerical package and subgrid scale model work in harmony. Currently, the implicit LES approach places the modelling role entirely with the scheme, and the classical explicit LES approach places the subgrid scale modelling role totally with a model. However, there could perhaps be potential for flexibility and hybridizations. Care would be needed to ensure that any hybridizations are rationally based.

The filter used for eddy-resolving simulations is vexing. There are wide ranges of filter definitions, many of which are likely to have as big an impact as the LES model. Strictly, the filter should also have a temporal and not just spatial component. Perhaps the latter should be further considered.

For many multiscale multi-physics problems, subgrid scale models could be developed that go beyond just accounting for the dissipation of turbulence kinetic energy. They could, for example, be used to account for lost acoustic content. With noise predictions for many high-speed aerospace applications, there are serious frequency resolution limits. With eddy-resolving grids, for many decades, there may well be the problem of being unable to predict the high acoustic frequencies needed by industry. Acoustic subgrid scale models could alleviate this. However, typically in source zones for jet noise eddy resolving simulations sufficient frequency content is probably resolved. The loss in frequency comes at the integral surfaces (where the grid is coarser) used to propagate the sound to the far field. Also, for combustors there are massive ranges in scales that need to be resolved and advanced subgrid scale models could help in this area also to characterize the multiphase/physics processes taking place at the small scales (see Bisetti et al. 2014). More work is also needed dealing with the inapplicability of LES models around shocks and also their conditional use (see Hadjadj 2012).

It seems hard to imagine that RANS, after so many decades of intensive research, will greatly move forwards. However, energy budgets and the more deep flow physics insights provided by eddy-resolving simulations will doubtless enable improvements.

8.6 Advanced Simulation

8.6.1 *Simulation Documentation, Pre- and Post-Processing and Data Handling*

Capabilities for the management of large databases are required and there is the need for methods to merge data from various sources such as measurements, high fidelity CFD and low-order simulation results. Flow feature identification is currently a highly manual (through threshold tuning) process involving potentially massive data sets. Hence, more reliable flow feature identification methods are needed.

Expert systems are needed to help an engineer engaged in complex, multidisciplinary, design tasks to choose the best methods. This could also be supplemented by best practice guidelines updated to account for more modern CFD techniques. The best practices needed for engineers performing eddy-resolving simulations will be much different to those used for RANS. Brief documents that detail established processes for specific problem types can be very useful and ensure results repeatability and a better audit trail. The NAFEMS 'How To' documents are examples of this type of literature.

8.6.2 *Design Optimization*

This is a critical area for providing rationally verifiable attempts to optimize the design landscape. The noise from eddy-resolving simulations presents great challenges for classical gradient-based design optimization approaches. Larsson and Wang (2014) discuss these challenges and present what is called a shadowing technique (involving linearizing the governing equations) to address them. Adjoint methods are also challenged for unsteady flows. Indeed there is a strong incompatibility with the use of expensive high-fidelity unsteady CFD and the process of design optimization, which necessitates a high volume of calculations. This pushes towards strategies that hybridize many low fidelity calculations with carefully selected high-fidelity simulations in a rational fashion. As with general CFD there are a wide range of design optimization strategies (see Chapter 6, Section 6.2). Some of these are most optimal when combined in hybrid forms (see Section 6.2.11). This again makes it difficult for an engineer focused on a complex design task to be certain of the best method(s).

8.6.3 *Multi-physics and Multi-scale Problems*

As discussed in Chapter 6, there is the increasing need to deal with multi-physics calculations. Also, the use of flow control appears a necessary element to reducing the environmental impact of future aircraft. Over 50% of drag on modern aircrafts is due to skin friction. Current flow control features, such as vortex generators, globally energize the boundary layer. However, the streaks that so severely hamper eddy-resolving simulation grid demands play a critical skin friction and turbulence generation role. Components that interact at this scale are needed and this is a tremendous challenge. For multi-scale models low order models can be helpful and these can be informed by high-fidelity eddy-resolving simulations. However, the low order models need to be selected with care. Hence, again there is the need for documentation and knowledge capture. Section 6.3.3.2 of Chapter 6 discussed an example of a multi-physics and multi-scale problem, where plasma actuation was used for boundary layer control on an airfoil in an eddy-resolving simulation framework. The use of MEMS (microelectromechanical systems) devices for streak control (see Carpenter et al. 2007) has also been considered. Piezoelectric devices and electro osmosis (see Section 2.12) can also be used for flow actuation purposes. As noted by Fujii (2014), such simulations potentially moves us beyond geometry driven design to 'device design'. This also presents new challenges for design optimization.

Nanoscale flows are an important area. Their modelling can often need to be considered as part of larger scale systems. The study of nanoscale flows is well suited to atomistic approaches such as classical molecular dynamics (MD) and Direct Simulation Monte-Carlo (DSMC). However, these approaches places severe constraints on the scale of the system that can be considered. Hence, to allow the simulation of such extremes in scales, Werder et al. (2005) combine an atomistic approach with a continuum approach.

8.6.4 *Multi-fidelity Modelling*

Ultimately, as aircrafts develop the engines, the airframes will become increasingly integrated. Indeed, as the general need to increase simulation reality continues, increasingly coupled simulations are needed. This growth in problem scale is incompatible with the use of eddy-resolving simulations. Hence, low-order models will still be needed. However, there seems great potential to calibrate these with data from high fidelity eddy-resolving simulations. Also, hybrid simulations are likely to be performed with extensive RANS zones coupled to LES. Potentially, simulations might encompass a very wide range of fidelity levels in different zones of a single simulation. The interfacing of low and high order models raises further research challenges. Multi-fidelity modelling could also be considered in the context where one-dimensional model calculations closely interact with high fidelity, three-dimensional eddy resolving simulations. The latter being used to calibrate the former through some automated process at selected design points.

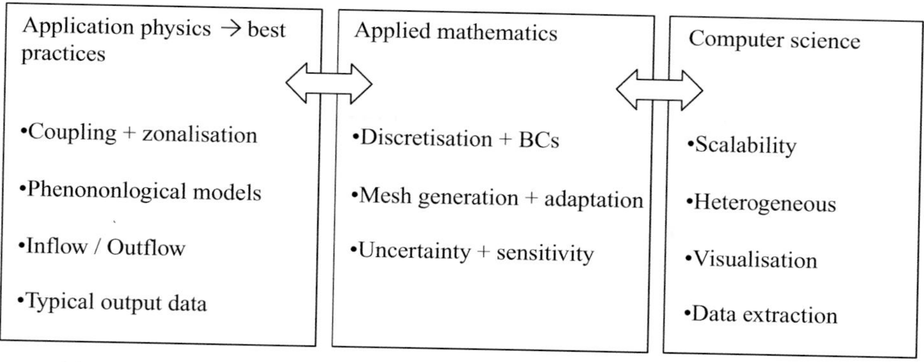

Figure 8.5. The range of fields that need to interact to advance CFD (used with the kind permission of the ASME – from Tyacke and Tucker 2015).

8.7 New Aerodynamic Technologies

The use of eddy-resolving simulations allows deeper exploration of flow control. The studies, when combined with advanced post-processing techniques such as outlined in Chapter 7 should allow substantial improvements in drag (see Fujii (2014)). Acoustics (see Epstein et al. 1989 and Williams 2002) offers the potential for flow control. Eddy-resolving simulations in compressible mode naturally capture all the acoustic information. This gives the potential to optimize through simulation both aerodynamics and aeroacoustics and potentially use aeroacoustics (even though the energy content in the acoustics component is typically relatively modest) to control the aerodynamic flow. Hence, the use of eddy-resolving simulations opens up new exciting research vistas. Surface roughness plays a critical role in generating and sometimes reducing aerodynamic drag. Different roughness topologies can have dramatically different aerodynamic impacts. The physics is highly complex and well beyond that which can be usefully studied using RANS. Hence, this is another critical future role of eddy-resolving simulations. These can be used to understand the aerodynamic impact of in-service degradation of surfaces. As noted by Fujii (2014), the use of eddy-resolving simulations could transform some areas of aerodynamic design through the use of multi-scale and multi-physics calculations.

8.8 Modern and Future CFD

8.8.1 *The Multi-Faceted Nature of Modern CFD*

Figure 8.5 summarizes the range of fields that need to interact to advance CFD. Also, there are substantial computational science challenges with respect to code scalability and so forth. Hence, for an aerospace company to have the optimal CFD resource (developed in house) needs people to collaborate/work across a wide range of academic research areas. Cleary, effectively making massive, coupled multi-physics simulations is a quest at the forefront of human intellectual endeavour.

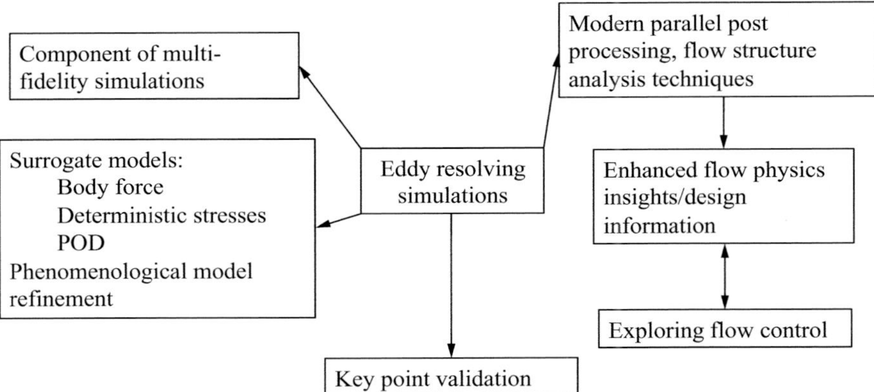

Figure 8.6. Potential future uses of eddy-resolving simulations (used with the kind permission of the ASME – from Tyacke and Tucker 2015).

Attention needs to be paid to a wide range of immature technologies such as quantum and molecular computing.

Figure 8.6 summarizes some potential future uses of eddy-resolving simulations as outlined in this text.

8.8.2 *Future Vistas*

To lighten and reduce aircraft emissions, flow control might involve highly deflecting or moving surfaces. The latter, if used to alleviate load, could promote highly separated flows. To reduce risk on product performance, as increasing use is made of simulation, software qualification is likely to become a bigger issue.

Successful porting of unstructured solvers to GPUs are already beginning to emerge (see Corrigan et al. 2010).

For structured solvers, this is relatively straightforward – see, for example, Brandvik and Pullan (2010), who achieved substantial (around a factor of 30) speedups. With structured solvers, it is possible to ensure that cells close in space are also close in memory. For unstructured solvers (see Figure 8.7), this is not so easy to guarantee. Hence special processes are needed to optimize this aspect. Also, for unstructured solvers, standard processes, such as the use of sliding planes in turbomachinery (see Chapter 6) are challenging. The grids are essentially spilt. This complicates load balancing. Also, there is relative movement between domains and this complicates data transfer with some form of searching process needed. Granted k-d trees can offer efficient interpolation.

As ever, there is a need for improved validation data. Such aspects were briefly discussed at the end of Chapter 7. Eddy-resolving simulations provide a wealth of high-order correlations, including spectral data. Hence, ideally, measurements at the high speeds found in industrial systems are needed to make data comparisons at this level.

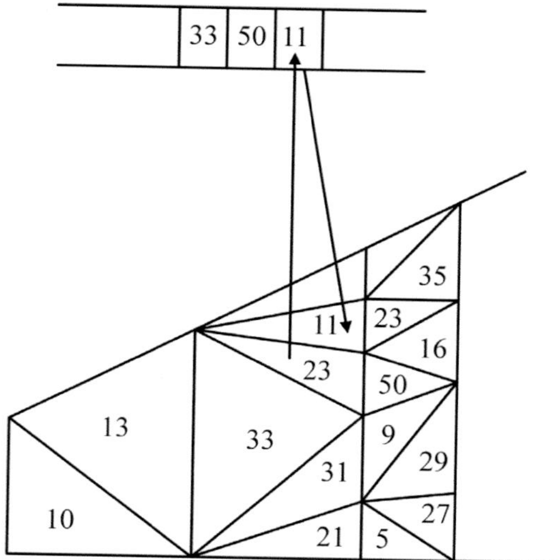

Figure 8.7. Difficulty of un-coalesced memory access (adapted from Corrigan et al., 2010).

As noted, for many of the above areas best practices are needed. For example, with fluid structure interactions a user needs to know when one or two way coupling is needed, loose or strong coupling and implicit or the simpler explicit coupling. Also, for multi-physics problems multiple codes might need to coupled and load balanced. For example, with an aircraft one might want to couple electromagnetics, aeroacoustics, non-linear FE composite structure, airframe CFD and engine CFD codes along with fatigue and weight analysis codes with some model for the aircraft pilot interaction. As well as presenting code coupling challenges with a range of differing physical time scales, human communication (granted this must to a degree already take place in aerospace) and collaboration perhaps becomes a challenge as different numerical analysis partners need to work closely together. Biomechanics problems can also offer problems with a high degree of coupling. For example, electromagnetics can be used to actuate chemotherapeutics, thus again coupling the fluid and electromagnetic fields. These large scale, multi-physics coupled problems might also be combined with a design optimization process with multi-fidelity surrogate models, creating highly complex simulations that present substantial computational science issues. Validation and verification of such simulations also perhaps poses new challenges.

REFERENCES

ADDAD, Y., GAITONDE, U., LAURENCE, D. & ROLFO, S. 2008. Optimal unstructured meshing for large eddy simulations. In *Quality and Reliability of Large-Eddy Simulations*, Springer.

BISETTI, F., ATTILI, A. & PITSCH, H. 2014. Advancing predictive models for particulate formation in turbulent flames via massively parallel direct numerical simulations.

Philosophical Transactions of the Royal Society A: Mathematical, Physical and Engineering Sciences, 372, 20130324.

BRANDVIK, T. & PULLAN, G. 2010. SBLOCK: A framework for efficient stencil-based PDE solvers on multi-core platforms. IEEE 10th International Conference on Computer and Information Technology (CIT). Paper No. 1181–1188.

CARPENTER, P. W., KUDAR, K. L., ALI, R., SEN, P. K. & DAVIES, C. 2007. A deterministic model for the sublayer streaks in turbulent boundary layers for application to flow control. *Phil. Trans. R. Soc.* A 365, 2419–2441. doi:10.1098/rsta.2007.2016.

CORRIGAN, A., CAMELLI, F., LÖHNER, R. & MUT, F. 2010. Porting of FEFLO to GPUs. 5th European Conference on Computational Fluid Dynamics ECCOMAS CFD, J. C. F. Pereira and A. Sequeira (Eds.), Lisbon, Portugal, 14–17 June.

DECK, S., GAND, F., BRUNET, V. & KHELIL, S. B. 2014. High-fidelity simulations of unsteady civil aircraft aerodynamics: stakes and perspectives. Application of zonal detached eddy simulation. *Philosophical Transactions of the Royal Society A: Mathematical, Physical and Engineering Sciences*, 372, 20130325.

EPSTEIN, A., FFOWCS WILLIAMS, J. & GREITZER, E. 1989. Active suppression of aerodynamic instabilities in turbomachines. *Journal of Propulsion and Power*, 5, 204–211.

FUJII, K. 2014. High-performance computing-based exploration of flow control with micro devices. *Philosophical Transactions of the Royal Society A: Mathematical, Physical and Engineering Sciences*, 372, 20130326.

GILES, M. & REGULY, I. 2014. Trends in high-performance computing for engineering calculations. *Philosophical Transactions of the Royal Society A: Mathematical, Physical and Engineering Sciences*, 372, 20130319.

GILES, M. B., MUDALIGE, G. R., SHARIF, Z., MARKALL, G. & KELLY, P. H. 2011. Performance analysis and optimization of the OP2 framework on many-core architectures. *The Computer Journal*, 55(2), 168–180.

GOURDAIN, N., SICOT, F., DUCHAINE, F. & GICQUEL, L. 2014. Large eddy simulation of flows in industrial compressors: A path from 2015 to 2035. *Philosophical Transactions of the Royal Society A: Mathematical, Physical and Engineering Sciences*, 372, 20130323.

HADJADJ, A. 2012. Large-eddy simulation of shock/boundary-layer interaction. *AIAA Journal*, 50, 2919–2927.

JAMESON, A. 2008. Formulation of kinetic energy preserving conservative schemes for gas dynamics and direct numerical simulation of one-dimensional viscous compressible flow in a shock tube using entropy and kinetic energy preserving schemes. *Journal of Scientific Computing*, 34, 188–208.

LARSSON, J. & WANG, Q. 2014. The prospect of using large eddy and detached eddy simulations in engineering design, and the research required to get there. *Philosophical Transactions of the Royal Society A: Mathematical, Physical and Engineering Sciences*, 372, 20130329.

LELE, S. K. & NICHOLS, J. W. 2014. A second golden age of aeroacoustics? *Philosophical Transactions of the Royal Society A: Mathematical, Physical and Engineering Sciences*, 372, 20130321.

LÖHNER, R. & BAUM, J. D. 2013. Handling tens of thousands of cores with industrial/legacy codes: Approaches, implementation and timings. *Computers & Fluids*, 85, 53–62.

MORTON, S. A., CUMMINGS, R. M. & KHOLODAR, D. B. 2007. High resolution turbulence treatment of F/A-18 tail buffet. *Journal of Aircraft*, 44, 1769–1775.

REYNOLDS-BARREDO, J. M., NEWMAN, D. E., SANCHEZ, R., SAMADDAR, D., BERRY, L. A. & ELWASIF, W. R. 2012. Mechanisms for the convergence of time-parallelized, parareal turbulent plasma simulations. *Journal of Computational Physics*, 231, 7851–7867.

SLOTNICK, J. P., KHODADOUST, A., ALONSO, J. J., DARMOFAL, D. L., GROPP, W. D., LURIE, E. A., MAVRIPLIS, D. J. & VENKATAKRISHNAN, V. 2014. Enabling the environmentally clean air transportation of the future: a vision of computational fluid dynamics in 2030. *Philosophical Transactions of the Royal Society A: Mathematical, Physical and Engineering Sciences*, 372, 20130317.

TUCKER, P. & DEBONIS, J. 2014. Aerodynamics, computers and the environment. *Philosophical Transactions of the Royal Society A: Mathematical, Physical and Engineering Sciences*, 372, 20130331.

TYACKE, J. C, & TUCKER, P. G., 2015. Future use of large eddy simulation in aeroengines, *Journal of Turbomachinery*, 137(8), 081005.

VOIGT, C., FREY, C. & KERSKEN, H. 2010. Development of a generic surface mapping algorithm for fluid-structure-interaction simulations in turbomachinery. 5th European Conference on Computational Fluid Dynamics ECCOMAS CFD, J. C. F. Pereira and A. Sequeira (Eds.), Lisbon, Portugal, 14–17 June.

WERDER, T., WALTHER, J. H. AND KOUMOUTSAKOS P. 2005. Hybrid atomistic–continuum method for the simulation of dense fluid flows, *Journal of Computational Physics*, 205(1), 373–390.

WILLIAMS, J. F. 2002. Noise, anti-noise and fluid flow control. *Philosophical Transactions of the Royal Society A: Mathematical, Physical and Engineering Sciences*, 360, 821–832.

Appendices

Basic Finite Element Method

This appendix illustrates a very basic finite element discretization of the diffusion equation

$$\Gamma\nabla \cdot (\nabla\phi) = \partial\phi/\partial t \tag{A.1}$$

The discretization of Equation (A.1) is considered using triangular elements with linear shape functions. The derivation of the matrix equations for a linear element, using Galerkin's method, is discussed. The derivation here is made slightly easier because only Dirichlet and adiabatic boundary conditions are considered. If an approximate solution, at a point in the region, is substituted into Equation (A.1), there will be a residual R. This is because the approximate solution does not satisfy Equation (A.1). Therefore the residual Equation (A.2) can be written.

$$0 \neq R(r, z, t) = \Gamma\nabla \cdot (\nabla\phi) - \partial\phi/\partial t \tag{A.2}$$

For an axisymmetric volume, the weighted residual method requires

$$2\pi \int_A rWf_{(r,z,t)}R_{(r,z,t)}dA = 0 \tag{A.3}$$

where A is the cross-sectional area of the volume considered in the r-z plane and Wf is a weighting function. Note that for a constant radius, $r = 1$, the following equations will correspond to a Cartesian system.

Substituting Equation (A.2) into (A.3) and omitting the subscripts r, z and t for brevity gives

$$2\pi \int_A r\Gamma\nabla \cdot (\nabla\phi)WfdA = F \tag{A.4}$$

where $F = 2\pi \int_A r\partial\phi/\partial tdA$. Equation (A.4) can be rearranged to give

$$2\pi \boldsymbol{r} \int_A [\Gamma\nabla \cdot (Wf\nabla\phi) - (\nabla Wf) \cdot (\Gamma\nabla\phi)]dA = 0 \tag{A.5}$$

where the integral F is ignored for the moment (set $F = 0$), that is, the variation of ϕ with time is not considered. As can be seen, the constant 2π can be ignored. Also, (see Owen and Hinton 1980), if the area to be integrated over is sufficiently small, r can be redefined as a mean radius $\boldsymbol{r}$. The form of $\boldsymbol{r}$ for a triangular element will be defined later. Since $\boldsymbol{r}$ (the mean radius) is used, $\nabla = [\partial/\partial z, \partial/\partial r]$ and $\nabla^2 = [\partial^2/\partial z^2, \partial^2/\partial r^2]$, and so the equations become the same as if working in a Cartesian coordinate system except for the $\boldsymbol{r}$ term.

Green's theorem is applied to Equation (A.5). This reduces the order of the differentiation and gives a separate integral relating to the region's surface. Therefore,

$$0 = 2\pi \boldsymbol{r} \int_S (Wf\Gamma\nabla\phi) \cdot nds - 2\pi r \int_A (\nabla Wf) \cdot (\Gamma\nabla\phi)dA = 0 \qquad \text{(A.6)}$$

For Dirichlet boundary conditions, there is no unknown associated with the boundary point. Therefore, the first integral in Equation (A.6) can be ignored.

Next we consider a triangular region where it is assumed the ϕ varies linearly. The equation to express the ϕ variation over the region can be written in the form of Equation (A.7), where Wf_i, Wf_j and Wf_k are Lagrange interpolating polynomials (see Segerlind 1984). The subscripts i, j and k refer to the apexes of the triangle,

$$\phi = Wf_i\phi_i + Wf_j\phi_j + Wf_k\phi_k \qquad \text{(A.7)}$$

where $Wf_i = (a_i + b_i z + c_i r)/2A_e$ (Note that A_e is the element area) and the coefficients are as follows: $a_i = z_j\, r_k$, $b_i = r_j - r_k$, $c_i = z_k - z_i$. The coefficients for the weighting functions Wf_j and Wf_k can be obtained by moving the subscripts of the given expressions in a cyclic fashion.

The Galerkin method uses the values Wf_i, Wf_j and Wf_k as the weighting functions Wf in Equation (A.3), which is convenient. This can be summarised as follows: The weighting function for the ith node, Wf_i, consists of the Lagrange interpolating polynomial associated with the ith node, and so forth.

Considering the solution of the ϕ field over a single triangular element and restating Equation (A.7), we get

$$\phi = \sum_{j=1,3} Wf_j\phi_j \qquad \text{(A.8)}$$

An equation for node 'i' of the triangular element can be made by substituting Equation (A.8) into Equation (A.6) to give

$$0 = 2\pi \boldsymbol{r}\boldsymbol{\Gamma} \int_{Ae} \sum_{j=1,3} ((\nabla Wf_i) \cdot (\nabla Wf_j))\phi_j dA \qquad \text{(A.9)}$$

which can be simplified to

$$0 = 2\pi \boldsymbol{r}\Gamma \int_{Ae} \sum_{j=1,3} ((\nabla Wf_i) \cdot (\nabla Wf_j))\phi_j dA \qquad \text{(A.10)}$$

Evaluation of Equation (A.10) yields the nodal equation which can be used in a recursive manner to generate a set of three equations (one for each apex of the triangular element). The elements of what is normally described as the stiffness array can be written as

$$k_{ij} = 2\pi \boldsymbol{r}\Gamma \int_{Ae} ((\nabla Wf_i) \cdot\ (\nabla Wf_j))\, dA \qquad \text{(A.11)}$$

where $\boldsymbol{r}$ is given by

$$\boldsymbol{r} = (r_i + r_j + r_k)/3 \tag{A.12}$$

The element stiffness arrays can be summed to give the global stiffness array $[K]_g$ using the so-called direct stiffness procedure, as described by Davies (1980). This involves summing all the element stiffness values with the subscripts, say p, q, and storing them in the global stiffness array in the location p, q, for example.

For a solution of the unsteady equation, the integral F must be considered in the discretization, and this term will now be dealt with. Before the integral F can be evaluated, an assumption of how the rate of change of ϕ with respect to time varies spatially over the element must be made. It can be assumed that it varies linearly over the element between node points (piecewise-linear). However, this has disadvantages that will be mentioned later. A better method is to assume that the rate of change of ϕ with respect to time varies in a step-wise fashion. This is normally referred to as the 'lumped formulation' (see Segerlind 1984).

To use this method, new step-wise functions Wf^s_{ijk} are defined, where,

$$\partial\phi/\partial t = Wf^S_i \partial\phi_i/\partial t + Wf^S_j \partial\phi_j/\partial t + Wf^S_k \partial\phi_k/\partial t \tag{A.13}$$

An equation for node i of an element is given by

$$C_{ij} = 2\pi \boldsymbol{r} \int_{Ae} \sum_{j=1,3} (Wf^S_i Wf^S_j dA) \tag{A.14}$$

Evaluation of Equation (A.14) will give what is normally described as a capacitance array $[C]$ given by Equation (A.15). This array, like the stiffness array, is assembled using the direct stiffness procedure to give $[C]_g$ (see Segerlind 1984).

$$[C] = \frac{2\pi \boldsymbol{r} A_e}{3} \begin{vmatrix} 1 & 0 & 0 \\ 0 & 1 & 0 \\ 0 & 0 & 1 \end{vmatrix} \tag{A.15}$$

Combining $[C]_g$ and $[K]_g$ gives the first order differential equation

$$[C]_g[\dot{\phi}] + [K]_g[\phi] = 0 \tag{A.16}$$

As shown in Chapter 4, the Equation (A.16) can be integrated in time as

$$([C]_g + \lambda \Delta t [K]_g)\phi^{new} = ([C]_g - (1-\lambda)\Delta t [K]_g)\phi^{old} \tag{A.17}$$

where $0 \leq \lambda \leq 1$. Whatever value of λ is used, Equation (A.17) can be cast into the form of

$$[D][\phi]^{new} = [P][\phi]^{old} = [S] \tag{A.18}$$

The lumped formulation has the advantage that larger time-steps can be used without numerical instabilities arising. Also, the full range of the parameter λ can be used. To ensure physical reality of the solution (Segerlind 1984, Patankar 1980), all coefficients off the diagonal of array $[D]$ must be negative and all on the diagonal positive. This criterion is violated if any of the interior angles of a triangular element are greater than 90° (Segerlind 1984). For a right triangular element with two sides of equal length, the maximum allowable time-step is given by

$$\Delta t < (2A_e)/(9\Gamma(1-\lambda)) \tag{A.19}$$

REFERENCES

DAVIES, A. J. 1980. *The finite element method*, Oxford University Press.

OWEN, D. R. J. & HINTON, E. 1980. *A simple guide to finite elements*, Pineridge Press Swansea.

PATANKAR, S. 1980. *Numerical heat transfer and fluid flow*, CRC Press.

SEGERLIND, L. J. 1984. *Applied finite element analysis*, Wiley.

Discretization of the Equations for a Simple Pressure Correction Solver

B.1 General Equation Discretization

As with Appendix A, all the equations in this appendix relate to a cylindrical polar coordinate system. Transformation to a Cartesian system involves setting radial coordinates to unity ($r = 1$). Also, the following substitutions should be made, $\Delta x = \Delta z$, $\Delta y = \Delta r$, $u = w$ and $v = u$.

The general transport equation can be written in the form of Equation (B.1), where the time derivative $\partial(\rho\phi)/\partial t$ is incorporated in the source term S_ϕ.

$$\frac{1}{r}\frac{\partial(F_r)}{\partial r} + \frac{\partial(F_z)}{\partial z} = S_\phi \tag{B.1}$$

The terms F_r and F_z can be considered to be total fluxes (convection + diffusion) given by Equations (B.2) and (B.3).

$$F_r = \rho r u\phi - \Gamma_r r\frac{\partial\phi}{\partial r} \tag{B.2}$$

$$F_z = \rho w\phi - \Gamma_z\frac{\partial\phi}{\partial z} \tag{B.3}$$

Equation (B.1) can be rewritten as

$$[F_n - F] + [F_e - F_w] = S_\phi \tag{B.4}$$

where

$$F_n = \frac{1}{r_P\Delta r}\left[\rho_n r_n u_n\phi_n - \Gamma_{rn}r_n\left.\frac{\partial\phi}{\partial r}\right|_n\right] \tag{B.5}$$

$$F_s = \frac{1}{r_P\Delta r}\left[\rho_s r_s u_s\phi_s - \Gamma_{rs}r_s\left.\frac{\partial\phi}{\partial r}\right|_s\right] \tag{B.6}$$

$$F_e = \frac{1}{\Delta z}\left[\rho_e w_e\phi_e - \Gamma_{ze}\left.\frac{\partial\phi}{\partial z}\right|_e\right] \tag{B.7}$$

$$F_w = \frac{1}{\Delta z}\left[\rho_w w_w\phi_w - \Gamma_{zw}\left.\frac{\partial\phi}{\partial z}\right|_w\right] \tag{B.8}$$

The following convection coefficients can be defined:

$$C_n = \frac{\rho_n r_n u_n}{r_p \Delta r} \tag{B.9}$$

$$C_s = \frac{\rho_s r_s u_s}{r_p \Delta r} \tag{B.10}$$

$$C_e = \frac{\rho_e w_e}{\Delta z} \tag{B.11}$$

$$C_w = \frac{\rho_w w_w}{\Delta z} \tag{B.12}$$

The following diffusion coefficients are also defined:

$$D_n = \frac{\Gamma_{rn} r_{rn}}{\delta r_n r_p \Delta r} \tag{B.13}$$

$$D_s = \frac{\Gamma_{rs} r_s}{\delta r_s r_p \Delta r} \tag{B.14}$$

$$D_e = \frac{\Gamma_{ze}}{\delta z_e \Delta z} \tag{B.15}$$

$$D_w = \frac{\Gamma_{zw}}{\delta z_w \Delta z} \tag{B.16}$$

The convection coefficients C multiply ϕ at the control volume face itself. Therefore a value for ϕ must be interpolated from the main grid points onto the respective control volume faces. Numerous schemes for doing this are available. For the simpler schemes, an interpolation involving the two neighbouring grid points to either side of the control volume face is used. So, for the general variable ϕ at the 'w' control volume face, the following interpolation is used:

$$\phi_w = Wf_W \phi_W + Wf_p \phi_p \tag{B.17}$$

Using the central difference ($Wf = 1/2$) scheme results in the following main grid point coefficients:

$$A_N = D_n - \frac{C_n}{2} \tag{B.18}$$

$$A_S = D_s + \frac{C_s}{2} \tag{B.19}$$

$$A_E = D_e - \frac{C_e}{2} \tag{B.20}$$

$$A_W = D_w + \frac{C_w}{2} \tag{B.21}$$

The following discretization is then obtained:

$$\begin{aligned} A_{P,\phi}\phi_P = {} & A_{N,\phi}\phi_N + A_{S,\phi}\phi_S + A_{W,\phi}\phi_W + A_{E,\phi}\phi_E \\ & + S_{1,\phi}\phi_P + S_{2,\phi}\phi_P \end{aligned} \tag{B.22}$$

The central grid point coefficients for axial velocity, w, and radial velocity, u, are as follows:

$$A_{P,w} = \sum_{NB} A_{NB} + \frac{\rho_P^{old}}{\Delta t} \tag{B.23}$$

$$A_{P,u} = \sum_{NB} A_{NB} + \frac{2\mu_P}{r_P^2} + \alpha_G \left| \frac{v_P}{r_P \rho_P} \right| + \frac{\rho_P^{old}}{\Delta t} \tag{B.24}$$

Note that in the above is the factor of α is a tuneable parameter arising from the factor of Gosman et al. (1976) described in Section 4.6.3. The source terms are given by the following equations:

$$\begin{aligned} S_{1,w} &= \frac{p_w - p_e}{\Delta z} + \frac{\lambda_e(r_n u_{ne} - r_s u_{se})}{r_P \Delta z \Delta r} - \frac{\lambda_w(r_n u_{nw} - r_s u_{sw})}{r_P \Delta z \Delta r} \\ &+ \frac{\mu_n r_n (u_{ne} - u_{nw})}{r_P \Delta z \Delta r} - \frac{\mu_s r_s (u_{se} - u_{sw})}{r_P \Delta z \Delta r} + \frac{\rho_P^{old} w_P^{old}}{\Delta t} \end{aligned} \tag{B.25}$$

$$S_{2,w} = 0 \tag{B.26}$$

$$\begin{aligned} S_{1,u} &= \frac{p_s - p_n}{\Delta r} + \frac{\mu_e(w_{ne} - w_{se})}{\Delta z \Delta r} - \frac{\mu_w(w_{nw} - w_{sw})}{\Delta z \Delta r} \\ &+ \frac{\lambda_n(w_{ne} - w_{nw})}{\Delta z \Delta r} - \frac{\lambda_s(w_{se} - w_{sw})}{\Delta z \Delta r} \\ &+ \frac{\rho_p v_p}{r_p} + \frac{\rho_p^{old} v_p^{old}}{\Delta t} + 2\rho_p v \Omega + \alpha_G \left| \frac{v_p}{\rho_p r_p} \right| u_p \end{aligned} \tag{B.27}$$

$$S_{2,u} = \frac{1}{\Delta r} \left(\frac{\lambda_n}{r_n} - \frac{\lambda_s}{r_s} \right) \tag{B.28}$$

Note that these sources also include the factor of Gosman et al. (1976) for coupling tangential and radial momentum. The symbol Ω represents the coordinate system angular velocity. For steady state flows, under certain conditions $\sum_{NB} A_{NB} = 0$, which will result in a division by zero. Therefore, the following term is added to the left and right hand side of Equation (B.22):

$$|C_{n,\phi} - C_{s,\phi} + C_{e,\phi} - C_{w,\phi}| \phi_P \tag{B.29}$$

This term will also increase diagonal dominance (the quotient $A_P / \sum A_{NB}$) of Equation (B.22), which improves convergence.

B.1.1 *Coefficients and Source Terms for Pressure Correction Equation*

The pressure field is computed iteratively by solving an equation to predict pressure corrections, p', to an initial guessed pressure field. The form of the pressure correction equation is the same as Equation (B.22), with the coefficients and source terms defined as follows:

$$A_{N,p'} = \frac{\rho_n d_{n.u} r_n}{r_P \Delta r \delta r_n} \tag{B.30}$$

$$A_{S,p'} = \frac{\rho_s d_{s.u} r_s}{r_P \Delta r \delta r_s} \tag{B.31}$$

$$A_{E,p'} = \frac{\rho_e d_{e,w}}{\Delta z \delta z_e} \tag{B.32}$$

$$A_{W,p'} = \frac{\rho_w d_{w,w}}{\Delta z \delta z_w} \tag{B.33}$$

$$A_{P,p'} = \sum_{NB} A_{NB,p'} \tag{B.34}$$

$$S_{1,p'} = \frac{\rho_s u_s r_s - \rho_n u_n r_n}{r_P \Delta r} + \frac{\rho_w w_w - \rho_e w_e}{\Delta z} + \frac{\rho_P^{old} - \rho_P^{new}}{\Delta t} \tag{B.35}$$

$$S_{2,p'} = 0. \tag{B.36}$$

REFERENCE

GOSMAN, A., KOOSINLIN, M., LOCKWOOD, F. & SPALDING, D. 1976. Transfer of heat in rotating systems. Gas Turbine Conference and Products Show, ASME Paper No. 76-GT-25.

TDMA Simultaneous Equation Solvers

Two subroutines follow. The first one is for a normal system with standard boundary conditions defined with coefficients and the right-hand side. The second is for periodic boundary conditions.

```
!****************************************************************
!****************************************************************
!****************************************************************
subroutine tdmauy(ny,ap,ac,am,fj)
implicit none
! - -- - -- - -- - -- - -- ---
! Input/Output variables
! - -- - -- - -- - -- - -- ---
  integer:: ny
  real:: ap(ny), ac(ny), am(ny), fj(ny), uj(ny)
! - -- - -- - -- - -- ---
! Internal variables
! - -- - -- - -- - -- ---
  integer:: n, j
  real:: cm
  n = ny-1
  do j = 2,n
    cm = am(j)/ac(j-1)
          ac(j) = ac(j)-cm*ap(j-1)
          fj(j) = fj(j)-cm*fj(j-1)
  enddo
  uj(n) = fj(n)/ac(n)
  do j = n-1,1,-1
    uj(j) = (fj(j)-ap(j)*uj(j+1))/ac(j)
  enddo
  do j = 1,n
    fj(j) = u(j)
  enddo
end subroutine tdmauy
```

```
!***************************************************************
!***************************************************************
!***************************************************************

subroutine tdpauz(ji,jf,ap,ac,am,fi)
implicit none
! - -- - -- - -- - -- - -- ---
! Input/Output variables
! - -- - -- - -- - -- - -- ---
  integer:: ji, jf
  real:: ap(jf+1), ac(jf+1), am(jf+1), fi(jf+1)
! - -- - -- - -- - -- ---
! Internal variables
! - -- - -- - -- - -- ---
  integer:: i, j, ja, jj
  real:: fnn, pp
  real:: qq(jf+1), ss(jf+1), fei(jf+1)
  ja = ji+1
  jj = ji+jf
  qq(ji) = -ap(ji)/ac(ji)
  ss(ji) = -am(ji)/ac(ji)
  fnn = fi(jf)
  fi(ji) = fi(ji)/ac(ji)
! forward elimination sweep
! -- - -- - -- - -- - -- - -- - -- -
  do j = ja,jf
    pp = 1.0d0/(ac(j)+am(j)*qq(j-1))
             qq(j) = -ap(j)*pp
             ss(j) = -am(j)*ss(j-1)*pp
             fi(j) = (fi(j)-am(j)*fi(j-1))*pp
  enddo
! backward pass
! -- - -- - -- - -- -
  ss(jf) = 1.0d0
  fei(jf) = 0.0d0
  do i = ja,jf
    j = jj-i
             ss(j) = ss(j)+qq(j)*ss(j+1)
             fei(j) = fi(j)+qq(j)*fei(j+1)
  enddo
  fi(jf) = (fnn-ap(jf)*fei(ji)-am(jf)*fei(jf-1))/ &
  & (ap(jf)*ss(ji)+am(jf)*ss(jf-1)+ac(jf))
```

```
! backward substitution
! -- - -- - -- - -- - -- - -- -
  do i = ja,jf
    j = jj-i
         fi(j) = fi(jf)*ss(j)+fei(j)
  enddo
end subroutine tdpauz
```

Index

Figures and tables are denoted in bold typeface